D1253541

CAMBRIDGE MONOGRAPHS ON APPLIED AND COMPUTATIONAL MATHEMATICS

Series Editors
M. J. ABLOWITZ, S. H. DAVIS, E. J. HINCH, A. ISERLES,
J. OCKENDON, P. J. OLVER

19 Matrix Preconditioning Techniques and Applications

The *Cambridge Monographs on Applied and Computational Mathematics* reflects the crucial role of mathematical and computational techniques in contemporary science. The series publishes expositions on all aspects of applicable and numerical mathematics, with an emphasis on new developments in this fast-moving area of research.

State-of-the-art methods and algorithms as well as modern mathematical descriptions of physical and mechanical ideas are presented in a manner suited to graduate research students and professionals alike. Sound pedagogical presentation is a prerequisite. It is intended that books in the series will serve to inform a new generation of researchers.

Also in this series:

1. A Practical Guide to Pseudospectral Methods, *Bengt Fornberg*
2. Dynamical Systems and Numerical Analysis, *A. M. Stuart and A. R. Humphries*
3. Level Set Methods and Fast Marching Methods, *J. A. Sethian*
4. The Numerical Solution of Integral Equations of the Second Kind, *Kendall E. Atkinson*
5. Orthogonal Rational Functions, *Adhemar Bultheel, Pablo González-Vera, Erik Hendiksen, and Olav Njåstad*
6. The Theory of Composites, *Graeme W. Milton*
7. Geometry and Topology for Mesh Generation *Herbert Edelsbrunner*
8. Schwarz–Christoffel Mapping *Tofin A. Driscoll and Lloyd N. Trefethen*
9. High-Order Methods for Incompressible Fluid Flow, *M. O. Deville, P. F. Fischer and E. H. Mund*
10. Practical Extrapolation Methods, *Avram Sidi*
11. Generalized Riemann Problems in Computational Fluid Dynamics, *Matania Ben-Artzi and Joseph Falcovitz*
12. Radial Basis Functions: Theory and Implementations, *Martin Buhmann*
13. Iterative Krylov Methods for Large Linear Systems, *Henk A. van der Vorst*
14. Simulating Hamiltonian Dynamics, *Benedict Leimkuhler and Sebastian Reich*
15. Collocation Methods for Volterra Integral and Related Functional Equations, *Hermann Brunner*
16. Topology for Computing, *Afra J. Zomorodian*
17. Scattered Data Approximation, *Holger Wendland*

Matrix Preconditioning Techniques and Applications

KE CHEN
Reader in Mathematics
Department of Mathematical Sciences
The University of Liverpool

CAMBRIDGE UNIVERSITY PRESS
Cambridge, New York, Melbourne, Madrid, Cape Town, Singapore, São Paulo

Cambridge University Press
The Edinburgh Building, Cambridge CB2 2RU, UK

www.cambridge.org
Information on this title: www.cambridge.org/9780521838283

© Cambridge University Press 2005

This book is in copyright. Subject to statutory exception
and to the provisions of relevant collective licensing agreements,
no reproduction of any part may take place without
the written permission of Cambridge University Press.

First published 2005

Printed in the United Kingdom at the University Press, Cambridge

A catalogue record for this book is available from the British Library

ISBN-13 978-0-521-83828-3 hardback
ISBN-10 0-521-83828-2 hardback

MATLAB® is a trademark of The MathWorks, Inc. and is used with permission. The MathWorks does not warrant the accuracy of the text or exercises in this book. This book's use or discussion of MATLAB® software or related products does not constitute endorsement or sponsorship by The MathWorks of a particular pedagogical approach or particular use of the MATLAB® software.

Cambridge University Press has no responsibility for the persistence or accuracy of URLs for external or third-party internet websites referred to in this book, and does not guarantee that any content on such websites is, or will remain, accurate or appropriate.

Dedicated to
Zhuang and Leo Ling Yi
and the loving memories of my late parents Wan-Qing and Wen-Fang

In deciding what to investigate, how to formulate ideas and what problems to focus on, the individual mathematician has to be guided ultimately by their own sense of values. There are no clear rules, or rather if you only follow old rules you do not create anything worthwhile.

Sir Michael Atiyah (FRS, Fields Medallist 1966). What's it all about? *UK EPSRC Newsline Journal – Mathematics* (2001)

Contents

Preface

> The experiences of Fox, Huskey, and Wilkinson [from solving systems of orders up to 20] prompted Turing to write a remarkable paper [in 1948]... In this paper, Turing made several important contributions... He used the word "*preconditioning*" to mean improving the condition of a system of linear equations (a term that did not come into popular use until 1970s).
>
> NICHOLAS J. HIGHAM. *Accuracy and Stability of Numerical Algorithms.* SIAM Publications (1996)

Matrix computing arises in the solution of almost all linear and nonlinear systems of equations. As the computer power upsurges and high resolution simulations are attempted, a method can reach its applicability limits quickly and hence there is a constant demand for new and fast matrix solvers. Preconditioning is the key to a successful iterative solver. It is the intention of this book to present a comprehensive exposition of the many useful preconditioning techniques.

Preconditioning equations mainly serve for an iterative method and are often solved by a direct solver (occasionally by another iterative solver). Therefore it is inevitable to address direct solution techniques for both sparse and dense matrices. While fast solvers are frequently associated with iterative solvers, for special problems, a direct solver can be competitive. Moreover, there are situations where preconditioning is also needed for a direct solution method. This clearly demonstrates the close relationship between a direct and an iterative method.

This book is the first of its kind attempting to address an active research topic, covering these main types of preconditioners.

Type 1	Matrix splitting preconditioner	FEM setting
Type 2	Approximate inverse preconditioner	FEM setting
Type 3	Multilevel (approximate inverse) preconditioner	FEM setting
Type 4	Recursive Schur complements preconditioner	FEM setting
Type 5	Matrix splitting and Approximate inverses	Wavelet setting
Type 6	Recursive Schur complements preconditioner	Wavelet setting
Type 7	Implicit wavelet preconditioner	FEM setting

Here by 'FEM setting', we mean a usual matrix (as we found it) often formed from discretization by finite element methods (FEM) for partial differential equations with piecewise polynomial basis functions whilst the 'Wavelet setting' refers to wavelet discretizations. The iterative solvers, often called accelerators, are selected to assist and motivate preconditioning. As we believe that suitable preconditioners can work with most accelerators, many other variants of accelerators are only briefly mentioned to allow us a better focus on the main theme. However these accelerators are well documented in whole or in part in the more recent as well as the more classical survey books or monographs (to name only a few)

- Young, D. M. (1971). *Iterative Solution of Large Linear Systems*. Academic Press.
- Hageman A. L. and Young D. M. (1981). *Applied Iterative Methods*. Academic Press.
- McCormick S. F. (1992). *Multilevel Projection Methods for Partial Differential Equations*. SIAM Publications.
- Barrett R., *et al.* (1993). *Templates for the Solution of Linear Systems: Building Blocks for Iterative Methods*. SIAM Publications.
- Axelsson O. (1994). *Iterative Solution Methods*. Cambridge University Press (reprinted by SIAM Publications in 2001)
- Hackbusch W. (1994). *Iterative Solution of Large Sparse Systems*. Springer-Verlag.
- Kelly C. T. (1995). *Iterative Methods for Solving Linear and Nonlinear Equations*. SIAM Publications.
- Smith B., *et al.* (1996). *Domain Decomposition Methods*. Cambridge University Press.
- Golub G. and van Loan C. (1996). *Matrix Computations*, 3rd edn. Johns Hopkins University Press.
- Brezinski C. (1997). *Projection Methods for Systems of Equations*. North-Holland.
- Demmel J. (1997). *Applied Numerical Linear Algebra*. SIAM Publications.

- Greenbaum A. (1997). *Iterative Methods for Solving Linear Systems*. SIAM Publications.
- Trefethen N. and Bau D. (1997). *Numerical Linear Algebra*. SIAM Publications.
- Dongarra J., *et al.* (1998). *Numerical Linear Algebra on High-Performance Computers*. SIAM Publications.
- Briggs W., *et al.* (2000). *A Multigrid Tutorial*, 2nd edn. SIAM Publications.
- Varga R. (2001). *Matrix Iteration Analysis*, 2nd edn. Springer.
- Higham N. J. (2002). *Accuracy and Stability of Numerical Algorithms*, 2nd edn. SIAM Publications.
- van der Vorst H. A. (2003). *Iterative Krylov Methods for Large Linear Systems*. Cambridge University Press
- Saad Y. (2003). *Iterative Methods for Sparse Linear Systems*. PWS.
- Duff I. S., *et al.* (2006). *Direct Methods for Sparse Matrices*, 2nd edn. Clarendon Press.
- Elman H., *et al.* (2005) *Finite Elements and Fast Solvers*. Oxford University Press.

Most generally applicable preconditioning techniques for unsymmetric matrices are covered in this book. More specialized preconditioners, designed for symmetric matrices, are only briefly mentioned; where possible we point to suitable references for details. Our emphasis is placed on a clear exposition of the motivations and techniques of preconditioning, backed up by MATLAB®[1] Mfiles, and theories are only presented or outlined if they help to achieve better understanding. Broadly speaking, the convergence of an iterative solver is dependent of the underlying problem class. The robustness can often be improved by suitably designed preconditioners. In this sense, one might stay with any preferred iterative solver and concentrate on preconditioner designs to achieve better convergence.

As is well known, the idea of providing and sharing software is to enable other colleagues and researchers to concentrate on solving newer and harder problems instead of repeating work already done. In the extreme case, there is nothing more frustrating than not being able to reproduce results that are claimed to have been achieved. The MATLAB Mfiles are designed in a friendly style to reflect the teaching of the author's friend and former MSc advisor Mr Will

[1] MATLAB is a registered trademark of MathWorks, Inc; see its home page `http://www.mathworks.com`. MATLAB is an easy-to-use script language, having almost the full capability as a *C* programming language without the somewhat complicated syntax of *C*. Beginners can consult a MATLAB text e.g. [135] from `http://www.liv.ac.uk/maths/ETC/matbook` or any tutorial document from the internet. Search `http://www.google.com` using the key words: **MATLAB tutorial**.

McLewin (University of Manchester) who rightly said: 'in Mathematics, never use the word 'obviously'.' A simple and useful feature of the supplied Mfiles is that *typing* in the file name invokes the help page, giving working examples. (The standard MATLAB reply to such a usage situation is *??? Error... Not enough input arguments.*)

The book was born mainly out of research work done in recent years and partly out of a need of helping out graduate students to implement a method following the not-always-easy-to-follow descriptions by some authors (who use the words 'trivial', 'standard', 'well-known', 'leave the reader to work it out as an exercise' in a casual way and in critical places). That is to say, we aspire to emphasize the practical implementation as well as the understanding rather than too much of the theory. In particular the book is to attempt a clear presentation and explanation, with the aid of illustrations and computer software, so that the reader can avoid the occasional frustration that *one must know the subject already before one can really understand and appreciate a beautiful mathematical idea or algorithm* presented in some (maybe a lot of) mathematical literature.

▶ **About the solvers and preconditioners.**

Chapter 1. (Introduction) defines the commonly used concepts; in particular the two most relevant terms in preconditioning: condition number and clustering. With non-mathematics majors' readers in mind, we give an introduction to several discretization and linearization methods which generate matrix equations – the idea of mesh ordering affecting the resulting matrix is first encountered. Examples of bounding conditioned numbers by considering norm equivalence (for symmetric systems) are given; these appealing theories are not a guarantee for fast convergence of iterative solvers. Both the fast Fourier transforms (FFT) and fast wavelet transforms (FWT) are introduced here (mainly discrete FWT and the continuous to come later); further discussions of FFT and FWT are in Chapters 2, 4 and 8.

Chapter 2. (Direct methods) discusses the direct Gaussian elimination method and the Gauss–Jordan and several variants. Direct methods are on one hand necessary for forward type preconditioning steps and on the other hand provide various motivations for designing an effective preconditioner. Likewise, for some ill-conditioned linear systems, there is a strong need for scaling and preconditioning to obtain accurate direct solutions – a much less addressed subject. Algorithms for inverting several useful special matrices are then given; for circulant matrices diagonalization by Fourier transforms is explained before considering Toeplitz matrices. Block Toeplitz matrices are considered later in Chapter 13. Algorithms for graph nodal or mesh (natural graph) orderings by reverse Cuthill–McKee method (RCM), spiral and domain decomposition methods (DDM) are given. The Schur complements and partitioned LU decompositions are presented together; for symmetric positive definite (SPD) matrices, some Schur properties are discussed. Overall, this chapter contains most of the ingredients for implementing a successful preconditioner.

Chapter 3. (Iterative methods) first discusses the classical iterative methods and highlights their use in multigrid methods (MGM, Chapter 6) and DDM. Then we introduce the topics most relevant to the book, conjugate gradient methods (CGM) of the Krylov subspace type (the complex variant algorithm does not appear in the literature as explicitly as presented in Section 3.7). We elaborate on the convergence with a view on preconditioners' design. Finally the popular fast multipole expansion method (along with preconditioning) is introduced. The mission of this chapter is to convey the message that preconditioning is relatively more important than modifying existing or inventing new CGM solvers.

Chapter 4. (Matrix splitting preconditioners: Type 1) presents a class of mainly sparse preconditioners and indicates their possible application areas, algorithms and limitations. All these preconditioners are of the forward type, i.e. $M \approx A$ in some way and efficiency in solving $Mx = b$ is assured. The most effective and general variant is the incomplete LU (ILU) preconditioner with suitable nodal ordering. The last two main sections (especially the last one) are mainly useful for dense matrix applications.

Chapter 5. (Approximate inverse preconditioners: Type 2) presents another large class of sparse approximate inverse preconditioners for a general sparse matrix problem, with band preconditioners suitable for diagonally dominant matrices and near neighbour preconditioners suitable for singular operator equations. All these preconditioners are of the backward type, i.e. $M \approx A^{-1}$ in some way and application of each sparse preconditioner M requires a simple multiplication.

Chapter 6. (Multilevel methods and preconditioners: Type 3) gives an introduction to geometric multigrid methods for partial differential equations (PDEs) and integral equations (IEs) and algebraic multigrid method for sparse linear systems, indicating that for PDEs, in general, smoothing is important but can be difficult while for IEs operator compactness is the key. Finally we discuss multilevel domain decomposition preconditioners for CG methods.

Chapter 7. (Multilevel recursive Schur preconditioners: Type 4) surveys the recent Schur complements based recursive preconditioners where matrix partition can be based on functional space nesting or graph nesting (both geometrically based and algebraically based).

Chapter 8. (Sparse wavelet preconditioners: Type 5) first introduces the continuous wavelets and then considers to how construct preconditioners under the wavelet basis in which an underlying operator is more amenable to approximation by the techniques of Chapters 4–7. Finally we discuss various permutations for the standard wavelet transforms and their use in designing banded arrow (wavelet) preconditioners.

Chapter 9. (Wavelet Schur preconditioners: Type 6) generalizes the Schur preconditioner of Chapter 7 to wavelet discretization. Here we propose to combine the non-standard form with Schur complement ideas to avoid finger-patterned matrices.

Chapter 10. (Implicit wavelet preconditioners: Type 7) presents some recent results that propose to combine the advantages of sparsity of finite elements, sparse approximate inverses and wavelets compression. Effectively the wavelet theory is used to justify the a priori patterns that are needed to enable approximate inverses to be efficient; this strategy is different from Chapter 9 which does not use approximate inverses.

▶ About the selected applications.

Chapter 11. (Application I) discusses the iterative solution of boundary integral equations reformulating the Helmholtz equation in an infinite domain modelling the acoustic scattering problem. We include some recent results on high order formulations to overcome the hyper-singularity. The chapter is concluded with a discussion of the open challenge of modelling high wavenumber problems.

Chapter 12. (Application II) surveys some recent work on preconditioning coupled matrix problems. These include Hermitian and skew-Hermitian splitting, continuous operators based Schur approximations for Oseen problems, the block diagonal approximate inverse preconditioners for a coupled fluid structure interaction problem, and FWT based sparse preconditioners for EHL equations modelling the isothermal (two dependent variables) and thermal (three dependent variables) cases.

Chapter 13. (Application III) surveys some recent results for iterative solution of inverse problems. We take the example of the nonlinear total variation (TV) equation for image restoration using operator splitting and circulant preconditioners. We show some new results based on combining FWT and FFT preconditioners for possibly more robust and faster solution and results on developing nonlinear multigrid methods for optimization. Also discussed is the 'matrix-free' idea of solving an elliptic PDE via an explicit scheme of a parabolic PDE, which is widely used in evolving level set functions for interfaces tracking; the related variational formulation of image segmentation is discussed.

Chapter 14. (Application IV) shows an example from scientific computing that typifies the challenge facing computational mathematics – the bifurcation problem. It comes from studying voltage stability in electrical power transmission systems. We have developed two-level preconditioners (approximate inverses with deflation) for solving the fold bifurcation while the Hopf problem remains an open problem as the problem dimension is 'squared'!

Chapter 15. (Parallel computing) gives a brief introduction to the important subject of parallel computing. Instead of parallelizing many algorithms, we motivate two fundamental issues here: firstly how to implement a parallel algorithm in a step-by-step manner and with complete MPI Fortran programs, and secondly what to consider when adapting a sequential algorithm for parallel computing. We take four relatively simple tasks for discussing the underlying ideas.

The Appendices give some useful background material, for reference purpose, on introductory linear algebra, the Harwell–Boeing data format, a MATLAB tutorial, the supplied Mfiles and Internet resources relevant to this book.

► **Use of the book.** The book should be accessible to graduate students in various scientific computing disciplines who have a basic linear algebra and computing knowledge. It will be useful to researchers and computational practitioners. It is anticipated that the reader can build intuition, gain insight and get enough hands on experience with the discussed methods, using the supplied Mfiles and programs from

http : //www.cambridge.org/9780521838283
http : //www.liv.ac.uk/maths/ETC/mpta

while reading. As a reference for researchers, the book provides a toolkit and with it the reader is expected to experiment with a matrix under consideration and identify the suitable methods first before embarking on serious analysis of a new problem.

▶ **Acknowledgements.** Last but not least, the author is grateful to many colleagues (including Joan E. Walsh, Siamiak Amini, Michael J. Baines, Tony F. Chan and Gene H. Golub) for their insight, guidance, and encouragement and to all my graduate students and research fellows (with whom he has collaborated) for their commitment and hard work, on topics relating to this book over the years. In particular, Stuart Hawkins and Martyn Hughes have helped and drafted earlier versions of some Mfiles, as individually acknowledged in these files. Several colleagues have expressed encouragement and comments as well as corrected on parts of the first draft of the manuscript – these include David J. Evans, Henk A. van der Vorst, Raymond H. Chan, Tony F. Chan, Gene H. Golub and Yimin Wei; the author thanks them all. Any remaining errors in the book are all mine. The handy author index was produced using the authorindex.sty (which is available from the usual LaTeX sites) as developed by Andreas Wettstein (ISE AG, Zurich, Switzerland); the author thanks him for writing a special script for me to convert bibitem entries to a bibliography style. The CUP editorial staff (especially Dr Ken Blake), the series editors and the series reviewers have been very helpful and supportive. The author thanks them for their professionalism and quality support. The continual funding in recent years by the UK EPSRC and other UK funding bodies for various projects related to this book has been gratefully received.

▶ **Feedback.** As the book involves an ambitious number of topics with preconditioning connection, inevitably, there might be errors and typos. The author is happy to hear from any readers who could point these out for future editions. Omission is definitely inevitable: to give a sense of depth and width of the subject area, a search on `www.google.com` in April 2004 (using keywords like 'preconditioned iterative' or 'preconditioning') resulted in hits ranging from 19000 to 149000 sites. Nevertheless, suggestions and comments are always welcome. The author is also keen to include more links to suitable software that are readable and helpful to other researchers, and are in the spirit of this book. Many thanks and happy computing.

Ke Chen
Liverpool, September 2004
EMAIL = k.chen@liverpool.ac.uk
URL = http://www.liv.ac.uk/~cmchenke

Nomenclature

All the beautiful mathematical ideas can be found in Numerical Linear Algebra. However, the subject is better to be enjoyed by researchers than to teach to students as many excellent ideas are often buried in the complicated notation. A researcher must be aware of this fact.

GENE H. GOLUB. Lecture at University of Liverpool (1995)

Throughout the book, capital letters such as A denote a rectangular matrix $m \times n$ (or a square matrix of size n), whose (i, j) entry is denoted by $A(i, j) = a_{ij}$, and small letters such as x, b denote vectors of size n unless stated otherwise i.e. $A \in \mathbb{R}^{m\times n}$ and $x, \ b \in \mathbb{R}^n$.

Some (common) abbreviations and notations are listed here

$\mathbb{C}^n$ $\rightarrow$ the space of all complex vectors of size n

$\mathbb{R}^n$ $\rightarrow$ the space of all real vectors of size n
[note $\mathbb{R}^n \subset \mathbb{C}^n$ and $\mathbb{R}^{n\times n} \subset \mathbb{C}^{n\times n}$)]

$\|A\|$ $\rightarrow$ A norm of matrix A (see §1.5)

$|A|$ $\rightarrow$ The matrix of absolute values of A i.e.
$(|A|)_{ij} = |A(i, j)| = |a_{ij}|$.

A^T $\rightarrow$ The transpose of A i.e. $A^T(i, j) = A(j, i)$.
[A is symmetric if $A^T = A$]

A^H $\rightarrow$ The transpose conjugate for complex A i.e. $A^H(i,j) = \overline{A(j, i)}$.
[A is Hermitian if $A^H = A$. Some books write $A^* = A^H$]

$det(A)$ $\rightarrow$ The determinant of matrix A

$\text{diag}(\alpha_j)$ $\rightarrow$ A diagonal matrix made up of scalars α_j

$\lambda(A)$ $\rightarrow$ An eigenvalue of A

$A \oplus B$ $\rightarrow$ The direct sum of orthogonal quantities A, B

$A \odot B$ $\rightarrow$ The biproduct of matrices A, B (Definition 14.3.8)

$A \otimes B$ $\rightarrow$ The tensor product of matrices A, B (Definition 14.3.3)

$\sigma(A)$	$\rightarrow$	A singular value of A
$\kappa(A)$	$\rightarrow$	The condition number $cond(A)$ of A (in some norm)
$\Lambda(A)$	$\rightarrow$	Eigenspectrum for A
$\Lambda_\epsilon(A)$	$\rightarrow$	ϵ-Eigenspectrum for A
$\Sigma(A)$	$\rightarrow$	Spectrum of singular values of A
$\mathcal{Q}_k$	$\rightarrow$	Set of all degree k polynomials with $q(0) = 1$ for $q \in \mathcal{Q}_k$
$\mathcal{W}(A)$	$\rightarrow$	Field of values (FoA) spectrum
AINV	$\rightarrow$	Approximate inverse [55]
BEM	$\rightarrow$	Boundary element method
BIE	$\rightarrow$	Boundary integral equation
BCCB	$\rightarrow$	Block circulant with circulant blocks
BTTB	$\rightarrow$	Block Toeplitz with Toeplitz blocks
BPX	$\rightarrow$	Bramble–Pasciak–Xu (preconditioner)
CG	$\rightarrow$	Conjugate gradient
CGM	$\rightarrow$	CG Method
CGN	$\rightarrow$	Conjugate gradient normal method
DBAI	$\rightarrow$	Diagonal block approximate inverse (preconditioner)
DDM	$\rightarrow$	Domain decomposition method
DFT	$\rightarrow$	Discrete Fourier transform
DWT	$\rightarrow$	Discrete wavelet transform
FDM	$\rightarrow$	Finite difference method
FEM	$\rightarrow$	Finite element method
FFT	$\rightarrow$	Fast Fourier transform
FFT2	$\rightarrow$	Fast Fourier transform in 2D (tensor products)
FMM	$\rightarrow$	Fast multipole method
FoV	$\rightarrow$	Field of values
FSAI	$\rightarrow$	Factorized approximate inverse (preconditioner) [321]
FWT	$\rightarrow$	Fast wavelet transform
GMRES	$\rightarrow$	Generalised minimal residual method
GJ	$\rightarrow$	Gauss–Jordan decomposition
GS	$\rightarrow$	Gauss–Seidel iterations (or Gram–Schmidt method)
HB	$\rightarrow$	Hierachical basis (finite elements)
ILU	$\rightarrow$	Incomplete LU decomposition
LU	$\rightarrow$	Lower upper triangular matrix decomposition
LSAI	$\rightarrow$	Least squares approximate inverse (preconditioner)
MGM	$\rightarrow$	Multigrid method
MRA	$\rightarrow$	Multilresolution analysis
OSP	$\rightarrow$	Operator splitting preconditioner
PDE	$\rightarrow$	Partial differential equation

PSM → Powers of sparse matrices
QR → Orthogonal upper triangular decomposition
SDD → Strictly diagonally dominant
SOR → Successive over-relaxation
SSOR → Symmetric SOR
SPAI → Sparse approximate inverse [253]
SPD → Symmetric positive definite matrix ($\lambda_j(A) > 0$)
SVD → Singular value decomposition
WSPAI → Wavelet SPAI

1
Introduction

To obtain improved convergence rates for the methods of successive displacement we require the coefficient matrix to have a P-condition number as small as possible. If this criterion is not satisfied, then it is advisable to prepare the system or 'precondition' it beforehand.

D. J. EVANS. *Journal of the Institute of Mathematics and Applications,* (1968)

In devising a preconditioner, we are faced with a choice between finding a matrix M that approximates A, and for which solving a system is easier than solving one with A, or finding a matrix M that approximates A^{-1}, so that only multiplication by M is needed.

R. BARRETT, *et al.* The *Templates* book. SIAM Publications (1993)

This book is concerned with designing an effective matrix, the so-called preconditioner, in order to obtain a numerical solution *with more accuracy or in less time*. Denote a large-scale linear system of n equations, with $A \in \mathbb{R}^{n\times n}$, $b \in \mathbb{R}^n$, by

$$Ax = b \tag{1.1}$$

and one simple preconditioned system takes the following form

$$MAx = Mb. \tag{1.2}$$

(Our primary concern is the real case; the complex case is addressed later.) We shall present various techniques of constructing such a preconditioner M that the preconditioned matrix $A_1 = MA$ has better matrix properties than A. As we see, preconditioning can strike the balance of success and failure of a numerical method.

This chapter will review these introductory topics (if the material proves difficult, consult some suitable textbooks e.g. [80,444] or read the Appendix of [299]).

Here Sections 1.2–1.5 review some basic theories, Sections 1.6 and 1.7 some numerical tools with which many matrix problems are derived from applied mathematics applications, and Section 1.8 on general discussion of preconditioning issues. Finally Section 1.9 introduces several software Mfiles which the reader can download and use to further the understanding of the underlying topics.

1.1 Direct and iterative solvers, types of preconditioning

There are two types of practical methods for solving the equation (1.1): the direct methods (Chapter 2) and the iterative methods (Chapter 3). Each method produces a numerical solution x, that should approximate the analytical solution $x^* = A^{-1}b$ with a certain number of accurate digits. Modern developments into new solution techniques make the distinction of the two types a bit blurred, because often the two are very much mixed in formulation.

Traditionally, a direct method refers to any method that seeks a solution to (1.1) by simplifying A explicitly

$$Ax = b \qquad \Longrightarrow \qquad A_j x = b_j \qquad \Longrightarrow \qquad Tx = c, \tag{1.3}$$

where T is a much simplified matrix (e.g. T is ideally diagonal) and $c \in \mathbb{R}^n$. The philosophy is essentially *decoupling* the interactions of components of $x = [x_1, \ldots, x_n]^T$ in a new system. One may say that the 'enemy' is A. A somewhat different approach is taken in the Gauss–Purcell method Section 15.5 that views x from a higher space $\mathbb{R}^{n+1}$.

On the other hand, without modifying entries of A, an iterative method finds a sequence of solutions $x_0, x_1, \ldots, x_k, \ldots$ by working closely with the residual vector

$$r = r_k = b - Ax_k, \tag{1.4}$$

which can not only indicate how good x_k is, but also may extract analytical information of matrix A. One hopes that an early termination of iterations will provide a sufficient accurate solution, which is cheaper to obtain than a direct solver. Indeed it is the analytical property of A that determines whether an iterative method will converge at all. See Chapter 3.

A preconditioner may enter the picture of both types of methods. It can help a direct method (Chapter 2) to achieve the maximum number of digits allowable in full machine precision, whenever conditioning is an issue. For an iterative method (Chapter 3), when (often) convergence is a concern, preconditioning is expected to accelerate the convergence to an approximate solution quickly [48,464].

Although we have only listed one typical type of preconditioning in (1.2) namely the *left inverse preconditioner* as the equation makes sense only when $M \approx A^{-1}$ in some sense. There are several other types, each depending on how one intends to approximate A and whether one intends to transform A to $\widetilde{A}$ in a different space before approximation.

Essentially all preconditioners fall into two categories.

Forward Type:	aiming $M \approx A$	
	I (left)	$M^{-1}Ax = M^{-1}b$
	II (right)	$AM^{-1}y = b,\ x = M^{-1}y$
	III (mixed)	$M_2^{-1}AM_1^{-1}y = M_2^{-1}b,\ x = M_1^{-1}y$
Inverse Type:	aiming $M \approx A^{-1}$	
	I (left)	$MAx = Mb$
	II (right)	$AMy = b,\ x = My$
	III (mixed)	$M_2AM_1y = M_2b,\ x = M_1y.$

Clearly matrix splitting type (e.g. incomplete LU decomposition [28] as in Chapter 4) preconditioners fall into the **Forward Type** as represented

$$M^{-1}Ax = M^{-1}b \tag{1.5}$$

while the approximate inverse type preconditioners (e.g. AINV [57] and SPAI [253] as in Chapter 5) fall into the **Inverse Type** and can be represented by (1.2) i.e.

$$MAx = Mb. \tag{1.6}$$

Similarly it is not difficult to identify the same types of preconditioners when A is first transformed into $\widetilde{A}$ in another space, as in Chapters 8–10.

Remark 1.1.1. Here we should make some remarks on this classification.

(1) As with all iterative solution methods, explicit inversion in the **Forward Type** is implemented by a direct solution method e.g. implement $z = M^{-1}w$ as solving $Mz = w$ for z (surely one can use another iterative solver for this).
(2) Each preconditioner notation can be interpreted as a product of simple matrices (e.g. factorized form).
(3) In the context of preconditioning, the notation $M \approx A$ or $M^{-1} \approx A$ should be broadly interpreted as approximating either A (or A^{-1}) directly or simply its certain analytical property (e.g. both M and A have the same small eigenvalues).

1.2 Norms and condition number

The magnitude of any scalar number is easily measured by its modulus, which is non-negative, i.e. $|a|$ for $a \in \mathbb{R}$. The same can be done for vectors in $\mathbb{R}^n$ and matrices in $\mathbb{R}^{m\times n}$, through a non-negative measure called the *norm*. The following definition, using three norm axioms, determines if any such non-negative measure is a norm.

Definition 1.2.2. *Let V be either $\mathbb{R}^n$ (for vectors) or $\mathbb{R}^{m\times n}$ (for matrices). A measure $\|u\|$ of $u \in V$, satisfying the following* **Norm axioms**, *is a valid norm:*

- $\|u\| \geq 0$ *for any u and $\|u\| = 0$ is and only if $u = 0$,*
- $\|\alpha u\| = |\alpha|\|u\|$ *for any u and any $\alpha \in \mathbb{R}$,*
- $\|u + v\| \leq \|u\| + \|v\|$ *for any $u, v \in V$.*

Remark that the same axioms are also used for function norms.
One can verify that the following are valid vector norms [80], for $x \in \mathbb{R}^n$,

$$\|x\|_p = \begin{cases} \left(\sum_{i=1}^{n} |x_i|^p \right)^{1/p}, & \text{if } 1 \leq p < \infty, \\ \max_{1\leq i\leq n} |x_i|, & \text{if } p = \infty. \end{cases} \tag{1.7}$$

and similarly the following are valid matrix norms, for $A \in \mathbb{R}^{m\times n}$,

$$\|A\|_p = \sup_{x\neq 0\in\mathbb{R}^n} \frac{\|Ax\|_p}{\|x\|_p} = \sup_{\|x\|_p=1} \|Ax\|_p, \tag{1.8}$$

where 'sup' denotes 'supremum' and note $Ax \in \mathbb{R}^m$.

While the formulae for vector norms are easy, those for matrices are not. We need to take some specific p in order to present computable formulae from (1.8)

$$
\begin{aligned}
p = 1: & \qquad \|A\|_1 = \max_{1 \le j \le n} \sum_{i=1}^{m} |a_{ij}| = \max_{1 \le j \le n} \|a_j\|_1, \\
p = 2: & \qquad \|A\|_2 = \rho(A^T A)^{1/2} = \max_{1 \le j \le n} \lambda_j (A^T A)^{1/2} = \sigma_{\max}(A) \\
p = \infty: & \qquad \|A\|_\infty = \max_{1 \le i \le m} \sum_{j=1}^{n} |a_{ij}| = \max_{1 \le i \le m} \|\widetilde{a}_i\|_1,
\end{aligned}
\tag{1.9}
$$

where $A \in \mathbb{R}^{m \times n}$ is partitioned in columns first $A = [a_1, \ldots, a_n]$ and then in rows $A = [\widetilde{a}_1^T, \ldots, \widetilde{a}_m^T]^T$, λ is the notation for eigenvalues,[1] ρ denotes the spectral radius and $\sigma_{\max}(A)$ denotes the maximal singular value.[2] Notice that the matrix 2-norm (really an eigenvalue norm) does not resemble the vector 2-norm. The proper counterpart is the following matrix Frobenius norm:

$$
\begin{aligned}
\|A\|_F &= \sqrt{\sum_{i=1}^{n} \sum_{j=1}^{n} a_{ij}^2} = \sqrt{\sum_{j=1}^{n} \|a_j\|_2^2} = \sqrt{\sum_{i=1}^{m} \|\widetilde{a}_i\|_2^2} \\
&= \left[\mathrm{trace}(A^T A)\right]^{1/2} = \left[\mathrm{trace}(A A^T)\right]^{1/2} = \left[\sum_{j=1}^{n} \lambda_j (A A^T)\right]^{1/2}.
\end{aligned}
\tag{1.10}
$$

A closely related norm is the so-called Hilbert–Schmidt ‘weak’ norm

$$
\|A\|_{HS} = \frac{\|A\|_F}{\sqrt{n}} = \sqrt{\frac{1}{n} \sum_{i=1}^{n} \sum_{j=1}^{n} a_{ij}^2}.
$$

We now review some properties of norms for vectors and square matrices.

(1) If $M^{-1} A M = B$, then A, B are similar so $\lambda_j(B) = \lambda_j(A)$, although this property is not directly useful to preconditioning since the latter is supposed to change λ.

Proof. For any j, from $A x_j = \lambda(A) x_j$, we have (define $y_j = M^{-1} x_j$)

$$
M^{-1} A M M^{-1} x_j = \lambda_j(A) M^{-1} x_j \implies B y_j = \lambda_j y_j.
$$

Clearly $\lambda_j(B) = \lambda_j(A)$ and the corresponding jth eigenvector for B is y_j.

[1] Recall the definition of eigenvalues λ_j and eigenvectors x_j of a square matrix $B \in \mathbb{R}^{n \times n}$:

$$
B x_j = \lambda_j x_j, \qquad \text{for } j = 1, \cdots, n,
$$

with $\rho(B) = \max_j |\lambda_j|$. Also $det(B) = \prod_{j=1}^{n} \lambda_j$ and $trace(B) = \sum_{j=1}^{n} B(j, j) = \sum_{j=1}^{n} \lambda_j$.

[2] Recall that, if $A = U \Sigma V^T$ with $\Sigma = \mathrm{diag}(\sigma_j)$ for orthogonal U, V, then σ_j's are the singular values of A while U, V contain the left and right singular vectors of A respectively. Denote by $\Sigma(A)$ the spectrum of singular values.

(2) One of the most useful properties of the 2-norm and F-norm is the so-called *norm invariance* i.e. an orthogonal matrix $Q \in \mathbb{R}^{n\times n}$ (satisfying $Q^T Q = QQ^T = I$ or $Q^T = Q^{-1}$) does not change the 2-norm of a vector or both the 2-norm and the F-norm of a matrix.
The *proof* is quite simple for the vector case: e.g. let $y = Qx$, then $\|y\|_2^2 = y^T y = (Qx)^T Qx = x^T Q^T Qx = x^T x = \|x\|_2^2$.
For matrices, we use (1.8) to prove $\|PAQ\| = \|A\|$ for F and 2-norms.
Proof. (i) 2-norm case: **(Norm invariance)**
Noting that $\lambda(Q^T A^T AQ) = \lambda(A^T A)$ since Q serves as a similarity transform, then

$$\|PAQ\|^2 = \rho((PAQ)^T\ PAQ) = \rho(Q^T A^T AQ) = \rho(A^T A) = \|A\|^2.$$

(ii) F-norm case: Let $AQ = W = [w_1,\ w_2,\ \ldots,\ w_n]$ (in columns). We hope to show $\|PW\|_F = \|W\|_F$ first. This follows from a column partition of matrices and the previous property for vector norm invariance: $\|PAQ\|_F^2 = \|PW\|_F^2 = \sum_{j=1}^n \|Pw_j\|_2^2 = \sum_{j=1}^n \|w_j\|_2^2 = \|W\|_F^2 = \|AQ\|_F^2$. Next it remains to show that $\|AQ\|_F^2 = \|A\|_F^2$ from a row partition $A = \left(a_1^T\ a_2^T \ldots a_n^T\right)^T$ and again the vector norm invariance:

$$\|AQ\|_F^2 = \sum_{i=1}^n \|a_i Q\|_2^2 = \sum_{i=1}^n \|Q^T a_i^T\|_2^2 = \sum_{i=1}^n \|a_i^T\|_2^2 = \|A\|_F^2.$$

The same result holds for the complex case when Q is unitary.[3]

(3) The spectral radius is the lower bound for all matrix norms: $\rho(B) \le \|B\|$. The proof for p-norms is quite easy, as (define $Bx_k = \lambda_k x_k$)

$$\begin{aligned}\rho(B) &= \max_{1\le j\le n} |\lambda_j(B)| = |\lambda_k(B)| = \frac{|\lambda_k(B)| \cdot \|x_k\|}{\|x_k\|} \\ &= \frac{\|Bx_k\|}{\|x_k\|} \le \sup_{x\ne 0} \frac{\|Bx\|_p}{\|x\|_p} = \|B\|_p.\end{aligned}$$

For the F-norm, one can use the Schur unitary decomposition

$$B = UTU^H, \qquad \text{with triangular} \qquad T = \Lambda + N, \qquad (1.11)$$

where U^H is the conjugate transpose of the matrix U and $\Lambda = \operatorname{diag}(T)$ contains the eigenvalues, so[4] $\|B\|_F^2 = \|T\|_F^2 = \|\Lambda\|_F^2 + \|N\|_F^2$. Hence $\rho(B) \le \sqrt{\sum_{j=1}^n |\lambda_j|^2} \le \|T\|_F$.

[3] In contrast to $Q^T Q = I$ for a real and orthogonal matrix Q, a complex matrix Q is called *unitary* if $Q^H Q = I$.
[4] This equation is used later to discuss non-normality. For a normal matrix, $\|N\| = 0$ as $N = 0$.

(4) If A is symmetric, then $A^TA = A^2$ so the 2-norm is simply: $\|A\|_2 = \rho(A^2)^{1/2} = \rho(A)$.
The proof uses the facts: $\lambda_j(A^2) = \lambda_j(A)^2$ and in general $\lambda_j(A^k) = \lambda_j(A)^k$ for any matrix A which is proved by repeatedly using $Ax_j = \lambda x_j$ (if $\lambda_j \neq 0$ and $\forall$ integer k).

(5) Matrix ∞-norm (1.9) may be written into a product form

$$\|A\|_\infty = \max_{1\le i\le m} \|\widetilde{a}_i\|_1 = \|\,|A|\mathbf{e}\|_\infty, \tag{1.12}$$

with $\mathbf{e} = (1, \ldots, 1)^T$, and similarly

$$\|A\|_1 = \max_{1\le j\le n} \|a_j\|_1 = \left\|\,|A^T|\mathbf{e}\right\|_\infty = \left\|\left(\mathbf{e}^T|A|\right)^T\right\|_\infty,$$

i.e.

$$\|\underbrace{A}_{\text{matrix}}\|_\infty = \|\underbrace{|A|\mathbf{e}}_{\text{vector}}\|_\infty, \qquad \text{and} \qquad \|\underbrace{A}_{\text{matrix}}\|_1 = \|\underbrace{|A^T|\mathbf{e}}_{\text{vector}}\|_\infty.$$

(6) Besides the above 'standard matrix norms', one may also view a $n \times n$ matrix $A \in \mathbb{R}^{n\times n}$ as a member of the long vector space $A \in \mathbb{R}^{n^2}$. If so, A can be measured in vector p-norms directly (note: $\|A\|_{2,v} = \|A\|_F$ so this explains how F-norm is invented)

$$\|A\|_{p,v} = \left(\sum_{j=1}^{n}\sum_{k=1}^{n} |a_{jk}|^p\right)^{1/p}.$$

(7) Different norms of the same quantity can only differ by at most a constant multiple that depends on the dimension parameter n. For example,

$$\|A\|_1 \le n\|A\|_\infty, \quad \|A\|_\infty \le n\|A\|_1, \qquad \text{and} \tag{1.13}$$

$$\|A\|_2 \le \sqrt{n}\|A\|_1, \quad \|A\|_1 \le \sqrt{n}\|A\|_2, \tag{1.14}$$

from the vector norm inequalities

$$\frac{\|x\|_1}{\sqrt{n}} \le \|x\|_2 \le \|x\|_1. \tag{1.15}$$

Proof. To prove (1.13), use (1.8) and matrix partitions in (1.9): for any $x \neq 0 \in \mathbb{R}^n$,

$$\frac{\|Ax\|_1}{\|x\|_1} = \frac{\sum_{i=1}^n |\widetilde{a}_i x|}{\sum_{i=1}^n |x_i|} \le \frac{n\max_i |\widetilde{a}_i x|}{\sum_{i=1}^n |x_i|} \le \frac{n\max_i |\widetilde{a}_i x|}{\max_i |x_i|} = \frac{n\|Ax\|_\infty}{\|x\|_\infty},$$

$$\frac{\|Ax\|_\infty}{\|x\|_\infty} = \frac{n\max_i |\widetilde{a}_i x|}{n\max_i |x_i|} \le \frac{n\max_i |\widetilde{a}_i x|}{\sum_{i=1}^n |x_i|} \le \frac{n\sum_{i=1}^n |\widetilde{a}_i x|}{\sum_{i=1}^n |x_i|} = \frac{n\|Ax\|_1}{\|x\|_1}.$$

Thus the proof for (1.13) is complete from (1.8). To prove the left inequality (since the second one is easy) in (1.15), we can use induction and the Cauchy–Schwarz inequality

$$|x \cdot y| \leq \|x\|_2 \|y\|_2, \quad ||x_1| + |x_2|| \leq \sqrt{2}\sqrt{x_1^2 + x_2^2}.$$

Using (1.15), inequalities (1.14) follow from (1.8). ■

Further results will be given later when needed; consult also [80,229,280].

(8) The condition number of a nonsingular matrix $A \in \mathbb{R}^{n \times n}$ is defined as

$$\kappa(A) = \|A\| \|A^{-1}\|. \tag{1.16}$$

From $\|AB\| \leq \|A\| \|B\|$ and $\|I\| = 1$, we have

$$\kappa(A) \geq 1. \tag{1.17}$$

Here certain matrix norm is used and sometimes for clarity we write explicitly $\kappa_\ell(A)$ if the ℓ-norm is used (e.g. $\ell = 1, 2$ or $\ell = F$).

Although widely publicized and yet vaguely convincing, a condition number measures how well-conditioned a matrix is. However, without involving a particular context or an application, the meaning of $\kappa(A)$ can be seen more precisely from the following two equivalent formulae [280]:

$$\kappa(A) = \lim_{\epsilon \to 0} \sup_{\|\Delta A\| \leq \epsilon \|A\|} \frac{\|(A + \Delta A)^{-1} - A^{-1}\|}{\epsilon \|A^{-1}\|}, \tag{1.18}$$

$$\kappa(A) = \frac{1}{\text{dist}(A)} \quad \text{with} \quad \text{dist}(A) = \min_{A + \Delta A \text{ singular}} \frac{\|\Delta A\|}{\|A\|}. \tag{1.19}$$

If A is 'far' from its nearest singular neighbour, the condition number is small.

Two further issues are important. Firstly, if A is nonsingular and P, Q are orthogonal matrices, then $\kappa(PAQ) = \kappa(A)$ for F and 2-norms.

Proof. This result is a consequence of matrix norm invariance:

$$\kappa(PAQ) = \|(PAQ)^{-1}\| \, \|PAQ\| = \|Q^T A^{-1} P^T\| \, \|A\| = \|A^{-1}\| \, \|A\| = \kappa(A).$$

Secondly, if A is symmetric and nonsingular, then eigensystems are special so

$$\kappa_2(A) = \|A\|_2 \|A^{-1}\|_2 = \rho(A)\rho(A^{-1}) = \frac{|\lambda|_{\max}}{|\lambda|_{\min}}. \tag{1.20}$$

For SPD matrices, as $\lambda_j(A) > 0$,

$$\kappa_2(A) = \frac{\lambda_{\max}}{\lambda_{\min}}. \tag{1.21}$$

Example 1.2.3. *The following* 3×3 *matrix*

$$A = \begin{pmatrix} 12 & 1 & -17 \\ 5 & 4 & -9 \\ 7 & 1 & -10 \end{pmatrix}$$

has three eigenvalues $\lambda = 1, 2, 3$, *from solving* $|A - \lambda I| = \lambda^3 - 6\lambda^2 + 11\lambda - 6 = 0$, *corresponding to three eigenvectors* $x_1 = \begin{pmatrix} 3 \\ 1 \\ 2 \end{pmatrix}$, $x_2 = \begin{pmatrix} 5 \\ 1 \\ 3 \end{pmatrix}$ *and* $x_3 = \begin{pmatrix} 2 \\ -1 \\ 1 \end{pmatrix}$.

Clearly the spectral radius of A *is* $\rho(A) = 3$, $\|A\|_1 = 36$ *and* $\|A\|_\infty = 30$. *Finally putting the eigensystems* $Ax_j = \lambda_j x_j$ *into matrix form, noting* $X = [x_1\ x_2\ x_3]$ *and* $AX = [Ax_1\ Ax_2\ Ax_3]$, *we obtain* $AX = XD$ *with* $D = diag(1, 2, 3)$. *From* $A^{-1} = \begin{pmatrix} -31/6 & -7/6 & 59/6 \\ -13/6 & -1/6 & 23/6 \\ -23/6 & -5/6 & 43/6 \end{pmatrix}$, *we can compute that* $\kappa_1(A) = 36\,(125/6) = 750$.

Remark 1.2.4. Although this book will focus on fast solvers for solving the linear system (1.1), the solution of the eigenvalue problems may be sought by techniques involving indefinite and singular linear systems and hence our preconditioning techniques are also applicable to 'preconditioning eigenvalue problems' [36,58,360]. Refer also to [486].

1.3 Perturbation theories for linear systems and eigenvalues

A computer (approximate) solution will not satisfy (1.1) exactly so it is of interest to examine the perturbation theory.

Theorem 1.3.5. *Let* $Ax = b$ *and* $(A + \Delta A)(x + \Delta x) = b + \Delta b$. *If* $\|A^{-1}\|\|\Delta A\| < 1$, *then*

$$\frac{\|\Delta x\|}{\|x\|} \le \frac{\kappa(A)}{1 - \|A^{-1}\|\|\Delta A\|}\left(\frac{\|\Delta A\|}{\|A\|} + \frac{\|\Delta b\|}{\|b\|}\right). \tag{1.22}$$

The *proof* of this result is easy once one notes that

$$\Delta x = (A + \Delta A)^{-1}(-\Delta A x + \Delta b) = \left(I + A^{-1}\Delta A\right)^{-1} A^{-1}(-\Delta A x + \Delta b),$$

$\| \left(I + A^{-1}\Delta A\right)^{-1} \| \le 1/(1 - \|A^{-1}\| \|\Delta A\|), \ \|b\| \le \|A\| \|x\|.$

Here for our application to iterative solution, ΔA is not a major concern but Δb will be reflected in the usual stopping criterion based on residuals. Clearly the condition number $\kappa(A)$ will be crucial in determining the final solution accuracy as computer arithmetic has a finite machine precision.

Eigenvalue perturbation. We give a brief result on eigenvalues simply for comparison and completeness [218].

Theorem 1.3.6. (Bauer–Fike). *Let $A, X \in \mathbb{R}^{n\times n}$ and X be nonsingular. For $D = diag(\lambda_i)$ with $\lambda_i = \lambda_i(A)$, let $\|.\|$ be one of these norms: 1, 2 or ∞, such that $\|D\| = \max_{1\le i\le n} |\lambda_i|$. If $B = A + \Delta A$ is a perturbed matrix of A and $X^{-1}AX = D$, then all eigenvalues of B are inside the union of n discs $\bigcup_{i=1}^{n} \Omega_i$, where the discs are defined by*

General case: 1, 2, ∞ *norm and any matrix X:* $\quad cond(X) = \|X\| \, \|X^{-1}\| \ge 1$

$$\Omega_i = \{z \in C \quad | \quad |z - \lambda_i| \le cond(X)\|\Delta A\|\}.$$

Special case: 2-norm and orthogonal matrix X: $\quad cond(X) = \|X\|_2 \, \|X^{-1}\|_2 = 1$

$$\Omega_i = \{z \in C \quad | \quad |z - \lambda_i| \le \|\Delta A\|_2\}.$$

Proof. Define $K = D - \mu I$. If μ is an eigenvalue of B, then $B - \mu I = A + \Delta A - \mu I$ is singular; we may assume that this particular $\mu \ne \lambda_i(A)$ for any i (otherwise the above results are already valid as the difference is zero). That is, we assume K is nonsingular. Also define $W = K^{-1}X^{-1}\Delta A X$. Now consider decomposing

$$\begin{aligned} B - \mu I = A + \Delta A - \lambda I &= X D X^{-1} + \Delta A - \mu I \\ &= X\left[D + X^{-1}\Delta A X - \mu I\right] X^{-1} \\ &= X\left[(D - \mu I) + X^{-1}\Delta A X\right] X^{-1} \\ &= X\left[K + X^{-1}\Delta A X\right] X^{-1} \\ &= X K\left[I + K^{-1}X^{-1}\Delta A X\right] X^{-1} \\ &= X K\,[I + W]\, X^{-1}. \end{aligned}$$

Clearly, by taking determinants both sides, matrix $I + W$ must be singular as $B - \mu I$ is, so W has the eigenvalue -1. Further from the property

$\rho(W) \le \|W\|$,

$$1 = |\lambda(W)| \le \rho(W) \le \|W\| = \|K^{-1}X^{-1}\Delta A X\|$$
$$\le \|K^{-1}\|\,\|X^{-1}\|\,\|\Delta A\|\,\|X\| = \|K^{-1}\|\,\kappa(X)\|\Delta A\|.$$

As K is a simple diagonal matrix, we observe that

$$K^{-1} = \operatorname{diag}(1/(\lambda_1 - \mu), 1/(\lambda_2 - \mu) \ldots, 1/(\lambda_n - \mu))_{n\times n}\,.$$

Therefore for 1, 2, ∞ norm, $\|K^{-1}\| = \max_i \left|\dfrac{1}{\lambda_i - \mu}\right| = \dfrac{1}{\min_i |\lambda_i - \mu|} = \dfrac{1}{|\lambda_k - \mu|}$. Thus the previous inequality becomes

$$|\lambda_k - \mu| = 1/\|K^{-1}\| \le \kappa(X)\|\Delta A\|.$$

Furthermore, if $X^T = X^{-1}$ for 2-norm, $\kappa(X) = 1$, the above result reduces to a simpler inequality. ■

Remark 1.3.7.

(1) Clearly from the above proof, if the F-norm must be used, $\|K^{-1}\| = \sqrt{\sum_{j=1}^n 1/(\lambda_j - \mu)^2}$ and a slightly more complex formula can still be given.
(2) If A is the numerical matrix and $\lambda_j(A)$'s are the numerical eigenvalues, this perturbation result says that the distance of these numerical values from the true eigenvalues for the true matrix $B = A + \Delta A$ depends on the conditioning of matrix X (i.e. the way $\lambda_j(A)$'s are computed!).
(3) Clearly the perturbation bound for $\lambda(A)$ is quite different from the linear system case as the latter involves $\kappa(A)$ rather than $\kappa(X)$.

1.4 The Arnoldi iterations and decomposition

The Arnoldi decomposition for matrix A is to achieve $Q^T A Q = H$ or $AQ = QH$, where Q is orthogonal and H is upper Hessenberg

$$H = \begin{pmatrix} h_{1,1} & h_{1,2} & h_{1,3} & \cdots & h_{1,n-1} & h_{1,n} \\ h_{2,1} & h_{2,2} & h_{2,3} & \cdots & h_{2,n-1} & h_{2,n} \\ & h_{3,2} & h_{3,3} & \cdots & h_{3,n-1} & h_{3,n} \\ & & \ddots & \ddots & \vdots & \vdots \\ & & & \ddots & h_{n-1,n-1} & h_{n-1,n} \\ & & & & h_{n,n-1} & h_{n,n} \end{pmatrix}. \tag{1.23}$$

There are three methods of realizing this decomposition and we shall use the third method quite often.

♦ **Method 1 — Givens rotation.** The Givens matrix has four main nonzeros that are in turn determined by an angle parameter θ

$$P(i,j) = \begin{bmatrix} 1 & & & & & & & & \\ & \ddots & & & & & & & \\ & & 1 & & & & & & \\ & & & \cos(\theta) & \cdots & \sin(\theta) & & & \\ & & & \vdots & \ddots & \vdots & & & \\ & & & \sin(\theta) & \cdots & -\cos(\theta) & & & \\ & & & & & & 1 & & \\ & & & & & & & \ddots & \\ & & & & & & & & 1 \end{bmatrix} \overset{\text{for short}}{=} \begin{bmatrix} c & s \\ s & -c \end{bmatrix}. \tag{1.24}$$

Note $P^T = P = P^{-1}$.

We shall use Givens rotations to zero out these positions of A: $(j,i) = (3,1),\cdots,(n,1),\ (4,2),\cdots,(n,n-2)$. Each of these (j,i) pairs defines a Givens rotation $P(i+1,j)$ and the crucial observation is that the target (j,i) position of matrices $P(i+1,j)A$ and $P(i+1,j)AP(i+1,j)$ are identical (this would be false if matrix $P(i,j)$ is used). Therefore, to specify $P(i+1,j) = P(j,i+1)$, we only need to consider $P(i+1,j)A$ whose (j,i) position can be set to zero:

$$\widetilde{a}_{ji} = a_{i+1,i}\sin(\theta_{i+1,j}) - a_{ji}\cos(\theta_{i+1,j}) = 0,$$

giving

$$\begin{aligned}\theta_{i+1,j} &= \tan^{-1}\frac{a_{ji}}{a_{i+1,i}} = \arctan\frac{a_{ji}}{a_{i+1,i}} = \arctan\frac{\text{Lower position}}{\text{Upper position}} \\ &= \arctan\left(ratio_{ji}\right).\end{aligned}$$

From properties of trigonometric functions, we have

$$\sin(\theta) = \frac{r}{\sqrt{1+r^2}}, \qquad \cos(\theta) = \frac{1}{\sqrt{1+r^2}} \qquad \text{with } r = \frac{a_{ji}}{a_{i+1,i}}. \tag{1.25}$$

That is to say, it is not necessary to find θ to work out $P(i+1,j)$. Here and in this section, we assume that A is over-written $A = P(i+1,j)AP(i+1,j)$ (by intermediate and partially transformed matrices). The final Q matrix comes from the product of these rotation matrices:

$$Q = \underbrace{P(2,3)\cdots P(2,n)}_{\text{Column } 1} \cdots \underbrace{P(n-2,n-1)\,P(n-2,n)}_{\text{Column } n-3}\underbrace{P(n-1,n)}_{\text{Column } n-2}.$$

♦ **Method 2 — Householder transform.** For any nonzero vector v, the following Householder matrix is symmetric and orthogonal, and it can be used to design our method 2: (note: ℓ can be any integer so $e_1, v \in \mathbb{R}^\ell$ and $P, vv^T \in \mathbb{R}^{\ell\times\ell}$)

$$P = P(v) = P_{\ell\times\ell} = I - 2\frac{vv^T}{v^Tv}. \tag{1.26}$$

Defining $w = \frac{v}{\|v\|_2}$ (unit vector) and $\beta = \frac{2}{v^Tv} = \frac{2}{\|v\|_2^2}$, the above form can be written equivalently as

$$P = P(w) = P_{\ell\times\ell} = I - 2ww^T = I - \beta vv^T.$$

For a vector $x = x_{\ell\times 1}$, here is the magic choice of v,[5]

$$v = v_{\ell\times 1} = x + \alpha e_1$$

with $\alpha = S(x_1)\|x\|_2$ ensuring that $Px = -\alpha e_1$. Here $S = SIGN = \pm 1$ denotes the usual sign function. The full vector (basic) case is summarised as:

$$\boxed{\begin{aligned} v &= [\, x_1 + S(x_1)\|x\|_2,\ x_2, \cdots, x_\ell]^T \\ Px &= [\quad -S(x_1)\|x\|_2,\ 0, \cdots, 0]^T \end{aligned}}$$

The most useful form of Householder transform is the following:

$$\boxed{\begin{aligned} x &= [\ \underline{x_1, x_2, \cdots, x_{k-1}, \qquad x_k,}\quad \underline{x_{k+1}, \cdots, x_n}\ ]^T \\ &= [\qquad\qquad \widetilde{x}^T \qquad\qquad\qquad \overline{x}^T \qquad]^T \\ v &= [\ 0,\ 0, \cdots, 0,\ \ x_k + S(x_k)\|\bar{x}_{k,n}\|_2,\ \underline{x_{k+1}, \cdots, x_n}\ ]^T \\ &= [\qquad\qquad \underline{\mathbf{0}^T} \qquad\qquad\qquad \overline{v}^T \qquad]^T \\ P_k x &= [\ x_1, x_2, \cdots, x_{k-1},\ -S(x_k)\|\bar{x}_{k,n}\|_2,\ \ 0, \cdots, 0]^T \end{aligned}}$$

where $\overline{x} = \overline{x}_{k,n} = x(k:n) = [x_k,\ x_{k+1},\ \cdots,\ x_n]^T$. Here the MATL4B® notation[6] is used; more uses can be found in Section 1.6. The following diagram

[5] The derivation [229] uses the properties that $P(\gamma v) = P(v)$ for any scalar γ and if $Px = (1 - W)x + W\beta e_1 = -\alpha e_1$, then $W = 1$.

[6] In MATLAB, if x is an $n \times 1$ vector, then $y = x(3:6)$ extracts positions 3 to 6 of x. Likely $z = x([1:2\ 7:n])$ extracts a new vector excluding positions $3:6$.

summarizes the above procedure $[\bar{P} = I - 2\bar{v}\bar{v}^T/(\bar{v}^T\bar{v})]$

$$x = \begin{pmatrix} \times \\ \times \\ \times \\ \hline \times \\ \times \\ \times \end{pmatrix} = \begin{bmatrix} \widetilde{x} \\ - \\ \overline{x} \end{bmatrix} \overset{v=\begin{pmatrix}\mathbf{0}\\ \overline{v}\end{pmatrix}}{\Longrightarrow} P_k x = \begin{bmatrix} \widetilde{x} \\ \hline \\ -\alpha \bar{e}_1 \end{bmatrix} = \begin{pmatrix} \times \\ \times \\ \times \\ \hline -\alpha \\ 0 \\ 0 \end{pmatrix}, \; P_k = \begin{pmatrix} I_{k-1} & \\ & \bar{P} \end{pmatrix}. \tag{1.27}$$

To realize $A = QHQ^T$, at step $i = 1, 2, \cdots, n-2$, denote the $(i+1)$th column of A by $x = \begin{pmatrix} \widetilde{x} \\ \overline{x} \end{pmatrix}$, we take $\overline{x} = A((i+1) : n, i)$ to define matrix P_{i+1}. The final Q matrix is

$$Q = P_2 P_3 \cdots P_{n-1}.$$

Remark 1.4.8. Our Givens formula (1.24) for $P_{i,j}$ differs from almost all textbooks by (-1) in row j in order to keep it symmetric! One useful consequence is that when v has only two nonzero positions, both Methods 1 and 2 define the identical transform matrix.

♦ **Method 3 — the Arnoldi method.** This is based on the well-known Gram–Schmidt orthogonalization process [229]. The method reduces to the Lanczos method in the symmetric case. The big idea in this famous Gram–Schmidt method is to make full use of vector orthogonalities in a direct manner. The Gram–Schmidt method is also heavily used in other subjects involving *orthogonal polynomials* or *basis*.

Let $Q^T A Q = H$ or $AQ = QH$. Expanding both sides of the latter form gives

$$\begin{array}{l|lllllll}
\text{Column} \longmapsto 1 & Aq_1 & = q_1 h_{\mathbf{11}} & + q_2 h_{21} & & & \\
\text{Column} \longmapsto 2 & Aq_2 & = q_1 h_{12} & + q_2 h_{\mathbf{22}} & + q_3 h_{32} & & \\
\text{Column} \longmapsto 3 & Aq_3 & = q_1 h_{13} & + q_2 h_{23} & + q_3 h_{\mathbf{33}} & + q_3 h_{43} & \\
 & & \vdots & & & & \\
\text{Column} \longmapsto n-1 & Aq_{n-1} & = q_1 h_{1,n-1} & + q_2 h_{2,n-1} & + q_3 h_{3,n-1} & + \cdots & + q_n h_{n,n-1} \\
\text{Column} \longmapsto n & Aq_n & = q_1 h_{1,n} & + q_2 h_{2,n} & + q_3 h_{3,n} & + \cdots & + q_n h_{\mathbf{n,n}}, \\
\hline
 & & \uparrow & \uparrow & \uparrow & & \uparrow \\
\text{Observe } h_{i,j} & \Longrightarrow & [\text{row } 1] & [\text{row } 2] & [\text{row } 3] & & [\text{row } n]
\end{array}$$

where equation j involves column j elements of H. Using the orthogonality relation $q_i^T q_j = \delta_{ij}$, we obtain the recursive formula (the Arnoldi method).[7]

Algorithm 1.4.9.

(1) *Let q_1 be given ($q_1^T q_1 = 1$).*
(2) **for** $j = 1, 2, \cdots, n$

$$h_{k,j} = q_k^T A q_j (\text{for } k = 1, \cdots, j)$$

$$r_j = Aq_j - \sum_{k=1}^{j} q_k h_{k,j}, \quad h_{j+1,j} = \|r_j\|_2 = \sqrt{r_j^T r_j}$$

if $h_{j+1,j} \neq 0$, *then* $q_{j+1} = r_j / h_{j+1,j}$, **otherwise stop** *with* $m \equiv j$ *(rank of Krylov matrix $\mathcal{K}_n(A, q_1)$ found §3.4)*

end for $j = 1, 2, \cdots, n$

(3) *H_j may be applied with say QR method to find the eigenvalues of A.*

The algorithm is implemented in `hess_as.m`, which will produce a tridiagonal matrix H if A is symmetric (and then the Arnoldi method reduces to the Lanczos method [229]).

Although Methods 1, 2 can always find a Hessenberg matrix of size $n \times n$ with Q having a unit vector e_1 at its column 1, Method 3 can have a 'lucky' breakdown whenever r_j is 0 or too small. When this happens, due to some special choice of q_1, $\mathcal{K}_j(A, q_1) \equiv \mathcal{K}_n(A, q_1)$ and $\mathcal{K}_j$ defines an invariant subspace in $\mathbb{R}^n$ for A:

$$A(Q_j)_{n\times j} = Q_j (H_j)_{j\times j}, \qquad \text{or} \quad H_j = Q_j^T A Q_j.$$

Regardless of breakdown or not, at step j,

$$AQ_j = Q_j H_j + h_{j+1,j} q_{j+1} e_j^T, \tag{1.28}$$

the special matrix H_j has interesting properties resembling that of A (e.g. eigenvalues). In particular, the eigenvalues θ_k and eigenvectors s_k of H_j are usually called the *Ritz values* and *Ritz vectors* respectively:

$$H_j s_k = \theta_k s_k. \tag{1.29}$$

If $h_{j+1,j}$ information is also used, the Harmonic Ritz values $\hat{s}_k$ and vectors $\hat{\theta}_k$ are defined similarly by

$$\left(H_j^T H_j + h_{j+1,j}^2 e_j e_j^T\right)\hat{s}_k = \hat{\theta}_k H_j^T \hat{s}_k. \tag{1.30}$$

[7] The definition for Krylov subspace $\mathcal{K}$ is $\mathcal{K}_\ell(A, q) = \mathtt{span}\left(q, Aq, \ldots, A^{\ell-1}q\right)$. See 3.4 and 3.6.

As these Ritz values approximate eigenvalues and the approximation improves with increasing size j, there have been a lot of recent work on these Ritz-related methods for many purposes; see [331,305,359]. For symmetric matrices, the Arnoldi method reduces to the Lanczos method with H_j replaced by a symmetric tridiagonal T_j and those $\lambda(T_j)$ can approximate the extreme eigenvalues of A very well; see [383].

We make an early remark on the eigenvalue connection to the context of preconditioning via deflation. As the approximations from Ritz values are not accurate, it is not entirely clear how to make use this information to design preconditioners for (1.2); a direct application via deflation may not be appropriate [138].

1.5 Clustering characterization, field of values and ϵ-pseudospectrum

In the theory of iterative methods and preconditioners, we often use different measures of the eigenspectrum or the singular value spectrum to indicate levels of expected fast convergence. The key measure is clustering.

By 'cluster' or 'clustering' we mean that there are a large number of eigenvalues (or singular values) that are inside a small interval [384] or close to a fixed point [28]. If we define, for any $\mu_1 \le \mu_2$, a complex row vector set by

$$\Upsilon_{[\ell,\mu_2]}^{[n_1,\mu_1]} = \left\{ \mathbf{a}^T \;\middle|\; \mathbf{a} = \begin{pmatrix} a_1 \\ \vdots \\ a_n \end{pmatrix} \in \mathbb{C}^n, \quad \begin{matrix} |a_j - \ell| \le \mu_2 \text{ for } j \ge 1 \text{ and} \\ |a_k - \ell| \le \mu_1 \text{ for } k \ge n_1 \end{matrix} \right\},$$

then a more precise statement can be made as follows.

Definition 1.5.10. *Given a square matrix $A_{n\times n}$, if $\Lambda(A) \in \Upsilon_{[\ell,\mu_2]}^{[n_1,\mu_1]}$ for some relatively small n_1 (with respect to n), we say $\Lambda(A)$ is clustered at point ℓ with a cluster size μ_1 and cluster radius μ_2.*

Here μ_2 is the radius of a disc, centering at ℓ, containing all the eigenvalues and μ_1 is the radius of a smaller disc that contains most of the eigenvalues (i.e., all eigenvalues except the first $n_1 - 1$).

Remark 1.5.11. As far as convergence of conjugate gradients methods is concerned, point clusterings imply that at step n_1 the underlying approximation in $\mathtt{span}\left(q_1, \ldots, q_{n_1}\right)$ (in Section 3.6 and Section 3.5) is almost as accurate as in $\mathtt{span}(q_1, \ldots, q_n)$. In this sense, both condition number estimates (popular in the literature) and interval clusterings are not as effective as point clusterings (Definition 1.5.10) in measuring convergence.

In view of [364,297], it is might be more suitable but not difficult to define a concept of multiple clustering for spectrum $\Lambda(A)$ or $\Sigma(A)$.

Compactness and eigenvalue clustering. We recall that for a compact operator defined in a Hilbert space, its eigenvalues cluster at zero. For compact $\mathcal{K}$, the adjoint $\mathcal{K}^*$ and product $\mathcal{K}^*\mathcal{K}$ are also compact. Therefore in this case for A, $\lambda(A) \in \Upsilon^{[n_1,\mu_1]}_{[\alpha,\mu_2]}$ and $\lambda(A^*A) \in \Upsilon^{[n_1,\mu_1]}_{[|\alpha|^2,\mu_2]}$, i.e., clustered at $\alpha = 0$ and $|\alpha|^2 = 0$, respectively, with μ_1 arbitrarily small for some suitable and fixed n_1. The use of this operator idea will be picked up again in Chapter 4. Further the superlinear convergence of both CGN and GMRES has been established in [493] and [358], respectively.

Before we complete this section, we must comment on the suitability of eigenvalue clustering (singular value clustering is always meaningful and suitable regardless of A's normality). This is because, in addition to condition number $\kappa(A)$, eigenvalues with eigenspectrum $\Lambda(A)$ and singular values with spectrum $\Sigma(A)$, two other properties of a matrix A namely field of values (FoV) with their spectrum denoted by $\mathcal{W}(A)$ and ϵ-pseudospectra denoted by $\Lambda_\epsilon(A)$, are useful in the context of iterative methods and preconditioning techniques. We here define and discuss these concepts to conclude that eigenvalue clustering may only be meaningful in preconditioners' design if non-normality is first tested.

Definition 1.5.12. (Field of values). *The field of values of a matrix $A \in \mathbb{C}^{n\times n}$ is defined as the set of all Rayleigh quotients*

$$\mathcal{W}(A) = \left\{ \frac{x^H Ax}{x^H x} \;\middle|\; x \neq 0 \in \mathbb{C}^n \right\}$$

with the numerical radius $\mu(A) = \max\{|z| \mid z \in \mathcal{W}(A)\}$.

Observe that

$$\Lambda(A) \subset \mathcal{W}(A) \tag{1.31}$$

so $\rho(A) \leq \mu(A)$. From the following

$$\left|\frac{x^H Ax}{x^H x}\right| \leq \frac{\|x\|_2\|Ax\|_2}{x^H x} = \frac{\|Ax\|_2}{\|x\|_2} = \frac{(x^H A^H Ax)^{1/2}}{\|x\|_2} \leq \rho(A^H A)^{1/2}\frac{(x^H x)^{1/2}}{\|x\|_2}$$
$$= \|A\|_2$$

we see that $\mu(A) \leq \|A\|_2$. Further one can prove [243,256] that $\|A\|_2 \leq 2\mu(A)$ so we have

$$\rho(A) \leq \mu(A) \leq \|A\|_2 \leq 2\mu(A). \tag{1.32}$$

If A is diagonalizable i.e. $A = V \Lambda V^{-1}$, then

$$\mu(A) \leq \|V\|_2 \|V^{-1}\|_2 \|\Lambda\|_2 = \kappa_2(V) \rho(A). \tag{1.33}$$

As is known, for unsymmetric and highly non-normal matrices, eigenvalues may not represent the correct behaviour of matrix A as they become seriously unstable e.g. with $\mathbf{n}$ a 'random' matrix with entries in $(0, 0.01)$, for

$$A = \begin{pmatrix} 2 & 1 & 0 & 0 \\ 0 & 1 & 20 & 0 \\ 0 & 0 & 2 & 300 \\ 0 & 0 & 0 & 4 \end{pmatrix},$$

$$B = A + \mathbf{n} = \begin{pmatrix} 2.0090 & 1.0041 & 0.0016 & 0.0014 \\ 0.0043 & 1.0013 & 20.0007 & 0.0078 \\ 0.0014 & 0.0089 & 2.0037 & 300.0046 \\ 0.0095 & 0.0009 & 0.0025 & 4.0035 \end{pmatrix}, \tag{1.34}$$

see the remarkable differences in the eigenvalues:

$$\begin{aligned} \lambda(A) &= [\quad 2 \quad\quad 1 \quad\quad 2 \quad\quad 4\ ] \\ \lambda(B) &= [\ 5.5641,\ 2.0129 + 2.4928i,\ 2.0129 - 2.4928i,\ -0.5725\ ]. \end{aligned}$$

The dilemma is that, when what you have obtained is just $\lambda(B)$, one could not use it to describe the behaviour of matrix A (or vice versa). Of course, singular values would give correct information as

$$\begin{aligned} \sigma(A) &= [\ 300.0334,\ 20.0246,\ 2.2355,\ 0.0012\ ] \\ \sigma(B) &= [\ 300.0380,\ 20.0253,\ 2.2454,\ 0.0025\ ]. \end{aligned}$$

Here we present another eigenvalues related concept, the so-called ϵ-pseudospectrum [456,366,457], which is a suitable description of eigensystems (especially when A is non-normal).

Definition 1.5.13. (ϵ-pseudospectra). *The 2-norm ϵ-pseudospectra of a matrix $A \in \mathbb{C}^{n \times n}$ is the following*

$$\Lambda_\epsilon(A) = \left\{ z \;\middle|\; z \in \mathbb{C}^n,\ \left\|(zI - A)^{-1}\right\|_2 \geq \frac{1}{\epsilon} \right\}$$

Note that if $z = \lambda(A)$, then $det(zI - A) = 0$ and $\|(zI - A)^{-1}\|_2 = \infty$ so $\lambda(A) \in \Lambda_\epsilon(A)$ and $\Lambda(A) \subset \Lambda_\epsilon(A)$ for any $\epsilon > 0$; therefore $\Lambda_\epsilon(A)$ will be able to contain all other values of z satisfying $det(zI - A) \approx 0$.[8] This connection to nearby eigenvalues of A can be formally established [457] to give an alternative

[8] Here one should not use the unscaled determinant $det(B)$ to judge the magnitude of B as this is highly unreliable. e.g. $det(B) = 10^{-n}$ if $B =$ diag([0.1, ..., 0.1]). Refer to Chapter 14.

to Definition 1.5.13:

$$\Lambda_\epsilon(A) = \{z \mid z \in \Lambda(A+E), \ \|E\|_2 \le \epsilon\}.$$

For non-normal matrices, eigenvectors play a dominating role! Even if $\Lambda(A)$ appears to be 'nicely' distributed (or clustered Section 1.5), the matrix V of A's eigenvectors may have a large condition number e.g. $cond(V) = \infty$ for the example in (1.34). If $\Lambda_\epsilon(A)$ is capable of representing A, it should be related to $cond(V) = \kappa(V)$ somehow; indeed one such result has been given in [457]:

$$\Lambda_\epsilon(A) \subseteq \Lambda(A) \bigcup \Delta_r, \tag{1.35}$$

where $r = \epsilon\kappa_2(V)$ and $\Delta_r = \{z \mid |z| \le r, \ z \in \mathbb{C}^n\}$. It is also interesting to point out that the ϵ-pseudospectrum is also closely related to the FoV spectrum [198,457]:

$$\Lambda_\epsilon(A) \subseteq \mathcal{W}(A) \bigcup \Delta_\epsilon. \tag{1.36}$$

Clearly from (1.31), (1.35) and (1.36), for a given unsymmetric matrix MA in (1.2), the eigenspectrum is only 'trustful' if its distance from both $\mathcal{W}(MA)$ and $\Lambda_\epsilon(MA)$ is not large. One hopes that a good preconditioner might do that.

1.6 Fast Fourier transforms and fast wavelet transforms

Both the Fourier analysis and wavelet analysis provide us a chance to transform a given problem (usually defined in a space of piecewise polynomials) to a new problem in a different functional space [482,481]. At a matrix level, however, both the fast Fourier transforms and the fast wavelet transforms are indispensable tools for certain applications. Here we give a short introduction.

1.6.1 The fast Fourier transform

The fast Fourier transform (FFT) represents a fast method of implementing the same discrete Fourier transform (DFT) [465,310]. The DFT comes about from an attempt of representing a general (and maybe nonperiodic) function $f(t)$, only available at n equally-spaced discrete points $\{f_k\}_0^{n-1} = \{f(kL)\}_0^{n-1}$, by trigonometric functions just as continuous periodic functions are represented by Fourier series or nonperiodic ones by Fourier transforms [300].

One may either use the so-called semi-discrete Fourier transforms by embedding these points into an infinite and periodic sequence, or simply use the

Fourier transforms for the sampled function: $f_s(t) = \sum_{k=0}^{N-1} f(kL)\delta(t - kL)$, where the usual Delta function δ indicates a local pulse. Then the Fourier transform for $f_s(t)$ is

$$F_s(\omega) = \int_{-\infty}^{\infty} f_s(t)e^{-i\omega t}dt = \sum_{k=0}^{n-1} f_k e^{-ik\omega L},$$

where $i = \sqrt{-1}$. Note that $F_s(\omega)$ is periodic with period $2\pi/L$ (or periodic in variable ωL with period 2π) with n coefficients so it is sufficient to consider n equally-spaced samples of $F_s(\omega)$ in interval $[0, 2\pi/L]$ i.e. take $\{F_j\}_{j=0}^{n-1} = \{F_s(j\Delta\omega)\}_{j=0}^{n-1}$ with

$$n\Delta\omega = \frac{2\pi}{L}, \qquad \text{i.e.} \qquad (\omega_j L) = j\Delta\omega L = \frac{2jL\pi}{nL} = \frac{2j\pi}{n}.$$

Therefore the Discrete Fourier Transform (DFT) for sequence $\{f_k\}_0^{n-1}$ is

$$F_j = \sum_{k=0}^{n-1} f_k e^{-\frac{2\pi kji}{n}} = \sum_{k=0}^{n-1} f_k \exp\left(-\frac{2\pi kji}{n}\right), \qquad \text{for } j = 0, 1, \ldots, n-1. \tag{1.37}$$

To put (1.37) into matrix form, we define the usual notation (note that $\omega_n^n = 1$)

$$\omega_n = \exp(-2\pi i/n) = \cos(2\pi i/n) - i\sin(2\pi i/n).$$

Therefore the DFT equation (1.37) becomes

$$\mathbf{g} = \begin{pmatrix} F_1 \\ F_1 \\ \vdots \\ F_n \end{pmatrix} = \begin{pmatrix} 1 & 1 & 1 & \cdots & 1 \\ 1 & \omega_n & \omega_n^2 & \cdots & \omega_n^{(n-1)} \\ 1 & \omega_n^2 & \omega_n^4 & \cdots & \omega_n^{2(n-1)} \\ \vdots & \vdots & \vdots & \ddots & \vdots \\ 1 & \omega_n^{(n-1)} & \omega_n^{2(n-1)} & \cdots & \omega_n^{(n-1)^2} \end{pmatrix} \begin{pmatrix} f_1 \\ f_1 \\ \vdots \\ f_n \end{pmatrix} = \mathbf{F}_n\mathbf{f}. \tag{1.38}$$

Clearly carrying out a DFT amounts to a matrix vector multiplication which takes $2n^2$ operations. Noting that the pth row of $\mathbf{F}_n^H$ and the qth column of $\mathbf{F}_n$ are respectively (with $\bar{\omega}_n = 1/\omega_n = \exp(2\pi i/n)$)

$$\left[1,\ \bar{\omega}_n^{p-1},\ \bar{\omega}_n^{2(p-1)},\ \bar{\omega}_n^{2(p-1)},\ \cdots,\ \bar{\omega}_n^{(n-1)(p-1)}\right],$$

$$\left[1,\ \omega_n^{q-1},\ \omega_n^{2(q-1)},\ \omega_n^{2(q-1)},\ \cdots,\ \omega_n^{(n-1)(q-1)}\right],$$

one can show that the DFT matrix $\mathbf{F}_n/\sqrt{n}$ is unitary via $\mathbf{F}_n^H\mathbf{F}_n = nI_n$ i.e. $\mathbf{F}_n^{-1} = \mathbf{F}_n^H/n$. This implies that to obtain the inverse DFT, we set $\omega_n = \exp(2\pi i/n)$

and divide the transformed result by n i.e.

$$f_k = \frac{1}{n}\sum_{j=0}^{n-1} F_j \exp\left(\frac{2\pi kji}{n}\right), \qquad \text{for } j = 0, 1, \ldots, n-1. \tag{1.39}$$

One may also use the idea of a sampled function $F_s(\omega)$ to derive (1.39) [300].

To introduce the FFT, consider $n = 8$ case and focus on its odd and even columns separately (noting $\omega_8^2 = \omega_4$ or more generally $\omega_n^{2k} = \omega_{n/2}^k$).

$$\mathbf{F}_8 = \begin{pmatrix} 1 & 1 & 1 & 1 & 1 & 1 & 1 & 1 \\ 1 & \omega_8^1 & \omega_8^2 & \omega_8^3 & \omega_8^4 & \omega_8^5 & \omega_8^6 & \omega_8^7 \\ 1 & \omega_8^2 & \omega_8^4 & \omega_8^6 & \omega_8^8 & \omega_8^{10} & \omega_8^{12} & \omega_8^{14} \\ 1 & \omega_8^3 & \omega_8^6 & \omega_8^9 & \omega_8^{12} & \omega_8^{15} & \omega_8^{18} & \omega_8^{21} \\ 1 & \omega_8^4 & \omega_8^8 & \omega_8^{12} & \omega_8^{16} & \omega_8^{20} & \omega_8^{24} & \omega_8^{28} \\ 1 & \omega_8^5 & \omega_8^{10} & \omega_8^{15} & \omega_8^{20} & \omega_8^{25} & \omega_8^{30} & \omega_8^{35} \\ 1 & \omega_8^6 & \omega_8^{12} & \omega_8^{18} & \omega_8^{24} & \omega_8^{30} & \omega_8^{36} & \omega_8^{42} \\ 1 & \omega_8^7 & \omega_8^{14} & \omega_8^{21} & \omega_8^{28} & \omega_8^{35} & \omega_8^{42} & \omega_8^{49} \end{pmatrix}$$

$$= \begin{pmatrix} 1 & 1 & 1 & 1 & 1 & 1 & 1 & 1 \\ 1 & \omega_8^1 & \omega_4^1 & \omega_8^3 & \omega_4^2 & \omega_8^5 & \omega_4^3 & \omega_8^7 \\ 1 & \omega_8^2 & \omega_4^2 & \omega_8^6 & \omega_4^4 & \omega_8^{10} & \omega_4^6 & \omega_8^{14} \\ 1 & \omega_8^3 & \omega_4^3 & \omega_8^9 & \omega_4^6 & \omega_8^{15} & \omega_4^9 & \omega_8^{21} \\ 1 & \omega_8^4 & \omega_4^4 & \omega_8^{12} & \omega_4^8 & \omega_8^{20} & \omega_4^{12} & \omega_8^{28} \\ 1 & \omega_8^5 & \omega_4^5 & \omega_8^{15} & \omega_4^{10} & \omega_8^{25} & \omega_4^{15} & \omega_8^{35} \\ 1 & \omega_8^6 & \omega_4^6 & \omega_8^{18} & \omega_4^{12} & \omega_8^{30} & \omega_4^{18} & \omega_8^{42} \\ 1 & \omega_8^7 & \omega_4^7 & \omega_8^{21} & \omega_4^{14} & \omega_8^{35} & \omega_4^{21} & \omega_8^{49} \end{pmatrix}$$

Firstly the odd columns present two 4×4 blocks (noting $\omega_4^4 = \omega_4^8 = \omega_4^{12} = 1$):

$$\mathbf{F}_8(\text{odd columns}) = \begin{pmatrix} 1 & 1 & 1 & 1 \\ 1 & \omega_4^1 & \omega_4^2 & \omega_4^3 \\ 1 & \omega_4^2 & \omega_4^4 & \omega_4^6 \\ 1 & \omega_4^3 & \omega_4^6 & \omega_4^9 \\ 1 & \omega_4^4 & \omega_4^8 & \omega_4^{12} \\ 1 & \omega_4^5 & \omega_4^{10} & \omega_4^{15} \\ 1 & \omega_4^6 & \omega_4^{12} & \omega_4^{18} \\ 1 & \omega_4^7 & \omega_4^{14} & \omega_4^{21} \end{pmatrix} = \begin{pmatrix} 1 & 1 & 1 & 1 \\ 1 & \omega_4^1 & \omega_4^2 & \omega_4^3 \\ 1 & \omega_4^2 & \omega_4^4 & \omega_4^6 \\ 1 & \omega_4^3 & \omega_4^6 & \omega_4^9 \\ 1 & 1 & 1 & 1 \\ 1 & \omega_4^1 & \omega_4^2 & \omega_4^3 \\ 1 & \omega_4^2 & \omega_4^4 & \omega_4^6 \\ 1 & \omega_4^3 & \omega_4^6 & \omega_4^9 \end{pmatrix} = \begin{pmatrix} \mathbf{F}_4 \\ \mathbf{F}_4 \end{pmatrix}$$

Secondly noting $\omega_8^8 = 1$, $\omega_8^4 = \omega_8^{12} = \omega_8^{20} = \omega_8^{28} = -1$ and $\omega_8^5 = -\omega_8$, $\omega_8^6 = -\omega_8^2$ etc:

$$\mathbf{F}_8(\text{even columns}) = \begin{pmatrix} 1 & 1 & 1 & 1 \\ \omega_8^1 & \omega_8^3 & \omega_8^5 & \omega_8^7 \\ \omega_8^2 & \omega_8^6 & \omega_8^{10} & \omega_8^{14} \\ \omega_8^3 & \omega_8^9 & \omega_8^{15} & \omega_8^{21} \\ \omega_8^4 & \omega_8^{12} & \omega_8^{20} & \omega_8^{28} \\ \omega_8^5 & \omega_8^{15} & \omega_8^{25} & \omega_8^{35} \\ \omega_8^6 & \omega_8^{18} & \omega_8^{30} & \omega_8^{42} \\ \omega_8^7 & \omega_8^{21} & \omega_8^{35} & \omega_8^{49} \end{pmatrix} = \begin{pmatrix} 1 & 1 & 1 & 1 \\ \omega_8^1 & \omega_8^3 & \omega_8^5 & \omega_8^7 \\ \omega_8^2 & \omega_8^6 & \omega_8^{10} & \omega_8^{14} \\ \omega_8^3 & \omega_8^9 & \omega_8^{15} & \omega_8^{21} \\ -1 & -1 & -1 & -1 \\ -\omega_8^1 & -\omega_8^3 & -\omega_8^5 & -\omega_8^7 \\ -\omega_8^2 & -\omega_8^6 & -\omega_8^{10} & -\omega_8^{14} \\ -\omega_8^3 & -\omega_8^9 & -\omega_8^{15} & -\omega_8^{21} \end{pmatrix}$$

$$= \begin{pmatrix} \mathbf{1} &&&&&&& \\ & \omega_8 &&&&&& \\ && \omega_8^2 &&&&& \\ &&& \omega_8^3 &&&& \\ &&&& \mathbf{-1} &&& \\ &&&&& -\omega_8 && \\ &&&&&& -\omega_8^2 & \\ &&&&&&& -\omega_8^3 \end{pmatrix} \begin{pmatrix} 1 & 1 & 1 & 1 \\ 1 & \omega_4^1 & \omega_4^2 & \omega_4^3 \\ 1 & \omega_4^2 & \omega_4^4 & \omega_4^6 \\ 1 & \omega_4^3 & \omega_4^6 & \omega_4^9 \\ 1 & 1 & 1 & 1 \\ 1 & \omega_4^1 & \omega_4^2 & \omega_4^3 \\ 1 & \omega_4^2 & \omega_4^4 & \omega_4^6 \\ 1 & \omega_4^3 & \omega_4^6 & \omega_4^9 \end{pmatrix}$$

$$= \begin{pmatrix} \Omega_4 \mathbf{F}_4 \\ -\Omega_4 \mathbf{F}_4 \end{pmatrix},$$

where Ω_4 is a 4×4 diagonal matrix

$$\Omega_4 = \begin{pmatrix} 1 &&& \\ & \omega_8 && \\ && \omega_8^2 & \\ &&& \omega_8^3 \end{pmatrix} \quad \text{defined in terms of } \omega_8, \text{ not of } \omega_4.$$

Assuming Π_8 is a permutation matrix swapping odd even columns, then

$$\mathbf{F}_8 \Pi_8 = [\mathbf{F}_8(\text{odd columns}) \;\; \mathbf{F}_8(\text{even columns})] = \begin{pmatrix} I_4 & \Omega_4 \\ I_4 & -\Omega_4 \end{pmatrix} \begin{pmatrix} \mathbf{F}_4 & \\ & \mathbf{F}_4 \end{pmatrix},$$

where, using the MATLAB notation,[9] $\Pi_8 = I_8(:, r_8 + 1)$ with I_8 is the 8×8 identity matrix and $r_8 = [0 : 2 : 6 \;\; 1 : 2 : 7]$. That is, the DFT matrix can be

[9] In MATLAB, if A is an $n \times n$ matrix, then $B = A(:, 1 : 3)$ extracts the all the rows and only columns 1 to 3 of A. Likely $C = A(1 : 4, 2 : 2 : 10)$ extracts a block matrix of rows 1 to 4 and columns 2, 4, 6, 8, 10 of matrix A.

decomposed

$$\mathbf{F}_8 = \begin{pmatrix} I_4 & \Omega_4 \\ I_4 & -\Omega_4 \end{pmatrix} \begin{pmatrix} \mathbf{F}_4 & \\ & \mathbf{F}_4 \end{pmatrix} \Pi_8^T.$$

As the above process is completely general, the same result holds for a general n:

$$\mathbf{F}_n = \begin{pmatrix} I_{n/2} & \Omega_{n/2} \\ I_{n/2} & -\Omega_{n/2} \end{pmatrix} \begin{pmatrix} \mathbf{F}_{n/2} & \\ & \mathbf{F}_{n/2} \end{pmatrix} \Pi_n^T, \tag{1.40}$$

where $\Pi_n = I_n(:, r_n + 1)$, the order $r_n = [0 : 2 : n-2 \ 1 : 2 : n-1]$ and $\Omega_n = diag(1, \omega_{2n}, \ldots, \omega_{2n}^{n-1})$. Evidently $I_1 = 1$, $\Omega_1 = 1$, $r_2 = [0\ 1]$ so $\Pi_2 = I_2$ and $\mathbf{F}_1 = 1$. When $n = 2$, the above formula becomes

$$\mathbf{F}_2 = \begin{pmatrix} I_1 & \Omega_1 \\ I_1 & -\Omega_1 \end{pmatrix} \begin{pmatrix} \mathbf{F}_1 & \\ & \mathbf{F}_1 \end{pmatrix} \Pi_2^T = \begin{pmatrix} 1 & 1 \\ 1 & -1 \end{pmatrix}.$$

It only remains to find a more compact notation. The first notation is for a Kronecker product (or tensor product in Definition 14.3.3) $A \otimes W$ of two matrices, $A \in \mathbb{C}^{p\times q}$ and $W \in \mathbb{C}^{m\times n}$, is defined by

$$(A \otimes W)_{pm\times qn} = \begin{bmatrix} a_{0,0}W & a_{0,1}W & \cdots & a_{0,q-1}W \\ a_{1,0}W & a_{1,1}W & \cdots & a_{1,q-1}W \\ \vdots & \vdots & \ddots & \vdots \\ a_{p-1,0}W & a_{p-1,1}W & \cdots & a_{p-1,q-1}W \end{bmatrix}. \tag{1.41}$$

Observe that the Kronecker product $A \otimes W$ is essentially an $p \times q$ block matrix with its ij 'element' being $a_{i,j}W$. The second notation, for the above 2×2 block matrix with simple blocks of size $k/2 \times k/2$, is the butterfly matrix of size $k \times k$

$$B_k = \begin{pmatrix} I_{k/2} & \Omega_{k/2} \\ I_{k/2} & -\Omega_{k/2} \end{pmatrix} = \left(\begin{array}{cccc|cccc} 1 & & & & 1 & & & \\ & 1 & & & & \omega_k & & \\ & & \ddots & & & & \ddots & \\ & & & 1 & & & & \omega_k^{k/2-1} \\ \hline 1 & & & & -1 & & & \\ & 1 & & & & -\omega_k & & \\ & & \ddots & & & & \ddots & \\ & & & 1 & & & & -\omega_k^{k/2-1} \end{array}\right).$$

With these two notations and noting for square matrices $I \otimes (ABC) = (I \otimes A)(I \otimes B)(I \otimes C)$, we can rewrite equation (1.40) recursively as

$$
\begin{aligned}
\mathbf{F}_n \Pi_n &= B_n \begin{pmatrix} \mathbf{F}_{n/2} & \\ & \mathbf{F}_{n/2} \end{pmatrix} = B_n \Big(I_2 \otimes \mathbf{F}_{n/2} \Big) \\
&= B_n \Big[I_2 \otimes \Big(B_{n-1} (I_2 \otimes \mathbf{F}_{n/4}) \Pi_{n-1}^T \Big) \Big] \\
&= B_n \Big(I_2 \otimes B_{n/2} \Big) \Big(I_4 \otimes \mathbf{F}_{n/4} \Big) (I_2 \otimes \Pi_{n-1}^T) \\
&= B_n \Big(I_2 \otimes B_{n/2} \Big) \Big(I_4 \otimes B_{n/4} \Big) \Big(I_8 \otimes \mathbf{F}_{n/8} \Big) (I_4 \otimes \Pi_{n-2}^T)(I_2 \otimes \Pi_{n-1}^T) \\
&= \cdots,
\end{aligned} \tag{1.42}
$$

giving rise to the following (note $\mathbf{F}_2 = B_2$)

Theorem 1.6.14. (Cooley–Tukey FFT factorization).
If $n = 2^t$ (power of 2), then the DFT matrix $\mathbf{F}_n$ can be factorized as

$$
\begin{aligned}
\mathbf{F}_n P_n &= (I_{2^0} \otimes B_n)(I_{2^1} \otimes B_{n/2})(I_{2^2} \otimes B_{n/2^2}) \cdots (I_{2^{t-1}} \otimes B_{n/2^{t-1}}) \\
&= (I_1 \otimes B_n)(I_2 \otimes B_{n/2})(I_4 \otimes B_{n/4}) \cdots (I_{n/4} \otimes B_4)(I_{n/2} \otimes B_2),
\end{aligned}
$$

where the permutation matrix P_n is

$$
P_n = (I_1 \otimes \Pi_n)(I_2 \otimes \Pi_{n/2})(I_{2^2} \otimes \Pi_{n/2^2}) \cdots (I_{2^{t-1}} \otimes \Pi_{n/2^{t-1}}).
$$

Remark 1.6.15.

(1) Note that $\mathbf{F}_k \Pi_k$ *swaps the odd-even columns of* $\mathbf{F}_k$ *while* $\Pi_k^T \mathbf{F}_k$ *swaps the odd-even rows of* $\mathbf{F}_k$. In practice the transpose of the above permutation is used when using the FFT i.e. DFT (1.38)

$$
P_n^T = (I_{2^{t-1}} \otimes \Pi_{n/2^{t-1}})^T \cdots (I_{2^2} \otimes \Pi_{n/2^2})^T (I_2 \otimes \Pi_{n/2})^T (I_1 \otimes \Pi_n)^T,
$$

which can be done in the initial **permutation phase** while the main Kronecker products are implemented in the second **FFT phase**. In particular, $\Pi_{n/2^{t-1}} = \Pi_2 = I_2$ as there is no need to permute to odd-evens for size 2 vectors.

(2) Here the butterfly matrix B_k is more logic, with its elements essentially defined by $\omega_k = \exp(-2\pi i/k)$, than Ω_k awkwardly defined by ω_{2k}.

(3) Matrix P_n is essentially determined by a permutation vector r_n of size n i.e. $P_n = I(:, r_n + 1)$ or $P_n^T = I(r_n + 1, :)$. For a small n, one can afford to implement each permutation step by step to find r_n. However,

the better method for getting r_n is the so-called bit reversal algorithm [465]. Let $(b_0\ b_1\ \ldots\ b_{t-1})_2$ be the binary representation of integer $k \in [0, n-1]$. Here exactly t binary digits are needed, due to the fact $(111\ldots11)_2 = 2^0 + 2^1 + \cdots + 2^{t-1} = 2^t - 1 = n - 1$. Then by induction [465], one can show that the kth index in the permutation vector r_n satisfies

$$r_n(k) = (b_{t-1}b_{t-2}\cdots b_1 b_0)_2. \tag{1.43}$$

For example, when $n = 2^t = 2^5 = 32$, the 5th index $k = 5 = (00101)_2$ corresponds to the 5th index $r_{32}(5) = (10100)_2 = 20$. Similarly $r_{32}(11) = r_{32}\big((01011)_2\big) = (11010)_2 = 26$ and $r_{32}(14) = r_{32}\big((01110)_2\big) = (01110)_2 = 14$. Note the whole vector is

$$\begin{aligned} r_{32} = [0\ \ 16\ \ 8\ \ 24\ \ 4\ \ 20\ \ 12\ \ 28\ \ 2\ \ 18\ \ 10\ \ 26\ \ 6\ \ 22\ \ 14\ \ 30 \\ 1\ \ 17\ 9\ \ 25\ \ 5\ \ 21\ \ 13\ \ 29\ \ 3\ \ 19\ \ 11\ \ 27\ \ 7\ \ 23\ \ 15\ \ 31]. \end{aligned}$$

MATLAB can easily convert a decimal number a to a binary number b (and back)

```
>> a = 13,   b = dec2bin(a,4),   a_back = bin2dec(b)
   % Here b = (1101)
```

Therefore to implement an explicit bit reversal operation to $k = 13$, do the following

```
>> b=dec2bin(k,4), b1=b(4:-1:1), b2=bin2dec(b1)
   % b1=(1011) and b2=11
```

In Maple,[10] similar operations can be done as follows

```
>  a := 13;   b := convert(a,binary,decimal);
   # Here b = (1101)
>        a_back := convert(b,decimal,binary);
   # a_back = 13
>            a1 := convert(1011,decimal,binary);
   #      a1 = 11
>  #------------Cautions! [1,0,1,1] means 1101 and [3,1]
   means 13-----
>  b1 := convert(13,base,2);
   # giving [1,0,1,1]
>  b2 := convert([1,1,0,1],base,2,10);
```

[10] ©Waterloo Maple Software, inc. may be found at the Web site: http://www.maplesoft.on.ca/. See also the Appendix §C.8.

```
     # giving [1,1] => 11
>  b3 := convert([1,0,1,1],base,2,10);
     # giving [3,1] => 13
>  b4 := convert([1,0,1,0],base,2,10);
     # giving [5]   => 5
```

(4) Finally, despite the somewhat complicated mathematics associated with FFT, the implementation in MATLAB cannot be simpler:

```
>>  F = fft(f),   g = ifft(F)
    %% Vector f case
>>  B = transpose( fft(transpose(fft(A))) )
    %% Matrix A case
>>  g = transpose( ifft(transpose(ifft(B))) )
    %% (transpose NOT ')
```

Guess what g is in each case.

(5) We also remark that our discussion is on the standard FFT (the complex form that effectively involves both sine and cosine (trigonometric) functions). It is sometimes desirable to develop real trigonometric transforms; there exist four such transforms: The Discrete Sine Transform (DST), The Discrete Cosine Transform (DCT), The Discrete Sine Transform-II (DST-II), The Discrete Cosine Transform-II (DCT-II). See [441,465]. It is of interest to point that all fast algorithms of these transforms are based on the standard (complex) FFT.

Before we show an example of using DFT and FFT, we comment on complexity.

We can see clearly from Theorem 1.6.14 that, apart from P_n^T, $\mathbf{F}_n$ has t terms. Each term applying to a vector requires three arithmetic operations (ignoring the powers of ω_k). Therefore the overall cost from the **FFT phase** is $O(n)t = cn \log n$ with $c \approx 3$.

If the above mentioned bit reversal is directly used for the **permutation phase**, there is an additional cost of $O(n \log n)$. However, there are many competing algorithms that can work out r_n in $O(n)$ operations; here we mention two such algorithms. In each case, we try to focus on the manners in which the number of steps $t = \log n$ is cleverly avoided. Let $R[a]$ denote the bit reversal operator e.g. $(R[(10100)])_2 = (00101)_2$.

♦ **Method 1 — dimension doubling.** Observe that for $i = (b_{t-1}b_{t-2}\ldots b_1 b_0)_2$ and $r(i) = r_i = (b_0 b_1 \ldots b_{t-2} b_{t-1})_2$, the first two indices are quite easy i.e. $r(0) = 0 = (0\ldots0)_2$ and $r(1) = r_1 = (R[(0\ldots01)])_2 = (10\ldots0)_2 = 2^{t-1}$. The rest are arranged in $t-1$ sequential groups G_i, $i = 1, 2, \ldots, t-1$ of size $s_i = 2^i$ – hence the name *dimension doubling*. A similar method was independently proposed in [402]. The crucial result is the following

Theorem 1.6.16. *The bit reversal process can arranged in t groups, starting from group $G_0 = \{0, 1\}$ with $r(0) = 0$ and $r(1) = 2^{t-1}$. Then the indices of group G_k, $k = 1, 2, \ldots, t-1$, can be recursively generated from*

$$r(i + 2^k) = r(i) + 2^{t-1-k}, \qquad \textit{for } i = 0, 1, 2, \ldots, 2^k - 1.$$

Consequently the algorithm costs $O(n)$ operations from the estimate

$$\sum_{k=1}^{t-1} 2^k = 2(2^t - 2) = 2n - 4.$$

Proof. Note that for $i = 0, 1, 2, \ldots, 2^k - 1$,

$$\begin{aligned} i + 2^k = 2^k + (b_{t-1}\, b_{t-2} \cdots b_1\, b_0)_2 &= (\underbrace{0 \cdots 0}_{(t-k-1)\ digits}\ \underbrace{0\, b_{k-1} \cdots b_1\, b_0}_{(k+1)\ digits})_2 + 2^k \\ &= (\underbrace{0 \cdots 0}_{(t-k-1)\ digits}\ \underbrace{1\, b_{k-1} \cdots b_1\, b_0}_{(k+1)\ digits})_2. \end{aligned}$$

Refer to Table 1.1. So we have the required result

$$\begin{aligned} r(i + 2^k) = R[(r + 2^k)] &= (\underbrace{b_0\, b_1 \cdots b_{k-1}\, 1}_{(k+1)\ digits}\ \underbrace{0 \cdots 0}_{(t-k-1)\ digits})_2 \\ &= (\underbrace{b_0\, b_1 \cdots b_{k-1}\, 0}_{(k+1)}\underbrace{0 \cdots 0}_{(t-k-1)})_2 + (\underbrace{0 \cdots 0\, 1}_{(k+1)}\underbrace{0 \cdots 0}_{(t-k-1)})_2 \\ &= r(i) + 2^{t-k-1}. \end{aligned}$$

■

♦ **Method 2 — dimension squaring.** The second algorithm is due to [203]; see also [313]. Observe that the t digits in $i = (b_{t-1}\, b_{t-2} \cdots b_1\, b_0)_2$ can be equally divided into three parts (or two parts if t is even): the front, middle and end parts, i.e. $i = (f\ m\ e)_2$. Then all numbers in $i = 0, 1, \ldots, 2^t - 1$ can be split sequentially into $N = \sqrt{n/2} = 2^{(t-1)/2}$ groups (or simply $N = \sqrt{n} = 2^{t/2}$ if t is even), where in each group the f part is identical in all $\tau = n/N$ members and the e part is identical across all groups. Refer to Table 1.1. Moreover let *root*, of size N, denote the set of all e parts in group 1; one can check that the f part is indeed drawn from the set *root*.

Putting these observations and the fact $R[(i)_2] = R[(f\, m\, e)_2] = (R[e]\ m\ R[f])_2$ together, one can generate the new order $w = r_N$ for *root*,

Table 1.1. *Bit reversal methods 1 (left) and 2 (right) group indices differently to achieve efficiency, for $n = 2^4 = 16$ and $n = 2^5 = 32$. Here 'Bin' stands for 'binary' $[f\ m\ e]$ and 'Rev' for 'reverse binary $[R(e)\ m\ R(f)]$, where there is no m for $n = 16$.*

i	Bin	Rev	$r(i)$
0	000 0	0 000	0
1	000 1	1 000	8
2	001 0	0 100	4
3	001 1	1 100	12
4	01 00	00 10	2
5	01 01	10 10	10
6	01 10	01 10	6
7	01 11	11 10	14
8	1 000	000 1	1
9	1 001	100 1	9
10	1 010	010 1	5
11	1 011	110 1	13
12	1 100	001 1	3
13	1 101	101 1	11
14	1 110	011 1	7
15	1 111	111 1	15

i	Bin	Rev	$r(i)$
0	00 00	0000	0
1	00 01	1000	8
2	00 10	0100	4
3	00 11	1100	12
4	01 00	0010	2
5	01 01	1010	10
6	01 10	0110	6
7	01 11	1110	14
8	10 00	0001	1
9	10 01	1001	9
10	10 10	0101	5
11	10 11	1101	13
12	11 00	0011	3
13	11 01	1011	11
14	11 10	0111	7
15	11 11	1111	15

i	Bin	Rev	$r(i)$
0	0000 0	00000	0
1	0000 1	10000	16
2	0001 0	01000	8
3	0001 1	11000	24
4	001 00	00100	4
5	001 01	10100	20
6	001 10	01100	12
7	001 11	11100	28
8	01 000	00010	2
9	01 001	10010	18
10	01 010	01010	10
11	01 011	11010	26
12	01 100	00110	6
13	01 101	10110	22
14	01 110	01110	14
15	01 111	11110	30
16	1 0000	00001	1
17	1 0001	10001	17
18	1 0010	01001	9
19	1 0011	11001	25
20	1 0100	00101	5
21	1 0101	10101	21
22	1 0110	01101	13
23	1 0111	11101	29
24	1 1000	00011	3
25	1 1001	10011	19
26	1 1010	01011	11
27	1 1011	11011	27
28	1 1100	00111	7
29	1 1101	10111	23
30	1 1110	01111	15
31	1 1111	11111	31

i	Bin	Rev	$r(i)$
0	00 0 00	00 0 00	0
1	00 0 01	10 0 00	16
2	00 0 10	01 0 00	8
3	00 0 11	11 0 00	24
4	00 1 00	00 1 00	4
5	00 1 01	10 1 00	20
6	00 1 10	01 1 00	12
7	00 1 11	11 1 00	28
8	01 0 00	00 0 10	2
9	01 0 01	10 0 10	18
10	01 0 10	01 0 10	10
11	01 0 11	11 0 10	26
12	01 1 00	00 1 10	6
13	01 1 01	10 1 10	22
14	01 1 10	01 1 10	14
15	01 1 11	11 1 10	30
16	10 0 00	00 0 01	1
17	10 0 01	10 0 01	17
18	10 0 10	01 0 01	9
19	10 0 11	11 0 01	25
20	10 1 00	00 1 01	5
21	10 1 01	10 1 01	21
22	10 1 10	01 1 01	13
23	10 1 11	11 1 01	29
24	11 0 00	00 0 11	3
25	11 0 01	10 0 11	19
26	11 0 10	01 0 11	11
27	11 0 11	11 0 11	27
28	11 1 00	00 1 11	7
29	11 1 01	10 1 11	23
30	11 1 10	01 1 11	15
31	11 1 11	11 1 11	31

i.e. that for all $R[e]$; this index vector w may be generated by any other bit reversal algorithm for up to the cost $O(\sqrt{n}\log\sqrt{n}) \ll O(n)$ operations. So the idea is to use the new index vector $w = r_N$ for one group only, to work the rest of r_n. For both examples in Table 1.1, only the explicit bit reversal $R[e]$ for the root set $e = [00\ 01\ 10\ 11]$ is needed and the rest is worked out accordingly.

In details, for group 1 as $f = (0\ldots0)$, the order indices are obtained by simply multiplying w (for $R[e]$) by N since

$$W(i) = R[(i)_2] = (\underbrace{root(i)}_{e}\ \underbrace{m\ 0}_{m\ f})_2 = w(i)N \quad \text{for } i = 0, 1, \ldots, \tau - 1.$$

Once this is done, for any other group $k \geq 2$, it will be an adding job

$$R[(i)_2] = (\underbrace{root(i')}_{e}\ \underbrace{m\ root(k)}_{m\ f})_2 = W(i') + w(k),$$

where $i = (k-1)\tau + i',\ i' = 0, 1, \ldots, \tau - 1$. Overall, the computational cost is again remarkably about $2\tau N = 2n$ operations, where again N is the number of groups and τ is the number of members in each group; a computer code is listed in [203].

In summary, here is how the standard bit reversal algorithm, costing $O(n\ t) = O(n \log n)$ operations, is compared with Methods 1 – 2 having only $O(n)$:

$$\begin{aligned}
\text{Standard}:&\ \underbrace{O(n)}_{\text{Step } 1} + \underbrace{O(n)}_{\text{Step } 2} + \cdots + \underbrace{O(n)}_{\text{Step } t} &&= O(nt) = O(n\log n),\\
\text{Method 1}:&\ \underbrace{O(2)}_{\text{Step } 1} + \underbrace{O(2^2)}_{\text{Step } 2} + \cdots + \underbrace{O(2^{t-1})}_{\text{Step } t-1} &&= O(2^t) = O(n),\\
\text{Method 2}:&\ \underbrace{O(\sqrt{n})}_{\text{Step } 1} + \underbrace{O(\sqrt{n})}_{\text{Step } 2} + \cdots + \underbrace{O(\sqrt{n})}_{\text{Step } N} &&= O(n).
\end{aligned}$$

Example 1.6.17. (Computation of $\mathbf{g}_{16} = \mathbf{F}_{16}\mathbf{f}_{16}$ by FFT).
For a detailed example, consider $n = 16$ *and*

$$\mathbf{f}_{16} = \begin{pmatrix} 4\ 1\ 11\ 11\ 15\ 6\ 8\ 13\ 1\ 1\ 8\ 11\ 0\ 6\ 1\ 7 \end{pmatrix}^T$$

with $\omega_{16} = \exp(-2\pi i/n) = 0.9239 - 0.3827i$. *We again compute* $\mathbf{g}_{16} = \mathbf{F}_{16}\mathbf{f}_{16}$ *using the direct DFT and the fast FFT methods.*

Firstly, for the direct DFT, we form the 16×16 *matrix* $\mathbf{F}_{16}$ *and then compute* $\mathbf{g}_{16} = \mathbf{F}_{16}\mathbf{f}_{16}$. *To display* $\mathbf{F}_{16}$, *we define constants:* $k_1 = (1/2 - 1/2i)\sqrt(2)$, $k_2 = (-1/2 - 1/2i)\sqrt(2)$, $k_3 = (-1/2 + 1/2i)\sqrt(2)$, $k_4 = (1/2 + 1/2i)\sqrt(2)$. *Then the results are shown in Tables 1.2, 1.3.*

Table 1.2. *Example 1.6.17: analytical DFT matrix* $\mathbf{F}_{16}$.

$$
\mathbf{F}_{16} = \begin{pmatrix}
1 & 1 & 1 & 1 & 1 & 1 & 1 & 1 & 1 & 1 & 1 & 1 & 1 & 1 & 1 & 1 \\
1 & e^{-1/8i\pi} & k_1 & e^{-3/8i\pi} & -i & e^{-5/8i\pi} & k_2 & e^{-7/8i\pi} & -1 & e^{7/8i\pi} & k_3 & e^{5/8i\pi} & i & e^{3/8i\pi} & k_4 & e^{1/8i\pi} \\
1 & k_1 & -i & k_2 & -1 & k_3 & i & k_4 & 1 & k_1 & -i & k_2 & -1 & k_3 & i & k_4 \\
1 & e^{-3/8i\pi} & k_2 & e^{7/8i\pi} & i & e^{1/8i\pi} & k_1 & e^{-5/8i\pi} & -1 & e^{5/8i\pi} & k_4 & e^{-1/8i\pi} & -i & e^{-7/8i\pi} & k_3 & e^{3/8i\pi} \\
1 & -i & -1 & i & 1 & -i & -1 & i & 1 & -i & -1 & i & 1 & -i & -1 & i \\
1 & e^{-5/8i\pi} & k_3 & e^{1/8i\pi} & -i & e^{7/8i\pi} & k_4 & e^{-3/8i\pi} & -1 & e^{3/8i\pi} & k_1 & e^{-7/8i\pi} & i & e^{-1/8i\pi} & k_2 & e^{5/8i\pi} \\
1 & k_2 & i & k_1 & -1 & k_4 & -i & k_3 & 1 & k_2 & i & k_1 & -1 & k_4 & -i & k_3 \\
1 & e^{-7/8i\pi} & k_4 & e^{-5/8i\pi} & i & e^{-3/8i\pi} & k_3 & e^{-1/8i\pi} & -1 & e^{1/8i\pi} & k_2 & e^{3/8i\pi} & -i & e^{5/8i\pi} & k_1 & e^{7/8i\pi} \\
1 & -1 & 1 & -1 & 1 & -1 & 1 & -1 & 1 & -1 & 1 & -1 & 1 & -1 & 1 & -1 \\
1 & e^{7/8i\pi} & k_1 & e^{5/8i\pi} & -i & e^{3/8i\pi} & k_2 & e^{1/8i\pi} & -1 & e^{-1/8i\pi} & k_3 & e^{-3/8i\pi} & i & e^{-5/8i\pi} & k_4 & e^{-7/8i\pi} \\
1 & k_3 & -i & k_4 & -1 & k_1 & i & k_2 & 1 & k_3 & -i & k_4 & -1 & k_1 & i & k_2 \\
1 & e^{5/8i\pi} & k_2 & e^{-1/8i\pi} & i & e^{-7/8i\pi} & k_1 & e^{3/8i\pi} & -1 & e^{-3/8i\pi} & k_4 & e^{7/8i\pi} & -i & e^{1/8i\pi} & k_3 & e^{-5/8i\pi} \\
1 & i & -1 & -i & 1 & i & -1 & -i & 1 & i & -1 & -i & 1 & i & -1 & -i \\
1 & e^{3/8i\pi} & k_3 & e^{-7/8i\pi} & -i & e^{-1/8i\pi} & k_4 & e^{5/8i\pi} & -1 & e^{-5/8i\pi} & k_1 & e^{1/8i\pi} & i & e^{7/8i\pi} & k_2 & e^{-3/8i\pi} \\
1 & k_4 & i & k_3 & -1 & k_2 & -i & k_1 & 1 & k_4 & i & k_3 & -1 & k_2 & -i & k_1 \\
1 & e^{1/8i\pi} & k_4 & e^{3/8i\pi} & i & e^{5/8i\pi} & k_3 & e^{7/8i\pi} & -1 & e^{-7/8i\pi} & k_2 & e^{-5/8i\pi} & -i & e^{-3/8i\pi} & k_1 & e^{-1/8i\pi}
\end{pmatrix} = \begin{pmatrix} \mathbf{F}_a \\ \mathbf{F}_b \end{pmatrix}
$$

Table 1.3. *Example 1.6.17: numerical DFT matrix* $\mathbf{F}_{16} = (\mathbf{F}_a^H \ \mathbf{F}_b^H)^H$.

where

$$\mathbf{F}_a = \begin{pmatrix} 1 & 1 & 1 & 1 & 1 & 1 & 1 & 1 \\ 1 & 0.9239-0.3827i & 0.7071-0.7071i & 0.3827-0.9239i & -i & -0.3827-0.9239i & -0.7071-0.7071i & -0.9239-0.3827i \\ 1 & 0.7071-0.7071i & -i & -0.7071-0.7071i & -1 & -0.7071+0.7071i & i & 0.7071+0.7071i \\ 1 & 0.3827-0.9239i & -0.7071-0.7071i & -0.9239+0.3827i & i & 0.9239+0.3827i & 0.7071-0.7071i & -0.3827-0.9239i \\ 1 & -i & -1 & i & 1 & -i & -1 & i \\ 1 & -0.3827-0.9239i & -0.7071+0.7071i & 0.9239+0.3827i & -i & -0.9239+0.3827i & 0.7071+0.7071i & 0.3827-0.9239i \\ 1 & -0.7071-0.7071i & i & 0.7071-0.7071i & -1 & 0.7071+0.7071i & -i & -0.7071+0.7071i \\ 1 & -0.9239-0.3827i & 0.7071+0.7071i & -0.3827-0.9239i & i & 0.3827-0.9239i & -0.7071+0.7071i & 0.9239-0.3827i \\ 1 & -1 & 1 & -1 & 1 & -1 & 1 & -1 \\ 1 & -0.9239+0.3827i & 0.7071-0.7071i & -0.3827+0.9239i & -i & 0.3827+0.9239i & -0.7071-0.7071i & 0.9239+0.3827i \\ 1 & -0.7071+0.7071i & -i & 0.7071+0.7071i & -1 & 0.7071-0.7071i & i & -0.7071-0.7071i \\ 1 & -0.3827+0.9239i & -0.7071-0.7071i & 0.9239-0.3827i & i & -0.9239-0.3827i & 0.7071-0.7071i & 0.3827+0.9239i \\ 1 & i & -1 & -i & 1 & i & -1 & -i \\ 1 & 0.3827+0.9239i & -0.7071+0.7071i & -0.9239-0.3827i & -i & 0.9239-0.3827i & 0.7071+0.7071i & -0.3827+0.9239i \\ 1 & 0.7071+0.7071i & i & -0.7071+0.7071i & -1 & -0.7071-0.7071i & -i & 0.7071-0.7071i \\ 1 & 0.9239+0.3827i & 0.7071+0.7071i & 0.3827+0.9239i & i & -0.3827+0.9239i & -0.7071+0.7071i & -0.9239+0.3827i \end{pmatrix}$$

$$\mathbf{F}_b = \begin{pmatrix} 1 & 1 & 1 & 1 & 1 & 1 & 1 & 1 \\ -1 & -0.9239+0.3827i & -0.7071+0.7071i & -0.3827+0.9239i & i & 0.3827+0.9239i & 0.7071+0.7071i & 0.9239+0.3827i \\ 1 & 0.7071-0.7071i & -i & -0.7071-0.7071i & -1 & -0.7071+0.7071i & i & 0.7071+0.7071i \\ -1 & -0.3827+0.9239i & 0.7071+0.7071i & 0.9239-0.3827i & -i & -0.9239-0.3827i & -0.7071+0.7071i & 0.3827+0.9239i \\ -1 & 0.3827+0.9239i & 0.7071-0.7071i & -0.9239-0.3827i & i & 0.9239-0.3827i & -0.7071-0.7071i & -0.3827+0.9239i \\ 1 & -i & -1 & i & 1 & -i & -1 & i \\ 1 & -0.7071-0.7071i & i & 0.7071-0.7071i & -1 & 0.7071+0.7071i & -i & -0.7071+0.7071i \\ -1 & 0.9239+0.3827i & -0.7071-0.7071i & 0.3827+0.9239i & -i & -0.3827+0.9239i & 0.7071-0.7071i & -0.9239+0.3827i \\ 1 & -1 & 1 & -1 & 1 & -1 & 1 & -1 \\ -1 & 0.9239-0.3827i & -0.7071+0.7071i & 0.3827-0.9239i & i & -0.3827-0.9239i & 0.7071+0.7071i & -0.9239-0.3827i \\ 1 & -0.7071+0.7071i & -i & 0.7071+0.7071i & -1 & 0.7071-0.7071i & i & -0.7071-0.7071i \\ -1 & 0.3827-0.9239i & 0.7071+0.7071i & -0.9239+0.3827i & -i & 0.9239+0.3827i & -0.7071+0.7071i & -0.3827-0.9239i \\ 1 & i & -1 & -i & 1 & i & -1 & -i \\ -1 & -0.3827-0.9239i & 0.7071-0.7071i & 0.9239+0.3827i & i & -0.9239+0.3827i & -0.7071-0.7071i & 0.3827-0.9239i \\ 1 & 0.7071+0.7071i & i & -0.7071+0.7071i & -1 & -0.7071-0.7071i & -i & 0.7071-0.7071i \\ -1 & -0.9239-0.3827i & -0.7071-0.7071i & -0.3827-0.9239i & -i & 0.3827-0.9239i & 0.7071-0.7071i & 0.9239-0.3827i \end{pmatrix}$$

$$\mathbf{g}_{16} = (\mathbf{g}_a \quad \mathbf{g}_b)^T$$

$$\mathbf{g}_a = (104 \quad -5.3717-24.3672i \quad -18.4853-4.3431i \quad 3.5323+2.3857i \quad -8+28i \quad 8.1245-13.4722i \quad -1.5147+15.6569i \quad 5.7149+19.7750i)$$

$$\mathbf{g}_b = (-8 \quad 5.7149-19.7750i \quad -1.5147-15.6569i \quad 8.1245+13.4722i \quad -8-28i \quad 3.5323-2.3857i \quad -18.4853+4.3431i \quad -5.3717+24.3672i) .$$

Secondly, to compare with the above DFT results, we next compute the same transform using the step-by-step FFT algorithm. As $n = 16 = 2^4$, after the initial permutation, we only need $t = \log_2(n) = 4$ steps.

Permutation phase. *Directly starting from the natural sequence* $0:15$, *we first obtain in turn*

$$\begin{aligned} r_2 &= [0\ \ 2\ \ 4\ \ 6\ \ 8\ \ 10\ \ 12\ \ 14\ \ 1\ \ 3\ \ 5\ \ 7\ \ 9\ \ 11\ \ 13\ \ 15] \\ r_4 &= [0\ \ 4\ \ 8\ \ 12\ \ 2\ \ 6\ \ 10\ \ 14\ \ 1\ \ 5\ \ 9\ \ 13\ \ 3\ \ 7\ \ 11\ \ 15] \\ r_{16} &= [0\ \ 8\ \ 4\ \ 12\ \ 2\ \ 10\ \ 6\ \ 14\ \ 1\ \ 9\ \ 5\ \ 13\ \ 3\ \ 11\ \ 7\ \ 15]. \end{aligned}$$

To illustrate the above Method 1 for getting $r = r_n = r_{16}$, *we start with* $t = 4$, $r(0) = 0$ *and* $r(1) = n/2 = 8$:

$$\begin{array}{rl} \text{Group } k = 1: & r(2) = r(0+2^k) = r(0) + 2^{t-1-1} = 0+4 \ = 4 \\ & r(3) = r(1+2^k) = r(1) + 2^{t-1-1} = 8+4 \ = 12 \\ \hline \text{Group } k = 2: & r(4) = r(0+2^k) = r(0) + 2^{t-1-2} = 0+2 \ = 2 \\ & r(5) = r(1+2^k) = r(1) + 2^{t-1-2} = 8+2 \ = 10 \\ & r(6) = r(2+2^k) = r(2) + 2^{t-1-2} = 4+2 \ = 6 \\ & r(7) = r(3+2^k) = r(3) + 2^{t-1-2} = 12+2 = 14 \\ \hline \text{Group } k = 3 = t-1: & r(8) \ = r(0+2^k) = r(0) + 2^{t-1-3} = 0+1 \ = 1 \\ & r(9) \ = r(1+2^k) = r(1) + 2^{t-1-3} = 8+1 \ = 9 \\ & r(10) = r(2+2^k) = r(2) + 2^{t-1-3} = 4+1 \ = 5 \\ & r(11) = r(3+2^k) = r(3) + 2^{t-1-3} = 12+1 = 13 \\ & r(12) = r(2+2^k) = r(4) + 2^{t-1-3} = 2+1 \ = 3 \\ & r(13) = r(3+2^k) = r(5) + 2^{t-1-3} = 10+1 = 11 \\ & r(14) = r(2+2^k) = r(6) + 2^{t-1-3} = 6+1 \ = 7 \\ & r(15) = r(3+2^k) = r(7) + 2^{t-1-3} = 14+1 = 15, \end{array}$$

and obtain the same order $r = r_{16}$. *In this* r_{16} *order, the original vector* $\mathbf{f}_{16}$ *will be permuted to* $\tilde{\mathbf{f}}_{16} = \left(4\ 1\ 15\ 0\ 11\ 8\ 8\ 1\ 1\ 1\ 6\ 6\ 11\ 11\ 13\ 7\right)^T$.

FFT phase. — *computation of* $\mathbf{g}_{16} = (I_1 \otimes B_{16})(I_2 \otimes B_8)(I_4 \otimes B_4)(I_8 \otimes B2)\tilde{\mathbf{f}}_{16}$.

Step 1 should be easy to implement: we form the product $\mathbf{g}^{(1)} = (I_8 \otimes B_2)$ $\tilde{\mathbf{f}}_{16}$.

As $\mathbf{F}_2 = B_2 = \begin{pmatrix} 1 & \omega_2^0 \\ 1 & -\omega_2^0 \end{pmatrix} = \begin{pmatrix} 1 & 1 \\ 1 & -1 \end{pmatrix}$, *we obtain (Note:* $\omega_2 = \exp(-2\pi i/2) = -1$)

$$\mathbf{g}^{(1)} = f2\_vector = [5\ \ 3\ \ 15\ \ 15\ \ 19\ \ 3\ \ 9\ \ 7\ \ 2\ \ 0\ \ 12\ \ 0\ \ 22\ \ 0\ \ 20\ \ 6]^T.$$

In Step 2, we compute $\mathbf{g}^{(2)} = (I_4 \otimes B_4)\mathbf{g}^{(1)}$ *(note:* $\omega_4 = \exp(-2\pi i/4) = -i$*)*

$$B_4 = \begin{pmatrix} I_2 & \begin{pmatrix} 1 & \\ & \omega_4 \end{pmatrix} \\ I_2 & -\begin{pmatrix} 1 & \\ & \omega_4 \end{pmatrix} \end{pmatrix} = \begin{bmatrix} 1 & 0 & 1 & 0 \\ 0 & 1 & 0 & -i \\ 1 & 0 & -1 & 0 \\ 0 & 1 & 0 & i \end{bmatrix}, \; \mathbf{g}^{(2)} = f4\_vector$$

$$= \begin{bmatrix} B_4 & & & \\ & B_4 & & \\ & & B_4 & \\ & & & B_4 \end{bmatrix} \mathbf{g}^{(1)},$$

$$\mathbf{g}^{(2)} = \left[\, 20,\ 3+15i,\ -10,\ 3-15i,\ 28,\ 3+7i,\ 10,\ 3-7i,\ 14,\ 0,\ -10,\ 0,\ 42,\ 6i,\ 2,\ -6i \,\right].$$

In Step 3, compute $\mathbf{g}^{(3)} = (I_2 \otimes B_8)\mathbf{g}^{(2)}$ *(note:* $\omega_8 = \exp(-2\pi i/8) = (1-i)/\sqrt{2}$*)*

$$B_8 = \begin{pmatrix} I_4 & \begin{pmatrix} 1 & & & \\ & \omega_8 & & \\ & & \omega_8^2 & \\ & & & \omega_8^3 \end{pmatrix} \\ I_4 & -\begin{pmatrix} 1 & & & \\ & \omega_8 & & \\ & & \omega_8^2 & \\ & & & \omega_8^3 \end{pmatrix} \end{pmatrix}$$

$$= \left[\begin{array}{cccc|cccc} 1 & 0 & 0 & 0 & 1 & 0 & 0 & 0 \\ 0 & 1 & 0 & 0 & 0 & 0.7071(1-i) & 0 & 0 \\ 0 & 0 & 1 & 0 & 0 & 0 & -i & 0 \\ 0 & 0 & 0 & 1 & 0 & 0 & 0 & -0.7071(i+1) \\ \hline 1 & 0 & 0 & 0 & -1 & 0 & 0 & 0 \\ 0 & 1 & 0 & 0 & 0 & 0.7071(i-1) & 0 & 0 \\ 0 & 0 & 1 & 0 & 0 & 0 & i & 0 \\ 0 & 0 & 0 & 1 & 0 & 0 & 0 & 0.7071(1+i) \end{array}\right],$$

$$\mathbf{g}^{(3)} = f8\_vector = \begin{bmatrix} B_8 & \\ & B_8 \end{bmatrix} \mathbf{g}^{(2)} =$$

$[48, 0.1716 + 22.0711i, -10 + 10i, 5.8284 - 7.9289i, -8, 5.8284 + 7.9289i, \cdots - 10 - 10i, 0.1716 - 22.0711i, 56, -4.2426 + 4.2426i, -10 + 2i, \cdots 4.2426 + 4.2426i, -28, 4.2426 - 4.2426i, -10 - 2i, -4.2426 - 4.2426i]^T.$

Finally in Step 4, *compute from (note:* $\omega_{16} = \exp(-2\pi i/16) = 0.9239 - 0.3827i$*)*

$$\mathbf{g}_{16} = \mathbf{g}^{(4)} = f16\_vector = (I_1 \otimes B_{16})\mathbf{g}^{(3)} = B_{16}\mathbf{g}^{(3)}$$

$$= \left[\begin{array}{c|c} I_8 & \begin{pmatrix} 1 & & & & & & & \\ & \omega_{16} & & & & & & \\ & & \omega_{16}^2 & & & & & \\ & & & \omega_{16}^3 & & & & \\ & & & & \omega_{16}^4 & & & \\ & & & & & \omega_{16}^5 & & \\ & & & & & & \omega_{16}^6 & \\ & & & & & & & \omega_{16}^7 \end{pmatrix} \\ \hline I_8 & -\begin{pmatrix} 1 & & & & & & & \\ & \omega_{16} & & & & & & \\ & & \omega_{16}^2 & & & & & \\ & & & \omega_{16}^3 & & & & \\ & & & & \omega_{16}^4 & & & \\ & & & & & \omega_{16}^5 & & \\ & & & & & & \omega_{16}^6 & \\ & & & & & & & \omega_{16}^7 \end{pmatrix} \end{array}\right] \mathbf{g}^{(3)} = \begin{pmatrix} 104 \\ -5.3717 - 24.3672i \\ -18.4853 - 4.3431i \\ 3.5323 + 2.3857i \\ -8 + 28i \\ 8.1245 - 13.4722i \\ -1.5147 + 15.6569i \\ 5.7149 + 19.7750i \\ -8 \\ 5.7149 - 19.7750i \\ -1.5147 - 15.6569i \\ 8.1245 + 13.4722i \\ -8 - 28i \\ 3.5323 - 2.3857i \\ -18.4853 + 4.3431i \\ -5.3717 + 24.3672i \end{pmatrix}.$$

Clearly this vector $\mathbf{g}_{16}$ from FFT is identical to that from DFT (Table 1.3). The reader may use the supplied Mfile `exafft16.m` to check the details.

1.6.2 The fast wavelet transform

The wavelet transform is only part of a much larger subject of wavelet analysis [152,441]. In Section 8.1, we shall give a brief introduction. Here we consider the standard discrete wavelet transforms (DWTs), based on Daubechies' compactly supported orthogonal wavelets, for matrix A in (1.1). Refer to [60,269,390,441]. As wavelets form a basis of L_2, given a function $f \in L_2$, it can be written as an infinite linear combination of the wavelets and the wavelet coefficients uniquely determine the function; for smooth functions, most coefficients may be small and nearly zero.

For a given vector $\mathbf{f}$, from vector space $\mathbb{R}^n$, one may construct an infinite periodic sequence of period n and use it as coefficients of a scaling function

$f_L(x)$ in some fixed subspace $\mathbf{V}_L$

$$\mathbf{V}_L = \mathbf{V}_r \bigoplus \mathbf{W}_r \bigoplus \mathbf{W}_{r+1} \bigoplus \cdots \bigoplus \mathbf{W}_{L-1}, \tag{1.44}$$

of L_2 where L is an integer. (The case of $L = J, r = 0$ is considered in Chapter 8). Then the new coefficients of $f_L(x)$, when expressed in an equivalent wavelet basis, may be denoted by w. The wavelet transform $W : \mathbf{f} \mapsto \mathbf{g}$ (i.e. $\mathbf{g} = W\mathbf{f}$) is essentially a matrix mapping. To discuss W and specify the matrix structure of W, let $m = 2M = N + 1$ be the order of compactly supported wavelets with $m = N + 1$ lowpass filter coefficients $c_0, c_1, \cdots, c_N$ and $M = m/2$ vanishing moments. For clarity, we assume that

$$\sum_{k=0}^{m-1} c_k = \sqrt{2}. \tag{1.45}$$

The factor $\sqrt{2}$ is not difficult to understand if one thinks about the simpler case of a Harr wavelet with $N = 1$ and $m = 2$ (for $n = 2$)

$$W = \begin{bmatrix} \frac{\sqrt{2}}{2} & \frac{\sqrt{2}}{2} \\ \frac{\sqrt{2}}{2} & -\frac{\sqrt{2}}{2} \end{bmatrix} \tag{1.46}$$

where the filter matrix resembles a specific Givens rotation (1.24) and clearly (1.45) is satisfied. The orthogonality of W, due to the wavelets being orthogonal, is a direct consequence of the functional orthogonality reflected in the Fourier space (see Section 8.1 and [152, Ch.7]). Like the Fourier matrix $\mathbf{F}_n$ in DFT, the DWT matrix W is also in a factorized form and so W is not directly used. Therefore the so-called fast wavelet transform (FWT) refers to the algorithm of working out $\mathbf{g} = W\mathbf{f}$ without forming W. As the overall W is sparse, the FWT is 'automatic' (unlike the DFT case).

Assume $n = 2^t$ and r is an integer such that $2^r < m$ and $2^{r+1} \geq m$. Note $r = 0$ for $m = 2$ (Haar wavelets) and $r = 1$ for $m = 4$ (Daubechies order 4 wavelets); see [68,390,441]. Denote by $s = \mathbf{f}^{(t)}$ a column vector of A at the wavelet level t. Then the standard pyramidal algorithm transforms the vector $\mathbf{f}^{(t)}$ to

$$\mathbf{g} = \left[(\mathbf{f}^{(r)})^T \; (\mathbf{g}^{(r)})^T \; (\mathbf{g}^{(r+1)})^T \; \cdots \; (\mathbf{g}^{(t-1)})^T \right]^T$$

in a level by level manner, that is,

$$\begin{array}{llllllll} \texttt{to transform}: \mathbf{f}^{(t)} & \rightarrow \mathbf{f}^{(t-1)} & \rightarrow \mathbf{f}^{(t-2)} & \rightarrow \cdots & \rightarrow \mathbf{f}^{(\nu)} & \rightarrow \cdots & \rightarrow \mathbf{f}^{(r)} \\ & \searrow & \searrow & \searrow & \searrow & \searrow & \searrow \\ \texttt{to retain}: & \mathbf{g}^{(t-1)} & \mathbf{g}^{(t-2)} & \cdots & \mathbf{g}^{(\nu)} & \cdots & \mathbf{g}^{(r)} \end{array}$$

where the vectors $\mathbf{f}^{(\nu)}$ and $\mathbf{g}^{(\nu)}$ are of length 2^ν. Notice that the sum of these lengths is $n = 2^t$ since

$$2^t = 2^r + 2^r + 2^{r+1} + 2^{r+2} + \cdots + 2^{t-1}.$$

At a typical level ν, $\mathbf{f}^{(\nu)}$ and $\mathbf{g}^{(\nu)}$ are collections of scaling and wavelet coefficients respectively. In matrix form, $\mathbf{g}$ is expressed as

$$\mathbf{g} = P_{r+1}W_{r+1}\cdots P_{t-1}W_{t-1}P_tW_t\mathbf{f}^{(t)} \equiv W\mathbf{f}^{(t)}, \tag{1.47}$$

where

$$P_\nu = \begin{pmatrix} \overline{P}_\nu & \\ & J_\nu \end{pmatrix}_{n\times n}, \qquad W_\nu = \begin{pmatrix} \overline{W}_\nu & \\ & J_\nu \end{pmatrix}_{n\times n}$$

with $\overline{P}_\nu$ a permutation matrix of size $2^\nu = 2^t - k_\nu = n - k_\nu$, that is, $\overline{P}_\nu = I(1, 3, \cdots, 2^\nu - 1, 2, 4, \cdots, 2^\nu)$, and with $\overline{W}_\nu$ an orthogonal (sparse) matrix of size $2^\nu = 2^t - k_\nu = n - k_\nu$ and J_ν is an identity matrix of size k_ν. Here $k_t = 0$ and $k_\mu = k_{\mu+1} + 2^\mu$ for $\mu = t-1, \cdots, r+1$. The one level transformation matrix $\overline{W}_\nu$ is a compact quasi-diagonal block matrix, whose rows can be read from the following multiplication [152,§1.6] $y = \overline{W}_\nu x$ for $k = 1, \ldots, n$

$$y_k = \begin{cases} \displaystyle\sum_\ell c_{\ell-2\bar{k}}x_{\ell+1} = \sum_{\ell=0}^{m-1} c_\ell x_{\ell+1+2\bar{k}}, & \text{if } k \text{ is odd, } \bar{k} = (k-1)/2, \\ \displaystyle\sum_\ell d_{\ell-2\bar{k}}x_{\ell+1} = \sum_{\ell=0}^{m-1} d_\ell x_{\ell+1+2\bar{k}}, & \text{if } k \text{ is even, } \bar{k} = k/2 - 1, \end{cases} \tag{1.48}$$

implying from $y_3 = \sum_\ell c_\ell x_{\ell+3}$, $\bar{k} = 1$ that

$$(\overline{W}_\nu)_{33} = c_0,\ (\overline{W}_\nu)_{34} = c_1, \cdots, (\overline{W}_\nu)_{3,m+2} = c_{m-1},$$

and from $y_4 = \sum_\ell c_\ell x_{\ell+3}$, $\bar{k} = 1$ that

$$(\overline{W}_\nu)_{43} = d_0,\ (\overline{W}_\nu)_{44} = d_1, \cdots, (\overline{W}_\nu)_{4,m+2} = d_{m-1}.$$

For functions, this step corresponds to decomposing a fine level function into the sum of two coarse level functions.

For clarity, take the commonly used Daubechies' order $m = 4$ wavelets with $m/2 = 2$ vanishing moments and display $\overline{W}_\nu$ as

$$\overline{W}_\nu = \begin{pmatrix} c_0 & c_1 & c_2 & c_3 & & & & & & \\ d_0 & d_1 & d_2 & d_3 & & & & & & \\ & & c_0 & c_1 & c_2 & c_3 & & & & \\ & & d_0 & d_1 & d_2 & d_3 & & & & \\ & & & & \ddots & \ddots & \ddots & \ddots & & \\ & & & & \ddots & \ddots & \ddots & \ddots & & \\ & & & & & & c_0 & c_1 & c_2 & c_3 \\ & & & & & & d_0 & d_1 & d_2 & d_3 \\ c_2 & c_3 & & & & & & & c_0 & c_1 \\ d_2 & d_3 & & & & & & & d_0 & d_1 \end{pmatrix},$$

where the filtering coefficients $\{c_i, d_i\}$ are known to be $c_0 = (1 + \sqrt{3})/(4\sqrt{2})$, $c_1 = (3 + \sqrt{3})/(4\sqrt{2})$, $c_2 = (3 - \sqrt{3})/(4\sqrt{2})$, $c_3 = (1 - \sqrt{3})/(4\sqrt{2})$ and $d_k = (-1)^k c_{m-1-k}$ as usual. Note that (1.45) is satisfied. Letting $\widehat{W}_k = P_k W_k$, an alternative decomposition of the wavelet matrix W to (1.47) can be defined by

$$W = \widehat{W}_{r+1} \cdots \widehat{W}_{t-1} \widehat{W}_t \tag{1.49}$$

with

$$\widehat{W}_\nu = \begin{pmatrix} \breve{W}_\nu & \\ & J_\nu \end{pmatrix}_{n \times n} \quad \text{and}$$

$$\breve{W}_\nu = \left(\begin{array}{cccccccccc} c_0 & c_1 & c_2 & c_3 & & & & & & \\ & & c_0 & c_1 & c_2 & c_3 & & & & \\ & & & & \ddots & \ddots & \ddots & \ddots & & \\ & & & & & & c_0 & c_1 & c_2 & c_3 \\ c_2 & c_3 & & & & & & & c_0 & c_1 \\ \hline d_0 & d_1 & d_2 & d_3 & & & & & & \\ & & d_0 & d_1 & d_2 & d_3 & & & & \\ & & & & \ddots & \ddots & \ddots & \ddots & & \\ & & & & & & d_0 & d_1 & d_2 & d_3 \\ d_2 & d_3 & & & & & & & d_0 & d_1 \end{array}\right).$$

Note that such an alternative form can also be written for the general m.

Algorithm 1.6.18. *Forward DWT*	$x \to \widetilde{x}$.

(1) Set $\mathbf{s}^0 = x$.
(2) For level $k = 1, 2, \ldots, L$

(a) Let $\mathbf{s}_j^k = \sum_{l=0}^{m-1} c_l \mathbf{s}^{k-1}_{\langle l+2j-1\rangle_{n/2^{k-1}}}$, $\quad j = 1, 2, \ldots, n/2^k$.

(b) Let $\mathbf{d}_j^k = \sum_{l=0}^{m-1} d_l \mathbf{s}^{k-1}_{\langle l+2j-1\rangle_{n/2^{k-1}}}$, $\quad j = 1, 2, \ldots, n/2^k$.

(3) Accept $\widetilde{x} = (\mathbf{s}^L, \mathbf{d}^L, \mathbf{d}^{L-1}, \ldots, \mathbf{d}^1)$.

Here $\langle a\rangle_b = (a-1)\text{ Mod } b + 1 = \texttt{mod}(a-1, b) + 1$. See also (4.13). For example, $\langle 256\rangle_{256} = 256$ and $\langle 257\rangle_{256} = 1$.

Here a full DWT step for matrix A in (1.1), transforming column and row vectors, respectively, will be $\widetilde{A} = WAW^T$. However in a practical implementation, one should use the factorizations in (1.47) or (1.49) without forming W explicitly. Thus the DWT can be done in a fast way so it can 'automatically' become the FWT. For a dense vector, the cost of the FWT via (1.47) can be estimated by

$$T_v = \underbrace{mn}_{\text{Step 1}} + \underbrace{mn/2}_{\text{Step 2}} + \cdots + \underbrace{mn/2^{t-1}}_{\text{Step } t} \approx 2mn$$

and for a dense matrix

$$T_m = \underbrace{2mn^2}_{\text{Step 1}} + \underbrace{2mnn/2}_{\text{Step 2}} + \cdots + \underbrace{2mnn/2^{t-1}}_{\text{Step } t} \approx 4mn^2.$$

For the inverse FWT, from (1.49), we have

$$W^T = \widehat{W}_t^T \widehat{W}_{t-1}^T \cdots \widehat{W}_{r+1}^T \tag{1.50}$$
$$\widehat{W}_v^T = \begin{pmatrix} \breve{W}_v^T & \\ & J_v \end{pmatrix}_{n\times n}$$

and

$$\breve{W}_v^T = \left(\begin{array}{cccccc|cccccc}
c_0 & & & c_{m-2} & \cdots & c_2 & d_0 & & & d_{m-2} & \cdots & d_2 \\
c_1 & & & c_{m-1} & \ddots & c_3 & d_1 & & & d_{m-1} & \ddots & d_3 \\
c_2 & c_0 & & & \ddots & \vdots & d_2 & d_0 & & & \ddots & \vdots \\
c_3 & c_1 & & & & c_{m-2} & d_3 & d_1 & & & & d_{m-2} \\
\vdots & c_2 & & & & c_{m-1} & \vdots & d_2 & & & & d_{m-1} \\
c_{m-1} & \ddots & & & & & d_{m-1} & \ddots & & & & \\
& c_{m-2} & \ddots & & & & & d_{m-2} & \ddots & & & \\
& c_{m-1} & \ddots & \ddots & \ddots & & & d_{m-1} & \ddots & \ddots & \ddots & \\
& & & c_{m-4} & & c_0 & & & & d_{m-4} & & d_0 \\
& & & c_{m-3} & & c_1 & & & & d_{m-3} & & d_1
\end{array}\right).$$

For functions, this step corresponds to constructing a fine level function from two coarse level functions. Let $z = \widehat{W}_{\nu}^T x$ with the first half of x corresponding to the coefficients of a scaling function and the second half of x to that of a wavelet function (see [152, Section 1.6] and Section 8.1). Then, with $n_2 = n/2$ and $k(\ell) = (k + 1 - \ell)/2$, for $k = 1, \ldots, n$

$$z_k = \sum_{\ell} \left(c_{k-1-2\ell} x_{\ell+1} + d_{k-1-2\ell} x_{\ell+1+n_2} \right) = \sum_{\ell=0}^{m-1} \left(c_\ell x_{k(\ell)} + d_\ell x_{k(\ell)+n_2} \right), \tag{1.51}$$

where the notation $k(\ell)$, similar to (8.26), should be interpreted as a whole integer; a term does not exist (so is zero) if $k(\ell)$ is not an integer and a term should be added by n_2 if $k(\ell)$ is a negative integer or zero. More precisely, with $m_2 = m/2$,

$$z_k = \begin{cases} \displaystyle\sum_{j=0}^{m_2-1} \left(c_{2j} x_{(k+1-2j)/2} + d_{2j} x_{(k+1-2j)/2+n_2}, \right. & \text{if } k \text{ is odd,} \\ \displaystyle\sum_{j=0}^{m_2-1} \left(c_{2j+1} x_{(k-2j)/2} + d_{2j+1} x_{(k-2j)/2+n_2}, \right. & \text{if } k \text{ is even.} \end{cases}$$

For instance with $m = 4$, $m_2 = 2$,

$$\begin{aligned}
z_1 &= (c_0 x_1 + d_0 x_{1+n_2}) + (c_2 x_0 + d_0 x_{0+n_2}) = c_0 x_1 + d_0 x_{n_2+1} + c_2 x_{n_2} + d_0 x_n, \\
z_2 &= (c_1 x_1 + d_1 x_{1+n_2}) + (c_3 x_0 + d_3 x_{0+n_2}) = c_1 x_1 + d_1 x_{n_2+1} + c_3 x_{n_2} + d_3 x_n, \\
z_5 &= (c_0 x_3 + d_0 x_{3+n_2}) + (c_2 x_2 + d_2 x_{2+n_2}) = c_0 x_3 + d_0 x_{n_2+3} + c_2 x_2 + d_2 x_{n_2+3}, \\
z_6 &= (c_1 x_3 + d_1 x_{3+n_2}) + (c_3 x_2 + d_3 x_{2+n_2}) = c_1 x_3 + d_1 x_{n_2+3} + c_3 x_2 + d_3 x_{n_2+2}.
\end{aligned}$$

Algorithm 1.6.19. *Inverse DWT* $\widetilde{x} \to x$.

(1) Set $(\mathbf{s}^L, \mathbf{d}^L, \mathbf{d}^{L-1}, \ldots, \mathbf{d}^1) = \widetilde{x}$.

(2) For level $k = L, L-1, \ldots, 1$

(a) Let $\mathbf{s}_j^{k-1} = \sum_{l=0}^{m/2-1} c_{2l} \mathbf{s}^k_{\langle (j+1)/2 - l \rangle n/2^k} + \sum_{l=0}^{m/2-1} d_{2l} \mathbf{d}^k_{\langle (j+1)/2 - l \rangle n/2^k}$

for $j = 1, 3, \ldots, n/2^{k-1} - 1$.

(b) Let $\mathbf{s}_j^{k-1} = \sum_{l=0}^{m/2-1} c_{2l+1} \mathbf{s}^k_{\langle j/2 - l \rangle n/2^k} + \sum_{l=0}^{m/2-1} d_{2l+1} \mathbf{d}^k_{\langle j/2 - l \rangle n/2^k}$

for $j = 2, 4, \ldots, n/2^{k-1}$.

(3) Accept $x = \mathbf{s}^0$.

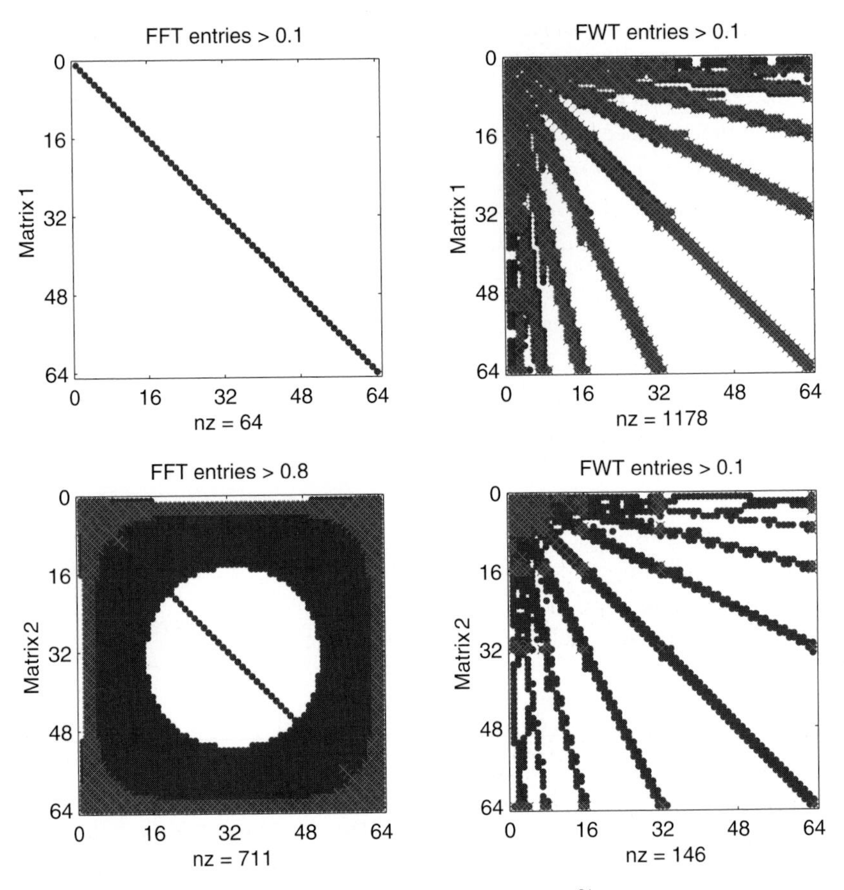

Figure 1.1. Comparison of FFT ($\widetilde{A} = \mathbf{F}A\mathbf{F}^H$) and FWT ($\widetilde{A} = WAW^T$) for compressing two test matrices. Clearly FFT is only good at circulant matrix 1 but FWT is more robust for both examples.

To try out a specific transform, one may use the supplied Mfiles `fwt.m` and `ifwt.m` as follows

```
>>  F = fwt(f,6),  g = ifwt(F,6)        %% Vector f case (Daub 6)
>>  B = fwt(A,4),  g = ifwt(B,4)        %% Matrix A case (Daub 4)
```

Guess again what g is in each case. For a matrix, it is known that $\widetilde{A} = WAW^T$ has a 'finger'-like sparsity pattern. This will be further considered in Chapter 8.

Finally, in Figure 1.1, we demonstrate the effectiveness of compression by the FFT and the FWT for two examples of a matrix A with $n = 64$

$$\begin{aligned} &\text{Test 1 } A_{ij} = \textit{circulant}(\mathbf{h}), \qquad \text{with } \mathbf{h} = (1, 2, \ldots, n)^T \\ &\text{Test 2 } A_{ij} = \textit{toeplitz}(\mathbf{h}), \qquad \text{with } \mathbf{h} = (1, 2, \ldots, n)^T \end{aligned} \tag{1.52}$$

where **h** is the root (column 1) for generating the Circulant and Toeplitz matrices Section 2.5.2. The reader should run the supplied Mfile `fft_fwt.m` to reproduce the result.

1.7 Numerical solution techniques for practical equations

As is known, linear systems such as (1.1) often arise from solving other equations. Here we briefly review a selection of solution techniques for partial differential equations (PDEs) and nonlinear systems.

Many PDEs of practical interest are of second order. Solution techniques typically (though not always) reduce one order of differentiation before discretization via Gauss' or Green's theorems or integration by parts. Below we discuss

(1) the finite element method (FEM)
(2) the boundary element method (BEM)
(3) the finite difference method (FDM)
(4) the finite volume method (FVM)
(5) the global element methods (GEMs).

As each topic involves a huge subject, the reader will be referred to detailed references (and therein) for special perusal. We shall consider the interior Helmholtz-like equation as our model PDE (with $p = (x, y) \in \Omega \subset \mathbb{R}^2$ and $\partial\Omega$ smooth)

$$\begin{cases} \mathbb{L}u \equiv -\Delta u - k^2 u = f(p), & p \in \Omega, \\ u|_{\partial\Omega} = g(p), & p \in \partial\Omega, \end{cases} \tag{1.53}$$

which is assumed to have a unique solution (for a suitable k).

1.7.1 The finite element method (FEM)

The FEM is the most widely used method for solving PDEs since it was invented in 1950s and matured in 1970s; see [308,282,499]. For (1.53), it does not look for a so-called classical solution $u \in C^2(\Omega)$ satisfying (1.53) partly because computer solutions are hardly in $C^2(\Omega)$ anyway. Instead it seeks a 'weak' solution $u = u(x, y) \in V \subset H^1(\Omega)$ such that

$$\int_\Omega (\mathbb{L}u - f)\, v d\Omega = 0, \qquad \forall v \in H^1(\Omega), \tag{1.54}$$

where

$$H^1(\Omega) = \left\{ w \;\middle|\; \int_\Omega (|w|^2 + |w_x|^2 + |w_y|^2)^{1/2} d\Omega < \infty, \; w \in L_2(\Omega) \right\}, \tag{1.55}$$

$$V = \left\{ w \;\middle|\; w \in H^1(\Omega), \, w|_{\partial\Omega} = g \right\}. \tag{1.56}$$

To reduce the second-order differentiation in $\mathbb{L}$ to first order, the analytical tool is the Green's first theorem (assume $x_1 = x, \; x_2 = y$)

$$\int_\Omega \frac{\partial u}{\partial x_i} v d\Omega = -\int_\Omega u \frac{\partial v}{\partial x_i} d\Omega + \int_{\partial\Omega} uvn_i dS \tag{1.57}$$

or (taking $\frac{\partial u}{\partial x_i}$s instead of u)

$$\int_\Omega \Delta uv d\Omega = -\int_\Omega \nabla u \cdot \nabla v d\Omega + \int_{\partial\Omega} v \frac{\partial u}{\partial \mathbf{n}} dS \tag{1.58}$$

where $\mathbf{n} = (n_1, n_2)^T$ denotes the unit outer normal to the boundary $\partial\Omega$ at $p = (x, y) \in \partial\Omega$ and $\frac{\partial u}{\partial \mathbf{n}} = \nabla u \cdot \mathbf{n} = \frac{\partial u}{\partial x} n_1 + \frac{\partial u}{\partial y} n_2$. Note if $\partial\Omega$ is parameterized as $x = x(t), \; y = y(t)$, then

$$n_1 = y'(t)/\sqrt{x'(t)^2 + y'(t)^2}, \qquad n_2 = -x'(t)/\sqrt{x'(t)^2 + y'(t)^2}. \tag{1.59}$$

In view of (13.14), if $\partial\Omega$ is implicitly defined as $G(x, y) = 0$, then the unit out normal is $\mathbf{n} = \nabla G / \|\nabla G\|_2$. Using (1.58), our main equation (1.54) becomes

$$\begin{aligned} 0 = \int_\Omega (\mathbb{L}u - f) v d\Omega &= -\int_\Omega \left(\Delta u + k^2 u + f\right) v d\Omega \\ &= \int_\Omega \nabla u \cdot \nabla v d\Omega - \int_{\partial\Omega} v \frac{\partial u}{\partial \mathbf{n}} dS - \int_\Omega \left(k^2 u + f\right) v d\Omega \\ &= a(u, v) - (f, v), \end{aligned} \tag{1.60}$$

where $v|_{\partial\Omega} = 0$ is assumed, the innocent looking notation $a(u, v)$ is the well-known bilinear functional, that proves useful in establishing the existence and uniqueness of u using the Lax–Milgram theorem, and is defined by

$$a(u, v) = \int_\Omega \nabla u \cdot \nabla v d\Omega - k^2 \int_\Omega uv d\Omega, \tag{1.61}$$

with $(f, v) = \int_\Omega f v d\Omega$. The minor issue $v|_{\partial\Omega} = 0$ calls for a new space notation $v \in H_0^1(\Omega)$, where $H_0^1(\Omega) = \left\{ w \;\middle|\; w \in H^1(\Omega), \, w|_{\partial\Omega} = 0 \right\}$ and in fact this turns out to be quite important as it manages to avoid the 'difficult' term $\partial u / \partial \mathbf{n}$ coming to the equations (this term will be a big issue for the BEM later).

Overall, the weak formulation of (1.60) as a basis for FEM is the following

$$\text{find } u \in V \subset H^1(\Omega): \qquad a(u,v) = (f,v), \qquad \forall v \in H_0^1(\Omega). \quad (1.62)$$

Remark 1.7.20.

(1) One commonly finds a weak formulation that appears to work for a homogeneous boundary condition only e.g. for, instead of (1.53),

$$\begin{cases} \mathbb{L}u \equiv -\Delta u - k^2 u = f(p), & p \in \Omega, \\ u|_{\partial\Omega} = 0, & p \in \partial\Omega. \end{cases}$$

This assumes that a function $u_0 \in V$, with V in (1.56), has been constructed first so it can be subtracted out from the main solution. Then $u, v \in H_0^1(\Omega)$. However, this is only for notational convenience as often one does not implement a FEM this way.

(2) The space $H_0^1(\Omega)$ is not required for the case of a Neumann's boundary condition i.e.

$$\begin{cases} \mathbb{L}u \equiv -\Delta u - k^2 u = f(p), & p \in \Omega, \\ \frac{\partial u}{\partial \mathbf{n}}|_{\partial\Omega} = g, & p \in \partial\Omega. \end{cases}$$

Then the weak formulation becomes

$$\text{find } u \in H^1(\Omega): \qquad a(u,v) = (f,v) + \int_{\partial\Omega} gv dS, \qquad \forall v \in H^1(\Omega).$$

(3) If the above procedure is repeated for a more general PDE (assume $x_1 = x,\ x_2 = y$)

$$\begin{cases} -\sum_{i,j=1}^{2} \frac{\partial}{\partial x_j}\left(a_{ij}(p)\frac{\partial u}{\partial x_j}\right) + \sum_{i=1}^{2} b_i(p)\frac{\partial u}{\partial x_i} + c(p)u = f(p), & p \in \Omega, \\ u|_{\partial\Omega} = g, & p \in \partial\Omega. \end{cases}$$

Then a similar weak formulation is obtained

$$\text{find } u \in V \subset H^1(\Omega): \qquad a(u,v) = (f,v), \qquad \forall v \in H_0^1(\Omega).$$

with $a(u,v) = \int_\Omega \left[\sum_{i,j=1}^2 a_{ij} \frac{\partial u}{\partial x_j}\frac{\partial v}{\partial x_i} + \sum_{i=1}^2 b_i(p)\frac{\partial u}{\partial x_i} v + c(p)uv\right] d\Omega$.

(4) The procedure we discussed is the Galerkin formulation. A related but equivalent form is the so-called Ritz formulation where we do not use the Green's theorem and (1.54). Assume (1.53) has a homogenous boundary condition $g = 0$. Define $J(v) = \frac{1}{2}a(v,v) - (f,v)$ with $a(v,v)$ as in (1.61) for (1.53). Then the Ritz formulation is

$$\text{find } u \in H^1(\Omega): \qquad J(u) = \min J(v), \qquad \forall v \in H_0^1(\Omega).$$

For the case of matrices, this method resembles the minimization formulation for a linear system, i.e. Theorem 3.4.6 in Section 3.4. There A must be a SPD; here likewise, $a(v, v)$ must satisfy the strongly elliptic condition

$$a(v, v) \geq \alpha \|v\|_2^2, \qquad \text{for } \alpha > 0, \ \forall v \in H_0^1(\Omega).$$

We are now to ready to introduce the FEM discretization. First partition a given domain Ω into E computationally convenient subdomains $\Omega_1, \ldots, \Omega_E$; in two dimensions (2D) the usual choice is to use triangles, as illustrated in Figure 1.2 for $E = 16$. Denote by n the total number of internal nodes, n_t the number of total nodes, and h the maximal side length of the triangulation. The domain Ω is approximated by

$$\Omega^h = \bigcup_{j=1}^{E} \Omega_j.$$

For $i = 1, \ldots, n_t$, a FEM will define *piecewise polynomials* $\phi_i(p)$, in Ω^h, of the interpolating type

$$\phi_i(p) = \begin{cases} 1, \text{ if } j = i, \\ 0, \text{ if } j \neq i. \end{cases} \tag{1.63}$$

Let $\mathbb{P}_\ell(\Omega_e)$ denote the set of all polynomials of degree ℓ defined in domain Ω_e. The commonly used piecewise linear functions can be denoted by

$$V^h = \left\{ u \;\middle|\; u \in C(\bar{\Omega}), \ u|_{\Omega_e} \in \mathbb{P}_1(\Omega_e), \ \text{ for } e = 1, \ldots, E \right\} \tag{1.64}$$

which approximates $H^1(\Omega)$ and similarly let $H_0^1(\Omega)$ be approximated by

$$V_0^h = \left\{ u \;\middle|\; u \in V^h \ \text{ and } u_{\partial\Omega \cup \Omega^h} = 0 \right\}. \tag{1.65}$$

Note that if we replace the above defined V_0^h subspace by another one made up of piecewise constants (i.e. not linear), we will obtain the so-called collocation approach (which is less popular for FEM but more with BEM discussed next).

To proceed with the Galerkin approach, one only needs to make one simple observation on basis sharing: $\{\phi_i\}_1^{n_t}$ form the basis for V^h while $\{\phi_i\}_1^n$ form the basis for V_0^h (note: $n < n_t$). Thus

$$u = \sum_{j=1}^{n_t} u_j \phi_j(p) \qquad \text{and} \qquad v = \sum_{j=1}^{n} v_j \phi_j(p). \tag{1.66}$$

Substituting (1.66) into (1.62) gives rise to (noting $\{u_j\}_{n+1}^{n_t}$ are known values)

$$\sum_{j=1}^{n} u_j a(\phi_j, v) = (f, v) - \sum_{j=n+1}^{n_t} u_j a(\phi_j, v).$$

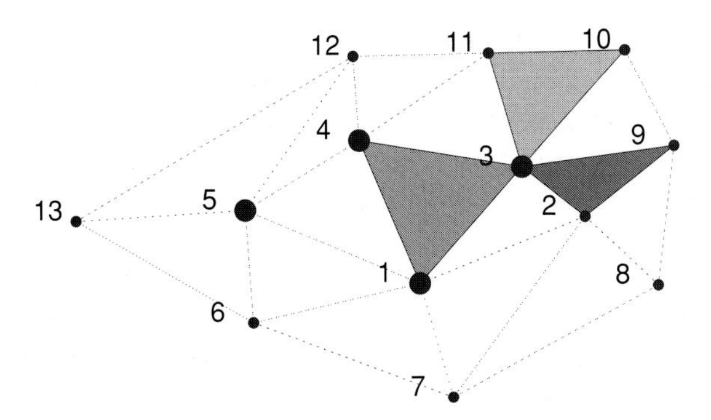

Figure 1.2. Sample FEM triangulation of a domain Ω ($n = 5$ internal points, $n_t = 13$).

As v must go through all members in V_0^h, it suffices to take all basis functions $v = \phi_i$ so we obtain the final linear system from solving (1.53):

$$\sum_{j=1}^{n} a(\phi_j, \phi_i) = (f, \phi_i) - \sum_{j=n+1}^{n_t} u_j a(\phi_j, \phi_i), \qquad \text{for } i = 1, \ldots, n. \quad (1.67)$$

Here it should be remarked that the above ϕ functions are actually sums of those $\hat{\phi}$ functions defined element-wise (i.e. each ϕ_i having a larger support than 1 element); for example, for node 3 in Figure 1.2, our ϕ_2 will denote the sum of six local basis functions from all neighbouring elements at node 3. This is the aspect of the local stiffness matrix. For the example in Figure 1.2 with $n = 5$ and $n_t = 13$, the local 3×3 stiffness out of the element enclosed by nodes 1, 3, 4, corresponding to three local basis functions $\hat{\phi}_1, \hat{\phi}_2, \hat{\phi}_3$, will enter the global 5×5 matrix A in these '$\times$' positions

$$\begin{pmatrix} a(\hat{\phi}_1, \hat{\phi}_1) & a(\hat{\phi}_2, \hat{\phi}_1) & a(\hat{\phi}_3, \hat{\phi}_1) \\ a(\hat{\phi}_1, \hat{\phi}_2) & a(\hat{\phi}_2, \hat{\phi}_2) & a(\hat{\phi}_3, \hat{\phi}_2) \\ a(\hat{\phi}_1, \hat{\phi}_3) & a(\hat{\phi}_2, \hat{\phi}_3) & a(\hat{\phi}_3, \hat{\phi}_3) \end{pmatrix} \quad \to \quad \begin{pmatrix} \times & \bigcirc & \times & \times \\ \bigcirc & \bigcirc & \bigcirc & \bigcirc \\ \times & \bigcirc & \times & \times \\ \times & \bigcirc & \times & \times \\ \bigcirc & \bigcirc & \bigcirc & \bigcirc \end{pmatrix}.$$

It should not come as a surprise that these are rows 1, 3, 4 and columns 1, 3, 4.

1.7.2 The boundary element method (BEM)

The boundary element method has a similar starting point to FEM in trying to find a weak formulation but it also aims to eliminate the domain integrals using different choices of v trial functions [159,16,277,301,159]. Admittedly, the standard BEM cannot eliminate the domain term (f, v) so we have to be content with solving a homogeneous problem (1.53) with $f = 0$ i.e. $(f, v) = 0$. Recently there are attempts to generalize the BEM to solve nonhomogeneous PDEs (even to general linear PDEs); see [385,428].

The first step in a BEM is to derive a boundary integral equation. To this end, we begin with the exact setup as in (1.58) and (1.60) with $q = (x, y)$:

$$
\begin{aligned}
0 &= \int_\Omega (\mathbb{L}u - f)\, v d\Omega = -\int_\Omega \left(\Delta u + k^2 u + f\right) v d\Omega \\
&= -\int_\Omega \left[u\Delta v + (k^2 u + f)v\right] d\Omega + \int_{\partial\Omega} \left(u\frac{\partial v}{\partial \mathbf{n}} - v\frac{\partial u}{\partial \mathbf{n}}\right) dS \\
&= -\int_\Omega u\left[\Delta v + k^2 v\right] d\Omega + \int_{\partial\Omega} \left(u\frac{\partial v}{\partial \mathbf{n}} - v\frac{\partial u}{\partial \mathbf{n}}\right) dS_q \\
&= u(p) + \int_{\partial\Omega} \left(u\frac{\partial v}{\partial \mathbf{n}} - v\frac{\partial u}{\partial \mathbf{n}}\right) dS_q,
\end{aligned}
\tag{1.68}
$$

i.e. with $G_k = G_k(p, q) = \frac{i}{4\pi}\mathbf{H}_0^{(1)}(kr) = \frac{i}{4\pi}\mathbf{H}_0^{(1)}(k|p - q|)$ and $q \in \partial\Omega$

$$
u(p) = \int_{\partial\Omega} \left(G_k \frac{\partial u}{\partial \mathbf{n}} - u\frac{\partial G_k}{\partial \mathbf{n}}\right) dS_q, \qquad \text{for } p \in \Omega, \tag{1.69}
$$

where we have made a particular choice of $v = v(q) = v_p(q) = G_k(p, q)$, $p \in \Omega$, with G_k being the so-called fundamental solution or free-space Green's function for (1.53) (note this v is highly dependent on the underlying operator)

$$
-\Delta G_k - k^2 G_k = \delta(p - q), \tag{1.70}
$$

and we have used the Green's second theorem, based on the first theorem in (1.57),

$$
\int_\Omega v\Delta u d\Omega = \int_\Omega u\Delta v d\Omega + \int_{\partial\Omega} \left(v\frac{\partial u}{\partial \mathbf{n}} - u\frac{\partial v}{\partial \mathbf{n}}\right) dS_q. \tag{1.71}
$$

Here from the properties of the Hankel functions (associated with the 2D Helmholtz)

(1) $\dfrac{i}{4}H_0^{(1)}(kr) = C - \dfrac{1}{2\pi}\log(kr) + \dfrac{k^2 r^2}{16\pi} + O(k^2 r^2)$ as $r \to 0$

(2) $\dfrac{dH_0^{(1)}}{dx}(x) = -H_1^{(1)}$ and $\dfrac{dH_1^{(1)}}{dx}(x) = H_0^{(0)} - H_1^{(1)}/x$,

one observes that the Helmholtz operator shares the same type of kernel singularities as the Laplacian since the latter has the fundamental solution

$$
G_0 = -\frac{1}{2\pi}\log(r). \tag{1.72}
$$

For the 3D case, the Helmholtz fundamental solution is $G_k(r) = \exp(ikr)/(4\pi r)$ with $G_0(r) = 1/(4\pi r)$ for the 3D Laplacian. The computation for $H0 = H_0^{(1)}(x) = J_0(x) + iY_0(x)$ and $H1 = H_1^{(1)}(x) = J_1(x) + iY_1(x)$ at $x = 0.234$, via Bessel functions, can be done in MATLAB by

```
>>  kind = 1                   % First kind Hankel function
>>  H0 = besselh(0, kind, x)   % giving 0.9864-0.9762i (order 0)
>>  H1 = besselh(1, kind, x)   % giving 0.1162-2.8732i (order 1)
```

and in Fortran using NAG routines[11]

```
info = 0                                       !Error indicator
H0 = DCMPLX(s17aEf(x,info),s17aCf(x,info)) !Give 0.9864-0.9762i
H1 = DCMPLX(s17aFf(x,info),s17aDf(x,info)) !Give 0.1162-2.8732i
```

From (1.69), one can see already that a BEM is advantageously avoiding the domain Ω as the solution u can be recovered from boundary data (the Cauchy data) of u and $\partial u/\partial \mathbf{n}$ alone. We now proceed to convert (1.69) into a boundary integral equation (BIE) which states a mapping relationship between the two kinds of boundary data: u and $\partial u/\partial \mathbf{n}$. For this reason, sometimes, a BIE is called a Dirichlet (i.e. u) and Neumann (i.e. $\partial u/\partial \mathbf{n}$) mapping.

To take the domain (observation) point p in (1.69) to the boundary $\partial\Omega$, we need to take the limit $p^- \to p \in \partial\Omega$; here the superscript indicates the direction approaching $\partial\Omega$ is from the interior Ω. For later use, define four BIE operators

$$\begin{cases} \text{Single layer} & (\mathcal{L}_k w)(p) = \displaystyle\int_{\partial\Omega} G_k(p,q) w(q) dS_q \\ \text{Double layer} & (\mathcal{M}_k w)(p) = \displaystyle\int_{\partial\Omega} \frac{\partial G_k}{\partial \mathbf{n}_q}(p,q) w(q) dS_q \\ \text{Transposed double layer} & (\mathcal{M}_k^T w)(p) = \displaystyle\frac{\partial}{\partial \mathbf{n}_p}\int_{\partial\Omega} G_k(p,q) w(q) dS_q \\ \text{Hyper-singular} & (\mathcal{N}_k w)(p) = \displaystyle\frac{\partial}{\partial \mathbf{n}_p}\int_{\partial\Omega} \frac{\partial G_k}{\partial \mathbf{n}_q}(p,q) w(q) dS_q. \end{cases} \tag{1.73}$$

Theorem 1.7.21. (Properties of layer operators [159,323]).

(1) From the interior domain to $\partial\Omega$,

$$\begin{aligned} \mathcal{L}_k^- w(p) &= \mathcal{L}_k w(p), && p \in \partial\Omega, \\ \mathcal{M}_k^- w(p) &= \mathcal{M}_k w(p) - \tfrac{1}{2} w(p), && p \in \partial\Omega, \\ (\mathcal{M}_k^T)^- w(p) &= \mathcal{M}_k^T w(p) + \tfrac{1}{2} w(p), && p \in \partial\Omega, \\ \mathcal{N}_k^- w(p) &= \mathcal{N}_k w(p), && p \in \partial\Omega. \end{aligned}$$

[11] ©Copyright Numerical Algorithms Group, Oxford, UK – http://www.nag.co.uk

(2) From the exterior domain to $\partial\Omega$,

$$\begin{aligned}
\mathcal{L}_k^+ w(p) &= \mathcal{L}_k w(p), & p \in \partial\Omega,\\
\mathcal{M}_k^+ w(p) &= \mathcal{M}_k w(p) + \tfrac{1}{2} w(p), & p \in \partial\Omega,\\
(\mathcal{M}_k^T)^+ w(p) &= \mathcal{M}_k^T w(p) - \tfrac{1}{2} w(p), & p \in \partial\Omega,\\
\mathcal{N}_k^+ w(p) &= \mathcal{N}_k w(p), & p \in \partial\Omega.
\end{aligned}$$

(3) Operators $\mathcal{L}_k : \ H^{1/2}(\partial\Omega) \to H^{1+1/2}(\partial\Omega)$ *and operators* $\mathcal{M}_k, \mathcal{M}_k^T, (\mathcal{N}_k - \mathcal{N}_0) : \ H^{1/2}(\partial\Omega) \to H^{1/2}(\partial\Omega)$ *are compact while operator* $\mathcal{N}_k : \ H^{1/2}(\partial\Omega) \to H^{-1/2}(\partial\Omega)$ *is bounded.*

Here the Sobolev space $H^{\tau-1/2}(\partial\Omega)$ contains functions of $H^{\tau}(\Omega)$ restricted to the boundary $\partial\Omega$, with

$$H^{\tau}(\Omega) = \left\{ w \;\middle|\; \|w\| = \int_{\Omega} \left(\sum_{j=0}^{\tau} \left|\mathbf{D}^j w\right|^2 \right)^{1/2} d\Omega < \infty, \ w \in L^2(\Omega) \right\}, \tag{1.74}$$

and $\mathbf{D}^j$ denoting the jth partial derivative [223,323]. So the spaces $H^{1/2}(\partial\Omega)$, $H^{-1/2}(\partial\Omega)$ will contain respectively piecewise linear and constant elements. Note the case of $\tau = 1$ has been introduced in (1.55).

With the operator notation, equation (1.69) can be written as

$$u(p) = \mathcal{L}_k \frac{\partial u}{\partial \mathbf{n}} - \mathcal{M}_k u \qquad \text{for } p \in \Omega, \tag{1.75}$$

which can be taken to the boundary using Theorem 1.7.21 to yield a BIE that is a Fredholm integral equation of the first kind

$$\mathcal{L}_k \frac{\partial u}{\partial \mathbf{n}} = \frac{1}{2} u(p) + \mathcal{M}_k u, \qquad \text{for } p \in \partial\Omega. \tag{1.76}$$

or recognizing $u|_{\partial\Omega} = g$ and letting $\tilde{g} = \frac{1}{2} g(p) + \mathcal{M}_k g$,

$$\mathcal{L}_k \frac{\partial u}{\partial \mathbf{n}} = \tilde{g}, \qquad \text{for } p \in \partial\Omega. \tag{1.77}$$

The good news is that we have obtained a simple BIE (1.77) with the boundary condition in (1.53) easily satisfied, and we shall discretize (1.77) to obtain a linear system and hence a numerical solution. But one has the option to reformulate the badly-conditioned problem (1.77) further as for some resonance wavenumbers k it does not have a unique solution (assuming g is already suitable for the solution to exist [159]). One solution [16] is to differentiate (1.76) (and then to combine the two equations)

$$\mathcal{M}_k^T \frac{\partial u}{\partial \mathbf{n}_q} = \frac{1}{2}\frac{\partial u}{\partial \mathbf{n}_p} + \mathcal{N}_k u. \tag{1.78}$$

Note that (1.78) alone can define a second kind equation in $\partial u/\partial \mathbf{n}$ to solve (1.53). However a linear combination of (1.77) and (1.78) gives a more robust formulation

$$-\frac{1}{2}\frac{\partial u}{\partial \mathbf{n}_p} + \mathcal{M}_k^T \frac{\partial u}{\partial \mathbf{n}_q} + \alpha_k \mathcal{L}_k \frac{\partial u}{\partial \mathbf{n}_q} = \mathcal{N}_k u + \alpha_k \tilde{g} = \mathcal{N}_k g + \alpha_k \tilde{g} \equiv \overline{g}, \tag{1.79}$$

or letting $\mathbf{q} = \frac{\partial u}{\partial \mathbf{n}}$

$$(I - \mathcal{K})\mathbf{q} = \widetilde{g}, \qquad \text{with } \mathcal{K} = \mathcal{M}_k^T + \alpha_k \mathcal{L}_k,\ \widetilde{g} = -2\overline{g}, \tag{1.80}$$

where α_k is a nonzero coupling parameter (usually taken as $-ik$). Here the BIE (1.80) is clearly a Fredholm integral equation of the second kind that is better conditioned; for convenience we shall write

$$(\mathcal{K}\psi)(p) = \int_{\partial\Omega} K(p,q)\psi(q)dS_q, \quad p \in \partial\Omega. \tag{1.81}$$

This idea first appeared in 1960s; see [16,159,83]. For the Neumann's boundary condition, similar results can be derived:

$$-\frac{1}{2}u + \mathcal{M}_k u + \alpha_k \mathcal{N}_k u = \left[\mathcal{L}_k + \alpha_k\left(\frac{1}{2}I + \mathcal{M}_k^T\right)\right]\frac{\partial u}{\partial \mathbf{n}_p} \equiv \overline{g} \text{ (known)}. \tag{1.82}$$

When $k = 0$, the Laplace case modelling the potential theory is often considered separately; see [504,306,488,222,261,238,262].

The above derivation, using Green's theorems explicitly, is the so-called direct BIE formulation. In fact, similar BIEs can be derived using the layer potentials (1.73) and their properties (Theorem 1.7.21), and directly matching up the boundary condition, since the single and double layer potentials satisfy (1.53) in Ω (and so does a combination of them as a hybrid potential).

We now come to our second step of a BEM to introduce discretizations into BIEs. For our concerned combined formulation (1.79), this amounts to specifying a computable subspace in $H^1(\partial\Omega)$; for the single layer equation one may choose a subspace in $H^0(\partial\Omega) = L^2(\partial\Omega)$. Divide the boundary domain $\partial\Omega$ into E subdomains $\partial\Omega_1, \ldots, \partial\Omega_E$; if $\Omega \subset \mathbb{R}^2$, $\partial\Omega$ is effectively one dimensional (1D) and if $\Omega \subset \mathbb{R}^3$, $\partial\Omega$ will be a 2D manifold. Thus for (1.53), the subdomains

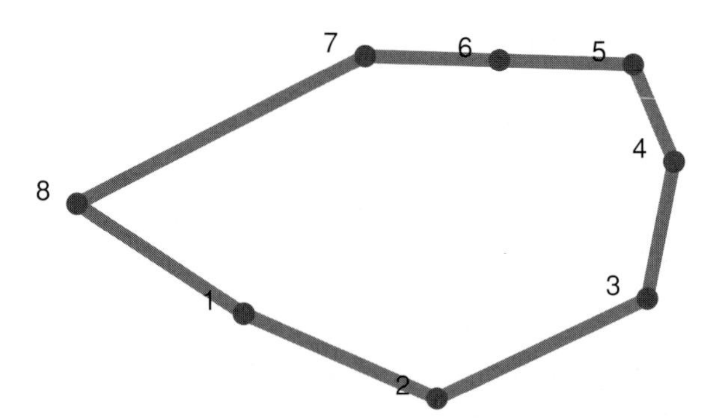

Figure 1.3. Sample BEM discretization of a domain Ω ($n = n_t = 8$ boundary points).

$\partial\Omega_j$ with $n = n_t$ end-points are simply segments along the boundary for the usual piecewise linear approximation (corresponding to the case of triangles for FEM) as illustrated in Figure 1.3 (compare to Fig 1.2). Here $n = E$. As with FEM, let $\phi_i \in \mathbb{P}_\ell(\partial\Omega^h)$ be similarly defined by (1.63) over the approximated domain of $\partial\Omega$

$$\partial\Omega^h = \bigcup_{j=1}^{E} \partial\Omega_j = \bigcup_{j=1}^{n} \partial\Omega_j.$$

Then our main approximation will be

$$\mathbf{q} = \sum_{j=1}^{n} \mathbf{q}_j \phi_j(p) \qquad \text{and} \qquad \overline{g} = \sum_{j=1}^{n} \overline{g}_j \phi_j(p). \tag{1.83}$$

Note that, different from the FEM case (1.62), the BEM weak formulation (1.79) has already done the choice of the trial function v. We need to invent another bilinear form (in a weak fashion) to seek a projection with a 'minimal' residual [24]

$$\begin{array}{llll} \text{Galerkin} & \text{find } u \in \mathbb{P}_1(\partial\Omega^h): & (r, v) = 0, & \forall v \in \mathbb{P}_1(\partial\Omega^h), \\ \text{Collocation} & \text{find } u \in \mathbb{P}_1(\partial\Omega^h): & (r, v) = 0, & \forall v \in \mathbb{P}_0(\partial\Omega^h), \end{array} \tag{1.84}$$

where r is the residual function from (1.80)

$$r = \widetilde{g} - (I - \mathcal{K})\,\mathbf{q}$$

and the inner product is now defined in $\partial\Omega^h$ (rather than in Ω)

$$(u, v) = \int_{\partial\Omega} u(q) v(q) dS_q.$$

Substituting (1.83) into (1.84) yields our linear system

$$\text{Galerkin:} \quad \sum_{j=1}^{n} \mathbf{q}_j \left((I - \mathcal{K})\phi_j, \phi_i \right) = (\widetilde{g}, \phi_i) \quad \text{for} \quad i = 1, \ldots, n,$$

$$\text{Collocation:} \quad \sum_{j=1}^{n} \mathbf{q}_j \left((I - \mathcal{K})\phi_j, \Phi_i \right) = (\widetilde{g}, \Phi_i) \quad \text{for} \quad i = 1, \ldots, n, \tag{1.85}$$

where the piecewise constant $\Phi_i(p) = 1$ if $p \in \partial\Omega_{i-1} \cup \partial\Omega_i$ with $\partial\Omega_{-1} = \partial\Omega_n$ (or similarly defined). One may develop a hybrid Galerkin method that is superior to a normal Galerkin method [238].

In addition to the above two methods, integral equations can be solved by the so-called Nyström quadrature method (see also Section 4.7)

$$(I - \mathcal{K}_n)\phi_n = \widetilde{g}, \tag{1.86}$$

where $\mathcal{K}_n$ approximates $\mathcal{K}$ and is obtained by direct discretization of $\mathcal{K}$ by a (composite) n-point quadrature rule, e.g. Trapezium or Gaussian rules [24].

Remark 1.7.22.

(1) The BEM clearly has the dimension reduction advantage over the FEM – the domain discretization is reduced to that of finite boundary part only. This is especially useful for exterior problems: the boundary condition at the exterior boundary is exactly met and there is no need to impose a large artificial boundary for approximate solution or coupling. There are also theoretical advantages regarding conditioning, as far as preconditioning and condition numbers are concerned, as operators from BIEs are usually bounded while FEM operators, still pro-differential, are unbounded. It should also be remarked that, for exterior problems, placing a specially-shaped artificial boundary can lead to interesting coupled operator equations [504] – combining the advantages of a FEM and a BEM.

(2) Of course, as stated, a BEM is somewhat restrictive. (i) It demands the knowledge of a fundamental solution to the underlying operator $\mathbb{L}$ in (1.53). Without it, a BEM does not apply. (ii) It cannot allow nonhomogeneous terms (i.e. f in (1.53)) to exist (this is a genuine requirement – no longer a convenient assumption). In these two cases, however, a generalized BEM may be tried [385,428].

1.7.3 The finite difference method (FDM)

The finite difference method is one of the oldest and most well-known methods for solving PDEs [354]. The method starts with a mesh and derives nodal equations in a direct manner. However, it should be noted that for rectangular domains and regular partitions, a FDM can produce the same linear system as a FEM with piecewise linear approximations [499].

First discretize a given domain Ω by placing a suitable mesh with internal nodes $p_1, \ldots, p_n$. Then apply the commonly used central finite differences, assuming h_x, h_y are the horizonal and vertical mesh sizes respectively at a node (x, y),

$$\frac{\partial u}{\partial x} \approx \frac{u(x+h_x) - u(x-h_x, y)}{2h_x}, \qquad \frac{\partial u}{\partial y} \approx \frac{u(x, y+h_y) - u(x, y-h_y)}{2h_y} \tag{1.87}$$

$$\begin{aligned} \frac{\partial^2 u}{\partial x^2} &\approx \frac{u(x+h_x, y) + u(x-h_x, y) - 2u(x)}{h_x^2} \\ \frac{\partial^2 u}{\partial y^2} &\approx \frac{u(x, y+h_y) + u(x, y-h_y) - 2u(x, y)}{h_y^2} \end{aligned} \tag{1.88}$$

to a PDE at each nodal point to derive the linear system. First derivatives can also be approximated by one-sided differences

$$\frac{\partial u}{\partial x} \approx \frac{u(x+h_x, y) - u(x, y)}{h_x} \equiv \frac{\delta_x^+ u(x, y)}{h_x}, \quad \frac{\partial u}{\partial y} \approx \frac{\delta_y^+ u(x, y)}{h_y}$$

$$\frac{\partial u}{\partial x} \approx \frac{u(x, y) - u(x-h_x, y)}{h_x} \equiv \frac{\delta_x^- u(x, y)}{h_x}, \quad \frac{\partial u}{\partial y} \approx \frac{\delta_y^- u(x, y)}{h_y}$$

therefore,

$$\begin{aligned} \frac{\partial}{\partial x}\left(D\frac{\partial u}{\partial x}\right) &\approx \frac{\delta_x^+\left(D(x-\frac{h}{2}, y)\delta_x^- u(x, y)\right)}{h_x^2} \\ &= \frac{D(x+\frac{h}{2}, y)(u(x+h, y) - u(x, y)) - D(x-\frac{h}{2}, y)(u(x, y) - u(x, y-h))}{h_x^2}, \\ \frac{\partial}{\partial y}\left(D\frac{\partial u}{\partial y}\right) &\approx \frac{\delta_y^+\left(D(x, y-\frac{h}{2})\delta_y^- u(x, y)\right)}{h_y^2} \\ &= \frac{D(x, y+\frac{h}{2})(u(x, y+h) - u(x, y)) - D(x, y-\frac{h}{2})(u(x, y) - u(x, y-h))}{h_y^2}. \end{aligned}$$

For the 2D case, one often uses the so-called stencil notation to denote a discrete equation at a grid point (i.e. the grid equation). Thus the above discretization of $\nabla(D\nabla u)$ may be denoted by the five-point stencil

$$\begin{bmatrix} & D(x,y+\frac{h}{2}) & \\ D(x-\frac{h}{2},y) & -\left(D(x,y+\frac{h}{2})+D(x,y-\frac{h}{2})+D(x-\frac{h}{2},y)+D(x+\frac{h}{2},y)\right) & D(x+\frac{h}{2},y) \\ & -D(x,y-\frac{h}{2}) & \end{bmatrix},$$

which reduces to the more familiar (discrete Laplacian) form when $D \equiv 1$

$$\begin{bmatrix} & 1 & \\ 1 & -4 & 1 \\ & 1 & \end{bmatrix}.$$

This essentially 2D notation can generalize to 3D if used as a vector of (stacking) 2D stencils [460].

For nonuniform tensor-product meshes (or discretization near nonrectangular boundaries), one may need to develop more accurate formulae using the Taylor expansions as only the low-order (one-sided) FD formulae are applicable. Given $f(x+h_2)$, $f(x-h_1)$, $f(x)$, the approximation to the first-order derivative by

$$\frac{f(x+h_2)-f(x-h_1)}{h_1+h_2} \approx f'(x) + \frac{h_1^2+h_2^2}{2(h_1+h_2)} f''(x) + \cdots \tag{1.89}$$

is only first-order accurate while a better approximation using three function values

$$\frac{h_1}{(h_1+h_2)h_2}[f(x+h_2)-f(x)] - \frac{h_2}{(h_1+h_2)h_1}[f(x-h_1)-f(x)]$$
$$\approx f'(x) + \frac{h_1h_2}{6} f'''(x) + \cdots$$

will be second-order accurate, derived from choosing w_1, w_2 in zeroing out the coefficient of the second derivative term in the Taylor expansion of $w_2\,[f(x+h_2)-f(x)] + w_1\,[f(x-h_1)-f(x)]$. (By the same idea, it turns out that we have to use four function values to design a FD formula of second-order accurate to approximate $f''(x)$).

Consider solving equation (1.53) on a rectangular domain Ω as illustrated in Figure 1.4 where we assume for simplicity $h = h_x = h_y = 1$ and $g = 0$. Then

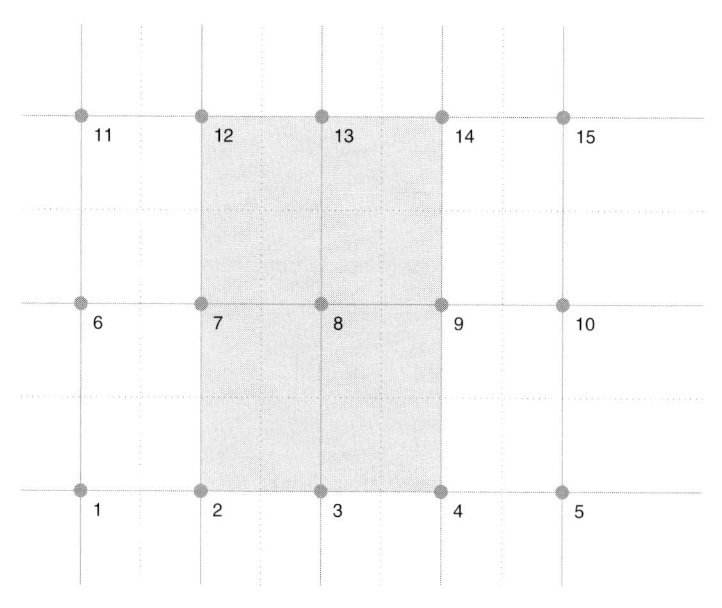

Figure 1.4. Sample FDM discretization of a domain Ω ($n = 15$ internal points).

the discrete equations for (1.53) centred at nodes 2, 8 are the following

$$
\begin{aligned}
\frac{2u_2 - u_1 - u_3}{h^2} + \frac{2u_2 - g(x_2, y_2 - h) - u_7}{h^2} - k^2 u_2 &= f(x_2, y_2) = f_2, \\
\frac{2u_8 - u_7 - u_9}{h^2} + \frac{2u_8 - u_3 - u_{13}}{h^2} - k^2 u_8 &= f(x_8, y_8) = f_8, \\
\text{i.e.} \qquad - u_1 + (4 - k^2)u_2 - u_7 &= f_2, \\
-u_3 - u_7 + (4 - k^2)u_8 - u_9 - u_{13} &= f_8.
\end{aligned}
\tag{1.90}
$$

Overall, the final linear system from solving (1.53) using the sample mesh in Figure 1.4 can be written as (1.1) with $b = f$ and

$$
A = \begin{bmatrix} T & -I & \\ -I & T & -I \\ & -I & T \end{bmatrix}, \text{ with } T = \begin{bmatrix} d & -1 & & & \\ -1 & d & -1 & & \\ & -1 & d & -1 & \\ & & -1 & d & -1 \\ & & & -1 & d \end{bmatrix} \text{ and } d = 4 - k^2.
\tag{1.91}
$$

The matrix structure of A can be visualized in MATLAB by (as shown in Figure 1.5)

```
>> spy(A,'s', 12)           % size 12 for larger symbol s='square'
```

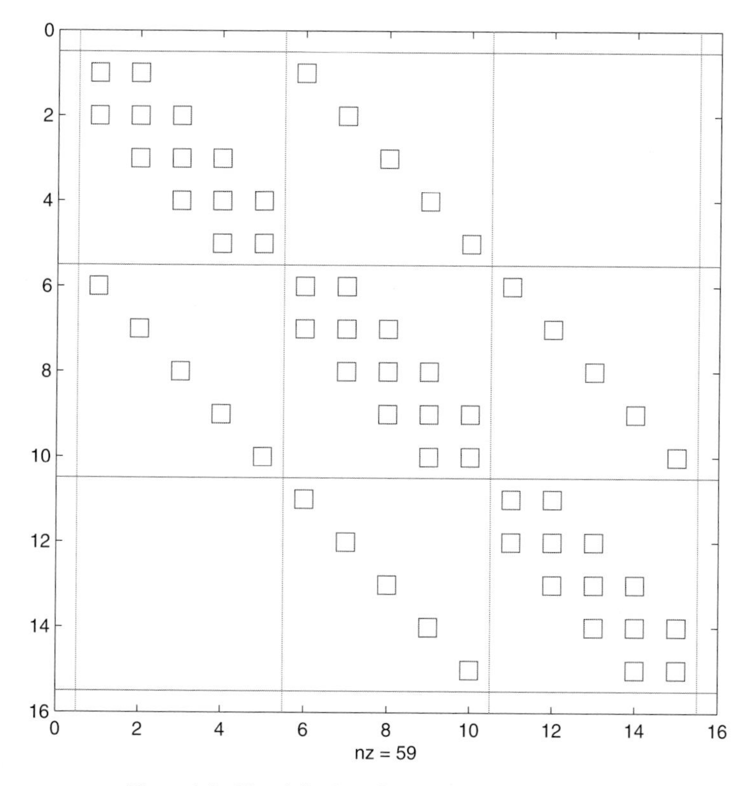

Figure 1.5. Visual display of a sparse matrix from FDM.

It remains to mention *upwinding schemes* for practical problems such as fluid flows where a central difference for first order terms may lead to oscillation when h is not small enough. Consider a modified PDE from (1.53)

$$-\Delta u + b(x, y)\frac{\partial u}{\partial x} - k^2 u = f(p), \quad p \in \Omega. \tag{1.92}$$

The usual central difference for $b\frac{\partial u}{\partial x}$ is

$$b\frac{\partial u}{\partial x} \approx b(x, y)\frac{u(x+h, y) - u(x-h, y)}{2h} = \frac{b(x, y)}{h}\frac{\delta_x^+ u + \delta_x^- u}{2},$$

where $u(x, y)$ is not influencing the result. The so-called upwind difference scheme is the following

$$b\frac{\partial u}{\partial x} \approx b\left[\frac{1 - sign(b)}{2}\frac{\delta_x^+ u(x, y)}{h_x} + \frac{1 + sign(b)}{2}\frac{\delta_x^- u(x, y)}{h_x}\right]. \tag{1.93}$$

From a matrix point, the scheme ensures that the overall matrix has taken a positive contribution from this first-order term to the diagonal entry (or towards

SDD Section 3.3.2)

$$A_{ii} = (4 - k^2) + |b(x_i, y_i)|.$$

This makes the upwinding formula easier to remember.

1.7.4 The finite volume method (FVM)

The finite volume method is like a hybrid method of FDM, FEM, and BEM, because it derives a nodal equation just like the FDM, and solves a PDE using a weak formulation with a weight and an integration domain resembling (but differing from) the FEM and BEM; see [473].

The first step is similar to a FDM. Place a discrete mesh over Ω with internal nodal points $p_1, \ldots, p_n$. The aim is also to produce a discrete equation at each p_j.

The second step resembles the FEM or BEM as we shall choose a simple trial function $v = 1$ locally in a piecewise manner. The key phrase is *control volume* V, which is a small and closed subdomain of Ω assigning to each nodal point p_i. Let Γ be the closed boundary of V and Ξ be the area of V (or the volume of V in three dimensions). Illustrated in Figure 1.6 is a simple example, where $\Gamma = \cup_j \Gamma_j$ denote the local (artificial) boundaries of V. Note that the union of all such control volumes V makes up the whole approximation domain Ω^h. Taking $v = 1$, we can deduce the following useful identities from the Green's first theorem (1.57)

$$\begin{cases} \displaystyle\int_V \frac{\partial u}{\partial x_i} d\Omega = \int_\Gamma u n_i dS, \\ \displaystyle\int_V \frac{\partial^2 u}{\partial x_i{}^2} d\Omega = \int_\Gamma \frac{\partial u}{\partial x_i} n_i dS, \\ \displaystyle\int_V \nabla \cdot (D\nabla u)\, d\Omega = \int_\Gamma D\nabla u \cdot \mathbf{n} dS. \end{cases} \tag{1.94}$$

Although a simple rectangular mesh is shown in Figure 1.6, one can equally use a triangulation just as in Figure 1.2 and this is partly why FVM has become increasingly popular with many research communities.

The FVM proposes to derive a nodal equation for the PDE (1.53) from the special weak formulation:

$$0 = \int_V (\mathbb{L}u - f)\, v d\Omega = \int_V (\mathbb{L}u - f)\, d\Omega = \int_V \left(-\Delta u - k^2 u - f\right) d\Omega. \tag{1.95}$$

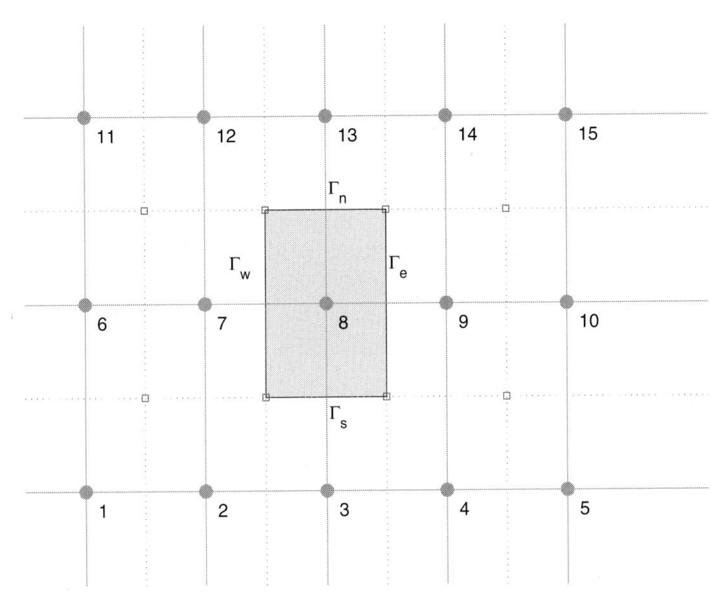

Figure 1.6. Sample FVM discretization of a domain Ω ($n = 15$ internal points).

Applying the above formulae (1.94) yields

$$-\int_{\Gamma} \nabla u \cdot \mathbf{n} dS - \int_{V} \left(k^2 u + f\right) d\Omega = 0.$$

We now consider the particular node $p_8 \in \Omega^h$ shown in Figure 1.6 with $\Gamma = \Gamma_e \cup \Gamma_n \cup \Gamma_w \cup \Gamma_s$. Note that the unit out normal $\mathbf{n}$ can be first computed at each Γ_j and denote by Ξ_j the length of Γ_j. Then (1.95) yields

$$\begin{aligned}
&-\int_{\Gamma} \nabla u \cdot \mathbf{n} dS - \int_{V} \left(k^2 u + f\right) d\Omega \\
&= \int_{\Gamma_w} \frac{\partial u}{\partial x} dS + \int_{\Gamma_s} \frac{\partial u}{\partial y} dS - \int_{\Gamma_e} \frac{\partial u}{\partial x} dS - \int_{\Gamma_n} \frac{\partial u}{\partial y} dS - \int_{V} \left(k^2 u + f\right) d\Omega \\
&\approx \Xi_w \frac{u_8 - u_7}{h_x} + \Xi_s \frac{u_8 - u_3}{h_y} - \Xi_e \frac{u_9 - u_8}{h_x} - \Xi_n \frac{u_{13} - u_8}{h_y} - \Xi \left(k^2 u_8 + f_8\right),
\end{aligned} \tag{1.96}$$

where the last step used the numerical middle-point quadrature rule. Thus from (1.96), we have obtained the nodal equation at p_8 as

$$-w_3 u_3 - w_7 u_7 + \left(w_3 + w_7 + w_9 + w_{13} - k^2\right) u_8 - w_9 u_9 - w_{13} u_{13} = f_8, \tag{1.97}$$

where $w_3 = \Xi_s/(h_x\Xi)$, $w_7 = \Xi_w/(h_y\Xi)$, $w_9 = \Xi_e/(h_x\Xi)$, $w_{13} = \Xi_n/(h_y\Xi)$. Clearly if $h = h_x = h_y$, then $\Xi_j = h$, $\Xi = h^2$ and $w_j = 1/h^2$ so the FVM produces the same equation as the FDM equation (1.90) at node 8. Applying the idea of (1.95) to all nodes in Ω^h will lead to the linear system for u.

Remark 1.7.23.

(1) The FVM appears to be a robust method. Its resemblance to FEM and FDM brings resembling advantages. On the FEM side, the FVM can work on any given mesh (not necessarily regular ones). On the FDM side, the FVM is node based so it is as easy to implement and to verify as the FDM, more convenient than the FEM in some sense. It should be remarked that if one insisted on using the point-wise Gauss–Seidel Newton methods, the FVM was considered as a 'matrix-free' iterative method by some research communities; clearly convergence cannot be very fast in general.

(2) A remarkable feature of a FVM is its treatment of convective terms, removing the need of directly discretizing the first derivatives all together! For the modified model (1.92), the nodal equation derivation similar to (1.95) will be

$$\begin{aligned} 0 &= \int_V \left(-\Delta u + b(x,y)\frac{\partial u}{\partial x} - k^2 u - f \right) d\Omega \\ &= -\int_\Gamma \nabla u \cdot \mathbf{n} dS + \int_\Gamma (ub)\, dS - \int_V \left(u\frac{\partial b}{\partial x} + k^2 u + f \right) d\Omega. \end{aligned} \tag{1.98}$$

Clearly $\partial u/\partial x$ is eliminated; *this is a step that the usual FEM does not do as nearly well as the elegant FVM.* This advantage is even clearer for constant coefficients e.g. the conservation laws in hyperbolic equations; see [473].

1.7.5 The global element methods (GEMs)

The GEMs refer to those FEM-like methods that use global functions as trial functions. These include spectral methods [91,458] and meshless methods [122]. We shall make a quick summary of them before concluding this chapter. As most of these methods employ basis functions that are more than twice differentiable, differentiating (rather than integrating) the trial functions is a common practice in derivation.

The general framework for such spectral methods to solve $\mathbb{L}u = f$ follows closely from the FEM

$$\begin{cases} u(p) = \sum_{j=1}^{n} \alpha_k \phi_k(p), \\ (\mathbb{L}u, \psi_i) = (f, \psi_i), \qquad \text{for } i = 1, \ldots, n - \ell, \end{cases}$$

where ℓ more equations are reserved to satisfy the boundary conditions; if such conditions are periodic, ℓ may be zero. Here the solution

$$u = u_n = \sum_{j=0}^{n_1} \sum_{\ell=0}^{n_2} \alpha_k \phi_k(p), \qquad k = \ell(n_1 + 1) + j + 1,$$

with the Fourier spectral methods choosing

$$\phi_k(p) = \phi_{j,\ell}(x_1, x_2) = \exp\left(i x_1 (j - \frac{n_1}{2})\right) \exp\left(i x_2 (\ell - \frac{n_2}{2})\right)$$

and the Chebyshev spectral methods choosing (see (3.45) for a definition)

$$\phi_k(p) = \phi_{j,\ell}(x_1, x_2) = T_j(x_1) T_\ell(x_2).$$

If ψ_i are chosen as piecewise constants, the resulting collocation algorithm is called the pseudo-spectral method.

Positively speaking, these methods offer a lot of advantages for several special classes of problems where the underlying domain is periodic. When applicable, a spectral method gives much greater accuracy than a FEM and the convergence rate can be exponential rather polynomial. It is known that matrices generated by the spectral methods from solving second order PDE's can be dense and ill-conditioned [231,458]. On the other hand, if a fixed accuracy is aimed at, linear systems generated from the highly accurate spectral methods are not too large scaled so iterative methods may not be needed but preconditioning techniques are needed to improve conditioning.

The so-called meshless methods [122], usually based on using radial basis functions (RBFs) [79], are easier methods to describe and apply than their spectral counterparts. However the power of these methods is different from (and a bit opposite to) that of spectral methods because the former, for scattered data interpolation and solving PDEs, is better suited to problems with a less regular mesh, and to low accuracy requirements. For instance, in oil exploration, the RBF may offer the only solution of 'prediction' of activities in the neighbouring areas if all you have obtained is a set of 10 measures dug at irregular places.

Let $\phi(r)$ be a RBF e.g. $\phi(r) = 1 + r$ with $r = r(p, q) = |p - q|$ the same as with the BEM, and $p_1, \ldots, p_n$ be a sequence of scattered data points in some closed domain Ω, on which we shall seek a RBF interpolation

$$s(p) = \sum_{j=1}^{n} \alpha_j \phi(r_j) \qquad \text{with} \qquad r_j = |p - p_j|, \tag{1.99}$$

for a given functions $f(p) = f(x, y)$

$$\text{find } \{\alpha_j\}'s : \quad s(p_i) = f(p_i), \quad \text{for } i = 1, \ldots, n.$$

Using (1.99) and $u = s(p)$, solving an operator equation $\mathbb{L}u = f$ yields a linear system (which is not difficult to derive). The difficulty lies in establishing invertibility and sparsifying the dense RBF matrix within the same numerical accuracy. Notice that the ordering of nodes p_j plays no part in the methods and so comes the name 'meshless methods'. See [8,122].

1.7.6 Linearization methods

We finally review a large and rich source of linear systems, which are obtained by linearizing a multidimensional nonlinear system. There are many ways of proposing a linearization [376]. We only discuss the Newton method; some more Newton variants can be found in Chapter 14.

Denote a nonlinear system in $\mathbb{R}^n$ by

$$\left.\begin{aligned} F_1(x_1, x_2, \ldots, x_n) &= 0, \\ F_2(x_1, x_2, \ldots, x_n) &= 0, \\ &\cdots \\ F_n(x_1, x_2, \ldots, x_n) &= 0, \end{aligned}\right\} \qquad \text{or} \qquad \mathbf{F}(\mathbf{x}) = 0. \tag{1.100}$$

Suppose $\mathbf{F}$ is differentiable (if not there are alternative techniques). Then starting from an initial guess $\mathbf{x}^{(0)}$, the Newton method computes the next iterate by

$$\mathbf{x}^{(j)} = \mathbf{x}^{(j-1)} - \boldsymbol{\Delta} x^{(j-1)}, \qquad \text{with} \qquad \underbrace{\mathbf{J}\boldsymbol{\Delta} x^{(j-1)} = \mathbf{F}(\mathbf{x}^{(j-1)})}_{\text{linear system}}, \tag{1.101}$$

where the Jacobian matrix, evaluated at $\mathbf{x}^{(j-1)}$, is

$$\mathbf{J} = \begin{bmatrix} \frac{\partial F_1}{\partial x_1} & \frac{\partial F_1}{\partial x_2} & \cdots & \frac{\partial F_1}{\partial x_n} \\ \frac{\partial F_2}{\partial x_1} & \frac{\partial F_2}{\partial x_2} & \cdots & \frac{\partial F_2}{\partial x_n} \\ \vdots & \vdots & \cdots & \vdots \\ \frac{\partial F_n}{\partial x_1} & \frac{\partial F_n}{\partial x_2} & \cdots & \frac{\partial F_n}{\partial x_n} \end{bmatrix}.$$

We wish to make an early remark on these linear systems. When the underlying nonlinear system (1.100) models some kind of stable equilibrium (e.g. in Chapter 14), to maintain stability of the nonlinear system, the eigenvalues of $\mathbf{J}$ are supposed to have negative real parts and this is about the worst scenario for an iterative method to work (Chapter 3) i.e. iterative solution of (1.101) requires preconditioning.

1.8 Common theories on preconditioned systems

We have mentioned that the purpose of preconditioning (1.1) by (1.2) or (1.5) is to ensure that the preconditioned system has better spectral properties, that are required by the iterative methods in later chapters.

Here we give a short list of common theories on preconditioned systems that a reader can find in the literature. These theories, while useful and mathematically interesting, are not always relevant to faster iterative solution.

For a symmetric matrix $A \in \mathbb{R}^{n\times n}$, any of the following proves that the symmetric M is an effective preconditioner.

(1) **Conditioned number estimates** for a SPD case – $\kappa(MA) \le O(1)$ for any n.

A common form of conditioned number estimates is presented by

$$c_1 x^T M x \le x^T A x \le c_2 x^T M x,$$

where $c_1, c_2 > 0$, and $M = M^{1/2}M^{1/2}$ and $M^{1/2}$ are both SPD. This implies that

$$\kappa_2(M^{-1/2}AM^{-1/2}) = \kappa_1(M^{-1}A) \le c = c_2/c_1.$$

To explain this result, first note that (setting $y = M^{1/2}x$ and $z = M^{-1/2}x$)

$M^{-1/2}AM^{-1/2}x = \lambda x$ implies $M^{-1/2}Az = \lambda M^{1/2}z$ and also $M^{-1}Az = \lambda z$

so $\lambda(M^{-1}A) = \lambda(M^{-1/2}AM^{-1/2})$. Second, the above inequalities may be written as

$$c_1 y^T y \le y^T M^{-1/2} A M^{-1/2} y \le c_2 y^T y, \qquad c_1 \le \frac{y^T M^{-1/2} A M^{-1/2} y}{y^T y} \le c_2.$$

Using the min–max theorem (on Rayleigh quotients) for symmetric matrices,

$$c_1 \le \lambda(M^{-1/2} A M^{-1/2}) \le c_2.$$

(2) **Bounding the largest and smallest eigenvalues.**

$$c_1 \le \lambda_{\min}(MA), \qquad \lambda_{\max}(MA) \le c_2,$$

where $c_1, c_2 > 0$.

(3) **Eigenvalue estimates**. There exists an fixed integer τ such that for $j \ge \tau$, $|\lambda_j(MA) - 1| < \epsilon$ where ϵ is some fixed small quantity.

For an unsymmetric matrix A, all of the above are doubtful as a test of the effectiveness of preconditioner M unless $MA \approx I$ or $\|MA - I\| \approx 0$. As remarked in Chapter 3, the preconditioned system MA must be close to normality to justify the validity of eigenvalue information.

1.9 Guide to software development and the supplied Mfiles

We first make some general remarks on the general and important issue of software. Then we comment on the supplied Mfiles for this book and finally list the specific files developed for this chapter.

♦ **Sources of software and development issues.** Producing **reliable**, **accurate** and **fast** numerical software for solving applied mathematical problems from any scientific computing discipline is the main aim of computational mathematics. However all three aspects must go through the usual process of research and development.

The software issue can be a big problem for many readers. Although there are many good examples and practices of published software, there exist two extreme practices that are not helpful to the general community and readers at large.

(1) On one hand, many high level, 'fast' and general purpose (some commercial) codes can solve various practical problems. Quite often, the exact descriptions of the mathematics and algorithms used are not always complete.

So it can be difficult to generalize, adapt or enhance these codes. Besides, some commercial codes may not have the latest methods available.

(2) On the other hand, increasingly, more and more new research methods and techniques come to the scene with the claim that they can achieve many wonderful aims. But, often, an average reader has no way of figuring out how the methods work exactly as a pseudo-code may not be given or a method may not be reproduced easily.

Therefore, the first and most crucial issue is to show transparency of a method or an algorithm so that a reader can either follow a published code or reproduce it. Sophistication and theory are secondary. The author strongly believes that developing quality software is easy if one really understands the mathematics behind it and algorithms in detail.

This book will aim to go beyond the 'usual practice' of showing ample experimental and theoretical results *without* giving full and useful details i.e. we illustrate ideas and methods using algorithms as well as pseudo-codes and inviting a reader to use existing (various pointers to such will be given) or to develop new codes hopefully with ease. Fortunately for the research community, the existence of the MATLAB[12] script language makes it possible for readers to share knowledge and algorithms in a portable and readable form. Moreover for later chapters, where necessary, we shall also give out our own programs for readers' benefit. All supplied Mfiles and programs will be available from

---

http : //www.cambridge.org/9780521838283
http : //www.liv.ac.uk/maths/ETC/mpta

♦ **Guide to the supplied Mfiles for this book.** A Mfile can also be written in such a sophisticated way that even an author may not understand it after a while! Bearing this mind, we have tried to stick to simplicity whenever possible. Where simplification at the price of sophistication to illustrate an idea is actually the main aim, this is clearly stated in the relevant Mfiles. There are three ways to find out the details of a Mfile (e.g. take `file.m` for example).

(1) In MATLAB, typing the name `file` without any parameters will invoke the self-help comments being shown (we realize that it is quite annoying for MATLAB to give you error messages on missing parameters). These self-help comments should explain how to run the Mfile `file`.

[12] MATLAB© is a registered trademark of MathWorks, Inc. See http://www.mathworks.com

(2) From the text, near an algorithm, we normally discuss the mathematics relevant to a Mfile. An index of the entry `file.m` can be found towards the end of the book in Appendix D.

(3) Check the last section of each chapter, where supplied Mfiles will be listed with short descriptions including their level of sophistication. For a Mfile classified as 'advanced', a novice reader should simply run it to further understanding only and is not expected to modify it.

♦ **Mfiles for Chapter 1.**

[1] hess_as.m – Implement the Arnoldi decomposition as in Algorithm 1.4.9 in Section 1.4 (via the Gram–Schmidt approach). The Mfile is moderate, not complicated. (Readers interested in the other two approaches for completing the Arnoldi decomposition should consult the Mfiles `givens.m` and `houses.m` as listed in Section 2.7.)

[2] exafft16.m – Compute the FFT of a vector as in Example 1.6.17 in a step-by-step manner.

[3] mygrid.m – Utility Mfile to supplement to exafft16.m. When used alone, it prints a mesh grid on an existing graph.

[4] mat_prt.m – Utility Mfile to supplement to exafft16.m. When used alone, it allows one to display a matrix or two matrices side-by-side without wrapping around.

[5] mat_prt4.m – Utility Mfile to supplement to exafft16.m. When used alone, it is similar to mat_prt.m but prints 4 digits.

[6] fwt.m – Compute the fast wavelet transform of a vector or a matrix, using a choice of Daubechies' compact wavelet filters. It implements the transform 'matrix'-free version; alternatively one can form the sparse transform matrices.

[7] ifwt.m – Compute the inverse fast wavelet transform of a vector or a matrix, reversing `fwt.m`. (Compare to `fwts.m` and `iwts.m` from Chapter 8).

[8] fft_fwt.m – Compare the two examples of FWT and FFT transforms in Figure 1.1.

♦ **Final remarks and main software sites.** Most numerical software in scientific computing can be obtained from the Internet in some form, as long as one knows where and what to look for. This book hopes to serve the purpose of helping readers who are not yet numerical experts in this direction. To summarize and re-iterate on the main message, the author believes that developing efficient software is not a hard job if one is in possession of a good numerical method and that often exposing the ideas behind methods is more important than going for the nitty-gritty implementation details. Nevertheless, in addition to developing a good practice in software writing, sharing software resources is a positive way forward. Fortunately one can always browse these places for software.

(1) **netlib** is like a gold mine for good numerical software that includes codes from ACM collection and LAPACK. Access via internet

http:/www.netlib.org

or by sending an email with the message **send index** to

netlib@ornl.org

(2) Professional and commercial packages:

Numerical Algorithm Groups (NAG) at: http://www.nag.co.uk/

and Mathematics and Statistics Libraries (IMSL) at: http://www.vni.com/

(3) Numerical recipes [390] programs (curtseys of Cambridge University Press)

Numerical Recipes at: http://www.nr.com/

(4) MATLAB exchanges (for techniques such the FEM)

http : //www.mathworks.com/matlabcentral/fileexchange/

Refer also to Appendix E for more lists.

2

Direct methods

How much of the matrix must be zero for it to be considered sparse depends on the computation to be performed, the pattern of the nonzeros, and even the architecture of the computer. Generally, we say that a matrix is sparse if there is an advantage in exploiting its zeros.

IAIN DUFF, *et al. Direct Methods for Sparse Matrices.* Clarendon Press (1986)

To be fair, the traditional classification of solution methods as being either direct or iterative methods is an oversimplification and is not a satisfactory description of the present state of affairs.

MICHELE BENZI. *Journal of Computational Physics,* Vol. 182 (2002)

A direct method for linear system $Ax = b$ refers to any method that seeks the solution x, in a finite number of steps, by simplifying the general matrix A to some special and easily solvable form (1.3), e.g. a diagonal form or triangular form. In the absence of computer roundoff, x will be the exact answer x^*; however unless symbolic computing is used, computer roundoff is present and hence conditioning of A will affect the quality of x. Often a direct method is synonymous with the Gaussian elimination method, which essentially simplifies A to a triangular form or equivalently decomposes matrix A into a product of triangular matrices. However one may also choose its closely related variants such as the Gauss–Jordan method, the Gauss–Huard method or the Purcell method especially when parallel methods are sought; refer to [143].

The purpose of this chapter is to mainly address the inversion step of a forward type preconditioner M i.e. solve $Mx = b$ for a given b. Of course, backward type (inverse based) preconditioners involve a simple matrix vector multiplication. As far as solving the original system $Ax = b$ is concerned, an inaccurate (but fast) solution procedure can also be adapted to produce a preconditioner. Thus many competing direct methods could potentially provide useful preconditioners; we give a brief but yet comprehensive account of these methods with these sections.

Section 2.1 The LU decomposition and variants
Section 2.2 The Newton–Schulz–Hotelling method
Section 2.3 The Gauss–Jordan decomposition and variants
Section 2.4 The QR decomposition
Section 2.5 Special matrices and their direct inversion
Section 2.6 Ordering algorithms for better sparsity
Section 2.7 Discussion of software and Mfiles

We first discuss the standard Gaussian elimination method in Section 2.1 and then variants of the Gauss–Jordan (GJ) method in Section 2.3. The material in Section 2.1 are perhaps well known to most readers while the material in Section 2.3 may not be familiar to many readers as the GJ method is often stated in passing in popular texts. The decomposition idea is later used in Section 5.6 in improving the approximate inverse preconditioner. We next briefly review special matrices that can be inverted analytically in a fast way and thus can be taken as a candidate for a preconditioner in suitable contexts. These include the FFT solution of the circulant and Toeplitz matrices. In matrix computing, the real challenge is to be able to identify these special matrices as part of A that can be used to justify their usefulness in preconditioning $\kappa(A)$ or its spectrum. Finally we consider ordering algorithms using graph theory before discussing the frequently-used Schur complements approaches which are essentially direct methods. These include the Duff's spiral ordering and the DDM ordering.

In matrix computation texts [229,280], the error analysis for Gaussian elimination with partial pivoting states that the computed solution $\hat{x}$ satisfies $(A+E)\hat{x} = b$ with

$$\|E\|_\infty \le \gamma_1(n)\,\rho\,\kappa_\infty(A)\,\mathbf{u}, \qquad \frac{\|x-\hat{x}\|_\infty}{\|x\|_\infty} \le \gamma_2(n)\rho\,\kappa_\infty(A)\,\mathbf{u}, \tag{2.1}$$

where $\mathbf{u}$ is the unit roundoff (machine precision), $\gamma_1(n)$ and $\gamma_2(n)$ are two low-order polynomials of n, ρ is the so-called growth factor and $\kappa_\infty(A)$ is the condition number of A. Note $\kappa_\infty(A) \le \sqrt{n}\kappa_2(A)$.

Clearly the two important quantities, ρ and $\kappa_2(A)$, determine the overall accuracy of these direct methods. Firstly, to reduce the effect of ρ, one can use the so-called **pivoting** strategies (partial pivoting or complete pivoting); see [229,280]. In any strategy, pivoting amounts to permuting the linear system (and possibly the variables). That is, apply permutation matrices P, Q to obtain either $PAx = Pb$ (partial pivoting case) or $PAQy = Pb$ with $x = Qy$ (complete pivoting case). Note that once pivoting is applied, the effect of ρ is reduced.

Therefore, we may assume that a pivoting permutation has been applied to our underlying linear system (without essential loss of generality).[1] Secondly, the condition number remains the same after pivoting as $\kappa_2(PA) = \kappa_2(PAQ) = \kappa_2(A)$ (using norm invariance of orthogonal matrices P, Q – see Section 1.5). The task of reducing this quantity further is usually called **scaling** and the commonly used approach for scaling is to multiply a suitable diagonal matrix to matrix A (e.g. use the diagonals of matrix A); for symmetric matrices one may multiply a diagonal matrix to A from both left and right hand sides [229]. This idea of scaling is a simple form of preconditioning for direct methods! Preconditioning a direct method is often necessary. One can find more examples of non-diagonal preconditioners in Chapters 4–10.

2.1 The LU decomposition and variants

The method of Gaussian elimination (GE) is well documented in all linear algebra texts. We shall present the algorithm for sake of later use in contexts of preconditioning. We pay special attention to those variants of the method that can be adapted for preconditioner designs. As mentioned, consult other texts e.g. [229,280] for results of theoretical and error analysis.

2.1.1 The standard LDM^T factorization

In the well known LU decomposition, $A = LU$, one of the lower triangular matrix L or the upper triangular matrix U has unit diagonals – a choice between Crout and Doolittle factorizations [23,229,280,298]. Here we present the LDM^T factorization of GE, allowing both L and M to have unit diagonals. Then the Crout and Doolittle factorizations $A = LU$ become respectively $L \Leftarrow L$ and $U \Leftarrow DM$, and $L \Leftarrow LD$ and $U \Leftarrow M$. Note that only in this section is M used as an upper triangular matrix (in order to be consistent with the general literature) while M refers to a preconditioner for matrix A elsewhere.

We first show the traditional kij form of GE method to simplify matrix A.

Algorithm 2.1.1. (GE method).

(1) Set $m = I$, $x = b$.
(2) for $k = 1, \ldots, n$, *do*
(3) *for* $i = k + 1, \ldots, n$, *do*

[1] There are two issues here: (1) in theory, a LU decomposition exists if all the principal submatrices are nonsingular. Note the largest principal submatrix is A itself. (2) In practice, our assumption also means that the growth factor will not be large for our permuted matrix A; see [280,281].

(4) $m_{ik} = a_{ik}/a_{kk}$
(5) $x_i = x_i - m_{ik}x_k$
(6) *for* $j = k, \ldots, n$, *do*
(7) $a_{ij} = a_{ij} - m_{ik}a_{kj}$
(8) *end*
(9) *end*
(10) *end*
(11) *For LU decomposition, accept* $L = m$, $U = A$;
(12) *For* LDM^T *decomposition, take* $L = m$, $D = diag(a_{11} \ \ldots \ a_{nn})$, $M^T = D^{-1}A$.

Readers interested in checking the details should use the supplied Mfiles `g_e.m` and `ge_all.m` as directed in Section 2.7. In Algorithm 2.1.1, at step k, all the multipliers $m_{ik} = m(i, k)$ can be collected to define the lower triangular matrix

$$M_k = \begin{bmatrix} I_{k-1} & & & & \\ & 1 & & & \\ & -m_{k+1,k} & 1 & & \\ & -m_{k+2,k} & & \ddots & \\ & \vdots & & & \ddots \\ & -m_{n,k} & & & & 1 \end{bmatrix} = I - m_k e_k^T \quad \text{with}$$

$$m_k = \begin{bmatrix} O_{k-1} \\ 0 \\ m_{k+1,k} \\ m_{k+2,k} \\ \vdots \\ m_{n,k} \end{bmatrix} \tag{2.2}$$

where e_k is the kth unit vector and O_{k-1} denotes a size $k-1$ zero vector. Note that the product $m_k e_k^T$ is only a notation to mean a matrix with an isolated vector m_k in its column k, although $m_k e_k^T$ is formally a matrix!

There are two amazing facts on this type of matrices. Firstly, the inverse of M_k is the same as M_k with its multiplier entries flipped signs i.e.

$$M_k^{-1} = (I - m_k e_k^T)^{-1} = \begin{bmatrix} I_{k-1} & & & & \\ & 1 & & & \\ & m_{k+1,k} & 1 & & \\ & m_{k+2,k} & & \ddots & \\ & \vdots & & & \ddots \\ & m_{n,k} & & & & 1 \end{bmatrix} = I + m_k e_k^T.$$

Actually there is nothing amazing about this result if you relate it to the Sherman–Morrison theorem in Section 2.5 with $A = I$, $u = m_k$ and $v = e_k$.

Secondly, one simply packs the multiplier entries if such a matrix with more entries multiplies with another one with fewer entries, e.g.

$$\begin{bmatrix} I_1 & & & & & \\ m_{2,1} & 1 & & & & \\ m_{3,1} & & 1 & & & \\ \vdots & & & \ddots & & \\ \vdots & & & & \ddots & \\ m_{n,1} & & & & & 1 \end{bmatrix} \begin{bmatrix} I_2 & & & & & \\ & 1 & & & & \\ & m_{4,3} & 1 & & & \\ & m_{5,3} & & \ddots & & \\ & \vdots & & & \ddots & \\ & m_{n,3} & & & & 1 \end{bmatrix}$$

$$= \begin{bmatrix} 1 & & & & & \\ m_{2,1} & 1 & & & & \\ m_{3,1} & & 1 & & & \\ m_{4,1} & & m_{4,3} & \ddots & & \\ \vdots & & \vdots & & \ddots & \\ m_{n,1} & & m_{n,3} & & & 1 \end{bmatrix}. \tag{2.3}$$

Multiplications in the reverse order are different (check this out using the Mfile `ge_all.m`!).

The whole GE process can be represented by

$$M_{n-1}M_{n-2}\cdots M_1 A = U, \qquad \text{or} \qquad A = LU = LDM, \tag{2.4}$$

where $D = \operatorname{diag}(u_{11}, \ldots, u_{nn})$, $M^T = D^{-1}U$ and

$$L = M_1^{-1} \cdots M_{n-1}^{-1} = \begin{bmatrix} 1 & & & & & \\ m_{2,1} & 1 & & & & \\ m_{3,1} & m_{3,2} & 1 & & & \\ m_{4,1} & m_{4,2} & m_{4,3} & \ddots & & \\ \vdots & \vdots & \vdots & \vdots & \ddots & \\ m_{n,1} & m_{n,2} & m_{n,3} & \cdots & m_{n,n-1} & 1 \end{bmatrix}.$$

One may observe that the GE process (Algorithm 2.1.1) literally implements the factorization $\hat{L}A = U$ with $\hat{L} = M_{n-1}M_{n-2}\cdots M_1$ a lower triangular matrix with unit diagonals and the commonly known $A = LU$ (or $A = LDM^T$) decomposition is only derived from the method.

Note that $M = L$ when A is symmetric so $A = LDM^T$ becomes the Cholesky decomposition $A = LDL^T$.

2.1.2 Schur complements and the partitioned LU factorization

For high performance computing, block forms of the LU decomposition can be similarly developed if one views new block entries as 'entries' in a scalar algorithm; see [184,229,280].

We first review the one level Schur complements idea of partitioning A into 2×2 blocks

$$A = \begin{bmatrix} A_{11} & A_{12} \\ A_{21} & A_{22} \end{bmatrix} \tag{2.5}$$

where $A_{11} \in \mathbb{R}^{r \times r}$; usually one takes $r = n/2$. Assuming the principal block submatrices A_{11} and A are both nonsingular, a block LU decomposition of A exists [229,280]. Then by equating corresponding blocks, we can easily find such a decomposition as follows

$$\begin{bmatrix} A_{11} & A_{12} \\ A_{21} & A_{22} \end{bmatrix} = \begin{bmatrix} I & \\ L_{21} & I \end{bmatrix} \begin{bmatrix} U_{11} & U_{12} \\ & U_{22} \end{bmatrix} = \begin{bmatrix} I & \\ A_{21}A_{11}^{-1} & I \end{bmatrix} \begin{bmatrix} A_{11} & A_{12} \\ & S \end{bmatrix}, \tag{2.6}$$

where $S = U_{22} = A_{22} - L_{21}A_{12} = A_{22} - A_{21}A_{11}^{-1}A_{12}$ is the so-called *Schur complement* of matrix A_{11} in A. From (2.6), we can find A^{-1} the inverse of A in terms of partitioned blocks

$$\begin{bmatrix} A_{11} & A_{12} \\ & S \end{bmatrix}^{-1} \begin{bmatrix} I & \\ A_{21}A_{11}^{-1} & I \end{bmatrix}^{-1} = \begin{bmatrix} \left(I + A_{11}^{-1}A_{12}S^{-1}A_{21}\right)A_{11}^{-1} & -A_{11}^{-1}A_{12}^{-1} \\ -S^{-1}A_{21}A_{11}^{-1} & S^{-1} \end{bmatrix}, \tag{2.7}$$

Clearly the assumption that all principal submatrices of A (including A_{11}) are nonsingular ensures that A_{11}^{-1} and S^{-1} exist since $det(A) = det(A_{11})det(S)$. However if A is positive definite (not necessarily symmetric), then much more can be said of the Schur complement S.

Theorem 2.1.2. (Schur complement). *If $A \in \mathbb{R}^{n \times n}$ is positive definite (PD), then the block LU decomposition exists for any r such that $A_{11} \in \mathbb{R}^{r \times r}$, and*

(1) the diagonal blocks A_{11}, A_{22} are also PD;
(2) the Schur complement S is PD;
(3) if A is additionally symmetric (i.e. SPD), then
  (a) $\lambda_{\min}(A) \le \min\left\{\lambda_{\min}(A_{11}),\ \lambda_{\min}(A_{22}),\ \lambda_{\min}(S)\right\}$,
  $\max\left\{\lambda_{\max}(A_{11}),\ \lambda_{\max}(A_{22}),\ \lambda_{\max}(S)\right\} \le \lambda_{\max}(A)$ for eigenvalues;
  (b) $\kappa_2(S) \le \kappa_2(A)$ for condition numbers.

Proof. There is only one main idea from Section 1.5 needed for this proof, namely, $x^T Ax > 0$ for any $x \neq 0 \in \mathbb{R}^n$. Suppose an index vector $p = [p_1, \ldots, p_r]$ corresponds with a particular principal matrix $A_p = A(p, p)$. For any $x_p \neq 0 \in \mathbb{R}^r$, map it to $x \neq 0 \in \mathbb{R}^n$ with $x(p) = x_p$ and set other components of x to 0. Then $x^T Ax = x_p^T A_p x_p > 0$ so A_p is also PD (and hence nonsingular). Therefore (2.6) exists for this r.

(1) Taking $p = [1, \ldots, r]$ and $p = [r+1, \ldots, n]$ respectively can show that A_{11}, A_{22} are also PD.

(2) From the existence of factorization (2.6), S^{-1} exists so from (2.7) we know that S^{-1} is a principal submatrix of A^{-1}. From §1.5, $\lambda(A^{-1}) = 1/\lambda(A)$ i.e. A^{-1} is PD. The result of part (1) immediately states that S^{-1} is PD and hence S is PD.

(3) We apply the above idea for $x^T Ax$ to the Rayleigh quotient $x^T Ax / x^T x$ (note A is now symmetric). Let $x^{(1)} \in \mathbb{R}^r$ be an eigenvector for $\lambda_{\min}(A_{11})$. Take $p = [1, \ldots, r]$ and map $x^{(1)}$ to $x \in \mathbb{R}^n$ with $x(p) = x^{(1)}$ and set other components of x to 0. Then

$$\lambda_{\min}(A) = \min_{y \neq 0} \frac{y^T Ay}{y^T y} \leq \frac{x^T Ax}{x^T x} = \frac{x^{(1)T} A_{11} x^{(1)}}{x^{(1)T} x^{(1)}} = \lambda_{\min}(A_{11}). \tag{2.8}$$

Similarly

$$\lambda_{\max}(A) = \max_{y \neq 0} \frac{y^T Ay}{y^T y} \geq \frac{x^T Ax}{x^T x} = \frac{x^{(1)T} A_{11} x^{(1)}}{x^{(1)T} x^{(1)}} = \lambda_{\max}(A_{11}).$$

The proof for A_{22} is similar if taking $p = [r+1, \ldots, n]$. Now consider S. The proof is again similar if we repeat the above argument for the PD matrix A^{-1}: as S^{-1} is a principal submatrix of A^{-1} from (2.7), then we have $\lambda_{\max}(S^{-1}) \leq \lambda_{\max}(A^{-1})$ [i.e. $\lambda_{\min}(S) \geq \lambda_{\min}(A)$] and $\lambda_{\min}(S^{-1}) \geq \lambda_{\min}(A^{-1})$ [i.e. $\lambda_{\max}(S) \leq \lambda_{\max}(A)$]. (Alternatively use $S = A_{22} - A_{21} A_{11}^{-1} A_{12}$, noting $A_{12}^T = A_{21}$, and take similar partitions on $x \in \mathbb{R}^n$). Finally $\kappa_2(S) = \lambda_{\max}(S)/\lambda_{\min}(S) \leq \kappa_2(A)$. ∎

Here in (2.6), as with most situations in scientific computing, one tries not to form A_{11}^{-1} explicitly unless r is small. In the context of preconditioning, as we shall see, the Schur complement is usually not formed and there are two acceptable ways to proceed with using this Schur approach in a multilevel fashion.

(1) Solve the Schur equation by a preconditioned iterative method. Then $A_{11}^{-1} v$, for a vector v, will be only a linear solution.

(2) Approximate A_{11}^{-1} by some simple matrix, e.g. of a diagonal or tridiagonal form, so that an approximate Schur can be formed explicitly and be partitioned recursively using more levels.

These ideas will be pursued further in Chapters 7, 9 and 12.

We now briefly review the so-called partitioned LU decomposition [229,280] in a very clear and elementary way. Similar to (2.6), we assume that the scalar LU decomposition exists, i.e.[2]

$$\begin{bmatrix} A_{11} & A_{12} \\ A_{21} & A_{22} \end{bmatrix} = \begin{bmatrix} L_{11} & \\ L_{21} & L_{22} \end{bmatrix} \begin{bmatrix} U_{11} & U_{12} \\ & U_{22} \end{bmatrix} \tag{2.9}$$

where $A_{11} \in \mathbb{R}^{r \times r}$, all L_{ij} and U_{ij} are all unknowns to be determined, L_{11} and L_{22} are lower triangular matrices with unit diagonals, and U_{11} and U_{22} are upper triangular matrices (of course, U_{12} and L_{21} can be unknown dense matrices). On equating the corresponding blocks, we obtain the following

$$\begin{aligned} L_{11}U_{11} &= A_{11},\ L_{11}U_{12} = A_{12}, \\ L_{21}U_{11} &= A_{21},\ L_{22}U_{22} = A_{22} - L_{21}U_{12}. \end{aligned}$$

Thus the first equation amounts to decomposing the submatrix A_{11}, the second and third to solving systems with a triangular coefficient matrix, and the last equation becomes $L_{22}U_{22} = B$ with $B = A_{22} - L_{21}U_{12}$ known after solving the first three equations. Then the whole procedure of partitions can be repeated for B.

One may deduce that if $r = 1$ is maintained, both the partitioned method (2.9) and the block Schur method (2.6) reverse back to the scalar Algorithm 2.1.1; however one may prefer to take a large r (say $r = n/2$) in order to maximize block matrix operations. When $r \geq 2$, the two methods (2.9) and (2.6) are different; however the former method (2.9) is always equivalent to the GE for any r.

2.1.3 The substitution free WAZ factorization

As the inverse of a triangular matrix is also of a triangular form, it is hardly surprising to rewrite $A = LU$ as $L^{-1}AU^{-1} = I$ or the WAZ form

$$W^T A Z = I \tag{2.10}$$

[2] Readers who are familiar with the Lanczos method (for tridiagonalization), or the Arnoldi method (for upper Hessenberg form Section 1.4) or the Gram–Schmidt method (for $A = QR$ decomposition Section 2.4) should agree that the underlying idea in all these four approaches is identical, i.e. the entire decomposition method (by way of 'method of under-determined coefficients') will come about as soon as an assumption of existence is made.

where $W^T = L^{-1}$ is a lower triangular matrix with unit diagonals and $Z = U^{-1}$ is an upper triangular matrix. The task of a WAZ decomposition is to compute W and Z without using the LU decomposition. It turns out that this task is not difficult as columns of W and Z can be worked out together in a manner similar to the Gaussian elimination.

The following algorithms show how to compute W and Z: in particular the first algorithm computes the biconjugation decomposition

$$WAZ = D, \qquad \text{or} \qquad w_i^T A z_j = D_{ij}\delta_{ij}, \tag{2.11}$$

as in [55,211] (similar to $A = LDU$ decomposition) while the second computes (2.10) as in [510]. The first approach will be discussed again when factorized approximate inverse preconditioners are introduced in Chapter 5.

Algorithm 2.1.3. ($WAZ = D$ method).

(1) *Set matrices* $W = Z = I$.
(2) *for* $i = 1, \ldots, n$, *do*
(3) $\quad\quad pq = W(:, i)^T A(:, i), \quad D_{ii} = pq,$
(4) $\quad$ *for* $j = i + 1, \ldots, n$, *do*
(5) $\quad\quad q_j = W(:, j)^T A(:, i)/pq$
(6) $\quad\quad p_j = Z(:, j)^T A(i, :)^T/pq$
(7) $\quad$ *end for* $j = i + 1, \ldots, n$
(8) $\quad$ *for* $j = i + 1, \ldots, n$, *do*
(9) $\quad\quad$ *for* $k = 1, \ldots, i$, *do*
(10) $\quad\quad\quad W(k, j) = W(k, j) - q_j W(k, i)$
(11) $\quad\quad\quad Z(k, j) = Z(k, j) - p_j Z(k, i)$
(12) $\quad\quad$ *end*
(13) $\quad$ *end*
(14) *end*

Algorithm 2.1.4. ($WAZ = I$ method).

(1) *Set* $W = Z = I$.
(2) *for* $i = 1, \ldots, n$, *do*
(3) $\quad\quad p_{ii} = W(:, i)^T A(:, i)$
(4) $\quad$ *for* $j = i + 1, \ldots, n$, *do*
(5) $\quad\quad q_{ij} = W(:, j)^T A(:, i)/p_{ii}$
(6) $\quad\quad p_{ij} = Z(:, j)^T A(i, :)^T/p_{ii}$
(7) $\quad\quad$ *for* $k = 1, \ldots, j - 1$, *do*

(8) $W(k, j) = W(k, j) - q_{ij} W(k, i)$
(9) $Z(k, j) = Z(k, j) - p_{ij} Z(k, i)$
(10) *end*
(11) *end*
(12) $Z(:, i) = Z(:, i)/p_{ii}$
(13) end

If all the multipliers p_{ij}, q_{ij} in Algorithm 2.1.4 are collected into suitable matrices, just as with (2.2) for the GE, we can obtain the so-called Zollenkopf bi-factorization [510]. Specifically letting

$$
\begin{aligned}
W_k &= \begin{bmatrix} I_{k-1} & & & & \\ & 1 & & & \\ & -q_{k+1,k} & 1 & & \\ & \vdots & & \ddots & \\ & -q_{n,k} & & & 1 \end{bmatrix}, \\
Z_k &= \begin{bmatrix} I_{k-1} & & & & \\ & 1/p_{kk} & -p_{k,k+1} & \cdots & -p_{k,n} \\ & & 1 & & \\ & & & \ddots & \\ & & & & 1 \end{bmatrix},
\end{aligned}
\tag{2.12}
$$

the Zollenkopf factorization becomes

$$
W^T A Z = I, \qquad W^T = W_{n-1} W_{n-2} \cdots W_2 W_1, \qquad Z = Z_1 Z_2 \cdots Z_{n-1} Z_n. \tag{2.13}
$$

Readers interested in trying the details out should use Mfiles `waz_fox.m` for $WAZ = D$, `waz_zol.m` for $WAZ = I$ and `waz_all.m` for (2.13) as shown in §2.7.

2.2 The Newton–Schulz–Hotelling method

We now introduce a direct method for computing an inverse of a matrix which has some potential applications to preconditioning of a linear system; see [59,202]. The method is not truly direct as it uses iterations!

This is the Schulz method [423], sometimes called the Hotelling–Bodewig method but more commonly known as the Newton–Schulz method in the literature possibly because the method may be 'viewed' as a quasi-Newton method for solving the nonlinear system $AX = I$:

$$X_{j+1} = X_j - X_j \left[AX_j - I \right]$$

or in the popular format (for $j = 0, 1, 2, \ldots$)

$$X_{j+1} = 2X_j - X_j A X_j. \tag{2.14}$$

Once convergence is reached, $X^* = \lim_{k\to\infty} X_k = A^{-1}$ and a linear system $Ax = b$ is solved from forming the product $x = x_k = X_k b$. As discussed in [227], to put its importance into the right prospective, this method was at one stage in 1940s believed to be much better than the GE following the paper by Hotelling [286] (before it was shown otherwise). We call the method Newton–Schulz–Hotelling to reflect the fact that it is referred to in differently combined names.

As a Newton method, it has a quadratic convergence rate provided the initial start is sufficiently close to the true solution (here the inverse A^{-1}). This can be simply demonstrated as follows (by first multiplying A and then subtracting I)

$$I - AX_{j+1} = I - 2AX_j + AX_j AX_j$$

i.e.

$$(I - AX_{j+1}) = (I - AX_j)^2. \tag{2.15}$$

Thus the convergence is assumed if the initial guess X_0 satisfies $\rho(I - AX_0) < 1$.

We remark that the method is adequate if dense matrices are used throughout. In this case, fast matrix–matrix multiplication algorithms should be used; see the Strassen's method in [280]. However for sparse matrices or for situations where only an approximate inverse is desired, one needs to consider if the intermediate iteration matrices can be kept sparse. Numerical experiments may be done using the supplied Mfile `nsh.m`.

2.3 The Gauss–Jordan decomposition and variants

The Gauss–Jordan (GJ) method should be very familiar with most readers as it is usually used to compute the row-echelon form of a matrix and the inverse of a square matrix using the method of augmented matrix (i.e. reduce $[A \mid I]$ by row operations to $[I \mid A^{-1}]$). However our emphasis here is on the less well-known aspect of matrix decomposition with a view to later applications to multilevel preconditioning Section 5.6.

2.3.1 The standard GJ factorization

The standard GJ method has been well documented in various texts e.g. [23,298,229,280]. Briefly, on top of the GE, it completes the back substitution step at the same time. In parallel to Algorithm 2.1.1, we show the following kij version of the GJ method.

Algorithm 2.3.5. (GJ method).

(1) *Set* $m = I$, $x = b$.
(2) *for* $k = 1, \ldots, n$, *do*
(3) *for* $i = 1, \ldots, k-1,\ k+1, \ldots, n$, *do*
(4) $m_{ik} = a_{ik}/a_{kk}$
(5) $x_i = x_i - m_{ik}x_k$
(6) *for* $j = k, \ldots, n$, *do*
(7) $a_{ij} = a_{ij} - m_{ik}a_{kj}$
(8) *end*
(9) *end*
(10) $m_{kk} = 1/a_{kk}$
(11) *for* $j = k, \ldots, n$, *do*
(12) $a_{kj} = m_{kk}a_{kj}$
(13) *end*
(14) $x_k = m_{kk}x_k$
(15) *end*
(16) *The solution* $x = A^{-1}b$ *is obtained.*

Readers interested in trying out the details should use Mfiles `g_j.m` and `gj_ek.m` as shown in Section 2.7.

Now we describe the slightly less well known GJ decomposition. Define a GJ elementary matrix (using the multiplier of Algorithm 2.3.5) by

$$E_k = \begin{bmatrix} 1 & & & -m_{1k} & & & \\ & \ddots & & -m_{2k} & & & \\ & & 1 & -m_{k-1,k} & & & \\ & & & m_{k,k} & & & \\ & & & -m_{k+1,k} & 1 & & \\ & & & \vdots & & \ddots & \\ & & & -m_{n,k} & & & 1 \end{bmatrix}. \tag{2.16}$$

Then the GJ decomposition is the following

$$E_n E_{n-1} \cdots E_2 E_1 A = I \qquad \text{or } A^{-1} = E_n E_{n-1} \cdots E_2 E_1. \tag{2.17}$$

One can verify, similar to inverting E_k for the GE, that

$$E_k^{-1} = \begin{bmatrix} 1 & & & -m_{1k}/m_{kk} & & & \\ & \ddots & & -m_{2k}/m_{kk} & & & \\ & & 1 & -m_{k-1,k}/m_{kk} & & & \\ & & & 1/m_{k,k} & & & \\ & & & -m_{k+1,k}/m_{kk} & 1 & & \\ & & & \vdots & & \ddots & \\ & & & -m_{n,k}/m_{kk} & & & 1 \end{bmatrix},$$

which defines $A = E_1^{-1}E_2^{-1}\cdots E_n^{-1}$. Remark that matrices that are a product of a few selected E_j type matrices can be inverted quickly; see one application in Section 5.6 and another one in [128,497].

Now consider further properties of E_k in relation to the GE decomposition (2.4). It turns out that the above multipliers m_{ik} are identical to those from Algorithm 2.1.1. Let us first decompose E_k into elementary GE matrices like (2.2)

$$\begin{aligned} E_k &= u_k \ell_k \\ &= \begin{bmatrix} 1 & & & -m_{1k} & & & \\ & \ddots & & \vdots & & & \\ & & 1 & -m_{k-1,k} & & & \\ & & & m_{k,k} & & & \\ & & & & 1 & & \\ & & & & & \ddots & \\ & & & & & & 1 \end{bmatrix} \begin{bmatrix} 1 & & & & & & \\ & \ddots & & & & & \\ & & 1 & & & & \\ & & & 1 & & & \\ & & & -m_{k+1,k} & 1 & & \\ & & & \vdots & & \ddots & \\ & & & -m_{n,k} & & & 1 \end{bmatrix} \end{aligned} \tag{2.18}$$

where ℓ_k and u_k commute: $E_k = \ell_k u_k$. Then the GJ decomposition (2.17) becomes

$$u_n \ell_n u_{n-1} \ell_{n-1} \cdots u_2 \ell_2 u_1 \ell_1 A = I. \tag{2.19}$$

Next observe an interesting property on commutability of u_j and ℓ_j i.e. for any $i,\ j = 1, \ldots, n$

$$u_j \ell_i = \ell_i u_j. \tag{2.20}$$

(However neither u_j's nor ℓ_j's can commute among themselves). Using (2.20), we can collect all u_j's in (2.19) to the left (though they cannot pass each other)

$$
\begin{aligned}
u_n u_{n-1} \cdots u_2 u_1 \ell_n \ell_{n-1} \cdots \ell_2 \ell_1 A = I, & \qquad \text{or} \\
\ell_n \ell_{n-1} \cdots \ell_2 \ell_1 A = u_1^{-1} u_2^{-1} \cdots u_{n-1}^{-1} u_n^{-1}, & \qquad \text{or} \\
\ell_n \ell_{n-1} \cdots \ell_2 \ell_1 A u_n u_{n-1} \cdots u_2 u_1 = I. &
\end{aligned} \tag{2.21}
$$

Clearly the above can be written as a LU decomposition

$$
A = LU \quad \text{with } L = \ell_1^{-1} \ell_2^{-1} \cdots \ell_{n-1}^{-1} \ell_n^{-1}, \quad U = u_1^{-1} u_2^{-1} \cdots u_{n-1}^{-1} u_n^{-1} \tag{2.22}
$$

where multiplications to obtain L and U again amount to a simple packing process as in (2.3). As the LU decomposition is unique, we conclude that the multipliers m_{ik} in Algorithm 2.3.5 are the same as those from Algorithm 2.1.1.

Likewise, one can also observe that the last equation of (2.21) is simply the WAZ decomposition as in Section 2.1.3. Hence the GJ method can reproduce both the GE and the WAZ decompositions.

2.3.2 A block variant of the GJ

Although the GE has two variants of block form Section 2.1.2, namely the Schur complements and the partitioned block methods, the GJ has one block variant, i.e. the counterparts of the two variants in Section 2.1.2 merge into one here.

We now derive the method using the motivation for (2.9), i.e. assume

$$
\begin{bmatrix} E_{11} & \\ E_{21} & I \end{bmatrix} A = \begin{bmatrix} E_{11} & \\ E_{21} & I \end{bmatrix} \begin{bmatrix} A_{11} & A_{12} \\ A_{21} & A_{22} \end{bmatrix} = \begin{bmatrix} I & U_{12} \\ & U_{22} \end{bmatrix}, \tag{2.23}
$$

where $A_{11} \in \mathbb{R}^{r \times r}$, all E_{ij} and U_{ij} are all unknowns (and possibly dense blocks). Equating corresponding blocks, we obtain

$$
\begin{aligned}
E_{11} A_{11} &= I, \quad E_{12} A_{11} = -A_{21}, \\
E_{11} A_{12} &= U_{12}, \; E_{12} A_{12} + A_{22} = U_{22}.
\end{aligned}
$$

Clearly this gives a neater solution than (2.9). This idea will be used again in Section 5.6. As stated earlier, the GJ decomposition is less well known and in particular the author has not seen any use of the block form (2.23) so far.

2.3.3 Huard and Purcell methods

There are perhaps two reasons why a GJ method is not widely used. Firstly the flops (floating point operations) count is 50% higher with the GJ than the GE for solving the same equation $Ax = b$. However there are better variants (i.e. Huard and Purcell [143]) of the GJ method. Secondly the error constant is believed to be larger when using the GJ with the usual column (partial) pivoting. However as shown in [176], this is no longer the case if row pivoting is used.

Both the Gauss–Huard and Gauss–Purcell methods are a special version of the block GJ method (2.23) that have a comparable flops count to the GJ. The unpivoted versions of Huard and Purcell are identical and the only difference lies in pivoting involving the right-hand side vector b in a Gauss–Purcell method where the augmented system $[A \mid b]\begin{pmatrix} x \\ -1 \end{pmatrix} = 0$ is solved; see [134,143]. Here we only review the Huard method and leave further discussion in Section 15.5.

Instead of (2.23), we consider a special block form

$$\begin{bmatrix} E_{11} & \\ & I \end{bmatrix} A = \begin{bmatrix} E_{11} & \\ & I \end{bmatrix} \begin{bmatrix} A_{11} & A_{12} \\ A_{21} & A_{22} \end{bmatrix} = \begin{bmatrix} I & U_{12} \\ A_{21} & A_{22} \end{bmatrix}, \tag{2.24}$$

where $A_{11} \in \mathbb{R}^{k\times k}$ at step k of the Gauss–Huard method. Clearly the matrix E_{11} is built up recursively to maintain efficiency. The following algorithm shows the computational details.

Algorithm 2.3.6. (GH method).

(1) Set $m = I$, $x = b$ and vector $s = 0 \in \mathbb{R}^{n-1}$.
    for steps $k = 1, \ldots, n$, do
        for rows $i = 1, \ldots, k-1$, do
(2) $m_{ik} = a_{ik}/a_{kk}$, $v_i = -m_{i,k}$
(3) $x_i = x_i - m_{ik} x_k$
(4)   for $j = k, \ldots, n$, do
                $a_{ij} = a_{ij} - m_{ik} a_{kj}$
            end j
        end i
(5) $m_{kk} = 1/a_{kk}$
(6)   for columns $j = k, \ldots, n$, do
                $a_{kj} = m_{kk} a_{kj}$
(7)   end j
(8) $x_k = m_{kk} x_k$
(9)   Set row index $r = k + 1$
(10)   if $r \le n$
(11)   for rows $i = 1, \ldots, k$, do
(12) $m_{ri} = a_{ri}$, $h_i = -m_{ri}$
(13) $x_r = x_r - m_{ri} x_i$
(14)   for columns $j = k, \ldots, n$, do
                    $a_{rj} = a_{rj} - m_{ri} a_{rj}$

(15) *end j*
(16) *for* $j = 1, \ldots, k$ *(row* $r = k + 1$*)*
$A(r, j) = 0$
(17) *end j*
(18) *Save the special product* $s_k = h^T v = \sum_{j=1}^{k} h_j v_j$ *for entry* (r, k).
(19) *end i*
(20) *end if*
end k
(21) The solution $x = A^{-1}b$ *is obtained.*

Readers interested in trying the details should use Mfiles `g_h.m` and `gh_all.m` as shown in Section 2.7.

Using the multipliers m_{ik} from Algorithm 2.3.6, we can present the underlying matrix decomposition. Define

$$H_k = \begin{bmatrix} 1 & & & -m_{1k} & & & \\ & \ddots & & \vdots & & & \\ & & 1 & -m_{k-1,k} & & & \\ & & & m_{k,k} & & & \\ -m_{k+1,1} & \cdots & -m_{k+1,k-1} & s_k & 1 & & \\ & & & & & \ddots & \\ & & & & & & 1 \end{bmatrix}. \tag{2.25}$$

Further the Gauss–Huard decomposition is the following

$$H_n H_{n-1} \cdots H_2 H_1 A = I, \qquad \text{or } A^{-1} = H_n H_{n-1} \cdots H_2 H_1. \tag{2.26}$$

On inspecting H_k, to connect the GE and GJ factorizations, we may further decompose H_k as

$$\begin{bmatrix} 1 & & & & & \\ & \ddots & & & & \\ & & 1 & & & \\ -m_{k+1,1} & \cdots & -m_{k+1,k} & 1 & & \\ & & & & \ddots & \\ & & & & & 1 \end{bmatrix} \begin{bmatrix} 1 & & & -m_{1k} & & & \\ & \ddots & & \vdots & & & \\ & & 1 & -m_{k-1,k} & & & \\ & & & m_{k,k} & & & \\ & & & & 1 & & \\ & & & & & \ddots & \\ & & & & & & 1 \end{bmatrix}. \tag{2.27}$$

We leave the reader to complete the study whenever such a need arises.

2.4 The QR decomposition

The QR decomposition of $A = QR$, where Q is orthogonal and R is upper triangular, is a well-documented topic [63,229,437,459,180,280]. Apart from many theoretical uses, it is an indispensable as well as useful method for least squares, rank deficient or eigenvalue type problems. Especially for the latter problem, the use of Q in a QR decomposition for transforming A leads to the well-known QR method for computing eigenvalues [229] (because, amongst similar transforms that can 'simplify' matrix A, the orthogonal matrices lead to least perturbation errors Section 1.3). We give a brief description of the QR decomposition and put a slight emphasis on the Givens approach which we have modified conveniently.

There are **three methods** of constructing Q for decomposing $A = QR$ with

$$R = \begin{pmatrix} u_{1,1} & u_{1,2} & u_{1,3} & \cdots & u_{1,n-1} & u_{1,n} \\ & u_{2,2} & u_{2,3} & \cdots & u_{2,n-1} & u_{2,n} \\ & & u_{3,3} & \cdots & u_{3,n-1} & u_{3,n} \\ & & & \ddots & \vdots & \vdots \\ & & & & u_{n-1,n-1} & u_{n-1,n} \\ & & & & & u_{n,n} \end{pmatrix},$$

while each Q is a product of several orthogonal matrices (in Methods 1, 2) or is formed directly in a column-by-column fashion (Method 3).

2.4.1 Method 1: Givens plane rotations

Using (1.24), consider to zero out the entry $\widetilde{A}_{ji}$ of $\widetilde{A} = P(i, j)A$ for $j = 1, 2, \cdots, n$ and $i < j$:

$$\widetilde{a}_{ji} = a_{ii}\sin(\theta_{ij}) - a_{ji}\cos(\theta_{ij}),$$

by choosing a suitable θ; here we hope to zero out these particular positions in turns: $(j, i) = (2, 1), \cdots, (n, 1), (3, 2), \cdots, (n, n-1)$.

Setting $\widetilde{a}_{ji} = 0$, from $a_{ii}\sin(\theta_{ij}) = a_{ji}\cos(\theta_{ij})$, gives us the required formula, similar to (1.25), i.e.

$$\sin(\theta) = \frac{r}{\sqrt{1+r^2}}, \qquad \cos(\theta) = \frac{1}{\sqrt{1+r^2}} \qquad \text{with } r = \frac{a_{ji}}{a_{ii}}.$$

The final Q matrix will be the product of these rotation matrices:

$$
\begin{aligned}
Q = & \underbrace{P(1,2)\cdots P(1,n)}_{\text{Column 1}}\,\underbrace{P(2,3)\cdots P(2,n)}_{\text{Column 2}} \\
& \cdots \underbrace{P(n-2,n-1)\;P(n-2,n)}_{\text{Column } n-2}\,\underbrace{P(n-1,n)}_{\text{Column } n-1}.
\end{aligned}
\tag{2.28}
$$

Use the Mfile `givens.m` for experiments and viewing individual $P(i,j)$.

2.4.2 Method 2: Householder transformations

Using (1.26), to realize $A = QR$, at step $k = 1, 2, \cdots, n-1$, denote the kth column of A by $x = \begin{pmatrix} \widetilde{x} \\ \overline{x} \end{pmatrix}$, we take $\overline{x} = A(k:n, 1:n)$. All we need to remember is where to start the active vector $\overline{x}$; refer to (1.27). Here we assume that A is over-written $A = P_j A$ (to denote intermediate and partially transformed matrices). Finally the Q matrix is

$$Q = P_1 P_2 \cdots P_{n-1}.$$

Use the Mfile `houses.m` for experiments.

2.4.3 Method 3: Gram–Schmidt

As in Section 1.4, using column vectors of A and Q, to find Q and R, equate columns of $A = QR$ as follows

$$
[a_1\; a_2\; \cdots\; a_n] = [q_1\; q_2\; \cdots\; q_n]
\begin{pmatrix}
u_{1,1} & u_{1,2} & u_{1,3} & \cdots & u_{1,n-1} & u_{1,n} \\
 & u_{2,2} & u_{2,3} & \cdots & u_{2,n-1} & u_{2,n} \\
 & & u_{3,3} & \cdots & u_{3,n-1} & u_{3,n} \\
 & & & \ddots & \vdots & \vdots \\
 & & & & u_{n-1,n-1} & u_{n-1,n} \\
 & & & & & u_{n,n}
\end{pmatrix},
$$

and

$$
\begin{array}{l|lllll}
\text{Column} \longmapsto 1 & a_1 & = q_1 u_{11} & & & \\
\text{Column} \longmapsto 2 & a_2 & = q_1 u_{12} & + q_2 u_{22} & & \\
\text{Column} \longmapsto 3 & a_3 & = q_1 u_{13} & + q_2 u_{23} & + q_3 u_{33} & \\
\vdots & & \vdots & & & \\
\text{Column} \longmapsto n-1 & a_{n-1} & = q_1 u_{1,n-1} & + q_2 u_{2,n-1} & + q_3 u_{3,n-1} & + \ddots \\
\text{Column} \longmapsto n & a_n & = q_1 u_{1,n} & + q_2 u_{2,n} & + q_3 u_{3,n} & + \cdots + q_n u_{n,n}
\end{array}
$$

The **basic Gram–Schmidt** method is the **column** approach (note $m = n$ is obtained if $det(A) \neq 0$):

(1) Since $r_1 = a_1$ is given, $u_{1,1} = \|r_1\|_2 = \|a_1\|_2$ and $q_1 = r_1/\|r_1\|_2$.

(2) for $j = 2, 3, \cdots, n$

$$u_{k,j} = q_k^T a_j \qquad \text{(for } k = 1, \cdots, j-1\text{)}$$

$$r_j = a_j - \sum_{k=1}^{j-1} q_k u_{k,j}, \quad \beta_j = \|r_j\|_2 = \sqrt{r_j^T r_j}$$

if $\beta_j \neq 0$, set $q_{j+1} = r_j/\|r_j\|_2$ **otherwise stop** and accept $m \equiv j$ (rank of A found)

end for $j = 2, 3, \cdots, n$

(3) $Q = [q_1\ q_2\ \cdots q_m]$ is the required orthogonal matrix.

Here note that the computation of q_j, a somehow easy task once u_{ij} are found, comes last at each step.

The modified Gram–Schmidt method is the row version as illustrated by the following

Step 1	(let $\Downarrow$ denote a u term being computed)
Column $\longmapsto$ 1	$a_1 = q_1 u_{11} \Downarrow$
Column $\longmapsto$ 2	$a_2 = q_1 u_{12} \Downarrow + q_2 u_{22}$
Column $\longmapsto$ 3	$a_3 = q_1 u_{13} \Downarrow + q_2 u_{23} + q_3 u_{33}$
$\vdots$	$\vdots$
Column $\longmapsto n-1$	$a_{n-1} = q_1 u_{1,n-1} \Downarrow + q_2 u_{2,n-1} + q_3 u_{3,n-1} + \ddots$
Column $\longmapsto n$	$a_n = q_1 u_{1,n} \Downarrow + q_2 u_{2,n} + q_3 u_{3,n} + \cdots + q_n u_{n,n}$

Step 2	
Column $\longmapsto$ 2	$a_2 = q_2 u_{22} \Downarrow$
Column $\longmapsto$ 3	$a_3 = q_2 u_{23} \Downarrow + q_3 u_{33}$
$\vdots$	$\vdots$
Column $\longmapsto n-1$	$a_{n-1} = q_2 u_{2,n-1} \Downarrow + q_3 u_{3,n-1} + q_4 u_{4,n-1} + \ddots$
Column $\longmapsto n$	$a_n = q_2 u_{2,n} \Downarrow + q_3 u_{3,n} + q_4 u_{4,n} + \cdots + q_n u_{n,n}$

Step 3	
Column $\longmapsto 3$	$a_3 = q_3 u_{33} \Downarrow$
Column $\longmapsto 4$	$a_4 = q_3 u_{34} \Downarrow + q_4 u_{44}$
$\vdots$	$\vdots$
Column $\longmapsto n-1$	$a_{n-1} = q_3 u_{3,n-1} \Downarrow + q_4 u_{4,n-1} + q_5 u_{5,n-1} + \ddots$
Column $\longmapsto n$	$a_n = q_3 u_{3,n} \Downarrow + q_4 u_{4,n} + q_5 u_{5,n} + \cdots + q_n u_{n,n}$

$\cdots$

Step $(n-1)$	
Column $\longmapsto n-1$	$a_{n-1} = q_{n-1} u_{n-1,n-1} \Downarrow$
Column $\longmapsto n$	$a_n = q_{n-1} u_{n,n-1} \Downarrow + q_n u_{nn}$

Step n	
Column $\longmapsto n$	$a_n = q_n u_{n,n} \Downarrow$

As the **row** version has better numerical stability, it is usually the preferred method:

(1) for $j = 1, 2, \cdots, n$

$$u_{j,k} = q_k^T a_j \qquad (\text{for } k = j, \cdots, n)$$

modify $a_k = a_k - q_j u_{j,k}$, (for $k = 2, \cdots, n$)

$$r_j = a_j, \quad \beta_j = \|r_j\|_2 = \sqrt{r_j^\top r_j}$$

if $\beta_j \neq 0$, set $q_{j+1} = r_j / \|r_j\|_2$ **otherwise stop** and accept $m \equiv j$ (rank of A found)

end for $j = 1, 2, \cdots, n$

(2) $Q = [q_1\ q_2\ \cdots q_m]$ is the required orthogonal matrix.

Note that the first column of Q, $q_1 = a_1/\|a_1\|_2$ computed at Step (1), does not need to be specified (compare to Arnoldi iterations, Section 1.4, where q_1 must be specified). Use the Mfile `mgs.m` for experiments.

As commented in [229], the QR decomposition (an essential tool for solving a least squares problem in Chapter 5) could be used for solving (1.1) reliably but it is normally not used as it is more expensive than the preferred LU approach Section 2.1.

2.5 Special matrices and their direct inversion

Returning from investigating general direct solvers, we wish to address the fast direct solution of a class of specially structured matrices.

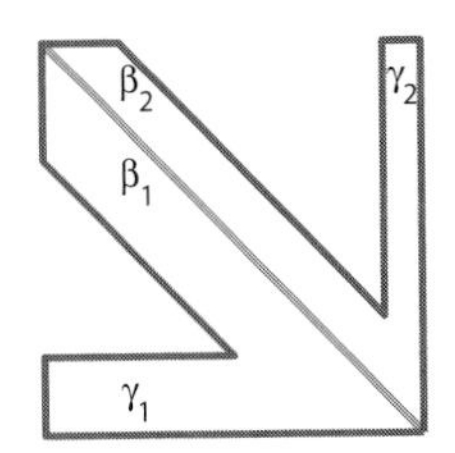

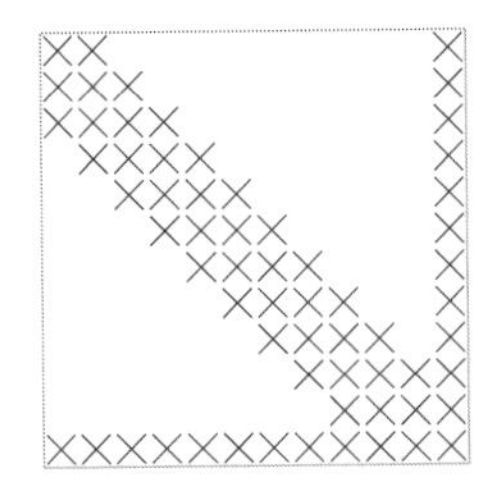
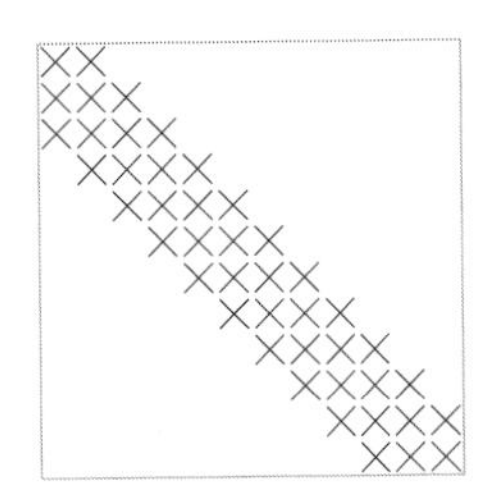

Figure 2.1. Graphical illustration of a banded arrow matrix of type **band**$(\beta_1, \beta_2, \gamma_1, \gamma_2)$ (left plot where entries in the closed domain are nonzero), an example matrix of type (2, 1, 1, 1) (middle plot — $\times$ means a nonzero) and an example matrix of type **band**(2, 1, 0, 0) (right plot i.e. a **band**(2, 1) matrix — $\times$ means a nonzero).

(1) As forward type preconditioners M i.e. $M \approx A$, fast solution procedures for these matrices are essential.
(2) In searching for inverse type preconditioners M i.e. $M \approx A^{-1}$, knowing the sparsity pattern of A^{-1} (even approximately) is of importance.

In most cases, we shall use the GE method. Among such special matrices, some are well known (e.g. banded ones) while others may not be so – in the latter case we try to give more background information for readers' benefit. If an exact inverse of a preconditioner can be obtained, the use of a theoretical dense matrix to precondition a sparse matrix becomes possible [255].

As also stated before, symmetric matrices deserve a special study; refer to those special techniques which can be found from [28,85,229,280,383] and the references therein.

A sparse matrix should be stored in a sparse storage scheme. The column-by-column row index scheme [189] is commonly used while the well-known Harwell–Boeing [347] storage format is very similar; we give an illustration in Appendix B using a 6×6 example.

2.5.1 Banded arrow matrices

Definition 2.5.7. (Banded arrow matrix). *A* **banded arrow matrix** $A \in \mathbb{R}^{n\times n}$ *denoted by type* **band**$(\beta_1, \beta_2, \gamma_1, \gamma_2)$ *satisfies* $a_{ij} = 0$, *under these conditions*

[1] $j + \beta_1 < i \le n - \gamma_1$ *(lower bandwidth* β_1 *and lower border* γ_1*);*
[2] $i + \beta_2 < j \le n - \gamma_2$ *(upper bandwidth* β_2 *and right border* γ_2*).*

When $\gamma_1 = \gamma_2 = 0$, *a banded arrow matrix reduces the more familiar* banded *matrix of type* (β_1, β_2), *denoted by* **band**(β_1, β_2).

These types of special matrices are illustrated in Figure 2.1.

In practice, as a sparse matrix, a banded arrow matrix should be stored in some suitable sparse format [189]. The reason is clearly evident from the following algorithm for computing the LU decomposition of a banded arrow matrix, adapted from Algorithm 2.1.1.

Algorithm 2.5.8. (Banded arrow GE method).

(1) *Set* $m = I$, $x = b$.
(2) *for* $k = 1, \ldots, n$, *do*
(3)   *for row* $i = k+1, \ldots k+\beta_1; (n+1-\gamma_1), \ldots, n$, *do*
(4) $m_{ik} = a_{ik}/a_{kk}$
(5) $x_i = x_i - m_{ik}x_k$
(6)   *for* $j = k, \ldots, k+\beta_2; (n+1-\gamma_2), \ldots, n$, *do*
(7) $a_{ij} = a_{ij} - m_{ik}a_{kj}$
(8)   *end*
(9)   *end*
(10) *end*
(11) *For LU decomposition, accept* $L = m$, $U = A$;

2.5.2 Circulant and Toeplitz matrices

Both circulant and Toeplitz matrices belong to a special class of matrices that have useful applications in matrix preconditioning for the simple reason that $Mx = b$ can be solved in a fast way using FFT Section 1.6. More references are from [104,465].

A Toeplitz matrix T_n may simply be considered given from some other applications – such a matrix is determined by its first row and first column only with each subdiagonal constant:

$$T_n = \begin{pmatrix} h_0 & h_{-1} & h_{-2} & \ddots & h_{2-n} & h_{1-n} \\ h_1 & h_0 & h_{-1} & \ddots & h_{3-n} & h_{2-n} \\ h_2 & h_1 & h_0 & \ddots & \ddots & h_{3-n} \\ \ddots & \ddots & \ddots & \ddots & h_{-1} & \ddots \\ h_{n-2} & h_{n-3} & \ddots & h_1 & h_0 & h_{-1} \\ h_{n-1} & h_{n-2} & h_{n-3} & \ddots & h_1 & h_0 \end{pmatrix}. \tag{2.29}$$

Here its generating vector is called root vector:

$$\mathbf{h} = [h_{1-n}, h_{2-n}, \ldots, h_{-1}, h_0, h_1, \ldots, h_{n-2}, h_{n-1}]^T.$$

A circulant matrix may be viewed as a Toeplitz matrix with $h_{-j} = h_{n-j}$ for $j = 1, 2, \ldots, n-1$

$$C_n = \begin{pmatrix} h_0 & h_{n-1} & h_{n-2} & \ddots & h_1 \\ h_1 & h_0 & h_{n-1} & \ddots & h_2 \\ h_2 & h_1 & h_0 & \ddots & h_3 \\ \ddots & \ddots & \ddots & \ddots & \ddots \\ h_{n-1} & h_{n-2} & h_{n-3} & \ddots & h_0 \end{pmatrix} \tag{2.30}$$

where one notes that each column is a copy of the previous one with entries down-shifted by one position in an overall 'circular' fashion. Matrices of type (2.30) are called **circulant** matrices where other columns are obtained via down-shifting from column one and so the root vector is simply this first column:

$$\mathbf{h} = [h_0, h_1, \ldots, h_{n-2}, h_{n-1}]^T.$$

Before we consider the mathematics of how to compute quickly $\mathbf{f} = C_n\mathbf{g}$ and $\mathbf{f} = T_n\mathbf{g}$, we briefly review the derivations of T_n and C_n in the context of convolutions as this aspect can often confuse readers.

♦ **Derivation I.** Consider the discretization of the convolution type integral operator

$$f(x) = \int_a^b \mathtt{k}(x-y)g(y)dy, \qquad \text{for } x \in [a,b], \; g \in C[a,b] \tag{2.31}$$

by the midpoint quadrature rule (taking uniformly distributed x_j as the midpoint of interval j for $j = 1, 2, \ldots, n$)

$$f_j = f(x_j) = \sum_{k=1}^{n} \mathtt{k}((j-k)\Delta x)g_k \qquad \text{for } j = 1, 2, \cdots, n. \tag{2.32}$$

Further, taking $h_\ell = \mathtt{k}(\ell\Delta x)$ for $\ell = 0, \pm 1, \pm 2, \ldots, \pm(n-1)$, we arrive at $\mathbf{f} = T_n\mathbf{g}$. If we redefine the kernel function $\mathtt{k}(x-y) = \mathtt{k}(|x-y|)$, then $\mathbf{f} = C_n\mathbf{g}$ is derived.

♦ **Derivation II.** Consider convoluting a periodic sequence of period n:

$$\cdots\ g_{n-1},\ g_0,\ g_1,\ \ldots,\ g_{n-1},\ g_0,\ \cdots$$

with a finite sequence of filters

$$h_{n-1}, h_{n-2}, \ldots, h_2, h_1, h_0, h_{-1}, h_{-2}, \ldots, h_{2-n}, h_{1-n}$$

centred at h_0. The first convolution sum is $f_1 = \sum_{k=0}^{n-1} h_{-k} g_k$. We can sketch the rule for this sum and the others in Table 2.1, where 'arrow ⇑' points to the corresponding filter coefficient in a sum. Clearly we see that relationship $\mathbf{f} = T_n \mathbf{g}$ holds, and it reads $\mathbf{f} = C_n \mathbf{g}$ if $h_{-j} = h_{n-j}$.

We first consider how to compute $\mathbf{f} = C_n \mathbf{g}$ quickly, It turns out that the fast computation becomes apparent once we express C_n as a sum of elementary permutation matrices.

Define a permutation matrix (for primarily linking columns 1 and 2 of C_n) by

$$R_n = \begin{pmatrix} 0 & 0 & \ddots & 0 & 1 \\ 1 & 0 & \ddots & 0 & 0 \\ 0 & 1 & \ddots & 0 & 0 \\ \ddots & \ddots & \ddots & 0 & 0 \\ 0 & 0 & \ddots & 1 & 0 \end{pmatrix} = \left[\begin{array}{cccc|c} 0 & 0 & \ddots & 0 & 1 \\ 1 & 0 & \ddots & 0 & 0 \\ 0 & 1 & \ddots & 0 & 0 \\ \ddots & \ddots & \ddots & 0 & 0 \\ 0 & 0 & \ddots & 1 & 0 \end{array}\right]$$

$$\text{and let } \mathbf{h} = \begin{pmatrix} h_0 \\ h_1 \\ h_2 \\ \vdots \\ h_{n-1} \end{pmatrix}.$$

Then one can observe that $R_n = (e_2, e_3, \ldots, e_n, e_1)$ is a simple permutation matrix and $R_n \mathbf{h} = C_n(:, 2)$ = 'column 2 of C_n'; that is, $R_n = I R_n$ and in words, R_n permutes the first column to the last column when multiplying to a matrix from the the right. Consequently, some amazing results follow this simple rule $R_n^k = I R_n \ldots R_n$:

$$R_n^k = \left[\begin{array}{ccc|ccc} \multicolumn{3}{c|}{\overbrace{\qquad\qquad}^{n-k}} & \multicolumn{3}{c}{\overbrace{\qquad\qquad}^{k}} \\ 0 & \cdots & 0 & 1 & \ddots & 0 \\ \ddots & \ddots & \ddots & \ddots & \ddots & \ddots \\ 0 & \ddots & 0 & 0 & 0 & 1 \\ 1 & \ddots & 0 & 0 & 0 & 0 \\ \ddots & \ddots & \ddots & \ddots & \ddots & \ddots \\ 0 & 0 & 1 & 0 & \ddots & 0 \end{array}\right] \quad \text{and} \quad \begin{aligned} R_n^k \mathbf{h} &= C_n(:, k+1) \\ &= \text{'column } k+1 \text{ of } C_n\text{'}, \\ C_n &= \begin{bmatrix} \mathbf{h} & R_n \mathbf{h} & R_n^2 \mathbf{h} & \cdots & R_n^{n-1} \mathbf{h} \end{bmatrix}. \end{aligned} \tag{2.33}$$

Table 2.1. *Illustration of convoluting sums with n filter coefficients. One may imagine a floating filter template for a fixed data set or moving data with a fixed filter template (below).*

Filters	h_{n-1}	h_{n-2}	h_{n-3}	$\cdots$	h_3	h_2	h_1	h_0	h_{-1}	h_{-2}	$\cdots$	h_{3-n}	h_{2-n}	h_{1-n}
Sum 1								$\Uparrow$ g_0	$\Uparrow$ g_1	$\Uparrow$ g_2	$\cdots$	$\Uparrow$ g_{n-1}	$\Uparrow$ g_{n-2}	$\Uparrow$ g_{n-1}
Sum 2							$\Uparrow$ g_0	$\Uparrow$ g_1	$\Uparrow$ g_2	$\Uparrow$ g_3	$\cdots$	$\Uparrow$ g_{n-2}	$\Uparrow$ g_{n-1}	
Sum 3						$\Uparrow$ g_0	$\Uparrow$ g_1	$\Uparrow$ g_2	$\Uparrow$ g_3	$\Uparrow$ g_4	$\cdots$	$\Uparrow$ g_{n-1}		
Sum 4					$\Uparrow$ g_0	$\Uparrow$ g_1	$\Uparrow$ g_2	$\Uparrow$ g_3	$\Uparrow$ g_4	$\vdots$	$\vdots$	$\cdots$		
$\vdots$				$\cdots$	$\vdots$	$\vdots$	$\vdots$	$\vdots$	$\vdots$	$\vdots$	$\cdots$			
Sum n-2			$\Uparrow$ g_0	$\Uparrow$ g_1	$\Uparrow$ g_2	$\cdots$	$\Uparrow$ g_{n-4}	$\Uparrow$ g_{n-3}	$\Uparrow$ g_{n-2}	$\Uparrow$ g_{n-1}				
Sum n-1		$\Uparrow$ g_0	$\Uparrow$ g_1	$\Uparrow$ g_2	$\cdots$	$\Uparrow$ g_{n-4}	$\Uparrow$ g_{n-3}	$\Uparrow$ g_{n-2}	$\Uparrow$ g_{n-1}					
Sum n	$\Uparrow$ g_0	$\Uparrow$ g_1	$\Uparrow$ g_2	$\cdots$	$\Uparrow$ g_{n-4}	$\Uparrow$ g_{n-3}	$\Uparrow$ g_{n-2}	$\Uparrow$ g_{n-1}						

Moreover using the notation of R_n^k, we can separate the full contribution of h_k to the circulant matrix C_n into the following elementary matrix sums

$$C_n = \begin{pmatrix} h_0 & & & & \\ & h_0 & & & \\ & & h_0 & & \\ & & & \ddots & \\ & & & & h_0 \end{pmatrix} + \begin{pmatrix} & & & & h_1 \\ h_1 & & & & \\ & h_1 & & & \\ & & \ddots & & \\ & & & h_1 & \end{pmatrix}$$

$$+ \begin{pmatrix} & & & h_2 & \\ & & & & h_2 \\ h_2 & & & & \\ & h_2 & & & \\ & & \ddots & & \\ & & & h_2 & \end{pmatrix} + \begin{pmatrix} & & & h_3 & & \\ & & & & h_3 & \\ & & & & & h_3 \\ h_3 & & & & & \\ & h_3 & & & & \\ & & \ddots & & & \\ & & & h_3 & & \end{pmatrix}$$

$$+ \cdots \quad + \begin{pmatrix} & h_{n-1} & & & \\ & & h_{n-1} & & \\ & & & \ddots & \\ & & & & h_{n-1} \\ h_{n-1} & & & & \end{pmatrix}$$

$$= h_0 I + h_1 R_n + h_2 R_n^2 + \cdots + h_{n-1} R_n^{n-1}$$

$$= \sum_{j=0}^{n-1} h_j R_n^j.$$

To study C_n and R_n, we first reveal a simple relationship between R_n and the DFT matrix $\mathbf{F}_n$ (Section 1.6). Denote the vector $\mathbb{I} = \mathbf{e} = (1, 1, \ldots, 1, 1)^T$ and $D_n = \text{diag}(d_n)$ with d_n the second column of $\mathbf{F}_n$ given by $d_n = (1, \omega_n, \omega_n^2, \ldots, \omega_n^{n-1})^T$, where $\omega_n = \exp(-2\pi i/n)$. As is known §1.6,

$$\begin{aligned} \mathbf{F}_n &= [\mathbb{I} \quad d_n \quad D_n d_n \; D_n^2 d_n \cdots D_n^{n-3} d_n \; D_n^{n-2} d_n] \\ &= [D_n^0 \mathbb{I} \; D_n^1 \mathbb{I} \; D_n^2 \mathbb{I} \quad D_n^3 \mathbb{I} \quad \cdots D_n^{n-2} \mathbb{I} \quad D_n^{n-1} \mathbb{I}]. \end{aligned} \tag{2.34}$$

Very crucially, note that $D_n^n = D_n^0 = \text{diag}(\mathbb{I})$ as $\omega_n^n = 1$. Therefore we have derived a powerful relationship by the following

$$\begin{aligned} D_n \mathbf{F}_n &= D_n [D_n^0 \mathbb{I} \; D_n^1 \mathbb{I} \; D_n^2 \mathbb{I} \; D_n^3 \mathbb{I} \cdots D_n^{n-2} \mathbb{I} \; D_n^{n-1} \mathbb{I}] \\ &= \quad [D_n^1 \mathbb{I} \; D_n^2 \mathbb{I} \; D_n^3 \mathbb{I} \; D_n^4 \mathbb{I} \cdots D_n^{n-1} \mathbb{I} \; D_n^n \mathbb{I}] \\ &= \quad [D_n^1 \mathbb{I} \; D_n^2 \mathbb{I} \; D_n^3 \mathbb{I} \; D_n^4 \mathbb{I} \cdots D_n^{n-1} \mathbb{I} \; D_n^0 \mathbb{I}] \\ &= \mathbf{F}_n R_n. \end{aligned} \tag{2.35}$$

Often this result is explained as follows: multiplying D_n to $\mathbf{F}_n$ is equivalent to shifting all columns to the left in a wrap-around manner.

Thus, clearly, $\mathbf{F}_n R_n = D_n \mathbf{F}_n$ implies

$$\mathbf{F}_n R_n \mathbf{F}_n^{-1} = D_n, \tag{2.36}$$

that is, the DFT matrix $\mathbf{F}_n$ can diagonalize the permutation matrix R_n. Therefore R_n has *known* eigenvalues and eigenvectors.

Since eigenvalues satisfy $\lambda(R_n^j) = \lambda(R_n)^j$, and R_n and R_n^j share eigenvectors,[3] along the similar lines of proving the Cayley–Hamilton theorem in linear algebra, we thus have proved the following theorem using (2.36).

Theorem 2.5.9. (Diagonalization of a circulant matrix). *The circulant matrix defined by (2.30) can be diagonalised by the DFT matrix* $\mathbf{F}_n$*:*

$$\begin{aligned}
\mathbf{F}_n C_n \mathbf{F}_n^{-1} &= \mathbf{F}_n \sum_{j=0}^{n-1} h_j R_n^j \mathbf{F}_n^{-1} \\
&= \sum_{j=0}^{n-1} h_j (\mathbf{F}_n R_n \mathbf{F}_n^{-1})^j = \sum_{j=0}^{n-1} h_j D_n^j = \ diag(\mathbf{F}_n \mathbf{h}).
\end{aligned} \tag{2.37}$$

Here the very last equality on the right-hand uses two small tricks: firstly from (2.34), $\mathbf{F}_n \mathbf{h} = \sum_{j=0}^{n-1} h_j D_n^j \mathbb{I}$; Secondly vector $D_n^j \mathbb{I}$ represents the diagonal of the diagonal matrix D_n^j. Equation (2.37) implies that the eigenvalues of circulant matrix C_n can be obtained by a FFT step to its first column $\mathbf{h}$.

Thus the convolution product $\mathbf{f} = C_n \mathbf{g}$ becomes

$$\mathbf{f} = C_n \mathbf{g} = \mathbf{F}_n^{-1} \ \text{diag}(\mathbf{F}_n \mathbf{h}) \ \mathbf{F}_n \mathbf{g} = \mathbf{F}_n^{-1} \Big[(\mathbf{F}_n \mathbf{h}). * (\mathbf{F}_n \mathbf{g}) \Big], \tag{2.38}$$

where the pointwise product $x. * y = \text{diag}(x) y = [x_1 y_1, \ x_2 y_2, \ \ldots, \ x_n y_n]^T$ and $\mathbf{F}_n^{-1}$ will denote an inverse FFT process (see Section 1.6 for FFT implementation details).

Example 2.5.10. (Fast computation of a circulant matrix product). *We now illustrate using FFT to compute the matrix vector product* $f = C_8 u$ *with*

$$C_8 = \begin{pmatrix} 2 & 5 & 8 & 4 & 3 & 7 & 6 & 1 \\ 1 & 2 & 5 & 8 & 4 & 3 & 7 & 6 \\ 6 & 1 & 2 & 5 & 8 & 4 & 3 & 7 \\ 7 & 6 & 1 & 2 & 5 & 8 & 4 & 3 \\ 3 & 7 & 6 & 1 & 2 & 5 & 8 & 4 \\ 4 & 3 & 7 & 6 & 1 & 2 & 5 & 8 \\ 8 & 4 & 3 & 7 & 6 & 1 & 2 & 5 \\ 5 & 8 & 4 & 3 & 7 & 6 & 1 & 2 \end{pmatrix}_{8\times 8}, \qquad u = \begin{pmatrix} 1 \\ 1 \\ 6 \\ 7 \\ 1 \\ 4 \\ 1 \\ 5 \end{pmatrix}, \qquad \mathbf{h} = \begin{pmatrix} 2 \\ 1 \\ 6 \\ 7 \\ 3 \\ 4 \\ 8 \\ 5 \end{pmatrix}.$$

[3] The proof of this result itself is effectively also following the similar lines of proving the Cayley–Hamilton theorem in linear algebra.

From the FFT diagonalization method, we first compute

$$FFT(\mathbf{h}) = F_8\mathbf{h} = \begin{pmatrix} 36 \\ -4.5355 + 2.7071i \\ -9 + 7i \\ 2.5355 - 1.2929i \\ 2 \\ 2.5355 + 1.2929i \\ -9 - 7i \\ -4.5355 - 2.7071i \end{pmatrix},$$

$$FFT(u) = F_8 u = \begin{pmatrix} 26 \\ -3.5355 - 4.2929i \\ -5 + 7i \\ 3.5355 + 5.7071i \\ -8 \\ 3.5355 - 5.7071i \\ -5 - 7i \\ -3.5355 + 4.2929i \end{pmatrix}.$$

Then after an inverse FFT, we obtain

$$f = IFFT\,[FFT(\mathbf{h}).*FFT(u)] = IFFT \begin{pmatrix} 936 \\ 27.6569 + 9.8995i \\ -4 - 98i \\ 16.3431 + 9.8995i \\ -16 \\ 16.3431 - 9.8995i \\ -4 + 98i \\ 27.6569 - 9.8995i \end{pmatrix} = \begin{pmatrix} 125 \\ 142 \\ 116 \\ 89 \\ 103 \\ 145 \\ 116 \\ 100 \end{pmatrix}.$$

Here FFT and $IFFT$ are notations for implementing $\mathbf{F}_n x$ and $\mathbf{F}_n^{-1} x$ steps; see Section 1.6.

It remains to discuss the fast solution of a circulant linear system $C_n\mathbf{g} = \mathbf{f}$ for preconditioning purpose. Assume C_n is nonsingular. From (2.37), we know that

$$\mathbf{F}_n C_n^{-1} \mathbf{F}_n^{-1} = \operatorname{diag}(1./(\mathbf{F}_n\mathbf{h})), \tag{2.39}$$

where './' refers to pointwise division (similar to notation '.∗'). Then we solve system $C_n\mathbf{g} = \mathbf{f}$ by this procedure

$$\mathbf{g} = C_n^{-1}\mathbf{f} = \mathbf{F}_n^{-1}\ \operatorname{diag}(1./(\mathbf{F}_n\mathbf{h}))\,\mathbf{F}_n\mathbf{f} = \mathbf{F}_n^{-1}\Big[(\mathbf{F}_n\mathbf{f}./(\mathbf{F}_n\mathbf{h})\Big]. \tag{2.40}$$

Clearly singularity of C_n is directly associated with any zero component of $\mathbf{F}_n\mathbf{h}$. Another way of checking the singularity of C_n is to study its eigenvalues from (2.37)

$$\lambda_k(C_n) = (\mathbf{F}_n\mathbf{h})_k = \sum_{j=0}^{n-1} h_j w_n^{kj} = \sum_{j=0}^{n-1} h_j \exp(-2\pi i k j/n),$$
$$\text{for } k = 0, 1, \cdots, n-1.$$

In general, the behaviour of these eigenvalues is determined by a real-valued generating function f; see [104] and the many references therein.

We now consider how to compute $\mathbf{f} = T_n\mathbf{g}$ quickly for a Toeplitz matrix T_n. As T_n cannot be diagonalised by $\mathbf{F}_n$, we associate a Toeplitz matrix with a circulant one. The common approach is to augment T_n so that it is embedded in a much larger circulant matrix C_m (here $m = 2n$) as follows:[4]

$$\left[\begin{array}{cccccc|cccccc}
a_0 & a_{-1} & a_{-2} & \ddots & a_{2-n} & a_{1-n} & 0 & a_{n-1} & a_{n-2} & \ddots & a_2 & a_1 \\
a_1 & a_0 & a_{-1} & \ddots & a_{3-n} & a_{2-n} & a_{1-n} & 0 & a_{n-1} & \ddots & a_3 & a_2 \\
a_2 & a_1 & a_0 & a_{-1} & \ddots & a_{3-n} & a_{2-n} & a_{1-n} & 0 & a_{n-1} & \ddots & a_3 \\
\ddots & \ddots & \ddots & \ddots & \ddots & \ddots & \ddots & \ddots & \ddots & \ddots & \ddots & \ddots \\
a_{n-2} & a_{n-3} & \ddots & a_1 & a_0 & a_{-1} & a_{-2} & \ddots & a_{2-n} & a_{1-n} & 0 & a_{n-1} \\
a_{n-1} & a_{n-2} & a_{n-3} & \ddots & a_1 & a_0 & a_{-1} & a_{-2} & \ddots & a_{2-n} & a_{1-n} & 0 \\
\hline
0 & a_{n-1} & a_{n-2} & \ddots & a_2 & a_1 & a_0 & a_{-1} & a_{-2} & \ddots & a_{2-n} & a_{1-n} \\
a_{1-n} & 0 & a_{n-1} & a_{n-2} & \ddots & a_2 & a_1 & a_0 & a_{-1} & \ddots & a_{3-n} & a_{2-n} \\
a_{2-n} & a_{1-n} & 0 & a_{n-1} & a_{n-2} & \ddots & a_2 & a_1 & a_0 & a_{-1} & \ddots & a_{3-n} \\
\ddots & \ddots & \ddots & \ddots & \ddots & \ddots & \ddots & \ddots & \ddots & \ddots & \ddots & \ddots \\
a_{-2} & a_{-3} & \ddots & a_{1-n} & 0 & a_{n-1} & a_{n-2} & a_{n-3} & \ddots & a_1 & a_0 & a_{-1} \\
a_{-1} & a_{-2} & a_{-3} & \ddots & a_{1-n} & 0 & a_{n-1} & a_{n-2} & a_{n-3} & \ddots & a_1 & a_0
\end{array}\right], \quad (2.41)$$

where we have padded one zero in column 1 (to make up an even number $m = 2n$). Further, any work on T_n will involve operations with C_m as the latter can be done very rapidly with FFT.

[4] The minimal m that can be allowed is $m = 2n - 1$. However such a choice will not lead to a convenient dimension $m = 2^\tau$. Similarly if m is too large (say $m = 4n$), more computations will have to be carried out later using C_m. Also the extra padded value 0 in (2.41) can be any other number as this will not enter later computation of $\mathbf{f} = T_n\mathbf{g}$.

To compute $\mathbf{f} = T_n\mathbf{g}$, we need to define $\widetilde{g}$ to match the dimension of C_m

$$\widetilde{g} = \begin{pmatrix} \mathbf{g} \\ \mathbf{0}_n \end{pmatrix}, \quad \widetilde{f} = C_m\widetilde{g} = \begin{pmatrix} T_n & T_a \\ T_b & T_c \end{pmatrix}\begin{pmatrix} \mathbf{g} \\ \mathbf{0}_n \end{pmatrix} = \begin{pmatrix} T_n\mathbf{g} \\ T_b\mathbf{g} \end{pmatrix} = \begin{pmatrix} \mathbf{f} \\ T_b\mathbf{g} \end{pmatrix},$$

where $\mathbf{0}_n$ denotes a zero vector of size n, and clearly $\mathbf{f} = T_n\mathbf{g}$ is precisely the first half of vector $\widetilde{f}$ once we have formed $\widetilde{f}$. The fast procedure to compute $\widetilde{f}$ using FFT is as before

$$\widetilde{f} = F_m^{-1}\left[(F_m\widetilde{g}).*(F_m\mathbf{h})\right], \tag{2.42}$$

where $\mathbf{h}$ denotes the first column of C_m.

Example 2.5.11. (Fast computation of a Toeplitz matrix product). *We now illustrate using FFT to compute the matrix vector product $\mathbf{f} = T_8 u$ with*

$$T = \begin{pmatrix} 2 & 1 & 6 & 7 & 3 & 4 & 8 & 5 \\ 4 & 2 & 1 & 6 & 7 & 3 & 4 & 8 \\ 8 & 4 & 2 & 1 & 6 & 7 & 3 & 4 \\ 9 & 8 & 4 & 2 & 1 & 6 & 7 & 3 \\ 3 & 9 & 8 & 4 & 2 & 1 & 6 & 7 \\ 7 & 3 & 9 & 8 & 4 & 2 & 1 & 6 \\ 4 & 7 & 3 & 9 & 8 & 4 & 2 & 1 \\ 7 & 4 & 7 & 3 & 9 & 8 & 4 & 2 \end{pmatrix}, \quad u = \begin{pmatrix} 0.6868 \\ 0.5890 \\ 0.9304 \\ 0.8462 \\ 0.5269 \\ 0.0920 \\ 0.6539 \\ 0.4160 \end{pmatrix}.$$

Firstly we construct the associated circulant matrix C_m that embeds T_n in its top left corner of C_{16}. Then we compute fast Fourier transforms of two vectors, with the root vector $\mathbf{h}$ denoting column 1 of C_{16}. Below we display C_{16}, $FFT(\mathbf{h}) = F_{16}\mathbf{h}$ and $FFT(u) = F_{16}u$ respectively:

$$\left[\begin{array}{cccccccc|cccccccc} 2&1&6&7&3&4&8&5&0&7&4&7&3&9&8&4 \\ 4&2&1&6&7&3&4&8&5&0&7&4&7&3&9&8 \\ 8&4&2&1&6&7&3&4&8&5&0&7&4&7&3&9 \\ 9&8&4&2&1&6&7&3&4&8&5&0&7&4&7&3 \\ 3&9&8&4&2&1&6&7&3&4&8&5&0&7&4&7 \\ 7&3&9&8&4&2&1&6&7&3&4&8&5&0&7&4 \\ 4&7&3&9&8&4&2&1&6&7&3&4&8&5&0&7 \\ 7&4&7&3&9&8&4&2&1&6&7&3&4&8&5&0 \\ \hline 0&7&4&7&3&9&8&4&2&1&6&7&3&4&8&5 \\ 5&0&7&4&7&3&9&8&4&2&1&6&7&3&4&8 \\ 8&5&0&7&4&7&3&9&8&4&2&1&6&7&3&4 \\ 4&8&5&0&7&4&7&3&9&8&4&2&1&6&7&3 \\ 3&4&8&5&0&7&4&7&3&9&8&4&2&1&6&7 \\ 7&3&4&8&5&0&7&4&7&3&9&8&4&2&1&6 \\ 6&7&3&4&8&5&0&7&4&7&3&9&8&4&2&1 \\ 1&6&7&3&4&8&5&0&7&4&7&3&9&8&4&2 \end{array}\right], \begin{pmatrix} 78 \\ -1.1395-5.1186i \\ -11.0711-6i \\ -6.7124-1.2918i \\ -18-2i \\ 7.8840-4.1202i \\ 3.0711+6i \\ 7.9680-7.9470i \\ -10 \\ 7.9680+7.9470i \\ 3.0711-6i \\ 7.8840+4.1202i \\ -18+2i \\ -6.7124+1.2918i \\ -11.0711+6i \\ -1.1395+5.1186i \end{pmatrix}, \begin{pmatrix} 4.7412 \\ 1.3307-2.8985i \\ 0.2071-0.9321i \\ -0.1394-1.1628i \\ -0.3707+0.5812i \\ 1.1218+0.0239i \\ 0.1126-0.3791i \\ 0.4339+0.3959i \\ 0.8549 \\ 0.4339-0.3959i \\ 0.1126+0.3791i \\ 1.1218-0.0239i \\ -0.3707-0.5812i \\ -0.1394+1.1628i \\ 0.2071+0.9321i \\ 1.3307+2.8985i \end{pmatrix}.$$

Then after an inverse FFT, we obtain

$$x_1^T = (369.8108 - 16.3529 - 3.5085i\ -7.8857 + 9.0770i\ -0.5668 + 7.9855i)$$
$$x_2^T = (7.8342 - 9.7207i\ 8.9431 - 4.4337i\ 2.6203 - 0.4888i\ 6.6035 - 0.2932i)$$
$$y_1^T = (-8.54956.6035 + 0.2932i\ 2.6203 + 0.4888i\ 8.9431 + 4.4337i)$$
$$y_2^T = (7.8342 + 9.7207i\ -0.5668 - 7.9855i\ -7.8857 - 9.0770i\ -16.3529 + 3.5085i)$$
$$\mathbf{f} = (22.7283\ 19.8405\ 17.9882\ 23.2110\ 26.1706\ 27.1592\ 23.5839\ 25.1406)^T$$

from using

$$\begin{pmatrix} \mathbf{f} \\ T_b u \end{pmatrix} = IFFT\,[FFT(\mathbf{h}).*FFT(u)] = IFFT \begin{pmatrix} x_1 \\ x_2 \\ y_1 \\ y_2 \end{pmatrix} = \begin{pmatrix} \mathbf{f} \\ T_b u \end{pmatrix}.$$

Finally we remark that the use of circulant preconditioners will be discussed in Chapter 4, and further examples including block forms will be studied in Chapter 13 for image restoration applications.

Readers who are interested in checking out the details of Examples 2.5.10 and 2.5.11 should use the supplied Mfile `circ_toep.m` as shown in Section 2.7.

We also remark that other circulant-like matrices may be diagonalized by other fast transform techniques. See [104] and refer also to Section 4.6.

2.5.3 Special matrices whose inverse can be analytically found

Matrices whose inverse can be analytically found can provide a better and more efficient preconditioner, as fast numerical solutions are not even needed.

2.5.3.1 (1) Special lower triangular matrices

As is known, the inverse of a lower triangular matrix is still lower triangular. We consider a case where the inverse is very sparse.

Let all components of two vectors $a = (a_j)_1^n,\ b = (b_j)_1^n \in \mathbb{R}^n$ be nonzero. Then

$$F_1^{-1} = \begin{bmatrix} a_1b_1 & & & & \\ a_2b_1 & a_2b_2 & & & \\ a_3b_1 & a_3b_2 & a_3b_3 & & \\ \vdots & \vdots & \vdots & \ddots & \\ a_nb_1 & a_nb_2 & a_nb_3 & \cdots & a_nb_n \end{bmatrix}^{-1} = \begin{bmatrix} \frac{1}{b_1a_1} & & & \\ -\frac{1}{b_2a_1} & \ddots & & \\ & \ddots & \frac{1}{b_{n-1}a_{n-1}} & \\ & & -\frac{1}{b_na_{n-1}} & \frac{1}{b_na_n} \end{bmatrix} \tag{2.43}$$

This result can be easily verified once we extract out two diagonal matrices diag(a), diag(b) and observe that

$$T = \begin{bmatrix} 1 & & & & \\ 1 & 1 & & & \\ 1 & 1 & 1 & & \\ \vdots & \vdots & \vdots & \ddots & \\ 1 & 1 & 1 & \cdots & 1 \end{bmatrix} = \begin{bmatrix} 1 & & & & \\ -1 & 1 & & & \\ & -1 & 1 & & \\ & & \ddots & \ddots & \\ & & & -1 & 1 \end{bmatrix}^{-1}. \tag{2.44}$$

2.5.3.2 (2) Special dense matrices

Let all components of vector $b = (b_j)_1^n \in \mathbb{R}^n$ be nonzero. Then

$$F_2^{-1} = \begin{bmatrix} b_1 & b_1 & b_1 & \cdots & b_1 \\ b_1 & b_1+b_2 & b_1+b_2 & \cdots & b_1+b_2 \\ b_1 & b_1+b_2 & b_1+b_2+b_3 & \cdots & b_1+b_2+b_3 \\ \vdots & \vdots & \vdots & \ddots & \vdots \\ b_1 & b_1+b_2 & b_1+b_2+b_3 & \cdots & b_1+b_2+\cdots+b_n \end{bmatrix}^{-1} \tag{2.45}$$

$$= \begin{bmatrix} 1/b_1+1/b_2 & -1/b_2 & & & \\ -1/b_2 & 1/b_2+1/b_3 & -1/b_3 & & \\ & -1/b_3 & \ddots & \ddots & \\ & & \ddots & 1/b_{n-1}+1/b_n & -1/b_n \\ & & & -1/b_n & 1/b_n \end{bmatrix}.$$

In the particular case, $b_j = 1,\ j = 1, \ldots, n$, we have

$$\begin{bmatrix} 1 & 1 & 1 & 1 & \cdots & 1 \\ 1 & 2 & 2 & 2 & \cdots & 2 \\ 1 & 2 & 3 & 3 & \cdots & 3 \\ 1 & 2 & 3 & 4 & \cdots & 4 \\ \vdots & \vdots & \vdots & \vdots & \cdots & \vdots \\ 1 & 2 & 3 & 4 & \cdots & n-2 \\ 1 & 2 & 3 & 4 & \cdots & n-1 \\ 1 & 2 & 3 & 4 & \cdots & n \end{bmatrix}^{-1} = \begin{bmatrix} 2 & -1 & & & & \\ -1 & 2 & -1 & & & \\ & -1 & 2 & \ddots & & \\ & & \ddots & \ddots & -1 & \\ & & & -1 & 2 & -1 \\ & & & & -1 & 1 \end{bmatrix}. \tag{2.46}$$

Remark that the matrix on the right-hand side might arise from discretizing a two-point boundary value problem in 1D with mixed Dirichlet/Neumann boundary conditions. The matrix is diagonalizable having known eigenvalues and eigenvectors as used in [389] – of course this matrix is nearly Toeplitz (or almost circulant, see Section 2.5.2 and Section 4.6).

To understand (2.45), one only needs to observe that (with T and T^{-1} from (2.44))

$$F_2 = T \begin{bmatrix} b_1 & & & \\ & b_2 & & \\ & & \ddots & \\ & & & b_n \end{bmatrix} T^T \qquad \text{and}$$

$$(T^{-1})^T \begin{bmatrix} a_1 & & & \\ & a_2 & & \\ & & \ddots & \\ & & & a_n \end{bmatrix} T^{-1} = \begin{bmatrix} a_1 + a_2 & -a_2 & & & \\ -a_2 & a_2 + a_3 & -a_3 & & \\ & -a_3 & \ddots & \ddots & \\ & & \ddots & a_{n-1} + a_n & -a_n \\ & & & -a_n & a_n \end{bmatrix}.$$

A more general type of symmetric dense matrices that have an exact tridiagonal inverse is the following: each row (from 2 to n) is proportional to row 1 pointwise; see [242, p. 179].

2.5.3.3 (3) Kroneker tensor products

The Kroneker tensor product $A \otimes B$ for two matrices A, B has been defined in Section 1.6. Here we highlight its inversion property which is potentially useful for preconditioning. The main result is that if A, B are nonsingular, then the product $A \otimes B$ is also nonsingular and one can verify that

$$(A \otimes B)^{-1} = A^{-1} \otimes B^{-1}. \tag{2.47}$$

For example, if $A = \begin{pmatrix} 2 & 2 \\ 1 & 3 \end{pmatrix}$ and if $B = \begin{pmatrix} 4 & \\ & 5 \end{pmatrix}$, then from $A^{-1} = \begin{pmatrix} 3/4 & -1/2 \\ -1/4 & 1/2 \end{pmatrix}$ and $B^{-1} = \begin{pmatrix} 1/4 & \\ & 1/5 \end{pmatrix}$, we obtain that

$$(A \otimes B)^{-1} = \begin{pmatrix} 8 & & 8 & \\ & 10 & & 10 \\ 4 & & 12 & \\ & 5 & & 15 \end{pmatrix}^{-1} = \begin{pmatrix} 3/16 & & -1/8 & \\ & 3/20 & & -1/10 \\ -1/16 & & 1/8 & \\ & -1/20 & & 1/10 \end{pmatrix}.$$

2.5.3.4 (4) Sherman–Morrison types

Theorem 2.5.12. (Sherman–Morrison–Woodburg). *Assume $A \in \mathbb{R}^{n \times n}$ is nonsingular and $(I + V^T A^{-1} U)^{-1}$ is also nonsingular for some $n \times m$ matrices U, V. Then*

$$(A + UV^T)^{-1} = A^{-1} - A^{-1}U(I + V^T A^{-1} U)^{-1} V^T A^{-1}. \tag{2.48}$$

Proof. Denote by B the right-hand side of (2.48). By direct multiplication,

$$\begin{aligned}(A+UV^T)B &= I + UV^TA^{-1} - (I+UV^TA^{-1})U(I+V^TA^{-1}U)^{-1}V^TA^{-1} \\ &= I + UV^TA^{-1} - U(I+V^TA^{-1}U)(I+V^TA^{-1}U)^{-1}V^TA^{-1} \\ &= I + UV^TA^{-1} - UV^TA^{-1} \; = \; I.\end{aligned}$$

Thus B is the inverse of $(A+UV^T)$. ■

The main formula can take on a more general form

$$(A+UGV^T)^{-1} = A^{-1} - A^{-1}U(G^{-1}+V^TA^{-1}U)^{-1}V^TA^{-1}. \tag{2.49}$$

Note that if $m = 1$ and $(1 + v^TA^{-1}u) \neq 0$ for vector $u, v \in \mathbb{R}^n$, the theorem becomes the simpler Sherman–Morrison formula

$$(A+uv^T)^{-1} = A^{-1} - A^{-1}\frac{uv^T}{(1+v^TA^{-1}u)}A^{-1}, \tag{2.50}$$

or in the special case $A = I$

$$(I+uv^T)^{-1} = I - \frac{uv^T}{(1+v^Tu)}. \tag{2.51}$$

Further, if taking $u = -2v/(v^Tv)$ for any $v \in \mathbb{R}^n$ (note the Sherman–Morrison condition is trivially met: $1 + v^Tu = 1 - 2 = -1 \neq 0$), the above becomes the familiar orthogonality property of the Householder transform [229] of the form $P = I - 2/(v^Tv)vv^T$ (which is symmetric; see also Section 2.4.2):

$$(I - \frac{2}{v^Tv}vv^T)^{-1} = I - \frac{uv^T}{(1+v^Tu)} = I + \frac{2}{v^Tv}\frac{vv^T}{(-1)} = I - \frac{2}{v^Tv}vv^T. \tag{2.52}$$

We finally remark that the Sherman–Morrison formula (2.50) can be flexibly used to define various easily invertible preconditioners e.g. taking $A = (a - b)I$, $v = (1, \ldots, 1)^T$ and $u = bv$ (assuming $(a-b) \neq 0,\ a + (n-1)b \neq 0$) gives

$$\begin{bmatrix} a & b & b & \cdots & b \\ b & a & b & \cdots & b \\ b & b & a & \cdots & b \\ \vdots & \vdots & \vdots & \ddots & \vdots \\ b & b & b & \cdots & a \end{bmatrix}^{-1} = (A+uv^T)^{-1} = \frac{I}{(a-b)} - \frac{uv^T/(a-b)}{a+(n-1)b}. \tag{2.53}$$

2.6 Ordering algorithms for better sparsity

Efficient treatment of sparse matrices is the key to scientific computing as differential equations are defined locally and the discretized equations (matrices) do not involve all variables in the whole domain.

A sparse matrix A with random nonzero locations may only be useful for working out Ax (matrix vector products) efficiently, as matrix vector products are the main operations with iterative methods Section 1.1. However for preconditioning purpose, sparsity itself is not enough and sparsity patterns are important in ensuring efficient direct solution.

Graph theory ideas can be used to pre-process a sparse matrix to arrange it in desirable patterns so that further computations are easy and fast. The subject can be both abstract and practical but the result is often amazingly good [189]. Although one may use graph theory to study unsymmetric matrices, techniques resulting from symmetric matrices are useful to general cases so we restrict our attention to symmetric matrices and undirected graphs.

Given a symmetric sparse matrix $A \in \mathbb{R}^{n\times n}$, a $\underline{\text{Node } d_i}$ is associated with each row and column ($i = 1, 2, \cdots, n$); an $\underline{\text{Edge } e_{ij}}$ from Node d_i to Node d_j is associated with a nonzero entry a_{ij}. Then nonzeros in matrix A can be represented by a sparse graph $G(A)$. On the other hand, once $G(A)$ is given, we can recover the sparsity pattern of A. On the left plot of Figure 2.2, we display a six-node graph $G(A_1)$; from $G(A_1)$, we can work out sparsity of A_1 in (2.54). On the other hand, given the following matrix A_2, we can work out $G(A_2)$ as shown on the right plot of Figure 2.2. Here

$$A_1 = \begin{pmatrix} 1 & 1 & & & 1 & 1 \\ 1 & 1 & & 1 & & \\ & & 1 & & 1 & 1 \\ & 1 & & 1 & 1 & \\ 1 & & 1 & 1 & 1 & \\ 1 & & 1 & & & 1 \end{pmatrix}, \quad A_2 = \begin{pmatrix} 2 & -1 & & & & -1 \\ -1 & 2 & -1 & & & \\ & -1 & 2 & -1 & & \\ & & -1 & 2 & -1 & \\ & & & -1 & 2 & -1 \\ -1 & & & & -1 & 2 \end{pmatrix}. \tag{2.54}$$

This section will review some re-ordering algorithms that are known to be effective for preconditioning purposes [48]. Many other ordering algorithms exist to assist direct solvers [189]; see also Chapter 7.

Definition 2.6.13. (Level sets of a sparse graph $G(A)$). *Given a graph $G(A)$ with n nodes $\{1, \cdots, n\}$, the j-th level set of* a fixed node *i is defined as*

$$L_i^{[j]} = L_{node\ i}^{[level\ j]} \Big\backslash L_{node\ i}^{[level\ j-1]}$$

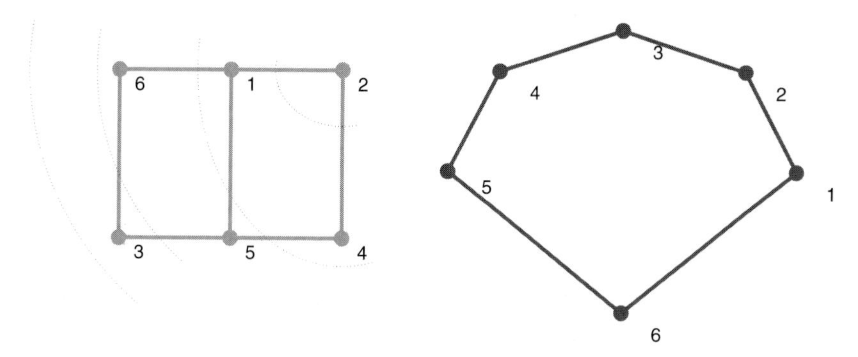

Figure 2.2. Illustration of sparse graphs: left: $G(A_1)$ is given, to find A_1; right: A_2 is given, to find $G(A_2)$.

which is assumed to have $m(i, j)$ members

$$\left\{L_i^{[j]} = L_i^{[j]}(1),\ L_i^{[j]}(2),\ \cdots,\ L_i^{[j]}(m(i, j))\right\},$$

where $1 \geq j \leq \ell(i)$ and $L_i^{[0]} = \{i\}$. Here $\ell(i)$ denotes the total number of level sets (w.r.t. node i) needed to exhaust all n nodes. Further, for graph $G(A)$, we call

- *$m(i, 1)$ the* **'Degree of node** *i'*, *which measures how many neighbours node i has;*
- *$\ell(i)$ the* **'Diameter of** $G(A)$ **with respect to** *i', which measures the total number of level sets (i.e. the depth).*

We can expand on the notation as follows

$$\begin{aligned}
&\text{for level 0,} && L_i^{[0]} = L_i^{[0]}(m(i,0)) = L_i^{[0]}(1) = \{i\}\\
&\text{for level 1,} && L_i^{[1]} = \left\{L_i^{[1]}(1),\ L_i^{[1]}(2),\ \cdots,\ L_i^{[1]}(m(i,1))\right\}\\
& && \quad = \left\{j : j \text{ is a neighbour of } L_i^{[0]}\right\} \setminus L_i^{[0]}\\
&\text{for level 2,} && L_i^{[2]} = \left\{L_i^{[2]}(1),\ L_i^{[2]}(2),\ \cdots,\ L_i^{[2]}(m(i,2))\right\}\\
& && \quad = \left\{j : j \text{ is a neighbour of } L_i^{[1]}\right\} \setminus L_i^{[1]}\\
& && \cdots \quad \cdots \quad \cdots \quad \cdots\\
&\text{for level } \ell(i), && L_i^{[\ell(i)]} = \left\{L_i^{[\ell(i)]}(1),\ L_i^{[\ell(i)]}(2),\ \cdots,\ L_i^{[\ell(i)]}(m(i,\ell(i)))\right\}\\
& && \quad = \left\{j : j \text{ is a neighbour of } L_i^{[\ell(i)-1]}\right\} \setminus L_i^{[\ell(i)-1]}.
\end{aligned}$$

Consider the example from the left plot of Figure 2.2 and take $i = 2$:
$m(2, 0) \equiv 1, m(2, 1) = 2, m(2, 2) = 2, m(2, 3) = 1$ with $\ell(i) = \ell(2) = 3$,

$$\text{Level } 0, \quad L_2^{[0]} = \{2\}$$
$$\text{Level } 1, \quad L_2^{[1]} = \{1, 4\}$$
$$\text{Level } 2, \quad L_2^{[2]} = \{5, 6\}$$
$$\text{Level } 3, \quad L_2^{[3]} = \{3\}.$$

Taking $i = 5$ gives $\ell(i) = \ell(5) = 2, m(5, 0) \equiv 1, m(5, 1) = 3, m(5, 2) = 2$ and

$$\text{Level } 0, \quad L_2^{[0]} = \{5\}$$
$$\text{Level } 1, \quad L_2^{[1]} = \{1, 3, 4\}$$
$$\text{Level } 2, \quad L_2^{[2]} = \{2, 6\}.$$

2.6.1 Reverse Cuthill–McKee algorithm for bandwidth reduction

We now introduce the reverse Cuthill–McKee (RCM) algorithm that intends to re-order the node order $r_{\text{old}} = \{1, 2, \ldots, n\}$ as $r_{\text{new}} = \{r_1, r_2, \ldots, r_n\}$ for a given sparse graph $G(A)$ so that the matrix A_{new} corresponding to r_{new} has a banded structure with a smaller bandwidth (Section 2.5.1).

Algorithm 2.6.14. (Reverse Cuthill–McKee Algorithm).

(1) Define the old node k as the new number 1 *node if*

$$m(k, 1) = \min_{1 \le i \le n} m(i, 1) \qquad \text{and} \qquad \ell(k) = \max_{1 \le i \le n} \ell(i).$$

Here the second equation implies that we have found the 'diameter' of the graph $G(A)$; often the first equation is easier to solve.

(2) Once the new node 1 *is identified, the rest of new nodes come from its level sets. Number the nodes in $L_k^{[j]}$ for $j = 1, \cdots, \ell(k)$ level-by-level. Within each level-set, number nodes in increasing degree. If two nodes have equal degrees, name first the node which is closer to the first node of previous level-set $L_k^{[j-1]}$.*

(3) Obtain the Cuthill–McKee ordering $r = r_n, r_{n-1}, \cdots, r_1$ first. Then RCM uses simply $r_{\text{new}} = r_1, r_2, \cdots, r_{n-1}, r_n$.

Take $G(A)$ from the left plot of Figure 2.2 for example. Four nodes 2, 3, 4, 6 satisfy that $\ell(i) = 2$ and $\ell(i) = 3$. We take $k = 2$ as the new node and obtain the new order $r = [2, 1, 4, 5, 6, 3]$. Thus an acceptable RCM order is $r_{\text{new}} = [3, 6, 5, 4, 1, 2]$:

$$\begin{aligned} A_{\text{new}} &= A_1(r_{\text{new}}, r_{\text{new}}) \\ &= PA_1P^T = \begin{pmatrix} 1 & 1 & 1 & & & \\ 1 & 1 & & & 1 & \\ 1 & & 1 & 1 & 1 & \\ & & 1 & 1 & & 1 \\ & 1 & 1 & & 1 & 1 \\ & & & 1 & 1 & 1 \end{pmatrix}, \quad P = \begin{pmatrix} & & 1 & & & \\ & & & & & 1 \\ & & & & 1 & \\ & & & 1 & & \\ 1 & & & & & \\ & 1 & & & & \end{pmatrix}. \end{aligned} \tag{2.55}$$

2.6.2 Duff's spiral ordering algorithm for closed surfaces

The RCM algorithm is often believed to be effective for 1D and 2D problems. However for 3D problems, the method has be found to be ineffective. While there exist other techniques [57], our recent experience [268] shows that the Duff's spiral ordering algorithm [192] is particularly suitable for 3D problems that involve a closed surface.

The algorithm, balancing $m(i, j)$ for each level j, can be presented as follows.

Algorithm 2.6.15. (Duff's spiral ordering algorithm).

(1) Define the old node k as the new number 1 *node if*

$$m(k, 1) = \min_{1 \le i \le n} m(i, 1) \qquad \textit{and} \qquad \ell(k) = \max_{1 \le i \le n} \ell(i),$$

similarly to Algorithm 2.6.14. From the new node 1 *(old k), the 'first' point under consideration, identify the corresponding 'last' point k^*,*

$$L_k^{[\ell(k)]}\left(m(k, \ell(k))\right),$$

in its last level $\ell(k)$.

(2) Find a middle node k_m between k and k^*.

(3) Re-order the graph following the level sets of the middle node k_m to obtain a spiral order: $r_{\text{spiral}} = r_1, r_2, \cdots, r_{n-1}, r_n$. (Note: $r_1 = k_m$).

For the example on the left plot of Figure 2.2, take $k = 2$ again. Then $k^* = 3$ so k_m may be either 5 or 1. Now select $k_m = 5$. Then a spiral order is $r_{\text{spiral}} = [5\ \ 4\ \ 1\ \ 3\ \ 2\ \ 4]$.

To see a more realistic example, refer to Figure 2.3. There nodes are presently ordered in a lexicographical manner. Among the four candidates $(1, 5, 21, 25)$ for the new node 1, we may take $k = 21$. Then $k^* = 5$ and $k_m = 13$. Therefore

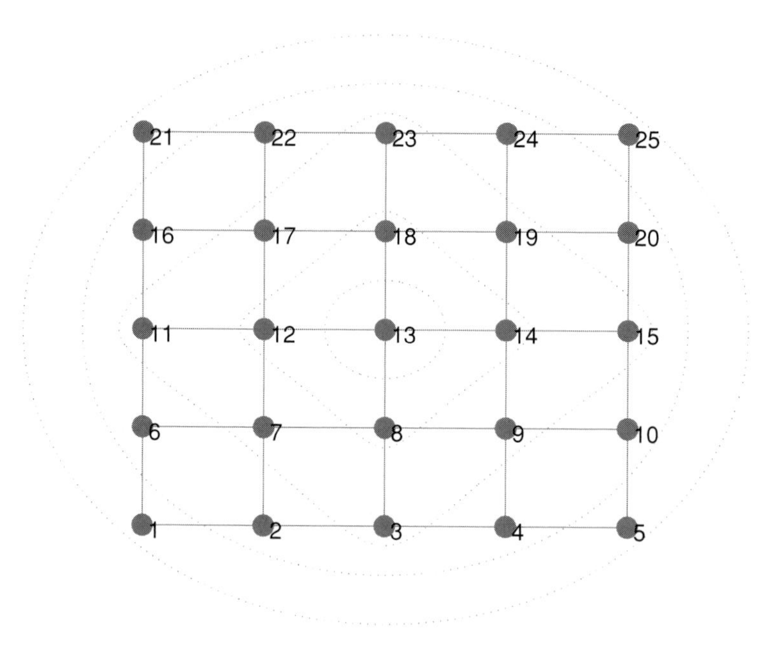

Figure 2.3. Spiral ordering illustration for a 25-node graph $G(A)$.

from the level sets of node $k_m = 13$ (spiralling anti-clockwise)

$$\begin{aligned}
\text{Level } 0, \quad & L_{13}^{[0]} = \{13\}, \\
\text{Level } 1, \quad & L_{13}^{[1]} = \{8, 14, 18, 12\}, \\
\text{Level } 2, \quad & L_{13}^{[2]} = \{3, 9, 15, 19, 23, 17, 11, 7\}, \\
\text{Level } 3, \quad & L_{13}^{[3]} = \{4, 10, 20, 24, 22, 16, 6, 2\}, \\
\text{Level } 4, \quad & L_{13}^{[4]} = \{5, 25, 21, 1\},
\end{aligned}$$

one obtains a spiral order as

$$r_{\text{spiral}} = [13, 8, 14, 18, 12, 3, 9, 15, 19, 23, 17, 11, 7, 4, 10, 20, 24, 22, 16, 6, 2, 5, 25, 21, 1].$$

2.6.3 Domain decomposition re-ordered block matrices

Domain decomposition method (DDM) is a well-established numerical technique for parallel solution of partial differential equations [432]. Here we discuss the use of non-overlapping (single level and multidomain) DDM ideas to work out a better graph ordering so that a preconditioner of an ideal sparsity structure can be constructed. That is, we are interested in the substructuring ordering (see Figure 7.2) which is closely related to another popular nested dissection ordering [214,215] (refer to Figures 7.3–7.4).

Consider the finite difference solution of some second-order operator equation (say the Poisson's equation i.e. (1.53) with $k = 0$) defined over a closed domain (with Dirichlet boundary conditions), using $n = 25$ internal nodes as shown on the left plot of Figure 2.4, which is lexicographically ordered. As is well known [355,413], the usual matrix is of a tridiagonal block structure (depending on the numerical schemes used)

$$A = \begin{pmatrix} B & X_2 & & \\ X_1 & B & \ddots & \\ & \ddots & \ddots & X_2 \\ & & X_1 & B \end{pmatrix},$$

where B is tridiagonal and X_j's are diagonal. However with the right plot of Figure 2.4, the DDM ordering defines a better block structure. To describe this

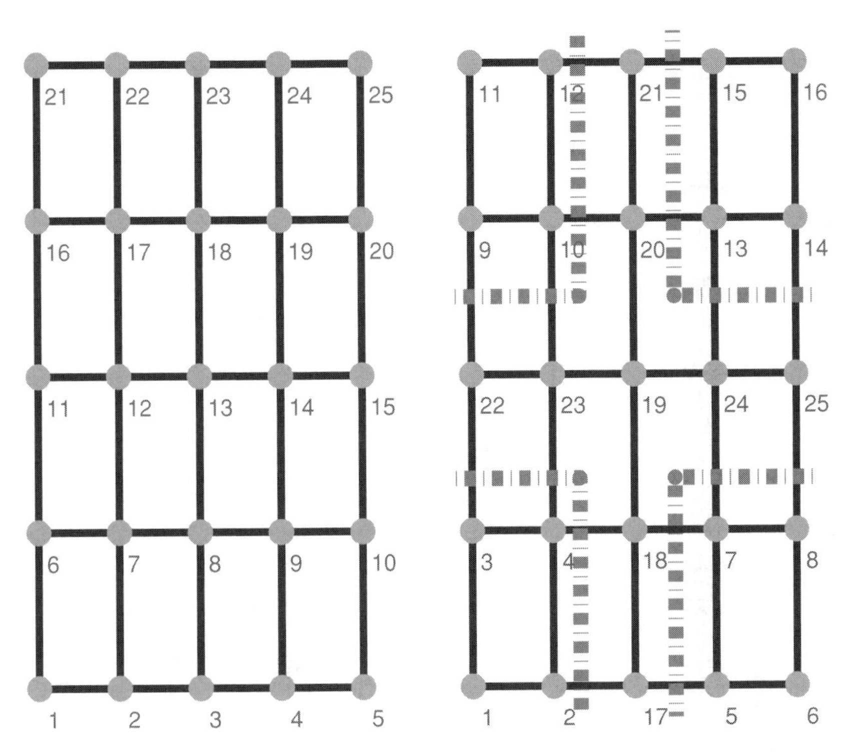

Figure 2.4. Illustration of graph re-ordering using domain decomposition: Left, $G(A)$ ordered lexicographically and Right, $G(A)$ ordered by DDM.

matrix, we first note the DDM ordering (in terms of geometry, not old ordering) is

$$r_{\rm ddm} = \begin{bmatrix} \Omega_1 & \Omega_2 & \Omega_3 & \Omega_4 & \Gamma_1 & \Gamma_2 & \Gamma_3 & \Gamma_4 & \Gamma_5 \end{bmatrix} \quad (2.56)$$

where in this particular domain partition

$$\Omega_1 = [1\ 2\ 3\ 4],\ \Omega_2 = [5\ 6\ 7\ 8],\ \Omega_3 = [9\ 10\ 11\ 12],\ \Omega_4 = [13\ 14\ 15\ 16],$$
$$\Gamma_1 = [17\ 18],\quad \Gamma_2 = 19,\quad \Gamma_3 = [20\ 21],\ \Gamma_4 = [22\ 23],\ \Gamma_5 = [24\ 25].$$

Further, in conjunction with the right plot of Figure 2.4, the corresponding matrix A is shown in Table 2.2.

In general, as $n \to \infty$, a DDM-ordered matrix approaches a large block diagonal matrix with small borders (representing interactions of internal boundaries) which is on one hand suited for parallel computing as well as direct solution, and on the other hand amenable to approximation by a banded block diagonal matrix with small borders for preconditioning purpose. Refer to [432].

Finally we remark that a very useful re-ordering algorithm due to [190] will be discussed in Chapter 5 in the context of scaling matrix A for approximate inverses while Chapter 7 will discuss some other ordering algorithms.

2.7 Discussion of software and the supplied Mfiles

This chapter has surveyed various useful direct solution techniques that are necessary tools for preconditioning implementations and for designing new preconditioners. We would encourage readers to try out these methods to gain further insight. Direct solution techniques have been implemented in virtually all software for scientific computing; see Appendix E for details.

For readers' benefit, we have developed several MATLAB® Mfiles for use to assist understanding of the material in this chapter. These are listed below with a short description of their use; we again remark that all Mfiles can be run by typing a Mfile's name, using a default setting without specifying any parameters. The following list also shows what values can be set before calling the Mfiles.

[1] `circ_toep.m` – Computes $f = Cu$ and $v = Tg$, and solves $Cf = u$ (see Section 2.5.2).

Optionally pre-set vectors r (row), c (column) for generating a Toeplitz matrix T. Pre-set c (column vector) to specify a user's own circulant matrix C. Similarly vectors u and g can be pre-set.

Table 2.2. *Illustration of matrix partition and node ordering (for the right plot of Figure 2.4).*

$$
\left[\begin{array}{cccc|ccccc}
A(\Omega_1,\Omega_1) & & & & A(\Omega_1,\Gamma_1) & & & A(\Omega_1,\Gamma_4) & \\
& A(\Omega_2,\Omega_2) & & & A(\Omega_2,\Gamma_1) & & & & A(\Omega_2,\Gamma_5) \\
& & A(\Omega_3,\Omega_3) & & & & A(\Omega_3,\Gamma_3) & A(\Omega_3,\Gamma_4) & \\
& & & A(\Omega_4,\Omega_4) & & & A(\Omega_4,\Gamma_3) & & A(\Omega_4,\Gamma_5) \\
\hline
A(\Gamma_1,\Omega_1) & A(\Gamma_1,\Omega_2) & & & A(\Gamma_1,\Gamma_1) & A(\Gamma_1,\Gamma_2) & & & \\
& & & & A(\Gamma_2,\Gamma_1) & A(\Gamma_2,\Gamma_2) & A(\Gamma_2,\Gamma_3) & A(\Gamma_2,\Gamma_4) & A(\Gamma_2,\Gamma_5) \\
& & A(\Gamma_3,\Omega_3) & A(\Gamma_3,\Omega_4) & & A(\Gamma_3,\Gamma_2) & A(\Gamma_3,\Gamma_3) & & \\
A(\Gamma_4,\Omega_1) & & A(\Gamma_4,\Omega_3) & & & A(\Gamma_4,\Gamma_2) & & A(\Gamma_4,\Gamma_4) & \\
& A(\Gamma_5,\Omega_2) & & A(\Gamma_5,\Omega_4) & & A(\Gamma_5,\Gamma_2) & & & A(\Gamma_5,\Gamma_5)
\end{array}\right]
$$

[2] `ge_all.m` – Computes all elementary matrices M_j of a LU decomposition
$M_{n-1}M_{n-2}\cdots M_1 A = U$ and solves $Ax = b$ (see Section 2.1.1). The Mfile is advanced and generates each individual $M_j = L_j^{-1}$.

Here the square matrix A and the right-hand side vector b may be pre-set if the default setting for a 7×7 matrix is not desired. (This style of generating named matrices via the MATLAB command `eval` will be used in several places, especially in the Mfile `mgm_2d.m` in Chapter 6. See also Appendix C.)

[3] `gh_all.m` – Computes all elementary matrices E_j and H_j of a Gauss–Huard (GH) decomposition $E_n H_{n-1}\cdots H_1 E_1 A = I$ and solves $Ax = b$ (see §2.3). The Mfile is advanced and generates each individual E_j, H_j.

Here the square matrix A and the right-hand side vector b may be pre-set if the default setting for a 7×7 matrix is not desired.

[4] `givens.m` – Use the Givens matrix (1.24) for either the QR decomposition or the Arnoldi decomposition of A. This is a simple Mfile.

[5] `houses.m` – Use the Householder matrix (1.26) for either the QR decomposition or the Arnoldi decomposition of A. This is another simple Mfile.

[6] `mgs.m` – Use the modified Gram–Schmidt method for the QR decomposition of A. (Compare to the Arnoldi decomposition by the Mfile `hess_as.m` in §1.9.)

[7] `gj_ek.m` – Experiments on the Gauss–Jordan transforms: products and inverses.

This Mfile is basic and the reader may modify it for more experiments.

[8] `g_e.m` – Computes L and U for $A = LU$ and solves $Ax = b$ (Algorithm 2.1.1).

Here the square matrix A and the right-hand side vector b may be pre-set if the default setting for a 7×7 matrix is not desired. This Mfile is fairly standard.

[9] `g_h.m` – Computes all elementary matrices H_j of a Gauss–Huard (GH) decomposition $H_n \cdots H_1 A = I$ and solves $Ax = b$ (see §2.3). The Mfile can be compared to `gh_all.m`.

Here the square matrix A and the right-hand side vector b may be pre-set if the default setting for a 7×7 matrix is not desired.

[10] `g_j.m` – Computes all elementary matrices E_j of a Gauss–Jordan (GJ) decomposition $E_n \cdots E_1 A = I$ and solves $Ax = b$ (see Section 2.3). The Mfile is also advanced. The reader is encouraged to compare with `g_h.m` and to check partial products e.g. $E_7 * E6 * E5$ and $E4 * E3 * E2 * E1$.

Here the square matrix A and the right-hand side vector b may be pre-set if the default setting for a 7×7 matrix is not desired.

[11] `mk_ek.m` – Experiments on the elementary Gauss transforms: products.
This Mfile is also basic and the reader may modify it for more experiments.

[12] `nsh.m` – Computes $X \approx A^{-1}$ and $x = Xb$ for $Ax = b$ using the Newton–Schultz–Hotelling method. This Mfile is simple and the reader may modify it.

[13] `waz_fox.m` – Computes $WAZ = D$ factorization (Algorithm 2.1.3). Here the square matrix A may be pre-set if the default setting for a 7×7 matrix is not desired.

[14] `waz_zol.m` – Computes the Zollenkopf factorization $WAZ = I$ (Algorithm 2.1.4).

[15] `waz_all.m` – Illustrates the Zollenkopf factorization

$$w_{n-1} \dots w_1 A z_1 \dots z_{n-1} z_n = I \text{ (see (2.13) and Algorithm 2.1.4).}$$

The Mfile is also advanced and generates each individual w_j, z_j. Here the square matrix A may be pre-set if the default setting for a 7×7 matrix is not desired.

3

Iterative methods

As we will see, iterative methods are not only great fun to play with and interesting objects for analysis, but they are really useful in many situations. For truly large problems they may sometimes offer the only way towards a solution.

HENK A. VAN DER VORST. *Iterative Krylov Methods for Large Linear Systems.* Cambridge University Press (2003)

A similar work [on the fast multipole method] was done in 3D by Rokhlin. As in 2D, the multistep algorithm was not properly explained.

ERIC DARVE. *Fast Multipole Method*, preprint, Paris, France (1997)

An iterative method for linear system $Ax = b$ finds an infinite sequence of approximate solutions $x^{(j)}$ to the exact answer x^*, each ideally with a decreased error, by using A repeatedly and without modifying it. The saving from using an iterative method lies in a hopefully early termination of the sequence as most practical applications are only interested in finding a solution x close enough to x^*. Therefore, it almost goes without saying that the essence of an iterative method is fast convergence or at least convergence. When this is not possible for (1.1), we shall consider (1.2) with a suitable M.

This chapter will review a selection of iterative methods for later use as building blocks for preconditioner designs and testing. No attempt is made to exhaust all the iterative methods as one can find them in many books and surveys (e.g. [41,464,416]) and most importantly we believe that, when a hard problem is solved, changing preconditioners is more effective than trying out a different iterative solver. One should be very sensitive about, and precise with, convergence requirements, and these should motivate preconditioning whenever convergence is a problem. Specifically, we shall review the subject in the following.

3.1 Solution complexity and expectations

It is instructive to state the 'obvious' question: what are we expecting to achieve with a fast solver for (1.1), either direct or iterative? If n is the number of unknowns, it turns out that an optimal method should produce an acceptable answer in $O(n)$ flops i.e. $O(1)$ flops per unknown. Reducing the constant in $O(1)$ further (to find the optimal method for a problem) is also interesting but a less urgent task.

For dense matrix problems, there is a general demand for fast solvers because this is the case where a direct solver normally needs

$\models$ to store $O(n^2)$ entries – a large task but often still bearable

$\models$ to carry out $O(n^3)$ flops – often prohibitive for $n \geq 10^5$.

An iterative solver (either the conjugate gradient type in this chapter or the multilevel type in Chapter 6) can 'easily' reduce the complexity to

$\models$ order $O(n^2)$ flops – more acceptable for moderately large n.

For idealized problems, the FFT Section 2.5.2, the FMM Section 3.8 and the fast wavelet methods Section 8.2 can reduce the overall complexity

$\models$ to $O(n)$ or $O(n \log n)$ flops – an optimal solution

which may be a driving motivation to develop these fast solvers further. Note that a dense matrix problem may be solved by an optimal method, after a massive flops reduction from $O(n^3)$ to $O(n)$.

For sparse matrix problems, the general demand for fast solvers is equally urgent for large scale computing. However, the expectations are similar but slightly different and occasionally confusing. We highlight these differences.

(i) Although a matrix with less than 50% nonzeros may be classified as sparse [492], matrices arising from many applications (e.g. PDEs) are often much sparser and commonly with only $O(n)$ nonzero entries. For instance, a second-order PDE in a rectangular 2D domain gives rise to only about $O(n) = 5n$ nonzero entries. This may also be part of the reason why many communities wish to compute large-scale systems with $n \gg 10^6$ unknowns (which is unthinkable with dense matrices in today's technology).

(ii) Unlike a dense matrix case, for a sparse matrix with $O(n)$ nonzeros, the direct solution of (1.1) typically requires up

$\models$ to $O(n^2)$ flops

and this implies that all fast solvers are 'optimal' if converging in a fixed number of iterations as they have thus reduced the usual complexity

$\models$ to order $O(n)$ flops.

Thus a typical sparse matrix problem may be solved by an optimal method, after a large flops reduction from $O(n^2)$ to $O(n)$.

(iii) Sparse matrices are more likely to be asymptotically ill-conditioned as $n \to \infty$, especially those arisen from a discretized PDE or PDE systems. Therefore a direct solver would require preconditioning to ensure solution accuracy and so does an iterative solver.

3.2 Introduction to residual correction

Many iterative methods can be described in the residual correction framework (or by repeated use of residual correction). Let $x^{(j-1)} = x^{\text{old}}$ be the current approximation (e.g. an initial guess when $j - 1 = 0$ or simply set $x^{(0)} = 0$). Then the residual vector of (1.1) is

$$r = r^{(j-1)} = b - Ax^{(j-1)}. \tag{3.1}$$

The residual correction idea seeks the next approximation $x^{(j)} = x^{\text{new}}$ from

$$x^{(j)} = x^{\text{new}} = x^{\text{old}} + e^{\text{new}} = x^{(j-1)} + e^{(j)} \tag{3.2}$$

by using the above residual r and the following correction $e^{(j)}$, with some *residual correction operator* $\mathbf{L} \approx A$,

$$\mathbf{L}e^{(j)} = \mathbf{L}e^{\text{new}} = r. \tag{3.3}$$

Here $x^{(j)}$ will be the exact solution x^* if $\mathbf{L} = A$ as one can verify that $Ax^{(j)} = Ax^{(j-1)} + Ae^{(j)} = Ax^{(j-1)} + r = b$. Eliminating r from (3.3) gives rise to the generalized Richardson iteration (setting $B = \mathbf{L}^{-1}$)

$$x^{(j)} = x^{(j-1)} + B\left(b - Ax^{(j-1)}\right), \tag{3.4}$$

where B is usually called the preconditioner (that is provided by the underlying method), although the original Richardson idea is to take $B = I$ (or $B = \omega_j I$ with ω_j a scalar parameter). The so-called iterative refinement technique [23,80] is also within this framework. In fact, different iterative methods only differ in choosing $\mathbf{L}$ (or B) – an operator that can be fixed and explicit, or variable and implicit.

We note that an alternative (but equivalent) form to (3.4) is the following

$$\mathbf{L}\frac{x^{(j)} - x^{(j-1)}}{\tau_j} + Ax^{(j-1)} = b,$$

where τ_j is some nonzero scalar, that might be 'absorbed' into $\mathbf{L}$.

3.3 Classical iterative methods

Many classical iterative methods, including Jacobi, Gauss–Seidel (GS) and successive over-relaxation (SOR), are routinely described in text books ever since the first edition of [467]; see also [264,287]. Although such methods are only useful for a restricted class of matrix problems, they become the standard smoothers for the modern multigrid method (Chapter 6). Moreover, in engineering computing, these classical methods are still in wide use [473], as they are simple.

It should be remarked that all the theories presented in this section are well known and can be found in these texts [23,80,180,260,264,287,413,467].

3.3.1 Richardson $A = M - N$

We first consider a general matrix splitting method and discuss its convergence requirements. Let $A = M - N$ and assume M is invertible. Then equation (1.1) takes, letting $T = M^{-1}N = I - M^{-1}A$ and $f = M^{-1}b$, a convenient format for stationary iterations:

$$Mx = Nx + b \qquad \text{or} \qquad x = Tx + f. \tag{3.5}$$

Naturally if M is easy to invert (i.e. solving $Mx = y$ is easy), we can form the iterations

$$Mx^{(j)} = Nx^{(j-1)} + b \qquad \text{or} \qquad \begin{aligned} x^{(j)} &= M^{-1}Nx^{(j-1)} + M^{-1}b \\ &= Tx^{(j-1)} + f. \end{aligned} \tag{3.6}$$

Here T is the *iteration matrix* for our generic splitting method. We can cast the second equation in (3.6) into the standard Richardson form (3.4)

$$\begin{aligned} x^{(j)} &= M^{-1}Nx^{(j-1)} + M^{-1}b \\ &= M^{-1}(M - A)x^{(j-1)} + M^{-1}b \\ &= x^{(j-1)} + M^{-1}\left(b - Ax^{(j-1)}\right). \end{aligned} \tag{3.7}$$

The *residual correction operator* is $\mathbf{L} = M$. To derive an iteration formula for the error $e^{(j)} = x^{(j)} - x$, take the difference of (3.6) and (3.5)

$$e^{(j)} = x^{(j)} - x = T\left(x^{(j)} - x\right) = Te^{(j-1)} = \cdots = T^j e^{(0)}. \tag{3.8}$$

Thus from $\|e^{(j)}\|_2 \le \|T\|_2^j \|e^{(0)}\|_2$, one can see that a sufficient condition for (3.6) to converge (i.e. $\|e^{(j)}\|_2 \to 0$ as $j \to \infty$) is $\|T\|_2 < 1$. Actually, a more precise statement can be made [23,80,180]:

Theorem 3.3.1. *The necessary and sufficient condition for $T^j \to 0$ as $j \to \infty$ is $\rho(T) < 1$, with $\rho(T) = \max |\lambda_j(T)|$ the spectral radius of T.*

Proof. Let λ_k be the kth eigenvalue and x_k the corresponding normalized eigenvector of T, for $k = 1, \ldots, n$. (Note $\|x_k\|_2 = 1$).

To prove the necessary condition, assume $\lim_{j\to\infty} T^j \to 0$. Then $T^j x_k = \lambda_k^j x_k$ and $x_k^H T^j x_k = \lambda_k^j x_k^H x_k = \lambda_k^j$ so $\lim_{j\to\infty} \lambda_k^j = \lim_{j\to\infty} x_k^H T^j x_k = 0$. Thus $|\lambda_k(T)| < 1$ and, as a result, $\rho(T) < 1$.

To prove the sufficient condition, assume that $\rho(T) < 1$. Then $|\lambda_k| < 1$ so $\lambda_k^j \to 0$ and $D^j \to 0$ as $j \to \infty$, where $D = \operatorname{diag}(\lambda_1, \ldots, \lambda_n)$. Let $T = XJX^{-1}$ be the Jordan canonical form of T with $J = D + U$ and U being strictly upper triangular (with some 1's on its super diagonal). Note[1] that $T^j = XJ^jX^{-1} = X\left[\operatorname{diag}(\lambda_k) + U\right]^j$ and $U^j = 0$ whenever $j \ge n$ (or more precisely when $j \ge \max \ell_k$ – the maximum algebraic multiplicity of T's eigenvalues).

Consider the binomial expansion of J^j, noting $U^\ell = 0$ if $\ell \ge n$,

$$\begin{aligned} J^j = (U + D)^j &= \tfrac{j!}{0!(j-0)!} U^0 D^j + \tfrac{j!}{1!(j-1)!} U^1 D^{j-1} + \tfrac{j!}{2!(j-2)!} U^2 D^{j-2} \\ &\quad + \cdots + \tfrac{j!}{(n-1)!(j-n+1)!} U^{n-1} D^{j-n+1} \end{aligned}$$

with $j \ge n$ (before $j \to \infty$). Here those U^ℓ terms for $\ell \ge n$, being 0, are not shown and the remaining n terms all contain high powers of D with entries $\operatorname{diag}(\lambda_k^{j-\ell})$ for $\ell = 0, 1, \ldots, n-1$. Since $\lim_{j'\to\infty} \lambda_k^{j'} \to 0$ this proves that $J^j \to 0$ and $T^j \to 0$. ■

Special cases (4.4) of A lead to stronger and more specific results [467,260] e.g.

Lemma 3.3.2. *If A is monotone (i.e. $A^{-1} \ge 0$) and the splitting $A = M - N$ is regular (i.e. $M^{-1} \ge 0$ and $N \ge 0$), then there holds*

$$\rho(T) = \rho(M^{-1}N) = \rho(I - M^{-1}A) = \frac{\rho(A^{-1}N)}{\rho(A^{-1}N) + 1} < 1.$$

[1] This is a simple trick commonly used in the context of similarity transforms and in working with powers of a matrix e.g. $T^2 = TT = XJ(X^{-1}X)JX^{-1} = XJ^2X^{-1}$. For the Jordan decomposition, refer to Section A.6.

3.3.2 Jacobi, GS and SOR

We now introduce the three well-known iterative methods, each choosing M, in the splitting $A = M - N$, differently. Assuming A has no zeros on its diagonal, we write

$$A = D - \widetilde{L} - \widetilde{U} = D(I - L - U), \tag{3.9}$$

where D is the diagonal of A, $-\widetilde{L}$ and $-\widetilde{U}$ are the strictly lower and upper triangular parts of A respectively, $L = D^{-1}\widetilde{L}$ and $U = D^{-1}\widetilde{U}$. As $L,\ U$ do not *share* nonzero entries, we see that $|L + U| = |L| + |U|$ after removing any negative signs. The notation can be illustrated for the following 4×4 case:

$$A = \begin{bmatrix} 14 & -3 & -2 & -4 \\ -4 & 11 & -2 & -2 \\ -1 & -4 & 15 & -9 \\ -3 & -3 & -4 & 12 \end{bmatrix},\ D = \begin{bmatrix} 14 & 0 & 0 & 0 \\ 0 & 11 & 0 & 0 \\ 0 & 0 & 15 & 0 \\ 0 & 0 & 0 & 12 \end{bmatrix},\ \widetilde{L} = \begin{bmatrix} 0 & 0 & 0 & 0 \\ 4 & 0 & 0 & 0 \\ 1 & 4 & 0 & 0 \\ 3 & 3 & 4 & 0 \end{bmatrix},$$

$$\widetilde{U} = \begin{bmatrix} 0 & 3 & 2 & 4 \\ 0 & 0 & 2 & 2 \\ 0 & 0 & 0 & 9 \\ 0 & 0 & 0 & 0 \end{bmatrix},\ L = \begin{bmatrix} 0 & 0 & 0 & 0 \\ 4/11 & 0 & 0 & 0 \\ 1/15 & 4/15 & 0 & 0 \\ 1/4 & 1/4 & 1/3 & 0 \end{bmatrix}, \tag{3.10}$$

$$U = \begin{bmatrix} 0 & 3/14 & 1/7 & 2/7 \\ 0 & 0 & 2/11 & 2/11 \\ 0 & 0 & 0 & 3/5 \\ 0 & 0 & 0 & 0 \end{bmatrix}.$$

To experiment on the three methods in this section, the reader may use the provided Mfile `iter3.m` as explained in Section 3.9.

♦ **Jacobi method.** The Jacobi method chooses $M = D,\ N = \widetilde{L} + \widetilde{U}$ for (3.6)

$$x^{(j)} = D^{-1}\left(\widetilde{L} + \widetilde{U}\right)x^{(j-1)} + D^{-1}b = (L + U)x^{(j-1)} + D^{-1}b \tag{3.11}$$

and can be written in the Richardson form (3.4) or (3.7):

$$x^{(j)} = x^{(j-1)} + D^{-1}\left(b - Ax^{(j-1)}\right) \tag{3.12}$$

with the *residual correction operator* $\mathbf{L} = M = D$ and the iteration matrix $T = T_J = L + U$.

To establish the convergence i.e. $\rho(T_J) < 1$ for a class of special matrices, we first recall the definition for a strictly diagonally dominant (SDD) matrix A:

$$|a_{ii}| > \sum_{j \neq i} |a_{ij}| \qquad \text{for } i = 1, \ldots, n. \tag{3.13}$$

Define $\mathbf{e} = (1, \ldots, 1)^T$ as the sized n vector of all ones. It is surprising as well as trivial to rewrite (3.13), in the notation of (3.9) and $\mathbf{e}$, as a vector inequality:

$$|D|\mathbf{e} > \left(|\widetilde{L}| + |\widetilde{U}|\right)\mathbf{e} \qquad \text{or} \qquad \mathbf{e} > \Big(|L| + |U|\Big)\mathbf{e} \tag{3.14}$$

From (1.12), $\|A\|_\infty = \| \; |A|\mathbf{e}\|_\infty$ for any matrix A. Hence (3.14) implies $\|L + U\|_\infty = \| \, (|L| + |U|)\,\mathbf{e}\|_\infty < \|\mathbf{e}\|_\infty = 1$. Thus $\rho(T_J) \le \|T_J\|_\infty < 1$ and the Jacobi method converges for SDD matrices. Note that SDD implies

$$(I - L)^{-1} = \sum_{i=0}^{\infty} L^i, \;\; (I - U)^{-1} = \sum_{i=0}^{\infty} U^i, \;\; \text{as } \|L\|_\infty < 1 \text{ and } \|U\|_\infty < 1. \tag{3.15}$$

If the diagonal dominance is only weak (when the '>' sign in (3.13) becomes '$\ge$'), matrix A must be additionally *irreducible* i.e. there exists no permutation matrix Q such that

$$QAQ^T = \begin{bmatrix} A_{11} & A_{12} \\ & A_{22} \end{bmatrix} \tag{3.16}$$

to prove $\rho(T_J) < 1$; see [467,180].

♦ **Gauss–Seidel method.** The GS method chooses $M = D - \widetilde{L}, \; N = \widetilde{U}$ for (3.6)

$$\begin{aligned} x^{(j)} &= \left(D - \widetilde{L}\right)^{-1} \widetilde{U} x^{(j-1)} + \left(D - \widetilde{L}\right)^{-1} b \\ &= (I - L)^{-1}\, U x^{(j-1)} + (I - L)^{-1}\, D^{-1} b \end{aligned} \tag{3.17}$$

and can be written in the Richardson form (3.4) or (3.7):

$$x^{(j)} = x^{(j-1)} + \left(D - \widetilde{L}\right)^{-1} \left(b - Ax^{(j-1)}\right) \tag{3.18}$$

with the *residual correction operator* $\mathbf{L} = M = D - \widetilde{L}$ and the iteration matrix $T = T_{\text{GS}} = (I - L)^{-1}\, U$.

Remark 3.3.3. The choice of M, N in (3.17) depends on A. For problems arising from a discretized PDE, a re-ordering of the nodes can fundamentally change the matrix splitting. Refer to [326,4] and Algorithm 4.3.3.

For SDD matrices, the proof of $\rho(T_{\text{GS}}) < 1$ is again based on using (3.14). Since $\|T_{\text{GS}}\|_\infty = \| \; |(I - L)^{-1}\, U|\mathbf{e}\|_\infty \le \| \; |(I - L)^{-1}| \; |U|\mathbf{e}\|_\infty$, we need to consider the term $|\,(I - L)^{-1}\,|$ in order to make use of (3.14). The trick lies in the sparsity structure of L that satisfies $L^j = 0$ and $|L|^j = 0$ if $j \ge n$. From (3.15), observe that both $(I - L)$ and $(I - |L|)$ are invertible because $\|L\| < 1$. So we have $(I - L)^{-1} = \sum_{i=0}^{\infty} L^i = \sum_{i=0}^{n-1} L^i$ and $(I - |L|)^{-1} = \sum_{i=0}^{\infty} |L|^i = \sum_{i=0}^{n-1} |L|^i$. Then from term by term comparison,

$$|\,(I - L)^{-1}\,| \; |U|\mathbf{e} \le (I - |L|)^{-1} \; |U|\mathbf{e}.$$

It remains to make use of the SDD property (3.14) i.e. $(I - |L| - |U|)\mathbf{e} > 0$ or

$$|L|(I - |L| - |U|)\mathbf{e} + |U|\mathbf{e} > |U|\mathbf{e}, \qquad \text{or} \qquad |U|\mathbf{e} < (I - |L|)\,(|L| + |U|)\mathbf{e}.$$

Therefore for any integer power i,

$$\begin{aligned} |L|^i|U|\mathbf{e} &< |L|^i(I - |L|)\,(|L| + |U|)\mathbf{e} \\ \text{i.e. } (I - |L|)^{-1}\,|U|\mathbf{e} &< (|L| + |U|)\mathbf{e} < \mathbf{e}. \end{aligned}$$

Thus we have proved that $\rho(T_{\text{GS}}) \le \|T_J\| < 1$, provided that A is a SDD matrix.

♦ **Successive over-relaxation (SOR) method.** The SOR method chooses $M = D/\omega - \widetilde{L}$, $N = D/\omega - D + \widetilde{U}$ (for some positive ω) for (3.6)

$$\begin{aligned} x^{(j)} &= \left(D/\omega - \widetilde{L}\right)^{-1}\left(D/\omega - D + \widetilde{U}\right)x^{(j-1)} + \left(D/\omega - \widetilde{L}\right)^{-1} b \\ &= (I/\omega - L)^{-1}\,(I/\omega - I + U)\,x^{(j-1)} + (I/\omega - L)^{-1}\,D^{-1}b \qquad (3.19) \\ &= (I - \omega L)^{-1}\left((1 - \omega)I + \omega U\right)x^{(j-1)} + \omega\,(I - \omega L)^{-1}\,D^{-1}b \end{aligned}$$

and can be written in the Richardson form (3.4) or (3.7):

$$x^{(j)} = x^{(j-1)} + \left(D/\omega - \widetilde{L}\right)^{-1}\left(b - Ax^{(j-1)}\right) \qquad (3.20)$$

with the *residual correction operator* $\mathbf{L} = M = D/\omega - \widetilde{L}$ and the iteration matrix $T = T_{\text{SOR}} = (I - \omega L)^{-1}\,((1 - \omega)I + \omega U)$. Here note that the inverse matrix $(I - \omega L)^{-1}$ (lower triangular with unit diagonal) always exists! A quick inspection can reveal that the SDD property (3.13) or (3.14) of A (as with GS) may no longer ensure the same SDD property for $(D/\omega - \widetilde{L})$ or $(I - \omega L)$ if $\omega > 1$, implying that SDD might not be the eventual sufficient condition for convergence.

Lemma 3.3.4. *If A is SPD, the necessary and sufficient condition for $\rho(T) = \rho(T_{SOR}) < 1$ is $0 < \omega < 2$.*

Proof. For the necessary condition, we assume $\rho(T_{\text{SOR}}) < 1$ and try to find a condition for ω. We shall use the property $\Pi_{j=1}^{n}\lambda_j(A) = \det(A)$ for any matrix A. Here $det(I - \omega L) = det\left((I - \omega L)^{-1}\right) = 1$ so $det(T_{\text{SOR}}) = det\left((I - \omega L)^{-1}\right) det((1 - \omega)I + \omega U) = (1 - \omega)^n$ since U is strictly lower triangular. Therefore $\rho(T_{\text{SOR}}) = |\lambda_k(T_{\text{SOR}})| \ge |\omega - 1|$ so $|\omega - 1| < 1$ i.e. $0 < \omega < 2$ is the necessary condition for SOR to converge.

To establish the sufficient condition, we assume $0 < \omega < 2$ (with A being SPD) and wish to show $\rho(T_{\text{SOR}}) < 1$. It turns out that the relationship between T_{SOR} and $M = D/\omega - \widetilde{L}$, via $T_{\text{SOR}} = (Q - I)(Q + I)^{-1}$ with $Q = A^{-1}(2M - A) = 2A^{-1}M - I$, is crucial. Let $\mu = \lambda(Q) = \Re + \Im i$ be an eigenvalue of Q with real part $\Re$ and imaginary part $\Im$ and it follows that $\lambda(Q + I) = \mu + 1$. Clearly $\mu \ne -1$ or $(\mu + 1) \ne 0$ because $A^{-1}M$ is non-singular. Hence from

$\lambda(T_{\mathrm{SOR}}) = (\mu - 1)/(\mu + 1)$, to prove

$$|\lambda(T_{\mathrm{SOR}})| = \left(\frac{(\Re - 1)^2 + \Im^2}{(\Re + 1)^2 + \Im^2} \right)^{1/2} < 1$$

amounts to showing that $\Re = \frac{\mu + \bar{\mu}}{2} > 0$ [165,180,298,467].

We note that $Qx = \mu x$ and $M + M^T - A = (2/\omega - 1)D$ imply several equalities:

$$(2M - A)x = \mu A x, \qquad x^H(2M - A)x = \mu x^H A x,$$
$$x^H(2M^T - A)x = \bar{\mu} x^H A x, \quad (\frac{2}{\omega} - 1)x^H D x = \frac{\mu + \bar{\mu}}{2} x^H A x.$$

Therefore using the assumptions proves $\Re > 0$ and completes the proof. ■

The SOR method can be modified so that the new iteration matrix is symmetric and hence has only real eigenvalues which may be useful for some applications (e.g. in combination with Chebyshev acceleration technique, by taking a linear combination of previous solutions, to speed up the convergence of the sequence $x^{(0)}, x^{(1)}, x^{(2)}, \ldots$). This will be the next iterative method.

♦ **Symmetric successive over-relaxation (SSOR) method.** The method is derived from applying a usual SOR step and a second SOR step in its reverse order (i.e. reverse the role of L and U). That is,

$$x^{(j+1/2)} = (I - \omega L)^{-1} [(1 - \omega)I + \omega U] x^{(j-1)} + b_{1/2},$$
$$x^{(j)} \quad = (I - \omega U)^{-1} [(1 - \omega)I + \omega L] x^{(j+1/2)} + b_1,$$

or combined into

$$x^{(j)} = T_{\mathrm{SSOR}} x^{(j-1)} + b_2,$$

with $b_2 = b_1 + (I - \omega U)^{-1} ((1 - \omega)I + \omega L) b_{1/2}$, $b_{1/2} = \omega (I - \omega L)^{-1} D^{-1} b$,

$$T_{\mathrm{SSOR}} = (I - \omega U)^{-1} [(1 - \omega)I + \omega L] (I - \omega L)^{-1} [(1 - \omega)I + \omega U]$$
$$= (I - \omega U)^{-1} (I - \omega L)^{-1} [(1 - \omega)I + \omega L] [(1 - \omega)I + \omega U].$$

One observes that the SSOR method chooses, with $0 < \omega < 2$,

$$M = \frac{\omega}{2 - \omega} \left(D/\omega - \widetilde{L}\right) D^{-1} \left(D/\omega - \widetilde{U}\right) = \frac{D}{(2 - \omega)\omega} (I - \omega L)(I - \omega U),$$
$$N = M - A = \frac{D}{(2 - \omega)\omega} [(1 - \omega)I + \omega L] [(1 - \omega)I + \omega U]$$

for equation (3.6)

$$\begin{aligned} x^{(j)} &= (I-\omega U)^{-1}(I-\omega L)^{-1}[(1-\omega)I+\omega L][(1-\omega)I+\omega U]x^{(j-1)} \\ &\quad + (2-\omega)\omega(I-\omega U)^{-1}(I-\omega L)^{-1}D^{-1}b \qquad (3.21) \\ &\equiv T_{\text{SSOR}}x^{(j-1)} + b_2 \end{aligned}$$

and can be written in the Richardson form (3.4) or (3.7):

$$\begin{aligned} x^{(j)} &= x^{(j-1)} + \frac{2-\omega}{\omega}\left(\frac{D}{\omega}-\widetilde{U}\right)^{-1} D \left(\frac{D}{\omega}-\widetilde{L}\right)^{-1}\left(b-Ax^{(j-1)}\right) \\ &= x^{(j-1)} + (2-\omega)\omega[D(I-\omega L)(I-\omega U)]^{-1}\left(b-Ax^{(j-1)}\right) \end{aligned} \qquad (3.22)$$

with the *residual correction operator* $\mathbf{L} = M = \frac{1}{(2-\omega)\omega}D(I-\omega L)(I-\omega U)$ and the iteration matrix $T = T_{\text{SSOR}}$ as shown above. Further discussions and theories (e.g. on choosing an optimal ω) can be found in [165,180,501,264].

Remark 3.3.5. Having written all three (Jacobi, GS and SOR) iterative methods in the Richardson form (3.7) i.e. with $\mathbf{L}e = r^{\text{old}}$ approximating $Ae = r^{\text{old}}$,

$$x^{\text{new}} = x^{\text{old}} + Br^{\text{old}}$$

one can see clearly that each method specifies $\mathbf{L} = M$ in $A = M - N$ (or $B = \mathbf{L}^{-1}$) differently and the new solution $x^{(j)}$ only involves the past solution $x^{(j-1)}$; these are simple single-step methods. Although sometimes B is referred as a preconditioner, there is not much we feel excited about mainly due to the somewhat stringent convergence requirements (SDD, SPD etc) of these methods. We believe a true preconditioner would make an otherwise divergent method converge. Note there are some papers that try to design preconditioners for these simple methods [174]. The usefulness of these methods to smooth out the residual vector r_k is a different matter which will be discussed in Chapter 6.

In the following sections, we shall review some more sophisticated iterative methods that either use information of *more than one previous solution* or involve a non-trivial choice of the residual correction operator $\mathbf{L}$. We shall call all these iterative methods **accelerator** techniques for (1.1), which the preconditioning methods presented in this book will try to speed up the iterations further.

3.4 The conjugate gradient method: the SPD case

Although the conjugate gradient (CG) method, originally proposed by [278] and popularized by [398], may be presented equivalently through the residual minimization [464] (in A^{-1}-norm) or Lanczos algorithms [180,228,429], the clearest and most convincing derivation comes from functional minimization as in [229]. Our main purpose of this section is to expose the convergence estimates to provide a concise motivation for preconditioning.

Let $A = A^T \in \mathbb{R}^{n\times n}$ be symmetric positive definite (SPD) and $b \in \mathbb{R}^n$. Then solving $Ax = b$ is equivalent to minimizing the following quadratic function

$$\phi(x) = \frac{1}{2}x^T Ax - x^T b. \tag{3.23}$$

More precisely, let $x^* = A^{-1}b$ be the exact solution.

Theorem 3.4.6. *The following equivalence holds:*

$$\min_{x\in\mathbb{R}^n} \phi(x) = \phi(x^*) \qquad \Longleftrightarrow \qquad Ax^* = b.$$

Proof. For any $h \in \mathbb{R}^n$, $\phi(x+h) - \phi(x) = \frac{1}{2}h^T Ah + h^T(Ax - b)$. For the sufficient condition, from $Ax^* = b$ and the assumption of SPD, $\phi(x^* + h) - \phi(x^*) = \frac{1}{2}h^T Ah > 0$ if $h \neq 0$. So x^* is the minimizer. For the necessary condition, assume some x is the minimizer i.e. $\frac{1}{2}h^T Ah + h^T(Ax - b) \geq 0$ for all h. Then we must have $Ax - b = 0$ or else we can find a particular h to make a contradiction (i.e. the left hand side negative). The proof is complete. ■

♦ **Minimization by line searches.** We now discuss the idea of line searches or one-dimensional minimization. Start from any initial guess $x_0 = x^{(0)} \in \mathbb{R}^n$. Suppose that a set of direction vectors $\{p_1, p_2, \cdots, p_k, \ldots\}$ are *available* and linearly independent. We construct a sequence of iterates that minimize ϕ along these direction vectors in turns (essentially we are doing dimension reduction).

Specifically, to set $x_k = x_{k-1} + \alpha p_k$ at point x_{k-1}, we solve for the best α from

$$\min_{\alpha\in\mathbb{R}} \phi(x_{k-1} + \alpha p_k). \tag{3.24}$$

To work out α, use the definition of function ϕ and the residual vector $r_j = b - Ax_j$:

$$\phi(x_{k-1} + \alpha p_k) = \phi(x_{k-1}) + \frac{\alpha^2}{2} p_k^T Ap_k - \alpha p_k^T r_{k-1}.$$

By minimizing the simple quadratic function for α,

$$\alpha = \alpha_{\min} = \frac{p_k^T r_{k-1}}{p_k^T Ap_k} \tag{3.25}$$

and $\phi(x_k) = \phi(x_{k-1}) - \dfrac{\alpha^2}{2} p_k^T Ap_k$.

♦ **Steepest descent algorithm.** One simple choice of p_k is from minimization along gradient directions. For any multi-variable function g, *locally*, the direction in which it decreases most rapidly is the so-called steepest descent direction, i.e. $p = -\nabla g$ (negative gradient)! Note the negative gradient of $\phi(x)$ at point x is the residual vector

$$p = -\nabla\phi(x) = -\left(\frac{\partial\phi}{\partial x_1}, \frac{\partial\phi}{\partial x_2}, \cdots, \frac{\partial\phi}{\partial x_n}\right)^T = b - Ax.$$

At point x_{k-1}, choosing $p_k = -\nabla\phi(x_{k-1}) = b - Ax_{k-1} = r_{k-1}$ as the next search direction gives rise to the well-known steepest descent method.

Algorithm 3.4.7. (Steepest descent).

To solve $Ax = b$, given x_0, $r_0 = b - Ax_0$, $k = 1$ and using $p_1 = r_0$,

(1) work out the steepest descent vector (residual) at step $k \ge 1$, $p_k = r_{k-1} = b - Ax_{k-1}$;
(2) minimize $\phi(x_{k-1} + \alpha p_k)$ with respect to α i.e. $\alpha_k = \dfrac{p_k^T r_{k-1}}{p_k^T A p_k} = \dfrac{r_{k-1}^T r_{k-1}}{r_{k-1}^T A r_{k-1}}$;
(3) update the solution $x_k = x_{k-1} + \alpha_k p_k$ and continue iterations with $k = k + 1$.

The fatal weakness of the steepest descent method is uncertainty associated with the linear independence of the generated search directions $p_1, p_2, \ldots$. Without linear independence, the method can be slowly converging because global minimization may not be achieved based on purely local information only.

♦ **The conjugate gradient algorithm.** The proper way to select suitable search directions p_j's is by achieving the global minimization [278,505]:

$$\underset{x = x_0 + \hat{x}, \hat{x} \in \text{span}\{p_1, \cdots, p_k\}}{\text{argmin}} \phi(x) = \underset{x = x_{k-1} + \alpha p_k, \alpha \in \mathbb{R}}{\text{argmin}} \phi(x), \tag{3.26}$$

where the usual notation argmin ('somewhat confusing') defines an argument e.g. $\text{argmin}[(x-2)^4 - 3] = 2$ as $x = 2$ is the minimizer of function $(x-2)^4 - 3$. Starting from x_0, suppose we have carried out $k - 1$ steps of a line search method to obtain

$$x_{k-1} = x_0 + \sum_{j=1}^{k-1} \alpha_j p_j \equiv x_0 + \overline{x}. \tag{3.27}$$

Here $\overline{x} \in \text{span}\{p_1, \cdots, p_{k-1}\}$. For any p_k, the next solution will be

$$x_k = x_{k-1} + \alpha_k p_k = x_0 + \overline{x} + \alpha_k p_k = x_0 + \sum_{j=1}^{k} \alpha_j p_j$$

and we consider if and when p_k can be selected such that the following problems are equivalent – this is the functional view as opposed to the argument view (3.26):

Problem I – global minimization in a dimension k subspace

$$\min_{x = x_0 + \hat{x},\, \hat{x} \in \text{span}\{p_1, \cdots, p_k\}} \phi(x),$$

Problem II – local minimization in a dimension 1 subspace

$$\min_{x = x_0 + \overline{x} + \hat{x},\, \hat{x} \in \text{span}\{p_k\}} \phi(x).$$

We expand Problem II and see how p_k would interact with other terms:

$$\begin{aligned}\phi(x) &= \phi(x_0 + \overline{x} + \alpha p_k)\\ &= \phi(x_0 + \overline{x}) + \tfrac{\alpha^2}{2} p_k^T A p_k - \alpha p_k^T r_{k-1} \qquad \text{(Note } x_{k-1} = x_0 + \overline{x}\text{)}\\ &= \phi(x_0 + \overline{x}) + \alpha p_k^T A \overline{x} + \tfrac{\alpha^2}{2} p_k^T A p_k - \alpha p_k^T r_0. \ \text{(Here } r_{k-1} = r_0 - A\overline{x}\text{)}\end{aligned} \tag{3.28}$$

Observe that, among the four terms, the first term does not interact with the third and fourth terms through the two unknown quantities α, p_k *if* the second term representing the interaction between the previous solution $\overline{x}$ and the next solution step disappears.

The superb idea of the CG method is decoupling of the previous solution $\overline{x}$ and the next solution step, which is achieved by setting $p_k^T A \overline{x} = 0$ or by killing off the second term. Examining the details of condition $p_k^T A \overline{x} = 0$ leads to A-conjugate directions. From (3.27), $A\overline{x} = \sum_{j=1}^{k-1} \alpha_j A p_j = A\mathcal{P}_{k-1} z$, where[2] $\mathcal{P}_{k-1} = [p_1, p_2, \ldots, p_{k-1}]$ and $z = [\alpha_1, \alpha_2, \ldots, \alpha_{k-1}]^T$. Then the above condition becomes $p_k^T A \overline{x} = p_k^T A \mathcal{P}_{k-1} z = 0$; if this holds for all possible z, then the condition is simply

$$p_k^T A \mathcal{P}_{k-1} = [0\ 0\ \cdots\ 0] \ \text{ i.e. } \ p_k^T A p_j = 0, \ \text{ for } j = 1, \ldots, k-1. \tag{3.29}$$

If $x^T A y = 0$, then x is A-conjugate to y. Here condition (3.29), implying

[2] This simple equation contains two fundamental tricks of linear algebra. Firstly, as in the context of defining linear independence, a linear combination of vectors can be written as a matrix (made up from the vectors packed into columns) and vector (made up of the coefficients) product; see Appendix A. Secondly, a matrix and matrix product can be written as a matrix multiplying individual columns (or rows) of the other matrix as in $AB = [Ab_1, Ab_2, \ldots, Ab_k]$ or $AB = [(a_1 B)^T, (a_2 B)^T, \ldots, (a_n B)^T]^T$ where $A = [a_1^T, a_2^T, \ldots, a_n^T]^T$ and $B = [b_1, b_2, \ldots, b_k]$ with a_j row vectors and b_j columns vectors.

$p_i^T A p_j = 0$ when $i \neq j$, states that $\{p_1, p_2, \cdots, p_k\}$ form a sequence of A-conjugate vectors.

Remark 3.4.8. The A-conjugate vectors are *weighted orthogonal* vectors. Since A is symmetric, let $PAP^T = D = \text{diag}(\lambda_1, \ldots, \lambda_n)$ with P orthogonal. The A-conjugate definition reads

$$x^T A y = x^T P^T D P y = (Px)^T D(Py) \equiv \widetilde{x}^T D \widetilde{y} = \sum_{j=1}^{n} \lambda_j \widetilde{x}_j \widetilde{y}_j = 0,$$

in contrast to the usual definition $x^T y = x^T P^T P y = (Px)^T (Py) = \widetilde{x}^T \widetilde{y} = \sum_{j=1}^{n} \widetilde{x}_j \widetilde{y}_j = 0$. Clearly when all eigenvalues are identical i.e. $\lambda_j \equiv \lambda$, then the two definitions are the same i.e. A-conjugation implies orthogonalization. This is why the inner product $(x, y)_A = x^T A y$ and the norm $(x, x)_A = \|x\|_A^2 = x^T A x$, based on the SPD matrix A, are widely used in the literature.

Once $p_k^T A \overline{x} = 0$, then the minimization of problem (3.28) becomes fully decoupled

$$\min \phi(x) = \min_{\bar{x} \in \text{span}\{p_1, \cdots, p_{k-1}\}} \phi(x_0 + \bar{x}) + \min_{\alpha \in \mathbb{R}^1} \left\{ \frac{\alpha^2}{2} p_k^T A p_k - \alpha p_k^T r_0 \right\} \quad (3.30)$$

or

$$\underbrace{\min_{x = x_0 + \hat{x}, \hat{x} \in \text{span}\{p_1, \cdots, p_k\}} \phi(x)}_{\text{Problem I — global}}$$
$$= \underbrace{\min_{x = x_0 + \overline{x}, \overline{x} \in \text{span}\{p_1, \cdots, p_{k-1}\}} \phi(x)}_{\text{Global — previous steps}} + \underbrace{\min_{\alpha \in \mathbb{R}^1} \left\{ \frac{\alpha^2}{2} p_k^T A p_k - \alpha p_k^T r_0 \right\}}_{\text{Problem II — Local step}}$$
$$= \phi(x_0) + \sum_{j=1}^{k} \min_{\alpha_j \in \mathbb{R}} \left\{ \frac{\alpha_j^2}{2} p_j^T A p_j - \alpha_j p_j^T r_0 \right\},$$

where the decoupling of global minimization into many single steps is justified i.e. the x_k from (3.26) will be global minimizer in the subspace $x_0 + \text{span}\{p_1, \cdots, p_k\}$ or equally

$$\phi(x_k) = \min_{x = x_0 + \hat{x}, \hat{x} \in \text{span}\{p_1, \cdots, p_k\}} \phi(x). \quad (3.31)$$

The construction of p_k will make use of the A-conjugate relationship (3.29) and the usual approach is to take (using $p_0 = r_0$)

$$p_{\text{new}} = p_k = r_{k-1} + \beta_k p_{k-1} = r_{\text{old}} + \beta_k p_{\text{old}}. \quad (3.32)$$

From (3.29), we immediately have $\beta_k = -\frac{p_{k-1}^T A r_{k-1}}{p_{k-1}^T A p_{k-1}}$. Once p_k is computed, the local minimization problem is identical to (3.25) i.e. $\alpha_k = \alpha_{\min} = \frac{p_k^T r_{k-1}}{p_k^T A p_k}$. Finally we set $x_k = x_{k-1} + \alpha_k p_k$ to complete step k. Putting all steps together, the conjugate gradient method (version 0) is obtained. However, the existing formulae for α_k and β_k can be much simplified as, currently, too many matrix vector products are used.

To further simplify α_k, β_k, we need (3.29) and two important relations:

$$r_k^T p_k = 0 \qquad \text{and} \qquad r_k^T r_{k-1} = 0, \tag{3.33}$$

which can be verified below. As $x = x_k = x_{k-1} + \alpha p_k$ is the global as well as the local minimizer, we must have $\frac{d\phi(x_k)}{d\alpha} = 0$ i.e. we have proved the first formula $\dfrac{d\phi(x_k)}{dx_k}\dfrac{dx_k}{d\alpha} = -r_k^T p_k = 0$. Now from $p_k = r_{k-1} + \beta_k p_{k-1}$, $r_k = r_{k-1} - \alpha_k A p_k$, and (3.29), we obtain the second formula $r_k^T r_{k-1} = r_k^T p_k - \beta_k r_k^T p_{k-1} = -\beta_k \left(r_{k-1}^T p_{k-1} - \alpha_k p_{k-1}^T A p_k \right) = 0$. Further with (3.33), we can reformulate

(1) from $p_k^T r_{k-1} = (r_{k-1}^T + \beta_k p_{k-1}^T) r_{k-1} = r_{k-1}^T r_{k-1}$,

$$\alpha_k = \frac{p_k^T r_{k-1}}{p_k^T A p_k} = \frac{r_{k-1}^T r_{k-1}}{p_k^T A p_k}. \tag{3.34}$$

(2) from $r_{k-1} = r_{k-2} - \alpha_{k-1} A p_{k-1}$,

$$\begin{aligned} r_{k-1}^T r_{k-1} &= r_{k-1}^T r_{k-2} - \alpha_{k-1} r_{k-1}^T A p_{k-1} = -\alpha_{k-1} r_{k-1}^T A p_{k-1} \\ &= -\alpha_{k-1} p_{k-1}^T A r_{k-1} \end{aligned}$$

and from $p_{k-1} = r_{k-2} + \beta_{k-1} p_{k-2}$,

$$\begin{aligned} r_{k-2}^T r_{k-2} &= r_{k-2}^T (r_{k-1} + \alpha_{k-1} A p_{k-1}) = \alpha_{k-1} (p_{k-1} - \beta_{k-1} p_{k-2})^T A p_{k-1} \\ &= \alpha_{k-1} p_{k-1}^T A p_{k-1}. \end{aligned}$$

Therefore,

$$\beta_k = -\frac{p_{k-1}^T A r_{k-1}}{p_{k-1}^T A p_{k-1}} = \frac{r_{k-1}^T r_{k-1}}{r_{k-2}^T r_{k-2}}. \tag{3.35}$$

Algorithm 3.4.9. (Conjugate gradient).

To solve $Ax = b$, *given* $x = x_0$, $r = b - Ax_0$, $k = 1$, $r_{new} = \|r\|_2^2 = r^T r$, $p = r$,

(1) minimize $\phi(x_k) = \phi(x_{k-1} + \alpha p_k)$ *with respect to* α *i.e. compute* α_k *using (3.34):* $\alpha_k = r_{new}/(p^T q)$ *with* $q = Ap$ *and update the solution* $x = x + \alpha_k p$*;*

(2) *work out the new residual vector at step* k, $r = b - Ax = r - \alpha_k q$ *and set* $r_{old} = r_{new}$.
(3) *compute* $r_{new} = r^T r$ *(exit if* r_{new} *is small enough) and* $\beta_k = r_{new}/r_{old}$.
(4) *update the search direction* $p = r + \beta_k p$ *and continue with step (1) for* $k = k + 1$.

Remark 3.4.10. We remark on the connection of the CG vectors with Lanczos iterations and the relationships between various notation for the CG analysis.

(1) The Lanczos method, as a special case of the Arnoldi method, Section 1.4, reduces a symmetric matrix A to a tridiagonal matrix T by orthogonal transforms: $U^T AU = T$ or $AU = UT$. Here we wish to use the already generated CG vectors r_j and p_j to produce U, T. Let $P = [p_1\ p_2\ \cdots\ p_n]$, $R = [r_0\ r_1\ \cdots\ p_{n-1}]$ and $D = \operatorname{diag}(\rho_1, \cdots, \rho_k)$ with $\rho_{i+1} = \sqrt{r_i^T r_i} = \|r_i\|_2$. Then from (3.33), $r_i^T r_j = 0$, and the fact $p_i^T A p_j = 0$ for $i \neq j$, we conclude that

$$R^T R = D^2, \qquad P^T AP = \operatorname{diag}(p_1^T A p_1, \cdots, p_n^T A p_n).$$

From (3.32),

$$\begin{cases} p_1 = r_0, \\ p_j = r_{j-1} + \beta_j p_{j-1},\ j = 2, 3, \cdots, n, \\ \quad \text{or} \\ r_0 = p_1, \\ r_j = p_{j+1} - \beta_{j+1} p_j,\ j = 1, 2, \cdots, n-1, \end{cases} \qquad R = PB,$$

$$B = \begin{bmatrix} 1 & -\beta_2 & & & \\ & 1 & -\beta_3 & & \\ & & 1 & \ddots & \\ & & & \ddots & -\beta_k \\ & & & & 1 \end{bmatrix}.$$

From the above definition, matrix $U = RD^{-1}$ is orthogonal as $U^T U = I$. Moreover, $U^T AU = D^{-1} B^T (P^T AP) B D^{-1} = T$ is a tridiagonal matrix.

Therefore, $U^T AU = T_{n\times n}$ is the tridiagonalization provided by the CG method. Moreover, if we use the Lanczos method with $q_0 = r_0/\rho_1$, the generated vectors $q_2, q_3, \cdots, q_k$ should lead to the same decomposition (the uniqueness of $U^T AU = T$ is determined by the signs of β_j's).

Observe the difference in deriving the above decomposition. The CG method requires that A be SPD (i.e. symmetric and positive definite) but it generates $T_{n\times n}$. The Lanczos method requires A be symmetric only but it can only manage to find $T_{m\times m}$ with $m \le n$ depending on the choice of the first column of U. Refer to [229,180,464].

(2) In the literature, the CG method has been described in several equivalent formulations. We show that the following three quantities are identical:

$$\Phi(x) \equiv \|r\|_{A^{-1}}^2 = \|x - x^*\|_A^2 = 2\phi(x) + b^T A^{-1} b$$

where $Ax^* = b$, $r = b - Ax$, $\|r\|_{A^{-1}} = r^T A^{-1} r$ and $\phi(x) = \frac{1}{2}x^T A x - x^T b$.

$$\begin{aligned}
\|r\|_{A^{-1}}^2 &\equiv r^T A^{-1} r = (b^T - x^T A)A^{-1}(b - Ax) \\
&= (b^T A^{-1} - x^T)(b - Ax) \\
&= x^T A x - 2x^T b + b^T A^{-1} b \\
&= 2\phi(x) + b^T A^{-1} b, \\
\|x^* - x\|_A^2 &= (x^* - x)^T A (x^* - x) = [(x^* - x)^T A] A^{-1} (Ax^* - Ax) \\
&= (Ax^* - Ax)^T A^{-1} (Ax^* - Ax) \\
&= r^T A^{-1} r \ = \ \|r\|_{A^{-1}}^2.
\end{aligned} \tag{3.36}$$

As A is SPD (and so is A^{-1}), $b^T A^{-1} b \ge 0$ and hence minimizations of $\|x - x^*\|_2$, $\|r\|_{A^{-1}}$ and $\phi(x)$ are identical problems. Recall that the presentation via (3.23) has been for the $\phi(x)$ minimization, whose global minimization issue was rigorously addressed.

Convergence rate. We next consider the all important issue of convergence rate by comparing $\phi(x_k)$ to $\phi(x_0)$. From Remark 3.4.10, comparing $\|r_k\|_{A^{-1}}$ to $\|r_0\|_{A^{-1}}$ from (3.36) is the same as comparing

$$\Phi(x_k) \equiv 2\phi(x_k) + b^T A^{-1} b$$

to $\Phi(x_0) = 2\left(\phi(x_0) + b^T A^{-1} b\right)$. To carry out this task, our main theory comes from the fact that x_k is the global minimizer of $\Phi(x) = \Phi(x_0 + \hat{x})$ with $\hat{x} \in \text{span}(p_1, \ldots, p_k)$. From the way p_j's are generated i.e. $p_j = r_{j-1} + \beta_j p_{j-1}$ and $p_0 = r_0$, we see that

$$\text{span}(r_0, r_1, \ldots, r_{k-1}) = \text{span}(p_1, p_2, \ldots, p_k)$$

and similarly from $x_j = x_{j-1} + \alpha_j p_j$ and $r_j = r_{j-1} - \alpha_j A p_j$,

$$\text{span}(r_0, r_1, \ldots, r_{k-1}) = \text{span}(r_0, Ar_0, \ldots, A^{k-1} r_0) \equiv \mathcal{K}_k(A, r_0)$$

The later subspace $\mathcal{K}_k(A, r_0)$ is called the Krylov subspace of order k for the matrix A and the generating vector r_0.

Observe that, although $\hat{x}, x_k - x_0 \in \mathcal{K}_k(A, r_0)$, $x_k \notin \mathcal{K}_k(A, r_0)$ if $x_0 \ne 0$; in this case one considers convergence rates for equation $Ae = r_0$ instead (since $x = x_0 + e$ satisfies $Ax = b$) and, for simplicity (and without loss of generality), one assumes that $x_0 = 0$ to avoid further notation. With $x_0 = 0$, $r_0 = b$ so the Krylov subspace for the solution x_k becomes

$$\mathcal{K}_k(A, b) = \text{span}(b, Ab, \ldots, A^{k-1}b). \tag{3.37}$$

Thus with $x_0 = 0$, we shall consider

$$\frac{\Phi(x_k)}{\Phi(x_0)} = \frac{\|x_k - x^*\|_A^2}{\|x_0 - x^*\|_A^2} = \frac{\|x_k - x^*\|_A^2}{\|x^*\|_A^2}. \tag{3.38}$$

The immediate task is to quantify (3.31). Now for any vector $z \in \mathcal{K}_k(A, b)$, we have $z = \sum_{j=0}^{k-1} \gamma_j A^j b$. Letting $g_{k-1}(\xi) = \sum_{j=0}^{k-1} \gamma_j \xi^j$ be a degree$(k-1)$ polynomial and noting that $b = Ax^*$, one obtains that $z = g_{k-1}(A)Ax^*$. Then we can expand $\Phi(z)$ in terms of $g_{k-1}(A)$ and x^* as follows

$$\begin{aligned}\Phi(z) = \|x^* - z\|_A^2 &= (x^* - g_{k-1}(A)Ax^*)^T \, A \, (x^* - g_{k-1}(A)Ax^*) \\ &= (x^*)^T q_k(A) A q_k(A) x^*,\end{aligned} \tag{3.39}$$

where $A = A^T$ is used and $q_k(\xi) = 1 - g_{k-1}(\xi)\xi$ satisfying $q_k(0) = 1$. Before detailing $\Phi(x_k)$, we note that $\Phi(x_0) = \Phi(0) = \|x^* - 0\|_A^2 = (x^*)^T A x^* = b^T A^{-1} b$. If $\mathcal{Q}_k$ denotes the set of all degree k polynomials, as each polynomial corresponds to a different minimizer, we can interpret our global minimizer $z = x_k$ in (3.31) as follows

$$\Phi(x_k) = \min_{z \in \mathcal{K}_k(A,b)} \Phi(z) = \min_{q_k \in \mathcal{Q}_k} (x^*)^T q_k(A) A q_k(A) x^*. \tag{3.40}$$

Here and throughout this chapter, the condition that $q(0) = 1$ if $q(x) \in \mathcal{Q}_k$ is imposed on the set $\mathcal{Q}_k$.

To simplify (3.40), we need to replace A by $A = QDQ^T$ (as A is symmetric, the eigenvector matrix Q must be orthogonal and $D = \text{diag}(\lambda_1, \lambda_2, \ldots, \lambda_n) \equiv \text{diag}(\lambda_j)$ contains eigenvalues $\lambda_j = \lambda_j(A)$). Then[3] we have $q_k(A) = q_k(QDQ^T) = Qq_k(D)Q^T \equiv Q \, \text{diag}(q_k(\lambda_j))Q^T$. Further, (3.40) can be simplified to

$$\begin{aligned}\Phi(x_k) &= \min_{q_k \in \mathcal{Q}_k} (x^*)^T Q \, \text{diag}(q_k(\lambda_j)) Q^T Q \, \text{diag}(\lambda_j) Q^T Q \, \text{diag}(q_k(\lambda_j)) Q^T x^* \\ &= \min_{q_k \in \mathcal{Q}_k} (x^*)^T Q \, \text{diag}\left(\lambda_j [q_k(\lambda_j)]^2\right) Q^T x^* \\ &= \min_{q_k \in \mathcal{Q}_k} y \, \text{diag}(\lambda_j [q_k(\lambda_j)]^2 y^T \qquad \text{(define } (x^*)^T Q = y) \\ &= \min_{q_k \in \mathcal{Q}_k} \sum_{j=1}^{n} \lambda_j [q_k(\lambda_j)]^2 y_j^2.\end{aligned} \tag{3.41}$$

[3] This trick has been used in the proof of Theorem 3.3.1, i.e. $A^j = A^{j-1}A = QD \ldots DQ^T = QD^j Q^T$.

Note that $\Phi(x_0) = (x^*)^T A x^* = (x^*)^T Q D Q^T x^* = y D y^T = \sum_{j=1}^n \lambda_j y_j^2$. Define

$$\mathcal{B}(\lambda_1, \lambda_2, \ldots, \lambda_n) \equiv \min_{q_k \in \mathcal{Q}_k} \max_{\lambda_j \in \Lambda(A)} \left| q_k(\lambda_j) \right|, \tag{3.42}$$

where $\Lambda(A) = \{\lambda_1, \ldots, \lambda_n\}$ is the eigenspectrum of matrix A.

Eigenspectrum bound. Now putting the above two equations together, we obtain

$$\Phi(x_k) \le \mathcal{B}(\lambda_1, \lambda_2, \ldots, \lambda_n)^2 \sum_{j=1}^n \lambda_j y_j^2 = [\mathcal{B}(\lambda_1, \lambda_2, \ldots, \lambda_n)]^2 \, \Phi(x_0),$$

that is,

$$\frac{\|r_k\|_{A^{-1}}}{\|r_0\|_{A^{-1}}} = \frac{\|x^* - x_k\|_A}{\|x^* - x_0\|_A} = \sqrt{\frac{\Phi(x_k)}{\Phi(x_0)}} \le \mathcal{B}(\lambda_1, \lambda_2, \ldots, \lambda_n). \tag{3.43}$$

Condition number bound. Further to simplify $\mathcal{B}$ to a much simpler upper bound, we shall use *cond*$(A) = \kappa(A) = \kappa \equiv \lambda_{\max}/\lambda_{\min}$, where interval $[\lambda_{\min}, \lambda_{\max}]$ contains $\Lambda(A)$. Replacing the *discrete* spectrum $\Lambda(A)$ by a *continuous* spectrum $[\lambda_{\min}, \lambda_{\max}]$ in $\mathcal{B}$ leads to the problem:

$$\min_{q_k \in \mathcal{Q}_k} \max_{\lambda \in [\lambda_{\min}, \lambda_{\max}]} |q_k(\lambda)| \tag{3.44}$$

and its solution

$$q_k = T_k \left(\frac{\lambda_{\max} + \lambda_{\min} - 2\lambda}{\lambda_{\max} - \lambda_{\min}} \right) \Big/ \; T_k(\mu) \qquad \text{with } \mu = \frac{\lambda_{\max} + \lambda_{\min}}{\lambda_{\max} - \lambda_{\min}},$$

which is a scaled Chebyshev polynomial T_k of degree k. Then one may use properties of Chebyshev polynomials [403]

$$\begin{aligned}
&|T_{k+1}(t)| = 2t T_k(t) - T_k(t), && T_0(t) = 1, \qquad T_1(t) = t, && \\
&T_k(a) = \cos(k \arccos(a)), && |T_k(a)| \le 1, && \text{if } |a| \le 1, \\
&T_k(a) = \frac{(a + \sqrt{a^2 - 1})^k + (a + \sqrt{a^2 - 1})^{-k}}{2} \ge \frac{(a + \sqrt{a^2 - 1})^k}{2}, && && \text{if } a \ge 1,
\end{aligned} \tag{3.45}$$

to derive that

$$\mathcal{B} \le \frac{1}{T_k(\mu)} \le 2 \left(\frac{1}{\mu + \sqrt{\mu^2 - 1}} \right)^k = 2 \left(\frac{\sqrt{\kappa} - 1}{\sqrt{\kappa} + 1} \right)^k. \tag{3.46}$$

This replaces (3.43) by the well-known (yet slightly over-used) result

$$\frac{\|x^* - x_k\|_A}{\|x^* - x_0\|_A} \le 2 \left(\frac{\sqrt{\kappa} - 1}{\sqrt{\kappa} + 1} \right)^k. \tag{3.47}$$

More results on convergence of the CG can be found in [28,303,180,229,413,

464]. In particular, one may sharpen the estimate (3.47) by considering special eigenvalue clustering patterns and super-linear convergence rates in certain cases; see [493,463,358,413].

Remark 3.4.11. We now make some remarks on convergence and preconditioning.

- If $x_0 \neq 0$ is used to start the iterations, then we have to re-define $x^* = x^* - x_0$, $x_k = x_k - x_0$, $b = b - Ax_0$, $r_k = r_k - r_0$ and re-set $x_0 = 0$ in order to validate (3.38), (3.43) and (3.47).
- The convergence result (3.47) is quite neat and elegant. However, in practice, it rarely represents the true convergence behaviour of the CG method unless for an easy problem with $\kappa \approx 1$. Although this observation is widely known, many researchers continue to use the reduction of $\kappa = cond(A)$ for (1.1) to a smaller $\widetilde{\kappa} = cond(PA)$ for a preconditioned system (1.2) as a main motivation for preconditioning. For, in many cases, it is difficult to propose a new theory other than to use the condition number.
- Nevertheless, we do not advocate using preconditioning to reduce the condition number alone as a good motivation. As we see, we encourage readers to consider using (3.43), rather than (3.47), to design new preconditioners whenever possible. This is where the real strength of preconditioning lies: we wish to achieve the idealized situations where a non-uniform distribution of eigenvalues would enable a hopefully much smaller $k \ll n$ to be sufficient for $\mathcal{B}$ to satisfy a given tolerance, i.e. a small number of CG steps is sufficient for any convergence requirement.
- Indeed, as is known, an efficient preconditioner can cluster eigenvalues to an ideal point 1 (refer to Section 1.5); see [303,463,14]. However from (3.43), one can expect fast convergence if the preconditioned spectrum is clustered at more than 1 points (yet still a fixed and small number of points); see [364,297] for work done this way. In particular, if a preconditioned matrix has only a small number $\mu \ll n$ distinct eigenvalues, the number of CG iterations is bounded by a function (almost linear) of μ, independent of n; see [464].

Symmetric preconditioner. This book addresses more of an unsymmetric matrix problem than a symmetric one because the latter has been better known; see e.g. [28]. Nevertheless, we hope to clarify two important points in applying a SPD preconditioner P to (1.1), with a SPD matrix A, before showing the so-called preconditioned conjugate gradient (PCG) algorithm for SPD systems.

The main point here is that the product matrix PA of two SPD matrices is positive definite (but unfortunately not symmetric). This is because PA is similar to a SPD matrix $\widetilde{A} = P^{1/2}AP^{1/2}$ i.e. $PA = P^{1/2}\widetilde{A}P^{-1/2}$, where the SPD matrix P admits the orthogonal decomposition $P = Q\Lambda Q^T$ and $P^{1/2} = Q\Lambda^{1/2}Q^T$

is another SPD matrix. Consequently, any ideal eigenspectrum of PA can be made full use of if we apply a CG method to $\widetilde{A}$ i.e. to system

$$\widetilde{A}y = \widetilde{b}$$

with $x = P^{1/2}y$ and $\widetilde{b} = P^{1/2}b$ (to reiterate the point, it is correct to use $\widetilde{A} = P^{1/2}AP^{1/2}$ but incorrect to relate $\widetilde{A} = P^{1/2}AP^{-1/2}$ to PA). The next technical point is to avoid forming $P^{1/2}$ explicitly by simply generating intermediate vectors differently, e.g. convert the step $y = y + \alpha\widetilde{p}$ in Step (1) of Algorithm 3.4.9 to $x = x + \alpha p$ (and effectively we have implemented $P^{1/2}y = P^{1/2}y + \alpha P^{1/2}\widetilde{p}$) – the same trick will be used later in CGNR Algorithm 3.5.13. Refer to [28,180,229,413].

Algorithm 3.4.12. (Preconditioned conjugate gradient).

To solve $Ax = b$ using preconditioner P, given $x = x_0$, $r = b - Ax_0$, $k = 1$, $p = Pr$, $y = Pr$, and using $r_{new} = y^T r$,

(1) minimize $\phi(x_k) = \phi(x_{k-1} + \alpha p_k)$ with respect to α, i.e. compute α_k using (3.34) i.e. $\alpha_k = r_{new}/(p^T q)$ with $q = Ap$ and update the solution $x = x + \alpha_k p$;

(2) work out the new residual vector at step k, $r = b - Ax = r - \alpha_k q$ and set $r_{old} = r_{new}$.

(3) compute $y = Pr$ and $r_{new} = y^T r$ (exit if r_{new} is small enough) and $\beta_k = r_{new}/r_{old}$.

(4) update the search direction $p = y + \beta_k p$ and continue with step (1) for $k = k + 1$.

3.5 The conjugate gradient normal method: the unsymmetric case

When A is not symmetric (no chance to ask for SPD), the CG method from previous section does not apply since Theorem 3.4.6 no longer holds. Actually this is where the fun starts! Many (completing) unsymmetric CG type iterative solvers exist [41] for solving (1.1), each trying to choose $\hat{x} \in \mathcal{K}_k(A, r_0)$ differently in forming the solution $x_k = x_0 + \hat{x}$. The subspace $\mathcal{K}_k$ naturally renames all such methods as the Krylov subspace methods. Among such solvers, we select only two representatives for subsequent discussion and testing: the CGN (conjugate gradient normal) and the GMRES (generalized minimal residual) methods. Further Krylov subspace methods can be found in [41,413,464]. Our intention in making this decision follows from years of research experience (see also [366,198]), which suggests that *developing better preconditioning techniques for an iterative solver is more productive than trying too many or developing new iterative solvers alone.*

We now describe the CGN method. There are two variants, not surprisingly, one solving the left normal equation (giving rise to the CGNL Algorithm 3.5.14)

$$A^T A x = A^T b \tag{3.48}$$

and the other solving the right normal equation (to the CGNR Algorithm 3.5.13)

$$A A^T y = b, \qquad \text{with} \quad x = A^T y. \tag{3.49}$$

Both equations, each having a SPD coefficient matrix whenever A is not singular, can be solved directly by Algorithm 3.4.9 without forming $A^T A$ or AA^T explicitly. We prefer the latter approach based on (3.49).

Firstly for (3.48), a direct application of Algorithm 3.4.9 would compute the residual $r_k = A^T(b - Ax_k)$ which can affect the quality of the search direction p_k. A better approach, named as CGLS by [381], is to reformulate the method so that the original residual $r_k = b - Ax_k$ and the residual for the normal equation $s_k = A^T r_k$ are both computed. Here we rename the CGLS as the conjugate gradient left normal algorithm (CGNL) for simplicity. To remove possibly surprising factors, Algorithm 3.5.14 lists the direct (naive) application of the CG method and the CGNL side-by-side.

Secondly for (3.49), Algorithm 3.4.9 can also be reformulated to work with x instead of the intermediate vector y. The resulting method is often associated with an adapted variant CGNE of the Craig's method [381]. A similar version for complex systems $AA^* y = b$ was given in [82] and used by [126]. As before, we give the conjugate gradient right normal algorithm (CGNR) and the direct application of CG side-by-side:

Owing to the squaring[4] of the 2-norm condition number $\kappa(A^T A) = \rho(A^T A) = \kappa(A)^2$ as $\rho\left((A^T A)^T (A^T A)\right)^{1/2} = \rho\left((A^T A)^2\right)^{1/2}$, the use of CGN methods is often not recommended in view of (3.47). However using (3.43), it is not difficult to find plenty of examples where the eigenspectrum of $A^T A$ (or AA^T), i.e. singular value distribution of A, has clustering patterns (Section 1.5) and CGN is at least as effective as other unsymmetric solvers; see also [366].

For later use, we now give a CGN algorithm that takes one or two preconditioners for (1.1):

$$M_2 A M_1 y = M_2 b, \qquad x = M_1 y. \tag{3.50}$$

This includes the case of (1.2). Here we only discuss the right normal approach:

$$M_2 A M_1 (M_1^T A^T M_2^T) z = M_2 b, \qquad x = M_1 M_1^T A^T M_2^T z. \tag{3.51}$$

[4] Note $\lambda(A^k) = \lambda(A)^k$ for any integer k (even negative k when $\lambda \neq 0$. This can be shown by using $Ax = \lambda x$ i.e. $A^2 x = AAx = A(\lambda x) = \lambda^2 x$ etc. The simple result is widely used e.g. in the proof of Theorem 3.3.1.

Using the above CGNL and CGNR ideas, we propose the following algorithms.

Algorithm 3.5.13. (CGNL).

To solve $Ax = b$, given $x = x_0$, $r = b - Ax_0$, $k = 1$ and $p = A^T r$,

(CGNL)

Set $s = p$, and $r_{new} = s^T s$

(1) $q = Ap$, (Note: $q^T q = p^T A^T A p$)
(2) $\alpha_k = r_{new}/(q^T q)$
(3) Update the solution $x = x + \alpha_k p$
(4) $r = b - Ax = r - \alpha_k q$ and set $r_{old} = r_{new}$
(5) Compute $s = A^T r$ and $r_{new} = s^T s$ (exit if r_{new} is small enough)
(6) $\beta_k = r_{new}/r_{old}$
(7) Update the search direction $p = s + \beta_k p$ and continue with step (1) for $k = k + 1$.

(naive CGNL)

Set $r = p$, and $r_{new} = r^T r$

(1) $q = A^T A p$, ($p^T q = p^T A^T A p$)
(2) $\alpha_k = r_{new}/(p^T q)$
(3) Update the solution $x = x + \alpha_k p$
(4) $r = b - Ax = r - \alpha_k q$ and set $r_{old} = r_{new}$
(5) Compute $r_{new} = r^T r$ (exit if r_{new} is small enough)
(6) $\beta_k = r_{new}/r_{old}$
(7) Update the search direction $p = r + \beta_k p$ and continue with step (1) for $k = k + 1$.

Algorithm 3.5.14. (CGNR).

To solve $Ax = b$, with $k = 1$,

(CGNR)

given $x = x_0$, $r = b - Ax_0$ and set initially $p = A^T r$, and $r_{new} = r^T r$

(1) $q = Ap$
(2) $\alpha_k = r_{new}/(p^T p)$
(3) Update the solution $x = x + \alpha_k p$
(4) $r = b - Ax = r - \alpha_k q$ and set $r_{old} = r_{new}$
(5) Compute $r_{new} = r^T r$ (exit if r_{new} is small enough)
(6) $\beta_k = r_{new}/r_{old}$
(7) Update the search direction $p = A^T r + \beta_k p$ and continue with step (1) for $k = k + 1$.

(naive CGNR)

given $y = y_0$, $r = b - AA^T y_0$ and set initially $p = r$, and $r_{new} = r^T r$

(1) $q = AA^T p$, ($p^T q = \|A^T p\|_2^2$)
(2) $\alpha_k = r_{new}/(p^T q)$
(3) Update the solution $y = y + \alpha_k p$
(4) $r = b - Ax = r - \alpha_k q$ and set $r_{old} = r_{new}$
(5) Compute $r_{new} = r^T r$ (exit $x = A^T y$ if r_{new} is small)
(6) $\beta_k = r_{new}/r_{old}$
(7) Update the search direction $p = r + \beta_k p$ and continue with step (1) for $k = k + 1$.

Algorithm 3.5.15. (CGNT). *(Here* **T** *stands for* **T**wo *preconditioners)*

To solve $M_2 A M_1 y = b$ *for* $x = M_1 y$ *through system (3.51), with* $k = 1$, $x = x_0$, $r = b - Ax_0$ *and set initially* $s = M_2 r$, $p = M_1^T A^T M_2^T s$, *and* $r_{new} = s^T s$,

(1) $u = M_1 p, \quad q = Au,$
(2) $\alpha_k = r_{new}/(p^T p)$
(3) *Update the solution* $x = x + \alpha_k u$
(4) $r = b - Ax = r - \alpha_k q$ *and set* $r_{old} = r_{new}$
(5) *Compute* $s = M_2 r$ *and* $r_{new} = s^T s$ *(exit if* r_{new} *is small enough)*
(6) *Compute* $\beta_k = r_{new}/r_{old}$ *and* $s = M_1^T A^T M_2^T s$
(7) *Update the search direction* $p = s + \beta_k p$ *and continue with step (1) for* $k = k + 1$.

It should be remarked that the above three algorithms CGNL, CGNR and CGNT are applicable to a complex linear system *if* all the transposes (e.g. x^T) are replaced by conjugate transposes (x^H).

We next introduce a method that does not require solving a normal equation. (As remarked several times before, there exist many competing iterative solvers [41,464]).

3.6 The generalized minimal residual method: GMRES

When matrix A in (1.1) is unsymmetric, the previous minimization of $\phi(x)$ in (3.23) and $\Phi(x)$ as in (3.38) is no longer meaningful as Theorem 3.4.6 does not hold. We need to set up a new functional to minimize first. Once a functional is decided, we have to think of a way to generate suitable search directions so that global minimization is achieved. We now introduce one such method, namely GMRES due to [415,413], where the minimizing functional in the Krylov subspace $\mathcal{K}_k(A, r_0)$ is chosen as the 2-norm of the residual r_k, similar to $\Phi(x)$ in (3.38) but with A replaced by I, and the search directions are implicitly chosen from the Arnoldi iterations (Section 1.4) with global minimization achieved by a least square minimization process. (Note that, to some extent, the GMRES can cope with a nearly singular matrix A [77], although we assume that A is non-singular here.) For other unsymmetric methods, refer to [41,464] and when A is symmetric but indefinite, refer to MINRES method due to [380].

We consider the restarted GMRES method, GMRES(k), to compute $x_k = x_0 + \hat{x}_k$ with $\hat{x}_k \in \mathcal{K}_k(A, r_0)$. To select $\hat{x}_k$, the first k orthogonal vectors q_j from Arnoldi iterations will be used so that $\hat{x}_k = \sum_{j=1}^{k} q_j y_j = [q_1, q_2, \ldots, q_k] y \equiv$

$Q_k y$ with $y \in \mathbb{R}^k$ and $\text{span}(q_1, q_2, \ldots, q_k) \in \mathcal{K}_k(A, r_0)$. Let $q_1 = r_0/\|r_0\|_2$ be the starting vector for the Arnoldi decomposition, Section 1.4. In full generality, allowing the existence of an invariant subspace [229] i.e. $\mathcal{K}_n(A, r_0) = \mathcal{K}_m(A, r_0)$, assume $H_m = Q_m A Q_m^T$ or $A Q_m = Q_m H_m$ is formed with $Q_m = Q_{n\times m} = [q_1, q_2, \ldots, q_m]$ orthogonal and $H_m = H_{m\times m}$ an upper Hessenberg matrix:

$$H_m = \begin{bmatrix} h_{11} & h_{12} & \cdots & h_{1,m-1} & h_{1m} \\ h_{21} & h_{22} & \cdots & h_{2,m-1} & h_{2m} \\ & h_{32} & \cdots & h_{3,m-1} & h_{3m} \\ & & \ddots & \ddots & \vdots \\ & & & h_{m,m-1} & h_{mm} \end{bmatrix}, \tag{3.52}$$

where $k \le m \le n$.[5] Also note, though $Q_m^T A Q_m = H_m$, $A \ne Q_m H_m Q_m^T$ if $m < n$. As $k \le m$, the generated k orthogonal vectors q_j satisfy

$$A_{n\times n}(Q_k)_{n\times k} = (Q_{k+1})_{n\times(k+1)}(H_{k+1})_{(k+1)\times k} = Q_k H_k + h_{k+1,k} q_{k+1} e_k^T, \tag{3.53}$$

where, with $\mathbf{h}_{k+1} = (0,\ 0,\ \ldots,\ h_{k+1,k})$,

$$H_{k+1} = \left[\begin{array}{ccccc|} h_{11} & h_{12} & \cdots & h_{1,k-1} & h_{1k} \\ h_{21} & h_{22} & \cdots & h_{2,k-1} & h_{2k} \\ & h_{32} & \cdots & h_{3,k-1} & h_{3k} \\ & & \ddots & \ddots & \vdots \\ & & & h_{k,k-1} & h_{kk} \\ \hline & & & & h_{k+1,k} \end{array}\right] = \begin{bmatrix} H_k \\ \mathbf{h}_{k+1} \end{bmatrix}. \tag{3.54}$$

As mentioned, the GMRES method minimizes the following functional in the subspace $x_0 + \mathcal{K}_k(A, r_0)$ (compare to (3.38)))

$$\begin{aligned} \Phi(x_k) = \|r_k\|_2^2 = \|b - Ax_k\|_2^2 &= \|b - A(x_0 + Q_k y)\|_2^2 \\ = \|r_0 - A Q_k y\|_2^2 &= \|r_0 - Q_{k+1} H_{k+1} y\|_2^2. \end{aligned} \tag{3.55}$$

[5] The restart parameter k cannot exceed $m \le n$. Many descriptions of the method assume $m = n$ which is not theoretically correct for Arnoldi iterations; in the very special case of $r_0 = e_1$, we can assume $m = n$ as the decomposition $A = QHQ^T$ can be obtained by Givens or Householder transforms Section 1.4. When $A = A^T$, the Arnoldi iterations reduce to Lanczos iterations and a similar argument follows i.e. $m \le n$.

So far only the Arnoldi algorithm is used to primarily generate orthogonal vectors q_j. An important observation made in [415] is that the selection of x_k can also make use of the upper Hessenberg matrix H_{k+1} produced during the same Arnoldi process to simplify (3.55). Noting $r_0 = q_1 \|r_0\|_2$, we need the orthogonal complement $\hat{Q}$ in $\mathbb{R}^n$ of Q_k to complete the presentation where $Q_{n\times n} = [Q_k \hat{Q}]$ is orthogonal and $\hat{Q}$ will not be needed in the GMRES method.

Further using the 2-norm invariance of Q, i.e. $\|Q^T x\|_2 = \|Qx\|_2 = \|x\|_2$, we have

$$\begin{aligned}\Phi(x_k) &= \|Q^T r_0 - Q^T Q_{k+1} H_{k+1} y\|_2^2 \\ &= \left\| \|r_0\|_2 e_1 - \begin{pmatrix} H_{k+1} y \\ 0 \end{pmatrix} \right\|_2^2 \\ &= \| \, \|r_0\|_2 \hat{e}_1 - H_{k+1} y\|_2^2,\end{aligned} \tag{3.56}$$

where $\hat{e}_1$ is the unit vector of dimension $k+1$ because the bottom rows are zero. To find the optimal vector y that solves (3.56), we solve the $(k+1)\times k$ least squares problem

$$H_{k+1} y = \|r_0\|_2 \hat{e}_1. \tag{3.57}$$

Remark 3.6.16. (The FOM). *A sub-optimal (but natural) solution to (3.57) is obtained from solving the first k equations (for the square matrix H_k)*

$$H_k y = \|r_0\|_2 \hat{e}_1(1:k).$$

This will give rise to the full orthogonalization method (FOM):

$$x_k = x_0 + Q_k y, \tag{3.58}$$

which is simpler but often less efficient than the GMRES.

The above problem (3.57) is usually solved by a QR decomposition method (see [63,229] and Section 2.4), say by Givens transforms to give

$$H_{k+1} \overset{\text{full QR}}{=} P_{k+1,k+1} \begin{pmatrix} R_{k\times k} \\ 0 \end{pmatrix} \overset{\text{reduced QR}}{=} P_{k+1,k} R_{k\times k}. \tag{3.59}$$

Then (3.57) can be solved to give

$$y = R_{k\times k}^{-1} \left[P_{k+1,k+1}^T \|r_0\|_2 \hat{e}_1 \right]_k = R_{k\times k}^{-1} P_{k+1,k}^T \|r_0\|_2 \hat{e}_1,$$

where $[x]_k = x(1:k)$ denote the vector of the first k components. Putting the above steps together, we obtain the GMRES(k) algorithm.

Algorithm 3.6.17. (GMRES(k)).

To solve $Ax = b$, with iter $= 0$, $e_1 = (e_1)_{n\times 1}$ and given an initial vector $x = x_0$,

(1) *Set $x_0 = x$, iter $=$ iter $+ 1$ and compute $r = b - Ax_0$;*

(2) *Generate the first vector $q_1 = r/\|r\|_2$ and the right-hand side vector* `rhs` $= \|r\|_2 e_1$;

for $i = 1 : k$,

(3) *Start step i of a modified Gran–Schmidt method for Arnoldi:* $w = Aq_i$;

(4) **for** $\ell = 1 : i$

$R(\ell, i) = w^T q_\ell; \quad w = w - R(\ell, i) q_\ell;$

end

(5) $R(i+1, i) = \|w\|_2$ *and* $q_{i+1} = w/R(i+1, i)$;

(6) *Apply the past Givens rotations to past rows of new column i:*

for $\ell = 1 : i - 1$,

$t = c_\ell R(\ell, i) + s_\ell R(\ell+1, i);$

$R(\ell+1, i) = s_\ell R(\ell, i) - c_\ell R(\ell+1, i);$

$R(\ell, i) = t;$

end

(7) *Compute the Givens rotations:*

if $|R(i+1, i)| \le 10^{-16}$ *set* $c_i = 1$ *and* $s_i = 0$,

else, *set* $c_i = 1/\sqrt{1+t^2}$ *and* $s_i = c_i t$, *with*

$t = R(i+1, i)/R(i, i)$, *if* $|t| \le 1$

or set for $t = R(i, i)/R(i+1, i)$, $s_i = 1/\sqrt{1+t^2}$ *and* $c_i = s_i t$.

(8) *Apply the rotations to the right-hand side*

$t = c_i$ `rhs`$_i$, `rhs`$_{i+1} = s_i$ `rhs`$_i$; `rhs`$_i = t$ *and*

then to rows i, $i+1$ of column i:

$R(i, i) = c_i R(i, i) + s_i R(i+1, i)$ *and* $R(i+1, i) = 0$;

(9) *Solve the triangular system for y:*

$R(1 : i, 1 : i) y =$ `rhs`$(1 : i)$;

(10) *Update the solution: $x = x_0 + Qy$ with $Q = [q_1, q_2, \ldots, q_i]$.*

(11) *Compute the current residual: $r = b - Ax$ and exit if $\|r\|_2$ is small enough.*

end i

(12) *Continue with Step (1).*

See the Mfile `gmres_k.m`.

We now consider preconditioned GMRES algorithms. For generality and simplicity, we design one GMRES(k) algorithm for problem (3.51) i.e.

$$M_2AM_1y = M_2b, \qquad\qquad x = M_1y.$$

Algorithm 3.6.18. (Preconditioned GMREST(k)). *(Here* **T** *stands for* **T***wo).*

To solve $M_2AM_1y = M_2b$ for $x = M_1y$ by GMRES(k), without forming M_2AM_1 explicitly, with $iter = 0$, $e_1 = (e_1)_{n\times 1}$ and given an initial starting vector $x = x_0$,

(1) *Set $x_0 = x$, $iter = iter + 1$. Compute $r = b - Ax_0$ and $r = M_2r$;*
(2) *Generate the first vector $q_1 = r/\|r\|_2$ and the right-hand-side vector* $\texttt{rhs} = \|r\|_2e_1$;
for $i = 1 : k$,
(3) *Start step i of a modified Gran–Schmidt method for Arnoldi:*
$w = M_2AM_1q_i$;
(4) **for** $\ell = 1 : i$
$R(\ell, i) = w^Tq_\ell; \quad w = w - R(\ell, i)q_\ell;$
end
(5) $R(i + 1, i) = \|w\|_2$ *and* $q_{i+1} = w/R(i + 1, i)$;
(6) *Apply the rotations to past rows of new column i:*
for $\ell = 1 : i - 1$,
$t = c_\ell R(\ell, i) + s_\ell R(\ell + 1, i); \; R(\ell + 1, i) = s_\ell R(\ell, i) - c_\ell R(\ell + 1, i); \; R(\ell, i) = t;$
end
(7) *Compute the Givens rotations:*
if $|R(i + 1, i)| \le 10^{-16}$, *set* $c_i = 1$ *and* $s_i = 0$,
else
if $|t| \le 1$ *for* $t = R(i + 1, i)/R(i, i)$, *set* $c_i = 1/\sqrt{1 + t^2}$, $s_i = c_it$,
else
compute $t = R(i, i)/R(i + 1, i)$ *and set* $s_i = 1/\sqrt{1 + t^2}$, $c_i = s_it$.
end
end
(8) *Apply the rotations to the right-hand side:* $t = c_i \; \texttt{rhs}_i$;
$rhs_{i+1} = s_i \; \texttt{rhs}_i$;
$\texttt{rhs}_i = t$ *and then to rows* $i, \; i + 1$ *of column* i:
$R(i, i) = c_iR(i, i) + s_iR(i + 1, i)$ *and* $R(i + 1, i) = 0$;
(9) *Solve the triangular system for y:* $\quad R(1 : i, 1 : i)y = \texttt{rhs}(1 : i)$;

(10) *Update the solution:* $x = x_0 + M_1 Q y$ *with* $Q = [q_1, q_2, \ldots, q_i]$.
(11) *Compute the current residual:* $r = b - Ax$ *and exit if* $\|r\|_2$ *is small enough.*
end i
(12) Continue with Step (1).

See the Mfile `gmrest_k.m`. Note in Chapter 5 (see Algorithm 5.7.16) we present the flexible GMRES to allow M_1 to be different per step. The algorithm can be adapted to using other types of preconditioners e.g.

$$M_2^{-1} A M_1^{-1} y = M_2^{-1} b, \qquad x = M_1^{-1} y,$$

or

$$M_2^{-1} M_1^{-1} A x = M_2^{-1} M_1^{-1} b,$$

or

$$A M_2^{-1} M_1^{-1} y = b, \qquad x = M_1^{-1} y.$$

Many other variants of the basic GMRES(k) algorithm can be found from [413,464].

Minimizing functionals. For the symmetric case, the main three minimizing functionals in (3.36) are equivalent. For our unsymmetric case, we still have three similar minimizing functionals that are equivalent. Firstly corresponding to the normal equation (3.48), define

$$\phi(x) = \frac{1}{2} x^T A^T A x - x^T A^T b. \tag{3.60}$$

Then (3.48) and $\min_x \phi(x)$ are equivalent. Then for the GMRES, we see that the following equalities hold (analogous to (3.36))

$$\begin{aligned}
\|r\|_2^2 &= (b^T - x^T A)(b - Ax) = x^T A^T A x - 2x^T A^T b + b^T b \\
&= 2\phi(x) + b^T b, \\
\|x^* - x\|_{A^T A}^2 &= (x^* - x)^T A^T A (x^* - x) = (Ax^* - Ax)^T (Ax^* - Ax) \\
&= (b - Ax)^T (b - Ax) \\
&= r^T r \ = \ \|r\|_2^2.
\end{aligned} \tag{3.61}$$

GMRES convergence rate. In the remainder of this section, we briefly review the convergence rate of the GMRES algorithm to provide some more motivating ideas for preconditioning. As explained before, the convergence of the CG method hinges on the global minimization issue. Here with GMRES, the main clue lies in the optimal choice of y directly (or $\hat{x}$, x_k indirectly) from

(3.57): $y = \sum_{j=1}^{k} y_j q_j \in \text{span}(q_1, \ldots, q_k)$. From the Arnoldi decomposition §1.4 i.e.

$$Aq_{j-1} = q_1 h_{1,j-1} + q_2 h_{2,j-1} + \ldots + q_{j-1} h_{j-1,j-1} + q_j h_{j,j-1}$$

with $q_1 = r_0/\|r_0\|_2$, we know that

$$\text{span}(q_1, \ldots, q_k) = \mathcal{K}_k(A, r_0).$$

Therefore the GMRES solution x_k satisfies

$$\Phi(x_k) = \|b - Ax_k\|_2^2 = \Phi(x_0 + y) = \min_{z \in \mathcal{K}_k(A, r_0)} \Phi(x_0 + z), \qquad (3.62)$$

which will be compared to $\Phi(x_0)$ for quantifying convergence. Following the similar arguments of (3.39)–(3.41), we take $x_0 = 0$.

♦ **I. Eigenvector bound.** For any $z \in \mathcal{K}_k(A, r_0) = \mathcal{K}_k(A, b)$, we obtain that $z = \sum_{j=0}^{k-1} \gamma_j A^j b = g_{k-1}(A)b$, where $g_{k-1}(\xi)$ is again the same degree $(k-1)$ polynomial determined by γ_j. Then

$$\begin{aligned} \Phi(z) &= \|b - Az\|_2^2 = \|(b - Ag_{k-1}(A)b)\|_2^2 \\ &= \|(I - g_{k-1}(A)A)b)\|_2^2 \\ &= \|q_k(A)r_0\|_2^2 \end{aligned} \qquad (3.63)$$

where $q_k(\xi) = 1 - g_{k-1}(\xi)\xi \in \mathcal{Q}_k$ satisfying $q_k(0) = 1$ as before. As we can no longer use $A = QDQ^T$, we assume that A is diagonalizable i.e. $A = XDX^{-1}$. As with (3.41), $q_k(A)$ can again be simplified

$$q_k(A) = q_k(XDX^{-1}) = Xq_k(D)X^{-1} = X \,\text{diag}(q_k(\lambda_j))X^{-1}. \qquad (3.64)$$

As with (3.42), define

$$\mathcal{B}(\lambda_1, \lambda_2, \ldots, \lambda_n) \equiv \min_{q_k \in \mathcal{Q}_k} \max_{\lambda_j \in \Lambda(A)} \left| q_k(\lambda_j) \right|.$$

Hence we can enlarge (3.63) as follows

$$\Phi(x_k) \le \mathcal{B}(\lambda_1, \lambda_2, \ldots, \lambda_n)^2 \|X\|_2^2 \|X^{-1}\|_2^2 \|r_0\|_2^2 = \kappa_2(X)^2 \mathcal{B}(\lambda_1, \lambda_2, \ldots, \lambda_n)^2 \Phi(x_0),$$

that is,

$$\frac{\|r_k\|_2}{\|r_0\|_2} = \frac{\|x^* - x_k\|_{A^TA}}{\|x^* - x_0\|_{A^TA}} = \sqrt{\frac{\Phi(x_k)}{\Phi(x_0)}} \le \mathcal{B}(\lambda_1, \lambda_2, \ldots, \lambda_n)\kappa_2(X). \qquad (3.65)$$

An immediate observation of (3.65) is that a small $\mathcal{B}$ (achievable if eigenvalues λ_j are clustered; see Section 1.5) is only useful for fast convergence of the GMRES *if* $\kappa_2(X)$ is relatively small i.e. the eigenvectors do not 'ruin' the eigenvalues. It turns out that this worst case can happen whenever matrix

A is highly non-normal [244], i.e $\|A^T A - AA^T\| \gg 0$. To tackle the problem directly, we have to reduce the effect of non-normality somehow by suitable preconditioning.

Various ways of deriving alternative bounds, to (3.65), have been attempted – the starting point is to work with (3.63) and enlarge it differently. One common approach is to consider bounding the ideal GMRES problem [245,243,198], without using $A = XDX^{-1}$ in (3.63),

$$\min_{q_k \in \mathcal{Q}_k} \|q_k(A)\|_2. \tag{3.66}$$

♦ **II. ϵ-pseudospectrum bound.** Note that the eigenspectrum $\Lambda(A)$ is contained in the ϵ-pseudospectrum $\Lambda_\epsilon(A)$ for $\epsilon \geq 0$, and for any polynomial $q_k \in \mathcal{Q}_k$

$$q_k(A) = \frac{1}{2\pi i} \int_\Gamma q_k(z)(zI - A)^{-1} dz,$$

where Γ is any finite union of Jordan curves containing $\Lambda(A)$ in its interior. Let the domain Ω_ϵ, with boundary $\partial\Omega_\epsilon$, enclose $\Lambda_\epsilon(A)$ tightly and denote by $\mathcal{L}(\partial\Omega_\epsilon)$ the contour length of $\partial\Omega_\epsilon$. Then an upper bound is obtained [456,198] from

$$\|q_k(A)\|_2 \leq \frac{1}{2\pi} \int_{\partial\Omega_\epsilon} |q_k(z)| \|(zI - A)^{-1}\|_2 |dz| \leq \frac{\mathcal{L}(\partial\Omega_\epsilon)}{2\pi\epsilon} \max_{z \in \partial\Omega_\epsilon} |q_k(z)|. \tag{3.67}$$

Therefore we have found another bound for (3.62), with $x_0 = 0$,

$$\frac{\|r_k\|_2}{\|r_0\|_2} \leq \frac{\mathcal{L}(\partial\Omega_\epsilon)}{2\pi\epsilon} \min_{q_k \in \mathcal{Q}_k} \max_{z \in \partial\Omega_\epsilon} |q_k(z)|. \tag{3.68}$$

In this bound, as ϵ is small, the size $\mathcal{L}$ (reflecting the sensitivity of eigenvalues) determines if eigenvalue distribution can provide useful convergence indication.

♦ **III. Field of values bound.** A further characterization of GMRES convergence [194] is based on field of values[6] (FoV) for definite problems satisfying $0 \notin \mathcal{W}(A)$:

$$\frac{\|r_k\|_2}{\|r_0\|_2} \leq 2 \min_{q_k \in \mathcal{Q}_k} \max_{z \in \mathcal{W}(A)} |q_k(z)|. \tag{3.69}$$

As is known [198,318,457], the FoVs almost contain the corresponding ϵ-pseudospectrum so the bound suggests fast convergence for GMRES only if the eigenvalues are not sensitive.

[6] Note $\Lambda(A) \subset \mathcal{W}(A)$ but bound (3.71) does not involve $\kappa(X)$. See §1.5.

Non-normality. Finally we comment on non-normality, which strongly influences the above three convergence results (3.65), (3.68), (3.69) for the GMRES. One early characterization was to measure the magnitude of the off-diagonal entries in the Schur decomposition [275]. As any matrix A possesses a Schur decomposition $A = UTU^*$, with U unitary i.e. $UU^* = I$, split $T = \Lambda + N$ where Λ is the diagonal of T containing all eigenvalues of A. Then the size

$$\|N\|_F = \sqrt{\|T\|_F^2 - \|\Lambda\|_F^2} = \sqrt{\|A\|_F^2 - \|\Lambda\|_F^2} \tag{3.70}$$

indicates how non-normal the matrix A is; if $\|N\|_F = 0$, the GMRES convergence bound (3.65) would be identical to the SPD case (3.43) as $\kappa_2(X) = \kappa_2(U) = 1$. *Therefore one anticipates that when* $\|N\|_F$ *is small,* $\kappa_2(X)$ *for the bound (3.65) will be small so that distribution of eigenvalues becomes relevant to fast convergence of GMRES.* For a diagonalizable matrix $A = XDX^{-1}$, the trouble being $XX^T \neq I$, one may decompose X by the QR method to give $X = QR$ and get $A = QRD(QR)^{-1} = Q(RDR^{-1})Q^T$. Then we can equally 'blame' the off-diagonal entries of R responsible for non-normality. Overall, it is our belief that a good preconditioner can improve on non-normality and thus speed up the GMRES method. Indeed the SPAI type preconditioners in Chapter 5 can reduce the non-normality measure $\|N\|_F$ to some extent.

3.7 The GMRES algorithm in complex arithmetic

So far A is assumed to be a real matrix and generalization of results to the complex case is usually straightforward. As far as computer programming and mathematics are concerned, there is no need to convert (1.1), with $A = A_1 + iA_2$, $x = x_1 + ix_2$ and $b = b_1 + ib_2$, to a real augmented system

$$\begin{bmatrix} A_1 & -A_2 \\ A_2 & A_1 \end{bmatrix} \begin{pmatrix} x_1 \\ x_2 \end{pmatrix} = \begin{pmatrix} b_1 \\ b_2 \end{pmatrix}, \tag{3.71}$$

as recommended by some books (though there may be situations where this is convenient). This is because a lot of mathematical formulae only require minor adjustments for a real algorithm to work in complex arithmetic, e.g. for the CGNT Algorithm 3.5.15 to work with complex arithmetic the required four changes will be at the initial setup, and steps (2), (5), (6) – essentially replacing transposes to conjugate transposes, e.g. replacing $r_{\text{new}} = s^T s$ by $r_{\text{new}} = s^H s$.

For the complex version of a GMRES method, the changes are also similar and for this reason the exact details (as shown below) may not be found in many references. Here we only concentrate on the Given rotations. Recall that a real

Given rotation matrix is defined by (1.24), in a shorthand notation,

$$P(j,k) = \begin{bmatrix} c & s \\ s & -c \end{bmatrix}. \tag{3.72}$$

Allowing θ to take complex values can satisfy $P(j,k)P(j,k) = P(j,k)^2 = I$ but does not make $P(j,k)$ an unitary matrix since

$$P(j,k)P(j,k)^H = \begin{bmatrix} c\bar{c} + s\bar{s} & c\bar{s} - s\bar{c} \\ s\bar{c} - c\bar{s} & s\bar{s} + c\bar{c} \end{bmatrix} \neq I \tag{3.73}$$

because, with $\theta = a + ib$, at least

$$c\bar{c} + s\bar{s} = |c|^2 + |s|^2 = (\cosh(b)^2 + \sinh(b)^2)(\sin(a)^2 + \cos(a)^2) = \cosh(2b) \neq 1,$$

though $c^2 + s^2 = \cos(\theta)^2 + \sin(\theta)^2 = (\cosh(b)^2 - \sinh(b)^2)(\sin(a)^2 + \cos(a)^2) = 1$ (e.g. let $\theta = 1/3 + 5/7i$, then $c \approx 1.196 - 0.254i$ and $s \approx 0.414 + 0.734i$ so $c^2 + s^2 = 1$ and $|c|^2 + |s|^2 = \cosh(2y) = \cosh(10/7) = 2.206$.) Let $z = x + yi \in \mathbb{C}$ with $x, y \in \mathbb{R}$ (so $\bar{z} = x - yi$) and set $a = \sqrt{1 + z\bar{z}} = \sqrt{1 + x^2 + y^2}$. A complex Givens rotation matrix can be defined as in Table 3.1. With the new definition, one can verify that $P(j,k)P(j,k)^H = I$ and, when $y = 0$, (3.74) reduces to the real case (3.72). We now consider the typical step (3.59) of decomposing an upper Hessenberg matrix H into QR form as in Table 3.1, by choosing z in $P(j, j+1)$ through[7]

$$\widetilde{h}_{j+1,j} = zh_{jj} - h_{j+1,j} = 0 \qquad \text{i.e.} \quad z = \frac{h_{j+1,j}}{h_{jj}}.$$

For readers' benefit, we give the complete complex GMRES(k) algorithm.

Algorithm 3.7.19. (Complex GMRES(k)).

To solve $Ax = b$, with iter $= 0$, $e_1 = (e_1)_{n\times 1}$ and given an initial vector $x = x_0$,

(1) Set $x_0 = x$, iter $=$ iter $+ 1$ and compute $r = b - Ax_0$;

(2) Generate the first vector $q_1 = r/\|r\|_2$ and the right-hand-side vector `rhs` $= \|r\|_2 e_1$;

for $i = 1 : k$,

(3) Start step i of a modified Gran–Schmidt method for Arnoldi: $w = Aq_i$;

(4) **for** $\ell = 1 : i$

$$R(\ell, i) = w^H q_\ell; \qquad w = w - R(\ell, i) q_\ell;$$

end

[7] Reset the matrix $P(j, j+1)$ to a permutation of two rows $j,\ j+1$ if $h_{jj} = 0$.

Table 3.1. *A complex Givens rotation matrix $P(j, k)$ and its use in a QR step.*

$$
\begin{aligned}
aP(j,k) &\equiv \begin{bmatrix} a &&&&&&&& \\ & \ddots &&&&&&& \\ && a &&&&&& \\ &&& 1 & \cdots & \bar{z} &&& \\ &&& \vdots & \ddots & \vdots &&& \\ &&& z & \cdots & -1 &&& \\ &&&&&& a && \\ &&&&&&& \ddots & \\ &&&&&&&& a \end{bmatrix} = \begin{bmatrix} a &&&&&&&& \\ & \ddots &&&&&&& \\ && a &&&&&& \\ &&& 1 & \cdots & x - yi &&& \\ &&& \vdots & \ddots & \vdots &&& \\ &&& x + yi & \cdots & -1 &&& \\ &&&&&& a && \\ &&&&&&& \ddots & \\ &&&&&&&& a \end{bmatrix}, \\
P(j,k) &=_{short}^{for} \frac{1}{a}\begin{bmatrix} 1 & \bar{z} \\ z & -1 \end{bmatrix}.
\end{aligned}
\tag{3.74}
$$

$$
\begin{aligned}
&P(j, j+1) \begin{bmatrix} r_{11} & \cdots & r_{1,j-1} & h_{1j} & \cdots & h_{1,k-1} & h_{1k} \\ & \ddots & \vdots & \vdots & \cdots & \vdots & \vdots \\ && r_{j-1,j-1} & h_{j-1,j} & \cdots & h_{j-1,k-1} & h_{j-1,k} \\ &&& \boxed{h_{jj}} & \cdots & h_{j,k-1} & h_{j,k} \\ &&& \boxed{h_{j+1,j}} & \ddots & \vdots & h_{j+1,k} \\ &&&& \ddots & \ddots & \vdots \\ &&&&& h_{k,k-1} & h_{kk} \end{bmatrix} \\
&= \begin{bmatrix} r_{11} & \cdots & r_{1,j-1} & h_{1j} && \cdots & h_{1,k-1} & h_{1k} \\ & \ddots & \vdots & \vdots && \cdots & \vdots & \vdots \\ && r_{j-1,j-1} & h_{j-1,j} && \cdots & \ddots & h_{j-1,k} \\ &&& \widetilde{h}_{jj} & \widetilde{h}_{j,j+1} && \ddots & \widetilde{h}_{j,k} \\ &&&& \widetilde{h}_{j+1,j+1} && \ddots & \widetilde{h}_{j+1,k} \\ &&&& h_{j+2,j+1} && \ddots & \vdots \\ &&&&&& \ddots & h_{kk} \end{bmatrix}.
\end{aligned}
\tag{3.75}
$$

Note that for a real matrix problem, the above Givens transform reduces to (1.24).

(5) $R(i+1,i) = \|w\|_2$ *and* $q_{i+1} = w/R(i+1,i)$;

(6) *Apply the past Givens rotations to past rows of new column* i:

for $\ell = 1 : i-1$,

$t = c_\ell R(\ell,i) + \bar{s}_\ell R(\ell+1,i)$;

$R(\ell+1,i) = s_\ell R(\ell,i) - c_\ell R(\ell+1,i)$;

$R(\ell,i) = t$;

end

(7) *Compute the Givens rotations: if* $|R(i,i)| \le 10^{-16}$ *set* $c_i = 0$ *and* $s_i = 1$, *else set* $c_i = 1/\sqrt{1+|t|^2}$ *and* $s_i = c_i t$, *with* $t = R(i+1,i)/R(i,i)$.

(8) *Apply the rotations to the right-hand side:*

$t = c_i\ \mathtt{rhs}_i$; $\mathtt{rhs}_{i+1} = s_i\ \mathtt{rhs}_i$;

$\mathtt{rhs}_i = t$ *and then to rows* $i,\ i+1$ *of column* i:

$R(i,i) = c_i R(i,i) + \bar{s}_i R(i+1,i)$ *and* $R(i+1,i) = 0$;

(9) *Solve the triangular system for* y: $R(1:i, 1:i)y = \mathtt{rhs}(1:i)$;

(10) *Update the solution:* $x = x_0 + Qy$ *with* $Q = [q_1, q_2, \ldots, q_i]$.

(11) *Compute the current residual:* $r = b - Ax$ *and exit if* $\|r\|_2$ *is small enough.*

end i

(12) *Continue with Step (1).*

3.8 Matrix free iterative solvers: the fast multipole methods

Before we present the increasingly popular idea of the fast multipole method (FMM), we must point out that the FMM is not an iterative method as quite a lot of readers seem to think. However, the wrong impression is inevitable since the use of an FMM for (1.1) often involves (and is embedded in) an iterative method; see [395,365,255,40,500,171].

The FMM is a fast method for computing a dense matrix vector product (or a large sum of N terms with structured components) whose coefficients can be associated with an analytical function that admits 'decaying' or 'smooth and easily computable' expansions. (A sparse matrix vector product is already fast so it is not applicable.)

The essential idea of an FMM is to turn element-to-element interactions to group-to-group interactions, or to put equally, to turn pointwise products into blockwise products. The somewhat mysterious word 'multipole' is indeed accurately reflecting the fact that complex analysis is used. Indeed, the underlying function (normally real-valued) is taken as the real part of a complex-valued analytical function and all function expansions are conveniently carried out with respect to poles in the complex plane (as we know from complex analysis on Taylor, Maclaurin, or Laurent series). Rearranging the way expansions are computed and avoiding repeated computations lead to a fast method of $O(N \log N)$ instead of $O(N^2)$ operations. This seems to suggest that the FMM resembles the FFT in that they both can provide a fast method for matrix vector products. In fact, for regularly distributed data in a special case, the underlying problem reduces to a Toeplitz matrix computation (Section 1.6) so a FFT [445,307] and an FMM can solve the problem in similar flops. However the beauty of an FMM lies in the fact that it can deal with irregular data and nontrigonometric functions.

Here for a real-valued problem, one may use real-valued expansions instead of involving complex ones at poles to derive an FMM-like method (see [220]), although the FMM formulation appears simpler and neater with the help of complex expansions. However, in three dimensions, such pole-like expansions are directly developed without connecting to theories of complex functions.

There exist other related fast methods such as the panel-clustering method [262]) and the H-matrix method [69], that are not discussed here.

We now list two commonly known problems which can be tackled by an FMM

$$
\begin{aligned}
&\text{compute for } j = 1, 2, \cdots, N : \varphi_j = \sum_{\substack{k=1 \\ k \neq j}}^{N} q_k \log |p_j - p_k|, \quad p_j \in \mathbb{R}^2, \\
&\text{compute for } j = 1, 2, \cdots, N : \varphi_j = \sum_{\substack{k=1 \\ k \neq j}}^{N} q_k \frac{n_k \cdot (p_k - p_j)}{|p_j - p_k|}, \quad p_j \in \mathbb{R}^2,
\end{aligned}
\tag{3.76}
$$

each of the matrix-vector product type, $\mathbf{q} = [q_1, q_2, \ldots, q_N]$,

$$
\varphi = A\mathbf{q}, \tag{3.77}
$$

where $n_k = (n_1, n_2)$ is the normal vector and q_k's are real. Here dense matrix problems (3.76) may be viewed as from modelling pairwise particle interactions [217,248] or from discretizing a BIE (Section 1.7).

A FMM uses a complex formulation, by 'equating' a point $p = (x, y)$ in $\mathbb{R}^2$ with the point $z = x + iy$ in the complex plane $\mathbb{C}^1$. For (3.76), the new formulae become

$$\text{compute for } j = 1, \cdots, N : \varphi_j = \Re\phi_j,\ \phi_j = \sum_{\substack{k=1\\k\neq j}}^{N} q_k \log(p_j - p_k),\ p_j \in \mathbb{C}^1$$

$$\text{compute for } j = 1, \cdots, N : \varphi_j = \Re\phi_j,\ \phi_j = \sum_{\substack{k=1\\k\neq j}}^{N} q_k \frac{n_k}{p_j - p_k}, \qquad p_j \in \mathbb{C}^1 \tag{3.78}$$

where $n_k = n_1 + i n_2$ may be absorbed into q_k for convenience.

Remark 3.8.20. Before we briefly discuss how an FMM works, some simple remarks and observations are made here.

(1) The key idea of an FMM is to accurately approximate the distant interactions (or far field contributions) relating to $O(N)$ work by a finite sum relating to $O(1)$ work. The technical points of a successful FMM lie in two acts

- *approximating* a local cluster of interacting particles .... act 1
- *flexible shifting* of a function expansion from one center z_0 to the next z_1 .... act 2

with minor updates on the coefficients and without changing the accuracy. Both acts jointly contribute to diminishing the N information. The convergence range is accordingly updated after a shifting. These expansion centres have nothing to do with the origin of a computational coordinate. Although the key FMM references [248,405] presented Lemmas of shifting an expansion apparently to the origin $z = 0$, the FMM algorithms require shifting to a centre away from the origin. The minor but crucially important point is that these Lemmas can be adapted before use; see below.

(2) In generalizing the FMM idea to solve a BIE in other cases (e.g. in three dimensions), often, one chooses to approximate the kernel by a separable function [395,406,40]. However, it remains to compare against the simple degenerate kernel method for BIEs [24] where no projections are actually needed to solve the BIE.

(3) In various informal discussions, helpful examples of the following types are frequently used to motivate the ideas of an FMM: compute $y = Ax$

with $x \in \mathbb{C}^N$ and

$$A = \begin{pmatrix} 0 & t & \cdots & t \\ t & 0 & \cdots & t \\ \vdots & \ddots & \ddots & \vdots \\ t & t & \cdots & t \end{pmatrix},$$

or

$$A = \begin{pmatrix} 0 & (p_1 - p_2)^2 & \cdots & (p_1 - p_N)^2 \\ (p_2 - p_1)^2 & 0 & \cdots & (p_2 - p_N)^2 \\ \vdots & \ddots & \ddots & \vdots \\ (p_N - p_1)^2 & (p_N - p_2)^2 & \cdots & 0 \end{pmatrix}. \tag{3.79}$$

Here the naive $O(N^2)$ work for computing y can be reduced to $O(N)$ by changing the order of computation, in a manner analogous to the Horner's method (7.14) for polynomials. The reformulated computation becomes respectively

$$\left.\begin{aligned} y_j &= \psi_1(x_j), \\ y_j &= \psi_2(p_j), \end{aligned}\right\} \quad \text{for } j = 1, \ldots, N$$

where the functions are ($a = \sum_{k=1}^N x_k,\ b = \sum_{k=1}^N p_k x_k,\ c = \sum_{k=1}^N p_k x_k^2$)

$$\begin{aligned} \psi_1(z) &= -tz + at, \\ \psi_2(z) &= az^2 - 2bz + c. \end{aligned} \tag{3.80}$$

Here the N-term summations are respectively replaced by calculations involving a *fixed* degree $\mathbf{p} = 1$ and degree $\mathbf{p} = 2$ polynomials. In the more general case, indeed, a *fixed* degree $\mathbf{p}$ polynomial will be used similarly but approximately. In a more sophisticated way, such a polynomial will be used for computing a partial sum involving a majority of terms! The interested reader can further modify the second example by replacing the power of 2 by other powers say 3 and see how a higher-order polynomial is found.

(4) Finally we remark that the usual convention on levels in FMM assumes that the full levels are used. However to be in line with the multigrid setting (Chapter 6) where any levels are beneficial and permitted, one only needs to re-define the number of boxes n_j as shown below in one and two levels.

To concentrate on the basic idea of FMM, we consider problem 1 of (3.78), i.e.

$$\phi(z) = \sum_{\substack{k=1 \\ k \neq j}}^{N} q_k \log(z - p_k) \tag{3.81}$$

in the 2D case without any boundary interactions, for $z = p_1, \ldots, p_N$. We shall look at $m = 1$ and 2 levels in details before presenting the main algorithm due to [248] for the general case. In modelling practices, physical particles are often allowed to be periodic while BIEs (Section 1.7) generate matrices with wrap around boundaries (associated with a closed surface); these will only require minor changes.

A description of the FMM is incomplete (or impossible) without introducing the underlying theory. Aiming to approximate distant interactions, such multipole expansions have been used by several authors; see [65,248] for historical details. However, the successful FMM will also have to address how to utilize such expansions. The following results are mainly based on [248]. Here we have replaced all references of the origin $z = 0$ in [248] by a new centre z_1 and added a meaningful heading for each result, so that less imagination is required of the reader to understand the subsequent FMM algorithm.

♦ **Some elementary expansions.** First we point out that the following elementary Taylor expansions will be extensively used:

$$\left.\begin{aligned}
\log(1-z) &= -\sum_{k=1}^{\infty} \frac{z^k}{k}, \qquad |z| < 1, \\
\frac{1}{(1-z)^k} &= \sum_{j=0}^{\infty} \begin{pmatrix} j+k-1 \\ k-1 \end{pmatrix} z^j = \sum_{j=k}^{\infty} \begin{pmatrix} j-1 \\ k-1 \end{pmatrix} z^{j-k} \\
&= \sum_{j=0}^{\infty} \begin{pmatrix} j-1 \\ k-1 \end{pmatrix} z^{j-k} \qquad |z| < 1, \\
(a+b)^n &= \sum_{k=0}^{n} a^k b^{n-k} C_n^k, \\
\sum_{k=0}^{n} a_k (z+z_0)^k &= \sum_{\ell=0}^{n} \left(\sum_{k=\ell}^{n} a_k C_k^{\ell} z_0^{k-\ell} \right) z^{\ell}.
\end{aligned}\right\} \tag{3.82}$$

where the binomial coefficient $C_k^j = \begin{pmatrix} k \\ j \end{pmatrix} = k!/(j!(k-j)!) = k(k-1)\cdots(k-j+1)/j!$ for $k \geq j$ $\begin{pmatrix} k \\ 0 \end{pmatrix} = 1$, and we define $C_j^k = \begin{pmatrix} j \\ k \end{pmatrix} = 0$ if $k > j$.

♦ **The basic far field expansion.** As mentioned, far fields relative to a particle refer to a cluster of particles far away from the underlying particle. Therefore the particle-to-cluster interaction may be represented by a series expansion. This idea is shown in the Theorem 2.1 of Greengard and Rokhlin [248, p. 282] which is the fundamental building block for the FMM.

Theorem 3.8.21. (Multipole expansion with a flexible expansion centre). *Suppose that τ charges of strengths $\{q_i, i = 1, \ldots, \tau\}$ are located at points $\{p_i, i = 1, \ldots, \tau\}$ with $|p_i - z_0| < r$ i.e. centred at z_0 and with a radius r. Then we can approximate $\phi_f(z)$ by*

$$\widetilde{\phi}_f(z) = a_0 \log(z - z_0) + \sum_{k=1}^{\mathbf{p}} \frac{a_k}{(z - z_0)^k}, \tag{3.83}$$

with $\mathbf{p} \geq 1$ and for any $z \in \mathbb{C}$ with $|z - z_0| \geq cr > 2r$,

$$a_0 = \sum_{i=1}^{\tau} q_i, \quad a_k = -\sum_{i=1}^{\tau} \frac{q_i}{k}(p_i - z_0)^k, \quad \left|\phi_f(z) - \widetilde{\phi}_f(z)\right| \leq \frac{B}{c-1}\left(\frac{1}{c}\right)^{\mathbf{p}}, \tag{3.84}$$

and $B = \sum_{i=1}^{\tau} |q_i|$.

Proof. Note that $\log(z - p_i) = \log(\ (z - z_0)\ +\ (z_0 - p_i)\ ) = \log(z - z_0) + \log(1 - y)$ with $y = (p_i - z_0)/(z - z_0)$ and $|y| \leq 1/c < 1$. Then using (3.82) completes the proof. Refer also to [248]. ■

♦ **The general case of ℓ-levels.** Let all $N = 4^L$ nodes ('particles') p_j be located in the square (computational) box $\Omega^{(0)}$ of level 0, which is further divided into $n_j \times n_j$ boxes $\Omega_k^{(j)}$ in level j for $j = 1, 2, \ldots, \ell$. The actual number of 'particles' in each box is not important; however for counting flops we assume a uniform distribution i.e. there are N/n_j^2 particles in each box on level j, with $n_j = 2^j$. Clearly

$$\Omega^{(0)} = \bigcup_{k=1}^{4^j} \Omega_k^{(j)}, \qquad j \leq L, \tag{3.85}$$

as shown in Figure 3.1.

For a given particle z, if the box $\Omega_k^{(m)}$ on level m contains z, then we write $\Omega_z^{(m)} = \Omega_k^{(m)}$. If the box $\Omega_k^{(m)}$ is nearest to box $\Omega_z^{(m)}$, then we call box $\Omega_k^{(m)}$ a near-field neighbour of z otherwise a far-field neighbour on level m. If the box $\Omega_k^{(m-1)}$ is a far-field neighbour of z i.e. of $\Omega_z^{(m-1)}$, then all four boxes on level m of this level-$(m-1)$ box are also far-field neighbours of z. We further denote the union of all level m near neighbours of the particle z, excluding $\Omega_z^{(m)}$, by $\Omega_{\text{near}(z)}^{(m)}$. The final notation we need [248] is the so-called *interaction list* of level

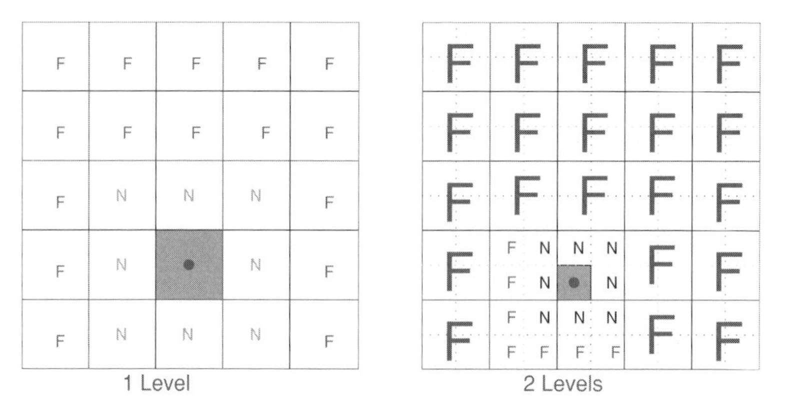

Figure 3.1. Illustration of near and far fields of two FMM schemes (2D) (z at $\bullet$).

m for box $\Omega_z^{(m)}$, which is an union of far-field boxes and is defined by

$$\Omega_{\text{intera}(z)}^{(m)} = \left\{ \bigcup_k \Omega_k^{(m)} \;\middle|\; \begin{array}{r} \Omega_k^{(m)} \text{ is a far-field box and } \Omega_k^{(m)} \subset \Omega_k^{(m-1)} \\ \text{with } \Omega_k^{(m-1)} \text{ a near-field box} \end{array} \right\}.$$

In words, the level m *interaction list* consists of the children boxes of all near neighbours of the *parent* box of $\Omega_z^{(m)}$. The union of all interaction lists makes up the so-called far field for particle z or box $\Omega_z^{(\ell)}$. The FMM will make full use of the decomposition property

$$\Omega^{(0)} = \Omega_z^{(\ell)} \bigcup \Omega_{\text{near}(z)}^{(\ell)} \bigcup \Omega_{\text{intera}(z)}^{(\ell)} \bigcup \Omega_{\text{intera}(z)}^{(\ell-1)} \bigcup \cdots \bigcup \Omega_{\text{intera}(z)}^{(2)}. \quad (3.86)$$

Clearly near and far fields complete a *covering* for the entire interaction list [262].

Once the interaction lists are set up, the evaluation of (3.81) can be divided into

$$\begin{aligned} \phi(z) &= \sum_{p_k \in \Omega^{(0)} \setminus \{z\}} q_k \log(z - p_k) \\ &= \underbrace{\sum_{p_k \in \{\Omega_z^{(\ell)} \setminus z\} \cup \Omega_{\text{near}(z)}^{(\ell)}} q_k \log(z - p_k)}_{\text{Near field terms to be evaluated directly}} + \\ &\quad \underbrace{\sum_{p_k \in \cup_{j=2}^{\ell} \Omega_{\text{intera}(z)}^{(j)}} q_k \log(z - p_k)}_{\text{Far fields to be approximated by } \mathbf{p}+1 \text{ terms.}} \end{aligned} \quad (3.87)$$

Here $\mathbf{p}$ will be the order of multipole expansions as in Theorem 3.8.21.

♦ **The special cases of one level and two levels.** Before we present the theory of multipole expansions of [248] and discuss the algorithmic details, we give an illustration of the simpler cases of one level and two levels. More precisely, we consider the partition of the whole domain $\Omega^{(0)}$ in (3.85) by

(1) the one-level case: the finest level $j = \ell$, n_ℓ boxes only;
(2) the two-level case: the fine level $j = \ell$ and the coarse level $j = (\ell - 1)$, with n_ℓ and $n_{\ell-1}$ boxes respectively.

Here the near and far fields are easier to demonstrate: Figure 3.1 depicts the cases of $n_\ell = 5$ (left plot for one level), and $n_\ell = 10$, $n_{\ell-1} = 5$ (right plot for two levels). In the left plot of Figure 3.1, the particular z has nine near-field neighbours (marked 'N') and 16 far-field level one neighbours (marked 'F'). In the right plot of of Figure 3.1, the set z (denoted by •) again has nine near-field neighbours and 28 far-field neighbours (21 level-one marked larger '**F**' and seven level-two marked smaller 'F').

According to (3.87), for Figure 3.1, each sum can be evaluated as follows

(1) the one-level case:

$$\phi(z) = \sum_{p_k \in \Omega^{(0)} \setminus \{z\}} q_k \log(z - p_k) = \sum_{p_k \in \cup N \setminus \{z\}} q_k \log(z - p_k) + \sum_{p_k \in \cup F} q_k \log(z - p_k).$$

(2) the two-level case:

$$\phi(z) = \sum_{p_k \in \Omega^{(0)} \setminus \{z\}} q_k \log(z - p_k) = \sum_{p_k \in \cup N \setminus \{z\}} q_k \log(z - p_k) + \sum_{p_k \in \cup F \cup \mathbf{F}} q_k \log(z - p_k).$$

Clearly from the above illustration and expectation, the treatment of far field interactions via multipole expansions will be a key development. We introduce the systematic method of shifting these expansions to derive an FMM algorithm.

♦ **Shifting local far field expansions to a coarse level centre of far fields.** We can adapt the Lemma 2.3 of Greengard and Rokhlin [248, p. 283] to the following

Theorem 3.8.22. (Multipole expansion fine-to-coarse shifting). *Suppose that a local multipole (approximate) expansion has been obtained for some*

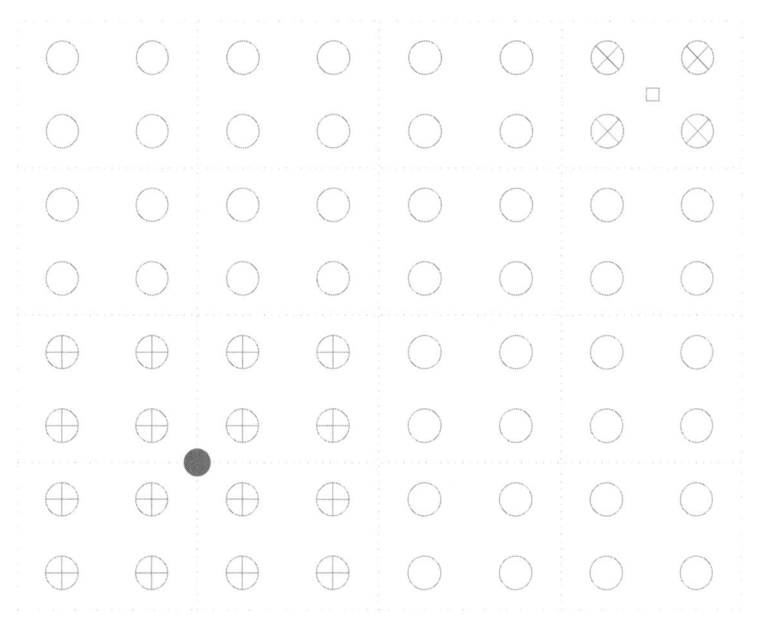

Figure 3.2. Illustration of shifting far-field centres. Case (i): shifting far expansions centred at $\oplus$ into the new far field centre at $\bullet$ in Theorem 3.8.22 for near field box $\square$. Case (ii): shifting far field expansions centred at $\bullet$ into the new near field centre at $\square$ in Theorem 3.8.23 for near field box $\square$.

local interacting box Ω_f, centered at z_0, as in Theorem 3.8.21

$$\widetilde{\phi}_f(z) = a_0 \log(z - z_0) + \sum_{k=1}^{\mathbf{p}} \frac{a_k}{(z - z_0)^k}.$$

If z_1 is a nearby far field centre of the next coarser level box (as shown in Figure 3.2), we can further approximate $\phi_f(z)$ and $\widetilde{\phi}_f(z)$ by

$$\hat{\phi}_f(z) = a_0 \log(z - z_1) + \sum_{\ell=1}^{\mathbf{p}} \frac{b_\ell}{(z - z_1)^\ell}, \tag{3.88}$$

for all $z \in \mathbb{C}$ with $|z - z_1| \geq r + |z_0 - z_1| \geq c_1 r > 2r$, where the modified coefficients can be generated from a_k's in $O(\mathbf{p}^2)$ flops

$$b_\ell = \left(\sum_{k=1}^{\ell} a_k (z_0 - z_1)^{\ell - k} \binom{k-1}{\ell-1} \right) - a_0 \frac{(z_0 - z_1)^\ell}{\ell}. \tag{3.89}$$

Here

$$\left| \phi_f(z) - \hat{\phi}_f(z) \right| \leq \frac{B}{c_1 - 1} \left(\frac{1}{c_1} \right)^{\mathbf{p}}.$$

Proof. As before, note that $(z - z_0) = (z - z_1) + (z_1 - z_0) = (z - z_1)(1 - y)$ with $y = (z_0 - z_1)/(z - z_1)$ and $|y| \le 1/c_1 < 1$. Then using (3.82) completes the proof. Note that in [248], Lemma 3.2 has a major typo in its formula for b_ℓ which should read $a_0 z_0^l/\ell$ instead of $a_0 z_0^l/0$. ∎

♦ **Far fields on an interaction list shifted to near fields on a fine level.** To enable the computed results of far field interactions to be used by near field boxes (and then by their particles), it is necessary to shift the expansions to the near field centres (without sacrificing the approximation accuracy). The exact progression across the levels will be done through the interactions lists.

Theorem 3.8.23. (Multipole expansion for coarse-to-fine shifting). *Suppose that a local multipole (approximate) expansion has been obtained for some far field interacting box Ω_f, centered at z_0 and with a radius R as in Theorem 3.8.21*

$$\widetilde{\phi}_f(z) = a_0 \log(z - z_0) + \sum_{k=1}^{\mathbf{p}} \frac{a_k}{(z - z_0)^k}.$$

If z_1 is a near field centre on the same level (see the illustration in Figure 3.2), we can describe the potential due to the above charges by a power series

$$\hat{\phi}_f(z) = \sum_{\ell=0}^{\mathbf{p}} b_\ell (z - z_1)^\ell, \tag{3.90}$$

for all $z \in \mathbb{C}$ in a circle centred at z_1 with radius R with $|z_0 - z_1| > (c_2 + 1)R > 2R$, where the modified coefficients can be generated from a_k's in $O(\mathbf{p}^2)$ flops

$$b_\ell = \left(\frac{1}{(z_0 - z_1)^\ell} \sum_{k=1}^{\mathbf{p}} \frac{a_k}{(z_0 - z_1)^k} \binom{\ell + k - 1}{k - 1} (-1)^k \right) - \frac{a_0}{(z_0 - z_1)^\ell \ell}, \quad \ell \ge 1,$$

$$b_0 = \sum_{k=1}^{\mathbf{p}} \frac{1}{(z_0 - z_1)^\ell} (-1)^k + a_0 \log(z_1 - z_0). \tag{3.91}$$

Here for some generic constant C and $\mathbf{p} \ge \max\{2, 2c_2/(c_2 - 1)\}$:

$$\left| \phi_f(z) - \hat{\phi}_f(z) \right| \le \frac{C}{c_2 - 1} \left(\frac{1}{c_2} \right)^{\mathbf{p}}.$$

Proof. As before, note that $(z - z_0) = (z - z_1) + (z_1 - z_0) = (z_1 - z_0)(1 - y)$ with $y = (z - z_1)/(z_0 - z_1)$ and $|y| \le 1/c_1 < 1$. Then use (3.82) to complete the proof. Refer to [248]. ∎

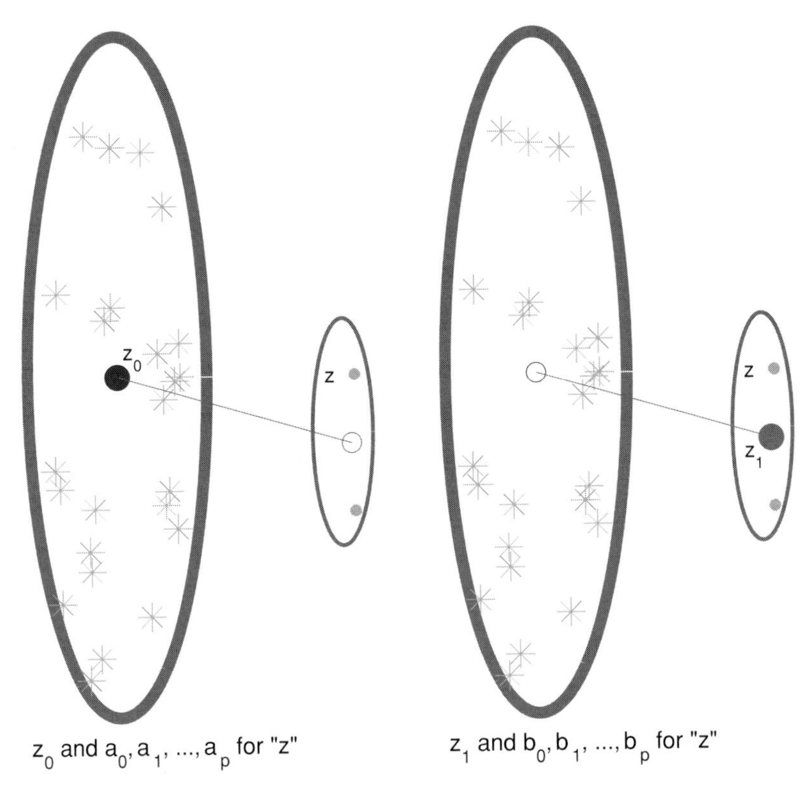

Figure 3.3. Illustration of the far field expansion manipulation in an FMM.

Up to now, we have presented all the three major stages of an FMM

- generate all local far field expansions on the finest level ℓ via Theorem 3.8.21;
- combine these local far field expansions for coarser levels via Theorem 3.8.22;
- shift far field expansions from coarser levels to all near fields on level ℓ via interaction lists and Theorem 3.8.23.

Here each initial (far field) expansion refers to an expansion centre z_0 and $\mathbf{p}+1$ coefficients $\{a_0, a_1, \ldots, a_{\mathbf{p}}\}$ as in Theorem 3.8.21 while the (far field) expansion in the last stage of an FMM refers to the new expansion centre z_1 and $\mathbf{p}+1$ coefficients $\{b_0, b_1, \ldots, b_{\mathbf{p}}\}$, as illustrated in Figure 3.3. The expansions represent the interaction of far field particles ('*' points) with the near field points ('z' points). To conform with the common terminology [248], *we shall call the above expansion with* $\mathbf{p}+1$ *terms the* **p**-*term expansion* for simplicity.

It remains to discuss the FMM algorithm in the general case of ℓ levels. We first give a simple one-level version of an FMM for computing (3.81) for $z = p_j$ with $j = 1, \ldots, N$ as follows.

Algorithm 3.8.24. (FMM in 1-level). *To compute* $y = Ax$*, with* A *as in (3.77) (with one level, Theorem 3.8.22 is not needed):*

Setting-up stage ....

(1) Divide the computational box $\Omega^{(0)}$ *with* N *particles on level* ℓ *into* $n_\ell \times n_\ell = 4^\ell$ *boxes.*

(2) Count the number of particles in each fine mesh box, work out its box centre.

• **Approximating stage** *on the fine mesh*..............................

(3) For each fine mesh box, use Theorem 3.8.21 to work a **p**-*term series expansion. Save all* n_ℓ^2 *sets of such coefficients.*

• **Shifting stage** *to the near fields* ....

(4) Shift all far field expansions to each near field box centre and add the corresponding coefficients using Theorem 3.8.23. Save all n_ℓ^2 *sets of coefficients.*

• **Summing stage** *for each particle on the fine mesh* ....

(5) compute its near field interactions with within all particles in the underlying box and the box's near-field level neighbours and add the far-field contributions implied in the associated expansion.

For the specific example in Figure 3.1 (left plot), we now estimate the flop counts using Algorithm 3.8.24 (assuming **p** is much smaller than N or n below). Let $n = N/n_\ell^2$ be the number of particles in each box. Then we have approximately

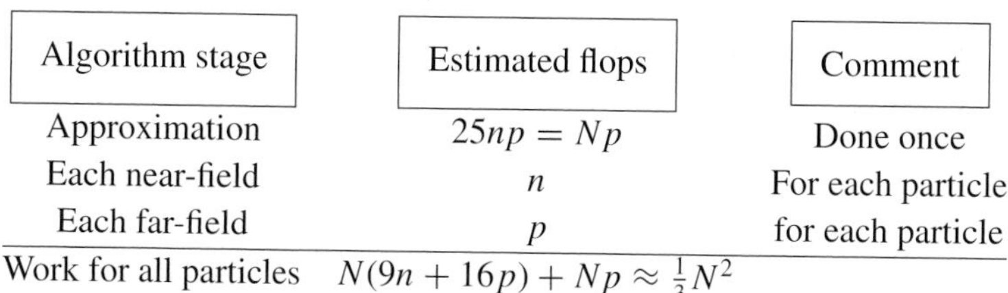

Algorithm stage	Estimated flops	Comment
Approximation	$25np = Np$	Done once
Each near-field	n	For each particle
Each far-field	p	for each particle
Work for all particles	$N(9n + 16p) + Np \approx \frac{1}{3}N^2$	

For the right plot in Figure 3.1, let $n = N/n_\ell^2$ be the number of particles in each box with $n_\ell = 10$. Then we can deduce these estimates.

Algorithm stage	Estimated flops	Comment
Approximation	$100np = Np$	Done once for F
Shifting	$25p^2$	Done once for $\mathbf{F}$
Each near-field	n	For each particle
Each far-field	p	for each particle
Work for all particles	$N(9n + 28p) + Np + 25p^2 \approx \frac{1}{11}N^2$	

Clearly for this single level algorithm, using only $n_\ell = 5$, the saving in flops is more than 60% while the advantage of using two levels is seen from a saving of about 90%. Within the context of a single level scheme, if the number of boxes is relatively small, the near field computation dominates. If one increases the number of boxes, the number of far fields begins to dominate; the optimal choice is about $n \approx \sqrt{N}$.

However, to the main point, optimal efficiency can be achieved from using multiple ℓ levels with $\ell = O(\log N)$. Then the reduction of $O(N^2)$ work to $O(N \log N)$ implies a saving close to 100%. This will ensure that each near field box contain $O(1)$ particles.

♦ **The FMM algorithm.** We are now ready to state the FMM algorithm. To clarify on the notation, we summarize the main definitions in Table 3.2. The crucial concept of interaction lists can be more clearly illustrated in Figures 3.4

Table 3.2. *Main definitions for the FMM.*

ℓ-levels	the computational box $\Omega^{(0)}$ is divided into 4^j smaller boxes $\Omega_k^{(j)}$ on level j, where $j = 0, 1, \ldots, \ell$ and $k = 1, 2, \ldots, 4^j$ for each j.
Interaction list (l, i)	the set of boxes on level l which are children of all nearest neighbours of box i's parent (on level $l - 1$, not including the parent itself) *and* which are well separated from box i at this level l.
$\Phi_{l,i}$	the **p**-term multipole expansion about the centre of box i at level l, describing the potential field created by the particles contained *inside* box i at this level.
$\Psi_{l,i}$	the **p**-term multipole expansion about the centre of box i at level l, describing the potential field created by the particles contained *outside* box i and its nearest neighbours at this level. (This is to be built up recursively from interaction lists).

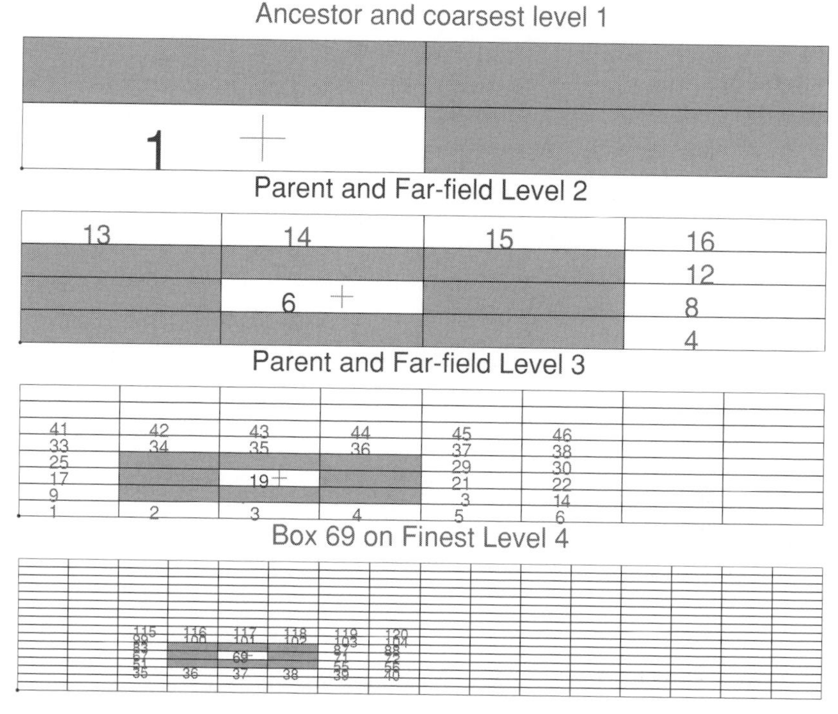

Figure 3.4. The FMM Illustration I of the interaction lists across all levels (for a specified box). Note the parent's near neighbours provide the interaction lists.

and 3.5, where we take $\ell = 4$ levels and consider the box numbered 69 on level 4. Its parent box is numbered as 19 on the coarser level 3, its grand-parent box as 6 on level 2 and the ancestor box as 1 on level 1. The two figures are respectively produced by the supplied Mfiles `intera.m` and `index2.m`.[8] Clearly one observes that the difference between a box's near field boxes and its parent (coarse level) box's near field boxes is precisely the interaction list.

Algorithm 3.8.25. (FMM in ℓ-levels). *To compute $y = Ax$ with A as in (3.77), given an accuracy tolerance ϵ and taking $\mathbf{p} = -\log_c(\epsilon)$, $c = 4/\sqrt{2} - 1$ as in [248,246], the FMM implements the following steps (via a two-pass process similar to the V-cycling of a MGM):*

Upward Pass – from the finest level ℓ to the coarsest level 1

[8] Both mfiles were developed based on several original codes, designed by a colleague Dr Stuart C. Hawkins (E-mail: stuarth@liv.ac.uk, University of Liverpool, UK). Here we adopt the lexicographical ordering but there was also a version with an interesting anti-clockwise ordering.

Parent and Far-field Level 1

3	4
1	2

Parent and Far-field Level 2

13	14	15	16
9	10	11	12
5	6	7	8
1	2	3	4

Parent and Far-field Level 3

57	58	59	60	61	62	63	64
49	50	51	52	53	54	55	56
41	42	43	44	45	46	47	48
33	34	35	36	37	38	39	40
25	26	27	28	29	30	31	32
17	18	19	20	21	22	23	24
9	10	11	12	13	14	15	16
1	2	3	4	5	6	7	8

Finest and Near-field Level 4

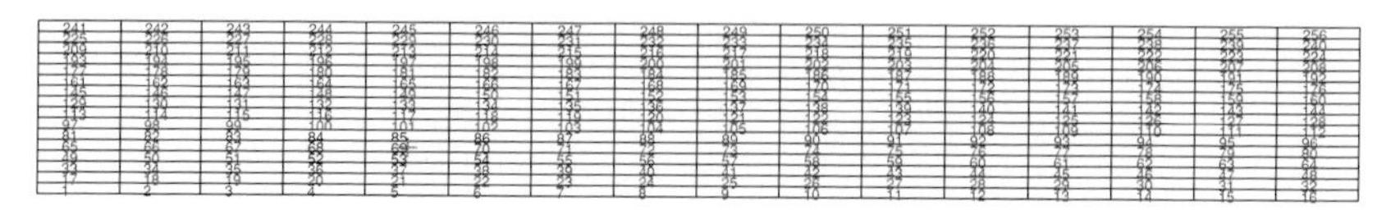

Figure 3.5. The FMM Illustration II of far fields for a specified box: watch the relationship between its far fields and the parent's near neighbours.

(1) • **Forming the multipole expansions at the finest level ℓ:** . . . .
for *ibox* $= 1, 2, \ldots, 4^{\ell}$

Form the **p**-*term multipole expansion* $\Phi_{\ell,ibox}$ *which represents the potential field due to particles in each box ibox* **end** *ibox*

(2) •**Forming the multipole expansions at all coarse levels:** . . . .
for $l = \ell - 1, \ldots, 1$

for *ibox* $= 1, 2, \ldots, 4^{l}$

Form the **p**-*term multipole expansion* $\Phi_{l,ibox}$ *by shifting the centre of each child box's expansion to the current box centre and adding the corresponding expansion coefficients together, which represents the potential field due to particles in each box ibox (from the* $\otimes$ *boxes to* $\square$ *in Figure 3.2)*

end *ibox*

end *l*

Downward Pass – from the coarsest level 1 *to the finest level* ℓ

(3) • **Combining all far field multipole expansions to local expansions up to all boxes on the finest level ℓ:** . . . .

Initialize *all expansion coefficients of* $\Psi_{l,i}$ *by zero i.e.* $\Psi_{l,i} = (0, \ldots, 0)$.

Assign *the level* 1 *expansion coefficients of* $\Phi_{1,i}$ *to* $\Psi_{1,i}$ *as these are the same.*

for $l = 1, 2, \ldots, \ell$

  if $l < \ell$ (exclude level ℓ)

  for $ibox = 1, 2, \ldots, 4^l$

    Create the **p***-term multipole expansion* $\Phi_{l+1,ibox}$ *for ibox's children by shifting the centre of expansions from the current box centre to its children's box centres, which represents the potential field due to all far field particles for the child box except the child's interaction list (refer to box* 6 *and its four children* 19, 20, 27, 28 *in Figure 3.5).*

  end *ibox*

  end if

  if $l > 1$ (exclude level 1)

  for $ibox = 1, 2, \ldots, 4^l$

    Update the **p***-term multipole expansion* $\Phi_{l,ibox}$ *by shifting the centre of expansions of each member on its interaction list to the current box centre and adding the corresponding expansion coefficients together, which represents the potential field due to particles in each box ibox's interaction lists and all of its parents (ancestors)' interaction lists (refer to Figure 3.3 and Figure 3.2 for the box* □*)*

  end *ibox*

  end if

end l

• **Evaluating separately the far field and the near field interactions for particles in each box on the finest level ℓ, before adding them:** . . . .

Evaluate *the far field expansion* $\Psi_{\ell,i}$ *and save it in a vector* F.

for $ibox = 1, 2, \ldots, 4^\ell$

  For each particle $z = p_j$ *in box ibox, evaluate the far field potential*

$$F_j = \Psi_{\ell,ibox}(z) = \sum_{k=0}^{\mathbf{p}} b_k^{ibox}(z - c_{ibox})^k \tag{3.92}$$

  where $c_{ibox} = c_{ibox}^{(\ell)}$ *is the centre of box ibox*

end *ibox*

Evaluate *the near field potential and save it in a vector* B.

for $ibox = 1, 2, \ldots, 4^\ell$

  For each particle $z = p_j$ *in box ibox, directly compute the near field potential due to all other particles in this box ibox and its nearest neighbours:*

$$B_j = \sum_{p_k \in \{\Omega_z^{(\ell)} \backslash z\} \cup \Omega_{near(z)}^{(\ell)}} q_k \log(z - p_k) \tag{3.93}$$

as in (3.87) (refer to box 69 *in Figure 3.5 whose eight nearest neighbours are* 52, 53, 54, 68, 70, 84, 85, 86*).*

end *ibox*

Complete *the FMM by adding the near and far field vectors together*

$$y = Ax = \Re(B + F). \tag{3.94}$$

Here we emphasize that each *far field expansion* $\Psi_{l,ibox}$ at a near field box centre $c_{ibox}^{(l)}$ has two contributions: the first one inherited indirectly from the grand-parent via its parent and the second one inherited directly from the parent. In [248,246], the first contribution was given a separate notation, namely, $\widetilde{\Psi}_{l,ibox}$. We have avoided this extra notation in the above description by using the word 'Create' for the first contribution and the word "Update" to imply the addition of the second contribution.

For highly nonuniform distributions of the particles, an adaptive algorithm has to be considered; refer to [246,500]. For application to BIE problems, the standard FMM algorithm as described has proved efficient.

Finally we show some test results using the FMM algorithm. The FMM toolbox[9] of Mfiles, downloaded from `http://www.madmaxoptics.com`, is run to produce the numerical tests. *It should be remarked that we are not implying that the madmaxoptics implementation is following Algorithm 3.8.25 in any way*. However, we believe the performance of all FMM variants should be similar for uniformly distributed source data i.e. p_j's. We have supplied a Mfile `ch3_fmm.m` containing these essential lines

```
>> x = rand(N,1); y = rand(N,1); q = rand(N,1)*N;
% Sample data
>> D = createsrc2d(x,y, q);
% Data formatting for the FMM toolbox
>> P_f = fmmcoul2d(D, x,y);
% The main step
```

which called two Mfiles `fmmcoul2d.m` and `createsrc2d.m` from the above toolbox. (The toolbox is comprehensive as it solves many other problems).

[9] The toolbox from ©MADMAX, Inc. is provided free of charge but the user must fill in their required registration form from `http://www.madmaxoptics.com` first to use the toolbox. The detailed usage is clearly explained in the manual document available in the software.

Firstly we show how the FMM toolbox works for a small problem with $N = 8$ source points $\mathbf{p} = \begin{bmatrix} p_1 & p_2 & p_3 & p_4 & p_5 & p_6 & p_7 & p_8 \end{bmatrix}$ and interaction strengths $\mathbf{q} = \begin{bmatrix} 6 & 4 & 6 & 2 & 5 & 1 & 3 & 7 \end{bmatrix}$ for problem (3.78):

$$\mathbf{p} = \begin{bmatrix} (7,7) \\ (4,3) \\ (6,6) \\ (4,1) \\ (0,6) \\ (2,3) \\ (0,1) \\ (7,4) \end{bmatrix},$$

$$A = \begin{bmatrix} 0.0000 & 1.6094 & 0.3466 & 1.9033 & 1.9560 & 1.8568 & 2.2213 & 1.0986 \\ 1.6094 & 0.0000 & 1.2825 & 0.6931 & 1.6094 & 0.6931 & 1.4979 & 1.1513 \\ 0.3466 & 1.2825 & 0.0000 & 1.6836 & 1.7918 & 1.6094 & 2.0554 & 0.8047 \\ 1.9033 & 0.6931 & 1.6836 & 0.0000 & 1.8568 & 1.0397 & 1.3863 & 1.4452 \\ 1.9560 & 1.6094 & 1.7918 & 1.8568 & 0.0000 & 1.2825 & 1.6094 & 1.9851 \\ 1.8568 & 0.6931 & 1.6094 & 1.0397 & 1.2825 & 0.0000 & 1.0397 & 1.6290 \\ 2.2213 & 1.4979 & 2.0554 & 1.3863 & 1.6094 & 1.0397 & 0.0000 & 2.0302 \\ 1.0986 & 1.1513 & 0.8047 & 1.4452 & 1.9851 & 1.6290 & 2.0302 & 0.0000 \end{bmatrix}.$$

With this datum set, the Mfile `ch3_fmm.m` produces the identical results from the direct product $P_d = A\mathbf{q}$ and the FMM $P_f = A\mathbf{q}$:

```
D =      Coords: [8x2 double]    % (x,y)
    Monocharge: [8x1 double]    % q vector
     Dipcharge: [8x1 double]    % ETC not used in this test
                                  example
```

$$P_d = \begin{bmatrix} 38.3150 & 40.0308 & 32.9442 & 48.8933 & 52.6448 & 46.5842 & 57.7231 & 36.5610 \end{bmatrix}^T,$$
$$P_f = \begin{bmatrix} 38.3150 & 40.0308 & 32.9442 & 48.8933 & 52.6448 & 46.5842 & 57.7231 & 36.5610 \end{bmatrix}^T.$$

Secondly we give a rough indication of performance by comparing the CPU; however, this CPU measure is not an entirely reliable way, as clearly shown below even with different versions of MATLAB® to judge performance (due to MATLAB platform and also CPU alone). Nevertheless, the improvements are quite evident from the comparison. We shall test the Mfile `ch3_fmm.m` for some larger runs on a Pentium IV (1.5 GHz) PC and present the experiments in Table 3.3. Clearly Table 3.3 demonstrates the superiority of the FMM for computing $y = Ax$ in a speedy way! Similar observations have been made in

Table 3.3. *Speedup of the FMM for dense matrix-vector products* $y = Ax$.

MATLAB	Size N	Direct `cputime`	FMM `cputime`	Ratio
version 6.5	128	0.200	0.291	0.69
	256	0.911	0.020	45.55
	512	5.037	0.050	100.74
	1024	29.002	0.080	362.53
	2048	186.007	0.110	1690.97
version 7.0β	128	0.0901	0.0200	4.50
	256	0.3305	0.0501	6.60
	512	2.6338	0.0401	65.75
	1024	19.0574	0.0701	271.86
	2048	159.8799	0.0801	1995.63

the literature in an abundant number of applications; see the recent work by [249,247,171,500] and the references therein.

Other matrix-free methods exist in the solution of PDEs in computational mathematics. Many of these may not be classified in the same category as the efficient FMM. For example, one may avoid forming large matrices by using Gauss–Seidel Newton methods and nonlinear iterations in CFD (computational fluid dynamics) problems, or by using explicit time-marching schemes in time-dependent PDEs, or by using finite differences to approximate the Jacobian matrix product $y = Jx = \nabla \mathbf{F} x$ for a nonlinear algebraic system.

Once a fast method for computing $y = Ax$ is available, an iterative solution of (1.1) will still require a suitable preconditioner. In normal circumstances, the near fields information of A are sufficient to find a good preconditioner but in some extreme cases, more entries of A may have to be available to construct a preconditioner. This will pose a computational challenge.

3.9 Discussion of software and the supplied Mfiles

The topic of iterative methods is vast. We only gave a selected introduction here, with a view to use these methods to test preconditioning later.

There are many rich sources of software available. We list a few of these.

(1) http://www.netlib.org/ contains a repository of various software including those from the template book [41] and the TOMS (Transactions on Mathematical Software).

(2) http://www-users.cs.umn.edu/∼saad/software/SPARSKIT/sparskit.html contains the Sparskit package for solving sparse linear systems.
(3) http://www.nag.co.uk/ contains some iterative solvers (e.g. in part D03).
(4) http://www.mathworks.com has adopted many iterative solvers in its MATLAB releases (e.g. the GMRES [413] and the BiCGSTAB [464]).
(5) http://www.mgnet.org/ contains many pointers to multigrid- and multilevel-related work and software.
(6) http://www.ddm.org/ contains many multidomain- and multilevel-related work and software.
(7) http://web.comlab.ox.ac.uk/projects/pseudospectra/ contains the Oxford pseudospectra gateway on Mfiles for pseudospectra computing.
(8) http://www.ma.umist.ac.uk/djs/software.html contains a suite of Mfiles for iterative solution of several types of fluid problems, using the preconditioned conjugate gradient and multigrid methods.
(9) http://www.madmaxoptics.com/ contains software for fast multipole methods for various simulations (e.g. (3.78)). (The developers are also the pioneers of FMM.) We have already mentioned the particular Mfiles in the FMM toolbox: `fmmcoul2d.m` and `createsrc2d.m` for solving the simple model (3.78).

This chapter is accompanied by several Mfiles for experimenting iterative methods.

[1] `iter3.m` – Implement the three well-known relaxation methods: Jacobi, GS and the SOR. This Mfile is quite elementary. As remarked in the preface, typing `inter3` alone will invoke the help and usage comments.
[2] `gmres_k.m` – Implement the basic GMRES(k) as in Algorithm 3.6.17. This is a moderate Mfile.
[3] `gmrest_k.m` – Implement the preconditioned GMRES(k) as in Algorithm 3.6.18. This is another moderate Mfile and it is similar to the MATLAB default Mfile `gmres.m`, although our Mfile `gmres_k` is simpler without the extra checking steps.
[4] `gmres_c.m` – Implement the complex arithmetic GMRES(k) as in Algorithm 3.7.19. This Mfile is new, as the author is not aware of a similar algorithm to Algorithm 3.7.19.
[5] `intera.m` – Illustrate how an interaction list for an FMM can be worked out for a 2D square domain (as noted, this Mfile is developed jointly with Dr Stuart C. Hawkins, University of Liverpool, UK).
[6] `index2.m` – Computes the index list for a node's near neighbours (suitable for an FMM in a 2D square domain); as with `intera.m`, this Mfile is

also developed jointly with Dr Stuart C. Hawkins, University of Liverpool, UK.

[7] `ch3_fmm.m` – A driver Mfile used to call the FMM package (the package itself is not supplied here) from `madmaxoptics`; the reader needs to register (free) directly to download the package: see the web address listed above.

4

Matrix splitting preconditioners [T1]: direct approximation of $A_{n \times n}$

The term **"preconditioning"** appears to have been used for the first time in 1948 by Turing [461],.... The first use of the term in connection with iterative methods is found in a paper by Evans [200] ... in 1968.

MICHELE BENZI. *Journal of Computational Physics*, Vol. 182 (2002)

For such problems, the coefficient matrix A is often highly nonsymmetric and non-diagonally dominant and hence many classical preconditioning techniques are not effective. For these problems, the circulant preconditioners are often the only ones that work.

RAYMOND CHAN and TONY CHAN. *Journal of Numerical Linear Algebra and Applications*, Vol. 1 (1992)

In ending this book with the subject of preconditioners, we find ourselves at the philosophical center of the scientific computing of the future... Nothing will be more central to computational science in the next century than the art of transforming a problem that appears intractable into another whose solution can be approximated rapidly. For Krylov subspace matrix iterations, this is preconditioning.

LLOYD NICHOLAS TREFETHEN and DAVID BAU III. *Numerical Linear Algebra*. SIAM Publications (1997)

Starting from this chapter, we shall first describe various preconditioning techniques that are based on manipulation of a given matrix. These are classified into four categories: direct matrix extraction (or operator splitting type), inverse approximation (or inverse operator splitting type), multilevel Schur complements and multi-level operator splitting (multilevel methods). We then discuss the similar sparse preconditioning techniques in the wavelet space.

This chapter will discuss the direct extraction techniques for constructing **Forward Type** preconditioners (or operator splitting type). Essentially the preconditioner M in (1.5), either dense (but structured Section 2.5) or sparse (and easily solvable), will arise from a direct splitting of A

$$M = A - C \qquad \text{or} \qquad A = M + C, \tag{4.1}$$

where M is obtained either algebraically or graphically; in the latter case non-zeros in M, C do not normally overlap. Specifically we present the following.

Section 4.1 Banded preconditioner
Section 4.2 Banded arrow preconditioner
Section 4.3 Block arrow preconditioner from DDM ordering
Section 4.4 Triangular preconditioner
Section 4.5 ILU preconditioner
Section 4.6 Fast circulant preconditioner
Section 4.7 Singular operator splitting preconditioners
Section 4.8 Preconditioning the fast multipole method
Section 4.9 Numerical experiments
Section 4.10 Discussion of software and Mfiles

4.1 Banded preconditioner

A banded matrix $\mathbf{band}(\alpha, \beta)$ Section 2.5.1, including the diagonal matrix $\mathbf{band}(0, 0)$, provides us with one of the easiest preconditioners. The idea is widely used. Also one can develop a block version, i.e. a block diagonal matrix of a banded form with diagonal blocks.

While a banded matrix as a preconditioner can be identified or constructed easily, some other forms that may be permuted to a banded matrix are also of interest but harder to identify. We consider a specific class of permutations that have applications for matrices arising from PDEs and leave the reader with a Mfile `bandb.m` for further experiments.

Lemma 4.1.1. *Let A be a dense $b \times b$ block matrix with $k \times k$ blocks. If all the block entries are in turn in block diagonal form with its blocks having identical sparsity structures i.e.*

$$diag(D_{k_1 \times k_1}, \cdots, D_{k_\ell \times k_\ell}),$$

then matrix A can be permuted to a block diagonal matrix and the permutation will naturally follow sequential ordering of the corresponding blocks i.e. $A(p, p)$ is block diagonal with

$$p = [1, 1+k, \cdots, 1+(b-1)k,\ 2, 2+k, \cdots, 2+(b-1)k,\ \cdots,\\ k, k+k, \cdots, k+(b-1)k].$$

The proof will follow from a direct construction. Note that this result is quite simple from a graphical point of view. However if the exact permutation is written down, the notation will confuse the ideas (see the quote on page xvi).

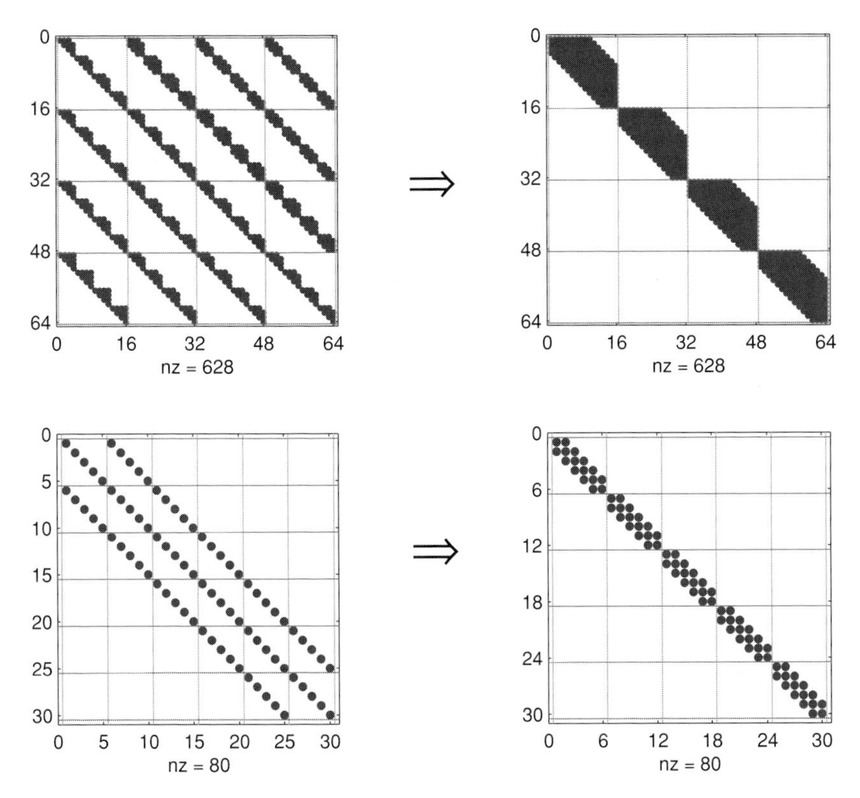

Figure 4.1. Illustration of Lemma 4.1.1 for two examples.

Figure 4.1 considers two examples (using `bandb.m`) where $k = 4, k_j \equiv 4$ in the first example with each small block of up to the type **band**$(1, 2)$ and $k = 6, k_j \equiv 1$ in the second example. Here the second example presents a typical matrix from using an ADI (alternating direction implicit) method for solving a PDE using a FDM [386].

The success of a banded preconditioner $M = B$ ultimately depends on the nature of A. If $\|A - B\|_\infty$ is small or B is dominant in A, such a preconditioner should be the first choice. For indefinite problems, this is unlikely to be the case and one good approach seems to achieve such banded dominance in another transformed space (Chapter 8) where off-diagonal elements (or information from $A - B$) are utilized.

4.2 Banded arrow preconditioner

Banded arrow preconditioners represent a slightly more sophisticated version of a banded preconditioner. As demonstrated in Section 2.5.1, this kind of preconditioners are easily invertible.[1] Strictly speaking, the type of matrices that have a dominant banded arrow part is quite limited. Here by a banded arrow preconditioner, we mean all matrices that can be permuted to possess a dominant banded arrow part.

Aiming to design a suitable M such that $\|A - M\|$ is 'small', for some problems with 'features', we can identify a number of rows and columns of A that must be included in M. Upon permutation, the resulting matrix will be a banded arrow matrix. In [205], we proposed a method that detects such nonsmooth local features. It is based on taking the differences of adjacent rows and columns and taking out those whose vectors norms are relatively larger than the average. An algorithm can be stated as follows.

Algorithm 4.2.2. (Banded arrow preconditioner).

(1) *For matrix $A_{n\times n}$, take the differences of adjacent rows and save the* 1*-norm of the different vectors (there are* $(n-1)$ *of them). Identify those vectors whose norms are some percentage (say* 10%*) larger than the average. Record the corresponding row indices* $r_1, r_2, \ldots, r_a$.
(2) *Similarly obtain column indices* $c_1, c_2, \ldots, c_b$ *from checking the differences of all columns.*
(3) *Optionally combine the row and column indices into a single list for symmetric permutation.*
(4) *Permute matrix* A *to* $\tilde{A}$ *so that the interested rows and columns are set to the end.*
(5) *Use the matrix* M *from splitting* **band**(α, β, a, b) *off* $\tilde{A}$ *as the banded arrow preconditioner.*

It turns out that the algorithm is suitable for at least two problems: (i) BIEs (Chapter 11) defined over a geometrically singular boundary where large column variations near the singularities can be detected. (ii) Elasto-hydrodynamic lubrication modelling (Section 12.5).

4.3 Block arrow preconditioner from DDM ordering

In Section 2.6.3, we have briefly discussed how a DDM ordering (on a single level and multi-domain discretization) can lead to a globally block

[1] In the context of designing preconditioners, one talks about 'inversion' in $x = M^{-1}y$ and in practice this should be understood as solving $Mx = y$ for x.

diagonal matrix A with arrow patterns. There are many applications of this idea. For example, one can design and study iterative substructuring methods if implementing a DDM approach by a stationary iteration scheme and using Schur complements (Chapter 7). See [111,432]. As better efficiency is achieved with multilevel DDM, study and implementation of reliable multilevel DDM preconditioners lead to many on-going[2] research results and challenges [111,494,432] when one goes much beyond the usual model problems.

For single level preconditioners' design, two approaches are of interest.

(1) Block diagonal preconditioners
(2) Block diagonal arrow preconditioners.

Here the first approach may be considered as a special case of (2). However this approach includes the famous example of red-black ordering and multicolour ordering methods for PDEs [413,189].

Algorithm 4.3.3. (Multicolour ordering algorithm).

Given a set of nodes $\{1, 2, \ldots, n\}$ *of a graph* $G(A)$ *for a sparse matrix* A and a maximal number ℓ of colours,

(1) Initialize the colour vector $c(1 : n) = 0$.
for $j = 1, 2, \ldots, n$
(2) Find the list L *of adjacent nodes to node* j *(from* $A_{ij} \neq 0$ *and* $A_{jk} \neq 0$*);*
(3) Find the existing colour list $C = c(L)$ *of the nodes in* L*;*
(4) Find the smallest colour index $l \in [1, \ell]$ *that is not in* C*;*
(5) Set $c(j) = l$*;*
end j
(6) Obtain the new order r *from ordering vector* c *sequentially.*

We have supplied a Mfile `multic.m` for experimenting this algorithm.

Example 4.3.4. *For the* 15×15 *matrix* A *(in Table 4.1), we may use* `multic.m` *to work out a new and re-ordered matrix* $A_1 = A(r, r)$ *that is more amenable to obtaining a block diagonal preconditioner as shown in Table 4.1.*

[2] The site http://www.ddm.org provides much useful information on the latest developments on the subject.

Table 4.1. *Multicolour ordering for a* 15×15 *matrix A (4 colours).*

$$A = \begin{bmatrix} 1 & 1 & 1 & 0 & 0 & 0 & 0 & 0 & 0 & 0 & 0 & 0 & 0 & 0 & 0 \\ 1 & 1 & 1 & 1 & 1 & 0 & 0 & 0 & 0 & 0 & 0 & 0 & 0 & 0 & 0 \\ 1 & 1 & 1 & 0 & 1 & 1 & 0 & 0 & 0 & 0 & 0 & 0 & 0 & 0 & 0 \\ 0 & 1 & 0 & 1 & 1 & 0 & 1 & 1 & 0 & 0 & 0 & 0 & 0 & 0 & 0 \\ 0 & 1 & 1 & 1 & 1 & 1 & 0 & 1 & 1 & 0 & 0 & 0 & 0 & 0 & 0 \\ 0 & 0 & 1 & 0 & 1 & 1 & 0 & 0 & 1 & 0 & 0 & 0 & 0 & 0 & 0 \\ 0 & 0 & 0 & 1 & 0 & 0 & 1 & 1 & 0 & 1 & 1 & 0 & 0 & 0 & 0 \\ 0 & 0 & 0 & 1 & 1 & 0 & 1 & 1 & 1 & 0 & 1 & 1 & 0 & 0 & 0 \\ 0 & 0 & 0 & 0 & 1 & 1 & 0 & 1 & 1 & 0 & 0 & 1 & 0 & 0 & 0 \\ 0 & 0 & 0 & 0 & 0 & 0 & 1 & 0 & 0 & 1 & 1 & 0 & 1 & 0 & 0 \\ 0 & 0 & 0 & 0 & 0 & 0 & 1 & 1 & 0 & 1 & 1 & 1 & 1 & 1 & 0 \\ 0 & 0 & 0 & 0 & 0 & 0 & 0 & 1 & 1 & 0 & 1 & 1 & 0 & 1 & 0 \\ 0 & 0 & 0 & 0 & 0 & 0 & 0 & 0 & 0 & 1 & 1 & 0 & 1 & 1 & 1 \\ 0 & 0 & 0 & 0 & 0 & 0 & 0 & 0 & 0 & 0 & 1 & 1 & 1 & 1 & 1 \\ 0 & 0 & 0 & 0 & 0 & 0 & 0 & 0 & 0 & 0 & 0 & 0 & 1 & 1 & 1 \end{bmatrix},$$

$$A_1 = \left[\begin{array}{|cccccc|cccc|ccc|cc|} \hline 1 & 0 & 0 & 0 & 0 & 0 & 1 & 0 & 0 & 0 & 1 & 0 & 0 & 0 & 0 \\ 0 & 1 & 0 & 0 & 0 & 0 & 1 & 1 & 0 & 0 & 0 & 1 & 0 & 1 & 0 \\ 0 & 0 & 1 & 0 & 0 & 0 & 0 & 0 & 1 & 0 & 1 & 0 & 0 & 1 & 0 \\ 0 & 0 & 0 & 1 & 0 & 0 & 0 & 1 & 0 & 1 & 0 & 0 & 0 & 0 & 1 \\ 0 & 0 & 0 & 0 & 1 & 0 & 0 & 0 & 1 & 0 & 0 & 1 & 1 & 0 & 1 \\ 0 & 0 & 0 & 0 & 0 & 1 & 0 & 0 & 0 & 1 & 0 & 0 & 1 & 0 & 0 \\ \cline{1-10} 1 & 1 & 0 & 0 & 0 & 0 & 1 & 0 & 0 & 0 & 1 & 0 & 0 & 1 & 0 \\ 0 & 1 & 0 & 1 & 0 & 0 & 0 & 1 & 0 & 0 & 0 & 1 & 0 & 0 & 1 \\ 0 & 0 & 1 & 0 & 1 & 0 & 0 & 0 & 1 & 0 & 0 & 1 & 0 & 1 & 0 \\ 0 & 0 & 0 & 1 & 0 & 1 & 0 & 0 & 0 & 1 & 0 & 0 & 1 & 0 & 1 \\ \cline{7-13} 1 & 0 & 1 & 0 & 0 & 0 & 1 & 0 & 0 & 0 & 1 & 0 & 0 & 1 & 0 \\ 0 & 1 & 0 & 0 & 1 & 0 & 0 & 1 & 1 & 0 & 0 & 1 & 0 & 1 & 1 \\ 0 & 0 & 0 & 0 & 1 & 1 & 0 & 0 & 0 & 1 & 0 & 0 & 1 & 0 & 1 \\ \cline{11-15} 0 & 1 & 1 & 0 & 0 & 0 & 1 & 0 & 1 & 0 & 1 & 1 & 0 & 1 & 0 \\ 0 & 0 & 0 & 1 & 1 & 0 & 0 & 1 & 0 & 1 & 0 & 1 & 1 & 0 & 1 \\ \hline \end{array}\right].$$

The order $r = \left[\boxed{1\ 4\ 6\ 10\ 12\ 15} \ \boxed{2\ 7\ 9\ 13} \ \boxed{3\ 8\ 14} \ \boxed{5\ 11} \right]$, *in* 4 *colour groups (as boxed), is obtained from the usage*

```
>> [r b c] = multic(A,4)
```

For the second approach, the arrow parts (bottom rows and far right columns) should be 'slim' enough to be comparable to the maximal block size in the main diagonal.

We remark that preconditioners resulting from single level domain decomposition are simple to use but apparently not optimal: if the number of subdomains is less, the block preconditioners may be efficient but the block size is large and it inherits the same (unfriendly) matrix structure as the original matrix. On the other hand, if the number of subdomains is large, one can see that the preconditioners will be cheaper but provide increasingly less effective preconditioning. If using multilevels, then the issues of effective relaxation and convergence have to be addressed. Although it has been widely recognized to use a multilevel DDM algorithm as a preconditioner (i.e. not to achieve any convergence), much research is still needed to assess the overall algorithmic issues in accelerating a Krylov subspace method.

4.4 Triangular preconditioners

As triangular matrices are easy to invert, using them to precondition (1.1) is a natural idea to consider. If such a triangular matrix M is extracted from A, the resulting algorithm is identical to solving the stationary iteration equation (3.5) (also Section 3.3.1) by another iterative solver (e.g. a Krylov subspace method)

$$(I - T)x = M^{-1}b, \qquad M = A - N. \tag{4.2}$$

Using such a M as a preconditioner for solving (1.1) was proposed some decades ago [201,44]. The proposal makes sense as $\|A - M\|$ can be 'minimized' if we choose, following (3.9), either $M = -\widetilde{L}$ or $M = -\widetilde{U}$. It is not difficult to propose even a mixed scheme where M is chosen to take values along a row or a column in a step-by-step and adaptive manner in order to minimize again $\|A - M\|$, i.e. M can take the following form

$$M = \begin{bmatrix} a_{11} & a_{12} & a_{13} & \cdots & a_{1n} \\ & a_{22} & & & \\ & a_{32} & a_{33} & \cdots & a_{3n} \\ & \vdots & & \ddots & \\ & a_{n2} & & \ddots & a_{nn} \end{bmatrix}, \tag{4.3}$$

which has a dense vector of a decreasing size along its row or column just like a triangular matrix. Indeed, M can be permuted to a triangular matrix and so M is easily invertible.

In the following example, we can split $A = M + R$ as follows

$$
\begin{bmatrix}
2 & 5 & 1 & 8 & 6 & 3 \\
1 & 7 & 1 & 1 & 1 & 1 \\
1 & -2 & 2 & 1 & 1 & 1 \\
1 & 1 & 4 & 4 & 3 & 8 \\
1 & 5 & 6 & 1 & 1 & 3 \\
1 & 6 & 5 & 1 & 1 & 3
\end{bmatrix}
=
\begin{bmatrix}
2 & 5 & 1 & 8 & 6 & 3 \\
 & 7 & & & & \\
 & -2 & 2 & & & \\
 & 1 & 4 & 4 & 3 & 8 \\
 & 5 & 6 & & 1 & 3 \\
 & 6 & 5 & & & 3
\end{bmatrix}
+
\begin{bmatrix}
 & & & & & \\
1 & & 1 & 1 & 1 & 1 \\
1 & & & 1 & 1 & 1 \\
1 & & & & & \\
1 & & & 1 & & \\
1 & & & 1 & 1 &
\end{bmatrix}
$$

$$
=
\begin{bmatrix}
\frac{1}{2} & -\frac{131}{42} & -\frac{11}{3} & -1 & & \frac{13}{6} \\
 & \frac{1}{7} & & & & \\
 & \frac{1}{7} & \frac{1}{2} & & & \\
 & \frac{73}{84} & \frac{37}{24} & \frac{1}{4} & -\frac{3}{4} & \frac{1}{12} \\
 & & -\frac{1}{2} & & 1 & -1 \\
 & -\frac{11}{21} & -\frac{5}{6} & & & \frac{1}{3}
\end{bmatrix}^{-1}
+ R,
$$

where one can check that $M(r, r)$ is triangular with the order $r = [1\ 4\ 5\ 6\ 3\ 2]$.

While these triangular preconditioners are simple and inexpensive to apply (even with a SSOR(ω)-like variant), they cannot give us robustness as it is not trivial to control $\|A - M\|$ unless A is 'simple'. A generalization of this idea of triangular preconditioners leads to a class of so-called incomplete LU (ILU) factorization methods, where we choose $M = LU \approx A$ but allow L, U to take values not just at nonzero entries of A directly.

4.5 ILU preconditioners

The ILU type preconditioners are widely used and newer and better variants frequently emerge [304,66,349,413]. Historically such preconditioners were proposed for specially structured and positive definite matrices, rather than for general matrices. Possibly for this reason, while useful, these methods still have some difficulties in providing robustness and reliability for general and especially indefinite problems.

♦ **A class of special matrices suited for ILU.** Without going into too much detail, we first present a brief review of classes of special matrices (SPD like)

- M-matrices
- Stieltjes matrices
- H-matrices
- Generalized diagonally dominant matrices (GDD)

for which ILU preconditioners are suitable and theories are well developed. Interested readers should consult [28,413,467] for further details of various sufficient conditions and properties.

Definition 4.5.5. (M-matrix). *A matrix $A = (a_{ij}) \in \mathbb{R}^{n\times n}$ is called a M-matrix if $a_{ij} \le 0$ for $i \ne j$ and $A^{-1} > 0$ (i.e. $(A^{-1})_{ij} > 0$).*

Definition 4.5.6. (Stieltjes matrix). *A matrix $A = (a_{ij}) \in \mathbb{R}^{n\times n}$ is called a Stieltjes matrix if $a_{ij} \le 0$ for $i \ne j$ and A is SPD (so $a_{ii} > 0$).*

Definition 4.5.7. (H-matrix). *A matrix $A = (a_{ij}) \in \mathbb{R}^{n\times n}$ is called a H-matrix if its comparison matrix $H = H(A) = (H_{ij})$ defined by*

$$H_{ij} = \begin{cases} |a_{ii}|, \; i = j, \\ -|a_{ij}|, \; i \ne j, \end{cases}$$

is a M-matrix (i.e. $H^{-1} > 0$).

Definition 4.5.8. (Generalized diagonally dominant matrix). *A matrix $A = (a_{ij}) \in \mathbb{R}^{n\times n}$ is called a generalized diagonally dominant matrix (GDD) if for some vector $x > 0 \in \mathbb{R}^n$ [or generalized strictly diagonally dominant matrix (GSDD) if valid with strict inequality]*

$$|a_{ii}|x_i \ge \sum_{j\ne i} |a_{ij}|x_j, \qquad i = 1, 2, \ldots, n.$$

We highlight some properties on their relationships [28,260]:

$$\left.\begin{array}{lll}
\bullet\; A \text{ is SDD via (3.13)} & \Longrightarrow & \text{GSDD} \\
\bullet\; A \text{ is SDD and } a_{ii} > 0\; \forall\, i & \Longrightarrow & A \text{ is SPD} \\
\bullet\; A \text{ is GSDD} & \Longleftrightarrow & A \text{ is a H-matrix} \\
\bullet\; A \text{ is monotone i.e. } A^{-1} \ge 0 & \Longleftrightarrow & Ax \ge 0 \text{ implies } x \ge 0 \\
\bullet\; a_{ij} \le 0 \text{ for } i \ne j \text{ and } A \text{ is GSDD} & \Longrightarrow & A \text{ is a M-matrix} \\
\bullet\; a_{ij} \le 0 \text{ for } i \ne j \text{ and } A \text{ is SPD} & \Longrightarrow & A \text{ is a Stieltjes matrix} \\
\bullet\; a_{ij} \le 0 \text{ for } i \ne j \text{ and } A \text{ is} & \Longrightarrow & \text{All principal minors of A} \\
\qquad\qquad\qquad\qquad \text{a M-matrix} & & \text{are positive} \\
\bullet\; A \text{ is a M-matrix in (2.6)} & \Longrightarrow & \text{The Schur complement S} \\
 & & \text{is also a M-matrix}
\end{array}\right\} \qquad (4.4)$$

To appreciate that these matrices are special, one may use one additional property of a M-matrix, that if $a_{ij} \le 0$ (for $i \ne j$) then A is a M-matrix if and only if $Re(\lambda(A)) > 0$, to see that indefinite matrices are nowhere near a M-matrix.

♦ **General ILU algorithms.** In an implicit manner, the ILU preconditioners attempt the matrix splitting M with sparse triangular matrices L, U ($L_{ii} = 1$)

$$M = LU = A - \widetilde{E} \tag{4.5}$$

with the aim of finding a lower triangular L and an upper triangular U such that

$$\min_{L,\,U} \|LU - A\|. \tag{4.6}$$

Although theoretically the optimal L, U may be dense, even for a general sparse matrix, we can impose some pattern restrictions on L, U to maintain efficiency. If A is a dense matrix, we may make a suitable threshold on entries of A before considering an ILU preconditioner.

In practice, one finds that for solving (1.1) using (4.6), it is sometimes beneficial to consider the LU factorization (approximation) of a permuted matrix PAQ and such permutations P, Q amount to finding a different nodal ordering (geometrically or algebraically using a matrix graph) for a sparse matrix; see [189,413].

One simple idea that works remarkably well for some problems is to let L, U take nonzero values only at the nonzero positions of A, i.e. keep the sparsity patterns of A and $L + U$ the same. This is the so-called ILU(0) method with no fill-ins allowed (0-level fill-in). Let $\mathcal{S}(A)$ denote the nonzero pattern of indices in A, e.g. all $(i, j) \in \mathcal{S}(A)$.

Algorithm 4.5.9. (ILU(0) preconditioner).

To find incomplete L, U factors for a sparse matrix $A \in \mathbb{R}^{n\times n}$, use the matrix B to keep L, U and set initially $b_{ij} = a_{ij}$

for $i = 2, \ldots, n$
  for $k = 1, \ldots, i-1$ *and* $(i, k) \in \mathcal{S}(A)$
(1)   *compute the multiplier* $b_{ik} = b_{ik}/b_{kk}$
    for $j = k+1, \ldots, n$ *and* $(i, j) \in \mathcal{S}(A)$
(2)   *set* $b_{ij} = b_{ij} - b_{ik}b_{kj}$
    end j
  end k
end i

For some tests, the reader may try the supplied Mfile `ilu_0.m`.

To allow a controlled increase of fill-ins, there exists the ILU(p) preconditioner where p specifies how many levels of fill-ins are permitted [413] but these methods, while interesting to consider and efficient to implement, do not offer robustness in term of reliability and fast convergence. Within the same category of methods, one finds MILU (modified ILU) where any discarded elements on a row are lumped together and added to the diagonal position of factor U; see [296,413] and the many references therein.

We now discuss the more robust ILU preconditioner based on thresholding namely the ILUT(p, τ) with p specifying a fixed number of nonzero elements and τ a threshold tolerance (for discarding small elements).

Algorithm 4.5.10. (ILU(p, τ) preconditioner).

To find the incomplete L, U factors for a sparse matrix $A \in \mathbb{R}^{n\times n}$, use the matrix B to contain L, U and set initially $b_{ij} = a_{ij}$

(1) **compute** *the 2-norms of all rows of matrix A and save them in vector v.*
  for $i = 1, \ldots, n$
(2) Take row i of A and save it in vector $w = B(i, :)$.
    for $k = 1, \ldots, i-1$ *and* $w_k \neq 0$
(3) *compute the multiplier $w_k = w_k/b_{kk}$*
(4) *Apply the L threshold: if $|w_k|/v_i < \tau$, set $w_k = 0$ for this k.*
(5) *Update row k if going ahead: if $w_k \neq 0$, set $w = w - w_k B(k, :)$ for all nonzeros.*
    end k
(6) Apply the U threshold: if $|w_j|/v_i < \tau$, set $w_j = 0$ for all j.
  Update $b_{ij} = w_j$ for up to p largest nonzeros w_j's in each of the first L range $j = 1, \ldots, (i-1)$ and the second U range $j = i, \ldots, n$.
  end i

If some diagonal elements become too small, it may be necessary to introduce partial pivoting into the algorithm. To test the algorithm using the supplied Mfile `ilu_t.m`, try $\tau = 0.1$ and $p = 3$.

There have been many recent results on developing block forms of the ILU type preconditioners suitable for parallel computing [311,67,394] and on more robust versions that are less dependent on parameters p, τ [66]. The work [66] is specially interesting as it links the usual idea (of regarding ILU as an operator splitting) to that of approximate inversions (in studying ILU for the inverse matrix; see next Chapter) in designing new methods for a suitable threshold parameter τ. The use of ILU in the Schur complements methods will be discussed in Chapter 7.

4.6 Fast circulant preconditioners

The FFT (Section 1.6) method provides one of the fastest techniques for solution of the Poisson's equation defined over a rectangular domain [180,Ch.6]. For other matrix problems from different applications, the same method may be considered by looking for a circulant approximation M out of matrix A (i.e. split $A = M - N$) and to use M as a 'dense' but structured preconditioner. No matter what a given matrix A is, a circulant matrix M is the easiest matrix to invert Section 2.5.2 as the complexity is low with FFT. As demonstrated in Section 2.5.2, circulant matrices (easily diagonalizable by FFT) are also Toeplitz matrices. For a complete account of fast solvers for Toeplitz problems, refer to [307].

♦ **Analytical properties of a Toeplitz matrix.** The analytical study (in the vast literature on the topic) of these matrices is usually based on investigating properties of its generating function for the Toeplitz coefficients (with $\theta \in [0, 2\pi]$)

$$f(\theta) = \sum_{k=-\infty}^{\infty} h_k \exp(ik\theta) = \lim_{n\to\infty} \sum_{k=1-n}^{n-1} h_k \exp(ik\theta) \tag{4.7}$$

where the function $f(\theta)$ fully defines the Toeplitz matrix with entries

$$h_k = \frac{1}{2\pi}\int_{-\pi}^{\pi} f(\theta)\exp(-ik\theta)d\theta,$$

using its first $(2n+1)$ coefficients. To constraint the infinite sequence $\{h_k\}$, we require it be square summable

$$\sum_{k=-\infty}^{\infty} |h_k|^2 < \infty$$

or be absolutely summable in a stronger assumption (or f in the Wiener class)

$$\sum_{k=-\infty}^{\infty} |h_k| < \infty, \tag{4.8}$$

(note $\sum |h_k|^2 < (\sum |h_k|)^2$). A Toeplitz matrix is Hermitian iff f is real or $h_{-k} = \bar{h}_k$. In the Hermitian case, there is a close relationship between $\lambda(T_n)$ and $f(\theta)$:

Lemma 4.6.11. ([241,104,250]). *Let T_n be any Toeplitz matrix with f from (4.7) being its generating function in the Wiener class. Then matrix T_n is uniformly bounded i.e.*

$$\|T_n\|_2 \le 2|f|_{\max}.$$

If T_n is additionally Hermitian, then

$$\lambda(T_n) \subset [f_{\min}, f_{\max}].$$

The proof of the first result requires the result (12.4) i.e. splitting a general matrix into a Hermitian part plus a skew-Hermitian part (with the latter part having a factor $-i = 1/i$) while the proof of the second result uses the Rayleigh quotient (2.8) and Fourier series properties.

♦ **Strang's circulant approximation of a Toeplitz matrix.** Strang [439] and Olkin [373] were one of the earliest to propose the use of a circulant preconditioner for a Toeplitz matrix (see (2.29) in Section 2.5.2) with elements from the top right to the bottom left

$$T_n = T(h_{1-n}, h_{2-n}, \ldots, h_{-1}, h_0, h_1, \ldots, h_{n-2}, h_{n-1}). \tag{4.9}$$

The Strang approximation of T_n is the following

$$c_j = \begin{cases} h_j, & j \le n/2, \\ h_{j-n}, & j > n/2, \end{cases} \qquad j = 0, 1, \ldots, n-1,$$

to construct the circulant matrix

$$C_n = \begin{pmatrix} c_0 & c_{n-1} & c_{n-2} & \cdots & c_1 \\ c_1 & c_0 & c_{n-1} & \ddots & c_2 \\ c_2 & c_1 & c_0 & \ddots & c_3 \\ \vdots & \ddots & \ddots & \ddots & \vdots \\ c_{n-1} & c_{n-2} & c_{n-3} & \ddots & c_0 \end{pmatrix} \tag{4.10}$$

Such a preconditioner C satisfies the optimality [104]

$$\min_{C:\text{circulant}} \|C - T\|_1 = \|C_n - T_n\|_1 \text{ and } \min_{C:\text{circulant}} \|C - T\|_\infty = \|C_n - T_n\|_\infty.$$

However the approximation using such a C to approximate T might be known much earlier [250,241] (though not used for preconditioning) but the precise theory for the preconditioned matrix $C^{-1}T$ (i.e. the eigenspectra of $C^{-1}T$ cluster at 1 for large n) was established much later (see the references in [104,250]).

♦ **T. Chan's circulant approximation of a Toeplitz matrix.** While the Strang's preconditioner C_n aims to approximate the middle bands of matrix T_n well (optimal in 1-norm and ∞-norm), the T. Chan's preconditioner achieves the F-norm optimality

$$\min_{C:\text{circulant}} \|C - T\|_F = \|C_n - T_n\|_F, \tag{4.11}$$

and involves all the subdiagonals

$$c_j = \frac{j h_{j-n} + (n-j) h_j}{n}, \qquad j = 0, 1, \ldots, n-1, \tag{4.12}$$

to construct the matrix C_n as in (4.10). Note (4.12) can be written in a convenient vector form (root vector of C_n)

$$\begin{bmatrix} c_0 \\ c_1 \\ \vdots \\ c_{n-1} \end{bmatrix} = \frac{1}{n} w \cdot \begin{bmatrix} 0 \\ a_{1-n} \\ a_{2-n} \\ \vdots \\ a_{-1} \end{bmatrix} + \frac{1}{n}(1-w) \cdot \begin{bmatrix} a_0 \\ a_1 \\ a_2 \\ \vdots \\ a_{n-1} \end{bmatrix}, \qquad w = \begin{bmatrix} 0 \\ 1 \\ 2 \\ \vdots \\ n-1 \end{bmatrix}.$$

It is worth noting that the T. Chan's circulant preconditioner, taking the form,

$$c_\ell = \frac{1}{n} \sum_{j=1}^{n} a_{[(j+\ell-1)\text{mod } n]+1, j} \qquad \text{for } \ell = 0, 1, \ldots, n-1 \tag{4.13}$$

is also optimal in F-norm approximation for a general *non-Toeplitz* matrix $A = (a_{kj})$. Here the operator mod, though simple at stand-alone, is often confusing in many mathematical contexts; here in MATLAB® command notation, the first index in (4.13) should read $\langle j + \ell - 1 \rangle_n + 1 = \texttt{mod}(j + \ell - 1, n) + 1$. When matrix A is Toeplitz, formula (4.13) reduces to (4.12). There exist a large collection of research papers that are devoted to developing circulant related dense preconditioners for matrices arising from discretizing a PDE (elliptic, parabolic and hyperbolic types); see [104] and the references therein. Of particular interest is the addition of a small quantity of magnitude ρn^{-2} onto the diagonals of the circulant approximation to improve the conditioning of the preconditioned system $C^{-1}A$ — a trick used by many numerical methods (modified ILU, stability control in time-stepping methods, viscosity methods for hyperbolic equations).

Both the Strang's and T. Chan's preconditioners are illustrated in the supplied Mfile `circ_pre.m`, which the reader may use to approximate other non-Toeplitz matrices as long as its first row and column are 'representative' of the overall matrix (e.g. when the underlying matrix A arises from discretization of a constant coefficient PDE). In the default examples of `circ_pre.m`, one can observe that the Strang's preconditioner C approximates A better than T Chan's in the Toeplitz example 1 but for the second PDE example, the T. Chan's preconditioner is better and the Strang's preconditioner happens to be singular for this case. Of course, the second application is far beyond its original usage for a Toeplitz context. A general theory with precise conditions for preconditioners (based on other norms) e.g.

$$\min_{C:\text{circulant}} \|I - C^{-1}T\|_F \qquad \text{or} \qquad \min_{C:\text{circulant}} \|I - C^{-1/2}TC^{-1/2}\|_F$$

are discussed in [104,289,462,310].

We also remark that while FFT can diagonalize a circulant matrix, other noncirculant (but circulant like) matrices must be diagonalized by other fast transforms $A_t = W^T A W$ including

$$\text{Fast sine transform (DST I)} \quad W_{jk} = \sqrt{\frac{2}{n+1}} \sin\left(\frac{jk\pi}{n+1}\right),$$

$$\text{Fast cosine transform (DCT II)} \quad W_{jk} = \sqrt{\frac{2c_k}{n}} \cos\left(\frac{(j-1/2)(k-1)\pi}{n}\right),$$

$$\text{Fast Hartley transform} \quad W_{jk} = \frac{1}{\sqrt{n+1}}\left[\sin\left(\frac{2(j-1)(k-1)\pi}{n+1}\right) + \cos\left(\frac{2(j-1)(k-1)\pi}{n+1}\right)\right], \tag{4.14}$$

where $1 \le j, k \le n$, $c_1 = 1/2$ and $c_k = 1$ for $k > 1$. Note that both since and cosine transforms have other variants [393,484,440] including

$$\text{(DCT I)} \ W_{jk} = \sqrt{\frac{2c_j c_k c_{n+1-j} c_{n+1-k}}{(n-1)}} \cos\left(\frac{(j-1)(k-1)\pi}{n-1}\right), \tag{4.15}$$

$$\text{(DST II)} \ W_{jk} = \sqrt{\frac{2c_{n+1-k}}{n}} \sin\left(\frac{(j-1/2)(k-1)\pi}{n}\right), \tag{4.16}$$

$$\text{(DCT III)} \ W_{jk} = \sqrt{\frac{2c_j}{n}} \cos\left(\frac{(j-1)(k-1/2)\pi}{n}\right), \tag{4.17}$$

$$\text{(DCT IV)} \ W_{jk} = \sqrt{\frac{2}{n}} \cos\left(\frac{(2j-1)(2k-1)\pi}{4n}\right). \tag{4.18}$$

Refer to [441,104,440,393] and experiment with the supplied Mfile `schts.m`.

♦ **Geometry-induced circulant matrix.** In the literature, often, the reader meets beginning statements like 'let us consider the solution of a Fredholm integral equation of the convolution type' (as in Section 2.5.2) and may feel helpless as any real life problem is harder than this idealized (or contrived) model problem. Indeed this is true. It is therefore instructive to look at specific situations where such an equation is 'born' naturally. These situations turn out to be associated with simple geometries and simple kernel functions and we shall make full use of this observation. For complex geometries, by using geometric embedding techniques, we try to design an operator splitting that gives rise to a convolution part for preconditioning purpose. For very complex kernel functions, we split the kernel function to produce a convolution operator again for preconditioning purpose. In either case, we arrive at a circulant preconditioner. We give a few examples below and the intention is that the reader will be sufficiently motivated to think about this clever technique whenever applicable.

Boundary integral operators. All boundary integral operators of practical use appear to possess a convolution kernel (of some function of $r = |p - q|$) in the physical domain but *not normally* the computational domain. In fact, both the Jacobian function J and the distance function (depending on the boundary geometry) can ruin the convolution setting: J is not of convolution type unless it is a constant and r is not of convolution type unless the boundary is special. Therefore one way to seek a circulant preconditioner is to take a nearby domain whose boundary can be mapped to an interval $[a, b]$ in 1D or $[a, b] \times [c, d]$ in 2D to ensure the Jacobian to be constant and the kernel to remain a convolution type. In 1D, the only boundary satisfying these requirements is the familiar circle Γ_c because with

$$\Gamma_c : \qquad \begin{cases} x = x(t) = a\cos(t), \\ y = y(t) = a\sin(t), \end{cases} \qquad 0 \le t \le 2\pi,$$

$J = \sqrt{x'(t)^2 + y'(t)^2} = a$ is a constant and the kernel stays convoluting

$$\begin{aligned} |p - q| &= a\sqrt{(\cos(s) - \cos(t))^2 + (\sin(s) - \sin(t))^2} = a\sqrt{2(1 - \cos(s - t))} \\ &= a\sqrt{4\sin^2(\frac{s-t}{2})} = 2a\left|\sin(\frac{s-t}{2})\right|. \end{aligned} \tag{4.19}$$

Although domain embedding (similar to the so-called capacitance matrix method [84,391]) is an analytical and numerical technique in its own right, we may use domain embedding in the general sense to approximate the boundary Γ. The mathematical statement is the following

$$\min_{\Gamma_c} \max_{p \in \Gamma} \left| \int_\Gamma k(p, q)u(q)dS(q) - \int_{\Gamma_c} k(p, q)u(q)dS(q) \right|$$

Here Γ_c does not have to enclose Γ (especially when Γ is polygonal). Once

Γ_c is identified, one may construct a circulant preconditioner from discretizing $\int_{\Gamma_c} k(p,q)u(q)dS(q)$ directly.

Kernel function splitting. Consider the example of the single layer operator (1.73) in 2D (for either the Laplacian (1.72) or the Helmholtz case) that admits

$$
\begin{aligned}
-2\pi \int_{\partial\Omega} K(p,q)\psi(q)dS_q &= \int_{\partial\Omega} \log|p-q|\psi(q)dS_q + \int_{\partial\Omega} R_1(p,q)\psi(q)dS_q \\
&= \bar{J}\int_0^{2\pi} \log|s-t|\psi(t)dt + \int_0^{2\pi} \log|s-t|(J(t)-\bar{J})\psi(t)dt \qquad (4.20) \\
&+ \int_0^{2\pi} \log\sqrt{(\frac{x(s)-x(t)}{s-t})^2 + (\frac{y(s)-y(t)}{s-t})^2}\, J(t)dt \\
&+ \int_0^{2\pi} R_1(s,t)\psi(t)J(t)dt = \int_0^{2\pi} \log|s-t|\psi(t)dt + \int_0^{2\pi} R_2(s,t)\psi(t)dt.
\end{aligned}
$$

Here $\bar{J}$ denotes a constant that approximates the Jacobian $J(t)$, R_1 and R_2 combined other terms together. Clearly the first part in the last equation of (4.20), of convolution form, will lead to a circulant matrix. In view of (4.19) for a circular boundary, one may rewrite the above splitting differently to associate with a circulant boundary case

$$
\begin{aligned}
-2\pi \int_{\partial\Omega} K(p,q)\psi(q)dS_q &= \underbrace{\bar{J}\int_0^{2\pi} \log\left|\sin\frac{s-t}{2}\right|\psi(t)\,dt}_{\text{mimic a circulant } \partial\Omega} \qquad (4.21) \\
&+ \bar{J}\int_0^{2\pi} \log\frac{|s-t|}{\left|\sin\frac{s-t}{2}\right|}\psi(t)dt + \int_0^{2\pi} R_3(s,t)\psi(t)dt,
\end{aligned}
$$

as done in [497] and the references therein.

Wiener–Hopf equations. The half line Wiener–Hopf integral equation

$$
u(s) + \int_0^{\infty} k(s-t)u(t)dt = f(s), \qquad 0 \le s < \infty, \qquad (4.22)
$$

with $k(t) \in L_1(\mathbb{R})$ and $f \in L_2(\mathbb{R}^+)$, arises from many applications. The usual discretization is on the finite section equation [121]

$$
u_\tau(s) + \int_0^{\tau} k(s-t)u_\tau(t)dt = f(s), \qquad 0 \le s \le \tau. \qquad (4.23)
$$

Here the approximated finite section operator can be written as

$$(\mathcal{A}_\tau u)(s) = \begin{cases} \int_0^\tau k(s-t)u(t)dt, & 0 \le s \le \tau, \\ 0, & t > \tau. \end{cases}$$

A circulant integral operator suitable for preconditioning is defined by

$$(\mathcal{C}_\tau u)(s) = \int_0^\tau c_\tau(s-t)u(t)dt, \qquad 0 \le s \le \tau, \tag{4.24}$$

where c_τ is periodic and conjugate symmetric (to be specified below)

$$c_\tau(s+\tau) = c_\tau(s) \quad \text{and} \quad c_\tau(-s) = \overline{c_\tau(s)}, \quad -\tau \le s \le \tau.$$

Then $\mathcal{C}_\tau$ is a compact and self-adjoint operator in $L_1[-\tau, \tau]$ and the following preconditioned equation may be proposed [225]

$$(\mathcal{I} + \mathcal{C}_\tau)^{-1}(\mathcal{I} + \mathcal{A}_\tau)u_\tau(s) = (\mathcal{I} + \mathcal{C}_\tau)^{-1} f(s), \qquad 0 \le s \le \tau, \tag{4.25}$$

with the spectra of the circulant preconditioned operator $(\mathcal{I} + \mathcal{C}_\tau)^{-1}(\mathcal{I} + \mathcal{A}_\tau)$ clustered at 1 due to the compactness of $\mathcal{A}_\tau$ and the convergence of $\mathcal{C}_\tau - \mathcal{A}_\tau \to 0$ as $\tau \to \infty$. (See Lemma 4.7.12).

The construction of $\mathcal{C}_\tau$ follows a continuous analog of the Strang and T. Chan Toeplitz preconditioners, yielding two methods

$$\begin{aligned} (\mathcal{C}_1 u)(s) &= \begin{cases} \displaystyle\int_0^\tau s_\tau(s-t)u(t)dt, & 0 \le s \le \tau, \\ 0, & t > \tau. \end{cases} \\ s_\tau(t) &= k(t), \qquad -\tau/2 \le t \le \tau/2, \\ (\mathcal{C}_2 u)(s) &= \begin{cases} \displaystyle\int_0^\tau c_\tau(s-t)u(t)dt, & 0 \le s \le \tau, \\ 0, & t > \tau. \end{cases} \\ c_\tau(t) &= \frac{\tau - t}{\tau} k(t) + \frac{t}{\tau} k(t-\tau), \qquad -\tau/2 \le t \le \tau/2. \end{aligned} \tag{4.26}$$

Similar development for constructing a circulant preconditioner has also been carried out for the case of non-convoluting kernels [104].

4.7 Singular operator splitting preconditioners

Circulant preconditioners can be computed or inverted in $O(n \log n)$ operations using FFT and they are the definite ones to consider if the underlying operator is weakly singular. For strongly singular operators or operators with geometric singularities or for higher dimensional problems (say 3D), approximation based circulant preconditioners will not work as singularities must be separated to recover compactness.

We now consider how to develop operator splitting preconditioners, that are simple and can be inverted in $O(n)$ operations, for singular integral equations. Operator splitting is a widely used technique in solving singular BIEs; see [127,128,285,497] and the references therein. There are two main approaches. Both make use of the following elementary lemma (from functional analysis).

Lemma 4.7.12. *Let linear operators $\mathcal{A}$ and $\mathcal{C}$ be defined in a normed space with $\mathcal{A}$ bounded and $\mathcal{C}$ compact. Then*

1. *operator $\mathcal{A}^{-1}\mathcal{C}$ is also compact;*
2. *operator $\mathcal{D} = \mathcal{A} - \mathcal{C}$ has a bounded inverse if $\mathcal{D}$ is injective.*
3. *from property 1, $\mathcal{D} = \mathcal{A}(I - \mathcal{A}^{-1}\mathcal{C})$ is bounded provided $\lambda(\mathcal{A}^{-1}\mathcal{C}) \neq 1$.*

The first splitting approach (similar to the previous section) is based on expanding the singular kernel into a principal term of simple forms and a smooth part of remaining terms, giving rise to two splitting operators: the latter part gives rise to a compact operator while the former to a bounded operator. Further because of the simple forms in the kernel, fast algorithms (e.g., the FFT; see [104] and [497]) are used to invert the former operator which serves as a preconditioner.

Here we apply the second idea of operator splitting, previously used in [127]. This is based on domain decomposition rather than domain embedding or kernel decomposition. Let $\Omega \in \mathbb{R}^2$ denote[3] a closed domain that may be interior and bounded, or exterior and unbounded, and $\Gamma = \partial\Omega$ be its (finite part) boundary that can be parameterized by $p = (x, y) = (x(s), y(s)), a \leq s \leq b$. Then a boundary integral equation that usually arises from reformulating a partial differential equation in Ω can be written as

$$\alpha u(p) - \int_\Gamma \bar{k}(p, q)u(q)d\Gamma = f(p), \qquad p \in \Gamma, \tag{4.27}$$

or

$$\alpha u(s) - \int_a^b k(s, t)u(t)dt = f(s), \qquad s \in [a, b], \tag{4.28}$$

i.e., simply,

$$(\alpha I - \mathcal{K})u = f. \tag{4.29}$$

[3] The 3D case can be described similarly; see [16,131,15].

Here u may be a density function; see [16,24]. We do not assume that $\alpha \neq 0$ so our methods will work for both first and second kind boundary integral equations of the Fredholm type. For the latter type, $\alpha = 1/2$ at smooth points on Γ. We assume that $\mathcal{K}$ is the full operator (not just a principal part); for the Helmholtz equation this refers to the unique formulation which is valid for all wavenumbers (see [16] and Chapter 11). To solve the above equation numerically, we divide the boundary Γ (interval $[a, b]$) into m boundary elements (nonintersecting subintervals $I_i = (s_{i-1}, s_i)$). On each interval I_i, we may either approximate the unknown u by an interpolating polynomial of order τ that leads to a collocation method or apply a quadrature method of τ nodes and weights w_i, that gives rise to the Nyström method. Both discretization methods approximate (4.29) by

$$(\alpha I - \mathcal{K}_n)u_n = f, \tag{4.30}$$

where we can write

$$\mathcal{K}_n u = \mathcal{K}_n u_n = \sum_{j=1}^{m} \left[\sum_{i=1}^{\tau} w_i k(s, t_{ji}) u_{ji} \right], \; u_n(t_{ji}) = u(t_{ji}) = u_{ji}, \text{ and } n = m\tau.$$

We use the vector $\underline{u}$ to denote u_{ji}'s at all nodes. By a collocation step in (4.30), we obtain a linear system of equations

$$(\alpha I - K)\underline{u} = \underline{f}, \qquad \text{or} \qquad A\underline{u} = \underline{f}, \tag{4.31}$$

where matrices K and A are dense and unsymmetric (in general). The conditioning of A depends on the smoothness of kernel function $k(s, t)$. A strong singularity (as $t \to s$) leads to noncompactness of operator $\mathcal{K}$ and consequently the iterative solution of (4.31) requires preconditioning. $[a, b] = \bigcup_{i=1}^{m} I_i$. Accordingly we can partition the variable u and vector $\underline{u}$ as follows: $u = (u_1, u_2, \ldots, u_m)^T$ and $\underline{u} = (\underline{u}_1, \underline{u}_2, \ldots, \underline{u}_m)^T$. Similarly writing the operator $\mathcal{A}$ in matrix form, we obtain the splitting $\mathcal{A} = \alpha\mathcal{I} - \mathcal{K} = \mathcal{D} - \mathcal{C}$ with $\mathcal{D} = \mathcal{I} + \overline{\mathcal{K}}$) and

$$\overline{\mathcal{K}} = \begin{pmatrix} \mathcal{K}_{1,1} & \mathcal{K}_{1,2} & & & \mathcal{K}_{1,m} \\ \mathcal{K}_{2,1} & \mathcal{K}_{2,2} & \mathcal{K}_{2,3} & & \\ & \mathcal{K}_{3,2} & \ddots & \ddots & \\ & & \ddots & \ddots & \mathcal{K}_{m-1,m} \\ \mathcal{K}_{m,1} & & & \mathcal{K}_{m,m-1} & \mathcal{K}_{m,m} \end{pmatrix}.$$

Observe that all singularities of $\mathcal{A}$ are contained in the above operator and so the smooth operator $\mathcal{C}$ is compact. Note also that, after discretization, the corresponding matrix out of K is

$$\overline{K} = \begin{pmatrix} K_{1,1} & K_{1,2} & & & K_{1,m} \\ K_{2,1} & K_{2,2} & K_{2,3} & & \\ & K_{3,2} & \ddots & \ddots & \\ & & \ddots & \ddots & K_{m-1,m} \\ K_{m,1} & & & K_{m,m-1} & K_{m,m} \end{pmatrix}.$$

Also define matrix $B = \alpha I - \overline{K}$ and $C = K - \overline{K}$. Then from the above lemma, it can be shown that the operator $\mathcal{D}$ is bounded. Since the operator $\mathcal{B}^{-1}\mathcal{C}$ is also compact, we can use $\mathcal{P} = \mathcal{D}$ as an operator preconditioner and $P = B$ as a matrix preconditioner. Note that in [127] we propose a small ϵ based decomposition to define a minimal operator splitting that corresponds to a diagonal preconditioner for special cases.

Thus the solution of $A\underline{u} = \underline{f}$ is reduced to that of $P^{-1}A\underline{u} = P^{-1}\underline{f}$, i.e., $[I - P^{-1}C]\underline{u} = P^{-1}\underline{f}$. Here B is in general a block quasi-tridiagonal matrix and the solution of $B\underline{x} = \underline{y}$ is via $B = LU$, where L, U are of the same sparsity structure as B apart from the last row of L and the last column of U; see [17,128].

We remark that our techniques of constructing P, for singular operators, may be viewed as efficient regularization methods. Therefore from properties of compact operators, the preconditioned matrix $(I - P^{-1}C)$ and its normal matrix should have most of its eigenvalues clustered at 1. Many other sparse preconditioners (e.g. approximate inverses [253] (or Chapter 5), ILU type preconditioners) do not possess the latter property of the normal matrix having clustered eigenvalues because of the unsymmetric nature of the matrix (where $\lambda(P^{-1}A)$ and $\sigma(P^{-1}A)$ are not related).

For the case of nonsmooth boundaries, one can adjust the operator splitting accordingly to include the singularities in $\mathcal{D}$; see [132,420].

4.8 Preconditioning the fast multipole method

The FMM as introduced in Section 3.8 provides a fast and matrix-free method to compute $y = Ax$. As matrix A is not accessible and only a 'banded' part of A (falling into near fields) are computed, the natural way to precondition equation (1.1) is to accept this natural splitting to define a preconditioner [240,365].

Fortunately for many problems where the FMM is applicable, the underlying kernel function is singular so the near fields defined operator splitting provides a sufficient preconditioner. In fact, the full near fields may not be needed; see Section 5.8.

However for non-singular kernels or even smooth kernels (associated with infinite problems) [96], it remains a challenge to design an efficient FMM while extracting enough far field preconditioning information (that is normally not available). A related situation is to how to design a suitable FMM with proper preconditioners for oscillating kernels when oscillating basis functions are used e.g. [121,387].

4.9 Numerical experiments

To demonstrate some of the algorithms presented, we set aside four test matrices (supplied as `matrix0.mat`, `matrix1.mat`, `matrix2.mat`, `matrix3.mat` respectively)

(1) **Matrix** 0 with $\gamma = 1/4$ and **Matrix** 1 with $\gamma = 1$: The problem arises from discretizing an indefinite PDE with coefficients similar to other 'hard' problems in [411]

$$-\left(\gamma + \frac{\sin 50\pi x}{2}\right) u_{xx} - \left(\gamma + \frac{\sin 50\pi x \sin 50\pi y}{2}\right) u_{yy}$$

$$+ 20 \sin 10\pi x \cos 10\pi y u_x - 20 \cos 10\pi x \sin 10\pi y u_y - 20u = f(x, y),$$
$(x, y) \in [0, 1]^2$.

(2) **Matrix** 2: This test matrix comes from solving the PDE

$$(a(x, y)u_x)_x + (b(x, y)u_y)_y + u_x + u_y = \sin(\pi x y),$$

with discontinuous coefficients defined as [116,483,107]:[4]

$$a(x, y) = b(x, y) = \begin{cases} 10^{-3} & (x, y) \in [0, 0.5] \times [0.5, 1] \\ 10^3 & (x, y) \in [0.5, 1] \times [0, 0.5] \\ 1 & \text{otherwise.} \end{cases}$$

(3) **Matrix** 3: This is a complex matrix arising from using piecewise constants to discretize the boundary integral equation (1.82) i.e.

$$-\frac{1}{2}u + \mathcal{M}_k u + \alpha_k \mathcal{N}_k u = \overline{g} \text{ (known)},$$

on a boundary (a generalized peanut shape) as described in [205].

[4] We thank W. L. Wan [483] for providing the original test data from the work [116], from which the second problem is extracted.

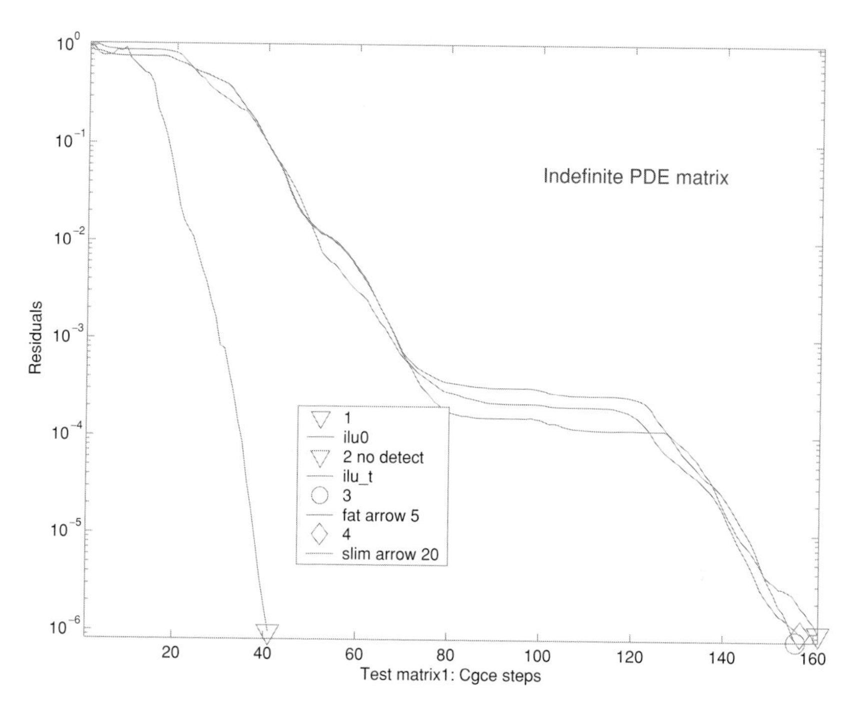

Figure 4.2. GMRES(50) results (`t_dete.m`) comparing banded arrow preconditioners with ILU preconditioners for Problem 1, which is mildly indefinite. With the help of `detect.m`, ILU(0) works fine. For `matrix0`, none of the four methods will work.

To test the effectiveness of Algorithms 4.2.2 and 4.5.10 when used with GMRES, we compare them with the unpreconditioned case and show some results in Figures 4.2–4.4. Clearly one observes that ILUT performs the best for problems which is not indefinite. The detection algorithm via `detect.m` may help an ILUT preconditioner as well as operator splitting types. The exact Mfiles used are given as `t_band.m` and `t_dete.m`.

4.10 Discussion of software and the supplied Mfiles

This chapter discussed the traditional forward type preconditioners with the operator splitting ones problem-dependent and ILU-related ones widely used. Many codes exist. The MATLAB default command `luinc` implements several variants of the ILU method. Efficient Fortran codes can be found in the following.

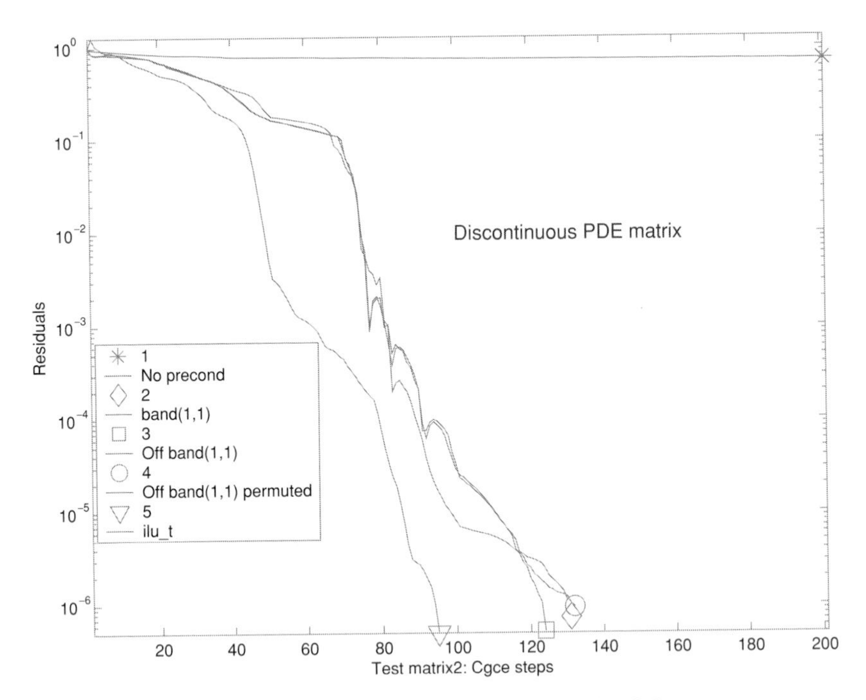

Figure 4.3. GMRES(50) results (t_band.m) comparing banded arrow preconditioners with ILU preconditioners for Problem 2. Discontinuity alone presents no problems with the ILUT preconditioner while no preconditioning ('*') does not work.

- **The Sparskit package**:

  http://www-users.cs.umn.edu/~saad/software/SPARSKIT/sparskit.html

  or directly `ftp : //ftp.cs.umn.edu/dept/sparse/`

- **The ILUM package**:

  http : //cs.engr.uky.edu/~jzhang/bilum.html

This book has supplied the following Mfiles for investigation.

[1] `banda.m` – Extract a banded arrow matrix for preconditioner. Both the diagonal and the bordering bands are allowed to vary.

[2] `bandb.m` – Extract a banded block matrix for preconditioner. Both the bandwidths and the block size are allowed to vary.

[3] `bandg.m` – Extract a far-field banded block matrix for preconditioner. (For the five-point Laplacian, this can extract the other far-field entries corresponding to the other derivative.)

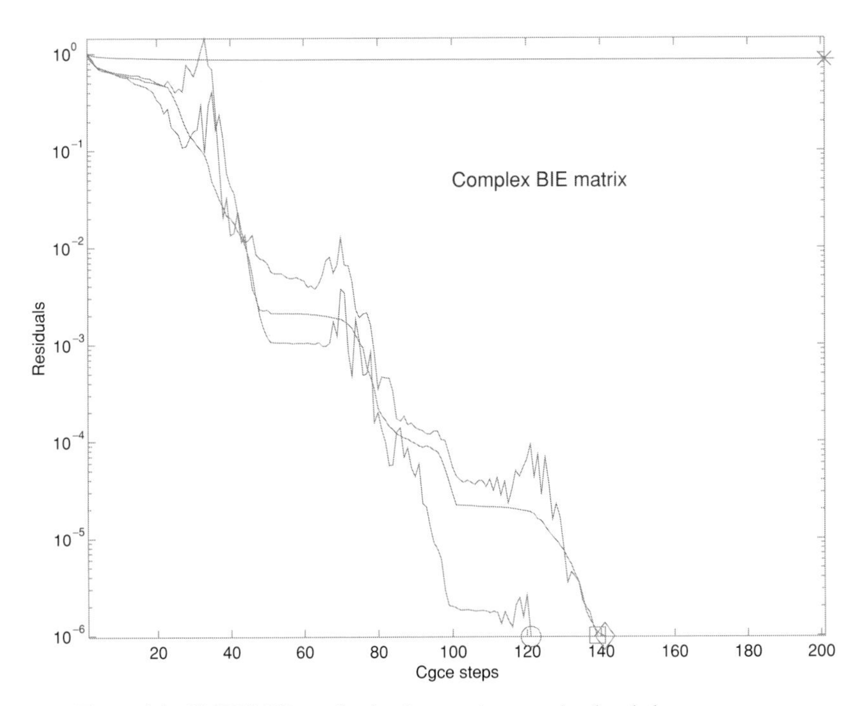

Figure 4.4. GMRES(50) results (t_dete.m) comparing banded arrow preconditioners with ILU preconditioners for Problem 3.

[4] circ_pre.m – Illustrate how the Strang and T Chan's approximations work for a Toeplitz matrix. The methods are also tested for a general matrix.

[5] detect.m – Detect locations of large features in a matrix for use in banda.m.

[6] t_band.m and t_dete.m – Two driver mfiles for testing the data files: matrix0.mat, matrix1.mat, matrix2.mat, matrix3.mat to display results in the format of Figures 4.2–4.4.

[7] ilu_0.m – Implement the Algorithm 4.5.9 for computing the ILU(0) preconditioner. (The same ILU factorization can also be obtained from the MATLAB command luinc.)

[8] ilu_t.m – Implement the Algorithm 4.5.10 for computing the ILU preconditioner by thresholding.

[9] multic.m – Implement the multicolour ordering Algorithm 4.3.3 for a permutation vector.

[10] schts.m – Illustrates various fast since/cosine transforms and their property of diagonalization of special matrices.

We comment that *if a reader has a different matrix to be tested, it is only a matter of commenting out the line relating to* `load` *of a matrix A to enable a new matrix to be used.*

Finally although ILU is the well-known choice for preconditioners, it should be remarked all *incomplete* factorizations might be developed for preconditioning purpose, e.g. the QR factorization [37] and the LQ factorization [411] (with LQ mimicking the transpose of QR).

5
Approximate inverse preconditioners [T2]: direct approximation of $A_{n\times n}^{-1}$

In the last few years we have studied preconditioning techniques based on sparse approximate inverses and have found them to be quite effective.

B. CARPENTIERI, *et al. SIAM Journal on Scientific Computing*, Vol. 25 (2003)

The objective is to remove the smallest eigenvalues of A which are known to slow down the convergence of GMRES.

JOCELYNE ERHEL, *et al. Journal of Computational and Applied Mathematics*, Vol. 69 (1996)

The most successful preconditioning methods in terms of reducing the number of iterations, such as the incomplete LU decomposition or symmetric successive relaxation (SSOR), are notoriously difficult to implement in a parallel architecture, especially for unstructured matrices.

MARCUS J. GROTE and THOMAS HUCKLE. *SIAM Journal on Scientific Computing*, Vol. 18 (1997)

This chapter will discuss the construction of **Inverse Type** preconditioners (or approximate inverse type) i.e. for equation (1.2)

$$MAx = Mb$$

and other types as shown on Page 3. Our first concern will be a theoretical one on characterizing A^{-1}. It turns out that answering this concern reveals most underlying ideas of inverse type preconditioners. We shall present the following.

Section 5.1 How to characterize A^{-1} in terms of A
Section 5.2 Banded preconditioner
Section 5.3 Polynomial $p_k(A)$ preconditioners
Section 5.4 General and adaptive SPAI preconditioners

5.1 How to characterize A^{-1} in terms of A

Computing A^{-1} exactly is a harder job than solving (1.1) in general. However, to solve (1.1) efficiently, we have to think about ways of estimating A^{-1} (as A^{-1} provides the ultimate preconditioner) while falling short of computing A^{-1} directly. We now address the question of characterizing A^{-1} in terms of the given matrix A, before considering any specific method. We first discuss the general approaches and then the subsequent applications in preconditioners' design. In line with the general notation in the literature, we shall use $\mathcal{S}(A)$ to denote the sparsity (nonzeros') pattern of matrix A and we are interested in $\mathcal{S}(A^{-1})$ or $\mathcal{S}(|A^{-1}| > \epsilon)$ for some ϵ if $\mathcal{S}(A^{-1})$ is dense.

♦ **Cayley–Hamilton theorem.** We first identify a spanned Krylov-like subspace where A^{-1} lies. Using the Cayley–Hamilton theorem, matrix A satisfies the characteristic polynomial

$$A^n + a_{n-1}A^{n-1} + \cdots + a_1 A + a_0 I = 0. \tag{5.1}$$

Hence, if A is non-singular, multiplying A^{-1} to (5.1) gives

$$A^{-1} \in \operatorname{span}(I,\ A,\ A^2, \ldots,\ A^{n-1}). \tag{5.2}$$

In terms of characterizing the sparsity pattern, we get

$$\mathcal{S}(A^{-1}) \subseteq \mathcal{S}\left((I + A)^{n-1}\right). \tag{5.3}$$

Similarly applying the same idea for the normal matrix $A^T A$ as in [290,147], yielding

$$A^{-1} \in \operatorname{span}\left(A^T,\ (A^T A)A^T,\ (A^T A)^2 A^T, \ldots,\ (A^T A)^{n-1} A^T\right). \tag{5.4}$$

The association of $\mathcal{S}(A^{-1})$ with $\mathcal{S}(A^T)$ may not always be a good idea because $\mathcal{S}(A^{-1}) = \mathcal{S}(A^T)$ if A is orthogonal (or diagonal) and $\mathcal{S}(A^{-1}) \neq \mathcal{S}(A^T)$ if A is triangular.

Having narrowed down the choices, the above results have not yet made the job of computing A^{-1} much easier. One hopes for such a subspace to be much smaller and definitely to remove the need of having to compute up to A^{n-1}.

♦ **Neumann's series approach.** An alternative way of expressing A^{-1} is to use the geometric series for $\rho(B) < 1$ (refer to Theorem 3.3.1)

$$(I - B)^{-1} = I + B + B^2 + B^3 + \cdots \tag{5.5}$$

assuming either

(1) $B = I - A,\ A^{-1} = (I - (I - A))^{-1} = (I - B)^{-1}$ or
(2) $B = I - A/\omega,\ A^{-1} = \omega(I - (I - A/\omega))^{-1} = (I - B)^{-1}$ for some scalar constant ω.

Here we note that to find sufficient conditions for the assumption $\rho(B) < 1$ to be valid, we require A be definite (i.e. $\Re(\lambda(A))$ *be of the same sign* which includes both positive definite or negative definite cases) due to the counter examples

$$A_1 = \begin{bmatrix} -0.5 & & \\ & 0.2 & \\ & & -0.5 \end{bmatrix} \text{ for (1) and } A_2 = \begin{bmatrix} 1 & -1 & \\ -1 & 1 & -1 \\ & -1 & 1 \end{bmatrix} \text{ for (2)},$$

because $\|A_1\| < 1,\ \rho(A_1) < 1$ but $\rho(B) = \rho(I \pm A_1) > 1$, and $\rho(A_2/100) < 1$ but $\rho(B) = \rho(I \pm A_2/100) > 1$.

Lemma 5.1.1. *If A is a definite matrix, then there exists a scalar constant ω such that $\rho(B) < 1$ for $B = I - A/\omega$.*

Proof. Without loss of generality, assume the real parts of eigenvalues of $A \in \mathbb{R}^{n\times n}$ are negative i.e. $\Re(\lambda(A)) < 0$. Let $A = U^T D U$ be the Schur decomposition with $D_{jj} = \lambda_j$. Consider bounding the eigenvalues of $\bar{B} = I - \frac{D}{\omega}$. Following the simple relationship ($\omega_j \in \mathbb{R}$)

$$\left| 1 - \frac{\Re(\lambda_j) + i\Im(\lambda_j)}{\omega_j} \right| < 1 \qquad \Longleftrightarrow \qquad \omega_j < -\frac{|\lambda_j|^2}{2|\Re(\lambda_j)|},$$

we choose $\omega = \min_j \omega_j$ to ensure that $\rho(\bar{B}) < 1$. Hence

$$B = I - \frac{A}{\omega} = U^T(I - \frac{D}{\omega})U = U^T \bar{B} U$$

satisfies $\rho(B) = \rho(\bar{B}) < 1$. ■

We emphasize the fact that whenever $\rho(I - A/\omega) \not< 1$ (even if $\rho(A) < 1$), one cannot rely on the Neumann's series to provide a way for polynomial preconditioning as (5.5) does not converge. This point is clearly stated in [375, p. 14]: *This (Neumann polynomial) method will be effective if the spectral radius of CD^{-1} is less than one.* Note that in [375] a diagonal scaling $D = \text{diag}(A)$ is first applied to A before using considering the Neumann's series; for the above example of A_2, such a scaling is not sufficient while it is effective for matrix A_1. A more general splitting of $A = M - N$ is studied in [187] where the Neumann's series is applied to $M^{-1}A = I - M^{-1}N$; however we must require $\rho(M^{-1}N) < 1$ for the series to converge. In theory, there exists M (e.g. $M \approx A$; see Remark 5.1.2) such that $\rho(M^{-1}N) < 1$ and both the inverse A^{-1} and its 'sparsity' pattern can be established from

$$A^{-1} = (I - M^{-1}N)^{-1}M^{-1} = \sum_{j=0}^{\infty} \left(M^{-1}N\right)^j M^{-1}. \tag{5.6}$$

As far as sparsity patterns are concerned, the pattern $\mathcal{S}(M^{-1})$ may be very close to be dense and higher powers of $M^{-1}N$ increase the density of nonzeros in $\mathcal{S}(A^{-1})$ e.g. if M is triangular, so is M^{-1} and if M is tridiagonal, then M^{-1} is dense.

In practice, the general difficulty of obtaining a converging splitting $A = M - N$, as shared by the task of constructing a preconditioner, lies in finding a simple and computationally efficient M. The idealized case is when $M^{-1}N$ is sparse while $(M^{-1}N)^j$ decays to zero very quickly as $j \to \infty$ so we can take a small number of terms in (5.6) in accurately approximating A^{-1} and equally we can take

$$\mathcal{S}(A^{-1}) \subset \mathcal{S}\left((I + M^{-1}N)^m\right), \qquad \text{with } m \text{ small, say } m = 2. \tag{5.7}$$

This section provides some guides on approximating A^{-1} as summarized below.

Remark 5.1.2.

(1) Polynomials up to high powers of A (nearly dense in general) always provide good approximations for A^{-1} and its 'sparsity pattern' – this property is not useful. To put in another way, the difficult preconditioning case will be when, for any small m,

$$\mathcal{S}(A^{-1}) \not\subseteq \mathcal{S}\left((I + A)^m\right).$$

(2) Polynomials up to low powers of A (nearly sparse in general even for dense matrices[1]) often provide good approximations for A^{-1} and its 'sparsity pattern' – this property is used in Section 5.3 for constructing a Chebyshev

[1] Assume a dense matrix is sparsified first. That is, we consider A_o^{-1} instead of A^{-1} with $A = A_o + E$ and E is 'small'. See [147].

polynomial preconditioner and in Section 5.4 for an approximate inverse preconditioner with polynomial guided patterns.

(3) The use of polynomials up to low powers of A to generate a priori sparsity patterns is known to be sensitive to scaling. This is much related to a possibly slow or no convergence of a Neumann's series. We anticipate that more work [190,191,340] towards scaling techniques will improve the inverse type preconditioners.

(4) The assumption of a convergent splitting $A = M - N$, with both M and M^{-1} sparse as well as $\rho(M^{-1}N) < 1$, may not be necessarily reasonable. To some extent, this is equivalent to assuming the existence of a suitable preconditioner of the forward type and then constructing an inverse type preconditioner!

Other methods of approximating A^{-1} will either be more problem dependent or arise from rewriting the forward types as inverse types i.e. from $PA = LU$ to $A^{-1}P^T = U^{-1}L^{-1}$ (§ 5.5).

5.2 Banded preconditioner

First we consider the special case where A is a block tridiagonal matrix with diagonal block dominance [179,351]. It is known that the inverse of such a matrix has off-diagonal elements exponentially decaying away from the main diagonal.

A banded matrix **band**(α, β) Section 2.5.1, including the tridiagonal matrix **band**$(1, 1)$, will be a suitable preconditioner of the inverse type. The procedure of computing M will follow from Section 5.4.2.

5.3 Polynomial preconditioner $p_k(A)$

We first remind ourselves that, unless a matrix A is suitably scaled, low-order polynomial preconditioners (based on the Neumann's series) do not necessarily work for any matrix. When a converging Neumann's series can be obtained (Section 5.1), we can relate it to low order polynomial preconditioners which in turn are linked to finding the best polynomial preconditioners. This is where Chebyshev polynomials come in. (A word of warning once again: as we have clearly stated, low order polynomials are not always the right thing to use so any blame on Chebyshev polynomials alone in this context is not justified.)

Once the best (low kth degree) polynomial of the inverse type $M = p_k(A)$ is targeted as our preconditioner i.e.

$$p_k(A)Ax = p_k(A)b \qquad \text{or} \qquad Ap_k(A)y = b, \; x = p_k(A)y, \tag{5.8}$$

with the preconditioner

$$p_k(A) = c_k A^k + c_{n-1} A^{k-1} + \cdots + c_1 A + c_0 I, \tag{5.9}$$

the selection problem for a diagonalizable matrix $A = XDX^{-1}$ (see (3.64)) becomes

$$\text{find } p_k \in \mathcal{P}_k : \quad \begin{cases} \min_M \|I - MA\|_\infty = \min_{p_k} \|I - p_k(A)A\|_\infty \\ \leq \kappa(X) \min_{p_k} \max_{\lambda \in \Lambda(A)} |1 - p_k(\lambda)\lambda| \\ = \kappa(X) \min_{q_{k+1} \in \mathcal{Q}_{k+1}} \max_{\lambda \in \Lambda(A)} |q_{k+1}(\lambda)|, \end{cases} \tag{5.10}$$

where $\mathcal{Q}_{k+1}$ is as defined previously in (3.40) and (3.64). It should be remarked that there do not appear yet to have existed any direct solutions of this min–max problem (as in Section 3.4).

For a fixed integer k, to use Chebyshev polynomials to find a solution to (5.10), we have to replace the discrete set $\Lambda(A)$ by a much larger and continuous set E. Although the Chebyshev approach can allow $\Lambda(A)$ to have a complex spectrum [257,309,410] (where we would enclose $\Lambda(A)$ by an ellipse domain), we restrict ourselves to the real case; let $\Lambda(A) \subset E = [\alpha, \beta]$. Then (5.10) is enlarged to

$$\text{find } q_{k+1} \in \mathcal{Q}_{k+1} : \quad \begin{cases} \min_M \|I - MA\|_\infty \\ \leq \kappa(X) \min_{q_{k+1} \in \mathcal{Q}_{k+1}} \max_{\lambda \in E} |q_{k+1}(\lambda)|, \end{cases} \tag{5.11}$$

which has the solution ($t \in [\alpha, \beta]$; see (3.45))

$$q_{k+1}(t) = T_{k+1}\left(\frac{\beta + \alpha - 2t}{\beta - \alpha}\right) \Big/ T_{k+1}\left(\frac{\beta + \alpha}{\beta - \alpha}\right),$$

if $\alpha > 0$ or A is SPD. Thus our polynomial preconditioner $p_k(A)$ will use the coefficients of

$$p_k(t) = \frac{1 - q_{k+1}(t)}{t}. \tag{5.12}$$

Then it is only a matter of algebraic manipulation to turn the three-term recursion (3.45) for T_k into one for generating p_k, q_{k+1} (or computing $y_j = p_k(A)Av_j$) recursively [410,413]. Clearly from the three-term recursion $T_{j+1} = 2tT_j(t) - T_{j-1}(t)$ with $T_0 = 1,\ T_1(t) = t$, we see that ($q_0(t) = 1$)

$$q_1(t) = \frac{\beta + \alpha - 2t}{\beta - \alpha} \Big/ \frac{\beta + \alpha}{\beta - \alpha} = 1 - \frac{2t}{\beta + \alpha} \text{ and } p_0(t) = \frac{1 - q_1(t)}{t} = \frac{2}{\beta + \alpha}. \tag{5.13}$$

Below we use the recursion of q_j's to derive similar formulae for $p_j = (1 - q_{j+1}(t))/t$. Defining $\theta = (\beta + \alpha)/2,\ \delta = (\beta - \alpha)/2$, we can rewrite q_{k+1} as

$$q_{k+1}(t) = T_{k+1}\left(\frac{\theta - t}{\delta}\right) \Big/ \sigma_k, \qquad \sigma_k = T_{k+1}\left(\frac{\theta}{\delta}\right). \tag{5.14}$$

Let $\rho_j = \sigma_j/\sigma_{j+1}$. We obtain ($\sigma_1 = \theta/\delta,\ \sigma_0 = 1,\ \rho_0 = 1/\sigma_1$)

$$\sigma_{j+1} = 2\frac{\theta}{\delta}\sigma_j - \sigma_{j-1}, \qquad \rho_j = \frac{\sigma_j}{\sigma_{j+1}} = \frac{1}{2\sigma_1 - \rho_{j-1}}, \qquad j \geq 1.$$

Hence from expanding T_{j+1} (identical to [413,(12.8)]) i.e.

$$\sigma_{j+1}q_{j+1}(t) = 2\frac{\theta - t}{\delta}\sigma_j q_j(t) - \sigma_{j-1}q_{j-1}(t)$$

we obtain the three-term recursion for the scaled Chebyshev polynomial q and $j \geq 1$ with $q_0(t) = 1$ and $q_1(t) = 1 - t/\delta$

$$q_{j+1}(t) = \rho_j\left[2\left(\sigma_1 - \frac{t}{\delta}\right)q_j(t) - \rho_{j-1}q_{j-1}(t)\right]. \tag{5.15}$$

To derive an explicit (recursive) formula for the preconditioning polynomial $p_j(t)$, we use (5.12) and (5.15) together with $p_{-1}(t) = 0,\ p_0(t) = 1/\theta$ or (note $\sigma_1\delta = \theta$)

$$p_1(t) = \frac{1 - q_2(t)}{t} = \frac{2\rho_1}{\delta}(2 - \frac{t}{\theta}), \quad \text{from} \quad q_2(t) = 1 + \frac{2\rho_1 t}{\delta\theta}\left(t - (\sigma_1\delta + \theta)\right).$$

We obtain the final formula for generating the required polynomial ($1 \leq j \leq k$)

$$\frac{1}{\rho_j}\frac{1 - q_{j+1}}{t} = (2\sigma_1 - \frac{2t}{\delta})\frac{1 - q_j}{t} + \frac{2}{\delta} - \rho_{j-1}\frac{1 - q_{j-1}}{t}, \quad \text{i.e.}$$

$$\boxed{p_j(t) = \rho_j\left[2\left(\sigma_1 - \frac{t}{\delta}\right)p_{j-1}(t) - \rho_{j-1}p_{j-2}(t) + \frac{2}{\delta}\right].} \tag{5.16}$$

Recall that our mission for finding $p_k(t)$ is to use it for solving (5.8); in the context of iterative solvers we require to form the matrix vector product

$$\mathbf{y} \equiv \mathbf{y}_k = MA\mathbf{v} = p_k(A)A\mathbf{v} = AM\mathbf{v} = Ap_k(A)\mathbf{v} = \mathbf{v} - q_{k+1}(A)\mathbf{v} \tag{5.17}$$

for any given $\mathbf{v} \in \mathbb{R}^n$ e.g. within a GMRES algorithm. However for either the left preconditioned or right preconditioned GMRES, we still need an explicit (recursive) formula to compute

$$\mathbf{x} \equiv \mathbf{x}_k = M\mathbf{v} = p_k(A)\mathbf{v}. \tag{5.18}$$

This is provided from (5.16)

$$\begin{cases} \theta = (\beta+\alpha)/2, \ \ \delta = (\beta-\alpha)/2, \ \sigma_1 = \theta/\delta, \\ \mathbf{x}_0 = \mathbf{v}/\theta, \ \rho_0 = 1/\sigma_1, \ \rho_1 = 1/(2\sigma_1 - \rho_0), \\ \mathbf{x}_1 = (2\rho_1/\delta)(2\mathbf{v} - A\mathbf{v}/\theta), \\ \begin{cases} \rho_j = 1/(2\sigma_1 - \rho_{j-1}), \\ \mathbf{x}_j = \rho_j \left[2\left(\sigma_1 \mathbf{x}_{j-1} - A\mathbf{x}_{j-1}/\delta\right) - \rho_{j-1}\mathbf{x}_{j-2} + 2\mathbf{v}/\delta \right], \ \text{for } j = 2, \ldots, k. \end{cases} \end{cases} \tag{5.19}$$

Algorithm 5.3.3. (Chebyshev polynomial preconditioner).

To compute $\mathbf{x}_k = p_k(A)\mathbf{v}$, *with* $\lambda(A) \in [\alpha, \beta]$ *and* $\alpha > 0$, *we do the following*

(1) set $\sigma_1 = \theta/\delta$, $\rho_0 = 1/\sigma_1$, $\rho = 1/(2\sigma_1 - \rho_0)$, $\mathbf{x}_0 = \mathbf{v}/\theta$ *and* $\mathbf{x}_k = 2\rho/\delta(2\mathbf{v} - A\mathbf{v}/\theta)$.
(2) update $\rho_0 = \rho$.
(3) **for** $j = 2, \ldots, k$
(4) *compute* $\rho = 1/(2\sigma_1 - \rho_0)$,
(5) *compute* $\mathbf{w} = A\mathbf{x}_k$,
(6) *update* $\mathbf{w} = \rho\,[2\,(\sigma_1\mathbf{x}_k - \mathbf{w}/\delta) - \rho_0\mathbf{x}_0 + 2\mathbf{v}/\delta]$,
(7) *update* $\mathbf{x}_0 = \mathbf{x}_k$, $\mathbf{x}_k = \mathbf{w}$, *and* $\rho_0 = \rho$,
(8) **end** j
(9) Accept $\mathbf{x} = \mathbf{x}_k$ *as the output.*

We have put details of this algorithm into a Mfile `cheb_fun.m` together with a driver Mfile `chebg.m` for readers' experiments of this method. Using the default example (in the Mfile) of a Laplacian PDE, one can see that increasing k from 1 to 8 would reduce the number of iteration steps steadily (and linearly). Actually with $k = -1, 0$, the iterations do not converge within the specified number of steps. While the ideas and the initial formulation are widely known, the specific algorithm is different from the usual form found in the literature where we focus on preconditioning rather than Chebyshev acceleration (as is often the case).

To achieve optimality in $\|I - p_k(A)A\|$ when $\alpha \le 0$, we have to change the norm used in (5.10) to the 2-norm with respect to chosen weight [413,337] and obtain a least squares polynomial. To allow complex eigenvalues (again $\Re(\lambda) > 0$), one may use a complex version of Chebyshev polynomials [413]. An alternative to using complex Chebyshev polynomials for real matrices is proposed in [216] where

$$q_{k+1}(t) = T_{k+1}\left(\frac{1-t}{\tilde{\rho}}\right) \Big/ \sigma_k, \qquad \sigma_k = T_{k+1}\left(1/\tilde{\rho}\right), \tag{5.20}$$

where $A = M - N$ is a convergent splitting for which $\rho(M^{-1}N) \leq \tilde{\rho} < 1$; as said, constructing such a M and then computing $\tilde{\rho}$ for a general matrix are not easy tasks. We shall not discuss these routes further. The work of [187,27] shows that the iteration gains are slower than the linear rate i.e. at most proportionally to the increased number of polynomial terms k; hence one may say that such preconditioners are not yet robust. Setting aside the issue of robustness, we recommend the use of a Chebyshev preconditioner for specialized problems where eigenvalue information is obtainable.

Nevertheless, as discussed in Section 5.1, there should be further scope for developing better polynomial-like preconditioners that may not use the continuous spectrum-based Chebyshev polynomials directly. For example, more work can be done towards getting a better distribution of discrete eigenvalue spectrum in association with operator preconditioning.

5.4 General and adaptive sparse approximate inverses

Instead of finding a polynomial $M = p_k(A)$ to approximate A^{-1}, the idea of computing a single and A^{-1}-like preconditioning matrix $M = P$ by minimizing $\|I - AM\|$ has been suggested since 1970s by several researchers; see [46,253, 312,320,321,48] and the references therein for historical notes. We shall mainly consider the right preconditioning for (1.1)

$$AMy = b, \quad x = My. \tag{5.21}$$

However the left preconditioning ($MAx = Mb$) and the double (two) sided preconditioning (5.64) will also be discussed.

5.4.1 Review of solution techniques for a least square problem

In this section, we shall make extensive use of solution techniques for a least square problem (when $m \geq n$)

$$Cz = f, \qquad C \in \mathbb{R}^{m \times n}, z \in \mathbb{R}^n, f \in \mathbb{R}^m. \tag{5.22}$$

Given an approximate solution z to (5.22), denote by $r = f - Cz$ the residual vector. Then noting $f^T Cz = z^T C^T f$, minimizing the quadratic functional (in $z \in \mathbb{R}^n$)

$$\Phi(z) = \|r\|_2^2 = r^T r = (f - Cz)^T (f - Cz) = z^T C^T Cz - 2z^T C^T f + f^T f$$

amounts to solving $\nabla\Phi(z) = 0$ and yielding [63,229]

$$C^T Cz = C^T f. \tag{5.23}$$

Assuming $rank(C) = n$, we obtain that

$$z = (C^T C)^{-1} C^T f \qquad \text{and} \qquad r = f - Cz = f - C(C^T C)^{-1} C^T f. \tag{5.24}$$

However the above result is only used as a theory. In practice, one solves (5.22) using the QR factorization of C (see Section 2.4). Let $C = QR$ be the reduced QR decomposition i.e. $Q \in \mathbb{R}^{m\times n}$, $R \in \mathbb{R}^{n\times n}$. Then $rank(C) = n$ implies that R^{-1} exists so the solution to (5.22) is given by

$$z = R^{-1} \hat{f}(1:n), \qquad \hat{f} = Q^T f, \tag{5.25}$$

and the residual vector takes on a very simple form

$$\begin{aligned} r = f - C(C^T C)^{-1} C^T f \; &= \; f - QR(R^T R)^{-1} R^T Q^T f \\ = f - QQ^T f \; &= \; (I - QQ^T) f \end{aligned} \tag{5.26}$$

where we used the facts $R(R^T R)^{-1} = (R^T R R^{-1})^{-1} = (R^T)^{-1}$, $Q^T Q = I$ but $QQ^T \neq I$ unless $m = n$. Again using $Q^T Q = I_{n\times n}$, we see that

$$Q^T r = Q^T f - Q^T Q Q^T f = 0. \tag{5.27}$$

Next we discuss how to compute an sparse approximate inverse (SPAI) preconditioner.

5.4.2 SPAI construction for a given pattern $\mathcal{S}$

Consider the right preconditioner M, given that the suitable M has a sparsity pattern $\mathcal{S}$ [162,321,232,150,253]. More precisely, let $\mathcal{N} = \{1, 2, \ldots, n\}$ and then $\mathcal{S}$ is a given set of index pairs (i, j). The SPAI preconditioner M will be sought as the best matrix that has the pattern $\mathcal{S}$ and minimizes the functional (in the Frobenius norm)

$$\min_M \|AM - I\|_F^2 = \min_M \|I - AM\|_F^2. \tag{5.28}$$

As studied by many researchers [46,147,148,162,232,253,290,320,321,451], the use of the F-norm naturally decouples (5.28) into n least squares problems

$$\min_{m_j} \|Am_j - e_j\|_2^2, \quad j = 1, \cdots, n, \tag{5.29}$$

where $M = [m_1, \ldots, m_n]$ and $I = [e_1, \ldots, e_n]$. Here we should note that (5.29) represents an usually small-sized least squares problem to solve (due to the sparsity of A), although n can be very large.

In detail, let the column vector m_j have k_j nonzeros with $k_j \ll n$ and let $\mathcal{J}$ denote the set of these k_j nonzero row indices in m_j. Then (5.29) is effectively reduced to a problem involving a submatrix A_j of A, consisted of the k_j columns of A corresponding to the nonzero positions of m_j. As both A and A_j are sparse, matrix A_j has only r_j (a small number) rows that are nonzero. Let $\mathcal{I}$ denote the set of the r_j row indices. Denote by $C_j = A(\mathcal{I}, \mathcal{J})$ such a $r_j \times k_j$ submatrix of A that is extracted from A. Assume $\hat{e}_j = e_j(\mathcal{I})$ is a reduced unit vector of e_j that is extracted corresponding to the r_j rows of A_j. Then (5.29) becomes

$$Am_j = e_j \quad \text{or} \quad A\hat{m}_j = e_j \quad \text{or} \quad C_j\hat{m}_j = \hat{e}_j, \quad j = 1, \cdots, n. \tag{5.30}$$

In MATLAB® notation, we have

```
>>  C_j = A(r, k),    e_hat = e(r)
```

if `r` denote the vector of the above r_j row indices and `k` the vector of the above k_j column indices. Here we claim that $r_j \geq k_j$ because we always insist on including the diagonal entries into the sparsity pattern $\mathcal{S}$. These diagonal entries in A_j ensure that C_j has at least k_j rows.

Using the QR approach of Section 2.4, problem (5.30) can be easily solved to give the solution

$$\hat{m}_j = R^{-1}\hat{c}(1 : k_j) \tag{5.31}$$

if $\hat{c}_{r_j \times 1} = Q^T \hat{e}_j$ and the QR decomposition of C_j is

$$C_j = Q\begin{bmatrix} R \\ 0 \end{bmatrix} = [Q_1\ Q_2]\begin{bmatrix} R \\ 0 \end{bmatrix} = Q_1 R. \tag{5.32}$$

The solution to (5.29) is given by $m_j(\mathcal{J}) = \hat{m}_j$ and $\|r\|_2 = \|\hat{c}(k_j + 1 : r_j)\|_2$ if the residual is denoted by

$$r = e_j - Am_j = e_j - A_j\hat{m}_j = e_j - A(:, \mathcal{J})\hat{m}_j. \tag{5.33}$$

This defines the right preconditioner M which is frequently used; construction of a left preconditioner can be similarly developed for the rows of M [232]. These SPAI preconditioners from (5.28) are essentially determined by the sparsity pattern $\mathcal{S}$ that approximates that of the inverse A^{-1}. However, there is no guarantee that $\|r\|_2 < \epsilon$ for small ϵ.

Remark 5.4.4. One interesting observation is the following. If $\mathcal{S}$ specifies a diagonal matrix, then matrix M is diagonal and contains the reciprocals of the 2-norms of all columns. Hence columns of the preconditioned matrix AM will be unit vectors.

Next we discuss the adaptive approaches of [232,253] that find a pattern $\mathcal{S}$ from any initial guess. See also [322] for another exposition. With adaptive approaches, we can guarantee that $\|r\|_2 < \epsilon$ for any small ϵ, provided that we are allowed to keep increasing k_j for column j. If very large k_j (say $k_j \approx n$) must be used i.e. when SPAI becomes ineffective, alternative methods in Section 5.6 are preferred.

5.4.3 Adaptive SPAI methods

Consider the solution of the least squares problem from (5.29) which can be solved exactly to produce a zero residual by

$$A\mathbf{m}_j = e_j, \tag{5.34}$$

for $j = 1, \ldots, n$ if $\mathbf{m}_j$ is allowed to be dense. Note $\mathbf{m}_j$ is the jth column of A^{-1}. However for practical purpose, we hope to achieve the following, using a pattern $\mathcal{S}$ as sparse as possible,

$$\|r\|_2 = \|Am_j - e_j\|_2 \leq \epsilon. \tag{5.35}$$

It remains to address how to find such an ideal pattern adaptively.

Assume that an initial pattern $\mathcal{S}$ is available (say using the diagonal matrix) but is not sufficient to ensure that the residual bound (5.35) is satisfied. *The idea of an adaptive approach is to enlarge the pattern $\mathcal{S}$ systematically* or to augment A_j by new columns of A (i.e. increase nonzero positions in m_j) adaptively to better approximate $\mathbf{m}_j$ so that (5.35) is eventually satisfied.

Each enlargement of $\mathcal{S}$ requires computing a new preconditioner M from solving (5.28). We hope to make use of the previously obtained information for sake of efficiency. Thus to add one new kth column $c = Ae_k$ to the existing matrix $C = A_j$, noting $A_j = A(:, \mathcal{J})$ and $C_j = A(\mathcal{I}, \mathcal{J})$, solve for vector z and scalar ξ

$$\sigma_{+c} = \min_{z \in \mathbb{R}^{k_j},\ \xi \in \mathbb{R}} \|Cz + \xi c - e_j\|_2^2 = \min_{z \in \mathbb{R}^{k_j},\ \xi \in \mathbb{R}} \left\| [C\ c] \begin{pmatrix} z \\ \xi \end{pmatrix} - e_j \right\|_2^2, \tag{5.36}$$

where $c = Ae_k$ is chosen [232] from the remaining columns of A that intersect with nonzero row indices of $[C\ e_j]$ (or of the residual vector $r = Am_j - e_j$) or more precisely

$$k \in \mathcal{J}_{\mathcal{N}} = \bigcup_{\ell \in L} \mathcal{N}_\ell$$

where $\boldsymbol{L}$ denotes the set of indices ℓ for which the current residual is nonzero i.e. $r(\ell) \neq 0$ and $\mathcal{N}_\ell$ is the column index set which corresponds to nonzero elements in row vector $A(\ell, :)$ and does not overlap with $\mathcal{J}$; see [253,232]. A simplified version [253] of (5.36) is the following approximation based on accepting $z = \hat{m}_j$ in (5.36) and solving for scalar ξ only

$$\sigma_{+c}^{\text{approx}} = \min_{\xi \in \mathbb{R}} \|Cz + \xi c - e_j\|_2^2 = \min_{\xi \in \mathbb{R}} \|\xi c - r\|_2^2, \ \xi = \frac{r^T c}{\|c\|_2^2}, \tag{5.37}$$

where $r = e_j - Cz = e_j - A_j \hat{m}_j$ is known.

Our objective here is to find the suitable column c to include in A_j so that the resulting residual will be smaller. This depends on what computable criteria are proposed to be used. At this stage, the solutions of these new least squares problems are not of interest yet.

♦ **Analytical solution of problem (5.36).** To propose a computable formula, we consider how to simplify (5.36) by re-using the QR decomposition of C and the known residual $r = e_j - C\hat{m}_j$. Note that both $C = A_j$ and $c = Ae_k$ are sparse so they can be reduced to compact forms. For brevity, we use the unreduced forms for C and c (of size n) and look for an update of the reduced QR decomposition:

$$[C\ c]_{n \times (k_1+1)} = [Y\ y] \begin{pmatrix} R & d \\ 0 & \rho \end{pmatrix}_{(k_j+1)\times(k_j+1)} \tag{5.38}$$

with $C = YR = Y_{n \times k_j} R_{k_1 \times k_j}$ the reduced QR decomposition of $C = A_j$. Assuming $C = A_j = YR$ is the previous QR decomposition, to compute y, d and ρ using the Gram–Schmidt idea, we equate both sides of (5.38)

$$c = Yd + \rho y.$$

Then using the orthogonality conditions $Y^T y = 0$ and $\|y\|_2 = 1$, we obtain

$$d = Y^T c, \quad \rho = \|c - Yd\|_2 = \|c - YY^T c\|_2 \quad \text{and} \quad y = (c - YY^T c)/\rho.$$

Note from $C_j = A(\mathcal{I}, \mathcal{J}) = QR = Q_1 R$ in (5.32) and $\hat{m}_j = R^{-1}\hat{c}(1 : k_j)$ from (5.31), we observe that

$$Y(\mathcal{I}, :) = Q_1 \quad \text{and} \quad Y^T e_j = Q_1^T e_j(\mathcal{J}) = (Q^T e_j(\mathcal{I}))(1 : k_j) = \hat{c}(1 : k_j)$$

and (5.36) becomes solving

$$\begin{cases} \begin{pmatrix} R & d \\ 0 & \rho \end{pmatrix} \begin{pmatrix} z \\ \xi \end{pmatrix} = \begin{pmatrix} \hat{c}(1 : k_j) \\ y^T e_j \end{pmatrix} \\ r_{\text{new}} = \left[I - (Y\ y) \begin{pmatrix} Y^T \\ y^T \end{pmatrix} \right] e_j = \left(I - YY^T - yy^T\right) e_j = r - yy^T e_j \end{cases} \tag{5.39}$$

where the second equation used (5.26) i.e. $r = e_j - C\hat{m}_j = (I - YY^T)e_j$.

We now concentrate on the quantity $\sigma_+ = \|r_{\text{new}}\|_2^2$ without solving individual least squares problem (5.39) for each candidate c. From the above derivation $y = (c - YY^T c)/\rho$ and the result (5.27) i.e. $Y^T r = 0$, we obtain that $e_j^T yyr = e_j^T yy^T e_j$, $y^T r = y^T d$ (due to $y^T Y = 0$) and

$$\sigma_{+c} = (r - yy^T e_j)^T (r - yy^T e_j) = r^T r - \|y^T r\|_2^2 = r^T r - \frac{\|c^T r\|_2^2}{\rho^2}.$$

♦ **Analytical solution of problem (5.37).** The least squares problem (5.37) for finding the best scalar $\xi \in \mathbb{R}$ can be solved directly by minimizing the quadratic function

$$\Phi(\xi) = \|\xi c - r\|_2^2 = (\xi c - r)^T(\xi c - r) = (c^T c)\xi^2 - 2(c^T r)\xi + r^T r \qquad (5.40)$$

that has a positive leading coefficient $c^T c = \|c\|_2^2$. From $\Phi'(\xi) = 0$, we obtain

$$\xi = c^T r/(c^T c) \qquad \text{and} \qquad \Phi = r^T r - (c^T r)^2/(c^T c)^2.$$

♦ **Adaptive SPAI strategies.** Thus the two minimization problems are solved to give the new residual norms

$$\sigma_{+c} = \sigma - \frac{(c^T r)^2}{\rho^2}, \qquad \sigma_{+c}^{\text{approx}} = \sigma - \frac{(c^T r)^2}{\|c\|_2^2} \qquad (5.41)$$

respectively, where $\sigma = \|r\|^2$ is the old residual norm and c is a candidate column of A. For an adaptive strategy, we shall loop through a list of candidate columns in the set $\mathcal{J}_N$ and select up to s columns that can make the new residual σ_{+c} (the Gould-Scott approach [232]) or $\sigma_{+c}^{\text{approx}}$ (the Grote–Huckle approach [253]) much smaller. Here for the former approach, one includes the new nonzero rows only rather than the full sized vector c. This is addressed below along with QR updates.

♦ **Adaptive QR updates from augmenting new columns.** Although we have already presented the Gram–Schmidt method for updating a *dense* QR factorization in minimizing σ_{+c}. Here we discuss the *sparse* versions using both the Gram–Schmidt and the Householder transform methods.

To improve the approximation of $\mathbf{m}_j$ by m_j that has a small number of nonzeros in set $\mathcal{J}$, in (5.34), we can use the above discussed adaptive strategies to augment set $\mathcal{J}$. Let $\widetilde{\mathcal{J}}$ denote the set of new row indices that are selected to add to m_j. Then we shall consider the submatrix $A(:, \mathcal{J} \cup \widetilde{\mathcal{J}})$ for a new least squares solution of type (5.29). Let $\widetilde{\mathcal{I}}$ be the set of new row indices, corresponding to the nonzero rows of $A(:, \mathcal{J} \cup \widetilde{\mathcal{J}})$ not already contained in $\mathcal{I}$. Denote by $\widetilde{r}_j$ and $\widetilde{k}_j = s$ the dimensions of sets $\widetilde{\mathcal{I}}$ and $\widetilde{\mathcal{J}}$ respectively. Then we consider how to update

$$C_j = A(\mathcal{I}, \mathcal{J}) = Q \begin{pmatrix} R \\ 0 \end{pmatrix} = [Q_1 \ Q_2] \begin{pmatrix} R \\ 0 \end{pmatrix} = Q_1 R \tag{5.42}$$

to find the QR decomposition for the augmented matrix

$$A_{\text{new}} = A(\mathcal{I} \cup \widetilde{\mathcal{I}}, \mathcal{J} \cup \widetilde{\mathcal{J}}) = \begin{pmatrix} A(\mathcal{I}, \mathcal{J}) & A(\mathcal{I}, \widetilde{\mathcal{J}}) \\ 0 & A(\widetilde{\mathcal{I}}, \widetilde{\mathcal{J}}) \end{pmatrix}, \tag{5.43}$$

where one notices that $A(\widetilde{\mathcal{I}}, \mathcal{J}) = 0$ as $A(\mathcal{I}, \mathcal{J})$ contains all nonzero rows already.

(1) **The Gram–Schmidt** approach. We shall carry out a column-by-column augmentation to find a reduced QR decomposition (see (5.38) for a single column case). For each new column $\ell = 1, \ldots, s$ (note $\widetilde{k}_j = s$), let the current matrix be $\overline{A}_\ell = Q_\ell R_\ell$ in a reduced QR form (to be augmented by a vector c_ℓ) and the new column vector c_ℓ brings $\hat{r}_\ell$ new nonzero rows. (As $\overline{A}_\ell$ is in sparse and compact form, it has be to augmented by $\hat{r}_\ell$ zero rows if $\hat{r}_\ell > 0$). Let $\overline{r}_{\ell+1} = \overline{k}_\ell + \hat{r}_\ell$ be the row dimension of matrix $\overline{A}_{\ell+1}$ for $\ell \geq 1$ with $\overline{r}_1 = r_j$, $\overline{r}_{\widetilde{k}_j+1} = r_j + \widetilde{r}_j$, and $\sum_{\ell=1}^{\widetilde{r}_j} \hat{r}_\ell = \widetilde{r}_j$ rows. Initially at $\ell = 1$, $\overline{A}_1 = C_j$ and $\widetilde{A}_1 = QR$.

Then the Gram–Schmidt approach proceeds as follows (for column $\ell = 1, \ldots, \widetilde{k}_j$)

$$\overline{A}_{\ell+1} = \begin{pmatrix} \overline{A}_\ell & c_\ell(1 : \overline{r}_\ell) \\ 0 & c_\ell(1 + \overline{r}_\ell : \overline{r}_{\ell+1}) \end{pmatrix}_{\overline{r}_{\ell+1} \times (k_j + \ell)} = \left(\begin{matrix} Q_\ell \\ 0 \end{matrix} \middle| y_\ell \right) \begin{pmatrix} R_\ell & d_\ell \\ 0 & \rho_\ell \end{pmatrix},$$

where the calculations were similar (5.38) (with Q augmented) i.e.

$$d_\ell = Q_\ell^T c_\ell(1 : \overline{r}_\ell), \quad \rho_\ell = \left\| \begin{pmatrix} (I - Q_\ell Q_\ell^T) c_\ell(1 : \overline{r}_\ell) \\ c_\ell(1 + \overline{r}_\ell) \end{pmatrix} \right\|_2, \quad \text{and}$$

$$y_\ell = \begin{pmatrix} (I - Q_\ell Q_\ell^T) c_\ell(1 : \overline{r}_\ell) \\ c_\ell(1 + \overline{r}_\ell) \end{pmatrix} \Big/ \rho_\ell.$$

(2) **The Householder** approach. Based on (5.42), the Householder approach (using the full QR decompositions) for (5.43) is simpler to describe.

$$A_{\text{new}} = \begin{pmatrix} \boxed{Q_1} & \boxed{Q_2} & \boxed{\begin{matrix} R \\ 0 \end{matrix}} & A(\mathcal{I}, \widetilde{\mathcal{J}}) \\ & 0 & & A(\widetilde{\mathcal{I}}, \widetilde{\mathcal{J}}) \end{pmatrix}$$

$$= \begin{pmatrix} Q_{r_j \times k_j} & \\ & I_{\widetilde{r}_j \times \widetilde{k}_j} \end{pmatrix} \begin{pmatrix} \boxed{\begin{matrix} R \\ 0 \end{matrix}} & \boxed{\begin{matrix} Q_1^T A(\mathcal{I}, \widetilde{\mathcal{J}}) \\ Q_2^T A(\mathcal{I}, \widetilde{\mathcal{J}}) \end{matrix}} \\ 0 & A(\widetilde{\mathcal{I}}, \widetilde{\mathcal{J}}) \end{pmatrix}$$

$$= \begin{pmatrix} Q_{r_j \times r_j} & \\ & I_{\widetilde{r}_j \times \widetilde{k}_j} \end{pmatrix} \begin{pmatrix} R & Q_1^T A(\mathcal{I}, \widetilde{\mathcal{J}}) \\ 0 & \boxed{\begin{matrix} Q_2^T A(\mathcal{I}, \widetilde{\mathcal{J}}) \\ A(\widetilde{\mathcal{I}}, \widetilde{\mathcal{J}}) \end{matrix}} \end{pmatrix} \quad \text{QR for the boxed matrix}$$

$$= \begin{pmatrix} Q & \\ & I \end{pmatrix} \begin{pmatrix} I_{k_j \times k_j} & \\ & \widetilde{Q}_{\ell \times \ell} \end{pmatrix} \begin{pmatrix} R & Q_1^T A(\mathcal{I}, \widetilde{\mathcal{J}}) \\ 0 & \boxed{\begin{matrix} \widetilde{R} \\ 0 \end{matrix}} \end{pmatrix}, \text{ with } \ell = (r_j - k_j) + \widetilde{r}_j.$$

♦ **The adaptive SPAI algorithm.** We have described all the ingredients for an adaptive algorithm. The algorithmic details can be summarized as follows.

Algorithm 5.4.5. (SPAI method [253]).

Given an initial sparsity pattern $\mathcal{S}$ (say from $diag(A)$), the following steps will work out an adaptive $\mathcal{S}$ and consequently an adaptive SPAI preconditioner M:

for *column $j = 1, \ldots, n$ of M*

(1) Let $\mathcal{J}$ be the present sparsity pattern of m_j for column j (set of nonzero row indices). Then $\mathcal{J}$ will determine the columns $\mathcal{J}$ of A for the least square problem (5.29).

(2) Find the set $\mathcal{I}$ of nonzero row indices in matrix $A(:, \mathcal{J})$.

(3) Compute the QR decomposition for $A(\mathcal{I}, \mathcal{J})$ as in (5.32).

(4) Find the least squares solution m_j from (5.31) and its residual r.

while $\|r\|_2 > \epsilon$

(5) *Let* **L** *be the set of indices* ℓ *for which* $r(\ell) \neq 0$.
(6) *Find the candidate index set* $\mathcal{J}_N$.
(7) *for each additional column* $\ell \in \mathcal{J}_N$*, find the new residual from (5.41).*
(8) *Decide on the s most profitable indices in* $\mathcal{J}_N$.
(9) *Determine the set of new indices* $\widetilde{\mathcal{I}}$ *and update the QR decomposition for the new sets* $\mathcal{I} = \mathcal{I} \cup \widetilde{\mathcal{I}}$ *and* $\mathcal{J} = \mathcal{J} \cup \widetilde{\mathcal{J}}$ *with* $A(\mathcal{I}, \mathcal{J})$ *of size* $r_j \times k_j$. *Obtain the new least squares solution* m_j *and the new residual* $r = e_j - Am_j$.

end while $\|r\|_2 > \epsilon$
end for *column* $j = 1, \ldots, n$ *of* M
(10) Accept the SPAI preconditioner $M = [m_1, \ldots, m_n]$.

For readers' convenience, we have provided the Mfile `spai2.m` to illustrate this algorithm; more optimal implementations are referred to in Section 5.10.

♦ **Some theoretical properties of the SPAI algorithm.** As shown in Chapter 3, the ultimate success of a preconditioner M lies in its ability to redistribute the spectrum of the preconditioned matrix AM (as far as an iterative solver is concerned). Here we summarize some results on the eigenspectrum $\lambda(AM)$ and the singular spectrum $\sigma(AM)$, following the work of [253] using the main SPAI residual bound (5.35) i.e. (note the number of nonzeros in m_j is k_j)

$$\|r\|_2 = \|\mathbf{r}_j\|_2 = \|Am_j - e_j\|_2 \leq \epsilon.$$

We now consider the overall approximation of M of A^{-1}.

Lemma 5.4.6. (Norm properties of SPAI [253]).
The SPAI residual error bound (5.35) implies that

(1) $\|AM - I\| \leq \sqrt{n}\epsilon$, $\|M - A^{-1}\| \leq \|A^{-1}\|_2\sqrt{n}\epsilon$ *in either F or 2-norm.*
(2) $\|AM - I\|_1 \leq \sqrt{p}\epsilon$, $\|M - A^{-1}\|_1 \leq \|A^{-1}\|_1\sqrt{p}\epsilon$,
where $p = \max_{1 \leq j \leq n}\{\text{number of nonzeros in } r_j\} \ll n$ *for a sparse matrix A.*

Proof. The proof essentially follows from definitions of norms, in particular[2]

$$\begin{cases} \|AM - I\|_F^2 = \displaystyle\sum_{j=1}^{n} \|(Am_j - e_j)\|_2^2, \\ \|AM - I\|_2 = \displaystyle\max_{\|x\|_2=1} \|(AM - I)x\|_2 == \max_{\|x\|_2=1} \left\| \sum_{j=1}^{n} (Am_j - e_j)x_j \right\|_2^2. \end{cases}$$

[2] Here the second line uses the usual relationship $Ax = [a_1, \ldots, a_n][x_1, \ldots, x_n]^T = \sum_{j=1}^{n} x_j a_j$.

Further enlarging the above quantities using (5.35) confirms the first two inequalities. To prove the first inequality in (2), we use the inequality (1.15); as r_j has at most p nonzeros, (1.15) becomes $\|r_j\|_1/\sqrt{p} \le \|r_j\|_2$ (demanding a similar simplification to (1.14)). Then

$$\|AM - I\|_1 = \max_{1\le j\le n} \|(Am_j - e_j)\|_1 \le \max_{1\le j\le n} \sqrt{p}\|(Am_j - e_j)\|_2 \le \sqrt{p}\epsilon,$$

proving the 1-norm result. Finally bounding $\|M - A{-}1\|$ comes from a simple rewriting: $M - A^{-1} = A^{-1}(AM - I)$. ■

Theorem 5.4.7. (Spectral properties of SPAI [253]).
Following Lemma 5.4.6, the SPAI method satisfying (5.35) has these properties

(1) the preconditioned matrix AM has a controllable departure of normality (3.70).
(2) the eigenvalues of AM, $\lambda(AM)$, are clustered at 1.
(3) the singular eigenvalues of AM, $\sigma(AM)$, are clustered at 1.

Proof. (1) Let $Q(AM)Q^T = \Lambda(AM) + N$ be a Schur decomposition of AM with $\Lambda(AM) = \text{diag}(\lambda_j(AM))$ and N upper triangular having zero diagonals. Then $Q(AM - I)Q^T = \Lambda(AM) - I + N$. From $\|AM - I\|_F^2 \le n\epsilon^2$, we see that $\|N\|_F^2 \le \|AM - I\|_F^2 \le n\epsilon^2$ and from (3.70) the preconditioned matrix AM has a controllable departure of normality that makes the study of $\lambda(AM)$ meaningful for the purpose of speeding up iterative solvers.

(2) As with (1), $\|\Lambda(AM) - I\|_F^2 \le \|AM - I\|_F^2 \le n\epsilon^2$. That is

$$\frac{1}{n}\sum_{j=1}^{n} \left|\lambda_j - 1\right|^2 \le \epsilon^2. \tag{5.44}$$

This implies that the eigenvalues are clustered at 1. Using Lemma 5.4.6 and an idea similar to Gerschgorin's theorem, we obtain, from $[(AM)^T - \lambda I]x = 0$ and $|x_k| = \max_j |x_j|$, that

$$\begin{aligned} |1-\lambda||x_k| &= \|(1-\lambda)x\|_\infty = \|(AM - I)^T x\|_\infty \qquad \text{let the max achieved at } \ell \\ &= |(Am_\ell - e_\ell)^T x| \le |x_k|\|Am_\ell - e_\ell\|_1, \quad \text{or} \end{aligned}$$

$$|1 - \lambda| \le \sqrt{p}\epsilon. \tag{5.45}$$

So the eigenvalues are clustered at 1 in radius $\sqrt{p}\epsilon$.

(3) Let the singular value be $\sigma = \sigma(AM)$ so $\sigma^2 = \lambda(AM(AM)^T)$. We consider

$$I - AM(AM)^T = (I - AM)^T + (I - AM)(AM)^T.$$

As transposes do not change 2-norms, we obtain the following

$$\begin{aligned}\|I - AM(AM)^T\|_2 &= \|(I - AM)^T + (I - AM)(AM)^T\|_2 \\ &\le \|(I - AM)\|_2 + \|(I - AM)\|_2\|AM\|_2 \\ &= \|(I - AM)\|_2 + \|(I - AM)\|_2\|(I - AM) - I\|_2 \\ &\le \sqrt{n}\epsilon\left(2 + \sqrt{n}\epsilon\right).\end{aligned}$$

Again using the Schur decomposition, we can obtain an identical equation to (5.44), proving that the singular values are clustered at 1. More precisely,

$$|1 - \sigma^2| \le \|I - AM(AM)^T\|_1 \le \sqrt{n}\|I - AM(AM)^T\|_2 \le n\epsilon\left(2 + \sqrt{n}\epsilon\right),$$

as with proving (5.45), where we have used (1.14). ■

Remark 5.4.8.

(1) If the quantity $\tau_\epsilon = \max_j k_j \le k_{\max} \ll n$ for a reasonably small ϵ (say 0.1), this indicates the success of Algorithm 5.4.5. That is to say, the algorithm is considered to have failed if $k_{\max} \approx n$ or ϵ is too large.
(2) The singular value clustering as shown in Theorem 5.4.7 suggests that CGN (Section 3.5) will be a serious method to consider when a SPAI preconditioner is developed.
(3) The adaptive methods described are not truly adaptive in a global sense. Theoretically minimizing σ_{+c} finds a better approximation than from $\sigma_{+c}^{\text{approx}}$, for the purpose of solving (5.28), at each step. However, there is no guarantee of a global minimizer for the essentially multi-dimensional problem – a situation somewhat mimicking the weakness of the steepest descent method [229]. The multi-dimensional problem may be posed as follows:

$$\sigma_{+C} = \min_{C;\, m_j} \|Cm_j - e_j\|_2^2 = \min_{c_k \in (a_1,\ldots,a_n);\xi_k \in \mathbb{R}} \|c_1\xi_1 + \cdots + c_\ell\xi_\ell - e_j\|_2^2$$

where a_k is column k of A, ξ_k's are the elements of $\hat{m}_j$ (i.e. the nonzero components of m_j) and hence C is chosen from any ℓ (a prescribed integer) columns of matrix A. Practically, for a class of sparse matrices, solving the multi-dimensional minimization is not necessary so both methods shown above (based on one-dimensional minimization) can work well. But it is not difficult to find surprising examples. For instance, the adaptive SPAI method of [253] applied to a triangular matrix A will not find a strictly triangular matrix as an approximate inverse unless one restricts the sparsity search pattern (Section 5.4.4) although this does not imply that SPAI will not lead to convergence of an iterative method [290]; of course, FSAI type methods will be able to return a triangular matrix as an approximation for this case. Although it remains to find an efficient way to solve the above multi-dimensional problem, narrowing down the choice for pattern $\mathcal{S}$ is regarded as an effective approach to speeding up the basic SPAI methods of type (5.28).

(4) The SPAI type may be improved using the polynomial preconditioners in the special case of a SPD matrix [98]. As stated before, there exist many other approaches for a SPD case [28].

5.4.4 Acceleration using a priori patterns

We now consider methods of selecting the initial sparsity pattern $\mathcal{S}$; this question was indirectly addressed in Section 5.1. There, we concluded that firstly the initial pattern (as well as the ideal pattern) $\mathcal{S}$ should be sought from the subspace as in (5.3) and secondly such a pattern can be accurately approximated by some low powers of A *if* A has been suitably scaled (which is in general a nontrivial task!).

For a class of problems, we hope that specifying suitable a priori patterns for the approximate inverse reduces or removes the need for any adaptive procedure and thus dramatically speeds up the SPAI preconditioner construction. With such a pattern $\mathcal{S}$, solving (5.28) yields the required preconditioner M. This is possible for many useful problems. For matrices arising from discretization of a class of partial differential equations, the so-called powers of sparsified matrices (PSM) methods have been found to give satisfactory and desirable patterns $\mathcal{S}$ [147,93,94,451]. The suggested procedure is as follows. Use 'drop' to denote a sparsification process; then $A_0 = drop(A)$. Here A_0 represents an initial sparsified matrix of A – necessarily for dense A and optionally for sparse A. As the number of nonzeros in high powers of A_0 can grow quickly to approach n^2, in practice, only low powers of A_0 are considered: use the pattern $\mathcal{S}$ defined by the graph of $A_i = A_0^{i+1}$ or, if less nonzeros are desired, $A_i = drop(A_{i-1}A_0)$. In this book we shall mainly use $i = 3$.

For matrices from boundary integral operators, the near neighbour patterns have been shown to be satisfactory [128,131,136,472]. The analytical approach of near neighbours is different from but related to the algebraic approach of PSM. That is to say, one can obtain approximately the pattern of near neighbours if adopting the (block-box) method of PSM. The coincidence again illustrates the usefulness of PSM.

Suitable scaling of a matrix is important before sparsification. In [147], symmetric scaling by diagonal matrices is suggested. However, our experience has shown that a better scaling method is the permutation and scaling method by Duff and Koster [190,191]. With this method, matrix A is permuted and scaled from both sides so that the resulting matrix has the largest entries in its rows and columns on the diagonal (there are related five algorithms). More precisely one variant of the method finds a maximum product transversal (permutation) that ensures the product of the diagonal elements

$$\prod_{j=1}^{n} |a_{p(j),j}|$$

is maximized; see also [53]. Using this method, the following matrix A can be scaled by two diagonal matrices and one column permutation to a better matrix B

$$A = \begin{pmatrix} 100 & 20 & & \\ 20 & 2 & -40 & \\ & 2 & 1 & 3 \\ & & 5 & 2 \end{pmatrix}, \qquad B = \begin{pmatrix} 1 & & 1 & \\ 1 & -1 & 1/2 & \\ & 1/20 & 1 & 1 \\ & 3/8 & & 1 \end{pmatrix}$$

$$= \begin{pmatrix} 1/10 & & & \\ & 1/2 & & \\ & & 1 & \\ & & & 3/2 \end{pmatrix} A \begin{pmatrix} 1/10 & & & \\ & 1/2 & & \\ & & 1/20 & \\ & & & 1/3 \end{pmatrix} \begin{pmatrix} 1 & & & \\ & & 1 & \\ & 1 & & \\ & & & 1 \end{pmatrix},$$

which is more amenable to sparsification by *global thresholding*. In fact, one can verify that the most important elements (the largest but always including the diagonals) of both matrix B^{-1} and B^3 follow some similar pattern $\mathcal{S}$. The software is available in Fortran; check the details of Harwell subroutine library (HSL) from `http://www.cse.clrc.ac.uk/nag/` and `http://www.numerical.rl.ac.uk/reports/reports.html`.

5.5 AINV type preconditioner

An alternative form of explicit sparse approximate inverse preconditioners to the single matrix M above is the factorized sparse approximate inverse (FSAI) preconditioner[3]

$$\min_{W,\ Z} \| W^T A Z - I \|_F^2, \tag{5.46}$$

which has a pair of upper triangular matrices $W,\ Z$, as proposed and studied in [320,321,319] for the symmetric case and [56,55,75] for the unsymmetric case where FSAI is more often referred to as AINV (sparse approximate inverse preconditioners). Here for symmetric A, $W = Z$ and the approach mimics the

[3] Although research papers with the abbreviations FSAI and AINV contain similar ideas and may be reviewed in the same framework, one notes that FSAI is synonymous with the authors of [320,321,319] and AINV with [56,55,50].

Cholesky factorization [229,280]; here we mainly consider the unsymmetric case. We remark that FSAI was originally proposed for preconditioning SPD matrices but can be adapted for unsymmetric cases, while AINV preconditioners were also designed for SPD matrices but later were generalized to unsymmetric cases [48].

In the unsymmetric case, the preconditioned system of (1.1) is the following

$$W^T AZy = W^T b, \quad x = Zy, \tag{5.47}$$

where $W^T AZ \approx I$. This preconditioning idea may be viewed as stemming from the bifactorization technique as discussed in Section 2.1.3 and [211,510]. As with ILU, we do not find the exact WAZ decomposition. The suggested approach is to develop an algorithm WAZT($\mathbf{p}$, τ), similar to ILUT(p, τ). Below we adapt Algorithm 2.1.3 for such a purpose.

Algorithm 5.5.9. (WAZT(p, τ) — $WAZ \approx I$ method).

(1) Set sparse matrices $W = Z = I$.
for $i = 1, \ldots, n$, do
(2) Set v to be the ∞-norm of row vector $A(i, :)$.
(3) $pq = W(:, i)^T A(:, i)$
for $j = i + 1, \ldots, n$, do
(4) $q_j = W(:, j)^T A(:, i)/pq$
(5) $p_j = Z(:, j)^T A(i, :)^T$
end j
for $k = 1, \ldots, i$
(6) $Z(k, i) = Z(k, i)/pq$
end k
for $j = i + 1, \ldots, n$, do
for $k = 1, \ldots, i$, do
(7) $W(k, j) = W(k, j) - q_j W(k, i)$
(8) Apply the W theshold: if $|W(k, j)|/v \le \tau$, set $W(k, j) = 0$.
(9) $Z(k, j) = Z(k, j) - p_j Z(k, i)$
(10) Apply the Z threshold: if $|Z(k, j)|/v \le \tau$, set $Z(k, j) = 0$.
end k
end j
(11) Update column $W(:, i)$ and $Z(:, i)$ by keeping only **p** *largest elements including the diagonal positions of W, Z.*
end i

For readers' convenience, we have developed the Mfile `waz_t.m` to illustrate this algorithm where we have included two default examples showing that the approximate WAZ decomposition is better than ILUT.

The AINV type preconditioners (Algorithm 5.5.9) have been compared with the SPAI (Algorithm 5.4.5) in [56] and also with the ILU type preconditioners (Algorithm 4.5.10) in [55]. The basic conclusions were that AINV offers more simplicity and robustness than ILU(0) but behaves similarly to ILUT. As with ILU, AINV preconditioners may not exist if breakdowns occur. One solution is to decompose $A + \alpha I$ instead of A; then one is not developing a preconditioner for A directly. Both ILU and AINV type preconditioners are cheaper than SPAI to construct, although the latter has much inherent parallelism to explore.

5.6 Multi-stage preconditioners

In this section, we consider one possible enhancement to SPAI preconditioners for problems where secondary preconditioners may prove useful. We shall mainly address the two-stage preconditioners as multi-stages may be prevented by sparsity requirements. Thus the setting is the following: given that M is a primary preconditioner for A, what can we do to further precondition matrix

$$A_1 = AM \tag{5.48}$$

assuming $AMy = b$ as in (5.21)) would benefit from further preconditioning.

Recall that Theorem 5.4.7 replies on the assumption of (5.35) i.e.

$$\|r\|_2 = \|Am_j - e_j\|_2 \leq \epsilon.$$

However (5.35) is not an easy task to achieve; as remarked in [253] such a bound is rarely satisfied in practice unless $k_{\max}$ (see Remark 5.4.8) is unrealistically close to n. In such a case of (5.35) not strictly satisfied, one can observe that most eigenvalues of $A_1 = AM$ are still clustered near 1 but there may be a few some small eigenvalues of A_1 that are very close to 0 and need to be deflated.

We now consider two types of two-stage methods for remedying this spectrum. We note in passing that (a) in general, A_1 has no large eigenvalues to be concerned about; (b) without the stage-1 preconditioner M, the following deflation method should be used with caution as $\lambda(A)$ may have eigenvalues on both sides of the imaginary axis (for indefinite problems) and a selective deflation may not be effective – this may explain the limited success of certain methods.

We first consider deflation techniques for matrix $A_1 = AM$, where M is the first stage preconditioner based on SPAI.

5.6.1 Deflating a single eigenvalue

Deflation of a single eigenvalue is a well-known idea which can be adapted for multiple eigenvalues. Briefly if u_1 is an unit eigenvector corresponding to eigenvalue $\lambda_1 = \lambda_1(A) \neq 0$, then the matrix

$$A_1 = AM_1 \qquad \text{with} \qquad M_1 = I + \sigma u_1 u_1^H \tag{5.49}$$

has all eigenvalues identical to A except one of them $\lambda_1(A_1) = \lambda_1 + \sigma\lambda_1$ e.g.

$$\begin{cases} \lambda_1(A_1) = 0, & \text{if } \sigma = -\lambda_1, \\ \lambda_1(A_1) = \lambda_1 + 1, & \text{if } \sigma = 1/\lambda_1, \\ \lambda_1(A_1) = 1, & \text{if } \sigma = \lambda_1^{-1} - 1, \end{cases}$$

To prove the result, the distinct eigenvalue λ_1 is easily dealt with as we can verify

$$A_1 u_1 = (\lambda_1 + \sigma\lambda_1)u_1, \tag{5.50}$$

while the others take some efforts. If A is symmetric, the task is relatively easy: since $u_1^H u_1 = u_1^T u_o = 0$ whenever $Au_o = \lambda_o u_o$ and $\lambda_o \neq \lambda_1$, proving that the other eigenvalues are identical follows easily from verifying $A_1 u_o = \lambda_o u_o$. However we must use unitary complements to u_1, in $\mathbb{C}^n$, for the unsymmetric case.

Lemma 5.6.10. (Deflation of a single eigenvalue [199]).
For any diagonalizable matrix A, the deflated matrix A_1 in (5.49) satisfies

$$A_1 u_o = \lambda_o u_o$$

whenever $Au_o = \lambda_o u_o$ and λ_o denotes any eigenvalue other than λ_1.

Proof. Let $Z = [u_1 \; W]$ be an unitary matrix i.e. $u_1^H W = 0$ and $u_1 \oplus W$ forming a basis matrix for $\mathbb{C}^n$ or W is the unitary complement of u_1 in $\mathbb{C}^n$. Then $A_1 W = AW$. Consider transforming matrices A, A_1, noting $u_1^H W = 0$, (5.50) and $Au_1 = \lambda_1 u_1$:

$$Z^H A Z = \begin{bmatrix} u_1^H \\ W^H \end{bmatrix} A[u_1 \ W] = \begin{bmatrix} u_1^H A u_1 & u_1^H A W \\ W^H A u_1 & W^H A W \end{bmatrix} = \begin{bmatrix} \lambda_1 & u_1^H A W \\ & W^H A W \end{bmatrix},$$

$$\begin{aligned} Z^H A_1 Z &= \begin{bmatrix} u_1^H \\ W^H \end{bmatrix} A_1[u_1 \ W] = \begin{bmatrix} u_1^H A_1 u_1 & u_1^H A_1 W \\ W^H A_1 u_1 & W^H A_1 W \end{bmatrix} \\ &= \begin{bmatrix} (\sigma+1)\lambda_1 & u_1^H A W \\ & W^H A W \end{bmatrix}. \end{aligned}$$

As Z is unitary (similar), the other $(n-1)$ eigenvalues of A and A_1 are identical; in particular they are the eigenvalues of matrix $W^H A W$. ■

Relevant to this work is adaption of (5.49) for deflating multiple eigenvalues of matrix $A_1 = AM$. Define $V = [v_1, \ldots, v_k]$ with $v_1, \ldots, v_k$ corresponding to k smallest eigenvalues $\lambda_1, \ldots, \lambda_k$ of matrix AM.

The generalization is only straightforward for the hypothetical case of matrix A_1 being symmetric

$$A_2 = AM_2 \qquad \text{with} \qquad M_2 = M + MVD_0V^H \tag{5.51}$$

where $D_0 = D_{k\times k} = \text{diag}(\lambda_1^{-1}, \ldots, \lambda_k^{-1})$. Then from $V^H V = I_{k\times k}$ and $V^H U = 0_{k\times(n-k)}$, for $U = [u_{k+1}, \ldots, u_n]$ with u_j's eigenvectors for other $(n-k)$ eigenvalues of AM, we can prove that

$$\lambda_j(A_2) = \begin{cases} \lambda_j(AM) + 1, & j = 1, \cdots, k \\ \lambda_j(AM), & j > k. \end{cases} \tag{5.52}$$

However, neither matrix A nor AM is in general symmetric. Hence, eigenvectors are *not* orthogonal to each other and one needs to construct an orthogonal basis in the subspace of the first k eigenvectors. We now consider two methods of deflating multiple eigenvalues in the unsymmetric case.

5.6.2 Deflation method 1: exact right eigenvectors

Suppose that we have computed the k normalized eigenvectors that correspond to the k smallest eigenvalues of $A_1 = AM$ and packed them as matrix V. Then

$$AMV = VD_1$$

with $D_1 = \text{diag}(\lambda_1, \ldots, \lambda_k)$. We propose to orthogonalize matrix $V = V_{n\times k}$ and decompose $V = UR$ with $U = U_{n\times k}$ orthogonal and $R = R_{k\times k}$ upper triangular. Then we have obtained a reduced Schur decomposition

$$AMU = UT \tag{5.53}$$

with $T = T_{k \times k} = R D_1 R^{-1}$ upper triangular. Clearly T and D_1 have the same eigenvalues. (Note that R would be an 'identity' matrix if A were symmetric).

To deflate the smallest k eigenvalues by shifting them by $+1$, we propose the following stage two preconditioner

$$A_2 = AM_2 \qquad \text{with} \qquad M_2 = M + MUT^{-1}U^H. \tag{5.54}$$

By comparing to (5.51), we observe that T^{-1} is upper triangular and

$$\operatorname{diag}(T^{-1}) = [\lambda_1^{-1}, \ldots, \lambda_k^{-1}].$$

From $U^H U = I_{k \times k}$ and

$$A_2 U = AMU + AMUT^{-1}U^H U = UT + UTT^{-1} = U(T + I),$$

we see that $\lambda_j(A_2) = \lambda_j(AM) + 1$ for $j = 1, \ldots, k$. To prove that the remaining eigenvalues are identical, one may use a unitary complements idea relating to an upper triangular matrix. That is, as in Lemma 5.6.10, let $Z = [U \ W]$ with $U^H W = 0$ and $U \oplus W$ forming a basis matrix for $\mathbb{C}^n$; e.g. see [199].

5.6.3 Deflation method 2: exact left and right eigenvectors

Here we consider a dual orthogonalization idea for deflation due to [95]. As the preconditioned matrix AM is unsymmetric, let $V_r = [x_1, \ldots, x_k]$ be the matrix of the k right eigenvectors and $V_l = [y_1, \ldots, y_k]$ be that of the k left eigenvectors, corresponding to the k smallest eigenvalues of AM. Then V_l and V_r are orthogonal to each other [492,95] i.e.

$$V_l^H V_r = \Lambda = \operatorname{diag}(y_1^H x_1, \cdots, y_k^H x_k) \tag{5.55}$$

and $S = V_l^H A V_r = V_l^H V_r D_1 = \Lambda D_1$ is a diagonal matrix (although $V_r^H V_r \neq I$). Note that the above equation reduces to the familiar result $V^T A V = D$ for a symmetric case (as $V_l = V_r = V$).

To deflate the smallest k eigenvalues by shifting them by $+1$, we can define a second stage two preconditioner (following [95])

$$A_2 = AM_2 \qquad \text{with} \qquad M_2 = M + MV_r S^{-1} V_l^H. \tag{5.56}$$

From $V_l^H V_r = S$ and

$$A_2 V_r = AMV_r + AMV_r S^{-1} V_l^H V_r = V_r D_1 + V_r D_1 S^{-1} \Lambda = V_r(D_1 + I),$$

we see that $\lambda_j(A_2) = \lambda_j(AM) + 1$ for $j = 1, \ldots, k$. To prove that the remaining eigenvalues are identical, one may use the matrix of remaining eigenvectors using the dual orthogonality i.e. the full form ($k = n$) of (5.55); see [95].

Remark 5.6.11. Finally, we comment on the implementation of the stage two preconditioner M_2 for both Method 1 and Method 2. As M_2 is an approximate inverse type preconditioner, it is not necessary to form M_2 explicitly and one simply uses M, T, S, U, V_l, V_r explicitly in matrix vector products.

5.6.4 Using approximate eigenvectors for deflation

The above two deflation methods can be efficient but at the same time expensive as computing quantities like T, S, U, V_l, V_r can be time-consuming unless k is kept small (say $k = 5$ or 10 as suggested in [95]).

Some approximation of these quantities can be provided by using the so-called Ritz or harmonic Ritz vectors [361]. Recall that after k steps of an Arnoldi method, we obtain from (1.28)

$$AQ_k = Q_k H_k + h_{k+1,k} q_{k+1} e_k^T.$$

Here the special matrix H_k, of size $k \times k$, has interesting properties (e.g. eigenvalues) resembling that of A. The *Ritz values* θ_k/*Ritz vectors* s_k and the *Harmonic Ritz values* $\hat{s}_k$/*Harmonic Ritz vectors* $\hat{\theta}_k$ are defined respectively by:

$$\begin{cases} H_j s_k = \theta_k s_k, \\ \left(H_j^T H_j + h_{j+1,j}^2 e_j e_j^T \right) \hat{s}_k = \hat{\theta}_k H_j^T \hat{s}_k, \end{cases}$$

as discussed in (1.29) and (1.30). Refer to [331,305,359].

For a relatively small k, computing these Ritz values/vectors is quite cheap. Therefore we can immediately develop a Ritz version of the deflation methods 1 and 2 as presented, viewing $\theta_k, \hat{\theta}_k \approx \lambda_k$. In fact this has been considered in [81]. Explicit deflation using Ritz vectors was also considered in [138] where restarted GMRES can be nearly as efficient as the full GMRES. Actually if a restarted GMRES (where Arnoldi factorizations are already available internally) is used, the deflation process can be combined with the solver itself; see [361,382] and the many references therein. The general result with these built-in (implicitly) deflation GMRES methods is the following: a restarted GMRES can be nearly as efficient as the full GMRES. Depending on one's interests and view points, this may be considered an interesting development but, equally, the full GMRES (even without counting the expensive work) may simply take many iterations for certain hard (say indefinite) problems – if so, different preconditioners may have be developed in addition to Ritz-based deflation methods. That is to say, *the so-called implicit deflation methods are only worthwhile if the full GMRES takes an acceptable number of iteration steps.*

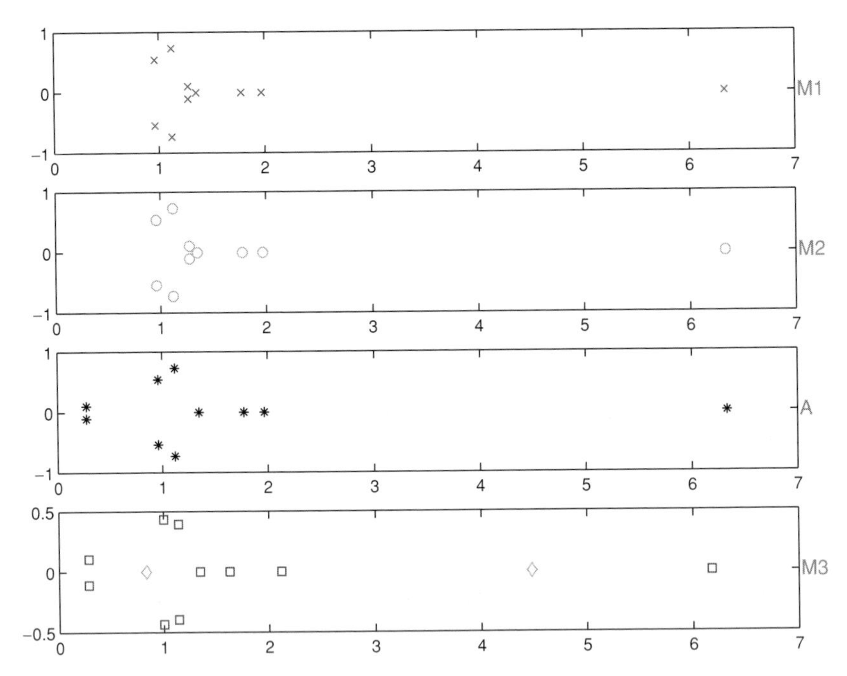

Figure 5.1. Exactly deflated eigenvalue spectra (Method 1 → plot 1 with symbol '×' and Method 2 → plot 2 with symbol '○') versus the approximated (Ritz vectors based) eigenvalue spectrum (plot 4 denoted by symbol '□'). Here symbol '◇' denotes the inaccurately approximated eigenvalues by the Ritz vectors.

When $A_1 = AM$ or A is unsymmetric, one inherent problem with the Ritz-type approach is that k must be reasonably large to see good approximation of *all* small eigenvalues. If k is small, isolated good approximations in spectrum will not fundamentally speed up an iterative solver such as GMRES. This observation can best be illustrated by Figure 5.1 with $k = 2$ and $n = 10$, where eigenvalues $\lambda(A)$ are shown in the third plot while preconditioned eigenvalues from Methods 1 and 2 are shown respectively in the first and second plots, and the fourth plot shows the preconditioned eigenvalues using the Ritz vectors as approximate eigenvectors. The reader can use the supplied Mfile `def_no.m` to see how this figure is generated. Clearly the Ritz case, the most unsuccessful, did shift some eigenvalues (actually the largest one instead of the smallest ones) but the poor performance is mainly due to the estimated eigenvalues (denoted by $\diamond$) are of poor quality. For this particular example, deflating more than $k = df = 4$ eigenvalues is not useful.

Deflating to a fixed value. If so desired one can adjust the above methods to shift k eigenvalues to α, for $\alpha > 0$ (e.g. $\alpha = 1$), rather than to $\lambda + 1$ i.e (5.54) and (5.56) become respectively

$$\begin{cases} \text{Method 1} \to A_2 = AM_2 \text{ with } M_2 = M + MU(\alpha T^{-1} - I)U^H, \\ \text{Method 2} \to A_2 = AM_2 \text{ with } M_2 = M + MV_r(\alpha S^{-1} - I)V_l^H, \end{cases} \tag{5.57}$$

where $U^H U = I_{k\times k}$ but $UU^T \neq I$ as $k < n$. We also note that [199] suggested to shift k eigenvalues to the largest eigenvalue λ_n i.e. $|\lambda(A_1)| \le |\lambda_n|$ and $\alpha = |\lambda_n|$:

$$\begin{cases} \text{Method 1} \to A_2 = AM_2 \text{ with } M_2 = M + MU(|\lambda_n|T^{-1} - I)U^H, \\ \text{Method 2} \to A_2 = AM_2 \text{ with } M_2 = M + MV_r(|\lambda_n|S^{-1} - I)V_l^H. \end{cases} \tag{5.58}$$

Here the shifting may not be useful if $\lambda_n \gg 1$.

Therefore we will not pursue this route further as not all of the deflated k eigenvalues (using Ritz vectors) are the smallest eigenvalues of matrix AM. Trying to resolve these issues will continue to be a future research direction.

5.6.5 Gauss–Jordan factorization-based stage-2 preconditioner

We now consider a method of designing a secondary (stage-2) preconditioner without involving the computation of explicit eigenvalues.

Our starting point is again (5.35): a SPAI preconditioner M in (5.28) approximates A^{-1} well if the underlying algorithm is successful in achieving (5.35), i.e. $\|Am_j - e_j\|_2 \le \epsilon$. If not, in general (apart from looking at smallest eigenvalues of $A_1 = AM$), the preconditioner M satisfies

$$AM = I + E = F \tag{5.59}$$

where F is approximately a special matrix with E having mostly zero columns except some dense ones. This matrix F resembles the product of some selected factoring matrices F_j from the Gauss–Jordan decomposition of matrix A (refer to Section 2.3, [42,143] or [298, p. 50]), where

$$F_n F_{n-1} \cdots F_1 A = I \tag{5.60}$$

and each F_j is an identity matrix with its column j containing a dense vector. We shall call a sparse matrix that is of the form of $F = I + E$, in (5.59), an elementary Gauss–Jordan matrix of order k if E has k dense columns.

One can verify that the inverse of an elementary Gauss–Jordan matrix retains the sparsity pattern of the original matrix. Moreover the essential work in computing the exact inverse of an elementary Gauss–Jordan matrix of order k is determined by the inverse of a $k \times k$ submatrix. We remark that, although the Gauss–Jordan method is well known [211,298], the Gauss–Jordan matrix decomposition implicitly defined is less commonly used. It is not difficult to see that any matrix A can be decomposed into a product of elementary Gauss–Jordan matrices whose orders can be summed up to n (assuming no partial pivoting is needed or some pre-permutation has been carried out). For instance, a simple form follows from rewriting (5.60) as $(F_j \ldots F_1)A(F_n \ldots F_{j+1}) = I$ or $A = (F_n^{-1} \ldots F_{j+1}^{-1})(F_j^{-1} \ldots F_1^{-1})$, where each bracket defines an elementary Gauss–Jordan matrix.

Example 5.6.12. (Partial Gauss–Jordan decomposition). *We give a simple example for $n = 6$ to demonstrate the partial decomposition of a matrix A using an order 4 elementary Gauss–Jordan matrix M_2 (to yield an order 2 matrix M_1^{-1}):*

$$M_2 A = \begin{bmatrix} 0.230 & -0.174 & 0.126 & -0.082 & 0 & 0 \\ -0.103 & 0.284 & -0.207 & 0.043 & 0 & 0 \\ -0.058 & -0.050 & 0.218 & -0.073 & 0 & 0 \\ -0.069 & 0.065 & -0.138 & 0.237 & 0 & 0 \\ -0.517 & 0.297 & -0.035 & -0.909 & 1 & 0 \\ -0.058 & -0.925 & 0.218 & 0.052 & 0 & 1 \end{bmatrix} \begin{bmatrix} 7 & 4 & 1 & 2 & 5 & 4 \\ 5 & 7 & 5 & 2 & 6 & 3 \\ 4 & 3 & 7 & 3 & 5 & 6 \\ 3 & 1 & 3 & 6 & 5 & 2 \\ 5 & 1 & 2 & 6 & 4 & 2 \\ 4 & 6 & 3 & 1 & 5 & 5 \end{bmatrix} = M_1^{-1}$$

$$= \begin{bmatrix} 1 & 0 & 0 & 0 & 0.329 & 0.993 \\ 0 & 1 & 0 & 0 & 0.371 & -0.716 \\ 0 & 0 & 1 & 0 & 0.136 & 0.783 \\ 0 & 0 & 0 & 1 & 0.539 & -0.435 \\ 0 & 0 & 0 & 0 & -1.520 & -1.200 \\ 0 & 0 & 0 & 0 & 0.511 & 3.410 \end{bmatrix} = \begin{bmatrix} 1 & 0 & 0 & 0 & 0.134 & -0.244 \\ 0 & 1 & 0 & 0 & 0.356 & 0.336 \\ 0 & 0 & 1 & 0 & 0.014 & -0.225 \\ 0 & 0 & 0 & 1 & 0.450 & 0.287 \\ 0 & 0 & 0 & 0 & -0.746 & -0.263 \\ 0 & 0 & 0 & 0 & 0.112 & 0.333 \end{bmatrix}^{-1}.$$

Here one observes that $M_2 A M_1 = I$ and the last matrix and its inverse are related through the smaller submatrix

$$\begin{bmatrix} -1.520 & -1.200 \\ 0.511 & 3.410 \end{bmatrix} = \begin{bmatrix} -0.746 & -0.263 \\ 0.112 & 0.333 \end{bmatrix}^{-1}.$$

The above example prompts us to consider situations where matrix AM_1 (or $M_2 A$) is approximately an elementary Gauss–Jordan matrix. If this is the case, we may naturally employ another elementary Gauss–Jordan matrix M_2 (or M_1) to achieve $M_2 A M_1 \approx I$. We shall use this idea to propose a two-stage preconditioner based on preconditioner M from (5.28) for (1.1).

A new two-stage preconditioner. We now formulate our two-stage preconditioner in detail. Consider the problem of post-processing the SPAI preconditioned linear system (rewritten from (5.21))

$$AM_1 y = b, \tag{5.61}$$

where the right preconditioner $M_1 = [m_1, \ldots, m_n]$ does not satisfy

$$\|Am_j - e_j\|_2 \le \epsilon \tag{5.62}$$

in k columns. That is to say,

$$AM_1 = I + E_1 + E_2, \tag{5.63}$$

where $\|E_2\|_F$ is very small and $I + E_1$ is an elementary Gauss–Jordan matrix with k dense columns (not necessarily small). We propose to further precondition (5.63) by $M_2 = (I + E_1)^{-1}$

$$M_2AM_1 y = M_2 b, \quad x = M_1 y. \tag{5.64}$$

The idea amounts to implementing a thresholding to the right preconditioned matrix $A_p = AM = AM_1$ and to seek a secondary left preconditioned matrix M_2 such that $M_2 A_p \approx I$ as with an usual SPAI preconditioner. This preconditioner will be effective as the preconditioned matrix

$$M_2AM_1 = (I + E_1)^{-1}(I + E_1 + E_2) = I + (I + E_1)^{-1}E_2$$

is expected to be a smaller perturbation of I than (5.63). For a simple case, we can establish this statement more precisely. To show that our new method (5.64) defines a better preconditioner than the standard SPAI method (5.63), we present the following.

Lemma 5.6.13. *Consider equations (5.61) and (5.64) in the simple case of* $\epsilon \le \|E_1\|_F < 1$ *and* $\|E_2\|_F < \epsilon < 1$ *with* $\|E_1\|_F + \|E_2\|_F < 1$. *Then the new two-stage preconditioner (5.64) satisfies*

$$\|M_2AM_1 - I\|_F < \|AM_1 - I\|_F < 1.$$

Proof. First of all, the standard SPAI preconditioner (5.63) satisfies

$$\|AM_1 - I\|_F = \|E_1 + E_2\|_F \le \|E_1\|_F + \|E_2\|_F.$$

Now from $\|E_1\|_F < 1$, we have $\|(I + E_1)^{-1}\|_F \le 1/(1 - \|E_1\|_F)$. Then our new preconditioner (5.64) satisfies

$$\|M_2AM_1 - I\|_F = \|(I + E_1)^{-1}E_2\|_F \le \frac{\|E_2\|_F}{1 - \|E_1\|_F}.$$

As $\|E_1\|_F + \|E_2\|_F < 1$, from

$$\frac{\|E_2\|_F}{1 - \|E_1\|_F} - \left(\|E_1\|_F + \|E_2\|_F\right) = \|E_1\|_F \frac{\|E_1\|_F + \|E_2\|_F - 1}{1 - \|E_1\|_F} < 0,$$

One sees that our new method (5.64) defines a better preconditioner than the standard SPAI method (5.63). ■

For instance, if $\|E_1\|_F = 0.9, \quad \|E_2\|_F = 0.001$, then

$$\|AM_1 - I\|_F \leq \|E_1\|_F + \|E_2\|_F = 0.901, \quad \|M_2AM_1 - I\|_F \leq \frac{\|E_2\|_F}{1 - \|E_1\|_F} = 0.01.$$

We now consider the implementation issue. Note that after permutation, the matrix $(I + E_1)$ can be written in the form

$$\begin{pmatrix} A_1 & 0 \\ A_2 & I_2 \end{pmatrix} \tag{5.65}$$

where A_1 is a matrix of size $k \times k$, A_2 of $(n-k) \times k$, I_2 of $(n-k) \times (n-k)$. The exact inverse of this matrix is

$$\begin{pmatrix} A_1^{-1} & 0 \\ -A_2A_1^{-1} & I_2 \end{pmatrix}.$$

This suggests that we approximate A_1^{-1} in order to work out the left approximate inverse M_2 in (5.64). The overall algorithm can be summarized as follows.

Algorithm 5.6.14. (Partial factorization based two-stage preconditioner).

(1) *For a tolerance tol and integer $k_{\max} = nzmax$ (the maximal number of nonzeros allowed per column such that $nzmax = NNZ/n$, where NNZ denotes the total number of nonzeros in a sparsified version of A), implement the SPAI algorithm, solving (5.28), to obtain the stage-1 right preconditioner M_1.*
(2) *Find the k sparsified columns of AM_1 that do not satisfy the tolerance tol.*
(3) *Compute the SPAI approximation for A_1.*
(4) *Assembly the stage-2 left preconditioner M_2.*
(5) *Solve the preconditioned system (5.64) by a Krylov subspace method.*

Here sparsification of AM_1 is important as it usually much less sparse than A and M_1 alone, and hence A_1 is implicitly sparsified. Note that in the adaptive SPAI approaches (Section 5.4.3), AM_1 is available in monitoring residual vectors r_j and hence our algorithm can be coupled naturally with an adaptive SPAI approach.

It should also be remarked that, instead of the two-sided scheme of (5.64), we can similarly propose the second stage preconditioner M_2 differently (from the right)

$$AM_1M_2y = b, \quad x = M_1M_2y. \tag{5.66}$$

However the computation of M_2 must be done as a left preconditioner for AM_1 in order to use a convenient sparsity pattern $\mathcal{S}$. All other discussions will follow as well. However we shall mainly study (5.64) in this work.

We now discuss the issue of complexity and the choice of A_1. Clearly the size k of matrix A_1 is an indication of the stage of difficulties in approximating A^{-1} by M_1. In most cases where our algorithm is particularly useful, we can assume that k is small. Then we may use a direct solver to compute A_1^{-1}. Thus the additional cost of using M_2 is simply $O(k^3) + O(nk^2) \approx 2k^3 + 2nk^2$. However for large k (e.g. $k = n$), we simply call an existing SPAI solver for the second time and the overall cost may be doubled. As with all SPAI preconditioners, in practical realizations, one should use parallel versions of Algorithm 5.6.14 to gain efficiency. Note that one may also take $M_2 = A_1^{-1}$ directly in (5.65) and implement the stage two preconditioner $y = M_2x$ from solving $A_1y = x$, and this gives rise to a mixed preconditioning strategy.

One simplification of A_1 may result from selecting at most a fixed number k_{fix} columns of AM_1 that have the largest least-squares-errors $\|Am_j - e_j\|_2$. For example, set $k_{\text{fix}} \le nzmax$. However for some extremely hard problems, this selection may not be sufficient.

Another possibility is to reset these identified $n - k$ columns of M_1 to unit vectors and then $AM_1 = I + E_1 + E_2$ becomes more pronounced as a Gauss–Jordan decomposition. A drawback of this approach is a possible scaling problem associated with matrix AM_1 thus complicating the further approximation by M_2.

For a general method, leaving out the issue of complexity, we expect a continuing repeated application of the SPAI idea will asymptotically generate an identity matrix

$$\cdots M_{2t} \cdots M_2 A M_1 \cdots M_{2t-1} \cdots = I.$$

In this case, intermediate products are not sparsified. Thus one envisages that a difficult matrix problem may need more than two preconditioning matrices. However we have not investigated this possibility further.

Remark 5.6.15. We have used the term 'two-stage', instead of 'two-level' [95], to differentiate it from the methods of the next chapter. In finding the approximate inverse of matrix (5.65), one might apply a SPAI method to the whole matrix (instead of applying to submatrix A_1). In this case, care must be taken to ensure that the zero positions in the right $(n - k)$ columns are not filled otherwise the stage 2 preconditioner M_2 will not be effective or in other words, the sparsity pattern should be restricted within the sparsity defined by $I + E_1$. Note also that this idea becomes less attractive for formulation (5.66), if taken as a secondary right preconditioner, because the second preconditioner may not be allowed to contain dense columns for the sake of efficiency; we recommend a secondary left preconditioner.

Although we are concerned with unsymmetric systems, similar two-stage preconditioning strategies based on triangular preconditioners for FSAI in solving symmetric systems have been suggested in [312,320,319]. There the choice of the second preconditioner M_2 is made to approximate a banded form of $M_1^T A M_1$. There does not appear to be any two-stage work generalizing the FSAI formulation (5.47) for unsymmetric systems.

5.7 The dual tolerance self-preconditioning method

A potentially useful preconditioning method is actually provided by the underlying problem itself. The idea, within the general framework of approximate inverse preconditioners, was suggested in [507] for the FOM (3.58) and also seen in the recent work of [96].

Consider the preconditioned equation for (1.1), of the type (1.2):

$$AMy = b, \qquad M = \tilde{A}^{-1} \approx A^{-1}, \; x = My, \tag{5.67}$$

and a typical preconditioning step $z = My$

$$\tilde{A}z = y. \tag{5.68}$$

As $\tilde{A} \approx A$, the solution of equations (5.67) and (5.68) can be sought from the same iterative solver with different tolerances ϵ_m, ϵ_p respectively (hence the name 'dual tolerance'). Clearly we take $\epsilon_p \gg \epsilon_m$ to focus on the main equation (5.67).

Assume the GMRES(k) method is used. Then for the main equation, effectively, we change the preconditioner per inner step j. Therefore we have to use the flexible GMRES (FGMRES) variant, which we now introduce, of Algorithm 3.6.18. This amounts to solving

$$AM_j w = b, \qquad x = M_j w, \tag{5.69}$$

with M_j depending the inner iteration index j. The Arnoldi decomposition (1.23), or (1.28), now aims to achieve

$$A\left[M_1q_1, M_2q_2, \cdots, M_kq_k\right] = \left[q_1, q_2, \cdots, q_k\right] H_k + h_{k+1,k}q_{k+1}e_k^T, \tag{5.70}$$

for the GMRES(k) method

$$w_k = w_0 + [q_1, q_2, \cdots, q_k]\, y = w_0 + \sum_{j=1}^{k} y_j q_j,$$

$$x_k = x_0 + [M_1 q_1, M_2 q_2, \cdots, M_k q_k]\, y = x_0 + \sum_{j=1}^{k} y_j M_j q_j,$$

with y computed from (3.57).

Algorithm 5.7.16. (Flexible GMRES(k)).

To solve $AM_j y = b$ for $x = M_j y$ for $j = 1, 2, \ldots, k$ by GMRES(k), with iter $=$ 0, $e_1 = (e_1)_{n\times 1}$ and given an initial starting vector $x = x_0$ and TOL,

(1) *Set $x_0 = x$, $iter = iter + 1$. Compute $r = b - Ax_0$;*

(2) *Generate the first vector $q_1 = r/\|r\|_2$ and the right-hand-side vector* $\texttt{rhs} = \|r\|_2 e_1$;

for $i = 1 : k$,

(3) *Start step i of a modified Gran–Schmidt method for Arnoldi: $z_i = M_i q_i$ and $w = Az_i$;*

(4) **for** $\ell = 1 : i$

$R(\ell, i) = w^T q_\ell; \qquad w = w - R(\ell, i) q_\ell;$

end

(5) *$R(i+1, i) = \|w\|_2$ and $q_{i+1} = w/R(i+1, i)$;*

(6) *Apply the rotations to past rows of new column i:*

for $\ell = 1 : i - 1$,

$t = c_\ell R(\ell, i) + s_\ell R(\ell + 1, i); \qquad R(\ell + 1, i) = s_\ell R(\ell, i) - c_\ell R(\ell + 1, i);\ R(\ell, i) = t;$

end

(7) *Compute the Givens rotations:*

if $|R(i+1, i)| \le 10^{-16}$, set $c_i = 1$ and $s_i = 0$,

else

if $|t| \le 1$ for $t = R(i+1, i)/R(i, i)$, set $c_i = 1/\sqrt{1+t^2}$, $s_i = c_i t$,

else

compute $t = R(i, i)/R(i+1, i)$ and set $s_i = 1/\sqrt{1+t^2}$, $c_i = s_i t$.

end

end

(8) *Apply the rotations to the right-hand side: $t = c_i\ \texttt{rhs}_i$; $\texttt{rhs}_{i+1} = s_i\ \texttt{rhs}_i$;*

$\texttt{rhs}_i = t$ *and then to rows i, $i + 1$ of column i:*

$R(i, i) = c_i R(i, i) + s_i R(i+1, i)$ and $R(i+1, i) = 0$;

(9) *Solve the triangular system for y:* $R(1:i, 1:i)y =$ `rhs`$(1:i)$;
(10) *Update the solution:* $x = x_0 + [z_1, z_2, \ldots, z_i]y$.
(11) *Compute the current residual:* $r = b - Ax$ *and exit if* $\|r\|_2 < TOL$.
end i
(12) *Continue with Step (1).*

Denote by $x = FGMRES(A, b, x_0, TOL, M_1, \ldots, M_k)$ the result of Algorithm 5.7.16. Then choose two tolerances ϵ_m, ϵ_p with ϵ_m a user-specified tolerance (say $\epsilon_m = 10^{-8}$) and $\epsilon_p \gg \epsilon_m$ (say $\epsilon_m = 10^{-1}$) for the preconditioning equation (5.67). Then the dual tolerance self-preconditioning algorithm can be described as follows.

Algorithm 5.7.17. (Self-preconditioning GMRES(k) method).

To solve $AM_jy = b$ *for* $x = M_jy$ *for* $j = 1, 2, \ldots, k$ *by GMRES(k), with* $iter = 0$, $e_1 = (e_1)_{n\times 1}$ *and given an initial starting vector* $x = x_0$ *and TOL=* ϵ_m,

(1) *Set* $x_0 = x$, $iter = iter + 1$. *Compute* $r = b - Ax_0$;
(2) *Generate the first vector* $q_1 = r/\|r\|_2$ *and the right-hand-side vector* `rhs` $= \|r\|_2 e_1$;
for $i = 1 : k$,
(3) *Start step i of a modified Gran–Schmidt method for Arnoldi:*
$z_i = FGMRES(A, q_i, 0, \epsilon_p, I, \ldots, I)$ *and* $w = Az_i$;
(4) **for** $\ell = 1 : i$
$R(\ell, i) = w^T q_\ell$; $w = w - R(\ell, i)q_\ell$;
end
(5) $R(i + 1, i) = \|w\|_2$ *and* $q_{i+1} = w/R(i + 1, i)$;
(6) *Apply the rotations to past rows of new column i:*
for $\ell = 1 : i - 1$,
$t = c_\ell R(\ell, i) + s_\ell R(\ell + 1, i)$; $R(\ell + 1, i) = s_\ell R(\ell, i) - c_\ell R(\ell + 1, i)$; $R(\ell, i) = t$;
end
(7) *Compute the Givens rotations:*
if $|R(i + 1, i)| \le 10^{-16}$, *set* $c_i = 1$ *and* $s_i = 0$,
else
if $|t| \le 1$ *for* $t = R(i + 1, i)/R(i, i)$, *set* $c_i = 1/\sqrt{1 + t^2}$, $s_i = c_i t$,
else

compute $t = R(i,i)/R(i+1,i)$ *and set* $s_i = 1/\sqrt{1+t^2}$, $c_i = s_i t$.
end

end

(8) *Apply the rotations to the right-hand side:* $t = c_i$ `rhs`$_i$; `rhs`$_{i+1} = s_i$ `rhs`$_i$;
`rhs`$_i = t$ *and then to rows* i, $i+1$ *of column* i:
$R(i,i) = c_i R(i,i) + s_i R(i+1,i)$ *and* $R(i+1,i) = 0$;

(9) *Solve the triangular system for y:* $R(1:i, 1:i)y =$ `rhs`$(1:i)$;

(10) *Update the solution:* $x = x_0 + [z_1, z_2, \ldots, z_i]y$.

(11) *Compute the current residual:* $r = b - Ax$ *and exit if* $\|r\|_2 < \epsilon_m$.

end i

(12) *Continue with Step (1).*

If other iterative solvers such as the FOM (3.58) rather than GMRES(k) are used, a similar algorithm to Algorithm 5.7.17 can be developed.

There is also certain natural connection of this method to the topic of the next Chapter, if one chooses to solve (5.68) on a different level from the current level.

5.8 Near neighbour splitting for singular integral equations

Singular BIEs represent a class of useful problems that can be solved by the boundary elements (Section 1.7) and accelerated by the fast multipole methods. As far as preconditioning is concerned, the so-called mesh near neighbour preconditioners due to [472] have been observed to be efficient. However there was no fundamental study on establishing the method so the work [131] has partially filled in such a gap by showing that the near neighbour preconditioner is equivalent to a diagonal operator splitting. The beauty of such a mesh near neighbour method is that it works very well for various 3D problems [136,268].

We shall call the mesh near neighbour method a diagonal block approximate inverse preconditioner (or DBAI). As it appears to be a hybrid method between an operator splitting method (OSP) and an approximate inverse preconditioner (denoted by LSAI below to emphasize the way the SPAI preconditioner is computed), we shall briefly discuss OSP and LSAI.

Firstly for singular BIEs, we have shown in Section 4.7 that for a suitable preconditioner $M^{-1} = B^{-1}$ (left or right), its inverse $M = B$ can be of the specific sparsity structure

$$\begin{pmatrix} \times & \times & & & & & & & \times \\ \times & \times & \times & & & & & & \\ & \times & \times & \times & & & & & \\ & & \ddots & \ddots & \ddots & & & & \\ & & & \ddots & \ddots & \ddots & & & \\ & & & & \times & \times & \times & & \\ & & & & & \times & \times & \times & \\ & & & & & & \times & \times & \times \\ \times & & & & & & & \times & \times \end{pmatrix}. \tag{5.71}$$

As the diagonal entries of B are large, numerical evidence suggests that the structure of the largest entries in B^{-1} is similar to the structure of B. In fact, if B is strictly diagonal dominant, B^{-1} can be shown to be exponentially decaying from the diagonal. In practice, this strong condition may not be strictly satisfied. For 3D problems, see Section 5.8.3 for further discussion.

However, it appears reasonable to seek a preconditioner M^{-1} of sparsity structure (5.71) that can be used as an approximate inverse of A. Thus $\mathcal{S}$ will represent all nonzero positions in (5.71).

5.8.1 Solution of the least squares problem

We now consider the solution of the least squares problem for finding the right preconditioner; the left preconditioner can be found similarly. Since matrix $M^{-1} = [m_1,\ m_2, \ldots, m_n]$ consists of column vectors, for each column j, the least squares problem is to solve

$$\min_{m_j \in \mathcal{G}\mathcal{S}_j} \|Am_j - e_j\|_2^2 = \min_{m_j \in \mathcal{G}\mathcal{S}_j} \|\hat{A}_j m_j - e_j\|_2^2$$

or

$$\begin{pmatrix} A_{1j_1} & A_{1j_2} & A_{1j_3} \\ \vdots & \vdots & \vdots \\ A_{j_1j_1} & A_{j_1j_2} & A_{j_1j_3} \\ A_{j_2j_1} & A_{j_2j_2} & A_{j_2j_3} \\ A_{j_3j_1} & A_{j_3j_2} & A_{j_3j_3} \\ \vdots & \vdots & \vdots \\ A_{nj_1} & A_{nj_2} & A_{nj_3} \end{pmatrix} \begin{pmatrix} M_{j_1j} \\ M_{j_2j} \\ M_{j_3j} \end{pmatrix} = \begin{pmatrix} 0 \\ \vdots \\ 0 \\ 1 \\ 0 \\ \vdots \\ 0 \end{pmatrix}, \tag{5.72}$$

or simply

$$\hat{A}_j \hat{m}_j = e_j,$$

where $j_2 = j$, $\hat{m}_j = [M_{j_1 j} \quad M_{j_2 j} \quad M_{j_3 j}]^T$, $m_j = [0^T \quad \hat{m}_j \quad 0^T]^T$, $j_1 = j-1$, $j_3 = j+1$ for $j = 2, \ldots, n$, $j_1 = n, j_3 = 2$ for $j = 1$, and $j_1 = n-1$, $j_3 = 1$ for $j = n$ due to the choice of $\mathcal{S}$ and the wrap-around nature of M^{-1}.

The least squares problem (5.72) may be solved by the QR method [63,229]. For the approximation using this specific pattern $\mathcal{S}$, we have the following theorem.

Theorem 5.8.18. *For the least squares problem* (5.72) *with $\hat{A}_j$ of size $n \times 3$,*

1. *the residual for the solution $\hat{m}_j$ satisfies $\|r_j\|_2 \le 1$ because the right-hand side of* (5.72) *is a unit vector;*
2. *problem* (5.72) *is equivalent in the least squares sense to the following problem with B_j of a smaller size (i.e., 4×3),*

$$B_j \hat{m}_j = \begin{pmatrix} b_{11} & b_{12} & b_{13} \\ b_{21} & b_{22} & b_{23} \\ & b_{32} & b_{33} \\ & & b_{43} \end{pmatrix} \begin{pmatrix} M_{j_1 j} \\ M_{j_2 j} \\ M_{j_3 j} \end{pmatrix} = \begin{pmatrix} 1 \\ 0 \\ 0 \\ 0 \end{pmatrix}.$$

Further, the residual for the solution $\hat{m}_j$ can be written more specifically as

$$r_j = [0 \ \bar{r}_j]^T \qquad \text{and} \qquad \bar{r}_j = -\sin\theta_1 \sin\theta_2 \sin\theta_3$$

for some θ_i's (so $\|r_j\|_2 < 1$ if $A_{1 j_1} \neq 0$).

Therefore the matrix residual for the approximate inverse M^{-1} will be $E = I - AM^{-1}$ and its F-norm satisfies $\|E\|_F^2 = \sum_{j=1}^n \|r_j\|_2^2 < n$ or $\|E\|_F < \sqrt{n}$.

Proof. For an orthogonal matrix $Q = [q_1, \ldots, q_n]$, let the QR-decomposition of $\hat{A}_j$ be

$$\hat{A}_j = Q^T \begin{pmatrix} R \\ 0 \end{pmatrix}$$

and let $Qe_j = q_j$. Define $q_j = [\hat{c}^T \quad \hat{d}^T]^T$ where $\hat{c}^T$ is of size 3 and $\hat{d}^T$ is of size $n-3$. Then (5.72) is equivalent in the least squares sense to $R\hat{m}_j = \hat{c}^T$. The solution is $\hat{m}_j = R^{-1}\hat{c}^T$. The residual error will be $r_j = e_j - \hat{A}_j \hat{m}_j = Q^T[0 \ \hat{d}^T]^T$.

The first result is trivial because q_j is a unit vector and so $\|r_j\|_2 = \|\hat{d}\|_2 \le \|q_j\|_2 = 1$. For the second result, we first multiply a permutation matrix $P_{1,j}$ (also orthogonal), permuting rows 1 and j, to (5.72). The resulting equation can be applied by three Householder transformations giving rise to the reduced 4×3 problem.

Further, a sequence of three successive Givens transformations

$$\begin{pmatrix} 1 & & & \\ & 1 & & \\ & & \cos\theta_3 & \sin\theta_3 \\ & & -\sin\theta_3 & \cos\theta_3 \end{pmatrix} \begin{pmatrix} 1 & & & \\ & \cos\theta_2 & \sin\theta_2 & \\ & -\sin\theta_2 & \cos\theta_2 & \\ & & & 1 \end{pmatrix} \begin{pmatrix} \cos\theta_1 & \sin\theta_1 & & \\ -\sin\theta_1 & \cos\theta_1 & & \\ & & 1 & \\ & & & 1 \end{pmatrix}$$

can reduce the 4×3 matrix B_j to an upper triangular matrix $(R^T \ \ 0^T)^T$ and the right-hand side to

$$\hat{c} = [\cos\theta_1 \quad -\sin\theta_1\cos\theta_2 \quad \sin\theta_1\sin\theta_2\cos\theta_3 \quad -\sin\theta_1\sin\theta_2\sin\theta_3]^T$$

for some θ_i's. Note that $|\bar{r}_j| \le 1$ but if $b_{11} = A_{1j_1} \ne 0$, $|\sin\theta_1| \ne 1$ so $\|r_j\|_2 < 1$. Thus the second result follows. ■

Remark 5.8.19. This theorem illustrates the accuracy of inverse approximation using structure (5.71). More general results of this type and on eigenvalue bounds can be found in [162,253] among others. In particular, note that the residual error $\|E\|$ is directly linked to the eigenspectrum $\lambda(AM^{-1})$. Using Definition 1.5.10, we may write

$$\lambda(AM^{-1}) \in \Upsilon^{[1,\mu_{\text{LSAI}}]}_{[1,\mu_{\text{LSAI}}]},$$

where μ_{LSAI} is generally small depending on the approximation accuracy. This behaviour of LSAI having the same (maybe small) cluster radius as the cluster size is different from OSP having a very small cluster size but not necessarily small cluster radius. We shall show that the DBAI is an interesting method that has both small cluster size and small cluster radius.

5.8.2 The DBAI preconditioner

In (5.72), we expect three rows $(j_1, \ j_2, \ j_3)$ of $\hat{A}_j$ to play a dominant role due to the singular nature of the original operator. Therefore we may approximately reduce (5.72) to a 3×3 system

$$\overline{A}_j \hat{m}_j = \begin{pmatrix} A_{j_1 j_1} & A_{j_1 j_2} & A_{j_1 j_3} \\ A_{j_2 j_1} & A_{j_2 j_2} & A_{j_2 j_3} \\ A_{j_3 j_1} & A_{j_3 j_2} & A_{j_3 j_3} \end{pmatrix} \begin{pmatrix} M_{j_1 j} \\ M_{j_2 j} \\ M_{j_3 j} \end{pmatrix} = \begin{pmatrix} 0 \\ 1 \\ 0 \end{pmatrix}, \qquad (5.73)$$

which of course makes sense from a computational point of view. This modified preconditioner M^{-1}, of form (5.71), is a DBAI preconditioner. This is the so-called method of mesh neighbours in [472]. The same idea was used in the local least squares inverse approximation preconditioner of [452], the truncated Green's function preconditioner of [240], and the nearest neighbour preconditioner of [365], among others.

While heuristically reasonable, computationally simple, and experimentally successful, the DBAI method has not been justified in theory. Here we present results on an analysis for the method before discussing the generalized version using more mesh neighbours.

To simplify the presentation, we first give two definitions and then a simple lemma.

Definition 5.8.20. (Band$^+(d_L, d_U, b_L, b_U)$). *A band matrix $A_{n\times n}$ with wrap-around boundaries is called* **Band**$^+(d_L, d_U, b_L, b_U)$ *if its lower and upper bandwidths are b_L and b_U, respectively, and if furthermore the first d_L bands below the main diagonal are all zeros and the first d_U bands above the main diagonal are also all zeros.*

Definition 5.8.21. (Band$^-(d_L, d_U, b_L, b_U)$). *A simple band matrix $A_{n\times n}$ (without wrap-around boundaries) is called* **Band**$^-(d_L, d_U, b_L, b_U)$ *if its lower and upper bandwidths are b_L and b_U, respectively, and if furthermore the first d_L bands below the main diagonal are all zeros and the first d_U bands above the main diagonal are also all zeros.*

Note that the first definition here is for matrices with wrap-around boundaries while the second is for simple band matrices without wrap-rounds. In both definitions, the parameters are non-negative integers, not exceeding $(n-1)$. Here if $d_L d_U \neq 0$, both **Band**$^+(d_L, d_U, b_L, b_U)$ and **Band**$^-(d_L, d_U, b_L, b_U)$ matrices have a zero diagonal. But if $d_L d_U = 0$ the diagonal information will be stated in the context. With $n = 6$ we may illustrate **Band**$^+(0, 1, 2, 1)$ and **Band**$^-(0, 1, 2, 1)$, respectively, by

$$\begin{bmatrix} 0 & 0 & \times & 0 & \times & \times \\ \times & 0 & 0 & \times & 0 & \times \\ \times & \times & 0 & 0 & \times & 0 \\ 0 & \times & \times & 0 & 0 & \times \\ \times & 0 & \times & \times & 0 & 0 \\ 0 & \times & 0 & \times & \times & 0 \end{bmatrix} \quad \text{and} \quad \begin{bmatrix} 0 & 0 & \times & 0 & 0 & 0 \\ \times & 0 & 0 & \times & 0 & 0 \\ \times & \times & 0 & 0 & \times & 0 \\ 0 & \times & \times & 0 & 0 & \times \\ 0 & 0 & \times & \times & 0 & 0 \\ 0 & 0 & 0 & \times & \times & 0 \end{bmatrix}.$$

Therefore, for matrices from section 4.7, B and $\overline{K}$ are **Band**$^+(0, 0, 1, 1)$, and C is **Band**$^-(1, 1, n-3, n-3)$. One may verify that, for example, **Band**$^+(0, 0, 2, 2) =$ **Band**$^+(0, 0, 1, 1) +$ **Band**$^+(1, 1, 1, 1)$, and **Band**$^+(d_L, d_U, b_L + 3, b_U + 4) =$ **Band**$^+(d_L, d_U, 3, 4) +$ **Band**$^+(d_L + 3, d_U + 4, b_L, b_U)$.

Lemma 5.8.22. (Multiplication of band matrices). *If matrix $A_{n\times n}$ is* **Band**$^+(0, 0, b_{L_1}, b_{U_1})$, *$B_{n\times n}$ is* **Band**$^+(0, 0, b_{L_2}, b_{U_2})$ *and $C_{n\times n}$ is* **Band**$^-(d_{L_3}, d_{U_3}, b_{L_3}, b_{U_3})$, *then*

1. AB *is* **Band**$^+(0, 0, b_{L_4}, b_{U_4})$*, with* $b_{L_4} = b_{L_1} + b_{L_2}$ *and* $b_{U_4} = b_{U_1} + b_{U_2}$*;*
2. AC *is* **Band**$^-(d_{L_5}, d_{U_5}, b_{L_5}, b_{U_5})$*, with* $d_{L_5} = \max(0, d_{L_3} - b_{L_1})$*,* $d_{U_5} = \max(0, d_{U_3} - b_{U_1})$*,* $b_{L_5} = b_{L_1} + b_{U_1} + b_{L_3}$*, and* $b_{U_5} = b_{L_1} + b_{U_1} + b_{U_3}$*.*

Proof. The proof is by simple inductions. ■

We are now in a position to study the singularity separation property of the preconditioned matrix AM^{-1}.

Theorem 5.8.23. *The DBAI preconditioner admits a diagonal operator splitting. Therefore for singular BIEs, the preconditioned matrix and its normal matrix have clustered eigenvalues.*

Proof. Partition matrix A as follows (as illustrated in Figure 5.2)

$$A = D + B_2 + C_2,$$

where D is the diagonal matrix of A, B_2 is **Band**$^+(0, 0, 2, 2)$ (with a zero diagonal), and C_2 is **Band**$^-(2, 2, n-5, n-5)$. That is,

$$B_2 = \begin{pmatrix} & A_{12} & A_{13} & & & & A_{1n-1} & A_{1n} \\ A_{21} & & A_{23} & A_{24} & & & & A_{2n} \\ A_{31} & A_{32} & & A_{34} & A_{35} & & & \\ & A_{42} & A_{43} & & A_{45} & \ddots & & \\ & & \ddots & \ddots & & \ddots & \ddots & \\ & & & \ddots & \ddots & & \ddots & A_{n-2n} \\ A_{n-11} & & & & \ddots & \ddots & & A_{n-1n} \\ A_{n1} & A_{n2} & & & & A_{nn-2} & A_{nn-1} & \end{pmatrix}.$$

First, from a similar matrix splitting of the operator $\mathcal{A}$, we can show that the off diagonal operators are compact due to smooth kernels. Therefore assuming the original operator $\mathcal{A}$ is bounded, using Lemma 4.7.12, we see that $\overline{A}_j$ has a bounded inverse and so M^{-1} is bounded.

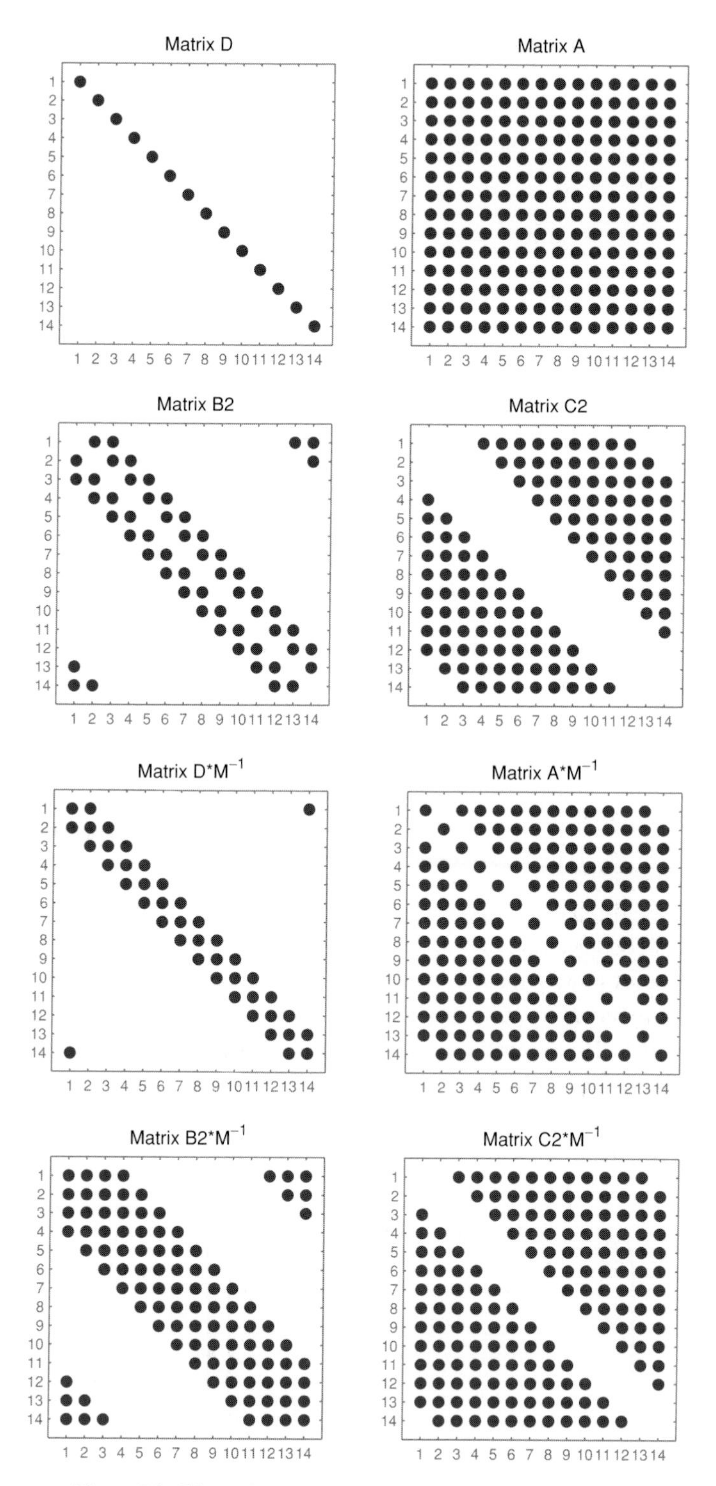

Figure 5.2. Illustration of operator splitting of DBAI ($n = 14$).

Secondly, to work out an explicit formula for $AM^{-1} - I$ in terms of D, B_2, C_2, we have

$$AM^{-1} = DM^{-1} + B_2M^{-1} + C_2M^{-1},$$

where DM^{-1} is $\mathbf{Band}^+(0, 0, 1, 1)$ as with M^{-1}. From Lemma 5.8.22, B_2M^{-1} is $\mathbf{Band}^+(0, 0, 3, 3)$ with a nonzero diagonal and C_2M^{-1} is $\mathbf{Band}^-(1, 1, n-3, n-3)$. Now do a simple splitting $B_2M^{-1} = B_2^{(1)} + B_2^{(2)}$ with $B_2^{(1)}$ as a $\mathbf{Band}^+(0, 0, 1, 1)$ matrix and $B_2^{(2)}$ as $\mathbf{Band}^+(1, 1, 2, 2)$. So defining $C_3 = B_2^{(2)} + C_2M^{-1}$ gives

$$AM^{-1} = DM^{-1} + B_2^{(1)} + C_3. \tag{5.74}$$

From the construction of M^{-1}, we see that

$$DM^{-1} + B_2^{(1)} = I.$$

Therefore the matrix D is implicitly inverted because (5.74) becomes

$$AM^{-1} = I + C_3. \tag{5.75}$$

In this formula, notice that C_3 is solely determined by terms B_2 and C_2 which correspond to compact operators (refer to Lemma 4.7.12). Thus matrix C_3 can be viewed as from a discretization of a compact operator and its eigenvalues and those of its normal matrix are thus clustered at 1. ■

The present DBAI is specified by the pattern in (5.71), that is, using the nearest neighbours. As is known from [452,240,365,147], one may use more than one level of neighbours. In our notation, this means that we use the new pattern of a $\mathbf{Band}^+(0, 0, s, s)$ matrix or a band $k = 2s + 1$ matrix instead of a $\mathbf{Band}^+(0, 0, 1, 1)$ matrix or a band 3 matrix. For brevity, we name such a preconditioner DBAI(k). Thus $s = 1$ (or $k = 3$) gives the same DBAI as before. We now consider $s > 1$ (or odd $k \geq 5$).

Then to solve for the jth column of M^{-1}, we solve a new $k \times k$ system

$$\overline{A}_j \hat{m}_j = \begin{pmatrix} A_{j_1 j_1} & \cdots & A_{j_1 j_{s+1}} & \cdots & A_{j_1 j_k} \\ \vdots & \cdots & \vdots & \cdots & \vdots \\ A_{j_{s+1} j_1} & \cdots & A_{j_{s+1} j_{s+1}} & \cdots & A_{j_{s+1} j_k} \\ \vdots & \cdots & \vdots & \cdots & \vdots \\ A_{j_k j_1} & \cdots & A_{j_k j_{s+1}} & \cdots & A_{j_k j_k} \end{pmatrix} \begin{pmatrix} M_{j_1 j} \\ \vdots \\ M_{j_{s+1} j} \\ \vdots \\ M_{j_k j} \end{pmatrix} = \begin{pmatrix} 0 \\ \vdots \\ 0 \\ 1 \\ 0 \\ \vdots \\ 0 \end{pmatrix}, \tag{5.76}$$

where $j_{s+1} = j$ always, $j_\ell = j + \ell - s - 1$ for $\ell = 1, \ldots, k$, and j_ℓ's take wrap-around index values outside the range of $[1, n]$ as with (5.72). Compare to (5.73).

It remains to identify the operator splitting implied in this DBAI(k). We can prove the following.

Theorem 5.8.24. *The DBAI(k) admits the same diagonal operator splitting as DBAI. Therefore for singular BIEs, the preconditioned matrix and its normal have clustered eigenvalues.*

Proof. Follow the similar lines of proving Theorem 5.8.23, partition matrix A as follows:

$$A = D + B_{2s} + C_{2s},$$

where B_{2s} is $\mathbf{Band}^+(0, 0, 2s, 2s)$ and C_{2s} is $\mathbf{Band}^-(2s, 2s, n - 2s - 1, n - 2s - 1)$, to complete the proof. ■

We have thus shown that DBAI is an OSP (having a small cluster size), although it appears more like an LSAI method (having a small cluster radius). So DBAI possesses advantages of both methods: inverse approximation (of LSAI) and operator splitting (of OSP). Using Definition 1.5.10, we may write for DBAI $\lambda(AM^{-1}) \in \Upsilon^{[]}_{[n_1, \mu_{\text{OSP}}]} 1, \mu_{\text{LSAI}}$, where the cluster radius μ_{LSAI} is related to the approximation error (that can be made smaller by increasing k) and μ_{OSP} is small due to operator splitting.

It remains to specify what k should be used. Since working out the preconditioner M^{-1} takes $O(k^3 n)$ operations, to ensure that this work does not exceed n^2 operations (one step of matrix vector multiplication), we suggest to choose k as an odd integer satisfying $3 \le k \le cn^{1/3}$ for some fixed constant c (say $c = 1$). This will be used in the experiments below.

5.8.3 Analysis of the 3D case

The analysis presented so far is mainly for two-dimensional (2D) problems. However, for 3D problems, a similar analysis can be done. For LSAI and DBAI, the essential difference is that the sparsity pattern $\mathcal{S}$ due to mesh neighbours, depending on the geometry of the surface and ordering, is more irregular and complex than that from (5.71). This is because the mesh neighbours are not always related to neighbouring entries in matrix A. In the 3D example of Section 5.9, the number of mesh neighbours varies from element to element (say one case with four neighbours and another with at least nine neighbours).

However, it is not difficult to understand why the analysis presented for DBAI can be generalized to this case in a similar way since all we need to do is to replace band matrices by pattern matrices. Let $\mathcal{S}$ denote the sparsity pattern of a mesh neighbouring strategy (see Section 5.9 for both edge and edge/vertex-based strategies). This includes the case of introducing levels of neighbours as in the 2D case.

Definition 5.8.25. *For any matrix B, given the sparsity pattern $\mathcal{S}$, define the pattern $\mathcal{S}$ splitting of B as*

$$B = Patt_{\mathcal{S}}(B) + Pato_{\mathcal{S}}(B),$$

where $Patt_{\mathcal{S}}(B)$ is the sparse matrix taking elements of B at location $\mathcal{S}$ and zeros elsewhere and $Pato_{\mathcal{S}}(B)$ is the complement matrix for B.

If we use M^{-1} to denote the DBAI preconditioner based on $\mathcal{S}$, then $M^{-1} = \text{Patt}_{\mathcal{S}}(M^{-1})$.

We can now establish that the DBAI preconditioner admits a diagonal splitting. As in the proof of Theorem 5.8.23, partition matrix A as follows:

$$A = D + C,$$

where $D = \text{diag}(A)$. Then

$$\begin{aligned} AM^{-1} &= DM^{-1} + CM^{-1} \\ &= \text{Patt}_{\mathcal{S}}(DM^{-1} + CM^{-1}) + \text{Pato}_{\mathcal{S}}(CM^{-1}) \\ &= I + \text{Pato}_{\mathcal{S}}(CM^{-1}) \\ &= I + C_3, \end{aligned}$$

because $\text{Patt}_{\mathcal{S}}$ is not affected by diagonal scaling and it also has the simple summation property. As with (5.75), matrix C_3 is solely determined by matrix C, which corresponds to a compact operator. Thus DBAI admits a diagonal operator splitting. Therefore the DBAI preconditioned matrix and its normal matrix have clustered eigenvalues at 1 with a small cluster size. Also from the approximation inversion property of DBAI, we know that the eigenvalues have a small cluster radius.

Remark 5.8.26. For both OSP- and LSAI-type methods, as is known, one may improve the eigenvalue clustering (in particular the cluster size for OSP and cluster radius for LSAI). However, as our analysis shows, the DBAI using a more complex sparsity pattern $\mathcal{S}$ does not imply a similar operator splitting beyond the diagonal splitting (i.e., one cannot observe much change in the cluster size) although the cluster radius will be reduced. It is straightforward to establish that a block matrix version of DBAI admits a block diagonal splitting. More work is needed to find a DBAI-like method admitting more than the block diagonal splitting (say, tridiagonal in two dimensions).

5.9 Numerical experiments

To illustrate the main inverse type preconditioners, we again take the test matrices as defined in Section 4.9.

For readers' benefit, we have shown the steps taken to test the methods in the driver Mfile `run_ai.m` with some results displayed in Figures 5.3 and 5.4. Clearly one observes that both variants of the approximate inverse preconditioner are effective. However the effectiveness is in practice dependent on the maximum number of nonzeros allowed per column, as remarked earlier, and for difficult situations deflation techniques have to be considered.

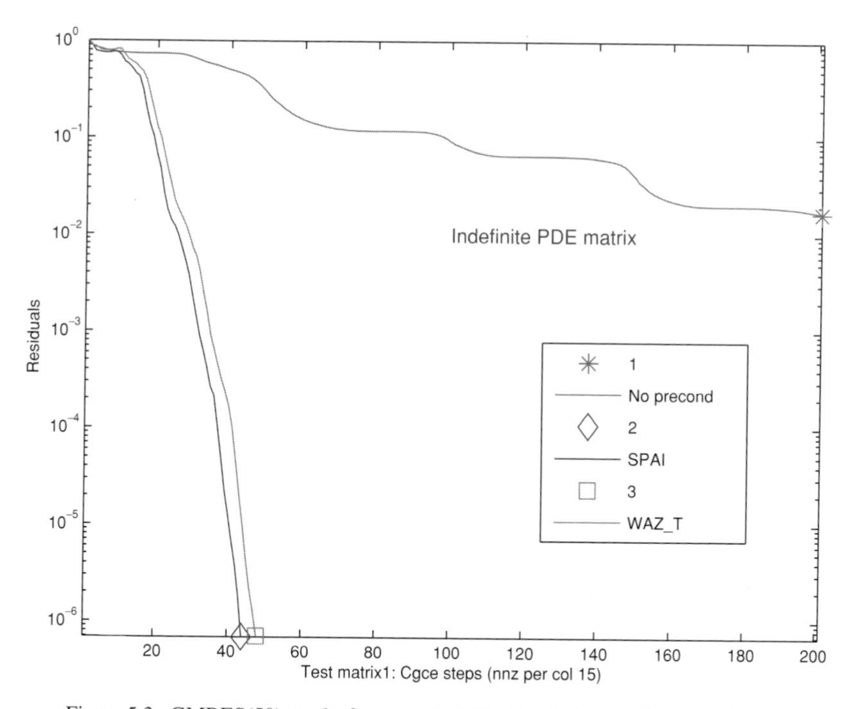

Figure 5.3. GMRES(50) results for `matrix1` (Section 4.9) using the approximate inverse preconditioners.

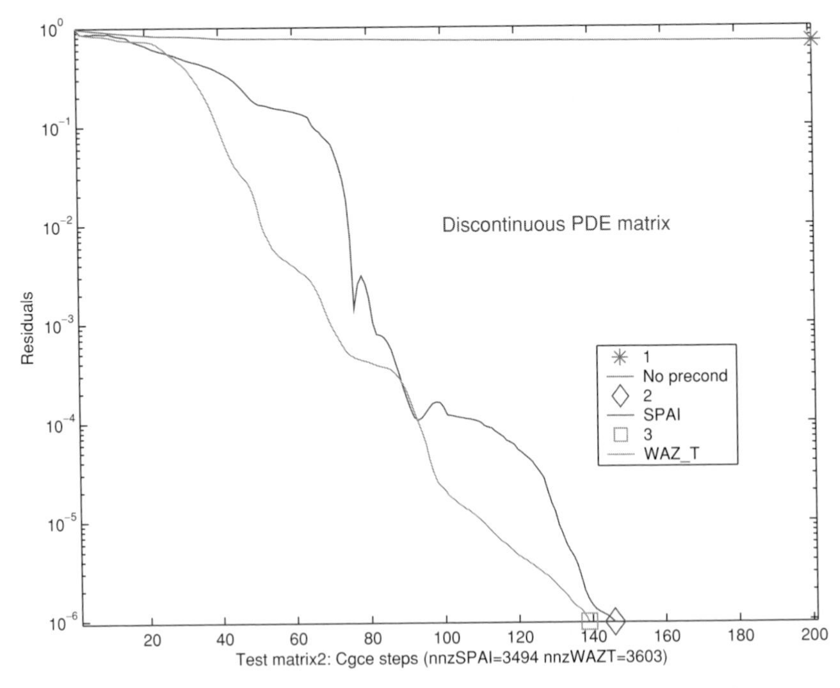

Figure 5.4. GMRES(50) results for `matrix2` (Section 4.9) using the approximate inverse preconditioners. Here '*' denotes the case of no preconditioning (and no convergence).

5.10 Discussion of software and the supplied Mfiles

There exist several pieces of software that implement the approximate inverse type preconditioners; we list a few of them.

(1) 'SPAI' (Marcus Grote and Thomas Huckle):

http://www.sam.math.ethz.ch/~grote/spai/

(2) 'SPARSLAB' (Michelle Benzi and Miroslav Tuma):

http://www.cs.cas.cz/~tuma/sparslab.html

(3) 'SAINV' preconditioner (Richard Bridson):

http://www.cs.ubc.ca/~rbridson

(4) 'HSL: MI12' (Nicholas Gould and Jennifer Scott):

http : //www.cse.clrc.ac.uk/nag/hsl/contents.shtml

At the level of research investigations, we have supplied the following Mfiles.

[1] chebg.m – Driver file for solving $Ax = b$ using the Chebyshev preconditioner via cheb_fun.m.
[2] cheb_fun.m – Implementation of Algorithm 5.3.3.
[3] spai2.m – Illustration of the SPAI algorithm 5.4.5
[4] waz_t.m – Illustration of FSAI/AINV algorithm 5.5.9.
[5] def_no.m – Illustration of two deflation methods and an approximate deflation (Ritz) method.
[6] run_ai.m – The main driver Mfile for illustrating spai2.m and waz_t.m using the test matrices as in Section 4.9.

6
Multilevel methods and preconditioners [T3]: coarse grid approximation

> Whenever both the multigrid method and the domain decomposition method work, the multigrid is faster.
>
> JINCHAO XU, *Lecture at University of Leicester EPSRC Numerical Analysis Summer School*, UK (1998)

> This paper provides an approach for developing completely parallel multilevel preconditioners.... The standard multigrid algorithms do not allow for completely parallel computations, since the computations on a given level use results from the previous levels.
>
> JAMES H. BRAMBLE, *et al.*
> Parallel multilevel preconditioners. *Mathematics of Computation*, Vol. 55 (1990)

> Multilevel methods [including multigrid methods and multilevel preconditioners] represent new directions in the recent research on domain decomposition methods ... they have wide applications and are expected to dominate the main stream researches in scientific and engineering computing in 1990s.
>
> TAO LU, *et al. Domain Decomposition Methods.*
> Science Press, Beijing (1992)

The linear system (1.1) may represent the result of discretization of a continuous (operator) problem over the finest grid that corresponds to a user required resolution. To solve such a system or to find an efficient preconditioner for it, it can be advantageous to set up a sequence of coarser grids with which much efficiency can be gained.

This sequence of coarser grids can be nested with each grid contained in all finer grids (in the traditional geometry-based multigrid methods), or non-nested with each grid contained only in the finest grid (in the various variants of the domain decomposition methods), or dynamically determined from the linear system alone in a purely algebraic way (the recent algebraic multigrid methods).

As it is impossible to include all such multilevel methods in the vast topic, this Chapter will highlight the main ideas and algorithms in a selective way.

6.1 Multigrid method for linear PDEs

We first describe the geometric multigrid method for solving a linear elliptic PDE

$$\begin{cases} \mathcal{L}u = f, & \Omega \subset \mathbb{R}^d, \\ u(p) = g(p), & p \in \Gamma = \partial\Omega. \end{cases} \tag{6.1}$$

In particular we shall illustrate the 1D case $d = 1$ graphically:

$$\begin{cases} \mathcal{L}u = -u'' = f, \quad \Omega = (0, 1) \subset \mathbb{R}, \\ u(0) = u(1) = 0, \end{cases} \tag{6.2}$$

and present multigrid algorithms for the 2D case $d = 2$. A multigrid method cleverly combines two separate (and old) mathematical ideas:

(1) *fine grid residual smoothing by relaxation*;
(2) *coarse grid residual correction*.

Although the first multigrid appeared in early 1960s, the actual efficiency was first realized by A. Brandt in 1973 and W. Hackbusch in 1976 independently; see [72,185,263,259,495,490] and the many references therein. A multigrid algorithm using the above two ideas has three components, namely:

(1) *Relaxation step*;
(2) *Restriction step*;
(3) *Interpolation (or prolongation) step*;

of which the first step is the most important.

6.1.1 Relaxation and smoothing analysis

Several relaxation methods were analyzed in Chapter 3 using matrix analysis. The so-called smoothing property of a relaxation method refers to its ability to smooth out the residual after a small number of iterations. In this sense, the residual vector $r = [1\ 1 \ldots 1]^T$ is perfectly smooth (but not small) while the residual vector $r = [1\ 1 \ldots 1\ 0 \ldots\ 0]^T$ is less smooth (though relatively smaller in norm). Mathematically speaking, a function is called smooth if its Fourier coefficients are decaying, i.e. it is essentially in a span of some low frequency Fourier basis functions (for vectors we use the grid function Fourier series). A relaxation method is called a good smoother if the smoothing rate (measured by the maximal ratio by which the high frequency coefficients of the error is reduced per step) is less than 1. Thus the task of a smoothing analysis is to establish if a relaxation method for a given problem is an adequate smoother. In this context, any convergence of a relaxation method is not relevant.

Let a multiple level of grids be denoted by $\mathcal{T}_k$ with $k = 1, 2, \ldots, J$ and $\mathcal{T}_J$ be the finest grid (directly associated with the system (1.1)) as shown in Figure 6.1 for $n = 15$ and $k = 3$ levels. Here J can be flexibly chosen to allow the coarsest grid $\mathcal{T}_1$ to have a suitable number of grid points – for linear problems $\mathcal{T}_1$ may have $n_1 = 1$ point but $n_1 \gg 1$ for nonlinear problems. With uniform refinement, one may define the grid $\mathcal{T}_{k-1}$ to obtain the grid points of $\mathcal{T}_k$ as follows (note: $n = n_J - 1$ on $\mathcal{T}_J$)

$$x_i^k = ih_k, \qquad i = 1, 2, \ldots, n_k - 1, \qquad h_k = \frac{1}{n_k} = \frac{2^{J-k}}{n_J} = \frac{2^{J-k}}{n+1}, \tag{6.3}$$

with $n_k = (n+1)/2^{J-k} = n_J/2^{J-k}$. Here $\mathcal{T}_k$ represents the k-th level discretization of the entire domain Ω and hence $\mathcal{T}_k$ can be analogously defined in higher dimensions [354,29]. Then on a typical grid $\mathcal{T}_k$, using the FDM (or the FEM with piecewise linear elements [308]) leads to the familiar linear

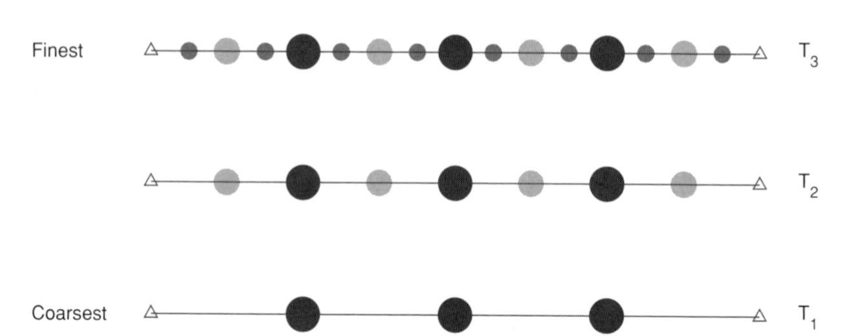

Figure 6.1. Illustration of the set-up of multiple grids ($k = 3$ levels).

system

$$\mathcal{L}_k \mathbf{u}_k = \begin{bmatrix} 2 & -1 & & & \\ -1 & 2 & -1 & & \\ & -1 & 2 & \ddots & \\ & & \ddots & \ddots & -1 \\ & & & -1 & 2 \end{bmatrix} \begin{pmatrix} u_1^k \\ u_2^k \\ \vdots \\ u_{n_k-1}^k \end{pmatrix} = \begin{pmatrix} g_1^k \\ g_2^k \\ \vdots \\ g_{n_k-1}^k \end{pmatrix} = \mathbf{g}_k, \quad (6.4)$$

where we set $g_j = h_k^2 g(x_j^k)$. We now consider the damped Jacobi method and the Gauss–Seidel method for (6.4). Let $e_j = u_j - u_j^k$ be the solution error with $u_j = u(x_j^k)$ the exact solution of (6.4).

To avoid any confusion, we recall that a periodic grid function can be represented by a complex Fourier series [490,Ch.7] with $i = \sqrt{-1}$

$$e_j = \sum_{\alpha=1-n_k/2}^{n_k/2} c_\alpha e^{2ij\pi\alpha/n_k} = \sum_{\alpha=1-n_k/2}^{n_k/2} c_\alpha e^{i2\alpha\pi x_j} = \sum_{\alpha=1-n_k/2}^{n_k/2} c_\alpha e^{i\theta_\alpha \frac{x_j}{h_k}} \quad (6.5)$$

with $\theta_\alpha = 2\alpha\pi/n_k \in [-\pi, \pi]$, or a Fourier since series (for homogeneous Dirichlet boundary conditions)

$$e_j = \sum_{\alpha=1}^{n_k-1} c_\alpha \sin j\pi\alpha/n_k = \sum_{\alpha=1}^{n_k-1} c_\alpha \sin \alpha\pi x_j = \sum_{\alpha=1}^{n_k-1} c_\alpha \sin \theta_\alpha \frac{x_j}{h_k} \quad (6.6)$$

with $\theta_\alpha = \alpha\pi/n_k \in [0, \pi]$. Here we used the mesh size information $h = h_k = 1/n_k$. Observe that the following Fourier modes provide the basis for expansion in (6.6):

$$\begin{gathered} \{\sin \pi x,\ \sin 2\pi x,\ \sin 3\pi x, \cdots,\ \sin(n_k - 1)\pi x\} \qquad \text{or} \\ \left\{\sin \theta_1 \frac{x}{h},\ \sin \theta_2 \frac{x}{h},\ \sin \theta_3 \frac{x}{h}, \cdots,\ \sin \theta_{n_k-1} \frac{x}{h}\right\} \end{gathered} \quad (6.7)$$

and in (6.5)

$$\begin{gathered} \left\{1,\ e^{\pm i\pi x},\ e^{\pm 2i\pi x},\ e^{\pm 3i\pi x}, \cdots,\ e^{\pm(n_k-1)i\pi x},\ e^{n_k i\pi x}\right\} \qquad \text{or} \\ \left\{1,\ \exp \pm i\theta_1 x,\ \exp \pm i\theta_2 x,\ \exp \pm i\theta_3 x, \cdots,\ \exp \pm i\theta_{n_k-1} x,\ \exp i\theta_{n_k} x\right\}. \end{gathered} \quad (6.8)$$

Within each basis, one can see that it ranges from the mildly varying basis functions to the fast oscillating functions. In the multigrid context, we shall name these mildly varying functions (the first half in the basis)

$$\text{the } \textit{low frequency} \text{ functions} — \sin \alpha\pi x, \quad \text{or} \quad \exp(i\theta_\alpha x)$$
$$\theta_\alpha \in [-\pi/2, \pi/2], \qquad \alpha = 1, 2, \ldots, n_k/2,$$

and the rest of fast oscillating functions

$$\text{the } \textit{high frequency} \text{ functions} — \sin \alpha\pi x, \qquad \text{or} \qquad \exp(i\theta_\alpha x)$$
$$\theta_\alpha \in [-\pi, \pi]\backslash[-\pi/2, \pi/2], \quad \alpha = 1 + n_k/2, \ldots, n_k - 1.$$

Correspondingly, the expansion for the error e_j will have a decomposition of *low frequency* terms and *high frequency* terms.

Remark 6.1.1. If the grid function is not periodic, the same Fourier expansion can still be used for smoothing analysis *approximately*. See [72,490]. However, for our particular example (6.4), the Fourier series happens to provide the eigenfunctions but we shall not use this fact for sake of generality. A careful reader can see the similarity between this smoothing analysis and the von Neumann's stability analysis for time-marching schemes for parabolic problems [400]. The multidimensional series is similar [490]; incidentally the first uses of the Fourier formula in [490,pages 6/7] had the incorrect range for α.

♦ **The damped Jacobi method.** The typical grid equation on $\mathcal{T}_k$ is

$$u_j^{\text{new}} = \omega u_j^{\text{old}} + (1 - \omega)\frac{g^k + u_{j+1}^{\text{old}} + u_{j-1}^{\text{old}}}{2}$$

As the exact u_j for the discretized equation naturally satisfies the grid equation, i.e.

$$u_j = \omega u_j + (1 - \omega)\frac{u_{j+1} + u_{j-1}}{2},$$

we obtain that

$$e_j^{\text{new}} = \omega e_j^{\text{old}} + (1 - \omega)\frac{e_{j+1}^{\text{old}} + e_{j-1}^{\text{old}}}{2}. \tag{6.9}$$

In what follows, we shall replace all grid functions by their Fourier series and essentially consider the so-called *amplification factor* i.e. the ratio between c_α^{new} and c_α^{old} for each α. The largest of such ratios will be the convergence rate (of no interest here) while that of those ratios for large $\alpha \geq n_k/2$ (the high frequency range) will be the smoothing factor. Owing to linearity, we only need to concentrate on the c_α^{new} term after substituting (6.6) into (6.9):

$$\begin{aligned}
&c_\alpha^{\text{new}} \sin j\pi\alpha/n_k \\
&= c_\alpha^{\text{old}} \omega \sin j\pi\alpha/n_k + (1 - \omega)c_\alpha^{\text{old}} \frac{\sin(j + 1)\pi\alpha/n_k + \sin(j - 1)\pi\alpha/n_k}{2} \\
&= c_\alpha^{\text{old}} \sin j\pi\alpha/n_k \left[\omega + (1 - \omega)\cos \alpha\pi/n_k\right] \\
&= c_\alpha^{\text{old}} \sin j\pi\alpha/n_k \left[1 - 2(1 - \omega)\sin^2 \frac{\alpha\pi}{2n_k}\right],
\end{aligned}$$

where $\alpha = 1, 2, \ldots, n_k/2, \ldots, n_k - 1$. Clearly the amplification factor for α is

$$a_\alpha = \left| \frac{c_\alpha^{\text{new}}}{c_\alpha^{\text{old}}} \right| = \left| 1 - 2(1-\omega)\sin^2 \frac{\alpha\pi}{2n_k} \right|.$$

With $0 < \omega < 1$ and for $\alpha\pi/(2n_k) \in (0, \pi/2)$

$$\begin{cases} 0 < \sin^2 \dfrac{\alpha\pi}{2n_k} \le \sin^2(\pi/4) = \dfrac{1}{2}, \ \alpha = 1, 2, \ldots, n_k/2, \text{(low frequency)} \\ \dfrac{1}{2} = \sin^2(\pi/4) < \sin^2 \dfrac{\alpha\pi}{2n_k} < 1, \ \alpha = n_k/2 + 1, \ldots, n_k - 1, \text{(high frequency)}, \end{cases}$$

we obtain the smoothing factor (noting $\gamma_\omega < 1$), from high frequency α's, as

$$\gamma_\omega = \max_{\alpha > n_k/2} a_\alpha = \begin{cases} \omega, & \text{if } 1/3 < \omega < 1, \\ 1 - 2\omega, & \text{if } 0 < \omega \le 1/3. \end{cases} \tag{6.10}$$

Note that the low frequency range $\alpha = 1, 2, \ldots, n_k/2$ for a fine grid $\mathcal{T}_k$ coincides with the complete frequency range of its *next coarser grid* $\mathcal{T}_{k-1}$ because

$$\begin{aligned} e_j^{k-1} &= \sum_{\alpha=1}^{n_{k-1}-1} c_\alpha \sin j\pi\alpha/n_{k-1} \quad \text{(complete range for } \mathcal{T}_{k-1}) \\ &= \sum_{\alpha=1}^{n_k/2-1} c_\alpha \sin \frac{2j}{n_k}\pi\alpha \qquad \text{(low frequency range for } \mathcal{T}_k). \end{aligned}$$

Hence assuming that the high frequency components on $\mathcal{T}_k$ are diminished by an effective smoother, the $\mathcal{T}_k$ equation can be accurately solved on the coarser grid $\mathcal{T}_{k-1}$ which is much cheaper. The process can be repeated until we reach grid $\mathcal{T}_1$ where the solution takes little effort. This is the basis of a good coarse grid approximation for a fine grid.

♦ **The Gauss–Seidel method.** Similar to the Jacobi case, the $\mathcal{T}_k$ grid equation is

$$2u_j^{\text{new}} - u_{j-1}^{\text{new}} = g^k + u_{j+1}^{\text{old}}$$

so corresponding to (6.9) we obtain that

$$2e_j^{\text{new}} - e_{j-1}^{\text{new}} = e_{j+1}^{\text{old}}. \tag{6.11}$$

On a quick inspection, the use of (6.6) will not lead to separation of the sine terms so it is more convenient to use (6.5) by considering the Fourier mode $c_\alpha^{\text{old}} e^{2ij\alpha\pi/n_k}$. Hence we have

$$2c_\alpha^{\text{new}} e^{2ij\alpha\pi/n_k} - c_\alpha^{\text{new}} e^{2i(j-1)\alpha\pi/n_k} = c_\alpha^{\text{old}} e^{2i(j-1)\alpha\pi/n_k},$$

which, noting $\theta_\alpha = 2\alpha\pi/n_k$, leads to the following amplification factor

$$\begin{aligned}
a_\alpha &= \left|\frac{c_\alpha^{\text{new}}}{c_\alpha^{\text{old}}}\right| = \left|\frac{e^{-i\theta_\alpha}}{2 - e^{i\theta_\alpha}}\right| \\
&= \left|\frac{\left[\sin^2\theta_\alpha + (2-\cos\theta_\alpha)\cos\theta_\alpha\right] + i\left[\frac{1}{2}\sin(2\theta_\alpha) + (2-\cos\theta_\alpha)\sin\theta_\alpha\right]}{(2-\cos\theta_\alpha)^2 + \sin^2\theta_\alpha}\right| \\
&= \frac{1}{\sqrt{(2-\cos\theta_\alpha)^2 + \sin^2\theta_\alpha}} = \frac{1}{\sqrt{5 - 4\cos\theta_\alpha}}.
\end{aligned}$$

Note that $\theta_\alpha \in (-\pi, \pi]$ and the different

Low frequency α range: $\alpha = -\frac{n_k}{4}, \ldots, \frac{n_k}{4}, \quad \theta_\alpha = \frac{2\alpha\pi}{n_k} \in [-\frac{\pi}{2}, \frac{\pi}{2}]$
High frequency α range: $\alpha = 1 - \frac{n_k}{2}, \ldots, 1 - \frac{n_k}{4};$
$\frac{n_k}{4} + 1, \ldots, \frac{n_k}{2}, \ \theta_\alpha = \frac{2\alpha\pi}{n_k} \in (-\pi, \pi] \backslash [-\frac{\pi}{2}, \frac{\pi}{2}].$

correspond to the cases of $0 \le \cos(\theta_\alpha) < 1$ and $-1 < \cos(\theta_\alpha) < 0$, respectively. In the high frequency range, therefore, the smoothing factor (as the maximal amplification factor) follows from the upper bound

$$\gamma_{GS} = \max_{|\alpha| > n_k/4} a_\alpha \approx \frac{1}{\sqrt{5 - 4\cos\pi/2}} = \frac{1}{\sqrt{5}} = 0.45. \tag{6.12}$$

Note $\gamma_{GS} < 1$ implies that the Gauss–Seidel is an effective smoother for the model PDE. As with the Jacobi case, the above low frequency range for a fine grid $\mathcal{T}_k$ will be covered by the complete frequency of the next coarser grid $\mathcal{T}_{k-1}$ which is seen from

$$\begin{aligned}
e_j^{k-1} &= \sum_{\alpha=1-n_{k-1}/2}^{n_{k-1}/2} c_\alpha e^{2ij\pi\alpha/n_{k-1}} \quad \text{(complete range for } \mathcal{T}_{k-1}) \\
&= \sum_{\alpha=1-n_k/4}^{n_k/4} c_\alpha e^{2i\frac{2j}{n_k}\pi\alpha} \quad \text{(low frequency range for } \mathcal{T}_k).
\end{aligned}$$

Excluding $\alpha = 0$, one can use the low frequency range to estimate the overall convergence rate of both the Jacobi and the Gauss–Seidel methods (from taking the maximal amplification factor) but we are not interested in this task.

Remark 6.1.2. To generalize the smoothing analysis to $\mathbb{R}^d$, it is proposed [72,460] to apply the so-called local Fourier analysis (LFA), i.e. to compute the amplification factor, locally, using the Fourier mode $e^{ij\theta}$ as with $\theta = \frac{2\alpha}{n_k}\pi \in [-\pi, \pi]$ in 1D ($d = 1$).

(1) In $\mathbb{R}^d$, one uses the general notation similar to (6.5)

$$\phi(\boldsymbol{\theta}, \mathbf{x}) = e^{i\boldsymbol{\theta}.*\mathbf{x}./\mathbf{h}} = e^{i(\theta_1 x_1/h_1 + \theta_2 x_2/h_2 + \cdots + \theta_d x_d/h_d)}$$

where '.*' and './' denote the pointwise operations for vectors $\mathbf{x} = (x_1, x_2, \ldots, x_d)$, $\boldsymbol{\theta} = (\theta_1, \theta_2, \ldots, \theta_d)$, $\mathbf{h} = (h_1, h_2, \ldots, h_d)$. Here $\theta_\ell \in [-\pi, \pi]$. If in each dimension the nested coarser grids are defined by doubling the mesh size of a finer grid, this is what we call the standard coarsening – for anisotropic problems it may be useful to keep some dimensions not coarsened (semi-coarsening). If the standard coarsening is used, each low frequency corresponds to $\boldsymbol{\theta}_\ell \in [-\pi/2, \pi/2]^d$ while the high frequency to $\boldsymbol{\theta}_\ell \in [\pi, \pi]^d \backslash [-\pi/2, \pi/2]^d$.

(2) An important issue on LFA is the word 'local', implying that all nonlinear PDEs can be analysed locally as a linearized PDE. Consider the following PDE in $\mathbb{R}^2$

$$\begin{aligned} \nabla\ (D(u,x,y)\nabla u) + c_1(u,x,y)\frac{\partial u}{\partial x} + c_2(u,x,y)\frac{\partial u}{\partial y} + c_3(u,x,y)u \\ = g(x,y). \end{aligned} \tag{6.13}$$

Then linearization involves evaluating all the nonlinear coefficients D, c_ℓ's using some known past iteration (or simply approximate) function $u = \overline{u}(x,y)$ – 'freezing' the coefficients as stated by [72]. Once this linearization is done in (6.13), with the FDM, a typical discrete grid equation will be studied locally at each discrete mesh point (x_l, y_m) 'linearly' as if for a simple Laplacian

$$k_0 u_{lm} - k_1 u_{l+1,m} - k_2 u_{l-1,m} - k_3 u_{l,m+1} - k_4 u_{l,m-1} = g_{lm} \tag{6.14}$$

or if first-order terms are kept separately to be treated differently (e.g. using the upwinding idea (1.93)), by a smoothing scheme

$$\begin{aligned} k_0 u_{lm} \underbrace{-k_1 u_{l+1,m} - k_2 u_{l-1,m} - k_3 u_{l,m+1} - k_4 u_{l,m-1}}_{\text{second-order } D \text{ term}} \\ \underbrace{-k_5 u_{l+1,m} - k_6 u_{l-1,m} - k_7 u_{l,m+1} - k_8 u_{l,m-1}}_{\text{first-order } c_1,\, c_2 \text{ terms}} = g_{lm} \end{aligned}$$

where we have assumed the 'constant' coefficients k_ℓ and g_{lm} have absorbed the mesh size information h_1, h_2. Further the LFA of the Gauss–Seidel smoother for (6.14) amounts to studying the grid equation

$$k_0 e_{lm}^{(\ell+1)} - k_2 e_{l-1,m}^{(\ell+1)} - k_4 e_{l,m-1}^{(\ell+1)} = k_1 e_{l+1,m}^{(\ell)} + k_3 e_{l,m+1}^{(\ell)}. \tag{6.15}$$

The LFA would use $e_{lm}^{(\ell+1)} = c_\alpha^{\text{new}} e^{i\left(\theta_1 \frac{x_l}{h_1} + \theta_2 \frac{y_m}{h_2}\right)}$ and $e_{lm}^{(\ell)} = c_\alpha^{\text{old}} e^{i\left(\theta_1 \frac{x_l}{h_1} + \theta_2 \frac{y_m}{h_2}\right)}$ in (6.15) to compute the amplification factor

$$a_\alpha = a_\alpha(\theta_1, \theta_2) = \left| \frac{c_\alpha^{\text{new}}}{c_\alpha^{\text{old}}} \right|$$

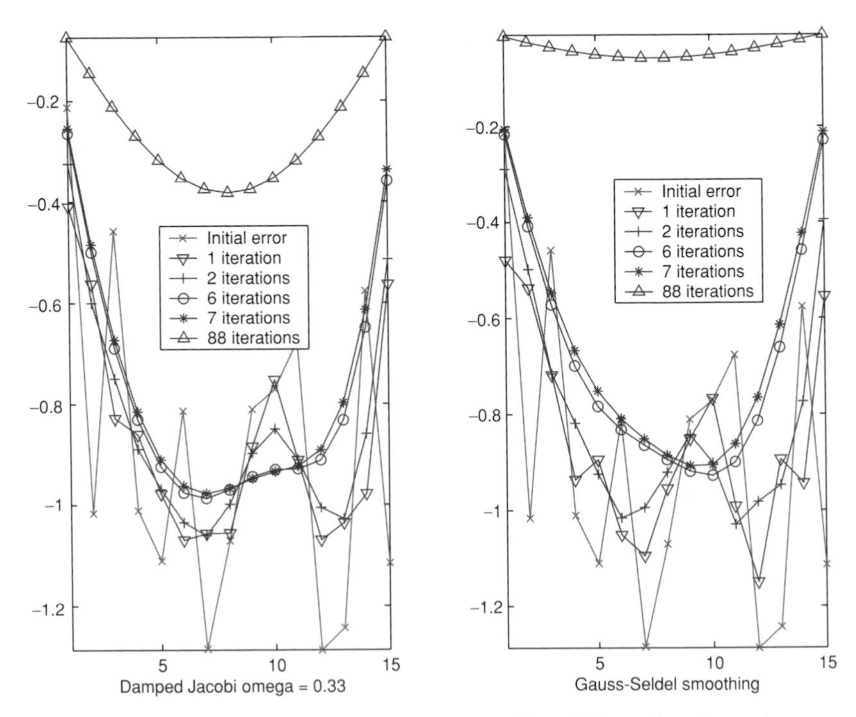

Figure 6.2. Illustration of error smoothing for (6.2) at different iterations using the damped Jacobi method (left plot $\omega = 1/3$) and the Gauss–Seidel method (right plot). Clearly after 88 iterations, the errors converge towards 0 but that is not needed by the multigrid method.

and the smoothing factor

$$\gamma_{\text{GS}} = \max_{\boldsymbol{\theta}_\ell=(\theta_1,\theta_2)\in[\pi,\pi]^2\setminus[-\pi/2,\pi/2]^2} a_\alpha(\theta_1, \theta_2).$$

(3) It goes without saying that whenever the smoothing rate of a particular smoother for a specific equation is not at least less than 1 (or less than 0.5 if more stringently), the overall multigrid method may not have convergence; if so, one has to find or design new and more appropriate smoothers.

In summary, the Fourier series-based smoothing factor analysis, as proposed by [72], provides an effective tool to show that many relaxation methods may be slow to converge but are always very fast in smoothing out the solution error. Figure 6.2 shows how the solution error to (6.2) behaves at different iterations using the damped Jacobi method (left plot) and the Gauss–Seidel method (right plot); the figure is generated by the Mfiles `ch6_mg2.m` and `ch6_gs.m`.

The method using grids $\mathcal{T}_k$ and $\mathcal{T}_{k-1}$ in any dimension d works as follows.

Algorithm 6.1.3. (The two-grid method).

(1) *Relax the fine grid equation:* $\mathcal{L}_k u_k = \mathbf{g}_k$ *on grid* $\mathcal{T}_k$ *for a small number of smoothing steps to obtain the approximation* u_k *such that the solution error* $e_k = u - u_k$ *is smooth. This implies this smooth error function (vector)* e_k *can be represented very accurately on a coarse grid* $\mathcal{T}_{k-1}$. *The set up (as shown in Chapter 3) is to solve the correction equation* $\mathcal{L}_k e_k = r_k \equiv \mathbf{g}_k - \mathcal{L}_k u_k$.
(2) *Restrict the residual function (vector) to grid* $\mathcal{T}_{k-1}$: $r_{k-1} = R_k^{k-1} r_k$.
(3) *Solve the coarse grid equation for the correction:* $\mathcal{L}_{k-1} e_{k-1} = r_{k-1}$.
(4) *Interpolation the coarse grid correction to obtain* $\hat{e}_k = P_{k-1}^k e_{k-1}$.
(5) *Add the fine grid correction to obtain the new approximation* $\hat{u}_k = u_k + \hat{e}_k$.
(6) *Return to step (1) and continue the iterations unless* $\|\hat{e}_k\|$ *is small.*

Here the transfer operators for restriction $R_k^{k-1} : \mathcal{T}_k \to \mathcal{T}_{k-1}$ and for prolongation $P_{k-1}^k : \mathcal{T}_{k-1} \to \mathcal{T}_k$ are usually linear mappings and described next.

6.1.2 Restriction and interpolation

The intergrid transfers are often based on standard choices and thus less demanding tasks than the smoothing step. Here we mainly discuss how the transfer operators work for the FDM setting – for the FEM setting the approach is similar but is applied to transfer coefficients of the elements rather than grid values (in fact, most FEMs use interpolatory elements and if so the coefficients are the grid values of a underlying function). For second-order differential equations such as (1.53), the standard choice is the full weighting for restriction and bilinear interpolation for interpolation [72,460].

♦ **Restriction.** Assume the fine grid values u_k over $\mathcal{T}_k$ have been obtained and we wish to restrict u_k to the coarse grid $\mathcal{T}_{k-1}$, as depicted by the left plot in Figure 6.3 ($\bullet \to \square$). The so-called full weighting (FW) restriction operator R_k^{k-1} makes use of all near neighbours of a coarse grid point, described by

$$\begin{aligned} R_k^{k-1} u_k(x, y) = \frac{1}{16}\Big[& u_k(x + h_k, y + h_k) + u_k(x + h_k, y - h_k) \\ & + u_k(x - h_k, y + h_k) + u_k(x - h_k, y - h_k) \\ & + 2u_k(x - h_k, y) + 2u_k(x - h_k, y) \\ & + 2u_k(x, y + h_k) + 2u_k(x, y - h_k) \\ & + 4u_k(x, y)\Big] \end{aligned} \tag{6.16}$$

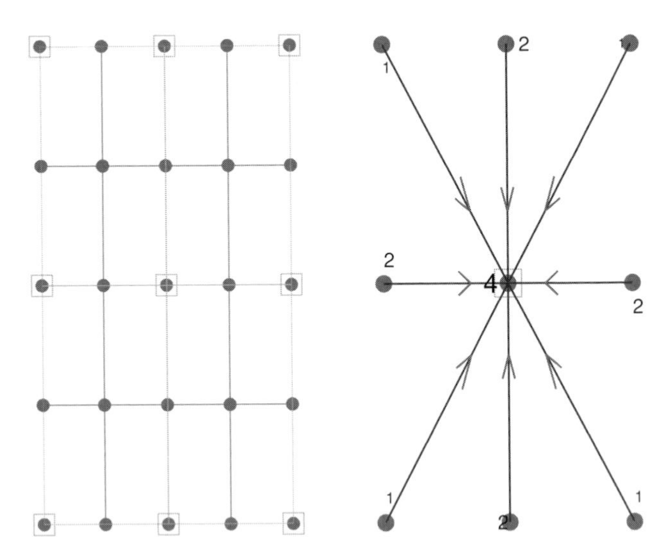

Figure 6.3. Full weighting restriction of fine grid • values to the coarse grid □ values (with integer k_f indicating the contribution to the weight $w_f = k_f / \sum k_f$).

with the stencil notation

$$u_{k-1} = R_k^{k-1} u_k(x, y) = \frac{1}{16} \begin{bmatrix} 1 & 2 & 1 \\ 2 & 4 & 2 \\ 1 & 2 & 1 \end{bmatrix}_k^{k-1} u_k.$$

Noting that the 2D stencil can be decomposed as two one-dimensional FW operators

$$\frac{1}{16} \begin{bmatrix} 1 & 2 & 1 \\ 2 & 4 & 2 \\ 1 & 2 & 1 \end{bmatrix} = \frac{1}{4} \begin{bmatrix} 1 \ 2 \ 1 \end{bmatrix} \otimes \frac{1}{4} \begin{bmatrix} 1 \\ 2 \\ 1 \end{bmatrix},$$

one deduces that tensor products may define the FW operator in any dimension d.

♦ **Prolongation.** Corresponding to the FW operator, the commonly used interpolation operator for prolongation from $\mathcal{T}_{k-1}$ to $\mathcal{T}_k$ is the bilinear (BL) operator as depicted by Figure 6.4 (□ → •). The BL operator again uses the immediate

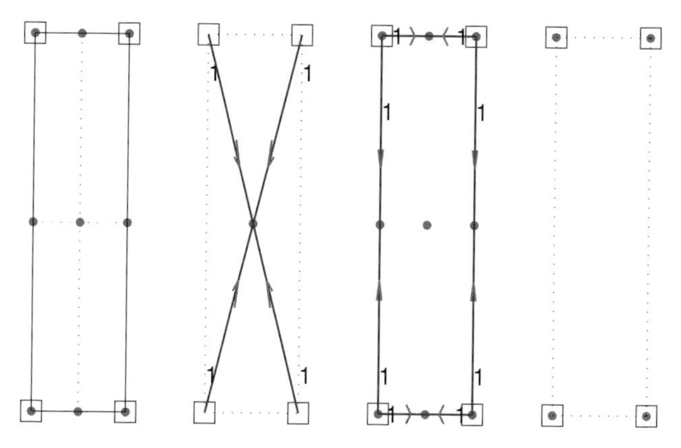

Figure 6.4. Bilinear interpolation of coarse grid □ values to the fine grid • values (with integer k_c indicating the weight $w_c = k_c / \sum k_c$).

coarse neighbours, defined by

$$P_{k-1}^k u_{k-1}(x, y) = \begin{cases} \frac{1}{4}\Big[u_{k-1}(x+h_k, y+h_k) + u_{k-1}(x+h_k, y-h_k) \\ \quad + u_{k-1}(x-h_k, y+h_k) + u_{k-1}(x-h_k, y-h_k)\Big] & \text{with four neighbours} \\ \frac{1}{2}\Big[u_{k-1}(x, y+h_k) + u_{k-1}(x, y-h_k)\Big] & \text{with two neighbours} \\ \frac{1}{2}\Big[u_{k-1}(x+h_k, y) + u_{k-1}(x-h_k, y)\Big] & \text{with two neighbours} \\ u_{k-1}(x, y) & \text{coinciding points} \end{cases} \tag{6.17}$$

with the (somewhat less intelligent) stencil notation

$$u_k = P_{k-1}^k u_{k-1}(x, y) = \frac{1}{4} \left]\begin{matrix} 1 & 2 & 1 \\ 2 & 4 & 2 \\ 1 & 2 & 1 \end{matrix}\right[_{k-1}^{k} u_{k-1}.$$

The reason why the two operators (FW and BL) are often used together is that they are adjoint to each other [490,460] in the sense that

$$\left(P_{k-1}^k u_{k-1}, v_k\right) = \left(u_{k-1}, R_k^{k-1} v_k\right), \qquad \forall v_k \in \mathcal{T}_k, \tag{6.18}$$

where $(,)$ denotes the usual inner product i.e. $(u, v) = u^T v = \sum_j u_j v_j$. Another pair of restriction and interpolation operators, popularized by [489,490], is the seven-point weighting operator as depicted by Figure 6.5. For convergence requirement [259,460,490], the sum of the orders of a pair of transfer operators should exceed the order of the differential operator. Here *the order of*

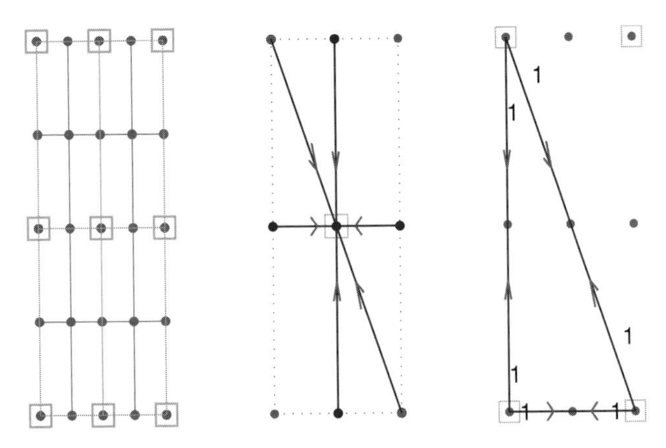

Figure 6.5. Wesseling's seven-point restriction operator (middle plot $\bullet \to \square$) and its adjoint operator for linear interpolation (right plot $\square \to \bullet$).

an interpolation operator is equal to $\ell + 1$ if the interpolation is exact for all polynomials of degree ℓ while the order of a restriction operator is equal to that of its transpose. As linear operators are often of order 2, therefore, the pairing of FW and BL will satisfy the convergence requirement for these orders if using to solve second order differential equations as $2 + 2 > 2$ (e.g. the Laplace and Helmholtz) but not for solving the biharmonic equation as $2 + 2 \not> 4$. For the same reason, the pairing of an injection operator (of order 0)

$$u_{k-1}(x, y) = R_k^{k-1} u_k(x, y) = u_k(x, y), \qquad \text{if} \quad (x, y) \in \mathcal{T}_{k-1} \subset \mathcal{T}_k \quad (6.19)$$

and the BL (bilinear) interpolation operator is not suitable for solving second-order PDEs as $2 + 0 \not> 2$.

6.1.3 Multigrid algorithms

A J-level multigrid algorithm, repeatedly using Algorithm 6.1.3, operates J grids:

$$\mathcal{T}_1 \subset \mathcal{T}_2 \subset \cdots \subset \mathcal{T}_J.$$

The aim is to solve the equations on the finest $\mathcal{T}_J$ iteratively, assuming that smoothing can be adequately done on all fine grids and we can afford to solve the coarsest grid equations on $\mathcal{T}_1$ directly (e.g. by Gaussian elimination as in Chapter 2).

Denote the grid $\mathcal{T}_k$ equations by system (such as (6.4))

$$\mathcal{L}_k \mathbf{u}_k = \mathbf{g}_k \tag{6.20}$$

where we assume that $\mathcal{L}_k$ is discretized directly from $\mathcal{L}$ in (6.1), although an alternative method is, by the so-called Galerkin approximation, to generate it recursively from the matrix $\mathcal{L}_J$ on the finest grid $\mathcal{T}_J$ i.e.

$$\mathcal{L}_k = R_{k+1}^k \mathcal{L}_{k+1} P_k^{k+1}.$$

Assume $\overline{\mathbf{u}}_k$ is the current approximation to (6.20). Then the task of a multi-grid method is to consider how to solve for the residual correction $\mathbf{v}_k$ efficiently

$$\mathcal{L}_k \mathbf{v}_k = r_k \equiv \mathbf{g}_k - \mathcal{L}_k \overline{\mathbf{u}}_k \tag{6.21}$$

with the aim of producing an improved solution, $\mathbf{u}_k = \overline{\mathbf{u}}_k + \mathbf{v}_k$ to (6.20).

Denote by $\mathbf{Relax}_k^{\nu}(u, f)$ the result of ν steps of some relaxation method for (6.20) with $\mathbf{g}_k = f$ and the initial guess $\mathbf{u}_k = u$.

Algorithm 6.1.4. (The multigrid method).

To solve the discretized PDE (6.20) on $\mathcal{T}_J$ with $k = J$, assume we have set up these multigrid parameters:

ν_1 *pre-smoothing steps on each level k (before restriction)*

ν_2 *post-smoothing steps on each level k (after interpolation)*

γ *the number of multigrid cycles on each level k or the cycling pattern ($\gamma = 1$ for V-cycling and $\gamma = 2$ for W-cycling – refer to Figure 6.6).*

(1) To obtain an initial guess $\mathbf{u}_J$ on $\mathcal{T}_J$, use the following FMG (full multigrid methods):

Solve on $\mathcal{T}_1$ for $\mathbf{u}_1$ exactly, $\mathcal{L}_1 \mathbf{u}_1 = \mathbf{g}_1$

for $k = 2, \ldots, J$, do

Interpolate to the next fine grid, $\mathbf{u}_k = P_{k-1}^k \mathbf{u}_{k-1}$,

if $k = J$, end the FMG step else continue,

Implement γ steps of $MGM(\mathbf{u}_k, \mathbf{g}_k, k)$,

end for $k = 2, \ldots, J$

(2) Based on the initial $\mathbf{u}_J$ to equation (6.20), to do ℓ steps of a γ-cycling multigrid, use

$$MGM(\mathbf{u}_J, \mathbf{g}_J, J).$$

(3) The general step of a γ-cycling multigrid $MGM(\mathbf{u}_k, \mathbf{g}_k, k)$ proceeds as follows

$$
\left\{
\begin{array}{l}
MGM(\mathbf{u}_k, \mathbf{g}_k, k): \\
\quad \textit{if } k = 1, \textit{ then} \\
\qquad\quad \textit{Solve on } \mathcal{T}_1 \textit{ for } \mathbf{u}_1 \textit{ exactly}, \ \mathcal{L}_1 \mathbf{u}_1 = \mathbf{g}_1 \\
\quad\ \textit{else, on grid } \mathcal{T}_k, \textit{ do} \\
\qquad\quad \textit{Pre-smoothing}: \mathbf{u}_k = \mathbf{Relax}_k^{\nu_1}(\mathbf{u}_k, \mathbf{g}_k), \\
\qquad\quad \mathbf{g}_{k-1} = R_k^{k-1}(\mathbf{g}_k - \mathcal{L}_k \mathbf{u}_k), \\
\qquad\quad \textit{Set the initial solution on } \mathcal{T}_{k-1} \textit{ to zero, } \mathbf{u}_{k-1} = 0, \\
\qquad\quad \textit{Implement } \gamma \textit{ steps of } MGM(\mathbf{u}_{k-1}, \mathbf{g}_{k-1}, k-1), \\
\qquad\quad \textit{Add the residual correction, } \mathbf{u}_k = \mathbf{u}_k + P_{k-1}^k \mathbf{u}_{k-1}, \\
\qquad\quad \textit{Post-smoothing}: \mathbf{u}_k = \mathbf{Relax}_k^{\nu_2}(\mathbf{u}_k, \mathbf{g}_k), \\
\quad \textit{end if } k \\
\quad \textit{end one step of } MGM(\mathbf{u}_k, \mathbf{g}_k, k).
\end{array}
\right\}
$$

The above algorithm using the self-loop idea is quite easy to implement if the reader intends to code it in C, Pascal, or Algol, or a MATLAB® script, since all these languages allow a subprogram to call itself. However, Fortran users may find it not helpful. Following the published code MG00D in [443,p.160], we may rewrite Algorithm 6.1.4 in a Fortran friendly way.

Algorithm 6.1.5. (The multigrid method (Fortran version)).

To solve the discretized PDE (6.20) on $\mathcal{T}_J$ with $k = J$, prepare as with Algorithm 6.1.4 and use the following non-recursive module for $MGM(\mathbf{u}_k, \mathbf{g}_k, k)$.

- *Set up an integer vector $ICGAM = ICGAM(1 : J)$ to control and count whether we have done γ steps on each grid $\mathcal{T}_k$.*
- *On the finest grid $\mathcal{T}_J$, assume $\mathbf{u}_J$ is the result of the initial guess from a FMG step. The general step of a γ-cycling multigrid $MGM(\mathbf{u}_k, \mathbf{g}_k, k)$ proceeds as in Table 6.1.*

Observe that both Algorithms 6.1.4 and 6.1.5 use $\mathbf{u}_j$ vectors for solution and correction i.e. they do not explicitly use the set of correction vectors $\mathbf{v}_j$ (to save storage); likewise the residual vectors r_j were replaced by the right-hand side vectors $\mathbf{g}_j$ i.e. once the FMG is done, $\mathbf{g}_j$ is over-written by residuals. Finally we remark that when the linear PDE (6.1) admits variable and highly nonsmooth coefficients (even with discontinuities or jumps), both the smoothing strategies and transfer operators may have to be adapted [460]. Refer also to the survey in [120].

Table 6.1. *Details of the MGM Algorithm 6.1.5 – non-self-recursive version.*

$MGM(\mathbf{u}_k, \mathbf{g}_k, k)$:
Label 10:
  $ICGAM(j) = 0, \; j = 1, \ldots, k$ (set the counters)
  Set the active level number, $\ell = k$
 if $\ell = 1$, goto Label 30 else continue
Label 20 :
 if $\ell < k$ and $ICGAM(\ell) = 0$,
  Set the initial solution on $\mathcal{T}_\ell$ to zero, $\mathbf{u}_\ell = 0$,
 end if $\ell < k$ and $ICGAM(\ell) = 0$
  Pre-smoothing : $\mathbf{u}_k = \mathbf{Relax}_k^{\nu_1}(\mathbf{u}_k, \mathbf{g}_k)$,
  Record one visit to $\mathcal{T}_\ell$, $ICGAM(\ell) = ICGAM(\ell) + 1$,
  Restrict $\mathbf{g}_{\ell-1} = R_k^{\ell-1}(\mathbf{g}_\ell - \mathcal{L}_\ell \mathbf{u}_\ell)$,
  Update the level number, $\ell = \ell - 1$,
 if $\ell > 1$, goto Label 20 else continue,
Label 30 :
  Solve on $\mathcal{T}_1$ for $\mathbf{u}_1$ exactly , $\mathcal{L}_1 \mathbf{u}_1 = \mathbf{g}_1$
 if $\ell = k$, goto Label 50 else continue,
Label 40 :
  Update the level number, $\ell = \ell + 1$,
  Add the residual correction, $\mathbf{u}_\ell = \mathbf{u}_\ell + P_{\ell-1}^{\ell} \mathbf{u}_{\ell-1}$,
  Post-smoothing : $\mathbf{u}_\ell = \mathbf{Relax}_\ell^{\nu_2}(\mathbf{u}_\ell, \mathbf{g}_\ell)$,
 if $\ell = k$, goto Label 50 else continue,
 if $ICGAM(\ell) < \gamma$, goto Label 20,
  else set the counter to zero, $ICGAM(\ell) = 0$,
 end if $ICGAM(\ell) < \gamma$,
 Continue from Label 20 on grid $\mathcal{T}_\ell$ (as not done γ steps yet)
Label 50 :
end one step of $MGM(\mathbf{u}_k, \mathbf{g}_k, k)$.

6.1.4 Convergence results

We now discuss the classical convergence result as shown in W. Hackbusch [258,259] for the multigrid Algorithm 6.1.4. Other results may be found in [503].

The convergence proof is consisted of two parts. Firstly relate the multigrid iteration to the two grid iteration (or if the two grid method converges, the multigrid would also converge). Secondly establish the convergence of the two-grid method from assumptions on the smoothing analysis and the properties of the transfer operators.

For solving the $\mathcal{T}_k$ equation (6.20), let the multigrid iteration matrix be $M_k^{\rm MG}$ defining the iteration from $\mathbf{u}_k^{(j)}$ to $\mathbf{u}_k^{(j+1)}$

$$\mathbf{u}_k^{(j+1)} = M_k^{\rm MG}\mathbf{u}_k^{(j)} + N_k^{\rm MG}\mathbf{g}_k \qquad \text{for } j = 0, 1, 2, \ldots, \tag{6.22}$$

and let $M_k^{\rm TG}$ be the two-grid iteration matrix (using the coarse grid $\mathcal{T}_{k-1}$). To characterize smoothing steps, let S_k^ν denote the smoothing matrix whose application corresponds to ν steps of a relaxation method. Recall that, between grids $\mathcal{T}_k$ and $\mathcal{T}_{k-1}$, Algorithm 6.1.4 implements (setting $\mathbf{v}_{k-1} = 0$, initially, $M_{k-1} = M_{k-1}^{\rm MG}$, $N_{k-1} = N_{k-1}^{\rm MG}$)

$$\begin{cases} \mathbf{u}_k = \mathbf{u}_k^{(j)} \rightsquigarrow \mathbf{u}_k^I : \nu_1 \text{ steps of } \mathbf{u}_k = S_k\mathbf{u}_k + [\ldots]\mathbf{g}_k, \\ \qquad\qquad\qquad \text{set } \mathbf{g}_{k-1} = R_k^{k-1}(\mathbf{g}_k - \mathcal{L}_k\mathbf{u}_k^I), \\ \mathbf{u}_k^I \rightsquigarrow \mathbf{v}_{k-1}^I : \gamma \text{ steps of } \mathbf{v}_{k-1} = M_{k-1}\mathbf{v}_{k-1} + N_{k-1}\mathbf{g}_{k-1}, \\ \mathbf{u}_k^I, \mathbf{v}_{k-1}^I \rightsquigarrow \hat{\mathbf{u}}_k : \ 1 \text{ step of } \mathbf{u}_k = \mathbf{u}_k^I + \mathbf{v}_k = \mathbf{u}_k^I + P_{k-1}^k\mathbf{v}_{k-1}^I, \\ \mathbf{u}_k = \hat{\mathbf{u}}_k \rightsquigarrow \mathbf{u}_k^{II} : \nu_2 \text{ steps of } \mathbf{u}_k = S_k\mathbf{u}_k + [\ldots]\mathbf{g}_k. \end{cases} \tag{6.23}$$

Each of the above five lines will be expanded in Table 6.2.

Here the notation [. . .] denotes terms that are not depending on **g** and **u** and it remains to determine $N_k^{\rm MG}$. To complete the task of relating the iteration matrix $M_k^{\rm MG}$ to $M_{k-1}^{\rm MG}$ (and $M_k^{\rm TG}$), we now consider how to express $N_{k-1}^{\rm MG}$ in terms of $M_{k-1}^{\rm MG}$ and $\mathcal{L}_{k-1}$.

Lemma 6.1.6. (Hackbusch [258]). *Let a linear and converging iteration for some linear system $A\mathbf{u} = \mathbf{f}$ such as (1.1) be denoted by*

$$\mathbf{u}^{(j+1)} = M\mathbf{u}^{(j)} + N\mathbf{f}, \qquad j = 0, 1, 2, \ldots \tag{6.26}$$

where M is the iteration matrix. If A is invertible, then N can be expressed by

$$N = (I - M)A^{-1}. \tag{6.27}$$

Proof. Let **u** be the exact solution to $A\mathbf{u} = \mathbf{f}$. Then $\mathbf{u} = M\mathbf{u} + N\mathbf{f}$ i.e. $(I - M)A^{-1}f = N\mathbf{f}$ for any given **f**. So the proof is complete. ■

Now applying Lemma 6.1.6 to $\mathbf{v}_{k-1}^{(j+1)} = M_{k-1}^{\rm MG}\mathbf{v}_{k-1}^{(j)} + N_{k-1}^{\rm MG}\mathbf{g}_{k-1}$, for $j = 0, 1, 2, \ldots,$ we obtain that

$$N_{k-1}^{\rm MG} = (I - M_{k-1}^{\rm MG})\mathcal{L}_{k-1}^{-1}. \tag{6.28}$$

Table 6.2. *Derivation of the convergence matrix for (6.22) and (6.23).*

Consider (6.23). The first line can be denoted by $\mathbf{u}_k^I = S_k^{\nu_1}\mathbf{u}_k^{(j)} + [\ldots]\mathbf{g}_k$. The (crucial) third line may be reformulated, using the second line $\mathbf{g}_{k-1} = R_k^{k-1}(\mathbf{g}_k - \mathcal{L}_k\mathbf{u}_k^I)$, as

$$\left.\begin{aligned}
&\text{Cycle 0}\ \ \mathbf{v}_{k-1} = 0\\
&\text{Cycle 1}\ \ \mathbf{v}_{k-1} = N_{k-1}\mathbf{g}_{k-1} = N_{k-1}\left(-R_k^{k-1}\mathcal{L}_k\right)\mathbf{u}_k^I + [\ldots]\mathbf{g}_k\\
&\text{Cycle 2}\ \ \mathbf{v}_{k-1} = M_{k-1}\mathbf{v}_{k-1} + N_{k-1}\mathbf{g}_{k-1}\\
&\qquad\qquad\quad = \sum_{j=0}^{1} M_{k-1}^j N_{k-1}\left(-R_k^{k-1}\mathcal{L}_k\right)\mathbf{u}_k^I + [\ldots]\mathbf{g}_k\\
&\text{Cycle 3}\ \ \mathbf{v}_{k-1} = M_{k-1}\mathbf{v}_{k-1} + N_{k-1}\mathbf{g}_{k-1}\\
&\qquad\qquad\quad = \sum_{j=0}^{2} = M_{k-1}^j N_{k-1}\left(-R_k^{k-1}\mathcal{L}_k\right)\mathbf{u}_k^I + [\ldots]\mathbf{g}_k\\
&\quad\vdots \qquad\qquad\quad \vdots\\
&\text{Cycle } \gamma\ \ \mathbf{v}_{k-1}^I = \mathbf{v}_{k-1} = M_{k-1}\mathbf{v}_{k-1} + N_{k-1}\mathbf{g}_{k-1}\\
&\qquad\qquad\quad = \sum_{j=0}^{\gamma-1} M_{k-1}^j N_{k-1}\left(-R_k^{k-1}\mathcal{L}_k\right)\mathbf{u}_k^I + [\ldots]\mathbf{g}_k.
\end{aligned}\right\} \tag{6.24}$$

Now we are ready to formulate the complete algorithm

$$\mathbf{u}_k^{(j+1)} = M_k^{\mathrm{MG}}\mathbf{u}_k^{(j)} + N_k^{\mathrm{MG}}\mathbf{g}_k = \mathbf{u}_k^{II}.$$

Noting that the fifth line is $\mathbf{u}_k^{(j+1)} = \mathbf{u}_k^{II} = S_k^{\nu_2}\hat{\mathbf{u}}_k + [\ldots]\mathbf{g}_k$, we obtain

$$\mathbf{u}_k^{(j+1)} = \mathbf{u}_k^{II} = S_k^{\nu_2}\hat{\mathbf{u}}_k + [\ldots]\mathbf{g}_k = S_k^{\nu_2}\underbrace{\left[\mathbf{u}_k^I + P_{k-1}^k\mathbf{v}_{k-1}^I\right]}_{\hat{\mathbf{u}}_k} + [\ldots]\mathbf{g}_k$$

$$\begin{aligned}
&= S_k^{\nu_2}\Big[I + P_{k-1}^k\sum_{j=0}^{\gamma-1} M_{k-1}^j N_{k-1}\left(-R_k^{k-1}\mathcal{L}_k\right)\Big]\mathbf{u}_k^I + [\ldots]\mathbf{g}_k\\
&= S_k^{\nu_2}\left[I - P_{k-1}^k\sum_{j=0}^{\gamma-1} M_{k-1}^j N_{k-1}R_k^{k-1}\mathcal{L}_k\right]\left(S_k^{\nu_1}\mathbf{u}_k^{(j)} + [\ldots]\mathbf{g}_k\right) + [\ldots]\mathbf{g}_k\\
&= S_k^{\nu_2}\Big[I - P_{k-1}^k\sum_{j=0}^{\gamma-1} M_{k-1}^j N_{k-1}R_k^{k-1}\mathcal{L}_k\Big]S_k^{\nu_1}\mathbf{u}_k^{(j)} + N_k^{\mathrm{MG}}\mathbf{g}_k\\
&= \underbrace{S_k^{\nu_2}\left[I - P_{k-1}^k\sum_{j=0}^{\gamma-1}(M_{k-1}^{\mathrm{MG}})^j N_{k-1}^{\mathrm{MG}}R_k^{k-1}\mathcal{L}_k\right]S_k^{\nu_1}}_{M_k^{\mathrm{MG}}}\mathbf{u}_k^{(j)} + N_k^{\mathrm{MG}}\mathbf{g}_k.
\end{aligned} \tag{6.25}$$

Substituting this into (6.25), we can simplify the multigrid iteration matrix to

$$\begin{aligned} M_k^{\rm MG} &= S_k^{\nu_2}\left[I - P_{k-1}^k \sum_{j=0}^{\gamma-1}(M_{k-1}^{\rm MG})^j \left(I - M_{k-1}^{\rm MG}\right)\mathcal{L}_{k-1}^{-1}R_k^{k-1}\mathcal{L}_k\right]S_k^{\nu_1} \\ &= S_k^{\nu_2}\left[I - P_{k-1}^k \left(\sum_{j=0}^{\gamma-1}(M_{k-1}^{\rm MG})^j \left(I - M_{k-1}^{\rm MG}\right)\right)\mathcal{L}_{k-1}^{-1}R_k^{k-1}\mathcal{L}_k\right]S_k^{\nu_1} \end{aligned}$$

That is,

$$M_k^{\rm MG} = S_k^{\nu_2}\Big[I - P_{k-1}^k \underbrace{\left(I - (M_{k-1}^{\rm MG})^\gamma\right)\mathcal{L}_{k-1}^{-1}}_{\text{from all coarse grids}} R_k^{k-1}\mathcal{L}_k\Big]S_k^{\nu_1}. \tag{6.29}$$

To simplify (6.29), we shall formulate the two-grid iteration matrix $M_k^{\rm TG}$, using

$$\mathbf{v}_k = \mathcal{L}_{k-1}^{-1}\mathbf{g}_{k-1},\ \mathbf{g}_{k-1} = R_k^{k-1}(\mathbf{g}_k - \mathcal{L}_k\mathbf{u}_k),\ \mathbf{u}_k = S_k^{\nu_1}\mathbf{u}_k^{(j)} + [\ldots]\mathbf{g}_k.$$

Again we use Lemma 6.1.6 to derive the following

$$\begin{aligned} \mathbf{u}_k^{(j+1)} &= M_k^{\rm TG}\mathbf{u}_k^{(j)} + N_k^{\rm TG}\mathbf{g}_k \\ &= S_k^{\nu_2}\left[\mathbf{u}_k + P_{k-1}^k\mathbf{v}_k\right] \\ &= S_k^{\nu_2}\left[I - P_{k-1}^k\mathcal{L}_{k-1}^{-1}R_k^{k-1}\mathcal{L}_k\right]S_k^{\nu_1}\mathbf{u}_k^{(j)} + N_k^{\rm TG}\mathbf{g}_k \end{aligned} \tag{6.30}$$

and, furthermore, to deduce from (6.29) that

$$\begin{cases} M_k^{\rm TG} = S_k^{\nu_2}\left[I - P_{k-1}^k\mathcal{L}_{k-1}^{-1}R_k^{k-1}\mathcal{L}_k\right]S_k^{\nu_1}, \\ M_k^{\rm MG} = M_k^{\rm TG} + S_k^{\nu_2}P_{k-1}^k(M_{k-1}^{\rm MG})^\gamma\mathcal{L}_{k-1}^{-1}R_k^{k-1}\mathcal{L}_kS_k^{\nu_1}. \end{cases} \tag{6.31}$$

This latter equation is the most important relation linking the multigrid convergence rate to the two-grid rate.

Thus the convergence analysis for the multigrid method [258,460] amounts to estimating the norm of the iteration matrix

$$\left\|M_k^{\rm MG}\right\| \le \left\|M_k^{\rm TG}\right\| + \left\|S_k^{\nu_2}P_{k-1}^k\right\|\left\|M_{k-1}^{\rm MG}\right\|^\gamma\left\|\mathcal{L}_{k-1}^{-1}R_k^{k-1}\mathcal{L}_kS_k^{\nu_1}\right\|. \tag{6.32}$$

Theorem 6.1.7. *Let σ be the upper bound for the convergence rate of two grid methods and the transfer operator-related terms in (6.32) are uniformly bounded i.e.*

$$\left\|M_k^{TG}\right\| \le \sigma \qquad \text{and} \qquad \left\|S_k^{\nu_2}P_{k-1}^k\right\|\left\|\mathcal{L}_{k-1}^{-1}R_k^{k-1}\mathcal{L}_kS_k^{\nu_1}\right\| \le C.$$

Then the sequence η_k, recursively defined by,

$$\eta_k = \sigma + C\eta_{k-1}^\gamma, \qquad \text{with} \qquad \eta_1 = \sigma,\ \ k = 2, \ldots, J,$$

provides the upper bound for $\|M_k^{MG}\|$ *i.e.*

$$\left\|M_k^{MG}\right\| \leq \eta_k.$$

If we assume the fast convergence of two-grid methods i.e. $\max(4C\sigma,\ 2\sigma) < 1$ *and* $\gamma = 2$, *then*

$$\left\|M_k^{MG}\right\| \leq \eta = \lim_{k\to\infty} \eta_k = \frac{1-\sqrt{1-4C\sigma}}{2C} = \frac{2\sigma}{1+\sqrt{1-4C\sigma}} \leq 2\sigma.$$

Proof. The limit equation is $C\eta^2 - \eta + \sigma = 0$ which is solvable if $4C\sigma < 1$, and $2\sigma < 1$ means that the multigrid method is convergent. See [258,259, 460]. ■

Further theories on verifying the satisfaction of the assumptions in Theorem 6.1.7 require specifying the exact problem classes [495,258,259,460,503]. Clearly the two extremes exist.

(1) If the good smoothers are hard to find, or if the two grid methods are not convergent, multigrid methods will not converge.
(2) If the operator $\mathcal{L}$ (or the matrix $\mathcal{L}_k$) is SPD, the multigrid convergence is guaranteed – this is the case (Chapter 3) when the Gauss–Seidel method is actually convergent (maybe slowly converging).

It may be remarked that, while it is justified to argue that multigrid methods offer a wonderful idea which is in danger of being spoiled by the traditional and restrictive relaxation methods (and hence one should try to stay away from them to achieve robustness), the multigrid convergence theory never specifies what smoothers should be used. In fact, a lot of problems beyond the strongly elliptic PDEs have been solved by other smoothers e.g. the node colouring GS smoothers [326,413,4], the ILU smoother [490] and the SPAI smoother [452]. See also [460] for discussion of other generalizing possibilities.

6.2 Multigrid method for nonlinear PDEs

We now describe the multigrid method for solving a nonlinear elliptic PDE

$$\begin{cases} \mathcal{N}u = f, & \Omega \subset \mathbb{R}^d, \\ u(p) = g(p), & p \in \Gamma = \partial\Omega. \end{cases} \tag{6.33}$$

Using the previous setting of J grids $\mathcal{T}_k$, $k = 1, \ldots, J$, we discretize (6.33) on $\mathcal{T}_k$ as

$$\mathcal{N}_k \mathbf{u}_k = \mathbf{g}_k. \tag{6.34}$$

For this nonlinear case, one solution strategy is to apply the linear multigrid method in the previous section after linearizing the the above nonlinear PDE by the Newton type global linearization techniques [72,460,316]. Here, instead, we mainly discuss the genuinely nonlinear multigrid method in the full approximation scheme (due to [72]) which provides an alternative and neater treatment of (6.33).

In the nonlinear case, the transfer operators may still be taken as those applicable to a linear PDE. However, we need to comment on nonlinear relaxation schemes as we must use a form of the Newton method to generalize a linear relaxation scheme. Take the Gauss–Seidel (GS) method as an example – there are two main variants depending on whether a Newton method is used first or second. Firstly in the *Newton–GS* method, we use the Newton method for global linearization (to obtain linear iterations involving the Jacobian matrix [316,460]) and then use the GS for the linear system as in the linear PDE case. Secondly in the *GS–Newton* method, we use the GS to reduce the dimension of each nonlinear equation to a single variable (nonlinear) case and then apply the one-dimensional Newton method for iterations. See also [124]. Similarly one can develop the Jacobi–Newton and SOR–Newton methods.

6.2.1 Full approximation schemes

The essence of the full approximation scheme (FAS) is to suggest a new way of computing the residual correction $\mathbf{v}_k$, i.e. consider how to derive a residual correction equation similar to (6.21) in the linear case.

It turns out that a similar equation to (6.21) is the following [72]

$$\mathcal{N}_k (\overline{\mathbf{u}}_k + \mathbf{v}_k) = r_k + \mathcal{N}_k \overline{\mathbf{u}}_k, \tag{6.35}$$

assuming $\overline{\mathbf{u}}_k$ is the current approximation to (6.33) on grid $\mathcal{T}_k$. Of course, the idea of a multigrid method is not to solve (6.35) for $\mathbf{v}_k$ on $\mathcal{T}_k$ but on the coarse grid $\mathcal{T}_{k-1}$

$$\mathcal{N}_{k-1} (\overline{\mathbf{u}}_{k-1} + \mathbf{v}_{k-1}) = r_{k-1} + \mathcal{N}_{k-1} \overline{\mathbf{u}}_{k-1}, \tag{6.36}$$

where $\overline{\mathbf{u}}_{k-1} = R_k^{k-1} \overline{\mathbf{u}}_k$, $r_{k-1} = R_k^{k-1} r_k$ and $\overline{\mathbf{u}}_{k-1} = R_k^{k-1} \overline{\mathbf{u}}_k$. Clearly if $\mathcal{N}$ is linear, all the above equations reduce to the linear ones that have been seen already in the previous section.

In realizing the solution of (6.36), we have to introduce an intermediate variable $\widetilde{\mathbf{u}}_{k-1}$ and solve

$$\mathcal{N}_{k-1}\widetilde{\mathbf{u}}_{k-1} = \underbrace{r_{k-1} + \mathcal{N}_{k-1}\overline{\mathbf{u}}_{k-1}}_{\text{known}},$$

before interpolating back to the fine grid: $\mathbf{v}_k = P_{k-1}^k(\widetilde{\mathbf{u}}_{k-1} - \overline{\mathbf{u}}_{k-1})$.

6.2.2 Nonlinear multigrid algorithms

We are now ready to state the FAS version of a nonlinear multigrid algorithm. Denote again by $\mathbf{Relax}_k^\nu(u, f)$ the result of ν steps of some relaxation method for (6.34) on grid $\mathcal{T}_k$ i.e. for

$$\mathcal{N}_k\mathbf{u}_k = \mathbf{g}_k \tag{6.37}$$

with the initial guess $\mathbf{u}_k = u$ and $\mathbf{g}_k = f$.

Algorithm 6.2.8. (The nonlinear MG).

To solve the discretized PDE (6.37) on $\mathcal{T}_J$ with level $k = J$, assume we have set up these multigrid parameters:

- ν_1 *pre-smoothing steps on each level k (before restriction)*
- ν_2 *post-smoothing steps on each level k (after interpolation)*
- γ *the number of multigrid cycles on each level k or the cycling pattern (usually $\gamma = 1$ for V-cycling and $\gamma = 2$ for W-cycling).*

(1) To obtain an initial guess $\mathbf{u}_J$ on $\mathcal{T}_J$, use the following FMG (full multigrid methods):

Solve on $\mathcal{T}_1$ for $\mathbf{u}_1$ accurately, $\mathcal{N}_1\mathbf{u}_1 = \mathbf{g}_1$

for $k = 2, \ldots, J$, do

Interpolate to the next fine grid, $\mathbf{u}_k = P_{k-1}^k\mathbf{u}_{k-1}$,

if $k = J$, end the FMG step else continue,

Implement γ steps of $FAS(\mathbf{u}_k, \mathbf{g}_k, k)$,

end for $k = 2, \ldots, J$

(2) Based on the initial guess $\mathbf{u}_J$ to equation (6.34), to do ℓ steps of a γ-cycling nonlinear multigrid, use

$$FAS(\mathbf{u}_J, \mathbf{g}_J, J).$$

(3) The general step of a γ-cycling nonlinear multigrid $FAS(\mathbf{u}_k, \mathbf{g}_k, k)$ proceeds as follows

$$\left\{\begin{array}{l} FAS(\mathbf{u}_k, \mathbf{g}_k, k): \\ \quad \textit{if } k = 1, \ \textit{then} \\ \qquad \textit{Solve on } \mathcal{T}_1 \textit{ for } \mathbf{u}_1 \textit{ accurately}, \ \mathcal{N}_1 \mathbf{u}_1 = \mathbf{g}_1 \\ \quad \textit{else, on grid } \mathcal{T}_k, \textit{ do} \\ \qquad \textit{Pre-smoothing}: \mathbf{u}_k = \mathbf{Relax}_k^{\nu_1}(\mathbf{u}_k, \mathbf{g}_k), \\ \qquad \textit{Restrict to the coarse grid, } \mathbf{u}_{k-1} = R_k^{k-1} \mathbf{u}_k \\ \qquad \textit{Compute } \mathbf{g}_{k-1} = R_k^{k-1}(\mathbf{g}_k - \mathcal{N}_k \mathbf{u}_k) + \mathcal{N}_{k-1} \mathbf{u}_{k-1}, \\ \qquad \textit{Set the initial solution on } \mathcal{T}_{k-1} \textit{ as } \widetilde{\mathbf{u}}_{k-1} = \mathbf{u}_{k-1}, \\ \qquad \textit{Implement } \gamma \textit{ steps of } FAS(\widetilde{\mathbf{u}}_{k-1}, \mathbf{g}_{k-1}, k-1), \\ \qquad \textit{Add the residual correction, } \mathbf{u}_k = \mathbf{u}_k + P_{k-1}^k (\widetilde{\mathbf{u}}_{k-1} - \mathbf{u}_{k-1}), \\ \qquad \textit{Post-smoothing}: \mathbf{u}_k = \mathbf{Relax}_k^{\nu_2}(\mathbf{u}_k, \mathbf{g}_k), \\ \quad \textit{end if } k \\ \quad \textit{end one step of } FAS(\mathbf{u}_k, \mathbf{g}_k, k). \end{array}\right\}$$

Similarly to Table 6.1 for a non-recursive FAS, we can develop a non-recursive FAS algorithm (especially suitable for Fortran implementation):

Algorithm 6.2.9. (The nonlinear MG (Fortran version)).

To solve the discretized PDE (6.34) on $\mathcal{T}_J$ with $k = J$, prepare as with Algorithm 6.2.8 and use the following non-recursive module for $FAS(\widetilde{\boldsymbol{u}}_k, \mathbf{g}_k, k)$.

(1) Set up an integer vector $ICGAM = ICGAM(1 : J)$ to control and count whether we have done γ steps on each grid $\mathcal{T}_k$.
(2) On the finest grid $\mathcal{T}_J$, assume $\widetilde{\boldsymbol{u}}_J$ is the result of the initial guess from a FMG step. Then the general step of a γ-cycling multigrid $FAS(\widetilde{\boldsymbol{u}}_k, \mathbf{g}_k, k)$, modifying Algorithm 6.1.5, proceeds as in Table 6.3.

To help the reader to understand when the main solution quantities $\widetilde{\boldsymbol{u}}_\ell$, $\mathbf{u}_\ell$ on each level $\mathcal{T}_\ell$ are updated, we show an illustration of a W-cycling (i.e. $\gamma = 2$) in Figure 6.6 of a $J = 4$ level method. There these two variables are marked as w, u respectively, with u indicating an update from restriction of the fine level variable and w indicating an update due to relaxation or interpolation. The figure is produced by the Mfile `ch6_w.m`.

Table 6.3. *Details of the FAS module in an alternative and non-self-recursive version as used by Algorithm 6.2.8*

$FAS(\widetilde{\mathbf{u}}_k, \mathbf{g}_k, k)$:
Label 10:
  $ICGAM(j) = 0, \; j = 1, \ldots, k$ (set the counters)
  Set the active level number, $\ell = k$
 if $\ell = 1$, goto Label 30 else continue
Label 20 :
  Pre-smoothing : $\widetilde{\mathbf{u}}_\ell = \mathbf{Relax}_\ell^{\nu_1}(\widetilde{\mathbf{u}}_\ell, \mathbf{g}_\ell)$,
  Restrict to the coarse grid, $\mathbf{u}_{\ell-1} = R_\ell^{\ell-1}\widetilde{\mathbf{u}}_\ell$
  Compute $\mathbf{g}_{\ell-1} = R_\ell^{\ell-1}(\mathbf{g}_\ell - \mathcal{N}_\ell\widetilde{\mathbf{u}}_\ell) + \mathcal{N}_{\ell-1}\mathbf{u}_{\ell-1}$,
  Update the level number, $\ell = \ell - 1$,
 if $\ell < k$ and $ICGAM(\ell) = 0$,
  Set the initial solution on $\mathcal{T}_\ell$, $\widetilde{\mathbf{u}}_\ell = \mathbf{u}_\ell$,
 end if $\ell < k$ and $ICGAM(\ell) = 0$
  Record one visit to $\mathcal{T}_\ell$, $ICGAM(\ell) = ICGAM(\ell) + 1$,
 if $\ell > 1$, goto Label 20 else continue,
Label 30 :
  Solve on $\mathcal{T}_1$ for $\widetilde{\mathbf{u}}_1$ exactly , $\mathcal{L}_1\widetilde{\mathbf{u}}_1 = \mathbf{g}_1$
 if $\ell = k$, goto Label 50 else continue,
Label 40 :
  Update the level number, $\ell = \ell + 1$,
  Add the correction, $\widetilde{\mathbf{u}}_\ell = \widetilde{\mathbf{u}}_\ell + P_{\ell-1}^{\ell}(\widetilde{\mathbf{u}}_{\ell-1} - \mathbf{u}_{\ell-1})$,
  Post-smoothing : $\widetilde{\mathbf{u}}_\ell = \mathbf{Relax}_\ell^{\nu_2}(\widetilde{\mathbf{u}}_\ell, \mathbf{g}_\ell)$,
 if $\ell = k$, goto Label 50 else continue,
 if $ICGAM(\ell) < \gamma$, goto Label 20,
  else set the counter to zero, $ICGAM(\ell) = 0$,
 end if $ICGAM(\ell) < \gamma$,
 Continute from Label 20 on grid $\mathcal{T}_\ell$ (as not done γ steps yet)
Label 50 :
 end one step of $FAS(\mathbf{u}_k, \mathbf{g}_k, k)$.

6.3 Multigrid method for linear integral equations

The multigrid method as described in Algorithm 6.1.4 can naturally be applied [22,24,259,274,420,132,125] to a linear integral equation such as (1.80)

$$(I - \mathcal{K})\mathbf{u} = \mathbf{g}, \qquad \text{with } (\mathcal{K}\psi)(p) = \int_{\partial\Omega} K(p, q)\psi(q)dS_q, \quad p \in \partial\Omega, \tag{6.38}$$

which may be discretized, on grid $\mathcal{T}_k$ that has n_k nodal points, as (see (1.85) and (1.86))

$$(I - \mathcal{K}_k)\mathbf{u}_k = \mathbf{g}_k. \tag{6.39}$$

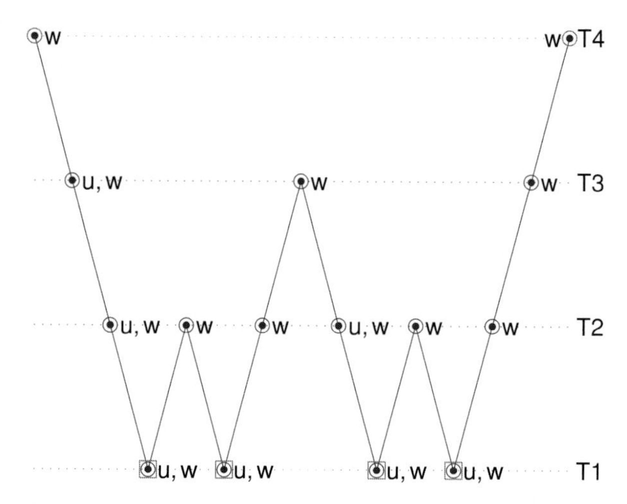

Figure 6.6. Illustration of a W-cycling (for one step of a $J = 4$ level multigrid method, $\gamma = 2$) and when the solution variables $w = \widetilde{\mathbf{u}}_\ell$, $u = \mathbf{u}_\ell$ are updated. $\mathcal{T}_1$ is the coarsest grid and $\square$ indicates an accurate solution on $\mathcal{T}_1$. If this figure is viewed for illustrating Algorithm 6.1.4, one can equate u to the action of initializing the correction vector $\mathbf{v}_\ell$ and w to the solution $\mathbf{u}_\ell$.

We assume that $n_{k-1} < n_k$ for $k = 2, 3, \ldots, J$; depending on the choice of the order of boundary elements and the dimension of Ω, the grids $\mathcal{T}_k$ do not have to be nested if we adopt the standard approach (shown below). Here in the 1D case, the boundary is a closed curve and the set up for the grid $\mathcal{T}_k$ is the same as the FDM or FEM while for the case of a 2D boundary (surface), the commonly used approach is based on triangulation for the surface $\partial\Omega$ (see [15,136]). For generality, we assume $\mathcal{K}$, $\mathbf{g}$, $\mathcal{K}_k$ are defined in some Banach space $X = X(\partial\Omega)$; for studying numerical properties of (6.39), more specific functional spaces such as $C^m(\partial\Omega)$ in [339] and $H^m(\partial\Omega)$ in [488] for piecewise smooth functions may be used as appropriate.

Although the standard transfer and smoothing operators for PDEs may equally be used for the integral equations, in practice, much easier options have been adopted. More specifically, the restriction is by the simple injection operator (6.19) if the grids are nested while the interpolation is by the Nystrom interpolation defined by

$$\mathbf{v}_k(p) = \mathcal{K}_{k-1}(p)\mathbf{v}_{k-1} + r_k, \qquad p \in \partial\Omega \backslash \mathcal{T}_{k-1} \tag{6.40}$$

which is essentially similar to the 'built-in' Picard iteration used for smoothing

$$\mathbf{u}_k(p) = \mathcal{K}_k(p)\mathbf{u}_k + \mathbf{g}_k, \qquad p \in \mathcal{T}_k \tag{6.41}$$

instead of considering the relaxation methods such as the Gauss–Seidel. If the grids are not nested, the convenient restriction operator is again by the Nystrom method (as in (6.40))

$$\mathbf{v}_{k-1}(p) = \mathcal{K}_k(p)\mathbf{v}_k + r_k, \qquad p \in \partial\Omega\backslash\mathcal{T}_k.$$

Here the collocation method (§1.7) provides the vital connection of a matrix equation

$$(I - \mathcal{K}_k(p))\,\mathbf{u}_k = \mathbf{g}_k(p), \qquad p \in \mathcal{T}_k,\ \mathbf{u}_k, \mathbf{g}_k \in \mathbb{R}^{n_k}$$

to an operator equation

$$(I - \mathcal{K}_k(p))\,\mathbf{u}_k(p) = \mathbf{g}_k(p), \qquad p \in \partial\Omega,\ \mathbf{u}_k(p) \in X$$

and the (Nystrom) interpolation

$$\mathbf{u}_k(p) = \mathcal{K}_k(p)\mathbf{u}_k + \mathbf{g}_k(p), \qquad p \in X\backslash\mathcal{T}_k,\ \mathbf{u}_k(p) \in X \tag{6.42}$$

where $\mathbf{u}_k$ is defined on grid $\mathcal{T}_k$. Clearly all coarse grid functions are also defined on fine grid $\mathcal{T}_k$. More formally, one needs to define a projection that takes a function from $\partial\Omega$ to a vector $\mathcal{T}_k$.

Note that for the linear system (1.1), the Picard iteration corresponds to the simple operator splitting (see §3.3.1) $A = M - N$ with $M = I$ and $N = (I - A)$.

6.3.1 The case of a second kind equation with a compact operator

The multigrid solution of the second kind equation (6.38) with a compact operator $\mathcal{K}$ was studied in [22,274,259,125] among others. The solution of the first kind equation (i.e. equation (6.38) without the term I) is harder; see [265,476]. As with the PDE case, definition of a two-grid method is essential.

As shown in [430,431], a compact operator can provide the natural smoother for a residual function i.e. $\mathcal{K}w$ is more smooth than w for any w. Based on (6.41) and (6.42), the first two-grid method (TG-1) by Atkinson [22] uses the residual correction idea: let $\overline{\mathbf{u}}_k = \mathbf{u}_k^{(j)}$ be the jth approximation of (6.39) and $r_k = \mathbf{g}_k - (I - \mathcal{K}_k)\mathbf{u}_k$. Then the residual equation $(I - \mathcal{K}_k)\mathbf{v}_k = r_k$ is solved on the coarse grid $\mathcal{T}_{k-1}$

$$(I - \mathcal{K}_{k-1})\mathbf{v}_{k-1} = r_k(p), \qquad p \in \mathcal{T}_{k-1}, \tag{6.43}$$

giving the overall method (with $j = 1, 2, \ldots$)

$$\begin{aligned} \mathbf{u}_k^{(j+1)} &= \mathbf{u}_k^{(j)} + \mathbf{v}_{k-1} = \mathbf{u}_k^{(j)} + (I - \mathcal{K}_{k-1})^{-1} r_k \\ &= (I - \mathcal{K}_{k-1})^{-1} (\mathcal{K}_k - \mathcal{K}_{k-1}) \mathbf{u}_k^{(j)} + (I - \mathcal{K}_{k-1})^{-1} \mathbf{g}_k. \end{aligned} \tag{6.44}$$

Although the method appears to be reasonable in the multigrid philosophy, its convergence is better seen from rewriting (6.44) as

$$\begin{aligned} \mathbf{u}_k^{(j+2)} &= (I - \mathcal{K}_{k-1})^{-1} (\mathcal{K}_k - \mathcal{K}_{k-1}) \mathbf{u}_k^{(j+1)} + (I - \mathcal{K}_{k-1})^{-1} \mathbf{g}_k \\ &= \left[(I - \mathcal{K}_{k-1})^{-1} (\mathcal{K}_k - \mathcal{K}_{k-1})\right]^2 \mathbf{u}_k^{(j)} + \\ &\quad \left[I + (I - \mathcal{K}_{k-1})^{-1} (\mathcal{K}_k - \mathcal{K}_{k-1})\right] (I - \mathcal{K}_{k-1})^{-1} \mathbf{g}_k \\ &= (\mathcal{L}_k - \mathcal{L}_{k-1})^2 \mathbf{u}_k^{(j)} + [I + (\mathcal{L}_k - \mathcal{L}_{k-1})] (I - \mathcal{K}_{k-1})^{-1} \mathbf{g}_k, \end{aligned} \tag{6.45}$$

where $\mathcal{L}_\ell = (I - \mathcal{K}_{k-1})^{-1} \mathcal{K}_\ell$ behaves like $\mathcal{K}_\ell$ if $\mathcal{K}$ is compact. As shown later, the convergence of (6.44) follows that of (6.45).

The second two-grid method (TG-2) of Atkinson [22] combines an explicit presmoothing step (6.41) with TG-1:

$$\begin{cases} \overline{\mathbf{u}}_k^{(j)} = \mathcal{K}_k \mathbf{u}_k^{(j)} + \mathbf{g}_k, \\ \mathbf{u}_k^{(j+1)} = \overline{\mathbf{u}}_k^{(j)} + (I - \mathcal{K}_{k-1})^{-1} \overbrace{\left[\mathbf{g}_k - (I - \mathcal{K}_k)\overline{\mathbf{u}}_k^{(j)}\right]}^{\text{residual for } \overline{\mathbf{u}}_k} \\ \qquad = (I - \mathcal{K}_{k-1})^{-1} (\mathcal{K}_k - \mathcal{K}_{k-1}) \mathbf{u}_k^{(j)} + [I + (I - \mathcal{K}_{k-1})^{-1} \mathcal{K}_k] \mathbf{g}_k, \end{cases} \tag{6.46}$$

giving (with $j = 1, 2, \ldots$)

$$\mathbf{u}_k^{(j+1)} = (\mathcal{L}_k - \mathcal{L}_{k-1}) \mathcal{K}_k \mathbf{u}_k^{(j)} + (I + \mathcal{L}_k) \mathbf{g}_k.$$

The convergence theory [22,274] for the two methods makes use of the results of [19], under mild assumptions for the compact operator $\mathcal{K}$, the boundedness of $(I - \mathcal{K})^{-1}$, and the discretizations $\mathcal{K}_k$,

$$\begin{cases} \lim_{k\to\infty} \|(\mathcal{K} - \mathcal{K}_k)M\| = 0, \\ \lim_{k\to\infty} \|(\mathcal{K} - \mathcal{K}_k)\mathcal{K}_k\| = 0, \\ \lim_{k\to\infty} \|(\mathcal{K} - \mathcal{K}_k)x\| = 0, \quad x \in X, \\ \lim_{k\to\infty} \|\mathcal{K} - \mathcal{K}_k\| \neq 0, \end{cases} \tag{6.47}$$

where $M : X \to X$ is any compact operator. Using (6.47), it is not difficult to see that both TG-1 and TG-2 converge asymptotically since $\mathcal{L}_k$ is also compact (from Lemma 4.7.12) and

TG-1	$\lim_{k\to\infty} \left\|(\mathcal{L}_k - \mathcal{L}_{k-1})^2\right\| = \lim_{k\to\infty} \|[(\mathcal{L} - \mathcal{L}_{k-1}) - (\mathcal{L} - \mathcal{L}_k)] (\mathcal{L}_k - \mathcal{L}_{k-1})\| = 0,$
TG-2	$\lim_{k\to\infty} \|(\mathcal{L}_k - \mathcal{L}_{k-1})\mathcal{K}_k\| = \lim_{k\to\infty} \|[(\mathcal{L} - \mathcal{L}_{k-1}) - (\mathcal{L} - \mathcal{L}_k)]\mathcal{K}_k\| = 0,$

where $\mathcal{L} = (I - \mathcal{K}_{k-1})^{-1}\mathcal{K}$.

Clearly, if n_k, n_{k-1} are large enough, we do not need many iterations of TG-1 or TG-2 because the convergence rates are not only less than 1 but also amazingly towards 0 (differently from the PDE case earlier).

Generalization of the two-grid methods to a multigrid case follows closely the usual multigrid set up as in Algorithm 6.1.4. To summarize, the TG-1 based method will have: (a) no explicit smoothing step; (b) restriction by injection or the Nystrom method (if not nested); (c) interpolation by the Nystrom method and the TG-2 based method will have: (i) one step of a Picard smoothing iteration; (ii) restriction by injection or the Nystrom method (if not nested); (iii) interpolation by the Nystrom method. For simplicity, we only show the TG-2 based multigrid algorithm below.

Algorithm 6.3.10. (The two-grid method for integral equations).

To solve the discretized second kind integral equation (6.39) on $\mathcal{T}_J$ with $k = J$, assume P^k_{k-1} denotes the Nystrom interpolation and we have set up the multigrid parameter:

γ *the number of multigrid cycles on each level k or the cycling pattern (usually $\gamma = 1$ for V-cycling and $\gamma = 2$ for W-cycling).*

(1) To obtain an initial guess $\mathbf{u}_J$ on $\mathcal{T}_J$, use the following FMG:

Solve on $\mathcal{T}_1$ for $\mathbf{u}_1$ exactly, $(I - \mathcal{K}_1)\mathbf{u}_1 = \mathbf{g}_1$

for $k = 2, \ldots, J$, do

Interpolate to the next fine grid, $\mathbf{u}_k = P^k_{k-1}\mathbf{u}_{k-1}$,

if $k = J$, end the FMG step else continue,

Implement γ steps of $MIE(\mathbf{u}_k, \mathbf{g}_k, k)$,

end for $k = 2, \ldots, J$

(2) Based on the initial guess $\mathbf{u}_J$ to equation (6.20), to do ℓ steps of a γ-cycling multigrid, use

$$MIE(\mathbf{u}_J, \mathbf{g}_J, J).$$

(3) The general step of a γ*-cycling multigrid* $MIE(\mathbf{u}_k, \mathbf{g}_k, k)$ *proceeds as follows*

$$\left\{\begin{array}{l} MIE(\mathbf{u}_k, \mathbf{g}_k, k): \\ \quad \textit{if } k = 1, \textit{ then} \\ \qquad \textit{Solve on } \mathcal{T}_1 \textit{ for } \mathbf{u}_1 \textit{ exactly}, \ (I - \mathcal{K}_1)\mathbf{u}_1 = \mathbf{g}_1 \\ \quad \textit{else, on grid } \mathcal{T}_k, \textit{ do} \\ \qquad \textit{Pre-smoothing}: \mathbf{u}_k = \mathcal{K}_k\mathbf{u}_k + \mathbf{g}_k, \\ \qquad \mathbf{g}_{k-1} = \mathbf{g}_k - \mathcal{L}_k\mathbf{u}_k, \\ \qquad \textit{Set the initial solution on } \mathcal{T}_{k-1} \textit{ to zero, } \mathbf{u}_{k-1} = 0, \\ \qquad \textit{Implement } \gamma \textit{ steps of } MIE(\mathbf{u}_{k-1}, \mathbf{g}_{k-1}, k-1), \\ \qquad \textit{Add the residual correction, } \mathbf{v}_k = P_{k-1}^k\mathbf{u}_{k-1}, \ \mathbf{u}_k = \mathbf{u}_k + \mathbf{v}_k, \\ \quad \textit{end if } k \\ \quad \textit{end one step of } MIE(\mathbf{u}_k, \mathbf{g}_k, k). \end{array}\right\}$$

The convergence analysis Algorithm 6.3.10 is analogous to the linear PDE case: from (6.46),

$$\mathbf{u}_k^{(j+1)} = M_k^{\mathrm{TG}}\mathbf{u}_k^{(j)} + N_k^{\mathrm{TG}}\mathbf{g}_k, \tag{6.48}$$

with $M_k^{\mathrm{TG}} = (I - \mathcal{K}_{k-1})^{-1}(\mathcal{K}_k - \mathcal{K}_{k-1})\mathcal{K}_k$ and from Lemma 6.1.6, $N_k^{\mathrm{TG}} = (I - M_k^{\mathrm{MG}})(I - \mathcal{K}_k)^{-1}$. Similarly for the multigrid case, note (with $\mathbf{v}_{k-1} = 0$ initially)

$$\mathbf{v}_{k-1}^{(j+1)} = M_{k-1}^{\mathrm{MG}}\mathbf{v}_{k-1}^{(j)} + N_{k-1}^{\mathrm{MG}}\mathbf{g}_{k-1}, \tag{6.49}$$

where $N_{k-1}^{\mathrm{MG}} = (I - M_{k-1}^{\mathrm{MG}})(I - \mathcal{K}_{k-1})^{-1}$ and the restricted residual

$$\mathbf{g}_{k-1} = [\mathbf{g}_k - (I - \mathcal{K}_k)\underbrace{(\mathcal{K}_k\mathbf{u}_k^{(j)} + \mathbf{g}_k)}_{\text{Picard smoother}}].$$

Thus the multigrid algorithm from repeating (6.49) γ times (refer to (6.25)) can be analysed precisely as in (6.29). That is, to formulate M_k^{MG}, we consider

$$\begin{aligned} \mathbf{u}_k^{(j+1)} &= M_k^{\mathrm{MG}}\mathbf{u}_k^{(j)} + N_k^{\mathrm{MG}}\mathbf{g}_k = \mathbf{u}_k + \mathbf{v}_k \\ &= \underbrace{\overbrace{\mathcal{K}_k\mathbf{u}_k^{(j)} + \mathbf{g}_k}^{\mathbf{u}_k}}_{\text{Pre-smoothing}} + \overbrace{\underbrace{\sum_{j=0}^{\gamma-1} M_{k-1}^j N_{k-1}}_{\text{iterate (6.49)}} \underbrace{\Big[\mathbf{g}_k - (I - \mathcal{K}_k)(\mathcal{K}_k\mathbf{u}_k^{(j)} + \mathbf{g}_k)\Big]}_{\text{residual } \mathbf{g}_{k-1}}}^{\text{correction } \mathbf{v}_k} + [\ldots]\mathbf{g}_k \\ &= \mathcal{K}_k\mathbf{u}_k^{(j)} + \sum_{j=0}^{\gamma-1} M_{k-1}^j \underbrace{(I - M_{k-1})(I - \mathcal{K}_{k-1})^{-1}}_{N_{k-1}^{\mathrm{MG}}}(\mathcal{K}_k - I)\mathcal{K}_k\mathbf{u}_k^{(j)} + [\ldots]\mathbf{g}_k \\ &= \Big[I + \big(I - M_{k-1}^{\gamma}\big)(I - \mathcal{K}_{k-1})^{-1}(\mathcal{K}_k - I)\Big]\mathcal{K}_k\mathbf{u}_k^{(j)} + [\ldots]\mathbf{g}_k. \end{aligned}$$

Therefore, noting $M_\ell = M_\ell^{\rm MG}$ (as with the linear PDE case) and

$$[I + (I - \mathcal{K}_{k-1})^{-1}(\mathcal{K}_k - I)]\mathcal{K}_k = (I - \mathcal{K}_{k-1})^{-1}(\mathcal{K}_k - \mathcal{K}_{k-1})\mathcal{K}_k = M_k^{\rm TG},$$

we obtain that

$$\begin{aligned} M_k^{\rm MG} &= \Big[I + \big(I - M_{k-1}^{\gamma}\big)(I - \mathcal{K}_{k-1})^{-1}(\mathcal{K}_k - I)\Big]\mathcal{K}_k \\ &= M_k^{\rm TG} + (M_{k-1}^{\rm MG})^{\gamma}\left(M_k^{\rm TG} - \mathcal{K}_k\right). \end{aligned} \tag{6.50}$$

Taking norms of both sides, one can see the similarity to (6.32) and develop a convergence result following Theorem 6.1.7. Clearly the multigrid convergence will follow that of the two-grid method (TG-2). We remark that there exists another version of multigrid methods in [274] that has even better convergence properties.

6.3.2 Operator splitting for a noncompact operator case

When operator $\mathcal{K}$ is noncompact, the main smoothing component of the multigrid method will fail to work because the Picard iteration will not be able to smooth out a function. In the previous Chapter, the same situation will also make Krylov type methods fail to work. It turns out that all these iterative methods will converge if we modify the smoothing steps by splitting the noncompact operator $\mathcal{K}$ into a bounded part $\mathcal{D}$ and a compact part $\mathcal{C}$ i.e.

$$\mathcal{K} = \mathcal{D} + \mathcal{C}, \qquad (I - \mathcal{K}) = (I - \mathcal{D}) + \mathcal{C}. \tag{6.51}$$

The theoretical basis for this splitting was provided by the theory of pseudo-differential operators [12,487,488]. Note that Lemma 4.7.12 assures that, in the Banach space X, the compactness of $\mathcal{C}$ will imply the boundedness of $(I - \mathcal{D})^{-1}$ if the original problem $(I - \mathcal{K})$ is invertible.

Once this splitting (6.51) is done, we can use it to solve a first kind integral equation via a second kind equation

$$\mathcal{K}\mathbf{u} = \mathbf{g} \qquad \text{by solving} \qquad (I + \underbrace{\mathcal{D}^{-1}\mathcal{C}}_{\text{compact}})\mathbf{u} = \mathcal{D}^{-1}\mathbf{g} \tag{6.52}$$

or a second kind integral equation via another second kind equation

$$(I - \mathcal{K})\mathbf{u} = \mathbf{g} \qquad \text{by solving} \qquad (I - \underbrace{(I - \mathcal{D})^{-1}\mathcal{C}}_{\text{compact}})\mathbf{u} = (I - \mathcal{D})^{-1}\mathbf{g}. \tag{6.53}$$

Here we assume that the inverses after discretization (as forward type preconditioners) are easy or efficient to implement. We shall mainly consider the second kind case (6.53). As the preconditioned equation has a compact operator, any theoretical discussion will follow the previous subsection of the compact case.

It remains to comment on the implementation. The use of this preconditioning idea for multigrid methods can be found in [420,125,132]. As operator $(I - \mathcal{D})^{-1}$ cannot be formed explicitly, numerical discretizations will be carried out for $\mathcal{D}$ and $\mathcal{C}$ in (6.51) individually. On the grids $\mathcal{T}_k$, $k = 1, 2, \ldots, J$, we denote our discretized equation of (6.53) as

$$(I - (I - \mathcal{D}_k)^{-1}\mathcal{C}_k)\mathbf{u}_k = (I - \mathcal{D}_k)^{-1}\mathbf{g}_k, \tag{6.54}$$

which corresponds to the previous equation (6.39). Thus Algorithm 6.3.10 will apply with appropriate adjustments e.g. the pre-smoothing step becomes

$$\mathbf{u}_k = (I - \mathcal{D}_k)^{-1}\mathcal{C}_k\mathbf{u}_k + (I - \mathcal{D}_k)^{-1}\mathbf{g}_k$$

which is implemented as

$$\text{solve } (I - \mathcal{D}_k)y = \mathcal{C}_k\mathbf{u}_k, \ (I - \mathcal{D}_k)z = \mathbf{g}_k \ \text{ and set } \ \mathbf{u}_k = y + z,$$

and the coarsest grid solver step becomes

$$(I - (I - \mathcal{D}_1)^{-1}\mathcal{C}_1)\mathbf{u}_1 = (I - \mathcal{D}_1)^{-1}\mathbf{g}_1 \qquad \text{solved as} \qquad (I - \mathcal{K}_k)\mathbf{g}_k = \mathbf{g}_1.$$

There exist several problem classes that can be tackled by the splitting (6.51); see [420,132] and Section 4.7.

6.4 Algebraic multigrid methods

The multigrid methods discussed so far are the so-called geometrical multigrid methods because they require (or rely on) the geometrically obtained grid $\mathcal{T}_k$'s. It must be said that these geometrical multigrids have successfully solved many PDE problems and integral equations. As also commented, whenever simple smoothers are not effective in smoothing out fine grid errors, geometrical multigrids will not work (e.g. for PDEs with strongly varying coefficients). In these difficult cases, various attempts have been made to modify the coarsening strategy i.e. the choice of coarse grids so that un-smoothed error components are better represented on the new coarse grids. Another approach [490,452] was to replace the classical relaxation-based smoothers all together by more modern smoothers such as the ones discussed in Chapters 4 and 5. See also [119]. However one robust approach is the algebraic multigrid method (AMG) [442].

The advantage of an AMG is that it only requires the input of a sparse matrix A to produce an efficient multilevel iterative solver. It appears completely general. The resulting algorithm works far beyond the model cases for which convergence proofs exist.

A more radical solution of the difficulties of geometrical multigrids is to abandon the reliance of the geometrical grids $\mathcal{T}_k$'s as they, in such cases, do not offer proper guides to smoothing and coarsening (though $\mathcal{T}_J$, equivalent to $G(A)$, cannot be abandoned as the underlying solution is sought on this finest grid). The result is the algebraic multigrid (AMG) methods [276,409,442,480]. With AMG, we fix the smoother first and then try to figure out which components (variables or 'nodes') are more representative of the finer levels and deserved to be on the coarser levels. It is this flexibility that makes AMGs robust (and always convergent if the coarser levels are allowed to be 'fine' enough).

To proceed, define $\mathcal{T}^J = \{1, 2, \ldots, n_J\}$ as the index set for the finest level (linear) system ($k = J$)

$$\mathcal{A}_k \mathbf{u}_k = \mathbf{g}_k \tag{6.55}$$

which is of the equation type (6.20) and similarly coarse grid equations will be defined as ($k = J - 1, \ldots, 1$)

$$\mathcal{A}_k \mathbf{u}_k = \mathbf{g}_k,$$

where $\mathcal{A}_k = R_{k+1}^k \mathcal{A}_{k+1} P_k^{k+1}$ is the Galerkin operator (once the transfer operators are defined). *Here we use the superscript for $\mathcal{T}$, in order to relate to the geometric grid $\mathcal{T}_k$ in previous sections and yet to show the difference.* In fact, the index sets for all levels $\mathcal{T}^k$ ($k = 1, 2, \ldots, J$) and the interpolation matrix P_{k-1}^k ($k = 2, \ldots, J$), of size $n_k \times n_{k-1}$, will be effectively determined from matrix $\mathcal{A}_J$ alone. The restriction will be taken as $R_k^{k-1} = (P_{k-1}^k)^T$ (once the rectangular matrix P_{k-1}^k is explicitly computed and stored).

6.4.1 Algebraic smoothness

As we are only given the sparse linear system (6.55), there is no geometrical information available to guide us on coarse grids (apart from the matrix graph $G(\mathcal{A}_J)$). It immediately becomes a theoretical question how to develop a purely algebraic method to represent the old ideas of smoothness and coarse level correction.

Although AMGs are applicable to more general linear systems, the development of and motivation for such a theory requires the assumption of $\mathcal{A}_J$ being a SPD matrix. (Potentially this is also the weakness of AMGs). Without loss of generality, we shall consider the two-level method on $\mathcal{T}^k$ with $k = J$ first and write $n = n_J$ for simplicity. Let $S = (I - Q^{-1}\mathcal{A})$ denote the smoother for a smoothing step for (6.55) on $\mathcal{T}^k$

$$\mathbf{u}^{\text{new}} = \mathbf{u}^{\text{old}} + Q^{-1}\left(\mathbf{g} - \mathcal{A}\mathbf{u}^{\text{old}}\right) = S\mathbf{u}^{\text{old}} + Q^{-1}\mathbf{g} \tag{6.56}$$

which implies $Q = D$ for the Jacobi iteration and $Q = D - L$ for the Gauss–Seidel iteration (Chapter 3), assuming $\mathcal{A} = \mathcal{A}_J = D - L - U$. If the present error is $e = \mathbf{u}^{\text{exact}} - \mathbf{u}^{\text{old}}$, the new error will be

$$e^{\text{new}} = \mathbf{u}^{\text{exact}} - \mathbf{u}^{\text{new}} = Se.$$

As smoothness in e^{old} implies $e \approx Se$, we shall characterize this fact by relating to dominant entries in $\mathcal{A}$ in order to propose a coarsening strategy.

Defining the discrete Sobolev semi-norms, based on $(u, v)_{\ell_2} = u^T v = \sum_j u_j v_j$,

$$\begin{aligned} (u, v)_0 &= (Du, v)_{\ell_2}, & \|u\|_0 &= \sqrt{(u, u)_0}, \\ (u, v)_1 &= (Au, v)_{\ell_2}, & \|u\|_1 &= \sqrt{(u, u)_1}, \\ (u, v)_2 &= (D^{-1}Au, Av)_{\ell_2}, & \|u\|_2 &= \sqrt{(u, u)_2}, \end{aligned}$$

the above smoother (either the Jacobi or Gauss–Seidel) satisfies the inequality [409,442], for some $\alpha > 0$,

$$\|Se\|_1^2 \le \|e\|_1^2 - \alpha\|e\|_2^2. \tag{6.57}$$

Therefore, any smoothness in e or slow convergence of (6.56) (corresponding to the geometric low frequency range reached and remained) in terms of $\|Se\|_1 \approx \|e\|_1$ or $\|e\|_2 \approx 0$ from (6.57) can be detected from checking

$$\|e\|_2^2 \ll \|e\|_1^2. \tag{6.58}$$

To simplify this condition, one uses the Cauchy–Schwarz inequality[1] to derive that

$$\|e\|_1^2 = (Ae, e)_{\ell_2} = (D^{-1/2}Ae, D^{1/2}e)_{\ell_2} \le \|D^{-1/2}Ae\|_{\ell_2}\|D^{1/2}e\|_{\ell_2} = \|e\|_2\|e\|_0,$$

as shown in [480,p.42]. Therefore, equation (6.57) becomes

$$\|e\|_1^2 \ll \|e\|_0^2 \quad \text{or} \quad \sum_{i,j} -a_{i,j}\frac{(e_i - e_j)^2}{2} + \sum_{i=1}^n \Big(\sum_{j=1}^n a_{i,j}\Big)e_i^2 \ll \sum_{i=1}^n a_{i,i}e_i^2. \tag{6.59}$$

We try to establish a local criterion, for each coarse level variable i, with which (6.59) helps to select as coarse level variables from those having many strong

[1] The Cauchy–Schwarz inequality $(u, v)_{\ell_2} \le \|u\|_{\ell_2}\|v\|_{\ell_2}$ holds for any two vectors. A similar inequality exists for two functions in the L_2 norm.

connections (with $a_{i,j}/a_{i,i}$ relatively large). We shall define a variable i to be strongly negatively coupled (or *strongly n-coupled*) to another variable j if

$$-a_{i,j} \geq \varepsilon \max_{a_{ik}<0} |a_{ik}| = \varepsilon \max_{k \neq i} -a_{ik} \tag{6.60}$$

and to be strongly positively coupled (or *strongly p-coupled*) to another j if

$$a_{i,j} \geq \varepsilon^+ \max_{k \neq i} |a_{ik}| \tag{6.61}$$

for some fixed $0 < \varepsilon, \varepsilon^+ < 1$ (say $\varepsilon = 0.25$, $\varepsilon^+ = 0.5$). This is where one can find many specific matrices to verify the heuristics.

The best illustration of (6.59) is for a M-matrix (Chapter 4) that has a weak diagonal dominance i.e. $\sum_{i \neq j} |a_{i,j}| \approx a_{i,i}$. Then, noting $a_{i,j} < 0$ for $i \neq j$ (i.e. $\sum_{j=1}^{n} a_{ij} = 0$), (6.59) becomes approximately

$$\sum_{i,j} -a_{i,j} \frac{(e_i - e_j)^2}{2} \ll \sum_{i=1}^{n} a_{i,i} e_i^2 \quad \text{or even} \quad \sum_{j \neq i} -a_{i,j} \frac{(e_i - e_j)^2}{2} \ll a_{i,i} e_i^2$$
$$\text{or} \quad \sum_{j \neq i} \frac{|a_{i,j}|}{a_{i,i}} \frac{(e_i - e_j)^2}{2e_i^2} \ll 1.$$

Clearly if $|a_{i,j}|/a_{i,i}$ is relatively large, then variables i, j belong to the same 'patch' because $e_i \approx e_j$. Therefore there is no need for both variables i, j to be coarse level variables – either one of them can be picked as a coarse level variable or both remain as fine level variables. See [442,480].

It should be remarked that the use of the M-matrix assumption is to aid derivation of the AMG components and, fortunately, the resulting AMG methods perform well for a much larger class of problems. For familiar equations that arise from discretized PDEs with a regular domain and gridding, AMGs can produce a coarsening and the transfer operator similar to the geometric multigrids. More astonishingly, AMGs can produce semi-coarsening automatically for special problems that are normally obtainable by a very elaborate geometrical coarsening strategy. The classical smoothers may be optimized in the AMG context [498] as was done in the traditional setting. We also note that the AMG has been applied to solve the challenging saddle point problem [330] (see Chapter 12). Finally as the sparse graph is mainly used in AMGs, the Vaidya's preconditioner [123] based on edge information may also be considered for smoothing purpose.

6.4.2 Choice of coarser levels

The purpose of a coarsening method is to identify a suitable list of indices from the current fine level set $\mathcal{T} = \mathcal{T}^J$. As a geometrical grid is not available, the study of nonzero connections in $\mathcal{A}$ falls into the topic of graph theory. We shall call $i \in \mathcal{T}$ coupled (i.e. connected) to point $j \in \mathcal{T}$ if $a_{i,j} \neq 0$ – a graph edge connects two coupled points. For each $i \in \mathcal{T}$, we shall refer the following index set as the neighbourhood of i

$$N_i = \{j \mid j \neq i, j \in \mathcal{T}, \ a_{i,j} \neq 0 (\text{all coupled points of } i)\}. \tag{6.62}$$

Our task is to split $\mathcal{T} = \mathcal{T}^J$ into two disjoint subsets $C, F, \mathcal{T} = C \cup F$ and $C \cap F = \emptyset$, with $C = \mathcal{T}^{J-1}$ for the coarse level variables (C-variables) and F the complementary fine level F-variables. This task is often called the C/F-splitting. Following the previous subsection, we shall denote all strong n-coupled variables of $i \in \mathcal{T}$ by set S_i

$$S_i = \{j \mid j \in N_i, \ \text{ and } i \text{ is strongly n-coupled to } j\}. \tag{6.63}$$

The transpose of S_i is

$$S_i^T = \{j \mid j \in \mathcal{T}, \ i \in S_j \ \text{ or } j \text{ is strongly n-coupled to } i\}. \tag{6.64}$$

Here as (6.60) uses essentially row vector information, symmetry in A (via column information) is not used so 'j is strongly n-coupled to i' is not the same as 'i is strongly n-coupled to j'.

The coarsening strategy will make use of the number of elements in these index sets (i.e. the number of strong connections matters). We use $|P|$ to denote the number of elements in P (i.e. the cardinality) and assign the set U to contain the current set of undecided variables that will be divided into C or F.

Then the standard coarsening strategy goes as follows.

Algorithm 6.4.11. (The coarsening method of an AMG).

Let $\mathcal{T}$ denote the index set of a fine level that is associated with matrix $\mathcal{A} = (a_{ij})$. Let $\{S_i\}$ be the sequence of neighbourhoods of all members in $\mathcal{T}$. To implement the C/F-splitting of $\mathcal{T} = C \cup F$,

(1) *Initialize $U = \mathcal{T}$ (all members of a fine level $\mathcal{T}$), and $C = F = \emptyset$.*
(2) *Compute, for all $i \in U$, the initial measure of how popular a variable is in terms of influences to others:*

$$\lambda_i = |S_i^T \cap U| + 2|S_i^T \cap F| \ = \ |S_i^T|. \tag{6.65}$$

(3) *If $U = \emptyset$, end the algorithm else continue.*
(4) *We name the variable i_c in U, that has the largest λ_i (i.e. the most needed variable in terms of influences to others), to be the next*

C-variable:

$$\lambda_{i_c} = \max_i \lambda_i. \tag{6.66}$$

Update $U = U \backslash \{i_c\}$.

(5) *Set all $j \in S_{i_c}^T \cap U$ as F-variables and then remove them from U accordingly.*

(6) *For all new members $j \in S_{i_c}^T$, update its neighbouring information i.e. update the weighted measure for all members in set $S_j^T \cap U$*

$$\lambda_k = |S_k^T \cap U| + 2|S_k^T \cap F|, \qquad k \in S_j^T \cap U, \tag{6.67}$$

where the factor 2 is weighted towards neighbouring points of F-variables in U, encouraging them to be coarse points.

(7) *Return to Step (3).*

To illustrate this algorithm, we have supplied a Mfile `cf_split.m` for the readers (its default examples produce the plots in Figure 6.7).

There are two other optional steps that might be combined to adjust the distribution of F/C variables, each with a different purpose (so using only one of them):

(a) More F-variables and less C-variables: using the so-called *aggressive coarsening*, in (5) above, we also set those variables indirectly (but closely) connected to i as F-variables.

(b) More C-variables and less F-variables: following [442], strong p-couplings mean locally oscillating errors and demand locally semi-coarsening. So in step (5) above, we test if there exists $j \in S_{i_c}^T$ that satisfies (6.61). If yes, j should be a C-variable rather than in F and hence two adjacent variables enter the set C – a typical case of a semi-coarsening.

6.4.3 Matrix-dependent transfer operators

As $R_k^{k-1} = (P_{k-1}^k)^T$ will be formed explicitly, we only need to consider how to construct P_{k-1}^k. The task of P_{k-1}^k is to work out the level point value for each $i \in F$, using the neighbouring coarse level values of variables in $P_i = C \cap S_i$.

The intention is to satisfy approximately the i-th equation of $\mathcal{A}e = 0$ i.e.

$$a_{ii}e_i + \sum_{j \in N_i} a_{ij}e_j = 0. \tag{6.68}$$

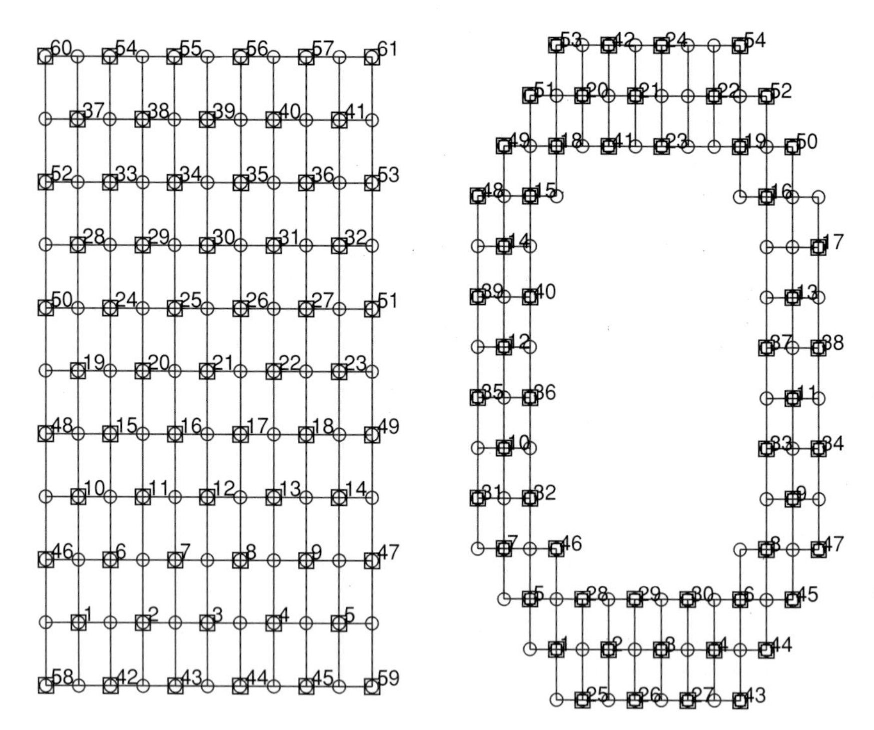

Figure 6.7. Illustration of automatic coarsening by an AMG method, with $o/\square$: fine grid points and $\square$: coarse grid points. (e.g. to produce the right plot, use `G=numgrid('A',n); A=delsq(G); lap_lab; xy=[x(:) y(:)]; [C F]=cf_split(A); gplot(A,xy,'bo-'), hold on, gplot(A(C,C),xy(C,:),'ks')`)

To cater for both negative and positive entries of a_{ij}, we use the notation [442]

$$a_{ij}^- = \begin{cases} a_{ij}, & \text{if } a_{ij} < 0, \\ 0, & \text{otherwise} \end{cases} \quad \text{and} \quad a_{ij}^+ = \begin{cases} a_{ij}, & \text{if } a_{ij} > 0, \\ 0, & \text{otherwise,} \end{cases} \tag{6.69}$$

and hence $a_{ij} = a_{ij}^- + a_{ij}^+$. For a fine level point i, its coupled coarse level points, $P_i = C \cap S_i$, will be used to provide an interpolation formula for e_i. Here as $S_i \subseteq N_i$, another reasonable choice for P_i is $P_i = C \cap N_i$. Because $N_i \backslash S_i$ involves only small entries of A, this alternative choice is unlikely to make a major difference.

To replace N_i in (6.68) by a small subset P_i, the somewhat analogous idea from approximating (with $w_0 = w_4 = 1/8$, $w_2 = 1/4$)

$$\int_a^b \frac{f(x)}{b-a} dx = w_0 f(a) + \frac{1}{4} f(\frac{3a+b}{4}) + w_2 f(\frac{a+b}{2}) + \frac{1}{4} f(\frac{a+3b}{4}) \\ + w_4 f(b),$$

by a small subset of nodes

$$\int_a^b \frac{f(x)}{b-a}dx = \frac{w_0}{w_0+w_2+w_4}f(a) + \frac{w_2}{w_0+w_2+w_4}f(\frac{a+b}{2}) + \frac{w_4}{w_0+w_2+w_4}f(b)$$
$$= \frac{1}{4}f(a) + \frac{1}{2}f(\frac{a+b}{2}) + \frac{1}{4}f(b)$$

is used here. For (6.68), we shall assume that

$$\frac{\sum_{k\in N_i} a_{ik}^- e_k}{\sum_{k\in N_i} a_{ik}^-} = \frac{\sum_{k\in P_i} a_{ik}^- e_k}{\sum_{k\in P_i} a_{ik}^-}, \qquad \frac{\sum_{k\in N_i} a_{ik}^+ e_k}{\sum_{k\in N_i} a_{ik}^+} = \frac{\sum_{k\in P_i} a_{ik}^+ e_k}{\sum_{k\in P_i} a_{ik}^+}.$$

Here and throughout this section, the corresponding terms do not exist if the subset of P_i for either a_{ik}^+ or a_{ik}^- is empty. With this assumption (or approximation), (6.68) leads to the matrix-dependent interpolation formula

$$\begin{aligned} e_i &= -\sum_{j\in N_i} a_{ij}e_j/a_{ii} = -\sum_{j\in N_i} a_{ij}^- e_j/a_{ii} - \sum_{j\in N_i} a_{ij}^+ e_j/a_{ii}, \\ &= \sum_{j\in P_i} w_{ik}e_k, \end{aligned} \tag{6.70}$$

for transferring variables between grids, where

$$w_{ik} = \begin{cases} -\dfrac{a_{ik}}{a_{ii}} \sum_{s\in N_i} a_{is}^- \Big/ \sum_{s\in P_i} a_{is}^-, & \text{if } a_{ik} < 0, \\ -\dfrac{a_{ik}}{a_{ii}} \sum_{s\in N_i} a_{is}^+ \Big/ \sum_{s\in P_i} a_{is}^+, & \text{if } a_{ik} > 0. \end{cases} \tag{6.71}$$

As mentioned, once P_{k-1}^k is formed, its transpose R_k^{k-1} is also defined.

Alternatively decompose the set $N_i = S_i \cup (N_i \backslash S_i)$. Then, in (6.68), approximate $e_j = e_i$ for $j \in N_i \backslash S_i$ so that the 'small' terms will be added to the diagonal entry a_{ii} [276]; accordingly (6.71) will be adjusted.

6.4.4 AMG algorithms and AMG preconditioners

We have presented the core components of an AMG based on the standard choices of the C/F-splitting and a matrix-dependent interpolation. It should be remarked that many individual steps may be carried out differently for a specific problem, fine-tuned for better performance. If some geometric information are available, combination of such information into the AMG can also be beneficial [442]. There exist a large literature on other variants of AMGs [73,371,466,399], some of which require less data storage and less computations at the price of reduced efficiency for very complex problems. However the true strength of an AMG lies in its ability to extend the applicability of geometric multigrid methods to complicated situations and also in its usefulness in providing a robust AMG preconditioner for Krylov subspace iterative solvers (Chapter 3). See [460,442,374].

As both the coarsening algorithm and the interpolation formula have been given, the presentation of an J level AMG is simply a matter of applying Algorithm 6.1.4.

Algorithm 6.4.12. (The AMG method).

To solve the given linear system (6.55) on $\mathcal{T}^J$ with $k = J$, assume we have set up these multigrid parameters:

ν_1 *pre-smoothing steps on each level k (before restriction)*

ν_2 *post-smoothing steps on each level k (after interpolation)*

γ *the number of multigrid cycles on each level k or the cycling pattern (usually $\gamma = 1$ for V-cycling and $\gamma = 2$ for W-cycling).*

Denote by $\textbf{Relax}_k^\nu(u, f)$ *the result of ν steps of some relaxation method for $\mathcal{A}_k \mathbf{u}_k = \mathbf{g}_k$ with $\mathbf{g}_k = f$ and the initial guess $\mathbf{u}_k = u$.*

(1) To set up the sequence of coarse levels $\mathcal{T}^k$, $k = J-1, \ldots, 1$, use Algorithm 6.4.11 repeatedly to obtain the sequences: $\{C_j\} = \mathcal{T}^{j-1}$, $\{F_j\}$ for $j = J, J-1, \ldots, 2$.
(Refer to the Mfile `cf_split.m`*).*

(2) **for** $k = J, J-1, \ldots, 2,$

work out the interpolation matrix $P_{k-1}^k = (w_{ij}^k)$
using the interpolation formula (6.70)–(6.71).
denote $R_k^{k-1} = (P_{k-1}^k)^T$ and
compute the Galerkin operator (matrix): $\mathcal{A}_{k-1} = R_k^{k-1} \mathcal{A}_k P_{k-1}^k$
initialize the new right-hand side: $\mathbf{g}_{k-1} = R_k^{k-1} \mathbf{g}_k$

end k

(3) To obtain an initial guess $\mathbf{u}_J$ on $\mathcal{T}_J$, use the following FMG (full multigrid methods):

Solve on $\mathcal{T}_1$ for $\mathbf{u}_1$ exactly, $\mathcal{L}_1 \mathbf{u}_1 = \mathbf{g}_1$

for $k = 2, \ldots, J,$

Interpolate to the next fine grid, $\mathbf{u}_k = P_{k-1}^k \mathbf{u}_{k-1}$,
if $k = J$, end the FMG step else continue,
Implement γ steps of $AMG(\mathbf{u}_k, \mathbf{g}_k, k)$,

end k

(4) *Based on the initial* $\mathbf{u}_J$ *to equation (6.55), to do* ℓ *steps of a* γ*-cycling multigrid, use*

$$AMG(\mathbf{u}_J, \mathbf{g}_J, J).$$

(5) *The general step of a* γ*-cycling multigrid* $AMG(\mathbf{u}_k, \mathbf{g}_k, k)$ *proceeds as follows*

$$\left\{\begin{array}{l} AMG(\mathbf{u}_k, \mathbf{g}_k, k): \\ \quad \textit{if } k = 1, \textit{ then} \\ \qquad \textit{Solve on } \mathcal{T}_1 \textit{ for } \mathbf{u}_1 \textit{ exactly}, \ \mathcal{L}_1 \mathbf{u}_1 = \mathbf{g}_1 \\ \quad \textit{else, on grid } \mathcal{T}_k, \textit{ do} \\ \qquad \textit{Pre-smoothing}: \mathbf{u}_k = \mathbf{Relax}_k^{\nu_1}(\mathbf{u}_k, \mathbf{g}_k), \\ \qquad \mathbf{g}_{k-1} = R_k^{k-1}(\mathbf{g}_k - \mathcal{L}_k \mathbf{u}_k), \\ \qquad \textit{Set the initial solution on } \mathcal{T}_{k-1} \textit{ to zero, } \mathbf{u}_{k-1} = 0, \\ \qquad \textit{Implement } \gamma \textit{ steps of } AMG(\mathbf{u}_{k-1}, \mathbf{g}_{k-1}, k-1), \\ \qquad \textit{Add the residual correction, } \mathbf{u}_k = \mathbf{u}_k + P_{k-1}^k \mathbf{u}_{k-1}, \\ \qquad \textit{Post-smoothing}: \mathbf{u}_k = \mathbf{Relax}_k^{\nu_2}(\mathbf{u}_k, \mathbf{g}_k), \\ \quad \textit{end if } k \\ \quad \textit{end one step of } AMG(\mathbf{u}_k, \mathbf{g}_k, k). \end{array}\right\}$$

6.5 Multilevel domain decomposition preconditioners for GMRES

A vast literature exists on multilevel methods (ML) and preconditioners, mostly based on ideas related to some variants of the domain decomposition methods (DDMs) [343,432]. Here ML methods include the geometric and algebraic multigrid methods. When one uses ML in conjunction with DDMs, one of the following scenarios emerges for a PDE such as (6.1) defined on some domain Ω.

$$\text{denoting} \qquad \Omega_k^{(j)} = \Omega_{k\text{-th subdomain}}^{(j\text{-th level})}$$

(i) Single level grid and multi-domains:

$$\Omega \approx \Omega^{(J)} = \Omega^{(1)} = \cup_{j=1}^J \Omega_j$$

(ii) Multilevel grids and single domain:

$$\Omega \approx \Omega_J = \Omega_1^{(J)}, \quad \Omega_1^{(1)} \subset \Omega_1^{(2)} \subset \cdots \subset \Omega_1^{(J)}$$

(iii) Multilevel grids and multi-domains: (level j having the maximal mesh size h_j)

$$\Omega \approx \Omega^{(j)} = \cup_{k=1}^{N_j} \Omega_k^{(j)}, \quad j = 1, \ldots, J \ \text{ with } \Omega_J = \Omega^{(J)},$$

where $h_J < h_{J-1} < \ldots < h_1$ (on the coarsest level), the grids may be optionally non-nested and the subdomains may be allowed to be nonoverlapping. Note that each new level corresponds to a different resolution with coarse grid points, approximating the whole domain Ω. Generally speaking, an increasing number of subdomains implies better parallel efficiency and slower overall convergence, and, vice versa, an increasing number of levels within each subdomain implies better sequential efficiency.

Although DDMs have been successfully applied to many classes of problems in scientific computing, the elegant convergence theory for these methods has mostly been established for a much narrower class of problems. The usual assumption is that $A_J = \mathcal{L}_J$ in (6.4) is SPD and this is when the (meaningful) bounded condition numbers are to be achieved. Nevertheless, the existing theory is so important that it may be stated that a book on iterative solvers is incomplete without mentioning the BPX preconditioner [71] and other works from their research group.

For unsymmetric problems, one can use one of these ML based DDM methods as preconditioners. One may also adapt the standard methods for specific situations. For instance, if there exists a SPD part that dominates the underlying operator, the preconditioner may primarily be based on the symmetric part [87,496,86]. Another possible approach is to construct the symmetric ML preconditioner using the symmetric part $\mathcal{H}$ (or even $\mathcal{H} + \alpha I$ with some $\alpha > 0$) from the Hermitian/skew-Hermitian splitting (12.4):

$$\mathcal{A} = \mathcal{H} + \mathcal{S}, \qquad \mathcal{H} = \frac{1}{2}\left(\mathcal{A} + \mathcal{A}^T\right), \qquad \mathcal{S} = \frac{1}{2}\left(\mathcal{A} - \mathcal{A}^T\right),$$

(as studied in other contexts [38,39]) and use it for preconditioning the whole unsymmetric matrix. A more direct use of the symmetric part $\mathcal{H}$ for preconditioning an unsymmetric system has been discussed in [161,491].

Below we mainly discuss the general case (iii) with non-nested multilevels $\Omega^{(j)}$ of grids each of which is divided into a variable number N_j of overlapping subdomains.

We now introduce the overlapping parallel and sequential preconditioners of multilevel grids and multi-domains [432,495]. Here the overlap between subdomains is to ensure the convergence of the Schwarz type iterations (additive or multiplicative variants as shown below) while the use of multilevels is to achieve efficiency.

For the bounded domain $\Omega \subset \mathbb{R}^d$, let $\{\Omega^{(j)}\}_{j=1}^{J}$ be a (possibly non-nested) sequence of discretizations each with the maximal mesh size h_j (or mesh diameter); for instance in $\mathbb{R}^2$ one may use triangulations and in $\mathbb{R}^3$ tetrahedrons. Let $A^{(j)}$ be the matrix obtained from discretizing a PDE such as (6.1) in $\Omega^{(j)}$ i.e.

$$A^{(j)}\mathbf{u}^{(j)} = \mathbf{g}^{(j)}, \qquad j = 1, 2, \ldots, J, \tag{6.72}$$

where $\mathbf{u}^{(j)}, \mathbf{g}^{(j)} \in \mathbb{R}^{n_j}$ with $n = n_J$ on the finest grid $\Omega^{(J)}$ with the aim of iterating

$$A^{(J)}\mathbf{u}^{(J)} = \mathbf{g}^{(J)}, \tag{6.73}$$

in order to provide a preconditioner for matrix $A^{(J)}$. We remark that the preconditioning algorithms (to be presented) may actually be converging for some problems and in such cases one may use these algorithms for solution of (6.73) directly.

Let each grid $\Omega^{(j)}$ be divided into $N^{(j)}$ overlapping subdomains

$$\Omega^{(j)} = \sum_{k=1}^{N^{(j)}} \Omega_k^{(j)} = \bigcup_{k=1}^{N^{(j)}} \Omega_k^{(j)}. \tag{6.74}$$

We need to define four kinds of restrictions for $j = 1, \ldots, J$:

$$\begin{aligned}
&\text{(1) the fine level } j+1 \text{ to a coarse level } j, && R^{(j)}\\
&\text{(2) the same level } j \text{ to its subdomain } k, && R_k^{(j)}\\
&\text{(3) the finest level } J \text{ recursively to} && \overline{R}^{(j)} =\\
&\qquad \text{a coarse level } j, && R^{(j)}R^{(j+1)}\cdots R^{(J-1)}\\
&\text{(4) the finest level } J \text{ recursively to} && \\
&\qquad \text{a coarse level } j\text{'s } k\text{-th subdomain} && \overline{R}_k^{(j)} = R_k^{(j)}\overline{R}^{(j)}.
\end{aligned} \tag{6.75}$$

Clearly $\overline{R}^{(J-1)} = R^{(J-1)}$; we reset $R^{(J)} = \overline{R}^{(J)} = I$ as they are not well defined otherwise. Then interpolations will be by the transposing the restrictions e.g. $P^{(j)} = R^{(j)T}$ is an interpolation from the level j back to its fine level $j+1$, for $j = 1, \ldots, J-1$. On the subdomain $\Omega_k^{(j)}$, within level j, the corresponding submatrix $A_k^{(j)} = R^{(j)}A^{(j)}R^{(j)T}$ is extracted from matrix $A^{(j)}$ for the entire level. Let $\mathbf{u}^{\text{old}} = (\mathbf{u}^{(J)})_\ell$ be the current approximation of $\mathbf{u}^{(J)}$ at iteration ℓ; initially one can use a FMG idea to obtain the first approximation at $\ell = 1$. Then the multilevel algorithms will be consisted of steps resembling the usual geometric multigrids:

- restriction to all subdomains at coarser grids;

- local solvers / smoothing at subdomains;
- prolongation to all fine levels to obtain new corrections.

6.5.1 The additive multilevel preconditioner

The first multilevel method, resembling the Jacobi iteration (3.11) for linear system (1.1), is of the so-called *additive Schwarz type*, and is suitable for parallel implementation. Let

$$r = \mathbf{g}^{(J)} - A^{(J)}\mathbf{u}^{\mathrm{old}} \tag{6.76}$$

be the current residual. Then the overall task is to solve the residual correction equation using all subdomains at coarse levels:

$$A_k^{(j)}\mathbf{v}_k^{(j)} = \mathbf{g}_k^{(j)} = \overline{R}_k^{(j)} r. \tag{6.77}$$

Algorithm 6.5.13. (Multilevel additive Schwarz preconditioner).

To solve (6.73) for the purpose of preconditioning, assume some initial guess $\mathbf{u}^{\mathrm{old}}$ *has been obtained from a FMG process. For a fixed number of MAXIT steps, repeat the following steps:*

(1) Compute the current residual r *from (6.76).*
(2) Restrict the residual r *on* $\Omega^{(J)}$ *to all subdomains at coarse levels:*

$$r_k^{(j)} = \mathbf{g}_k^{(j)} = \overline{R}_k^{(j)} r,$$

for $j = J-1, J-2, \ldots, 1$ *and* $k = 1, 2, \ldots, N^{(j)}$.
(3) Solve each residual correction equation as in (6.77) for $\mathbf{v}_k^{(j)} = A_k^{(j)^{-1}} r_k^{(j)}$.
(4) Set $\mathbf{v}^{(J)} = 0$.
(5) Interpolate all residual corrections back to the finest level on $\Omega^{(J)}$*:*

$$\mathbf{v}^{(J)} = \mathbf{v}^{(J)} + \overline{R}_k^{(j)^T} \mathbf{v}_k^{(j)}, \tag{6.78}$$

for $j = J-1, J-2, \ldots, 1$ *and* $k = 1, 2, \ldots, N^{(j)}$.
(6) Update the current solution $\mathbf{u}^{\mathrm{new}} = \mathbf{u}^{\mathrm{old}} + \mathbf{v}^{(J)}$.
(7) Update the old solution $\mathbf{u}^{\mathrm{old}} = \mathbf{u}^{\mathrm{new}}$ *and return to step (1) to continue or stop if MAXIT steps have been done.*

Writing the iteration method in the standard residual correction form (3.4),

$$
\begin{aligned}
\mathbf{u}^{\text{new}} &= \mathbf{u}^{\text{old}} + \mathbf{v}^{(J)} \\
&= \mathbf{u}^{\text{old}} + \sum_{j=1}^{J}\sum_{k=1}^{N^{(j)}} \overline{R}_k^{(j)T} \mathbf{v}_k^{(j)} \\
&= \mathbf{u}^{\text{old}} + \sum_{j=1}^{J}\sum_{k=1}^{N^{(j)}} \overline{R}_k^{(j)T} A_k^{(j)-1} r_k^{(j)} \\
&= \mathbf{u}^{\text{old}} + \sum_{j=1}^{J}\sum_{k=1}^{N^{(j)}} \overline{R}_k^{(j)T} A_k^{(j)-1} \overline{R}_k^{(j)} r \\
&= \mathbf{u}^{\text{old}} + B\left(\mathbf{g}^{(j)} - A^{(j)}\mathbf{u}^{\text{old}}\right),
\end{aligned}
\tag{6.79}
$$

where, as in (3.4), B is the additive multilevel preconditioner

$$
\begin{aligned}
B &= \sum_{j=1}^{J}\sum_{k=1}^{N^{(j)}} \overline{R}_k^{(j)T} A_k^{(j)-1} \overline{R}_k^{(j)} = \sum_{j=1}^{J} R^{(j)T}\left(\sum_{k=1}^{N^{(j)}} R_k^{(j)T} A_k^{(j)-1} R_k^{(j)}\right) R^{(j)} \\
&= \sum_{j=1}^{J-1} R^{(j)T}\left(\sum_{k=1}^{N^{(j)}} R_k^{(j)T} A_k^{(j)-1} R_k^{(j)}\right) R^{(j)} + \sum_{k=1}^{N^{(J)}} R_k^{(J)T} A_k^{(J)-1} R_k^{(J)},
\end{aligned}
\tag{6.80}
$$

since $R^{(J)} = \overline{R}^{(J)} = I$.

6.5.2 The additive multilevel diagonal preconditioner

The additive multilevel preconditioner just presented has one set of flexible parameters $N^{(J)}, N^{(J-1)}, \ldots, N^{(1)}$ which may be independent of the corresponding problem sizes $n_J, n_{J-1}, \ldots, n_1$ on all levels. Assuming each subdomain on level j has the same number of unknowns, then domain $\Omega_k^{(j)}$ should have at least $\tau_j = n_j/N^{(j)}$ unknowns (as overlaps are involved). The so-called additive multilevel diagonal preconditioner makes the simple choice of

$$
N^{(j)} = \begin{cases} n_j, & \text{if } 2 \le j \le J(\text{ or } \tau_j = 1), \\ 1, & \text{if } j = 1(\text{ or } \tau_1 = n_1), \end{cases}
$$

as the global coarsest level does not have too grid points. Further allow no overlaps to ensure $R_k^{(j)}$ is really simple.

Then each subdomain residual correction solver, at all levels, is to involve a 1×1 matrix which is the diagonal of matrix $A^{(j)}$ at level $j > 1$. Define

$D^{(j)} = \operatorname{diag}(A^{(j)})$. Then preconditioner B in (6.80) becomes

$$
\begin{aligned}
B = \sum_{j=2}^{J-1} R^{(j)^T} \underbrace{\left(\sum_{k=1}^{N^{(j)}} R_k^{(j)^T} A_k^{(j)^{-1}} R_k^{(j)} \right)}_{\text{diagonal matrix}} R^{(j)} + \underbrace{\sum_{k=1}^{N^{(J)}} R_k^{(J)^T} A_k^{(J)^{-1}} R_k^{(J)}}_{\text{diagonal matrix}} \\
+ \sum_{k=1}^{N^{(1)}} R_k^{(1)^T} A_k^{(1)^{-1}} R_k^{(1)} \qquad (6.81) \\
= R^{(1)^T} A^{(1)^{-1}} R^{(1)} + \sum_{j=2}^{J-1} R^{(j)^T} D^{(j)^{-1}} R^{(j)} + D^{(J)^{-1}},
\end{aligned}
$$

since $N^{(1)} = 1$ and so $R^{(1)} = R_k^{(1)}$.

6.5.3 The additive multilevel BPX preconditioner

The additive multilevel diagonal preconditioner might appear to be too simple to be simplified further. However, if we know enough information of an underlying operator, even the diagonal matrices $D^{(j)}$ can be replaced by constants. This was done in the BPX [71,495,432] preconditioner for a class of Laplacian type equations $-\nabla(a\nabla) = f$ in Ω whose standard finite element stiffness matrix has the diagonal elements of the form $O(h_j^{d-2})$ in $\Omega \subset \mathbb{R}^d$.

Therefore for this particular equation, we may modify (6.81) by replacing $D^{(j)^{-1}} = h_j^{2-d} I$, giving the BPX preconditioner

$$
\begin{aligned}
B &= R^{(1)^T} A^{(1)^{-1}} R^{(1)} + \sum_{j=2}^{J-1} R^{(j)^T} D^{(j)^{-1}} R^{(j)} + D^{(J)^{-1}} \\
&= R^{(1)^T} A^{(1)^{-1}} R^{(1)} + \sum_{j=2}^{J-1} h_j^{2-d} R^{(j)^T} R^{(j)} + h_J^{2-d} I.
\end{aligned}
\qquad (6.82)
$$

In implementation, step (3) of Algorithm 6.5.13 becomes the simple job of multiplying by a mesh constant. The reason that this seemingly simple BPX preconditioner has drawn much attention was not only because of its simplicity but also more importantly the proof of its optimality i.e. $\kappa(BA) = \kappa(BA^{(J)}) \le C$ with C independent of n_J. This optimality represents the best result one can achieve with a simple preconditioner for a SPD case. The result has inspired many researchers to devise other preconditioners in different contexts [436, 495].

6.5.4 The multiplicative multilevel preconditioner

The second multilevel method, resembling the Gauss–Seidel iteration (3.17) for linear system (1.1), is of the so-called *multiplicative Schwarz type*, and is suitable for sequential implementation. We now discuss the fully multiplicative variant (it is possible to propose a hybrid variant to parallelize some partial steps [432]). Note the so-called domain colouring idea (Algorithm 4.3.3), corresponding to some block Gauss–Seidel updating strategy, is one such hybrid variant.

The multiplicative methods converge, sequentially, faster than the additive methods but the parallel efficiency is not as good as the latter. As stated, our main purpose is to precondition the equation (6.73) so convergence (that may not a difficult task for general problems anyway) is not a main consideration. The sequence of operations will start the subdomains of the finest level J, then those of the coarser levels and finally interpolate from the coarsest level 1 back to the finest level J.

Algorithm 6.5.14. (Multilevel multiplicative Schwarz preconditioner)

To solve (6.73) for the purpose of preconditioning, assume some initial guess $\mathbf{u}^{\text{old}}$ *has been obtained from a FMG process. For a fixed number of MAXIT steps, repeat the following steps:*

(1) Compute the current residual $r^{(J)} = r$ *initially from (6.76).*
(2) **for level** $j = J, J-1, \ldots, 1,$
  for subdomain $k = 1, 2, \ldots, N^{(j)}$ *on level* j
  - *Restrict the residual to subdomain* $\Omega_k^{(j)}$, $r_k^{(j)} = R_k^{(j)} r^{(j)}$,
  - *Solve the residual correction equation as in (6.77),*
    $\mathbf{v}_k^{(j)} = A_k^{(j)^{-1}} r_k^{(j)}$,
  - *Interpolate the residual correction and add to the finer level,*

$$\mathbf{u}^{(J)} = \mathbf{u}^{(J)} + R^{(j)^T} \mathbf{v}_k^{(j)},$$

  - *Update the current residual on* $\Omega^{(j)}$ *at level* j,
    $r^{(j)} = r^{(j)} - A^{(j)} \mathbf{u}^{(J)}$.

  end for subdomain $k = 1, 2, \ldots, N^{(j)}$ *on level* j
  if $j = 1$*, finish the* j *loop else continue,*
  - *Restrict the residual to the next coarser level* $\Omega^{(j-1)}$,
    $r^{(j-1)} = R^{(j-1)} r^{(j)}$,
  - *Set the coarse grid correction* $\mathbf{u}^{(j-1)} = 0$,

end for *level* $j = J, J-1, \ldots, 1$

(3) *Interpolate all residual corrections back to the finest level on* $\Omega^{(J)}$*:*
for *level* $j = 2, \ldots, J-1, J$
- *Interpolate the residual correction and add to the finer level,*

$$\mathbf{u}^{(j)} = \mathbf{u}^{(j)} + R^{(j-1)^T} \mathbf{u}_k^{(j-1)}.$$

end for *level* $j = 2, \ldots, J-1, J$
(4) *Update the current solution,* $\mathbf{u}^{\text{new}} = \mathbf{u}^{(J)}$.
(5) *Update the old solution* $\mathbf{u}^{\text{old}} = \mathbf{u}^{\text{new}}$ *and return to step (1) to continue or stop if MAXIT steps have been done.*

6.6 Discussion of software and the supplied Mfiles

As the present chapter touches a topic that has been actively researched in recent years, various codes are available. Firstly we must mention the two important Web sites that contain many useful pointers and topical information:

- `http://www.mgnet.org` – Multigrid methods network
- `http://www.ddm.org` – Domain decomposition method network

Secondly we highlight these multilevel-related software from

(1) PETSc (Portable, Extensible Toolkit for Scientific Computation) (The PETSc project team):

http://www-unix.mcs.anl.gov/petsc/petsc-2/

(2) PLTMG (Piecewise linear triangular MG) (Radalph Bank):

http : //www.scicomp.ucsd.edu/~reb/software.html

(3) The MGM solvers from NAG (Numerical Algorithms Group, Ltd)

http : //www.nag.co.uk/

(4) SAMG (AMG for Systems, the algebraic multigrid methods, by Fraunhofer SCAI):

http : //www.scai.fraunhofer.de/samg.htm

(5) Algebraic Multigrid as a method or a preconditioner (Andreas Papadopoulos and Andrew Wathen):

http : //web.comlab.ox.ac.uk/oucl/work/andy.wathen/papadopoulos/soft.html

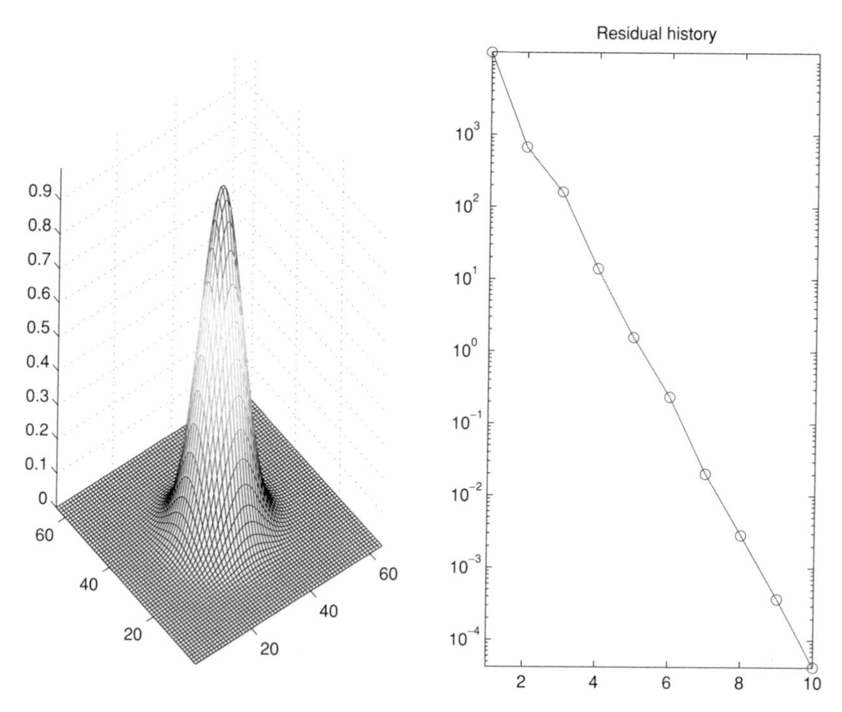

Figure 6.8. Illustration of the result from using `mgm_2d.m` to solve (6.83).

For research investigations, we have developed the following simpler Mfiles.

[1] `ch6_mg2.m` – Illustration of error smoothing at different iterations using the damped Jacobi method.
[2] `ch6_gs.m` – Illustration of error smoothing at different iterations using the Gauss–Seidel method.
[3] `ch6_w.m` – Illustration of multigrid cycling patterns and when main variables are updated.
[4] `mgm_2d.m` – The linear multigrid method (as in Algorithm 6.1.4) for solving

$$\operatorname{div}\mathbf{D} \cdot \operatorname{grad} u + c(x, y)u = d(x, y), \qquad \Omega = [0, 1]^2, \tag{6.83}$$

where $\mathbf{D} = \mathbf{D}(x, y) = (a, \ b)$. Two default examples (both using $a = 1 + x^2/100,\ b = 1 - y^2/100,\ c = -(x^2 + y^2)/100$) are given for tests, respectively using the following exact solution to supply the Dirichlet boundary conditions (and to work out d accordingly):

$$\text{iprob} = 1 : u = (x - \frac{1}{2})^3 + (y - \frac{1}{3})^4$$

$$\text{iprob} = 2 : u = \exp\left(-60(x - \frac{1}{2})^2 + (y - \frac{1}{1})^2\right).$$

The interested reader may adapt the Mfile for solving other PDEs; the result from solving the second case with $J = levs = 6$ is shown in Figure 6.8. It is particularly useful to examine how a MGM works by pausing this Mfile in selected places through adding the MATLAB command

```
>>  keyboard    % use dbcont or dbstop or return
    or CTRL-C to come back
```

The MGM is implemented with all vectors and matrices associated with the grid level number, rather than packing vectors from all levels to together as with other implementations in Fortran or C e.g. if $levs = 4$, $U4$ will be the solution vector on T_4 and likewise $R3$ is the residual vector on T_3. This makes it easier to display any quantity.

[5] `mgm_2f.m` – Similar to `mgm_2d.m`. In this version, we do not form the fine grid matrices to save storage. The relevant matrices when needed in smoothing and residual steps are computed entry-by-entry.

[6] `mgm_2s.m` – A short version of `mgm_2d.m` removing the comments and internal checks for fast running.

[7] `lap_lab.m` – Place numbering labels to a graph from `numgrid` (for Laplace's equation).

[8] `cf_split.m` – Compute the C/F-splitting for an AMG (to produce a list of coarse level nodes).

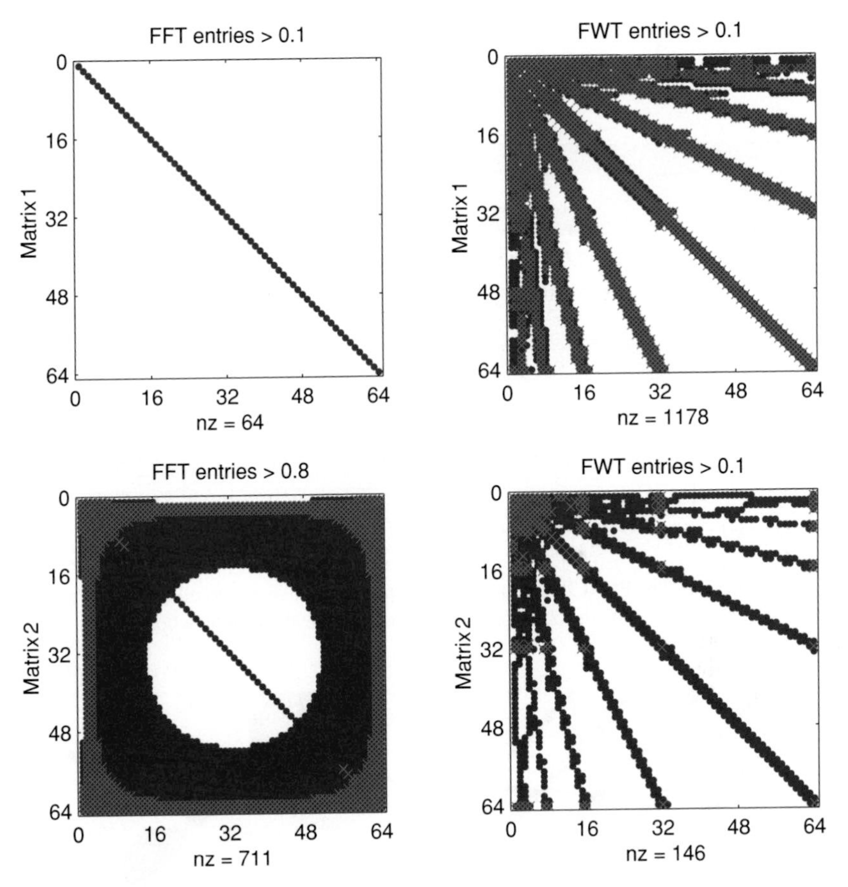

Figure 1.1. Comparison of FFT ($\widetilde{A} = \mathbf{F}A\mathbf{F}^H$) and FWT ($\widetilde{A} = WAW^T$) for compressing two test matrices. Clearly FFT is only good at circulant matrix 1 but FWT is more robust for both examples.

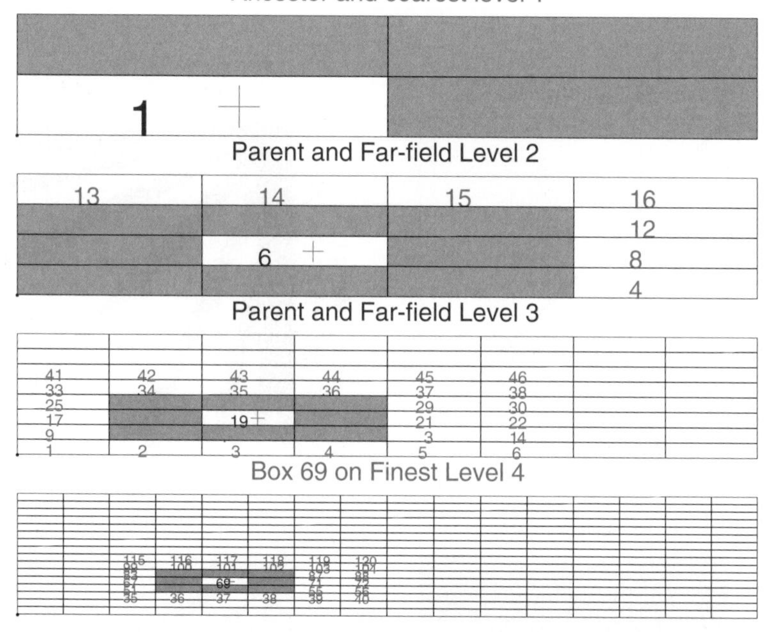

Figure 3.4. The FMM Illustration I of the interaction lists across all levels (for a specified box). Note the parent's near neighbours provide the interaction lists.

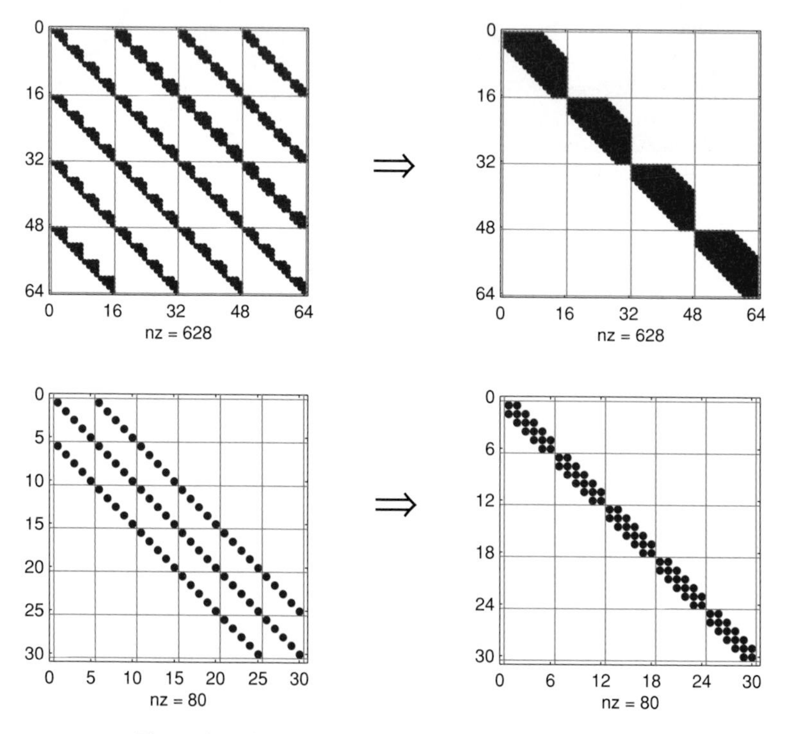

Figure 4.1. Illustration of Lemma 4.1.1 for two examples.

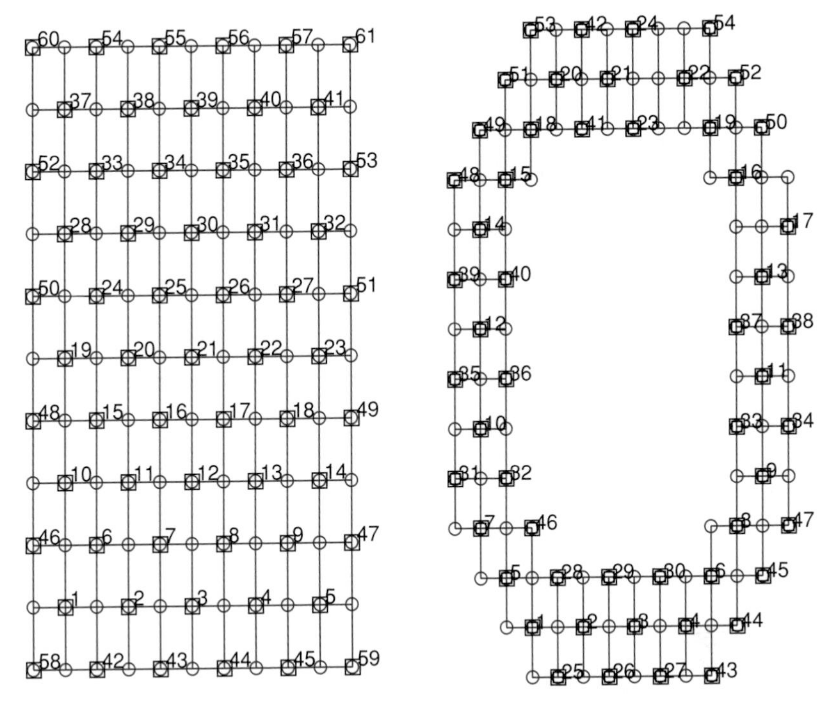

Figure 6.7. Illustration of automatic coarsening by an AMG method, with $o/\square$: fine grid points and $\square$: coarse grid points. (e.g. to produce the right plot, use `G=numgrid('A',n); A=delsq(G); lap_lab; xy=[x(:) y(:)]; [C F]=cf_split(A); gplot(A,xy,'bo-'), hold on, gplot(A(C,C),xy(C,:),'ks')`)

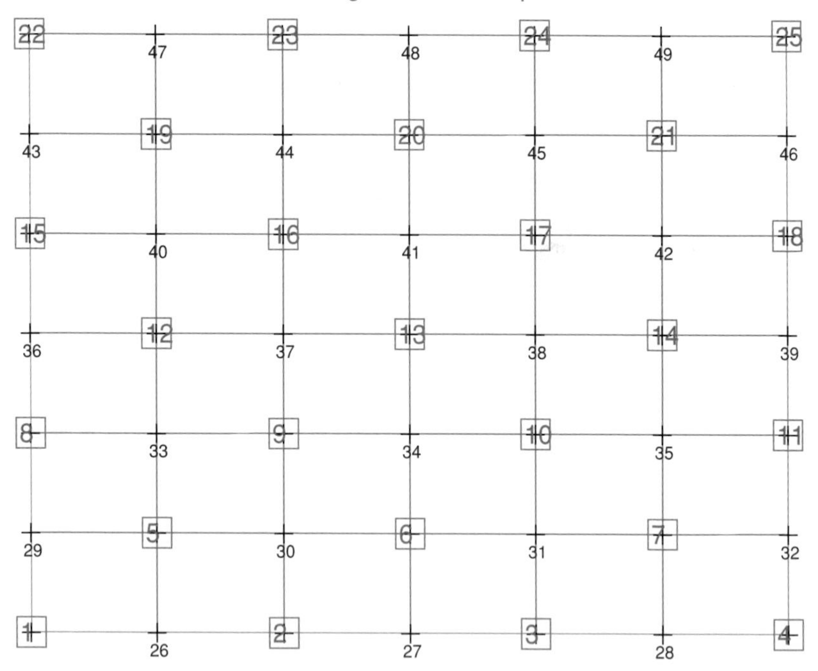

Figure 7.5. Illustration of a red-black ordering (with red in large fonts and black in small fonts, each forming an independent set of nodes). Check with the Mfile `multic.m`.

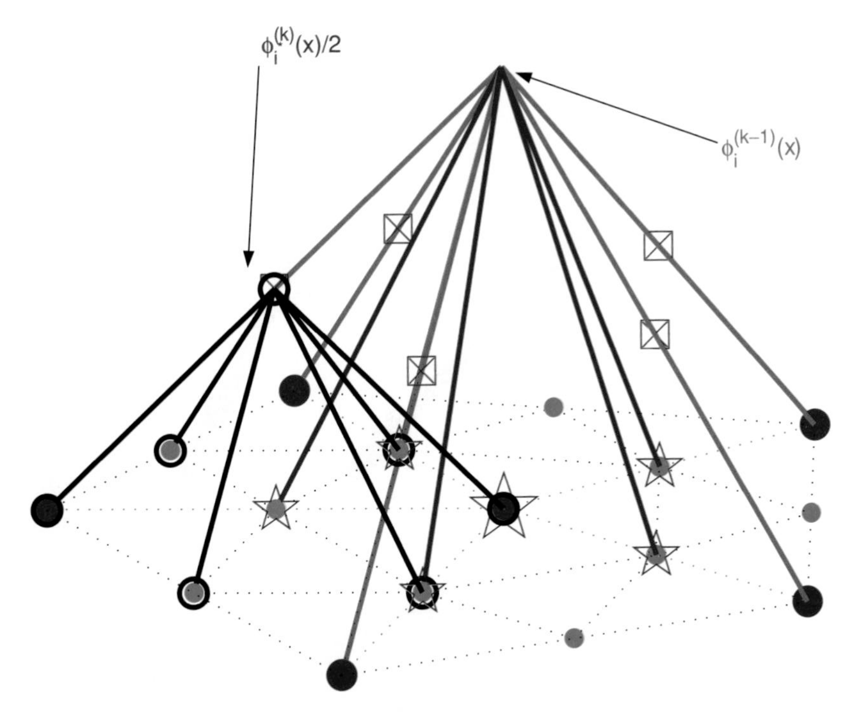

Figure 7.7. Illustration of the fundamental result (7.41) in $\mathbb{R}^2$ for the HB basis. Here large $\bullet$ points are the coarse points and small $\bullet$ ones are the fine level points. Observe the difference between $\phi^{(k)}$ and $\phi^{(k-1)}$ the same coarse point (large) $\star$. The $\square$ (and small $\star$) points indicate the position of neighbouring $\phi^{(k)}/2$.

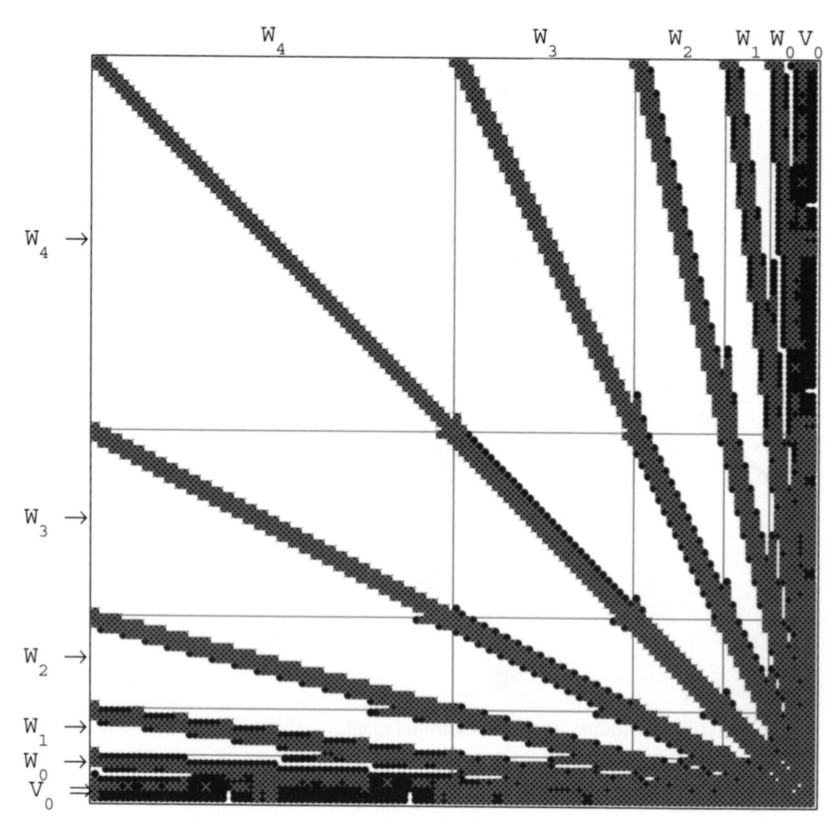

Figure 8.1. Illustration of the finger-like sparsity pattern in a wavelet basis ($J =$ 5 levels). Recall that $\mathbf{V}_5 = \mathbf{W}_4 + \mathbf{W}_3 + \mathbf{W}_2 + \mathbf{W}_1 + \mathbf{W}_0 + \mathbf{V}_0$.

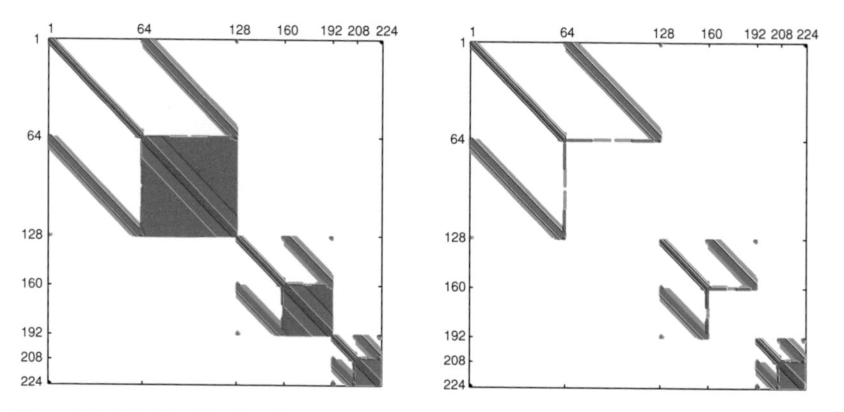

Figure 9.1. The level-by-level form (left) versus the nonstandard wavelet form [60] (right).

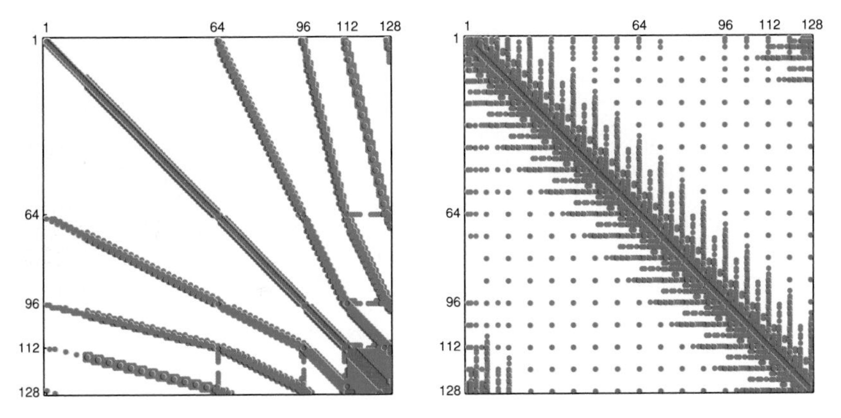

Figure 9.2. The standard wavelet form representation (left) versus an alternative centering form [130] (right) for the example in Figure 9.1.

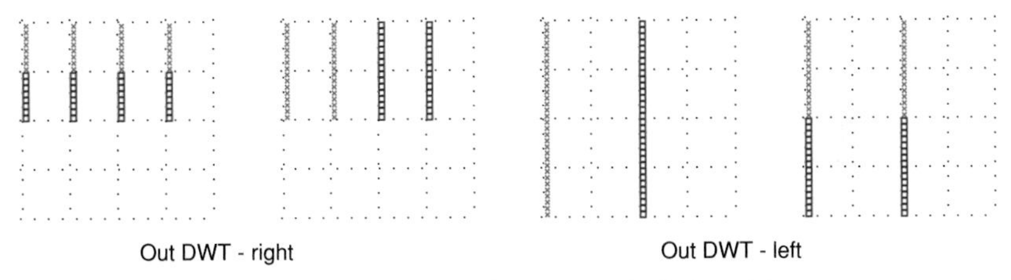

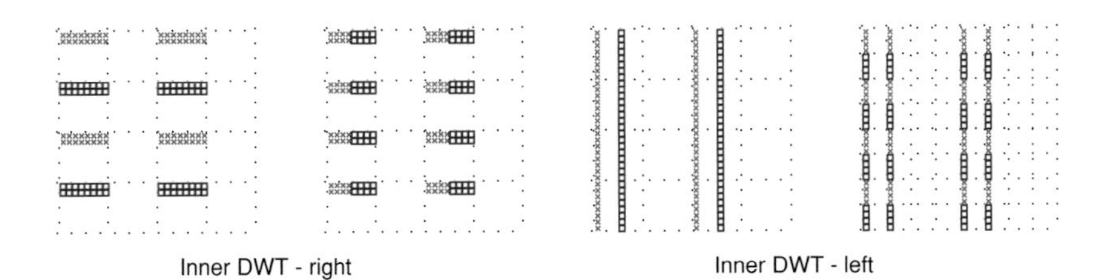

Figure 13.1. Graphical illustration of the proof of Lemma 13.2.6. The top plots show the outer block level transform $C_1 = (W \otimes I_n)C(W \otimes I_n)^T$ while the bottom plots show the inner block transform $\widetilde{C} = (I_n \otimes W)C_1(I_n \otimes W)^T = W_{2D}CW_{2D}^T$.

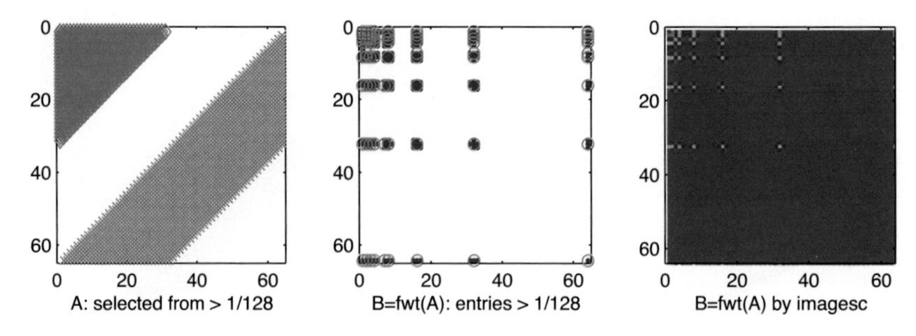

Figure C.2. MATLAB plots of matrices.

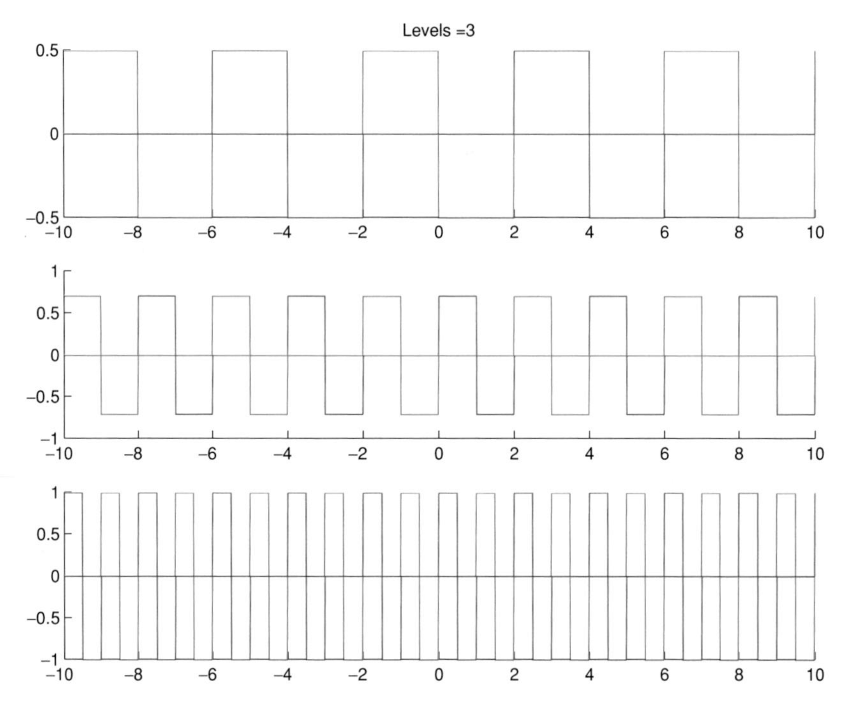

Figure C.3. Use of MATLAB functional Mfiles.

7

Multilevel recursive Schur complements preconditioners [T4]

Multilevel Preconditioners:... This new class of preconditioners can be viewed as one cycle of a standard multigrid method without the smoothing operations. They use the multigrid principle to capture the different length scales of the solution but rely on the conjugate gradient method to deal with other convergence difficulties. They offer the efficiency of multigrid methods and the robustness of the conjugate gradient method.

TONY F. CHAN. *Hierarchical Algorithms and Architectures for Parallel Scientific Computing.* CAM report 90-10 (1990)

In many applications, there is a natural partitioning of the given matrix... it is of importance to examine if the corresponding system can be solved more efficiently. For large-scale problems, iterative solution methods are usually more efficient than direct solution methods. The most important aspect of the iterative solution method is the choice of the preconditioning matrix.

OWE AXELSSON. *Iterative Solution Methods.* Cambridge University Press (1994)

As shown in Section 2.1.2, the Schur complement matrix

$$S = A_{22} - A_{21} A_{11}^{-1} A_{12} \tag{7.1}$$

obtained from eliminating the bottom left block of a 2×2 block matrix

$$A = \begin{bmatrix} A_{11} & A_{12} \\ A_{21} & A_{22} \end{bmatrix} = \begin{bmatrix} I & \\ A_{21} A_{11}^{-1} & I \end{bmatrix} \begin{bmatrix} A_{11} & A_{12} \\ & S \end{bmatrix} = \begin{bmatrix} A_{11} & \\ A_{21} & I \end{bmatrix} \begin{bmatrix} I & A_{11}^{-1} A_{12} \\ & S \end{bmatrix} \tag{7.2}$$

naturally occurs during the LU decomposition (2.6). When A is SPD, the condition number of S is less than A (Section 2.1.2) and also that of a block diagonal preconditioned matrix [345]. In the general case, computing an exact Schur complement amounts to implementing the LU decomposition while approximating a Schur complement is equivalent to obtaining a special ILU decomposition (preconditioner). However, powerful preconditioners can be obtained by using the Schur idea recursively in a multilevel setting.

This chapter considers various multilevel ideas of approximating S so as to develop an approximate Schur complement preconditioner for a sparse matrix A. We consider these selected approaches.

Section 7.1 Multilevel functional partition: AMLI approximated Schur
Section 7.2 Multilevel geometrical partition: exact Schur
Section 7.3 Multilevel algebraic partition: permutation based Schur
Section 7.4 Appendix – the FEM hierarchical basis
Section 7.5 Discussion of software and Mfiles

7.1 Multilevel functional partition: AMLI approximated Schur

The functional partition approach is usually considered in the context of a FEM solution of some elliptic PDE in $\mathbb{R}^d$. The approach is particularly useful for a self-adjoint second-order operator that leads to a SPD matrix $A = A_J$ on some finest grid $\mathcal{T}_J$, when solved by a finite element method [468,27,34,30,31,32]. The preconditioner proposed will be of the type of algebraic multilevel iterations (AMLI), where the essential idea is to approximate the Schur matrices in all levels in a recursive way. The preconditioner is obtained as a factorized LU decomposition. The idea is quite general but the applicability is normally to SPD type matrices. See [468,370,27,33,34,469].

We first need to introduce the functional partition. Following (1.60), consider the discretized equation

$$A^{(J)}\mathbf{u}_J = \mathbf{g}_J, \tag{7.3}$$

arising from the usual discretization of a variational formulation

$$a(u, v) = (f, v), \qquad u, v \in V \in H_0^1(\Omega) \tag{7.4}$$

for some second-order elliptic PDE (for instance (6.1)). Similar to the setting of the previous chapter on geometric multigrids, we assume that a sequence of refined grids $\mathcal{T}_k$ (e.g. triangulations in $\mathbb{R}^2$) have been set up, giving rise to the matrix $A^{(k)}$ of size $n_k \times n_k$, for each level $k = 1, 2, \ldots, J$ with $\mathcal{T}_1$ the coarsest grid and $\mathcal{T}_J$ the finest grid. *We are mainly interested in the algebraic* *construction i.e. only $A^{(J)}$ will be assumed given but other coarser grid matrices will be computed as approximation to the true Schur complement of a fine grid.* Therefore

$$n_1 < n_2 < \cdots < n_J.$$

Let N_k denote the set of all n_k nodes, $p_1, p_2, \ldots, p_{n_k}$, in grid $\mathcal{T}_k$. Then the nodal basis (e.g. piecewise linear polynomials) functions $\phi_j^{(k)}$ will satisfy

$$\phi_j(p_i)^{(k)} = \delta_{ij}.$$

The basis functions $\{\phi_j^{(k)}\}$ will be split accordingly to two nonoverlapping subsets of N_k:

$$N_k = (N_k \backslash N_{k-1}) \cup N_{k-1}. \tag{7.5}$$

The stiffness matrix $A^{(k)}$ on a coarser level may be defined and computed by different methods [469]. In particular the following four cases are of interest.

(1) **Nodal partition.** Given the stiffness matrix $A^{(J)}$ [34,370], we can use the nodal partition (7.5) and identify the matrix $A^{(k-1)}$ recursively from

$$A^{(k)} = \begin{bmatrix} A_{11}^{(k)} & A_{12}^{(k)} \\ A_{21}^{(k)} & A^{(k-1)} \end{bmatrix} \begin{matrix} \}(N_k \backslash N_{k-1}) \\ \}N_{k-1} \end{matrix} \tag{7.6}$$

where $A_{22}^{(k)} = A^{(k-1)}$ by the nodal ordering and $k = 2, 3, \ldots, J$.

(2) **Direct use of the hierarchical basis**. We assume that the nodal basis is replaced by the hierarchical basis [502] that defines

$$\begin{aligned} (A^{(k)})_{ij} &= a(\phi_i^{(k)}, \phi_j^{(k)}) \\ &= a({}^{HB}\phi_i^{(k)}, {}^{HB}\phi_j^{(k)}), \end{aligned} \tag{7.7}$$

where $i, j = 1, 2, \ldots, n_k$ and $k = 1, 2, \ldots, J-1$ and $\{{}^{HB}\phi_i^{(k)}\}$'s for a fixed k are defined at all nodes in $(N_k \backslash N_{k-1})$. Refer to Figure 7.1 and Section 7.4.

(3) **Indirect use of the hierarchical basis**. Let $A_{\text{NB}}^{(k)}$ denote the usual nodal basis matrix as defined by (7.6). A convenient way [34,468] to use a HB basis is to generate it from a nodal basis by a sparse transform $J = \begin{bmatrix} I & J_{12} \\ & I \end{bmatrix}$ in the form

$$A^{(k)} \equiv \hat{A}^{(k)} = J^T A_{\text{NB}}^{(k)} J. \tag{7.8}$$

As the hierarchical basis (HB) for a general differential operator may lead to less sparse stiffness matrices than the standard nodal basis, this transform idea may be considered as an effective and sparse implementation. Note

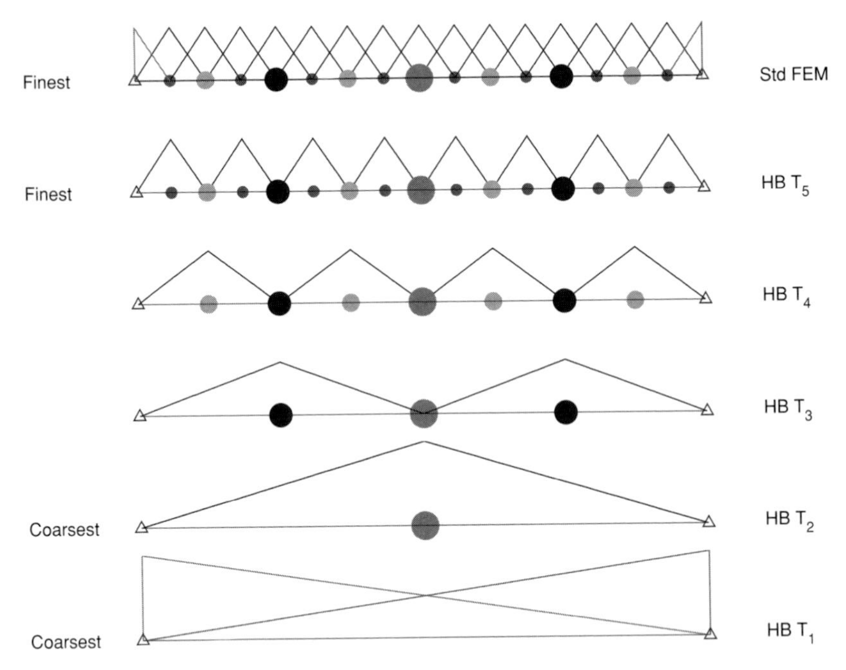

Figure 7.1. Comparison of the HB basis with $k = 5$ levels and the standard FEM linear basis. Note the HB ordering will start from the coarsest level and add (naturally) numbering until the finest.

that the HB basis can diagonalize the special Laplacian operator in $\mathbb{R}$ (1D); see the Mfile `HB1.m` and Section 7.4.

(4) **Direct use of the Schur matrix**. An alternative way [370,469] of defining $A^{(k)}$, for $k = 1, 2, \ldots, J-1$, is to let it be an approximate Schur matrix of (7.6)

$$A^{(k)} = A_{22}^{(k)} - A_{21}^{(k)} B_{11} A_{12}^{(k)} \approx A_{22}^{(k)} - A_{21}^{(k)} {A_{11}^{(k)}}^{-1} A_{12}^{(k)} = S, \tag{7.9}$$

where $B_{11} \approx {A_{11}^{(k)}}^{-1}$ may be considered to be some suitable SPAI matrix (see §5.4).

Once the coarse level matrices $\{A^{(k)}\}$ have been defined in one of the four cases, we are ready to specify the AMLI preconditioner $M^{(J)}$. Clearly on the coarsest level, the preconditioner $M^{(1)} = A^{(1)}$ is suitable and it remains to discuss other finer levels. The elementary version takes the following form for

$k = 2, 3, \ldots, J$:

$$M^{(k)} = \begin{bmatrix} A_{11}^{(k)} & \\ A_{21}^{(k)} & I \end{bmatrix} \begin{bmatrix} I & A_{11}^{(k)^{-1}} A_{12}^{(k)} \\ & M^{(k-1)} \end{bmatrix} = \begin{bmatrix} A_{11}^{(k)} & \\ A_{21}^{(k)} & M^{(k-1)} \end{bmatrix} \begin{bmatrix} I & A_{11}^{(k)^{-1}} A_{12}^{(k)} \\ & I \end{bmatrix}. \tag{7.10}$$

A more general version uses a polynomial $P_\nu(t)$ of degree ν to improve the approximation of the Schur matrix. If A is SPD [34,469,370], we may estimate the smallest eigenvalue of $M^{(k)^{-1}} A^{(k)}$ to define P_ν in terms of a scaled Chebyshev polynomial (as in (3.45)). Here we present the more generic choice:

$$P_\nu(t) = 1 - a_1 t - a_2 t^2 - \ldots - a_\nu t^\nu, \qquad \text{e.g.} \qquad P_2(t) = 1 - 2t + t^2.$$

Then the recursive AMLI preconditioner is defined, for $k = 2, 3, \ldots, J$, by

$$\begin{cases} S^{(k-1)} = A^{(k-1)} \left[I - P_\nu \left(M^{(k-1)^{-1}} A^{(k-1)} \right) \right]^{-1}, \\ M^{(k)} = \begin{bmatrix} A_{11}^{(k)} & \\ A_{21}^{(k)} & I \end{bmatrix} \begin{bmatrix} I & A_{11}^{(k)^{-1}} A_{12}^{(k)} \\ & S^{(k-1)} \end{bmatrix}, \end{cases} \tag{7.11}$$

where $S^{(1)} = A^{(1)}$ since $M^{(1)} = A^{(1)}$. Here the inverse of the top left block A_{11} across all levels can be optionally be approximated by a SPAI matrix as in (7.9) and likewise sparsification of $A^{(k)}$ may be also be adopted. Note (7.10) corresponds to the case of $P_1(t) = 1 - t$.

Applying (7.11) to the preconditioning step $M^{(k)}\mathbf{x} = \mathbf{y}$ amounts to a recursion

$$\begin{aligned} &\begin{bmatrix} A_{11}^{(k)} & \\ A_{21}^{(k)} & I \end{bmatrix} \begin{bmatrix} \mathbf{w}_1 \\ \mathbf{w}_2 \end{bmatrix} = \begin{bmatrix} \mathbf{y}_1 \\ \mathbf{y}_2 \end{bmatrix}, \quad \text{i.e.} \quad \begin{aligned} \mathbf{w}_1 &= A^{(k)^{-1}} \mathbf{y}_1, \\ \mathbf{w}_2 &= \mathbf{y}_2 - A_{21}^{(k)} \mathbf{w}_1, \end{aligned} \\ &\begin{bmatrix} I & A_{11}^{(k)^{-1}} A_{12}^{(k)} \\ & S^{(k-1)} \end{bmatrix} \begin{bmatrix} \mathbf{x}_1 \\ \mathbf{x}_2 \end{bmatrix} = \begin{bmatrix} \mathbf{w}_1 \\ \mathbf{w}_2 \end{bmatrix}, \quad \text{i.e.} \quad \begin{aligned} \mathbf{x}_2 &= S^{(k-1)^{-1}} \mathbf{w}_2, \\ \mathbf{x}_1 &= \mathbf{w}_1 - A^{(k)^{-1}} A_{12}^{(k)} \mathbf{x}_2, \end{aligned} \end{aligned} \tag{7.12}$$

where to find $\mathbf{x}_2$, we have to solve $S^{(k-1)}\mathbf{x}_2 = \mathbf{w}_2$. We finally discuss this most important substep of computing $\mathbf{x}_2$.

$$\begin{aligned} \mathbf{x}_2 = S^{(k-1)^{-1}} \mathbf{w}_2 &= \left[I - P_\nu \left(M^{(k-1)^{-1}} A^{(k-1)} \right) \right] A^{(k-1)^{-1}} \mathbf{w}_2 \\ &= \left[a_1 I + a_2 M^{(k-1)^{-1}} A^{(k-1)} + \cdots + a_\nu \left(M^{(k-1)^{-1}} A^{(k-1)} \right)^{\nu-1} \right] \\ &\quad M^{(k-1)^{-1}} \mathbf{w}_2, \end{aligned} \tag{7.13}$$

where we denote $P_\nu(t) = 1 - a_1 t - a_2 t^2 - \ldots - a_\nu t^\nu$. Recall that the Horner's rule [280] for computing $s = q_\nu(t) = a_0 + a_1 t + a_2 t^2 + \ldots + a_\nu t^\nu$ is the following:

$$
\begin{aligned}
s = q_\nu(t) &= a_0 + a_1 t + a_2 t^2 + \ldots + a_\nu t^\nu \\
&= (((\cdots((a_\nu)t + a_{\nu-1})t + \cdots)t + a_1)t + a_0),
\end{aligned}
\tag{7.14}
$$

which gives rise to the short recursive algorithm

$$
\begin{aligned}
&s = a_\nu \\
&\texttt{for } k = \nu - 1, \nu - 2, \ldots, 1, 0 \\
&\quad s = st + a_k \\
&\texttt{end } k.
\end{aligned}
$$

Adapting the Horner's rule to computing a matrix polynomial, with (7.13) in mind, leads to a similar method

$$
\begin{aligned}
\mathbf{x}_2 = q_\nu(W)\mathbf{z} &= a_1\mathbf{z} + a_2 W\mathbf{z} + a_3 W^2\mathbf{z} + \ldots + a_\nu W^{\nu-1}\mathbf{z} \\
&= (a_1\mathbf{z} + W(a_2\mathbf{z} + W(\cdots(+W(a_{\nu-1}\mathbf{z} + W(a_\nu\mathbf{z})))\cdots))),
\end{aligned}
\tag{7.15}
$$

with the associated algorithm

$$
\begin{aligned}
&\mathbf{x}_2 = a_\nu\mathbf{z} \\
&\texttt{for } k = \nu - 1, \nu - 2, \ldots, 1 \\
&\quad \mathbf{x}_2 = W\mathbf{x}_2 + a_k\mathbf{z} \\
&\texttt{end } k.
\end{aligned}
$$

Clearly on taking $\mathbf{z} = M^{(k-1)^{-1}}\mathbf{w}_2$ and $W = M^{(k-1)^{-1}}A^{(k-1)}$ in the above algorithm, one can rewrite (7.13) as $\mathbf{x}_2 = q_\nu(W)\mathbf{z}$ and compute it as follows [370]:

$$
\begin{aligned}
&\texttt{Solve } M^{(k-1)}\mathbf{x}_2 = a_\nu\mathbf{w}_2 \\
&\texttt{for } k = \nu - 1, \nu - 2, \ldots, 1 \\
&\quad \texttt{Solve } M^{(k-1)}\mathbf{x}_2 = A^{(k-1)}\mathbf{x}_2 + a_k\mathbf{w}_2 \\
&\texttt{end } k.
\end{aligned}
$$

Note ν plays the role of a cycling pattern in a multigrid algorithm (see Algorithm 6.1.4).

From a theoretical point of view, for SPD matrices, the AMLI algorithm as an algebraic multilevel preconditioner can provide stabilization for the HB method [502] as the latter does not have a simple and optimal preconditioner in three or more dimensions [495]. Another theoretically appealing stabilization method for the HB method was proposed by [471] and reviewed in [469,8], using the approximate wavelet-like L_2 projections (similar to [71]) to modify the HB basis. The BPX [71] preconditioner may be viewed as an optimal method for stabilizing the HB method for a special class of PDE's.

The AMLI type methods offer a large collection of excellent ideas for preconditioning and theories. However, the resulting algorithms are not immediately applicable to general sparse matrices. In the latter case, one may apply these preconditioners to the dominant part of a problem, as suggested in the remarks made in Section 6.5.

7.2 Multilevel geometrical partition: exact Schur

As mentioned in Section 6.5, a multilevel method may imply one of the following three scenarios, namely, (i) a single domain and multiple levels. (ii) a single level and multiple domains. (iii) multiple domains and multiple levels. The previous section falls into the first scenario just like the traditional multigrid method. Here we discuss the second scenario, i.e. the Schur complement method on a single level and multiple domains. This is technically the case of a geometrically driven matrix partition for Schur complements. In domain decomposition, we are entering a large topic of non-overlapping methods [111] and hence we only discuss some of the ideas and algorithms. The method is also called the substructuring method [432,352] as the idea originated from mechanical engineering. In graph theory (or finite element computation), the method is known as the nested dissection method [214,215].

Consider the solution of a PDE such as (6.1) in some domain $\Omega \subset \mathbb{R}^d$ and let the linear system after discretization by a domain method (see Section 1.7)

$$A\mathbf{u} = \mathbf{f}. \tag{7.16}$$

We shall first examine the direct substructuring method before discussing the iterative algorithms. For either method, the idea is to solve the global matrix problem by using local domain problems (in the form of a Schur decomposition) as preconditioners.

♦ **The direct substructuring method.** Although the domain Ω is divided into s nonoverlapping subdomains Ω_j and correspondingly the finest level mesh $\mathcal{T}$ is divided into patches of meshes $\mathcal{T}_j$ with mesh lines coinciding with the interfaces of the subdomains, in this section, we shall make a distinction of the mesh points (and subsequently the unknown variables) located at the interior and the boundary of a subdomain. See Figure 7.2 for an illustration of one-way domain partition or two ways. Here the substructuring method is related but different from the nested dissection method; the latter is illustrated in Figures 7.3 and 7.4 where we compare the use of 1-level dissection

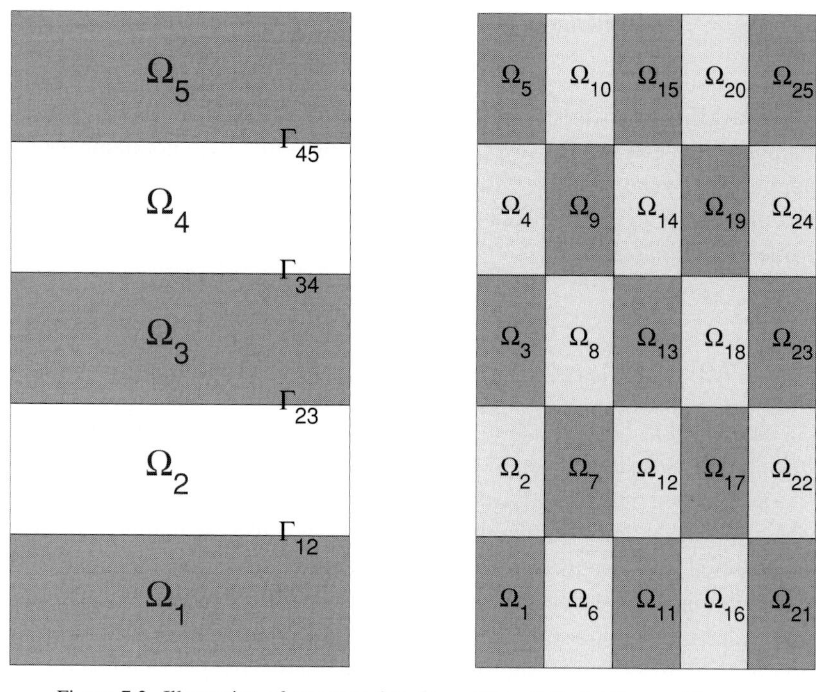

Figure 7.2. Illustration of one-way domain partition (left plot for $s = 5$) or two ways (right plot for $s = 25$). For general sparse matrices, one first constructs its graph and partitions its graph [215,352].

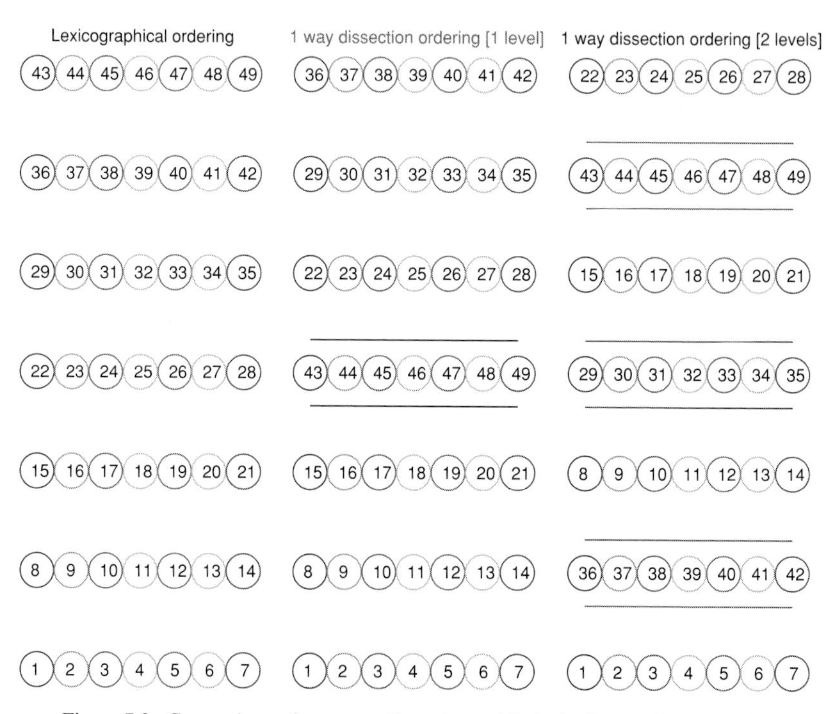

Figure 7.3. Comparison of one-way dissections with the lexicographical ordering.

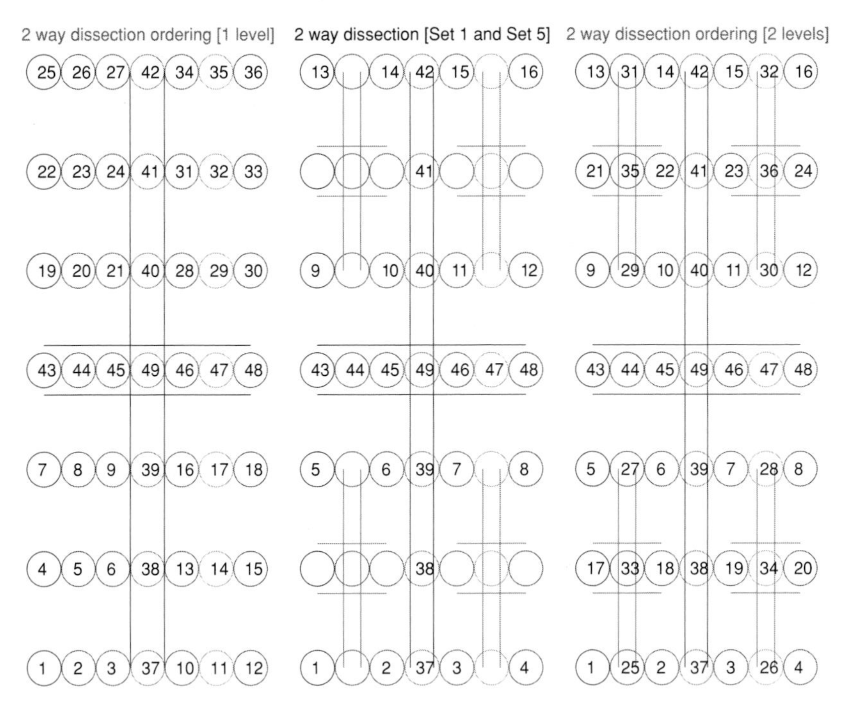

Figure 7.4. Illustration of two-way dissections (compare to Figure 7.2).

and 2 levels with the commonly used lexicographical ordering. From comparing to Figure 7.2, essentially, the substructuring method treats a multiple level dissection partition as a one-level dissection with interior domains interacting with interfaces. In the case of one way dissection, the two methods give the same result (i.e. the left Figure 7.2 is the 'same' as the far right plot of Figure 7.3 in terms of ordering).

Let $\mathbf{u}, \mathbf{f}$ be a vector defined over the entire mesh in $\mathcal{T} \subset \Omega$. Then assuming that all interior variables are ordered first and the boundary points second, both the vectors $\mathbf{u}, \mathbf{f}$ and the global matrix A admits the splitting into interior variables (I) and interface ones (B):

$$\mathbf{u} = \begin{bmatrix} u_I^{(1)} \\ u_I^{(2)} \\ \vdots \\ u_I^{(s)} \\ u_B \end{bmatrix}, \mathbf{f} = \begin{bmatrix} f_I^{(1)} \\ f_I^{(2)} \\ \vdots \\ f_I^{(s)} \\ f_B \end{bmatrix}, A = \left[\begin{array}{cccc|c} A_{II}^{(1)} & & & & A_{IB}^{(1)} \\ & A_{II}^{(2)} & & & A_{IB}^{(2)} \\ & & \ddots & & \vdots \\ & & & A_{II}^{(s)} & A_{IB}^{(s)} \\ \hline A_{BI}^{(1)} & A_{BI}^{(2)} & \cdots & A_{BI}^{(s)} & \sum_{j=1}^{s} A_{BB}^{(j)} \end{array}\right]. \tag{7.17}$$

Clearly this partitioned matrix A can be factorized in the block LU form

$$\left[\begin{array}{cccc|c} I & & & & A_{BI}^{(1)}A_{II}^{(1)^{-1}} \\ & I & & & A_{BI}^{(2)}A_{II}^{(2)^{-1}} \\ & & \ddots & & \vdots \\ & & & I & A_{BI}^{(s)}A_{II}^{(s)^{-1}} \\ \hline & & & & I \end{array}\right]^T \begin{bmatrix} A_{II}^{(1)} & & & & \\ & A_{II}^{(2)} & & & \\ & & \ddots & & \\ & & & A_{II}^{(s)} & \\ & & & & \sum_{j=1}^{s}\widetilde{S}^{(j)} \end{bmatrix}$$

$$\times \left[\begin{array}{cccc|c} I & & & & A_{II}^{(1)^{-1}}A_{IB}^{(1)} \\ & I & & & A_{II}^{(1)^{-1}}A_{IB}^{(2)} \\ & & \ddots & & \vdots \\ & & & I & A_{II}^{(1)^{-1}}A_{IB}^{(s)} \\ \hline & & & & I \end{array}\right], \tag{7.18}$$

where the Schur complement $\widetilde{S}^{(j)} = A_{BB}^{(j)} - A_{BI}^{(j)}A_{II}^{(j)^{-1}}A_{IB}^{(j)}$ and the transpose used in the first matrix is only for type setting purpose. Therefore using (7.18), the direct solution of (7.16) is given from solving

$$\begin{cases} \left(\sum_{j=1}^{s}\widetilde{S}^{(j)}\right)u_B = f_B - \sum_{j=1}^{s} A_{BI}^{(j)}A_{II}^{(j)^{-1}}f_I^{(j)}, \\ A_{II}^{(j)}u_I^{(j)} = f_I^{(j)} - A_{IB}^{(j)}u_B, \qquad \text{for } j = 1, 2, \ldots, s. \end{cases} \tag{7.19}$$

A possible confusion might arise, because the above equation is often not used in its present form. This is due to a need to further localize the set B of all internal interfaces (i.e. Γ_{ij} in Figure 7.2) to make each Schur matrix more sparse (with no approximation).

To this end, we define R_j to be a rectangular restriction matrix (consisted of ones and zeros) that restricts a global quantity to that of associated with the interface variables of Ω_j (or $\mathcal{T}_j$). That is, if $\mathbf{u} \in \mathbb{R}^n$ is defined on $\mathcal{T}$, then $u_B^{(j)} = R_j\mathbf{u}$ is a short vector extracted from $\mathbf{u}$ to associate with the interface variables of Ω_j. Conversely, $\overline{u} = R_j^T u_B^{(j)}$ is a global vector of size n that takes values of $u_B^{(j)}$ at the interface locations of Ω_j. Clearly $\widetilde{S}^{(j)} = R_j^T S^{(j)} R_j$,

$$\sum_{j=1}^{s}\widetilde{S}^{(j)} = \sum_{j=1}^{s} R_j^T S^{(j)} R_j, \tag{7.20}$$

and the quantity u_B is made up by assembling s contributions:

$$u_B = \sum_{j=1}^{s} R_j^T u_B^{(j)}, \tag{7.21}$$

where the reduced local Schur complement is

$$S^{(j)} = R_j \widetilde{S}^{(j)} R_j^T = R_j \left(A_{BB}^{(j)} - A_{BI}^{(j)} A_{II}^{(j)^{-1}} A_{IB}^{(j)} \right) R_j^T = A_{\Gamma\Gamma}^{(j)} - A_{\Gamma I}^{(j)} A_{II}^{(j)^{-1}} A_{I\Gamma}^{(j)}$$

and Γ is used to indicate all local interfaces of subdomain j. With the above notation to simplify (7.19), the direct solution of (7.16) is given from solving

$$\begin{cases} \left(\sum_{j=1}^{s} R_j^T S^{(j)} R_j \right) u_B = f_B - \sum_{j=1}^{s} R_j^T A_{\Gamma I}^{(j)} A_{II}^{(j)^{-1}} f_I^{(j)}, \\ \qquad A_{II}^{(j)} u_I^{(j)} = f_I^{(j)} - A_{I\Gamma}^{(j)} u_B^{(j)}, \qquad \text{for } j = 1, 2, \ldots, s. \end{cases} \tag{7.22}$$

Note that $A_{I\Gamma}^{(j)} u_B^{(j)} = A_{IB}^{(j)} u_B$ and all assembly and solution steps (except the solution of the first equation once assembled) can be implemented in parallel for all subdomains.

♦ **The iterative substructuring method.** The above direct method can be converted to an iterative method (using the GMRES for instance) instantly as the latter method, only needing matrix vector products, does not require to form any local Schur matrices. Thus the only remaining task of an iterative substructuring method is essentially to find a suitable preconditioner for the Schur complement (which is a sum of local Schur complements). There exists a vast literature on this topic [432,111,343,450]. We only present the balancing Neumann–Neumann method for the case of s subdomains.

To proceed, denote the first Schur equation in (7.22) for the interface variables by (note other equations are purely local)

$$S\mathbf{u} = \mathbf{g}, \tag{7.23}$$

where for simplicity $\mathbf{u} = \mathbf{u}_B \in \mathbb{R}^{n_B}$. A Krylov subspace iterative solver for (7.23) will make use of a preconditioner B defined by a s-subdomain algorithm of the type

$$\mathbf{u}^{\text{new}} = \mathbf{u}^{\text{old}} + B\left(\mathbf{g} - S\mathbf{u}^{\text{old}}\right), \tag{7.24}$$

which is iterated for a number of steps. There are two special features (coarse grid correction and diagonal scaling) that require additional notation:

- A_0 is the discretization matrix using a piecewise constant approximation over each subdomain.
- R_0 restricts a global vector on the entire mesh $\mathcal{T}$ to a global coarse grid space of piecewise constants.
- R_0^T interpolates a global coarse grid space of piecewise constants to the global vector on $\mathcal{T}$.

- $D = diag(1/b_\ell)$ is a global diagonal matrix of size $n_B \times n_B$ (the same as $\mathbf{u}_B$) where, for the ℓ-th diagonal, b_ℓ denotes the number of subdomains that share the ℓ-th node. (For instance, in $\mathbb{R}^2$, $b_\ell = 2$ for points in the middle of a subdomain edge).

Then the balancing Neumann–Neumann preconditioning method for (7.23) and (7.22) is to repeat the following (with $\mathbf{u}^0 = \mathbf{u}^{\text{old}}$)

$$\begin{cases} \mathbf{u}^{1/3} = \mathbf{u}^0 + R_0^T A_0^{-1} R_0 \left(\mathbf{g} - S\mathbf{u}^0\right), \\ \mathbf{u}^{2/3} = \mathbf{u}^{1/3} + D \sum_{j=1}^{s} R_j^T S^{(j)^{-1}} R_j D \left(\mathbf{g} - S\mathbf{u}^{1/3}\right), \\ \mathbf{u}^1 = \mathbf{u}^{2/3} + R_0^T A_0^{-1} R_0 \left(\mathbf{g} - S\mathbf{u}^{2/3}\right), \\ \mathbf{u}^0 = \mathbf{u}^1, \end{cases} \tag{7.25}$$

where the first step is optional for unsymmetric problems as it intends to maintain 'symmetry' in the preconditioner B (like the SSOR). To find the preconditioner B, we only need to combine all three steps into a single one in the form of (7.24). Such a matrix B is the following

$$B = \left(I - R_0^T A_0^{-1} R_0 S\right) D \Big(\sum_{j=1}^{s} R_j^T S^{(j)^{-1}} R_j\Big) D \left(I - S R_0^T A_0^{-1} R_0\right) + R_0^T A_0^{-1} R_0.$$

Note that the most important step two in (7.25) is done in parallel and moreover $S^{(j)}$ is of the reduced size – the number of interface mesh points associated with $\mathcal{T}_j$ (or Ω_j). For other Schur preconditioners, refer to [432,111,343,450].

7.3 Multilevel algebraic partition: permutation-based Schur

The previous preconditioners in the last two sections are respectively functional based and geometry driven. This section will discuss a further class of graph ordering (permutation)-based Schur complement methods, originated from the work of [412,413]. We remark that the similar use of graph theory in dissection methods for general sparse matrices was much earlier [214,215] but the work of [412] was designed for preconditioning rather than direct solution. Moreover, for more than two levels, the two methods are very different because the nested dissection replies on the exact graph of the original matrix while the new Schur method of [412] will use the graph of a Schur matrix (approximated). For related developments and extensions, refer to [506] and the references therein.

The main idea of this permutation-based Schur complement method is to ensure that the top left block A_{11} in (7.2) is occupied by a diagonal matrix (or a block diagonal matrix in the block version), achieved by permutation based on a

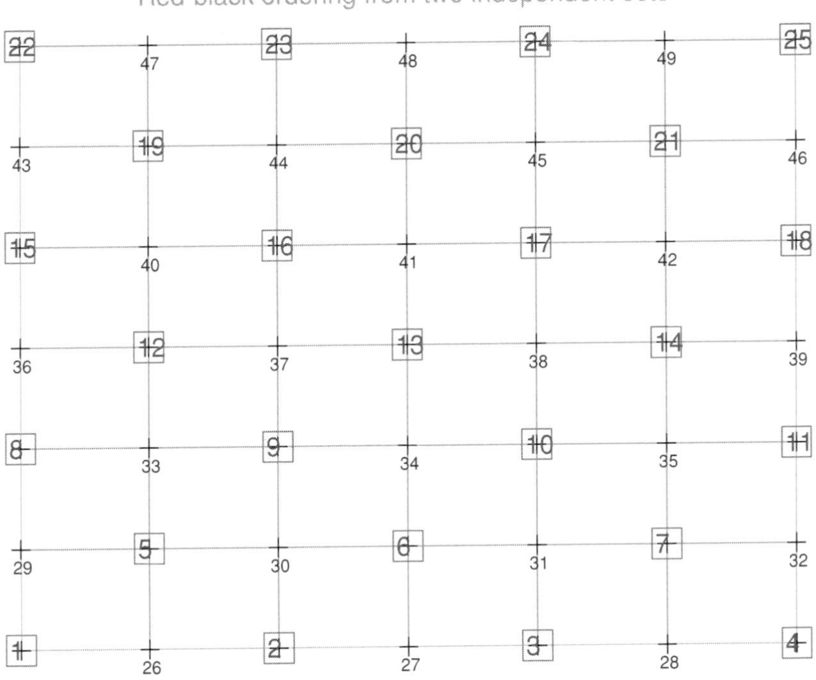

Figure 7.5. Illustration of a red-black ordering (with red in large fonts and black in small fonts, each forming an independent set of nodes). Check with the Mfile `multic.m`.

matrix graph. For a PDE discretized by a five-point FDM stencil, the resulting matrix A can be easily permuted (using the so-called red-black ordering as shown in Figure 7.5) to produce such a diagonal matrix $D = A_{11} = \text{diag}(d_j)$:

$$P_{rb} A P_{rb}^T = \begin{bmatrix} A_{11} & A_{12} \\ A_{21} & A_{22} \end{bmatrix} = \begin{bmatrix} \text{diag}(d_j) & F \\ E & C \end{bmatrix}, \tag{7.26}$$

which has a ‘computable’ Schur matrix $S = C - E\,\text{diag}(d_j)^{-1} F$; by ‘computable’ we mean that the resulting Schur matrix will most probably be sparse (instead of dense in the normal case). (Refer to Algorithm 4.3.3). Thus the idea is quite common for PDE problems that have a natural graph. For general sparse matrices, we need to use the matrix graph and introduce an algebraic tool for identifying an ordering for the nodes (or a permutation for a sparse matrix).

To realize (7.26), the suitable tool will be the independent set ordering. For symmetric matrices A, we have discussed its undirected graph $G(A)$ in Section 2.6. For a general sparse matrix, the directed graph $G(A)$ is denoted by $G(A) = (V, E)$ where V denotes the indices of vertices (nodes) and E the list of edges represented in the form of pairs of integers (the nonzero locations of A) e.g. for a 3×3 tridiagonal matrix A, its graph may be represented by $G(A) = (V, E)$ with

$$V = [1\ 2\ 3], \qquad E = [(1,1)\ (1,2)\ (2,1)\ (2,2)\ (2,3)\ (3,2)\ (3,3)]\,.$$

Definition 7.3.1. (The independent set). *Let $G(A) = (V, E)$ be the directed graph of a sparse matrix A. Then a subset S of V is called an independent set if any two distinct vertices in S are not connected i.e.*

$$\text{if } x, y \in S \text{ and } x \neq y, \text{ then } (x, y) \notin E \text{ and } (y, x) \notin E.$$

Two observations are in order. Firstly, one can see that the independent set is clearly geared up for fulfilling (7.26). Secondly, it is not difficult to envisage how to work out an independent set for a geometric graph arising from PDE discretization (e.g. from Figure 7.5). It is of interest to find the largest possible independent set in $G(A)$, although a reasonable size of such a set is practically sufficient – the size of this set will be the size of the corresponding block diagonal matrix. However, it is not all trivial to efficiently compute the largest possible independent set from $G(A)$ alone; see [332,333,412].

Practically there exist many algorithms for computing an independent set from $G(A) = (V, E)$. Here we only discuss one of them. As with the discussion of the RCM algorithm, there is this question of how to define the very first node (vertex) to join the set S. It makes sense not to start from a node that has too many connections (i.e. appearances in E). Similar to the symmetric case, the degree $deg(v)$ of a vertex v is the total number of edges that are adjacent to v. Assume that the vertices can be sorted by increasing degrees to obtain the initial ordering $i_1, i_2, \ldots, i_n$. Then we expect the vertex i_1 to be in set S.

Algorithm 7.3.2. (An independent set ordering).

Let the vertices V in a sparse graph $G(A) = (V, E)$ be initially ordered as $i_1, i_2, \ldots, i_n$ by increasing degrees. Let the set $S = \emptyset$.

for $j = 1, 2, \ldots, n$
  if vertex i_j is not yet marked
    Accept vertex i_j, $S = S \cup \{i_j\}$
    Mark vertex i_j and all its adjacent vertices (neighbours)
  end if vertex i_j is not yet marked
end j.

This algorithm produces the same subset of indices as Algorithm 4.3.3 with two colours. For the example in Figure 7.5, assuming only horizontal and vertical connections, one can generate the set S of all red nodes using Algorithm 7.3.2.

To estimate the size or cardinality $|S|$ of S, we need the information of the number ν that denotes the largest *unique* neighbours of all vertices and $\nu_{\max}$ that denotes the maximum degree of all nodes. By *unique*, assuming shared neighbours are equally divided, we mean the neighbouring vertices, at a fixed $v \in (V \backslash S)$, that are attributed to v and not to other vertices in S. Then we have (note $\nu \le \nu_{\max}$ and $|S| \le n$)

$$n - |S| \le \nu |S|, \quad \text{or} \quad |S| \ge \frac{n}{\nu + 1} \ge \frac{n}{\nu_{\max} + 1}.$$

For the graph in Figure 7.5, we have $\nu = 1$ and $\nu_{\max} = 4$ giving the prediction: $|S| \ge n/2 \ge n/5$ which is quite accurate. as the exact answer is $|S| = n/2$ (even n) or $n/2 + 1$ (odd n). We remark that a related method is the multi-colouring technique [412] that we do not pursue here while the colouring idea [432] in domain decomposition methods amounts to block relaxation (or other block operations).

We are now ready to introduce the Schur complement idea based on independent set orderings [412,506], for solving the sparse linear system (7.16) or (7.3) i.e.

$$A\mathbf{u} = \mathbf{f}. \tag{7.27}$$

To be consistent with previous ML methods, we shall consider J levels for solving (7.27) with $A^{(j)}$ the matrix obtained from the j-th step of a Schur reduction method for $j = J, J-1, \ldots, 1$. Here $A^{(J)} = A$ on the finest level J. For each matrix $A^{(j)}$, we shall apply Algorithm 7.3.2 to obtain an independent set S and with it to obtain an order $k_1, k_2 \ldots, k_n$ of the nodes, that orders S nodes first and the remaining nodes second. This ordering defines a permutation matrix P_j such that the top left block of $P_j A^{(j)} P_j^T$ is a diagonal matrix (by design):

$$\begin{aligned} P_j A^{(j)} P_j^T &= \begin{bmatrix} A_{11}^{(j)} & A_{12}^{(j)} \\ A_{21}^{(j)} & A_{22}^{(j)} \end{bmatrix} = \begin{bmatrix} D_j & A_{12}^{(j)} \\ A_{21}^{(j)} & C_j \end{bmatrix} \\ &= \begin{bmatrix} I & \\ A_{21}^{(j)} D_j^{-1} & I \end{bmatrix} \begin{bmatrix} D_j & A_{12}^{(j)} \\ & A^{(j-1)} \end{bmatrix} \end{aligned} \tag{7.28}$$

where the Schur complement

$$S^{(j)} = A^{(j-1)} = C_j - A_{21}^{(j)} D_j^{-1} A_{12}^{(j)}.$$

Observe that whenever $A_{21}^{(j)}, A_{12}^{(j)}$ are sparse, it becomes feasible to form the Schur complement $A^{(j-1)}$ explicitly. However if $j < J$, $A_{21}^{(j)}, A_{12}^{(j)}$ will be much less sparse than $A_{21}^{(J)}, A_{12}^{(J)}$ respectively. This is where one has to use the thresholding idea as in Algorithm 4.5.10; see also [506].

We are now in a similar (and slightly better) situation to that of (7.9) in constructing a sequence of coarse level approximations

$$P_j \bar{A}^{(j)} P_j^T = \begin{bmatrix} I & \\ \bar{A}_{21}^{(j)} \bar{D}_j^{-1} & I \end{bmatrix} \begin{bmatrix} \bar{D}_j & \bar{A}_{12}^{(j)} \\ & \bar{A}^{(j-1)} \end{bmatrix}, \qquad j = J, J-1, \ldots, 1$$

with $\bar{A}^{(J)} = A^{(J)}$ and for $k = J, J-1, \ldots, 2$

$$\bar{A}^{(k-1)} = \bar{C}_k - drop\left(\bar{A}_{21}^{(k)} \bar{D}_k^{-1} \bar{A}_{12}^{(k)}\right).$$

Here the 'drop' notation is the same as in Section 5.4.4 to indicate a step of thresholding to maintain sparsity.

On the coarsest level 1, we take $M^{(1)} = \bar{A}^{(1)}$ as the preconditioner. The finer level preconditioners are therefore ($j = 2, 3, \ldots, J$)

$$M^{(j)} = P_j^T \begin{bmatrix} I & \\ \bar{A}_{21}^{(j)} \bar{D}_j^{-1} & I \end{bmatrix} \begin{bmatrix} \bar{D}_j & \bar{A}_{12}^{(j)} \\ & M^{(j-1)} \end{bmatrix} P_j. \tag{7.29}$$

Applying (7.29) to the preconditioning step $P_j M^{(j)} P_j^T \mathbf{x} = \mathbf{y}$ amounts to a recursion in a V-cycling pattern

$$\begin{aligned}
&\begin{bmatrix} I & \\ \bar{A}_{21}^{(j)} \bar{D}_j^{-1} & I \end{bmatrix} \begin{bmatrix} \mathbf{w}_1 \\ \mathbf{w}_2 \end{bmatrix} = \begin{bmatrix} \mathbf{y}_1 \\ \mathbf{y}_2 \end{bmatrix}, \quad \text{i.e.} \quad \begin{array}{l} \mathbf{w}_1 = \mathbf{y}_1, \\ \mathbf{w}_2 = \mathbf{y}_2 - \bar{A}_{21}^{(j)} \bar{D}_j^{-1} \mathbf{w}_1, \end{array} \\
&\begin{bmatrix} \bar{D}_j & \bar{A}_{12}^{(j)} \\ & M^{(j-1)} \end{bmatrix} \begin{bmatrix} \mathbf{x}_1 \\ \mathbf{x}_2 \end{bmatrix} = \begin{bmatrix} \mathbf{w}_1 \\ \mathbf{w}_2 \end{bmatrix}, \quad \text{i.e.} \quad \begin{array}{l} \mathbf{x}_2 = {M^{(j-1)}}^{-1} \mathbf{w}_2, \\ \mathbf{x}_1 = \bar{D}_j^{-1} \left(\mathbf{w}_1 - \bar{A}_{12}^{(j)} \mathbf{x}_2\right), \end{array}
\end{aligned} \tag{7.30}$$

where to find $\mathbf{x}_2$, we have to solve $M^{(j-1)} \mathbf{x}_2 = \mathbf{w}_2$.

One important observation in [506] was that we do not need the last preconditioner $M^{(J)}$ for the whole matrix $A^{(J)} = A$ and rather it is better to use the preconditioner $M^{(J-1)}$ for the first Schur matrix $A^{(J-1)} = A$. This is because the system (7.27) is equivalent to

$$P_J A^{(J)} P_J^T P_J \mathbf{u} = P_J \mathbf{f},$$

or
$$P_J A^{(J)} P_J^T \widetilde{\mathbf{u}} = \widetilde{\mathbf{f}},$$

or
$$\begin{bmatrix} I & \\ A_{21}^{(J)} D_J^{-1} & I \end{bmatrix} \begin{bmatrix} D_J & A_{12}^{(J)} \\ & A^{(J-1)} \end{bmatrix} \begin{bmatrix} \widetilde{\mathbf{u}}_1 \\ \widetilde{\mathbf{u}}_2 \end{bmatrix} = \begin{bmatrix} \widetilde{\mathbf{f}}_1 \\ \widetilde{\mathbf{f}}_2 \end{bmatrix},$$

or
$$\begin{cases} \mathbf{x}_1 = \widetilde{\mathbf{f}}_1, \\ \mathbf{x}_2 = \widetilde{\mathbf{f}}_2 - A_{21}^{(J)} D_J^{-1} \mathbf{x}_1, \\ \widetilde{\mathbf{u}}_2 = A^{(J-1)^{-1}} \mathbf{x}_2, \\ \widetilde{\mathbf{u}}_1 = D_J^{-1} \left(\mathbf{x}_1 - A_{12}^{(J)} \mathbf{x}_2 \right). \end{cases} \tag{7.31}$$

where to find $\widetilde{\mathbf{u}}_2$, we use a Krylov subspace method using the preconditioner $M^{(J-1)}$ for

$$A^{(J-1)} \widetilde{\mathbf{u}}_2 = \mathbf{x}_2.$$

That is, we only need to solve the Schur equation iteratively (while all other equations involve no approximations)

$$\left(C_J - A_{21}^{(J)} D_J^{-1} A_{12}^{(J)} \right) \widetilde{\mathbf{u}}_2 = \mathbf{x}_2. \tag{7.32}$$

Once $\widetilde{\mathbf{u}}_2$ is solved, we use (7.31) to complete the solution for $\mathbf{u}$.

In summary, the last Schur method is designed for general sparse linear systems and appears to be a quite promising technique for future development.

7.4 Appendix: the FEM hierarchical basis

The FEM hierarchical basis proposed by [502] is a wonderful idea that deserves to be covered by all FEM texts. However this is not yet the case, partly because the presentations in the literature were often too advanced to follow for a general reader. Here we expose the essential ideas and facts so that further reading can be made easier.

The overview of the HB basis. For a given finest grid, the hierarchical basis [502] defines an equivalent subspace to the usual FEM space but has the useful property

$$\mathbf{V} = \mathbf{V}_1 \oplus \mathbf{V}_2 \oplus \cdots \oplus \mathbf{V}_J \tag{7.33}$$

where $\mathbf{V}_1$ is the coarsest level finite element subspace while $\mathbf{V}_k$ is the subspace corresponding to all nodes in $(N_k \backslash N_{k-1})$. For piecewise linear functions in $\mathbb{R}^2$, such a setup can be easily understood if one starts building the subspace $\mathbf{V}$ from adding new basis functions to $\mathbf{V}_1$ as shown in Figure 7.1. More precisely, let P_k be the piecewise linear interpolation operator satisfying

$$(P_k u)(x) = u(x), \qquad \in N_k.$$

Then $\mathbf{V}_k = \text{Range}(P_k - P_{k-1})$ for $k = 2, 3, \ldots, J$. Moreover, any function on the finest level J can be decomposed by a telescopic series (wavelets of Chapter 8 will define a similar series)

$$u = P_1 u + \sum_{k=2}^{J} (P_k - P_{k-1}) u. \tag{7.34}$$

Here function $(P_k - P_{k-1})u$ vanishes at nodes of N_{k-1} as they both take the value 1 there as an interpolation function.

However none of the above results seems to be convincing if we do not *explicitly* answer these questions?

(1) How to construct the HB basis from the standard FEM basis? and vice versa. One notes that [502] gives the detail algorithms of how to form a HB stiffness matrix from a standard FEM matrix. But the simple issue of basis construction is not discussed.
(2) Is the HB subspace identical to the standard FEM subspace for a given finest grid? If so, how to prove that?

Although these are simple questions, surprisingly, the full details cannot be found from the (somewhat large) literature.

The exact relationship of the HB basis and the standard FEM basis. Consider the 1D basis, as illustrated in Figure 7.1. The most important and yet simple fact is the following between basis functions of two adjacent levels:

$$\phi_i^{(k-1)}(x) = \phi_{2i-1}^{(k)}(x) + \frac{1}{2}\phi_{2i-2}^{(k)}(x) + \frac{1}{2}\phi_{2i}^{(k)}(x), \tag{7.35}$$

which is as shown in Figure 7.6; here we assume $\phi_j^{(k)} = 0$ if $j \le 0$ or $j \ge n_k + 1$ at the end nodes.

Clearly the standard linear FEM basis are

$$\{\phi_1^{(J)}, \phi_2^{(J)}, \cdots, \phi_{n_J}^{(J)}\} \qquad \text{on level } J, \tag{7.36}$$

and the HB basis (as in Figure 7.1) are

$$\{{}^{HB}\phi_1^{(J)}, {}^{HB}\phi_2^{(J)}, \cdots, {}^{HB}\phi_{n_J}^{(J)}\} \tag{7.37}$$

$$= \{ \underbrace{\phi_1^{(1)}, \phi_2^{(1)}}_{\text{Coarest } L1}; \underbrace{\phi_2^{(2)}}_{L2}; \underbrace{\phi_2^{(3)}, \phi_4^{(3)}}_{L3}; \underbrace{\phi_2^{(4)}, \phi_4^{(4)}, \ldots, \phi_{n_4-1}^{(4)}}_{L4}; \cdots; \underbrace{\phi_2^{(J)}, \phi_4^{(J)}, \ldots, \phi_{n_J-1}^{(J)}}_{LJ} \}.$$

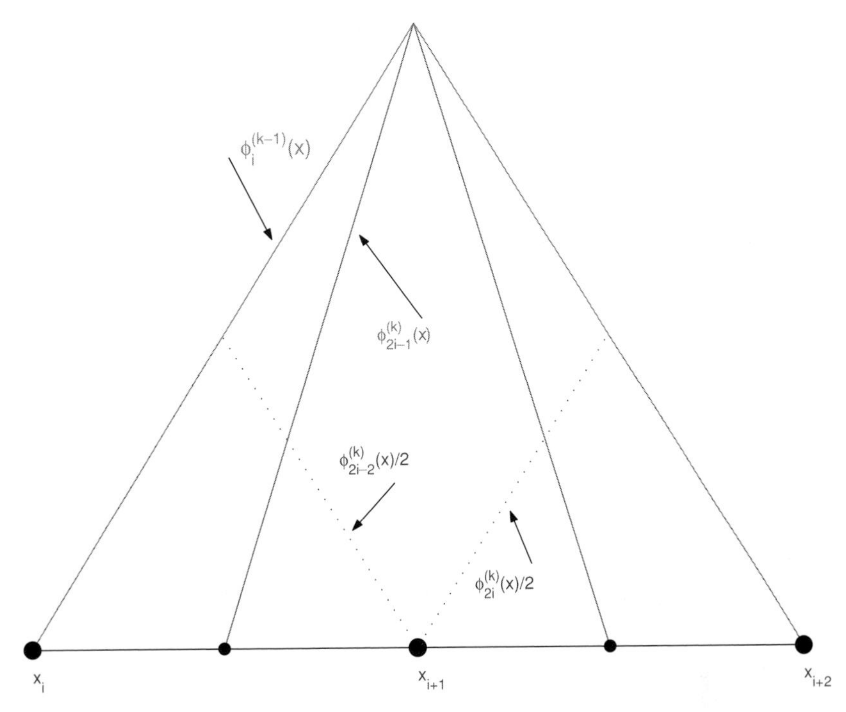

Figure 7.6. Illustration of the fundamental result (7.35) for the HB basis.

The identical representation of the two subspaces. From (7.35), one can verify that the above HB basis represents the same FEM subspace i.e.

$$\texttt{span}\left(\phi_1^{(J)}, \phi_2^{(J)}, \cdots, \phi_{n_J}^{(J)}\right) = \texttt{span}\left({}^{HB}\phi_1^{(J)}, {}^{HB}\phi_2^{(J)}, \cdots, {}^{HB}\phi_{n_J}^{(J)}\right). \quad (7.38)$$

With the aide of (7.35) i.e. $\phi_{2i-1}^{(k)}(x) = \phi_i^{(k)}(x) - \phi_{2i-2}^{(k)}(x)/2 - \phi_{2i}^{(k)}(x)/2$ and Figure 7.1, we see that

$$\begin{aligned}
&\texttt{span}\left(\phi_1^{(J)}, \phi_2^{(J)}, \cdots, \phi_{n_J}^{(J)}\right) \\
&= \texttt{span}\left(\phi_2^{(J)}, \phi_4^{(J)}, \cdots, \phi_{n_J-1}^{(J)}; \phi_1^{(J-1)}, \phi_2^{(J-1)}, \cdots, \phi_{n_{J-1}}^{(J-1)}\right) \\
&= \texttt{span}\left(\phi_2^{(J)}, \cdots, \phi_{n_J-1}^{(J)}; \phi_2^{(J-1)}, \cdots, \phi_{n_{J-1}-1}^{(J-1)}; \cdots; \phi_1^{(J-2)}, \cdots, \phi_{n_{J-2}}^{(J-2)}\right) \\
&= \qquad \cdots \\
&= \texttt{span}\left(\phi_2^{(J)}, \cdots, \phi_{n_J-1}^{(J)}; \cdots; \phi_2^{(3)}, \phi_{n_3-1}^{(3)}; \phi_2^{(2)}; \phi_1^{(1)}, \phi_2^{(1)}\right) \\
&= \texttt{span}\left({}^{HB}\phi_1^{(J)}, {}^{HB}\phi_2^{(J)}, \cdots, {}^{HB}\phi_{n_J}^{(J)}\right).
\end{aligned} \quad (7.39)$$

Higher dimensions. If we decompose the HB subspace by grouping its basis functions in (7.37) according to their level information, we obtain the space decomposition (7.33). Further one may use (7.33) to construct an additive multilevel preconditioner B for the underlying HB stiffness matrix A (assuming A is SPD) and show that [494,495] the preconditioner is optimal in 1D but not in higher dimensions $\mathbb{R}^d$ i.e.

$$\kappa(B^{-1}A) = \begin{cases} O(1) & \text{if } d = 1; \\ O(h|\log h|^2) & \text{if } d = 2; \\ O(h^{2-d}) & \text{if } d \geq 3. \end{cases} \tag{7.40}$$

In $\mathbb{R}^2$, we illustrate the generalized form of (7.35)

$$\phi_i^{(k-1)}(x) = \phi_{i^*}^{(k)}(x) + \frac{1}{2} \sum_{\ell \in \{\text{neighbouring fine level nodes}} \phi_\ell^{(k)}(x) \tag{7.41}$$

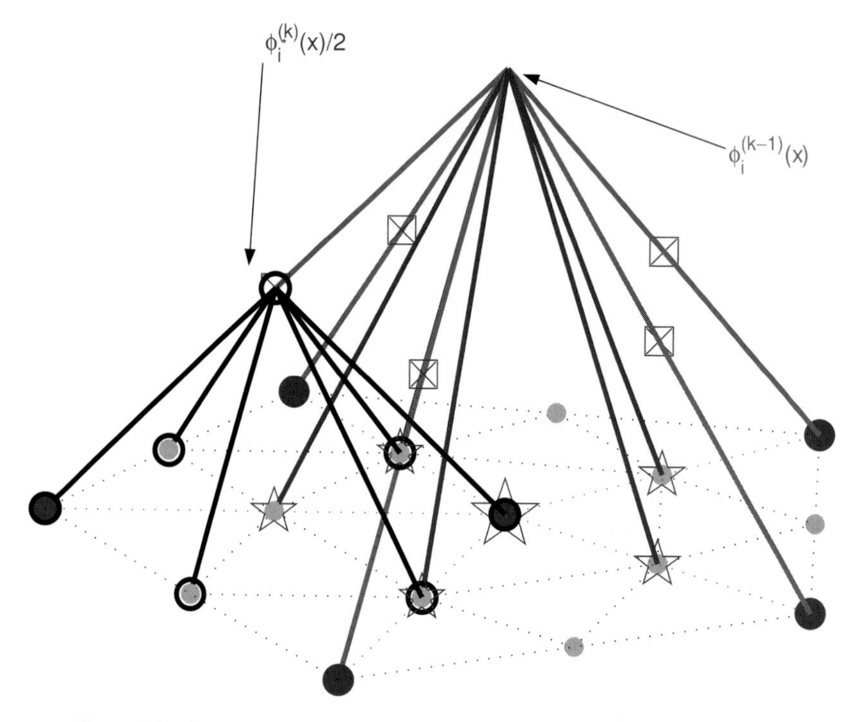

Figure 7.7. Illustration of the fundamental result (7.41) in $\mathbb{R}^2$ for the HB basis. Here large • points are the coarse points and small • ones are the fine level points. Observe the difference between $\phi^{(k)}$ and $\phi^{(k-1)}$ the same coarse point (large) ⋆. The □ (and small ⋆) points indicate the position of neighbouring $\phi^{(k)}/2$.

in Figure 7.7, where the particular node i at the centre has five neighbouring fine level nodes and i^* is the fine level k numbering given to the node i of level $(k-1)$. Clearly the hierarchical basis 'loses' its advantages visible in Figure 7.6, or the fine level basis functions in (7.41) (inside the sum $\sum_\ell$ are no longer orthogonal to each other!). It appears to be difficult to achieve the wonderful splitting (7.34) (associated with orthogonal basis) in higher dimensions.

7.5 Discussion of software and the supplied Mfiles

The Schur complements based preconditioners are widely used in coupled matrix problems (Chapter 12). For a single and sparse matrix, several well-written codes exist that use this kind of preconditioners.

(1) The AMLI package (Mfiles) for elliptic PDEs (Maya Neytcheva)

$$\text{http : //www.netlib.org/linalg/amli.tgz}$$

(2) The Finite Element ToolKit (FEtk) from

$$\text{http : //www.fetk.org/}$$

and its 2D MATLAB® (light) version

$$\text{http : //scicomp.ucsd.edu/}\sim\text{mholst/codes/mclite/index.html}$$

(3) The BILUM package (Jun Zhang and Youcef Saad):

$$\text{http : //cs.engr.uky.edu/}\sim\text{jzhang/bilum.html}$$

(4) The pARMS library (parallel solvers for distributed sparse linear systems of equations [336]):

$$\text{http : //www} - \text{users.cs.umn.edu/}\sim\text{saad/software/pARMS/}$$

For a sparse matrix, the use of `multic.m` (Chapter 4) may be used to develop a code making use of Schur complements for preconditioning. The use of HB ordering is an advanced proposal and below we supply a Mfile that illustrates it.

[1] HB1.m – Illustration of a standard FEM ordering and conversion to a HB ordering for a 1D example.

8

Sparse wavelet preconditioners [T5]: approximation of $\tilde{A}_{n\times n}$ and $\tilde{A}^{-1}_{n\times n}$

> The subject of "wavelets" is expanding at such a tremendous rate that it is impossible to give, within these few pages, a complete introduction to all aspects of its theory.
>
> RONALD A. DEVORE and BRADLEY J. LUCIER. Wavelets. *Acta Numerica* (1992)

> If A is a bounded operator with a bounded inverse, then A maps any orthogonal basis to a Riesz basis. Moreover, all Riesz bases can be obtained as such images of an orthogonal basis. In a way, Riesz bases are the next best thing to an orthogonal basis.
>
> INGRID DAUBECHIES. *Ten Lectures on Wavelets.* SIAM Publications (1992)

The discovery of wavelets is usually described as one of the most important advances in mathematics in the twentieth century as a result of joint efforts of pure and applied mathematicians. Through the powerful compression property, wavelets have satisfactorily solved many important problems in applied mathematics, such as signal and image processing; see [269,166,441,509] for a summary. There remain many mathematical problems to be tackled before wavelets can be used for solution of differential and integral equations in a general setting.

In this chapter, we aim to give an introduction to wavelet preconditioning and focus more on discrete wavelets. As far as the solution of operator equations is concerned, the construction of compactly supported and computable wavelet functions remains a challenge for the future. We discuss these issues in the following.

Section 8.1 Introduction to multiresolution and orthogonal wavelets
Section 8.2 Operator compression by wavelets and sparsity patterns

In this chapter, we shall use the tilde notation (e.g. $\widetilde{y}$) to mean a quantity in a wavelet space and the hat notation (e.g. $\widehat{y}$) for a function in the Fourier space. The bar notation is used for a complex conjugate in the context of Fourier analysis (e.g. $\overline{e^{-i\omega}} = \overline{\cos(\omega) - i\sin(\omega)} = \cos(\omega) + i\sin(\omega) = e^{i\omega}$).

8.1 Introduction to multiresolution and orthogonal wavelets

Although the subject area is large, we hope to summarize and capture enough main issues for a general reader to follow the argument even before diving into the subject. We pay particular attention to the success and failure of wavelets as far as applications are concerned. We refer to the interested reader to [152, 302,441,172,173] and many other wavelet books as listed on

`http://www.wavelet.org`

for further reading. To give a definition for a wavelet function in 1D, it is useful to mention the Riesz basis for the space $L_2(\mathbb{R})$ – a concept of stable basis, much more used in theoretical analysis [151].

Definition 8.1.1. *A functional basis $\{f_j(x)\}_{j=-\infty}^{\infty}$ for $L_2(\mathbb{R})$ is called a Riesz basis if its linear span is dense in $L_2(\mathbb{R})$ and there exists positive constants $A \le B < \infty$ (for any linear combination) such that*

$$A\left\|\{c_j\}\right\|_{l^2}^2 \le \left\|\sum_{j=-\infty}^{\infty} c_j f_j(x)\right\|_2^2 \le B\left\|\{c_j\}\right\|_{l^2}^2, \tag{8.1}$$

where

$$\|\{c_j\}\|_{l^2}^2 = \sum_{j=-\infty}^{\infty} |c_j|^2.$$

A function $\psi \in L_2(\mathbb{R})$ is called a $\mathcal{R}$-function, if the sequence defined by

$$\psi_{j,k}(x) = 2^{j/2}\psi(2^j x - k), \qquad j,k = 0, \pm 1, \pm 2, \ldots, \tag{8.2}$$

forms a Riesz basis in $L_2(\mathbb{R})$.

Definition 8.1.2. *A $\mathcal{R}$-function $\psi \in L_2(\mathbb{R})$ is called a wavelet (or $\mathcal{R}$-wavelet) if there exists a (dual) $\mathcal{R}$-function $\check{\psi} \in L_2(\mathbb{R})$ such that*

$$\left\langle \psi_{j,k}, \check{\psi}_{l,m} \right\rangle = \delta_{j,l}\delta_{k,m}, \qquad j,k,l,m = 0, \pm 1, \pm 2, \ldots, \tag{8.3}$$

with $\check{\psi}_{l,m}(x) = 2^{l/2}\check{\psi}(2^l x - m)$ and the inner product is defined as usual

$$\left\langle f, g \right\rangle = \int_{-\infty}^{\infty} f(x)\overline{g(x)}dx, \quad \text{with} \quad \|f\|_2 = \left\langle f, f \right\rangle^{1/2}. \tag{8.4}$$

Here the definition is for a general biorthogonal wavelet with (8.3) usually referred to as the biorthogonal property [446,447]. Of particular importance is the special case of orthogonal wavelets where $\psi(x) \equiv \check{\psi}(x)$ and then (8.3) becomes the orthogonal property

$$\left\langle \psi_{j,k}, \psi_{l,m} \right\rangle = \delta_{j,l}\delta_{k,m}, \qquad j,k,l,m = 0, \pm 1, \pm 2, \ldots. \tag{8.5}$$

8.1.1 Multiresolution analysis

All wavelets generate a direct sum of the $L_2(\mathbb{R})$ space. The partial sums of this direct sum will generate a function called the scaling function, characterized by the multiresolution analysis (MRA) property of a wavelet. Recall that a wavelet function (8.2) is associated with its basis functions $\psi_{j,k}(x) = 2^{j/2}\psi(2^j x - k)$, $j,k = 0, \pm 1, \pm 2, \ldots,$ where the first index j refers to the basis resolution (dilation of ψ) while the second index k implements the space covering (translation of ψ).

Let $\mathbf{W}_j$ be the subspace formed by those resolution j basis functions, more precisely, the closure of their linear span[1]

$$\mathbf{W}_j = \text{clos}_{L_2(\mathbb{R})}\left\{ \psi_{j,k} \mid k = 0, \pm 1, \pm 2, \ldots \right\}$$

Define the closed subspace (the partial sum of $\mathbf{W}_\ell$'s up to $j-1$) for any j

$$\mathbf{V}_j = \sum_{\ell=-\infty}^{j-1} \mathbf{W}_\ell = \cdots + \mathbf{W}_{j-2} + \mathbf{W}_{j-1}. \tag{8.6}$$

Then clearly (as all basis functions $\psi_{j,k}$ are included)

$$L_2(\mathbb{R}) = \sum_{j=-\infty}^{\infty} \mathbf{W}_j = \text{clos}_{L_2(\mathbb{R})}\left(\bigcup_{j=-\infty}^{\infty} \mathbf{V}_j \right).$$

[1] A closure of a set is roughly defined as the same set plus any limits of all possible sequences of this set. Consult any functional analysis book.

We now consider how $\mathbf{V}_j$ can be generated by some $\mathcal{R}$-function $\phi(x)$

$$\mathbf{V}_j = \text{clos}_{L_2(\mathbb{R})}\Big\{\phi_{j,k} \mid k = 0, \pm 1, \pm 2, \ldots.\Big\} \tag{8.7}$$

with $\phi_{j,k}(x) = 2^{j/2}\phi(2^j x - k),\ \ j, k = 0, \pm 1, \pm 2, \ldots.$ In particular, we shall focus on the subspace $\mathbf{V}_0$.

Definition 8.1.3. *A function $\phi \in L_2(\mathbb{R})$ is called a multiresolution analysis (MRA) and therefore a scaling function, if the sequence of subspaces $\mathbf{V}_j$ as from (8.7) satisfies*

(1) $\mathbf{V}_k \subset \mathbf{V}_{k+1}$ *i.e.* $\cdots \subset \mathbf{V}_{-1} \subset \mathbf{V}_0 \subset \mathbf{V}_1 \cdots$

(2) $L_2(\mathbb{R}) = clos_{L_2(\mathbb{R})}\Big(\bigcup_{k=-\infty}^{\infty} \mathbf{V}_k\Big).$

(3) $\bigcap_{k=-\infty}^{\infty} \mathbf{V}_k = \{0\}.$

(4) $f(x) \in \mathbf{V}_k \iff f(2x) \in \mathbf{V}_{k+1}.$

(5) $\{\phi_{0,k}\} = \{\phi(x-k)\}$ *forms a Riesz basis for the subspace* $\mathbf{V}_0$.

8.1.2 Two-scale equations, orthogonality and Fourier analysis

We shall briefly discuss the process of constructing the scaling and wavelet functions with desirable properties. This is where the real difficulty lies for a general setting.

The so-called two-scale equations arise from the simple observation that $\mathbf{V}_1 = \mathbf{V}_0 + \mathbf{W}_0$, spanned by basis functions $\phi_{1,k} = 2^{1/2}\phi(2x-k)$ with the scaling function $\phi(x) \in \mathbf{V}_0$ and the wavelet function $\psi(x) \in \mathbf{W}_0$. Therefore

$$\begin{cases} \phi(x) = \sum_{k=-\infty}^{\infty} \alpha_k \phi_{1,k}(x) = \sum_{k=-\infty}^{\infty} p_k \phi(2x-k), \\ \psi(x) = \sum_{k=-\infty}^{\infty} \beta_k \phi_{1,k}(x) = \sum_{k=-\infty}^{\infty} q_k \phi(2x-k), \end{cases} \tag{8.8}$$

for some sequences $\{p_k\}$, $\{q_k\}$. Eventually we hope for such sequences to be of finite length (small and compact support).

The task of constructing ϕ, ψ is greatly simplified by converting (8.8) to the Fourier space using the Fourier transform

$$\widehat{f}(\omega) = (\mathcal{F}f)(\omega) = \int_{\infty}^{\infty} e^{-i\omega x} f(x) dx. \tag{8.9}$$

One can verify that (letting $z = e^{-i\omega/2}$ and $P(z) = \frac{1}{2}\sum_{k=-\infty}^{\infty} p_k z^k$)

$$\begin{aligned}\int_{\infty}^{\infty} e^{-i\omega x} \sum_{k=-\infty}^{\infty} p_k \phi(2x-k)dx &= \frac{1}{2}\sum_{k=-\infty}^{\infty} p_k \int_{\infty}^{\infty} e^{-i\omega(y+k)/2}\phi(y)dy \\ &= \frac{1}{2}\sum_{k=-\infty}^{\infty} p_k e^{-\frac{ik\omega}{2}} \int_{\infty}^{\infty} e^{-\frac{i\omega y}{2}}\phi(y)dy \\ &= P(z)\widehat{\phi}(\tfrac{\omega}{2})\end{aligned}$$

and consequently (8.8) becomes

$$\widehat{\phi}(\omega) = P(z)\widehat{\phi}(\frac{\omega}{2}), \qquad \widehat{\psi}(\omega) = Q(z)\widehat{\phi}(\frac{\omega}{2}), \tag{8.10}$$

with the Laurent polynomial series given by (noting $z = e^{-i\omega/2}$ i.e. $|z| = 1$)

$$P(z) = \frac{1}{2}\sum_{k=-\infty}^{\infty} p_k z^k, \qquad Q(z) = \frac{1}{2}\sum_{k=-\infty}^{\infty} q_k z^k. \tag{8.11}$$

Further, requirements on ϕ, ψ will be imposed on the choice of $P(z)$, $Q(z)$

$$P(1) = 1, \qquad P(-1) = 1, \quad Q(1) = 0,$$

and more specifically on their coefficients $\{p_k\}$, $\{q_k\}$; see [152,173]. In particular, using the Parseval equality and the convolution theorem [226,455,324],

$$\begin{aligned}\left\langle f, g\right\rangle &= \frac{1}{2\pi}\left\langle \widehat{f}, \widehat{g}\right\rangle, \\ (f * g)(x) = \int_{-\infty}^{\infty} f(x-y)g(y)dy &= \frac{1}{2\pi}\int_{-\infty}^{\infty} \widehat{f}(\omega)\widehat{g}(\omega)e^{i\omega x}d\omega,\end{aligned} \tag{8.12}$$

the 'biorthogonality condition' $\left\langle \phi(\cdot - j), \psi(\cdot - k)\right\rangle = \delta_{jk}$ becomes

$$P(z)\overline{Q(z)} + P(-z)\overline{Q(-z)} = 1, \quad |z| = 1, \tag{8.13}$$

while the orthogonality condition $\left\langle \phi(\cdot - j), \phi(\cdot - k)\right\rangle = \delta_{jk}$ becomes[2]

$$|P(z)|^2 + |P(-z)|^2 = 1, \quad |z| = 1, \tag{8.14}$$

which follows from the above Parseval inequality.

To simplify the presentation, we focus on the construction of orthogonal wavelets in the Daubechies' family [173] with the compact support of length

[2] Note in some presentations [302,441] one defines $H(\omega) = P(e^{-i\omega}) = P(z)$ and $G(\omega) = Q(e^{-i\omega}) = Q(z)$ so a shift to ω by π results in $-z$ and hence the two orthogonal relations become respectively $H(\omega)\overline{G(\omega)} + H(\omega+\pi)\overline{G(\omega+\pi)} = 1$ and $|H(\omega)|^2 + |H(\omega+\pi)|^2 = 1$.

$N + 1$ on level 0 i.e.

$$\phi(x) = \sum_{k=0}^{N} p_k \phi(2x - k). \tag{8.15}$$

Here one solution to the orthogonality condition (8.13) is the following

$$Q(z) = -z\overline{P(-z)}, \quad |z| = 1, \tag{8.16}$$

from which one works out the sequence $\{q_k\}$, by equating the corresponding coefficients, and then the wavelet ψ as follows

$$\begin{aligned} q_n &= (-1)^n \overline{p}_{1-n} \quad \text{for} \quad n = 1-N, 2-N, \ldots, 0, 1 \text{ or} \\ q_k &= (-1)^k \overline{p}_{N-k} \quad \text{for} \quad k = 0, 1, 2, \ldots, N \\ \psi(x) &= \sum_{k=0}^{N} (-1)^k \overline{p}_{N-k} \phi(2x - k) \end{aligned} \tag{8.17}$$

since $(N + 1) = 2M$ is practically even. (Hence the famous Daubechies' order four wavelets corresponds to $N = 3$ and $M = 2$.)

It remains to find $P(z)$ or the coefficients $\{p_k\}$. We first remark that it is not difficult to find $P(z)$ such that (8.14) and $P(1) = 1$ are satisfied. We hope for $P(z)$ to possess other useful properties. Central to the success of wavelets' compression is the so-called property of *vanishing moments*:

$$\int_{-\infty}^{\infty} x^k \psi(x) dx = 0, \qquad k = 0, 1, \ldots, M - 1, \tag{8.18}$$

where $N = 2M - 1$ is odd (as before). If $p_\tau(x)$ is any polynomial with degree $\tau < M$, then $\left(p_\tau, \psi\right) = 0$. In the Fourier space, the above vanishing moment property takes the form (for $k = 0, 1, \ldots, M - 1$)

$$\int_{-\infty}^{\infty} x^k \psi(x) dx = \widehat{x^k \psi(x)}(2\pi) = i^k \frac{d^k \widehat{\psi}}{d\omega^k}(2\pi) = 0, \quad \text{or} \quad \frac{d^k P(z)}{dz^k}(-1) = 0. \tag{8.19}$$

Here the equality $\widehat{x^k f(x)} = i^k \frac{d^k \widehat{f}}{d\omega^k}$ is applied. Therefore the desirable $P(z)$ is expected to satisfy (8.19), (8.14) and $P(1) = 1$.

To satisfy (8.19), the clue comes from working with the cardinal B-spline $B_m(x)$ of order m

$$N_1(x) = \begin{cases} 1, \ 0 \le x \le 1, \\ 0 \ \text{ otherwise}, \end{cases}$$

$$N_m(x) = N_{m-1}(x) * N_1(x) = \int_{-\infty}^{\infty} N_{m-1}(x - y) N_1(y) dy = \int_0^1 N_{m-1}(x - y) dy.$$

With this scaling function in the context of a MRA

$$N_m(x) = \sum_{k=-\infty}^{\infty} p_k N_m(2x-k),$$

we can find more explicitly the result (note $P_m(-1)=0$)

$$\widehat{N_m}(\omega) = P_m(z)\widehat{N_m}(\frac{\omega}{2}) \quad \text{with} \quad P_m(z) = \frac{1}{2}\sum_{k=-\infty}^{\infty} p_k z^k = \left(\frac{1+z}{2}\right)^m. \tag{8.20}$$

This result can be used to define $P(z)$ to satisfy (8.19), if we choose

$$P(z) = P_M(z)S_{M-1}(z) = \left(\frac{1+z}{2}\right)^M S_{M-1}(z). \tag{8.21}$$

It is left to construct the degree $(M-1)$ polynomial $S_{M-1}(z)$ such that (8.14) and $P(1)=1$ are satisfied. Refer to [173,152,441].

The famous Daubechies' order $N+1=4$ wavelet with $M=2$ vanishing moments used the following choice

$$\begin{aligned} S_1(z) &= \frac{1}{\sqrt{2}}\sqrt{2-\sqrt{3}}\left((2+\sqrt{3}-z\right), \\ P(z) &= \left(\frac{1+z}{2}\right)^M S_1(z) \\ &= \frac{1}{2}\left(\frac{1+\sqrt{3}}{4} + \frac{3+\sqrt{3}}{4}z + \frac{3-\sqrt{3}}{4}z^2 + \frac{1-\sqrt{3}}{4}z^3\right). \end{aligned}$$

We finally clarify the notation for filter coefficients. In the above formulation, the partition of unity condition $P(1)=1$ means that

$$P(1) = \frac{1}{2}\sum_{k=0}^{\infty} p_k = 1 \quad \text{or} \quad \sum_{k=0}^{\infty} p_k = 2.$$

Also the condition $Q(1)=0$ means that

$$Q(1) = \frac{1}{2}\sum_{k=0}^{N} q_k = 0.$$

To be consistent with the notation of [441,390] and §1.6.2, we set

$$c_k = \frac{p_k}{\sqrt{2}}, \qquad d_k = \frac{q_k}{\sqrt{2}} = (-1)^k \frac{\overline{p_{N-k}}}{\sqrt{2}}$$

to ensure that the wavelet filter coefficients satisfy

$$\sum_{k=0}^{\infty} c_k = \sqrt{2}, \qquad \sum_{k=0}^{\infty} d_k = 0, \tag{8.22}$$

which is necessary for the wavelet transform matrix W (§1.6.2) to be orthogonal (think about the 2×2 matrix W in the Harr case with $N = 1$). An amazing and yet remarkable feature of this analysis is that the Fourier transform helped us to satisfy all requirements of a wavelet function $\psi(x)$, by finding the filter coefficients $\{p_k\}$, *without* formulating $\psi(x)$ directly.

Remark 8.1.4. This remark is obvious but may appear controversial. The commonly used wavelet construction in the Fourier space is clearly elegant in analysis and has achieved all the requirements of an orthogonal (or biorthogonal) wavelet. However, a major drawback of using the Fourier space is almost surely to restrict the usefulness of the resulting construction for the simple reason that the underlying functions are not explicitly available! This has partially contributed to the slow progress in applying wavelets to solve real life differential and integral equations. For some applications [61], it may still be useful to use the Parseval equality (8.12) to 'evaluate' the coefficients of a wavelet expansion in the Fourier space.

There exist other FEM basis like constructions without using the Fourier transforms but the wavelets or their dual wavelets are not as compact [166,397, 170,156].

8.1.3 Pyramidal algorithms

Following the previous subsection, as in [60,166,221,68,441], the filter coefficients c_j's and d_j's for the orthogonal wavelet setting define the scaling function $\phi(x)$ and the wavelet function $\psi(x)$. Further, dilations and translations of $\phi(x)$ and $\psi(x)$ define a multiresolution decomposition for L_2 in $\mathbf{d}$-dimensions, in particular,

$$\begin{aligned} L_2(\mathbb{R}^{\mathbf{d}}) &= \mathbf{V}_0 \bigoplus \mathbf{W}_0 \bigoplus \cdots \bigoplus \mathbf{W}_{J-1} \bigoplus \mathbf{W}_J \bigoplus \cdots \\ &= \mathbf{V}_0 \bigoplus_{j=0}^{\infty} \mathbf{W}_j, \end{aligned} \tag{8.23}$$

where the subspaces satisfy the relations

$$\begin{cases} \mathbf{V}_\ell \supset \mathbf{V}_{\ell-1} \supset \cdots \supset \mathbf{V}_1 \supset \mathbf{V}_0 \supset \mathbf{V}_{-1} \supset \cdots, \\ \mathbf{V}_{\ell+1} = \mathbf{V}_\ell \bigoplus \mathbf{W}_\ell. \end{cases}$$

In numerical realizations, we select a finite dimension space $\mathbf{V}_J$ (as the finest scale) as our approximation space to the infinite decomposition of L_2 in (8.23) i.e. effectively use

$$\mathbf{V}_0 \bigoplus_{j=0}^{J-1} \mathbf{W}_j = \mathbf{V}_0 \bigoplus \mathbf{W}_0 \bigoplus \mathbf{W}_1 \bigoplus \cdots \bigoplus \mathbf{W}_{J-1} \tag{8.24}$$

to approximate $L_2(\mathbb{R}^{\mathbf{d}})$; we mainly consider $\mathbf{d} = 1$.

The pyramidal algorithm [344], as used in Section 1.6.2, refers to the conversion of coefficients of each $f_{j+1}(x) \in \mathbf{V}_{j+1}$ to those of $f_{j+1}(x) = f_j(x) + g_j(x) \in \mathbf{V}_{j+1} = \mathbf{V}_j \oplus \mathbf{W}_j$ with

$$\left.\begin{aligned} f_{j+1}(x) &= \sum_m \mathbf{x}_m^{j+1} 2^{(j+1)/2} \phi(2^{j+1}x - m) &&\in \mathbf{V}_{j+1}, \\ f_j(x) &= \sum_m \mathbf{x}_m^{j} 2^{j/2} \phi(2^{j}x - m) &&\in \mathbf{V}_{j}, \\ g_j(x) &= \sum_m \mathbf{y}_m^{j} 2^{j/2} \psi(2^{j}x - m) &&\in \mathbf{W}_{j}. \end{aligned}\right\} \tag{8.25}$$

♦ **Wavelet decomposition** from $\mathbf{V}_{j+1}$ to $\mathbf{V}_j$ and $\mathbf{W}_j$. With the choice of q_k in terms of p_k from (8.17), one can verify that

$$\sum_k \left[p_{l-2k}\overline{p}_{m-2k} + q_{l-2k}\overline{q}_{m-2k} \right] = 2\delta_{l,m}$$

and furthermore

$$\begin{aligned} 2\phi(2x-m) &= \sum_k \left[\bar{p}_{m-2k}\phi(x-k) + \bar{q}_{m-2k}\psi(x-k) \right], \qquad \text{or} \\ 2^{1/2}\phi(2x-m) &= \frac{1}{\sqrt{2}} \sum_k \left[\bar{p}_{m-2k}\phi(x-k) + \bar{q}_{m-2k}\psi(x-k) \right]. \end{aligned}$$

Applying this to (8.25), we obtain (noting that c_k, d_k are real)

$$\begin{aligned} \mathbf{x}_k^j &= \frac{1}{\sqrt{2}} \sum_l \bar{p}_{l-2k}\mathbf{x}_l^{j+1} = \sum_l c_{l-2k}\mathbf{x}_l^{j+1} = \sum_{l=0}^{N} c_l \mathbf{x}_{l+2k}^{j+1}, \\ \mathbf{y}_k^j &= \frac{1}{\sqrt{2}} \sum_l \bar{q}_{l-2k}\mathbf{x}_l^{j+1} = \sum_l d_{l-2k}\mathbf{x}_l^{j+1} = \sum_{l=0}^{N} d_l \mathbf{x}_{l+2k}^{j+1}. \end{aligned}$$

The overall process can be dipicted by

$$\begin{array}{lccccccccccc} \text{to decompose:} & f_J(x) \rightarrow & f_{J-1}(x) & \rightarrow & f_{J-2}(x) & \rightarrow \cdots \rightarrow & f_j(x) & \rightarrow \cdots \rightarrow & f_0(x) \\ & \searrow & & \searrow & & \searrow \quad \searrow & & \searrow \quad \searrow & \\ \text{to retain:} & & g_{J-1}(x) & & g_{J-2}(x) & \cdots & g_j(x) & \cdots & g_0(x) \end{array}$$

♦ **Wavelet reconstruction** from $\mathbf{V}_j$ and $\mathbf{W}_j$ to $\mathbf{V}_{j+1}$. We use the two-scale equation (8.8) with a shifted value x i.e.

$$\begin{aligned} 2^{j/2}\phi(x-m) &= \frac{1}{\sqrt{2}} \sum_{k=0}^{N} p_k 2^{(j+1)/2}\phi(2(x-m)-k), \\ 2^{j/2}\psi(x-m) &= \frac{1}{\sqrt{2}} \sum_{k=0}^{N} q_k 2^{(j+1)/2}\phi(2(x-m)-k). \end{aligned}$$

The following is immediately obtained from equating (8.25)

$$\mathbf{x}_k^{j+1} = \frac{1}{\sqrt{2}} \sum_l \left[p_{k-2l} \mathbf{x}_l^j + q_{k-2l} \mathbf{y}_l^j \right] = \sum_{\ell=0}^{N} \left[c_\ell \mathbf{x}_{k(\ell)}^j + d_\ell \mathbf{y}_{k(\ell)}^j \right],$$

where the functional notation $k(\ell) = (k - \ell)/2$ should be interpreted as a whole integer; a term does not exist (so is zero) if $k(\ell)$ is not an integer i.e.

$$v_{k(\ell)} = \begin{cases} v_j, & \text{if } \mathtt{rem}(k-\ell, 2) = 0 \text{ and } j = k(\ell) = (k-\ell)/2, \\ 0, & \text{otherwise.} \end{cases} \tag{8.26}$$

The overall reconstruction process can be depicted by

$$\begin{array}{llllllll} \texttt{to decompose:} & f_J(x) \leftarrow & f_{J-1}(x) \leftarrow & f_{J-2}(x) \leftarrow & \cdots \leftarrow & f_j(x) \leftarrow & \cdots \leftarrow & f_0(x) \\ & \nwarrow & \nwarrow & \nwarrow & \nwarrow & \nwarrow & \nwarrow & \\ \texttt{to retain:} & & g_{J-1}(x) & g_{J-2}(x) & \cdots & g_j(x) & \cdots & g_0(x) \end{array}$$

8.1.4 Discrete wavelets as matrix computation tools

As remarked before, constructing user-friendly wavelets with compact supports and fewer requirements on the geometry of a domain continues to be a challenging and interesting topic. Presently the most useable wavelets are those tensor-products based biorthogonal ones that are defined by a collection of flexible patches covering a general domain [166,266].

Following the work of [60,397,169,479,266], it became clear that the tensor-product wavelets can offer remarkable improvements for solving the boundary integral equations. This is mainly due to the excellent compression rates that can be achieved. See Section 8.5.

However on solving a PDE, while some problems have been solved successfully, two comments may be made.

(i) most of the wavelet methods proposed so far have theoretical advantages over the traditional methods in terms of defining a Riesz basis and better conditioning of stiffness matrices for self-adjoint problems. The recent work of [155] even showed clear advantages of wavelets over the FEM on adaptive solution.

(ii) the wavelet methods, as practical methods, are not yet ready to compete with the traditional FEM approach. One reason was that the compactness of a wavelet, which defines the sparsity of the resulting stiffness matrix, is not easy to to realize, while the FEM can always produce a sparse matrix with more regular sparsity patterns.

Therefore for PDE problems, it is recommended to use first the FEM discretization to derive a sparse matrix problem and then to use the discrete wavelets for preconditioning. While this idea is also applicable to integral equations, whenever possible, we should try the optimal implementation of using wavelets as functions rather than as discrete wavelet transforms for dense matrices.

As mentioned in Section 1.6.2, when using a FWT based on pyramidal algorithms for a sparse matrix, one may or may not need to refer to the functional wavelet theory as this may not make sense for many cases. For pure matrix problems e.g. from signal and image processing, the wavelets can offer an excellent solution in most cases [441].

The advantages of using a DWT to solve a matrix problem may be explained using certain smoothness measure of the underlying matrix as done in [205] where we measure the norm of column and row differences. However a better method is given by [440] who interprets the functional vanishing moment conditions in terms of vanishing vector moments. This offers a purely algebraic way of using wavelets for a wide class of matrix computations. Below we shall discuss how this discrete approach can be used for preconditioning purposes.

8.2 Operator compression by wavelets and sparsity patterns

We are primarily interested in solving a pseudo-differential equation, which includes a large class of second-order PDEs of practical interest and integral equations [169,60]. The usual choices and questions are:

(i) which scaling and wavelet functions to use or construct?
(ii) do the wavelet functions process finite supports?
(iii) how to compute the wavelet matrix by suitable quadrature?
(iv) what sparsity patterns does the final matrix take?

Amazingly enough, the last question (iv) can be answered for a large class of pseudo-differential equations even before the three questions are answered as long as we make the usual assumption on the decaying of the distributional kernel of the operator and on the availability of some vanishing moments of the wavelet. The sparsity pattern concerned is the so-called 'finger-like' pattern as shown on the right plot of Figure 1.1. The first three questions (i–iii) are tough and it may take a few more years for the subject to become mature. *What is more*, as the theoretical inverse of a pseudo-differential operator is another pseudo-differential operator in a different order, we can predict the sparsity pattern of an inverse matrix in the wavelet space! This is the very property that is much needed to guarantee the efficiency of a SPAI type preconditioner (Chapter 5). In this viewpoint, the wavelet filters (rather than the underlying wavelet function explicitly) will play a major role for sparse matrices from PDEs; for dense matrices from integral equations, one may avoid transforming a dense matrix by using the implicit wavelet preconditioning (Chapter 10).

Let T be a pseudo-differential operator with the symbol $\sigma(x, \xi)$ of order λ. Then it can be written as an 'integral operator'

$$\begin{aligned} T(f)(x) = \sigma(x, D)f &= \int_{\mathbb{R}} e^{ix\xi}\sigma(x, \xi)\widetilde{f}(\xi)d\xi \\ &= \int_{\mathbb{R}} e^{ix\xi}\sigma(x, \xi)\frac{1}{2\pi}\int_{\mathbb{R}} e^{-iy\xi} f(y)dy\, d\xi \\ &= \int_{\mathbb{R}} K(x, y)f(y)dy, \qquad K(x, y) = \frac{1}{2\pi}\int_{\mathbb{R}} e^{i(x-y)\xi}\sigma(x, \xi)d\xi. \end{aligned} \tag{8.27}$$

Note the range of λ is usually $-1 \le \lambda \le 2$ with

(a) $\lambda = 2$ corresponding to a second-order differential operator (e.g. ∇^2)
(b) $\lambda = 1$ corresponding to a hyper-singular integral operator (e.g. $\mathcal{N}_k$)
(c) $\lambda = 0$ corresponding to a double-layer integral operator (e.g. $\mathcal{M}_k, \mathcal{M}_k^T$)
(d) $\lambda = -1$ corresponding to a single-layer integral operator (e.g. $\mathcal{L}_k$)

where the order of an operator is closely related to its mapping properties (e.g. refer to Theorem 1.7.21). The distributional kernel satisfies the normal condition

$$|\partial_x^l \partial_\xi^k \sigma(x, \xi)| \le C_{l,k}(1 + |\xi|)^{\lambda-k} \tag{8.28}$$

and $\sigma(x, \xi)$ has compact support in x; the result is also true in $\mathbb{R}^d$ for $d \ge 2$. For the 1D case in $[a, b]$, let $\psi(x)$ be some suitable wavelet function that has M vanishing moments

$$\int_a^b \psi(x)x^j dx = 0 \quad \text{for } j = 0, 1, 2, \ldots, M-1. \tag{8.29}$$

Suppose J levels of grids have been set up and we are interested in computing (or estimating) the particular matrix element

$$\begin{aligned} \alpha_{II'} &= \left\langle T\psi_I, \psi_{I'}\right\rangle = \left\langle \mathcal{K}\psi_I, \psi_{I'}\right\rangle \\ &= \int_a^b \int_a^b K(x, y)\psi_I(x)\psi_{I'}(y)dxdy = \int_I \int_{I'} K(x, y)\psi_I(x)\psi_{I'}(y)dxdy \end{aligned} \tag{8.30}$$

with the wavelet basis functions $\psi_I, \psi_{I'}$ supported on the intervals I (level j), I' (level j') respectively. Then to estimate $\alpha_{II'}$, assuming I, I' are not overlapping, we expand K into a Taylor series up to degree M around the center of $I \times I'$ and substitute into (8.30), using (8.29), to yield

$$|\alpha_{II'}| \le \frac{C_{M,\lambda,j,j'}}{\text{dist}(I, I')^{M+\lambda+1}}, \tag{8.31}$$

where the generic constant C may be further specified in particular situations. Refer to [60,166,167].

We now consider how (8.31) informs us of the overall sparsity pattern of the stiffness matrix $\widetilde{A} = (\alpha_{II'})$. Let the wavelet functions on level $j = J-1, \ldots, 1, 0$ be denoted by

$$\psi_1^j(x), \psi_2^j(x), \ldots, \psi_{n_j}^j(x),$$

and the scaling functions on level 0 by

$$\phi_1^0(x), \phi_2^0(x), \ldots, \phi_{n_0}^0(x).$$

Opposite to the right plots of Figure 1.1, we order these functions from the finest level to the coarsest level so the matrix $\widetilde{A}$ will arise from these blocks as follows

$$\left[\begin{array}{c|c|c|c|c}
\left\langle T\psi_\ell^{J-1}, \psi_k^{J-1}\right\rangle & \left\langle T\psi_\ell^{J-1}, \psi_k^{J-2}\right\rangle & \cdots & \left\langle T\psi_\ell^{J-1}, \psi_k^{0}\right\rangle & \left\langle T\phi_\ell^{J-1}, \phi_k^{0}\right\rangle \\ \hline
\left\langle T\psi_\ell^{J-2}, \psi_k^{J-1}\right\rangle & \left\langle T\psi_\ell^{J-2}, \psi_k^{J-2}\right\rangle & \cdots & \left\langle T\psi_\ell^{J-2}, \psi_k^{0}\right\rangle & \left\langle T\phi_\ell^{J-2}, \phi_k^{0}\right\rangle \\ \hline
\vdots & \vdots & \ddots & \vdots & \vdots \\ \hline
\left\langle T\psi_\ell^{0}, \psi_k^{J-1}\right\rangle & \left\langle T\psi_\ell^{0}, \psi_k^{J-2}\right\rangle & \cdots & \left\langle T\psi_\ell^{0}, \psi_k^{0}\right\rangle & \left\langle T\phi_\ell^{0}, \phi_k^{0}\right\rangle \\ \hline
\left\langle T\phi_\ell^{0}, \psi_k^{J-1}\right\rangle & \left\langle T\phi_\ell^{0}, \psi_k^{J-2}\right\rangle & \cdots & \left\langle T\phi_\ell^{0}, \psi_k^{0}\right\rangle & \left\langle T\phi_\ell^{0}, \phi_k^{0}\right\rangle
\end{array}\right], \tag{8.32}$$

where the pairing (ℓ, k) should go through all basis functions within each level. Applying (8.31) to (8.32), we can see that each wavelet–wavelet block (square or rectangular) has concentrated its large nonzeros near the main diagonal and small elements decaying away from the main diagonal whilst the wavelet-scaling blocks along the bottom row and right most column do not necessarily have as many small elements. This is clearly shown in Figure 8.1.

The idea of making use of the inverse matrix pattern in the wavelet matrix to propose a robust finger-patterned SPAI preconditioner was first recognized in [157,158], although mainly the Laplacian equation was considered. Unfortunately no comprehensive experiments were carried out and no numerical algorithms were presented to compute the approximate inverse for the wavelet matrix. In fact, as we see in Chapter 10, obtaining a finger-like patterned SPAI preconditioner is not computationally feasible due to high cost and alternative algorithms are needed.

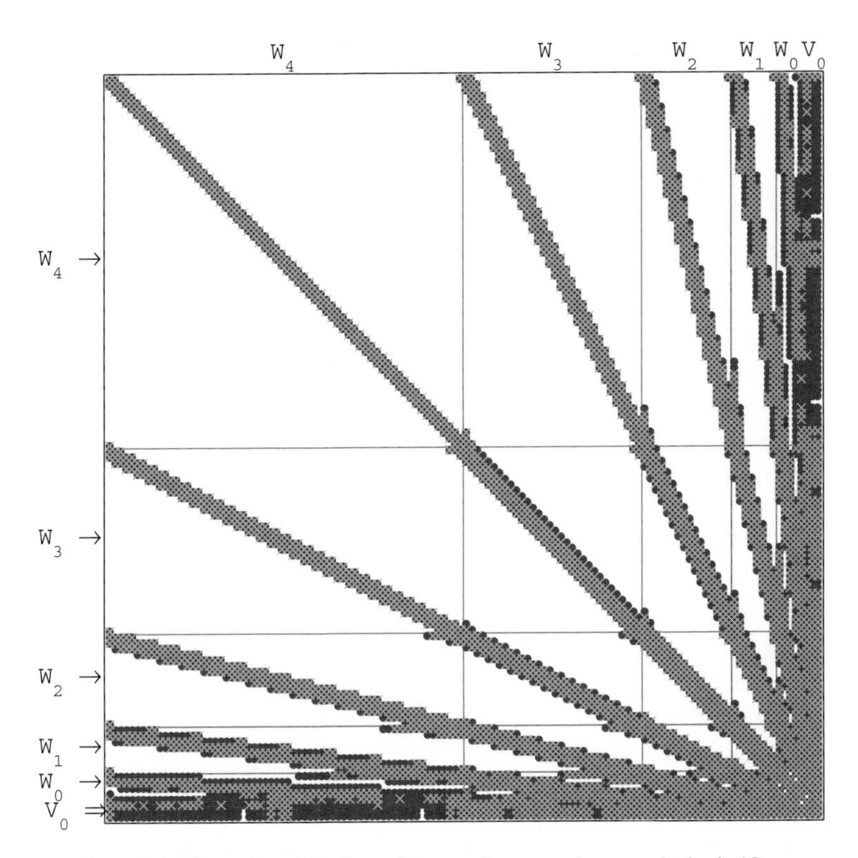

Figure 8.1. Illustration of the finger-like sparsity pattern in a wavelet basis ($J =$ 5 levels). Recall that $\mathbf{V}_5 = \mathbf{W}_4 + \mathbf{W}_3 + \mathbf{W}_2 + \mathbf{W}_1 + \mathbf{W}_0 + \mathbf{V}_0$.

8.3 Band WSPAI preconditioner

The first use of orthogonal wavelets to design a sparse preconditioner for sparse problems was made in [116,483], although the work was motivated by the general smoothness observation of some elliptic operator rather by the pseudo-differential argument. The resulting algorithm defines the wavelet sparse approximate inverse (WSPAI) preconditioner. The message was clear. We wish to explore the preconditioning techniques (as from Chapters 4 and 5) in a wavelet space.

As we intend to work with sparse matrices from FEM applications, it may not be necessary to use wavelet functions or invoke the wavelet compression theory directly. (However it is feasible to formulate the vanishing moment conditions in the discrete case [441].) As in Section 1.6.2, let W be an orthogonal transform matrix. Then the linear system (1.1) will become

$$\widetilde{A}\widetilde{x} = \widetilde{b} \quad \text{with } \widetilde{A} = WAW^T, \ x = W^T\widetilde{x}, \ \widetilde{b} = Wb, \tag{8.33}$$

since $WW^T = W^TW = I$. If a biorthogonal transform is used, the idea will be similar as one works with

$$\widetilde{A}\widetilde{x} = \widetilde{b} \quad \text{with } \widetilde{A} = WA\breve{W}, \ x = \breve{W}\widetilde{x}, \ \widetilde{b} = Wb,$$

since $W\breve{W} = I$. If A is sparse, $\widetilde{A}$ will also be sparse (even in the case of very few or no vanishing moments), though possibly denser than A if A is extremely sparse (e.g. tridiagonal).

We look for a SPAI type preconditioner $\widetilde{M}$ as in Chapter 5 for matrix $\widetilde{A}$. It turns out that such a preconditioner would correspond to a SPAI preconditioner M (which may or may not be found directly and easily) in the original space:

$$\begin{aligned} \min_M \|AM - I\|_F &= \min_M \|WAW^TWMW^T - I\|_F \\ &= \min_{\widetilde{M}} \|\widetilde{A}\widetilde{M} - I\|_F, \end{aligned} \tag{8.34}$$

where $\widetilde{A} = WAW^T$ and $\widetilde{M} = WMW^T$ are respectively the counterparts of A and M in the wavelet setting. Following Chapter 5, (8.34) is solved via n least squares problems

$$\min_{\widetilde{m}_j} \|\widetilde{A}\widetilde{m}_j - e_j\|_2, \qquad j = 1, 2, \ldots, n. \tag{8.35}$$

It only remains to specify a suitable sparsity pattern $\mathcal{S}$ for $\widetilde{M}$.

Algorithm 8.3.5. (WSPAI).

(1) Compute the transformed matrix, $\widetilde{A} = WAW^T$.
(2) Apply Algorithm 5.4.5 to obtain $\widetilde{M}$.
(3) Use $\widetilde{M}$ *to solve (8.33) by some iterative solver (e.g. Algorithm 3.6.18).*
(4) Apply the inverse wavelet transform to compute the solution, $x = W^T\widetilde{x}$.

As discussed in the previous section, the best pattern $\mathcal{S}$ is the finger-like pattern. However a direct use of such a $\mathcal{S}$ leads to high cost because there exist dense columns (see Figure 8.1). One solution is discussed in Chapter 10. The choice made in [116] is to use a banded sparsity pattern as this would balance the work required for each column.

The results are quite promising, although the WSPAI method cannot precondition PDE problems with jumps in coefficients. For such a case, an improved algorithm combining simple scaling (such as those from Chapter 4) is proposed in [107] in a two-stage setting

$$\begin{cases} D^{-1}Ax = D^{-1}b, \\ \widetilde{A_1}\widetilde{x} = \widetilde{b_1}, \end{cases}$$

where $\widetilde{A_1} = WA_1W^T = WD^{-1}AW^T$.

8.4 New centering WSPAI preconditioner

The first use of orthogonal wavelets to design a sparse preconditioner dense problems was in [130]. The motivation there was to improve on existing sparse preconditioners of the forward type (Chapter 4).

For dense matrices arising from the BEM discretization, two observations emerge. Firstly the conditioning alone is not a major issue as such matrices are less ill-conditioned than those sparse ones from FEM/PDE contexts. Secondly simple preconditioners based on the standard DWT do not work as they cannot preserve a diagonal operator splitting (necessary for eigenvalue clustering) in the original BEM space. It should be remarked that the eigenspectra of matrices A and $\widetilde{A}$ are identical, even though the latter appears approximately sparse (under a thresholding).

To introduce our new DWT, we first examine how the standard DWT works; we adopt the same notation used in Section 1.6.2. Table 8.1 shows a typical example of applying a three-level pyramidal algorithm to a vector $s = a = s^{(4)}$ of size $n = 2^L = 2^4$, where three forms (1 for components, 2 for vectors, 3 for matrices) of transform details are shown. Clearly local features in a are scattered in Wa, that is, the standard DWT is not centred.

Let $\overline{A}$ be such a 'finger'-like sparse matrix, truncated from $\widetilde{A}$ after some thresholding i.e. $\overline{A} = drop(\widetilde{A})$. For such matrices, matrix-vector products can be formed very efficiently. That is to say, if it is desirable to solve $\overline{A}u = z$ by some iterative methods without preconditioning, then as far as certain applications are concerned we have obtained an efficient implementation. Here $u = Wx$ and $z = Wb$.

However, approximately, $\overline{A}$ is spectrally equivalent to the original matrix A. Therefore, as the original problem requires preconditioning, we need to precondition $\overline{A}u = z$. Indeed, [422] reports that with GMRES, the number of iteration steps for $\widetilde{A}u = z$ and $\overline{A}u = z$ are the same.

Moreover, it is not an easy task to use $\overline{A}$ as a preconditioner because solving the linear system $\overline{A}u = z$ results in many fill-ins. We also note that such

Table 8.1. *Comparison of the standard DWT (left) and the New DWT (right) for a vector a of size* 16 ($m = 4$, $L = 4$, $r = 1$)

	Form	Lev 4	Lev 3		Lev 2	
1		a_1	×	×	×	×
2		a_2	×	0	×	0
3		0	0	0	0	×
4		0	0	0	0	×
5		0	0	0	×	×
6		0	0	0	×	0
7		0	0	×	×	×
8	1	0	0	0	×	×
9		0	0	×	×	×
10		0	0	0	0	0
11		0	0	0	0	0
12		0	0	0	0	0
13		0	0	0	0	0
14		0	0	0	0	0
15		0	×	0	0	0
16		0	×	×	×	×
	2	$s^{(4)}$	$\overline{W}_4 s^{(4)}$	$s^{(3)}$ $f^{(3)}$	$\overline{W}_3 s^{(3)}$ $f^{(3)}$	$s^{(2)}$ $f^{(2)}$ $f^{(3)}$
	3	$s^{(4)}$	$W_4 s^{(4)}$	$P_4 W_4 s^{(4)}$	$W_3 P_4 W_4 s^{(4)}$	$P_3 W_3 P_4 W_4 s^{(4)}$

	Form	Lev 4	Lev 3		Lev 2	
1		a_1	×	×	×	×
2		a_2	×	×	×	×
3		0	0	0	×	×
4		0	0	0	0	0
5		0	0	0	0	0
6		0	0	0	0	0
7		0	0	0	0	0
8	1	0	0	0	0	0
9		0	0	0	×	×
10		0	0	0	0	0
11		0	0	0	×	×
12		0	0	0	0	0
13		0	0	0	×	×
14		0	0	0	0	0
15		0	×	×	×	×
16		0	×	×	×	×
	3	$s^{(4)}$	$\widehat{W}_4 s^{(4)}$	$\widehat{W}_4 s^{(4)}$	$\widehat{W}_3 \widehat{W}_4 s^{(4)}$	$\widehat{W}_3 \widehat{W}_4 s^{(4)}$

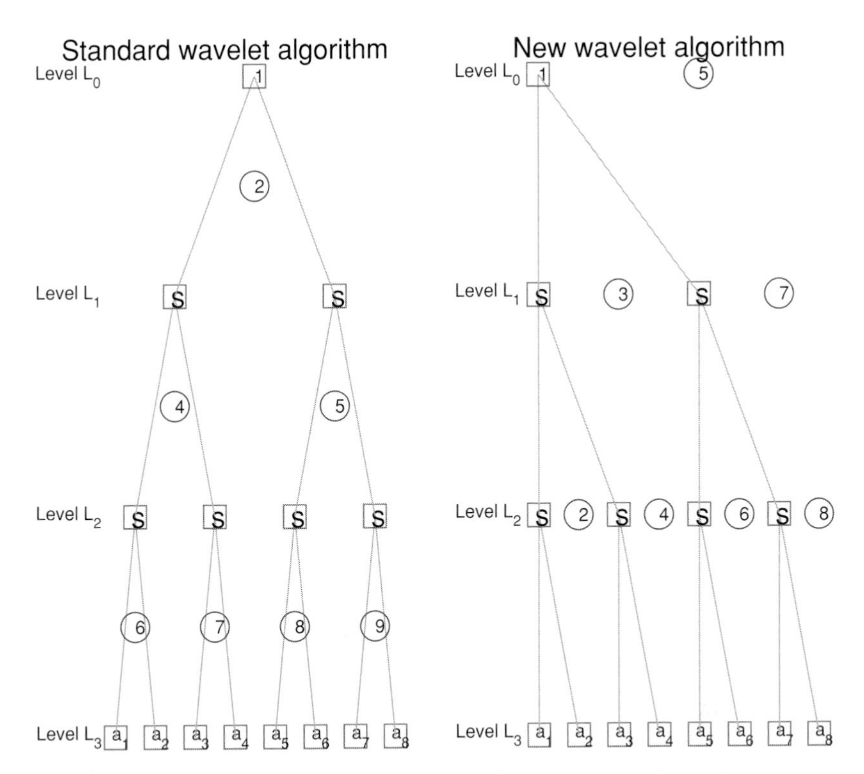

Figure 8.2. Comparison of the standard pyramidal algorithm (left) and the new wavelet ordering algorithm (right) of $J = 3$ levels for transforming a sized 8 vector $\mathbf{a} = (a_j)$. In both plots, the boxes □ denote the 'scaling' coefficients ('**S**ums') and the circles ○ the 'wavelet' coefficients ('differences'). The numbers indicate the final order for $\widetilde{\mathbf{a}}$.

a 'finger'-like matrix also gives problems in designing approximate sparse inverses; see [116]. Below we consider a new implementation of DWT's avoiding the generation of such 'finger'-like matrices.

8.4.1 A new nonstandard DWT

To begin with, we re-consider how a DWT should ideally deal with a local feature like that in the vector a shown in Table 8.1 (left) before introducing our proposed changes to the basic pyramidal algorithm. It turns out that a local feature preserving scheme is necessary and this amounts to removing the permutation matrices P_j's in the standard DWT. The difference in ordering the transformed quantity may be illustrated by Figure 8.2, where one can see that

Table 8.2. *The new centering DWT matrix [130].*

$$\widehat{W}_\nu = \begin{pmatrix}
c_0 \emptyset & c_1 & \emptyset & c_2 & \emptyset & \cdots & c_{m-1} & & & \\
\emptyset \mathcal{I} & \emptyset & \emptyset & \emptyset & \emptyset & \cdots & \emptyset & & & \\
d_0 \emptyset & d_1 & \emptyset & d_2 & \emptyset & \cdots & d_{m-1} & & & \\
\emptyset \emptyset & \emptyset & \mathcal{I} & \emptyset & \emptyset & \cdots & \emptyset & & & \\
 & & & c_0 & \emptyset & \ddots & \ddots & & & \\
 & & & \emptyset & \mathcal{I} & \ddots & \ddots & & & \\
\vdots \vdots & \vdots & \vdots & \cdots & \cdots & \ddots & \ddots & \ddots & \ddots & \\
c_2 \emptyset & \cdots & c_{m-1} & & & & c_0 & \emptyset & c_1 & \emptyset \\
\emptyset \emptyset & \cdots & \emptyset & & & & \emptyset & \mathcal{I} & \emptyset & \emptyset \\
d_2 \emptyset & \cdots & d_{m-1} & & & & d_0 & \emptyset & d_1 & \emptyset \\
\emptyset \emptyset & \cdots & \emptyset & & & & \emptyset & \emptyset & \emptyset & \mathcal{I}
\end{pmatrix}_{n\times n} . \tag{8.36}$$

the standard DWT algorithm orders vertically in a level by level fashion while our new DWT algorithm orders horizontally in an 'in-place' way.

We first define a new one-level DWT matrix in (8.36) of Table 8.2, similar to $\overline{W}_\nu$ in (1.47). Here $\mathcal{I}$ is an identity matrix of size $2^{(L-\nu)} - 1$ and Ø's are block zero matrices. For $\nu = L$, both $\mathcal{I}$ and Ø are of size 0 i.e. $\widehat{W}_L = \overline{W}_L = W_L$.

Further a new DWT (from (1.44)) for a vector $s^{(L)} \in \mathcal{R}^n$ can be defined by

$$\widehat{w} = \widehat{W} s^{(L)}$$

with

$$\widehat{W} = \widehat{W}_{r+1} \widehat{W}_{r+2} \cdots \widehat{W}_L \tag{8.37}$$

based on $(L + 1 - r)$ levels. For $L = 4$ and $m = 4$ (Daubechies' wavelets for $n = 2^L = 16$), Table 8.1 (right) shows details of a three-level transform using $\widehat{W}_4$ and $\widehat{W}_3$. One may verify that, unlike the left side of Table 8.1, the very last column ($\widehat{w}$) on the right side of Table 8.1 would possess a locally centered sparse structure depending only on the number of wavelet levels and not on size n.

For a matrix $A_{n\times n}$, the new DWT would give

$$\widehat{A} = \widehat{W} A \widehat{W}^\top.$$

Now to relate $\widehat{A}$ to $\widetilde{A}$ from a standard DWT, or rather $\widehat{W}$ to W, we define

$$P = P_L^\top P_{L-1}^\top \cdots P_{r+2}^\top P_{r+1}^\top, \tag{8.38}$$

which is a permutation matrix with P_k's from (1.47). Firstly, by induction, we can prove the following

$$\widehat{W}_k = \left(\prod_{\ell=1}^{L-k} P_{k+\ell}\right)^\top W_k \left(\prod_{\ell=1}^{L-k} P_{k+\ell}\right) \qquad \text{for } k = r+1, r+2, \cdots, L,$$

that is,

$$\begin{aligned} \widehat{W}_L &= W_L, \\ \widehat{W}_{L-1} &= P_L^\top W_{L-1} P_L, \\ &\vdots \\ \widehat{W}_{r+1} &= P_L^\top P_{L-1}^\top \cdots P_{r+2}^\top W_{r+1} P_{r+2} \cdots P_{L-1} P_L. \end{aligned}$$

Secondly, we can verify that

$$\begin{aligned} PW &= \left(P_L^\top P_{L-1}^\top \cdots P_{r+1}^\top\right)\left(P_{r+1} W_{r+1} \cdots P_L W_L\right) \\ &= \widehat{W}_{r+1}\left(P_L^\top P_{L-1}^\top \cdots P_{r+2}^\top\right)\left(P_{r+2} W_{r+2} \cdots P_L W_L\right) \\ &\vdots \\ &= \widehat{W}_{r+1} \widehat{W}_{r+2} \cdots \widehat{W}_{L-2} P_L^\top W_{L-1} \left(P_L W_L\right) \\ &= \widehat{W}_{r+1} \widehat{W}_{r+2} \cdots \widehat{W}_L \\ &= \widehat{W}. \end{aligned}$$

Consequently from $\widehat{W} = PW$, $\widehat{A} = P\widetilde{A}P^\top$ define our new DWT transform [130].

The practical implication of these relations is that the new DWT can be implemented in a level by level manner, either directly using $\widehat{W}_\nu$'s (via $\widehat{W}$) or indirectly using P_ν's (via P) after a standard DWT, and we obtain the same result.

To illustrate the new DWT, we apply the new DWT to a non-constant diagonal matrix A and the transformed matrix, which has an essentially band-like pattern, is shown in Figure 8.3 along with the standard DWT (the left plot). Clearly such a pattern (in the middle plot) may be used more advantageously than a finger-like one (for example, in the applications of [116]).

As far as preconditioning is concerned, to solve (1.1), we propose the following:

Algorithm 8.4.6.

1. *Apply the new DWT to* $Ax = b$ *to obtain* $\widehat{A}u = z$*;*
2. *Select a suitable band form* M *of* $\widehat{A}$*;*
3. *Use* M^{-1} *as a preconditioner to solve* $\widehat{A}u = z$ *iteratively.*

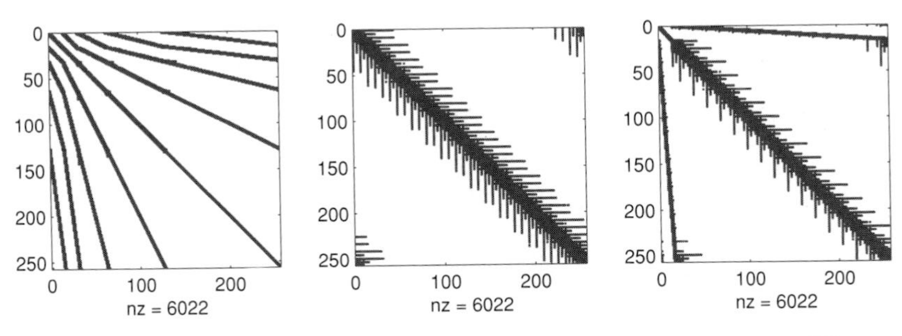

Figure 8.3. Illustration of the finger-like sparsity pattern (left) versus the new DWT pattern (middle) and its coarsest level version (right) for a diagonal matrix.

Here the band size of M determines the cost of a preconditioning step. If this size is too small, the preconditioner may not be effective. If the size is so large (say nearly $n-1$) that the preconditioner approximates $\widehat{A}^{-1}$ very accurately, then one may expect that one or two iterations are sufficient for convergence but each iteration is too expensive. Therefore, we shall next examine the possibility of constructing an effective preconditioner based on a relatively small band.

8.4.2 Band matrices under the new DWT

To discuss combining the new DWT with the operator splitting ideas in the next subsection, we first consider the process of transforming a band matrix. Let J denote the actual number of wavelet levels used $(1 \le J \le (L+1-r))$. Mainly we try to address this question: under what conditions does the new DWT transform a band matrix A into another band matrix $\widehat{A}$ (instead of a general sparse matrix)? Here, by a band matrix, we mean a usual band matrix with wrap-round boundaries. The correct condition turns out to be that J should be chosen to be less than $(L+1-r)$. This will be stated more precisely in Theorem 8.4.12.

To motivate the problem, we show in Figures 8.3 and 8.4 respectively the DWT of a diagonal matrix and a nondiagonal matrix, using $J=5$. Here with $m=4$ (so $r=1$) and $L=8$, a band structure in $\widehat{A}$ is achieved by not using the maximum level $J=L-r+1=8$. For a given band matrix A, to establish the exact band width for the transformed matrix $\widehat{A}$ under the new DWT, we need to view the one-step transformation matrix $\widehat{W}_\nu$ as a band matrix. For ease of presentation, we first introduce some definitions and a lemma before establishing the main theorem.

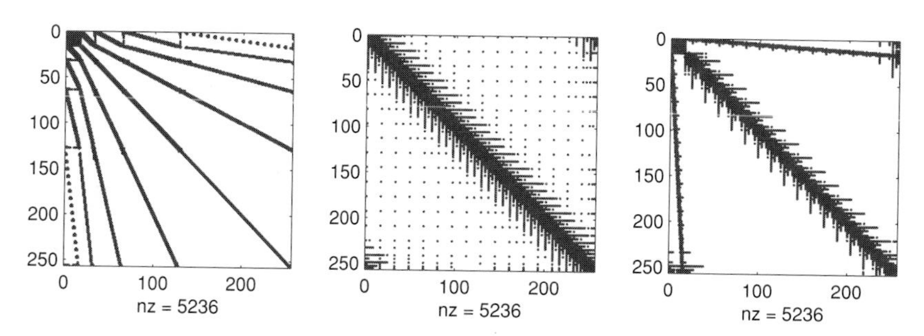

Figure 8.4. Illustration of the finger-like sparsity pattern (left) versus the new DWT pattern (middle) and its coarsest level version (right) for a Caldron–Zygmund matrix as from (8.39).

Definition 8.4.7. (Band(α, β, k)). *A block band matrix $A_{n\times n}$, with wrap-round boundaries and blocks of size $k \times k$, is called a Band(α, β, k) if its lower block band width is α and upper block band width β (both including but not counting the main block diagonal).*

Note that when $k = 1$, we write Band$(\alpha, \beta, 1) =$ Band(α, β).

Definition 8.4.8. (Band(α, β, k, τ)). *A Band(α, β, k) matrix $A_{n\times n}$ is called Band(α, β, k, τ) if each $k \times k$ block has a total band width of $2k - 1 - 2\tau$, that is, if there are τ bands of zeros at both ends of the anti-diagonal.*

Lemma 8.4.9. *For non-negative integers $\alpha, \beta, \gamma, \delta, k, \tau$, the following results hold:*

(1) Band(α, β)Band(γ, δ) = Band($\alpha + \gamma, \beta + \delta$);
(2) Band(α, β, k)Band(γ, δ, k) = Band($\alpha + \gamma, \beta + \delta, k$);
(3) Band(α, β, k) = Band($(\alpha + 1)k - 1, (\beta + 1)k - 1$);
(4) Band(α, β, k, τ) = Band($(\alpha + 1)k - 1 - \tau, (\beta + 1)k - 1 - \tau$).

Lemma 8.4.10. *With Daubechies' order m wavelets, the one-step transformation matrix $\widehat{W}_\nu$, for $n = 2^L$, is Band($0, (m/2 - 1), 2^{(L-\nu+1)}, 2^{(L-\nu)} - 1$) and is therefore Band($2^{(L-\nu+1)} - 2^{(L-\nu)}, 2^{(L-\nu)}m - 2^{(L-\nu)}$).*

Proof. Note that the band information of $\widehat{W}_\nu$ does not actually involve L or n, and the apparent involvement of L is due to index ν. It suffices to consider $\nu = L$. Then $\widehat{W}_L$ is consisted of 2×2 blocks with $m/2$ blocks on each row. That is, it is a Band$(0, m/2 - 1, 2, 0)$ matrix. Then use Lemma 8.4.9 to complete the proof. ■

Remark 8.4.11. Clearly Band$(\alpha, \beta, k) =$ Band$(\alpha, \beta, k, 0)$. However, for block matrices, it is often necessary to keep them as blocks until a final step in order to obtain improved results. For example, with Lemma 8.4.9.2–8.4.9.3,

Band(1, 1, 2)Band(1, 1, 2) = Band(2, 2, 2) = Band(5, 5), but with Lemma 8.4.9.3,

$$\text{Band}(1,1,2)\text{Band}(1,1,2) = \text{Band}(3,3)\text{Band}(3,3) = \text{Band}(6,6).$$

Similarly as blocks, Band(3, 2, 4)Band(1, 1, 4) = Band(4, 3, 4) = Band(19, 15) but as bands,

$$\text{Band}(3,2,4)\text{Band}(1,1,4) = \text{Band}(15,11)\text{Band}(7,7) = \text{Band}(22,18)$$

– an over-estimate! This suggests that if the band matrix A is Band($\alpha, \beta, 2$), then Theorem 8.4.12 below can be further improved.

Theorem 8.4.12. *Assume that $A_{n\times n}$ is a Band(α, β) matrix. Then the new DWT of ℓ levels, based on Daubechies' order m wavelets, transforms A into $\widehat{A}$ which is at most a Band(λ_1, λ_2) matrix with*

$$\lambda_1 - \alpha = \lambda_2 - \beta = m(2^{(\ell-1)} - 1).$$

Proof. For the new DWT with ℓ levels, the transform is $\widehat{A} = \widehat{W} A \widehat{W}^\top$

$$\widehat{W} = \widehat{W}_{L-\ell+2}\widehat{W}_{L-\ell+3}\cdots\widehat{W}_L.$$

From Lemma 8.4.10, the total lower and upper band widths of $\widehat{W}$ will be, respectively

$$low = \sum_{\nu=L-(\ell-2)}^{L} \left(2^{L-\nu+1} - 2^{L-\nu}\right) \qquad \text{and} \qquad up = \sum_{\nu=L-(\ell-2)}^{L} (m-1)2^{L-\nu}.$$

Therefore the overestimate for the lower band width of $\widehat{A}$ will be

$$\lambda_1 = \alpha + low + up = \alpha + \sum_{\nu=L-(\ell-2)}^{L} 2^{L-\nu} = \alpha + m(2^{(\ell-1)} - 1).$$

Similarly we get the result for λ_2 and the proof is complete. ∎

Note that as indicated before, parameters λ_1, λ_2 do not depend on the problem size n. When $\alpha = \beta$ for A, $\lambda_1 = \lambda_2$ for $\widehat{A}$. For a diagonal matrix A with distinct diagonal entries, for instance, a Band(0, 0) matrix, with $m = 4$, $J = \ell = 5$, and $n = 256$, the standard DWT gives a 'finger'-like pattern in $\widetilde{A}$ as shown in Figure 8.3 (left plot) while the new DWT gives a Band(46, 46) matrix in $\widehat{A}$ as shown in Figure 8.3 (middle plot). Here Theorem 8.4.12 gives over estimates $\lambda_1 = \lambda_2 = 4(2^{(5-1)} - 1) = 60$, a Band(60, 60) matrix.

As we shall see, preserving a banded matrix (or a diagonal matrix) under the new DWT is important for preconditioning a class of singular BIEs [130] using also banded preconditioners. However for other problems where a banded arrow matrix is more desirable and the coarsest level is not too 'coarse' ($\ell \ll L$), it may become desirable to order the 'sums' (scaling coefficients) on the coarsest level separately from the rest of wavelet coefficients. In this way, the resulting DWT (as a variation of [130]) has a tighter band (with added arrows) as shown in the right plots of Figures 8.3 and 8.4, where Figure 8.4 shows the results for the following matrix

$$(A)_{ij} = \left[\begin{array}{c} \dfrac{2}{1+|i-j|}, \ i \neq j \\ 0, \text{ otherwise.} \end{array} \right. \tag{8.39}$$

A similar idea was explored in [204], where all 'sums' related terms are ordered first. In fact, if ℓ levels are applied, the coarsest level nodes are located at $j = 1 : 2^{\ell} : n$ and this suggests that one should order nodes $j = 2 : 2^{\ell-k} : n$ first and the rest second in order to derive a new ordering (based on the centering DWT) that excludes $k+1$ levels in the centering scheme. We have coded the general idea in `perm0.m` that, taking the input $k \in [0, \ell]$, outputs the usual centering DWT if $k = \ell$ and the DWTPerMod [204] if $k = 0$. To see the changes to the pyramidal algorithm or rather the new centering algorithm [130], we refer to Figure 8.5.

8.4.3 Applications to preconditioning a linear system

Consider the transformed linear system $\widehat{A}y = \widehat{b}$ after a new DWT applied to $Ax = b$, where $\widehat{A} = \widehat{W}A\widehat{W}^{\top}$, $x = \widehat{W}^{\top}y$ and $\widehat{b} = \widehat{W}b$. We hope to select an efficient preconditioner M^{-1} to matrix $\widehat{A}$ based on operator (matrix) splitting, that is, $M = \widehat{D}$ and $\widehat{A} = \widehat{D} + \widehat{C}$.

The main issue to bear in mind is the following: any partition $\widehat{D}$ of $\widehat{A}$ corresponds to a partition D of matrix A (via an inverse DWT process). The latter partition, directly related to operators, must include all singularities; this will ensure a good eigenvalue distribution for the preconditioned matrix and its normal ([128]). Such a selection idea seems difficult to realize. Therefore we consider the reverse process.

The strategy that we take is to start with a preliminary partition $A = D + C$ and consider the linear relationships of A, D, C under a DWT, where we may assume as in Section 2.5.1 D is a Band(α, α) for some integer α (say $\alpha = 1$ for a tridiagonal matrix). First, apply the new DWT with a smaller number $\ell = J \leq (L + 1 - r) - 2 = L - r - 1)$ of wavelet levels, to give

$$\widehat{A}y = (\widehat{D} + \widehat{C})y = \widehat{b}, \tag{8.40}$$

where $\widehat{D} = \widehat{W}D\widehat{W}^{\top}$, $\widehat{C} = \widehat{W}C\widehat{W}^{\top}$ and $\widehat{b} = \widehat{W}b$. Now $\widehat{D}$ is also a band matrix (with wrap-around boundaries) and more specifically it is at most Band(λ, λ)

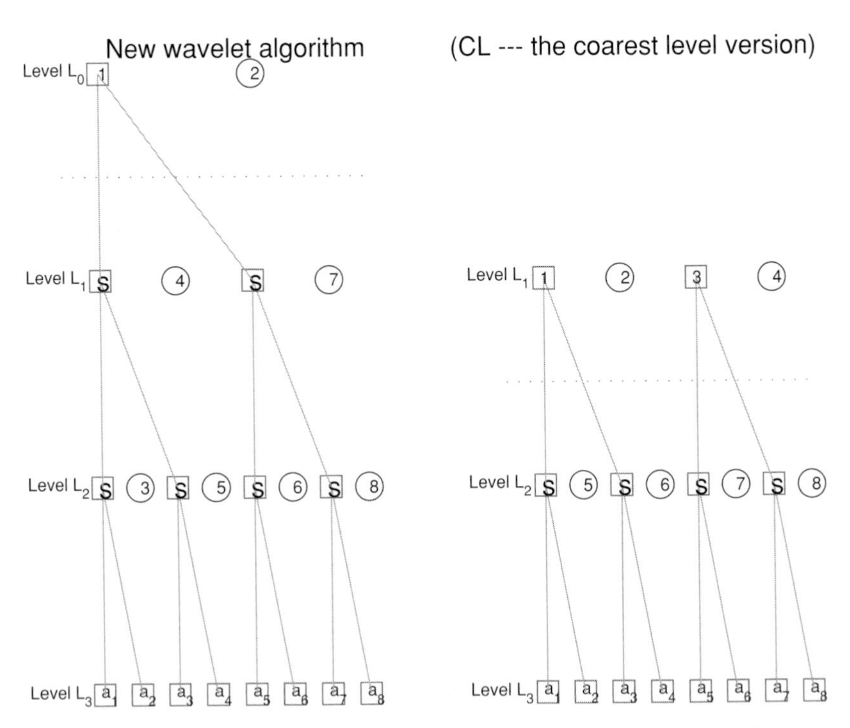

Figure 8.5. Illustration of the coarsest level version of the new wavelet ordering algorithm [130] for $J = 3$ levels (left) and for $J = 2$ levels (right plot). As in Figure 8.2, the boxes □ denote the 'scaling' coefficients ('Sums') and the circles ○ the 'wavelet' coefficients ('differences'). The numbering means that how the final sequence is ordered. The dotted red lines indicate separation of the coarsest level.

matrix with λ as predicted according to Theorem 8.4.12. Let B denote the Band(λ, λ) part of matrix $\widehat{A}$, and we can identify the composition of B in terms of $\widehat{D}, \widehat{C}$. Specifically $B = \widehat{D} + C_d$, where $\widehat{D}$ is enclosed in B and C_d is the band part of matrix $\widehat{C}$ that falls into the sparsity pattern of B. Secondly, partition matrix $\widehat{C} = C_d + C_f$ via B. That is, C_f contains the remaining elements of $\widehat{C}$. Finally, with $B = \widehat{D} + C_d$, we effectively partition the coefficient matrix of (8.40) by $(\widehat{D} + \widehat{C}) = (D_f + C_f) = B + C_f$, with $D_f = B = \widehat{D} + C_d$. Thus $M^{-1} = D_f^{-1}$ will be used as a preconditioner. Using inverse transforms, one can see that using the sparse matrix $M = D_f$ is spectrally equivalent to using a full matrix to precondition matrix A. So the use of DWT is a way to achieve this purpose efficiently.

It should be remarked that, in implementations, the wavelet transform is applied to A, not to D and C separately. The above discussion is mainly to identify the band structure B (say in the middle plot of Figure 8.3) and to explain the exact inclusion of $\widehat{W} D \widehat{W}^\top$ (or D) in matrix M. Thus a new algorithm can be stated as follows:

Algorithm 8.4.13.

(1) *Decide on an operator splitting $A = D + C$ with D a band matrix.*
(2) *Apply the new DWT to $Ax = b$ to obtain $\widehat{A}y = \widehat{b}$.*
(3) *Determine a band-width μ from Theorem 8.4.12 (to find B that encloses $\widehat{D}$).*
(4) *Select the preconditioner as the inverse of a band-width μ matrix of $\widehat{A}$ and use it to solve $\widehat{A}y = \widehat{b}$ iteratively.*

Here the band size $\mu = \lambda_1 + \lambda_2 + 1$ (total band width) is known in advance, once m (wavelet order) and ℓ (wavelet levels) have been selected, and generally small, with respect to problem size n. For example if D is Band(1, 1) (tridiagonal), with $m = 6$ and $\ell = 3$, the total band width for B will be $\mu = 2 \times 6 \times (2^{3-1} - 1) + 1 = 37$. In [129], we applied such an algorithm to a generalized BEM that involves the radial basis function interpolation [79] and obtained improvements to the Krylov solver used.

Note that Algorithm 2 includes Algorithm 1 as a special case if λ is chosen as a fixed integer in the second step.

Remark 8.4.14. Recall that wavelets compress well for smooth (or for smooth parts of) functions and operators. Here matrix C, after cutting off the non-smooth parts D of A, is smooth except on the bands near the cuts (artificially created). So most nonzeros in $\widehat{C}$ will be centered around these cuts and C_d will be significant.

8.5 Optimal implementations and wavelet quadratures

Optimal wavelet implementations refer to obtaining a sparse matrix from using quadratures involving the wavelet functions rather than DWTs. Although it is beyond this book to present wavelet quadratures, we wish to make these remarks.

- As far as the improvement for *conditioning* is concerned, a PDE can be solved by wavelets methods with the most advantage as conditioning, which is is normally extremely poor with FEM with piecewise polynomials, can be better controlled. However, wavelets compression cannot help PDEs since the usual FEM with piecewise polynomials leads to excellent sparsity while wavelets struggle on sparsity (depending on operators and domains).

- For an integral equation example, the opposite is true. Much can be gained from using wavelet functions for discretization since an approximate sparse matrix $\widetilde{A}$ (it does not matter how sparse it is) is generated and the usual BEM matrix from using piecewise polynomials gives the dense matrix. However, conditioning is not a major issue in the case of an integral equation.

We anticipate much progress to come in wavelets research.

8.6 Numerical results

To illustrate the DWT-based preconditioners, we have developed a Mfile `run8.m` to test the matrices as defined in Section 4.9. For simplicity, we have not transformed all solution quantities from $\widetilde{u}$ in the wavelet space to u in the normal space on the finest level; these tasks can be achieved by `ifwt.m` together with `perm2.m` (or `perm0.m` and `iperm0.m`) depending which permutations are involved.

In Figures 8.6–8.8, we display the convergence results for three of the test matrices with the scaling [107] applied to the second matrix before applying

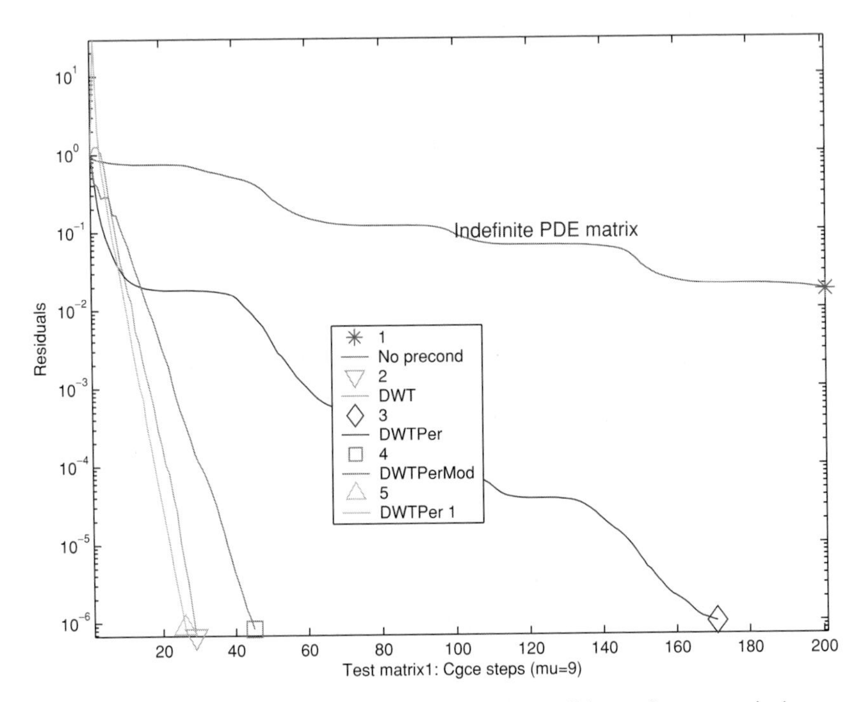

Figure 8.6. Convergence results of the wavelet preconditioners for test matrix 1 (Section 4.9).

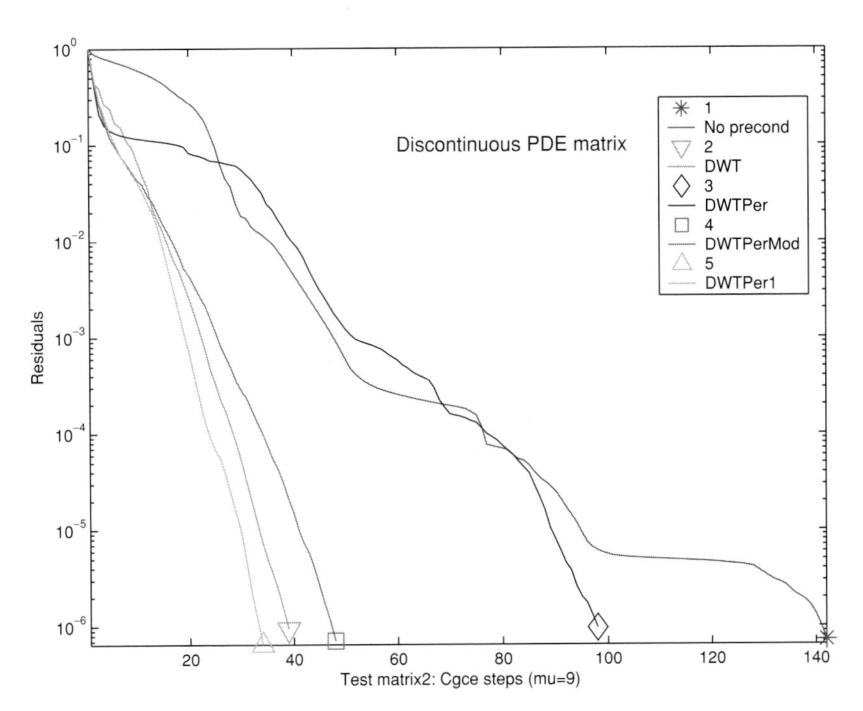

Figure 8.7. Convergence results of the wavelet preconditioners for test matrix 2 (Section 4.9).

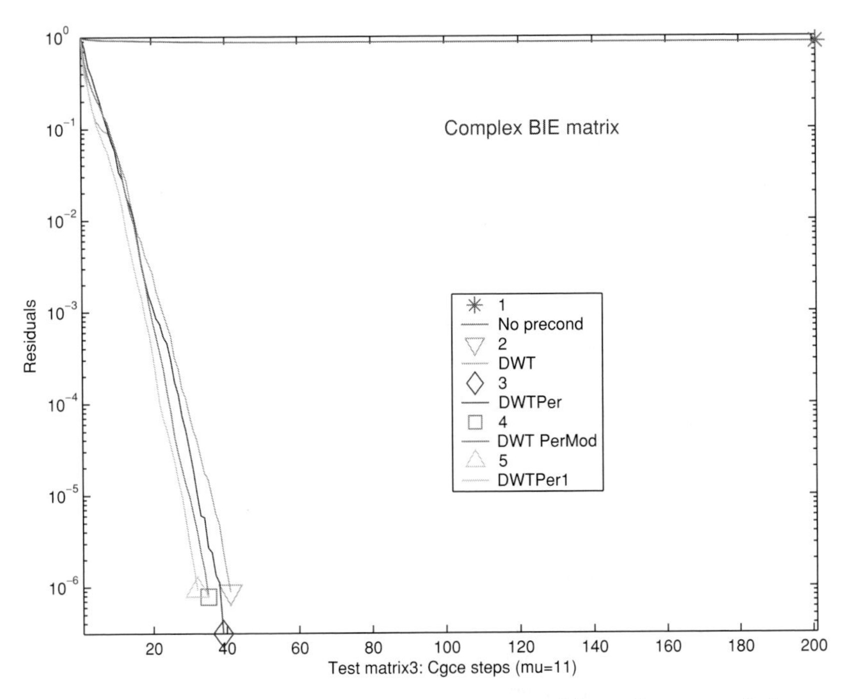

Figure 8.8. Convergence results of the wavelet preconditioners for test matrix 3 (Section 4.9).

the DWT. Clearly one observes that all these three examples are solved much faster than the previous methods in the original polynomial spaces. The hard example `matrix0` that cannot be solved easily by methods from Chapters 4 and 5 remains a challenge for the DWT methods.

8.7 Discussion of software and the supplied Mfiles

Several sources of software appear to exist for wavelet applications. None are completely general and suitable for solving operator equations, especially for preconditioning, reflecting the current state of affairs in wavelet research. Nevertheless, we recommend the following sites for general study.

(1) The MATLAB® wavelet toolbox

http : //www.mathworks.com/wavelet.html

(2) The wavelet digest (pointing to various software)

http : //www.wavelet.org/

(3) The Numerical Recipes [390]:

http : //www.nr.com/

(4) The TOMS (ACM Transactions on Mathematical Software):

http : //www.netlib.org/toms/

This book has supplied these Mfiles for investigating preconditioning related issues.

[1] `perm2.m` – Computes the new DWTper matrix for the wavelet matrix $\widetilde{A}$, that is obtained from using `fwt.m` (Chapter 1).
[2] `perm0.m` – Computes the new DWTPerc matrix, that includes the DWTPer and DWTPerMod as special cases.
[3] `fwts.m` – A sparse matrix version of `fwt.m` (see Section 1.9), which is identical to `fwt.m` for a dense matrix but is much faster for a sparse matrix due to MATLAB's sparse matrix facilities. (See also `iwts.m` for the corresponding inverse DWT.)
[4] `iperm2.m` – The inverse DWTPer transform (undo the work of `perm2` for recovering the original solution together with `ifwt.m` from Section 1.9).
[5] `iperm0.m` – The inverse DWTPer transform (undo the work of `perm0`).

[6] `spyc.m` – An utility Mfile for displaying a sparse matrix with colours (based on the `spy` command).
[7] `ch8.m` – A driver Mfile, illustrating the use of *perm2.m* and *perm0.m* as well as their use in developing preconditioners for GMRES.
[8] `cz.m` – Generation of a test matrix of the Calderon-Zygmund type.
[9] `run8.m` – The main driver Mfile for illustrating `fwt.m`, `perm2.m` and `perm0.m` using the test matrices in Section 4.9.

9

Wavelet Schur preconditioners [T6]

> There seem to be at least two important issues for which wavelet-like expansions have already proven to work with great success, namely *preconditioning* linear systems stemming from Galerkin approximations for elliptic problems and *compressing* full stiffness matrices arising in connection with integral or pseudodifferential operators, to facilitate nearly optimal complexity algorithms for the solution of the corresponding discrete problems.
>
> WOLFGANG DAHMEN, *et al.* Multiscale methods for pseudodifferential equations. *Recent Advances in Wavelet Analysis* (1994)

In the usual FEM setting, Schur complement methods from Chapter 7 perform the best if there is some kind of 'diagonal' dominance. This chapter proposes two related and efficient iterative algorithms based on the wavelet formulation for solving an operator equation with conventional arithmetic. In the new wavelet setting, the stiffness matrix possesses the desirable properties suitable for using the Schur complements. The proposed algorithms utilize the Schur complements recursively; they only differ in how to use coarse levels to solve Schur complements equations. In the first algorithm, we precondition a Schur complement by using coarse levels while in the second we use approximate Schur complements to construct a preconditioner. We believe that our algorithms can be adapted to higher dimensional problems more easily than previous work in the subject. The material is organized in the following

Section 9.1 Introduction
Section 9.2 Wavelets telescopic splitting of an operator
Section 9.3 An exact Schur preconditioner with level-by-level wavelets
Section 9.4 An approximate preconditioner with level-by-level wavelets
Section 9.5 Some analysis and numerical experiments
Section 9.6 Discussion of the accompanied Mfiles

9.1 Introduction

The motivation of this work follows from the observation that any one-scale compressed results (matrices) can be conveniently processed before applying the next scale. Our main algorithms will be designed from level-by-level wavelets, which is related to the BCR work[1] exploring fully the sparsity exhibited.

In this way, regular patterns created by past wavelet scales are not destroyed by the new scales like in the BCR work and unlike in the standard wavelet bases; we define the notation and give further details in Section 9.2. This radical but simple idea will be combined in Section 9.3 with the Schur complement method and Richardson iterations in a multi-level iterative algorithm. Moreover the Richardson iterations can be replaced by a recursive generalized minimal residuals (GMRES) method [415]. The essential assumption for this new algorithm to work is the invertibility of an approximate band matrix; in Subsection 9.5.2 we show that for a class of Calderon–Zygmund and pseudo-differential operators such an invertibility is ensured. In practice we found that our method works equally well for certain operators outside the type for which we can provide proofs. In Section 9.4, we present an alternative way of constructing the preconditioner by using approximate Schur complements. Section 9.5 discusses some analysis issues and several numerical experiments to illustrate the effectiveness of the present algorithms.

We remark that our first algorithm is similar to the framework of a non-standard (NS) form reformulation of the standard wavelets bases (based on the pyramid algorithm) but does not make use of the NS form itself, although our algorithm avoids a finger matrix (just like a NS form method) that could arise from overlapping different wavelet scales. As a by-product, the NS form reduces the flops from $O(n \log n)$ to $O(n)$. However, the NS form does not work with conventional arithmetic although operations with the underlying matrix (that has a regular sparse pattern) can be specially designed; in fact the NS form matrix itself is simply singular in conventional arithmetic. The recent work in [221] has attempted to develop a direct solution method based on the NS form that requires a careful choice of a threshold; here our method is iterative. In the context of designing recursive sparse preconditioners, it is similar to the ILUM type preconditioner to a certain extent [418]. Our second algorithm is similar to the algebraic multi-level iteration methods (AMLI, Section 7.1) that were developed for finite elements [30,34,469]; here our method uses wavelets and does not require estimating eigenvalues.

[1] The nonstandard (NS) form work by Beylkin, Coifman and Rokhlin [60] is often known as the BCR paper.

9.2 Wavelets telescopic splitting of an operator

This section will set up the notation to be used later and motivate the methods in the next sections. We first introduce the standard wavelet method. For simplicity, we shall concentrate on the Daubechies' order m orthogonal wavelets with low pass filters $c_0, c_1, \ldots, c_{m-1}$ and high pass filters $d_0, d_1, \cdots, d_{m-1}$ (such that $d_j = (-1)^j c_{m-1-j}$). In fact, the ideas and expositions in this paper apply immediately to the more general bi-orthogonal wavelets [166].

Following the usual setting of [60,166,221,68,441], the filter coefficients c_j's and d_j's define the scaling function $\phi(x)$ and the wavelet function $\psi(x)$. Further, dilations and translations of $\phi(x)$ and $\psi(x)$ define a multiresolution analysis for L_2 in **d**-dimensions, in particular (refer to (8.6)),

$$
\begin{aligned}
L_2(\mathbb{R}^{\mathbf{d}}) &= \bigoplus_{j=-\infty}^{\infty} \mathbf{W}_j = \mathbf{V}_0 \bigoplus_{j=0}^{\infty} \mathbf{W}_j \\
&= \mathbf{V}_0 \bigoplus \mathbf{W}_0 \bigoplus \mathbf{W}_1 \bigoplus \cdots \bigoplus \mathbf{W}_{\ell-1} \bigoplus \mathbf{W}_\ell \bigoplus \cdots,
\end{aligned}
\tag{9.1}
$$

where $\mathbf{V}_0$ will be used as a coarsest subspace and all subspaces satisfy the relations

$$
\begin{cases}
\cdots \supset \mathbf{V}_\ell \supset \mathbf{V}_{\ell-1} \supset \cdots \supset V_1 \supset V_0 \supset V_{-1} \supset \cdots, \\
\mathbf{V}_{j+1} = \mathbf{V}_j \bigoplus \mathbf{W}_j.
\end{cases}
$$

In numerical realizations, we select a finite dimension space $\mathbf{V}_\ell$ (in the finest scale) as our approximation space to the infinite decomposition of L_2 in (9.1) i.e. effectively

$$
\mathbf{V}_\ell = \mathbf{V}_0 \bigoplus_{j=0}^{\ell-1} \mathbf{W}_j \tag{9.2}
$$

is used to approximate $L_2(\mathbb{R}^{\mathbf{d}})$. Note that the wavelet approximation is for $\mathbb{R}^d$ while the HB basis (§7.4) is mainly exact for $\mathbb{R}$ (for $d > 1$, the HB will not define an orthogonal sum [494]). Consequently for a given operator $\mathcal{T} : L_2 \to L_2$, its infinite and exact operator representation in wavelet bases

$$
\begin{aligned}
\mathcal{T} &= \mathcal{P}_0 \mathcal{T} \mathcal{P}_0 + \sum_{j=0}^{\infty} \left(\mathcal{P}_{j+1} \mathcal{T} \mathcal{P}_{j+1} - \mathcal{P}_j \mathcal{T} \mathcal{P}_j \right) \\
&= \mathcal{P}_0 \mathcal{T} \mathcal{P}_0 + \sum_{j=0}^{\infty} \left(\mathcal{Q}_j \mathcal{T} \mathcal{Q}_j + \mathcal{Q}_j \mathcal{T} \mathcal{P}_j + \mathcal{P}_j \mathcal{T} \mathcal{Q}_j \right)
\end{aligned}
$$

is *approximated* in space $\mathbf{V}_\ell$ by

$$
\mathcal{T}_\ell = \mathcal{P}_\ell \mathcal{T} \mathcal{P}_\ell = \mathcal{P}_0 \mathcal{T} \mathcal{P}_0 + \sum_{j=0}^{\ell-1} \left(\mathcal{Q}_j \mathcal{T} \mathcal{Q}_j + \mathcal{Q}_j \mathcal{T} \mathcal{P}_j + \mathcal{P}_j \mathcal{T} \mathcal{Q}_j \right),
$$

where $\mathcal{P}_j : L_2 \to \mathbf{V}_j$ and $\mathcal{Q}_j = \mathcal{P}_{j+1} - \mathcal{P}_j : L_2 \to \mathbf{V}_{j+1} - \mathbf{V}_j \equiv \mathbf{W}_j$ are both projection operators. For brevity, define operators

$$\begin{aligned} \mathcal{A}_j &= \mathcal{Q}_j \mathcal{T} \mathcal{Q}_j : \mathbf{W}_j \to \mathbf{W}_j, \ \mathcal{B}_j = \mathcal{Q}_j \mathcal{T} \mathcal{P}_j : \mathbf{V}_j \to \mathbf{W}_j, \\ \mathcal{C}_j &= \mathcal{P}_j \mathcal{T} \mathcal{Q}_j : \mathbf{W}_j \to \mathbf{V}_j, \ \ \mathcal{T}_j = \mathcal{P}_j \mathcal{T} \mathcal{P}_j : \mathbf{V}_j \to \mathbf{V}_j. \end{aligned}$$

Then one can observe that

$$\mathcal{T}_{j+1} = \mathcal{A}_j + \mathcal{B}_j + \mathcal{C}_j + \mathcal{T}_j. \tag{9.3}$$

A further observation based on $\mathcal{T}_{j+1} : \mathbf{V}_{j+1} \to \mathbf{V}_{j+1}$ and $\mathbf{V}_{j+1} = \mathbf{V}_j \bigoplus \mathbf{W}_j$ is that the wavelet coefficients of $\mathcal{T}_{j+1}$ will be equivalently generated by the block operator

$$\begin{bmatrix} \mathcal{A}_j & \mathcal{B}_j \\ \mathcal{C}_j & \mathcal{T}_j \end{bmatrix} \quad : \quad \begin{pmatrix} \mathbf{W}_j \\ \mathbf{V}_j \end{pmatrix} \to \begin{pmatrix} \mathbf{W}_j \\ \mathbf{V}_j \end{pmatrix}.$$

Now we change the notation and consider the discretization of all continuous operators. Define matrix $T_\ell = A$ as the representation of operator $\mathcal{T}_\ell$ in space $\mathbf{V}_\ell$. Assume that A on the finest level (scale) $j = \ell$ is of dimension $\tau_\ell = n$. Then the dimension of matrices on a coarse level j is $\tau_j = \tau_\ell / 2^{\ell - j}$ for $j = 0, 1, 2, \ldots, \ell$. The operator splitting in (9.3) for the case of $\mathbf{d} = 1$ (higher dimensions can be discussed similarly [60,221]) corresponds to the two-dimensional wavelet transform

$$\widetilde{T}_{j+1} = W_{j+1} T_{j+1} W_{j+1}^\top = \begin{bmatrix} A_j & B_j \\ C_j & T_j \end{bmatrix}_{\tau_{j+1} \times \tau_{j+1}} \tag{9.4}$$

where the one level transform from $j+1$ to j (for any $j = 0, 1, 2, \cdots, \ell$) is

$$W_{j+1} = \left(W_{j+1} \right)_{\tau_{j+1} \times \tau_{j+1}} = \begin{bmatrix} P_j \\ Q_j \end{bmatrix},$$

$$\begin{aligned} A_j &= \left(A_j \right)_{\tau_j \times \tau_j} = Q_j T_{j+1} Q_j^\top, \quad B_j = \left(B_j \right)_{\tau_j \times \tau_j} = Q_j T_{j+1} P_j^\top, \\ C_j &= \left(C_j \right)_{\tau_j \times \tau_j} = P_j T_{j+1} Q_j^\top, \quad T_j = \left(T_j \right)_{\tau_j \times \tau_j} = P_j T_{j+1} P_j^\top, \end{aligned}$$

with rectangular matrices P_j and Q_j (corresponding to operators $\mathcal{P}_j$ and $\mathcal{Q}_j$) defined respectively as

$$P_j = \begin{bmatrix} c_0 & c_1 & \cdots & \cdots & c_{m-1} & & \\ & & c_0 & c_1 & \cdots & c_{m-1} & \\ & & & & \ddots & \ddots & \ddots \\ c_2 & c_3 & \cdots & c_{m-1} & & c_0 & c_1 \end{bmatrix}_{\tau_j \times \tau_{j-1}},$$

$$Q_j = \begin{bmatrix} d_0 & d_1 & \cdots & \cdots & d_{m-1} & & \\ & & d_0 & d_1 & \cdots & d_{m-1} & \\ & & & & \ddots & \ddots & \ddots \\ d_2 & d_3 & \cdots & d_{m-1} & & d_0 & d_1 \end{bmatrix}_{\tau_j \times \tau_{j-1}}.$$

For a class of useful and strongly elliptic operators i.e. Calderon–Zygmund and pseudo-differential operators, it was shown in [60] that matrices $A_j = (\alpha^j_{k,i})$, $B_j = (\beta^j_{k,i})$, $C_j = (\gamma^j_{k,i})$ are indeed 'sparse' satisfying the decaying property

$$\left|\alpha^j_{k,i}\right| + \left|\beta^j_{k,i}\right| + \left|\gamma^j_{k,i}\right| \le \frac{c_{m,j}}{(1 + |k - i|)^{m+1}}, \tag{9.5}$$

where $|k - i| \ge 2m$ and $c_{m,j}$ is a generic constant depending on m and j only. Refer also to (8.31).

To observe a relationship between the above level-by-level form and the standard wavelet representation, define a square matrix of size $\tau_\ell \times \tau_\ell = n \times n$ for any $j = 0, 1, 2, \ldots, \ell$

$$\overline{W}_j = \begin{bmatrix} I_{\nu_j} & \\ & W_j \end{bmatrix}, \tag{9.6}$$

where $\nu_j = n - \tau_j$; clearly $\nu_\ell = 0$ and $\overline{W}_\ell = W_\ell$. Then the standard wavelet transform can be written as

$$W = \overline{W}_1 \cdots \overline{W}_{\ell-1} \overline{W}_\ell, \tag{9.7}$$

that transforms matrix A into $\widetilde{A} = WAW^\top$.

Thus the diagonal blocks of $\widetilde{A}$ are the same as A_j's of a level-by-level form. However the off-diagonal blocks of the former are different from B_j and C_j of the latter. To gain some insight into the structure of the off-diagonal blocks of matrix $\widetilde{A}$ with the standard wavelet transform, we consider the following case of $\ell = 3$ (three levels) and $m = 4$ (order 4) wavelets. Firstly after level 1 transform, we obtain

$$\widetilde{A}_3 = \overline{W}_3 A \overline{W}_3^\top = W_3 T_3 W_3^\top = \begin{bmatrix} A_2 & B_2 \\ C_2 & T_2 \end{bmatrix}_{n \times n}.$$

Secondly after level two transform, we get

$$\widetilde{A}_2 = \overline{W}_2 \widetilde{A}_3 \overline{W}_2^\top = \begin{bmatrix} A_2 & B_2 W_2^\top \\ W_2 C_2 & \begin{bmatrix} A_1 & B_1 \\ C_1 & T_1 \end{bmatrix} \end{bmatrix}_{n \times n}.$$

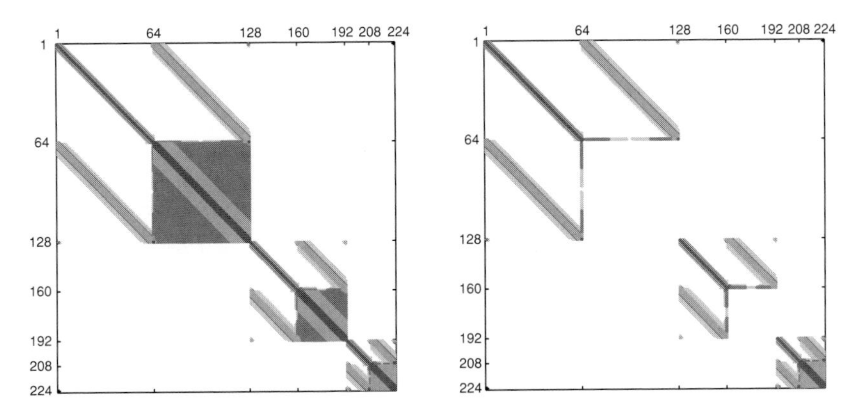

Figure 9.1. The level-by-level form (left) versus the nonstandard wavelet form [60] (right).

Finally after level three transform, we arrive at

$$\widetilde{A}_1 = \overline{W}_1 \widetilde{A}_2 \overline{W}_1^\top = \begin{bmatrix} A_2 & B_2 W_2^\top \begin{bmatrix} I_{3n/4} & \\ & W_1^\top \end{bmatrix} \\ \begin{bmatrix} I_{3n/4} & \\ & W_1 \end{bmatrix} W_2 C_2 & \begin{bmatrix} A_1 & B_1 W_1^\top \\ W_1 C_1 & \begin{bmatrix} A_0 & B_0 \\ C_0 & T_0 \end{bmatrix} \end{bmatrix} \end{bmatrix}_{n \times n}. \tag{9.8}$$

Clearly the off-diagonal blocks of $\widetilde{A}_1$ are perturbations of that of the level-by-level form off-diagonal blocks B_j and C_j; in fact the one-sided transforms for the off-diagonal blocks are responsible for the resulting (complicated) sparsity structure. This can be observed more clearly for a typical example with $n = 128$ (three levels and $m = 4$) in Figure 9.1 where the left plot shows the level-by-level representation set-up that will be used in this paper and in Figure 9.2 where the left plot shows the standard wavelet representation as in (9.8).

Motivated by the exposition in (9.8) of the standard form, we shall propose a preconditioning and iterative scheme that operates on recursive one-level transforms. Thus it will have the advantage of making full use of the NS form idea and its theory while avoiding the problem of a non-operational NS form matrix.

Remark 9.2.1. Starting from the level-by-level set-up, taking T_0 and the collection of all triplets $\{(A_j, B_j, C_j\}_{0 \le j \le \ell-1}$ as a sparse approximation to T_ℓ is the idea of the NS form [60,221]. By way of comparison, in Figure 9.1, the NS form representation versus the level-by-level form are shown. It turns out that

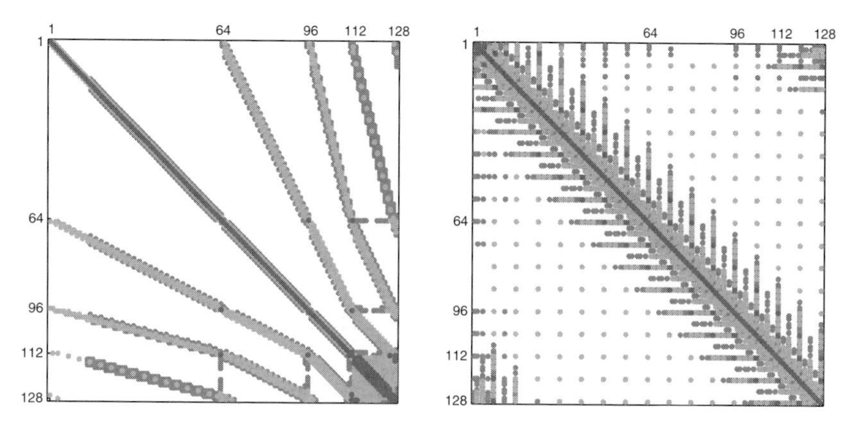

Figure 9.2. The standard wavelet form representation (left) versus an alternative centering form [130] (right) for the example in Figure 9.1.

this work uses the identical set-up to the NS form without using the NS form formulation itself because we shall not use the proposed sparse approximation. Note that the centering algorithm [130] (see the right plot in Figure 9.2) is designed as a permutation of the standard wavelet form (see the left plot of Figure 9.2) and is only applicable to a special class of problems where its performance is better.

9.3 An exact Schur preconditioner with level-by-level wavelets

We now present our first and new recursive method for solving the linear system $Ax = b$ defined on the finest scale $\mathbf{V}_\ell$ i.e.

$$T_\ell x_\ell = b_\ell, \tag{9.9}$$

where $T_\ell = A_\ell = A$ is of size $\tau_\ell \times \tau_\ell = n \times n$ as discussed in the previous section, and $x_\ell,\ b_\ell \in \mathbb{R}^n$. Instead of considering a representation of T_ℓ in the decomposition space (9.2) and then the resulting linear system, we propose to follow the space decomposition and the intermediate linear system in a level-by-level manner. A sketch of this method is given in Figure 9.3 (the left plot) where we try to show a relationship between the multiresolution (MR for wavelet representation) and the multi-level (ML for preconditioning via Schur) ideas from the finest level (top) to the coarsest level (bottom).

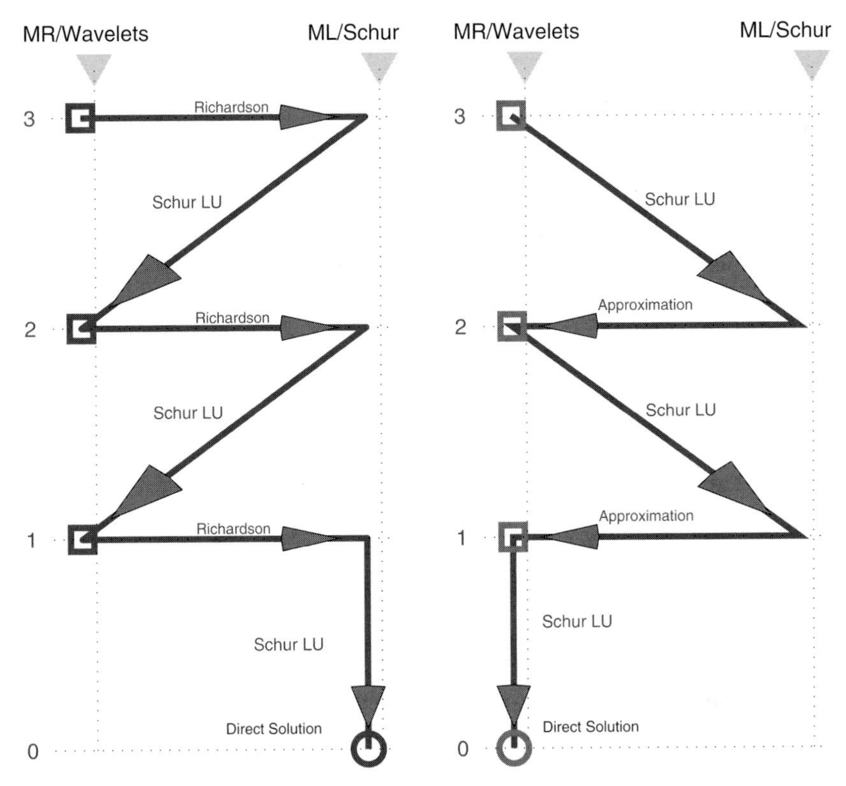

Figure 9.3. Illustration of Algorithms 9.3.2 (left) and 9.4.4 (right). Here we take $\ell = 3$ levels (3 indicates the finest level and 0 the coarsest level), use '□' to indicate a DWT step (one level of wavelets) and '○' to denote the direct solution process on the coarsest level. The arrows denote the sequence of operations (written on the arrowed lines) with each algorithm interacting the two states (two columns on the plots) of multiresolution wavelets and multi-level Schur decomposition. The left plot shows that for Algorithm 9.3.2, a Richardson step (or GMRES) takes the results of a DWT step to the next level via the Schur decomposition while the right plot shows that the Schur decomposition takes the results of a DWT step to the next level via a Schur approximation.

Firstly at level ℓ, we consider $\mathbf{V}_\ell = \mathbf{V}_{\ell-1} \bigoplus \mathbf{W}_{\ell-1}$ and the wavelet transform (9.4) yields

$$\widetilde{T}_\ell \widetilde{x}_\ell = \widetilde{b}_\ell \tag{9.10}$$

where $\widetilde{x}_\ell = W_\ell x_\ell$ and $\widetilde{b}_\ell = W_\ell b_\ell$. Since

$$\widetilde{T}_\ell = \begin{bmatrix} A_{\ell-1} & B_{\ell-1} \\ C_{\ell-1} & T_{\ell-1} \end{bmatrix}_{n\times n}, \tag{9.11}$$

following the general result in (9.5), it is appropriate to consider the approximation of $A_{\ell-1}, B_{\ell-1}, C_{\ell-1}$ by band matrices. To be more precise, let $\text{Band}_\mu(D)$ denote a banded matrix of D with semi-bandwidth μ

$$\left(\text{Band}_\mu(D)\right)_{ik} = \begin{cases} D_{ik}, & \text{if } |i-k| \le \mu, \\ 0, & \text{otherwise} \end{cases}$$

where integer $\mu \ge 0$. Define $\overline{A}_{\ell-1} = \text{Band}_\mu(A_{\ell-1})$, $\overline{B}_{\ell-1} = \text{Band}_\mu(B_{\ell-1})$, $\overline{C}_{\ell-1} = \text{Band}_\mu(C_{\ell-1})$ for some suitable μ (to be specified later). Then matrix $\widetilde{T}_\ell = \overline{T}_\ell - \overline{R}_\ell$ can be approximated by

$$\overline{T}_\ell = \begin{bmatrix} \overline{A}_{\ell-1} & \overline{B}_{\ell-1} \\ \overline{C}_{\ell-1} & T_{\ell-1} \end{bmatrix}_{n\times n}. \tag{9.12}$$

Or equivalently matrix

$$\overline{R}_\ell = \begin{bmatrix} \overline{A}_{\ell-1} - A_{\ell-1} & \overline{B}_{\ell-1} - B_{\ell-1} \\ \overline{C}_{\ell-1} - C_{\ell-1} & 0 \end{bmatrix}_{n\times n} \tag{9.13}$$

is expected to be small in some norm (refer to the Section 9.5.2). Write equation (9.10) as

$$\left(\overline{T}_\ell - \overline{R}_\ell\right)\widetilde{x}_\ell = \widetilde{b}_\ell. \tag{9.14}$$

Consequently we propose to use $M_\ell = \overline{T}_\ell$ as our preconditioner to equation (9.14). This preconditioner can be used to accelerate iterative solution; we shall consider two such methods: the Richardson method and the GMRES method [415].

The most important step in an iterative method is to solve the preconditioning equation:

$$\overline{T}_\ell y_\ell = r_\ell \tag{9.15}$$

or in a decomposed form

$$\begin{bmatrix} \overline{A}_{\ell-1} & \overline{B}_{\ell-1} \\ \overline{C}_{\ell-1} & T_{\ell-1} \end{bmatrix} \begin{pmatrix} y_\ell^{(1)} \\ y_\ell^{(2)} \end{pmatrix} = \begin{pmatrix} r_\ell^{(1)} \\ r_\ell^{(2)} \end{pmatrix}. \tag{9.16}$$

Using the Schur complement method we obtain

$$\begin{cases} \overline{A}_{\ell-1} z_{\ell-1} = r_\ell^{(1)} \\ z_2 = r_\ell^{(2)} - \overline{C}_{\ell-1} z_{\ell-1} \\ \left(T_{\ell-1} - \overline{C}_{\ell-1}\overline{A}^{-1}\overline{B}_{\ell-1}\right) y_\ell^{(2)} = z_2 \\ y_\ell^{(1)} = z_{\ell-1} - \overline{A}^{-1}\overline{B}_{\ell-1} y_\ell^{(2)}. \end{cases} \tag{9.17}$$

Here the third equation of (9.17), unless its dimension is small (i.e. when $\mathbf{V}_{\ell-1}$ is the coarsest scale), has to be solved by an iterative method with the preconditioner $T_{\ell-1}$; we shall denote the preconditioning step by

$$T_{\ell-1}x_{\ell-1} = b_{\ell-1}, \tag{9.18}$$

where $T_{\ell-1}$ is of size $\tau_{\ell-1} \times \tau_{\ell-1}$. This sets up the sequence of a multilevel method where the main characteristic is that each Schur complements equation in its exact form is solved iteratively with a preconditioner that involves coarse level solutions.

At any level j $(0 < j \leq \ell - 1)$, the solution of the following linear system

$$T_j x_j = b_j, \tag{9.19}$$

with T_j of size $\tau_j \times \tau_j$ and through solving

$$\widetilde{T}_j \widetilde{x}_j = \widetilde{b}_j,$$

can be similarly reduced to that of

$$T_{j-1}x_{j-1} = b_{j-1}, \tag{9.20}$$

with T_{j-1} of size $\tau_{j-1} \times \tau_{j-1}$. The solution procedure from the finest level to the coarsest level can be illustrated by the following diagram (for $j = 1, 2, \ldots, \ell - 1$):

$$\boxed{T_j x_j = b_j} \;\Longrightarrow\; \underset{\tau_j \times \tau_j}{\overset{\text{Transform}}{\Longrightarrow}} \;\Longrightarrow\; \boxed{\widetilde{T}_j \widetilde{x}_j = \widetilde{b}_j}$$

$$\Downarrow$$

$$\boxed{\left(\overline{T}_j - \overline{R}_j\right)\widetilde{x}_j = \widetilde{b}_j}$$

$$\Downarrow$$

$$\boxed{\overline{T}_j y_j = r_j}$$

$$\text{Schur} \Downarrow \text{Solver}$$

$$\boxed{\begin{aligned} \overline{A}_{j-1} z_1 &= r_j^{(1)} \\ z_2 &= r_j^{(2)} - \overline{C}_{j-1} z_1 \\ \left(T_{j-1} - \overline{C}_{j-1}\overline{A}_{j-1}^{-1}\overline{B}_{j-1}\right) y_j^{(2)} &= z_2 \\ y_j^{(1)} &= z_1 - \overline{A}_{j-1}^{-1}\overline{B}_{j-1} y_j^{(2)} \end{aligned}}$$

$$\boxed{T_{j-1}x_{j-1} = b_{j-1}} \;\Longleftarrow\; \underset{\tau_{j-1} \times \tau_{j-1}}{\overset{\text{Precondition}}{\Longleftarrow}} \;\Longleftarrow\; \text{(above box)}$$

$$\text{Set} \;\Uparrow\; j = j - 1 \quad \text{(from } T_{j-1}x_{j-1} = b_{j-1} \text{ back to } T_j x_j = b_j\text{)}$$

where as with (9.12) and (9.13)

$$\overline{T}_j = \begin{bmatrix} \overline{A}_{j-1} & \overline{B}_{j-1} \\ \overline{C}_{j-1} & T_{j-1} \end{bmatrix}_{n\times n} \quad \text{and} \quad \overline{R}_j = \begin{bmatrix} \overline{A}_{j-1} - A_{j-1} & \overline{B}_{j-1} - B_{j-1} \\ \overline{C}_{j-1} - C_{j-1} & 0 \end{bmatrix}_{n\times n}. \tag{9.21}$$

The coarsest level is $j = 0$, as set up in the previous section, where a system like (9.20) is solved by a direct elimination method. As with conventional multilevel methods, each fine level iteration leads to many coarse level iteration cycles. This can be illustrated in Figure 9.4 where $\ell = 3$ and $\mu = 2$ (top plot), 3 (bottom plot) are assumed and at the coarsest level ($\otimes$) a direct solution is used. In practice, a variable $\mu = \mu_j$ on level j may be used to achieve certain accuracy for the preconditioning step i.e. convergence up to a tolerance is pursued whilst by way of comparison smoothing rather convergence is desired in an usual multilevel method. Our experiments have shown that $\mu = 1,\ 2$ are often sufficient to ensure the overall convergence.

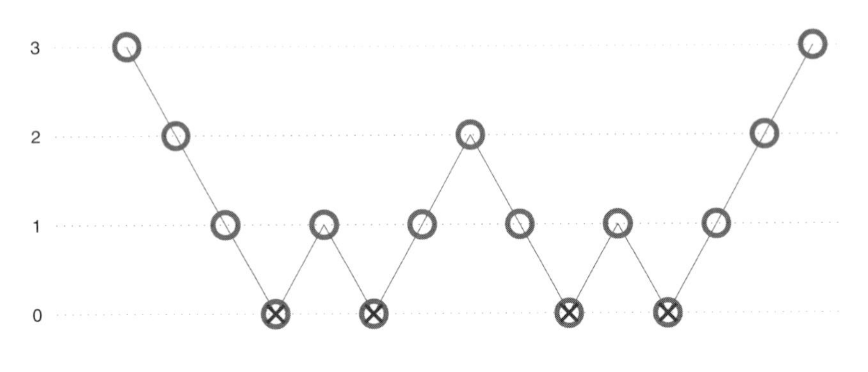

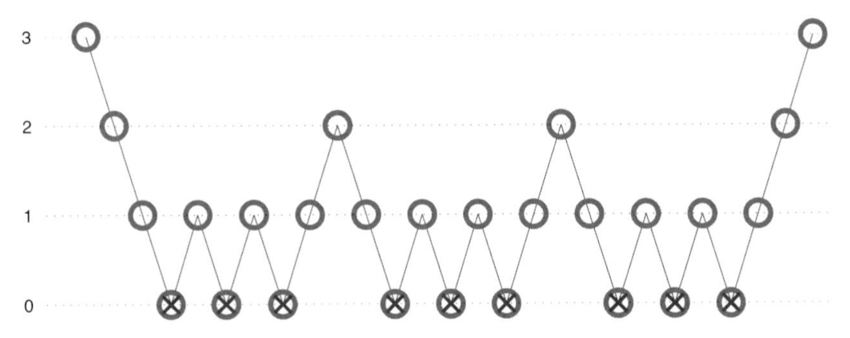

Figure 9.4. Iteration cycling patterns of Algorithm 9.3.2 with $\ell = 3$ levels: top for $\mu = 2$ and bottom for $\mu = 3$. In each case, one solves a fine level equation (starting from the finest level $\ell = 3$) by iteratively solving coarser level equations μ times; on the coarsest level $\otimes$ (here level 0) a direct solution is carried out.

We now summarize the formulation as an algorithm. The iterative solver for (9.19) at level j can be the Richardson method

$$\overline{T}_j \widetilde{x}_j^{(k)} = \overline{R}_j \widetilde{x}_j^{(k-1)} + \widetilde{b}_j \qquad \text{for } k = 1, 2, \cdots, \mu_j$$

or the GMRES method [415] for solving $\widetilde{T}_j \widetilde{x}_j = \widetilde{b}_j$ (or actually a combination of the two). For simplicity and generality, we shall use the word 'SOLVE' to denote such an iterative solver (either Richardson or GMRES).

Algorithm 9.3.2. (Recursive wavelet Schur I).

(1) *Set $j = \ell$ and start on the finest level.*
(2) *Apply one level DWT to $T_j x_j = b_j$ to obtain* $\tilde{T}_j \tilde{x}_j = \tilde{b}_j$.
(3) **Use** μ_j *steps of* SOLVE *for* $\tilde{T}_j \tilde{x}_j = \tilde{b}_j$.
(4) *In each step, implement the preconditioner $\overline{T}_j$: i.e. solve $\overline{T}_j y_j = r_j$ or*

Restrict *to the coarse level:*
$$\begin{cases} \overline{A}_{j-1} z_1 = r_j^{(1)} \\ z_2 = r_j^{(2)} - \overline{C}_{j-1} z_1 \\ \left(T_{j-1} - \overline{C}_{j-1} \overline{A}_{j-1}^{-1} \overline{B}_{j-1}\right) y_j^{(2)} = z_2 \\ y_j^{(1)} = z_1 - \overline{A}_{j-1}^{-1} \overline{B}_{j-1} y_j^{(2)} \end{cases}$$

(5) **Use** SOLVE *for the above third equation with the preconditioner T_{j-1} i.e. solve $T_{j-1} x_{j-1} = b_{j-1}$.*
(6) *Set $j := j - 1$.*
(7) *If $j = 0$ (on the coarsest level), apply a* **direct solver** *to* $T_j x_j = b_j$ *and proceed with Step 8; otherwise return to Step 2.*
(8) *Set $j := j + 1$.*
(9) **Interpolate** *the coarse level $j - 1$ solution to the fine level j:*

$$x_j^{(2)} = x_{j-1}, \qquad x_j^{(1)} = z_1 - \bar{A}_j^{-1} \bar{B}_j x_{j-1} \qquad i.e.$$

$$x_j = \begin{bmatrix} x_j^{(1)} \\ x_j^{(2)} \end{bmatrix} = \begin{bmatrix} x_j^{(1)} \\ x_{j-1} \end{bmatrix}.$$

(10) *Apply one level inverse DWT to $\widetilde{y}_j$ to obtain y_j.*
(11) *If $j = \ell$ (on the finest level), check the residual error – if small enough accept the solution x_0 and stop the algorithm. If $j < \ell$, check if μ_j steps (cycles) have been carried out; if not, return to Step 2 otherwise continue with Step 8 on level j.*

The rate of convergence of this algorithm depends on how well the matrix $\overline{T}_j$ approximates T_j and this approximation is known to be accurate for a suitable μ and for a class of Calderon–Zygmund and pseudo-differential operators [60]. For this class of problems, it remains to discuss the invertibility of matrix $\overline{A_j}$ which is done in Section 9.5.2; a detailed analysis on $\overline{T}_j \approx T_j$ may be done along the same lines as Lemma 9.5.7. For other problem classes, the algorithm may not work at all for the simple reason that $\overline{A_j}$ may be singular e.g. the diagonal of matrix $A = T_\ell$ in (9.9) may have zero entries. Some extensions based on the idea of [107] may be applied as discussed in [108].

It turns out that our recommended implementation is to specify SOLVE on the finest level by GMRES and to follow by simple Richardson iterations on coarse levels (see `gmres_nr.m`). In [208], an all GMRES version was considered.

Remark 9.3.3. We remark that for a class of general sparse linear systems, Saad, Zhang, Botta, Wubs *et al.* [412,70,417,418] have proposed a recursive multilevel preconditioner (named as ILUM) similar to this Algorithm 9.3.2. The first difference is that we need to apply one level of wavelets to achieve a nearly sparse matrix while these works start from a sparse matrix and permute it to obtain a desirable pattern suitable for Schur decomposition. The second difference is that we propose an iterative step before calling for the Schur decomposition while these works try to compute the exact Schur decomposition approximately. Therefore it is feasible to refine our Algorithm 9.3.2 to adopt the ILUM idea (using independent sets) for other problem types. However one needs to be careful in selecting the dimensions of the leading Schur block if a DWT is required for compression purpose.

9.4 An approximate preconditioner with level-by-level wavelets

In the previous algorithm, we use coarse level equations to precondition the fine level Schur complement equation. We now propose an alternative way of constructing a preconditioner for a fine level equation. Namely we approximate and compute the fine level Schur complement before employing coarse levels to solve the approximated Schur complement equation. A sketch of this method is shown in Figure 9.3 and demonstrates the natural coupling of wavelet representation (level-by-level form) and Schur complement. To differentiate from Algorithm 9.3.2, we change the notation for all matrices.

At any level k for $k = 1, 2, \ldots, \ell$, consider the solution (compare to (9.19))

$$A^{(k)} x_k = b_k. \tag{9.22}$$

Applying one-level of DWT, we obtain

$$\widetilde{A^{(k)}}\widetilde{x_k} = \widetilde{b_k} \quad \text{with} \quad \widetilde{A^{(k)}} = W_k A^{(k)} W_k^\top = \begin{bmatrix} A_{11}^{(k)} & A_{12}^{(k)} \\ A_{21}^{(k)} & A_{22}^{(k)} \end{bmatrix}. \tag{9.23}$$

Note that we have the block LU decomposition

$$A^{(k)} = \begin{bmatrix} A_{11}^{(k)} & 0 \\ A_{21}^{(k)} & I \end{bmatrix} \begin{bmatrix} I & A_{11}^{(k)^{-1}} A_{12}^{(k)} \\ 0 & \mathbf{S}^{(k)} \end{bmatrix}$$

where $\mathbf{S}^{(k)} = A_{22}^{(k)} - A_{21}^{(k)} A_{11}^{(k)^{-1}} A_{12}^{(k)}$ is the true Schur complement. To approximate this Schur complement, we must consider approximating the second term in the above $\mathbf{S}^{(k)}$. We propose to form band matrix approximations

$$\begin{aligned} B_{11} &= Band_\mu(A_{11}^{(k)}) \approx A_{11}^{(k)^{-1}}, \\ \overline{A_{12}} &= Band_\mu(A_{12}^{(k)}) \approx A_{12}^{(k)}, \\ \overline{A_{21}} &= Band_\mu(A_{21}^{(k)}) \approx A_{21}^{(k)}. \end{aligned}$$

For level $k = 0$, these approximations are possible for a small bandwidth μ; see Section 9.5.2. Seeking a band approximation to the inverse of $A_{11}^{(k)}$ makes sense because $A_{11}^{(k)} = A_k$ is expected to have a decaying property (refer to (9.5)). Let $\mathcal{S}$ denote the set of all matrices that have the sparsity pattern of a band μ matrix $Band_\mu(A_{11}^{(k)})$. The formation of a sparse approximate inverse (SPAI) is to find a band matrix $B_{11} \in \mathcal{S}$ such that

$$\min_{B \in \mathcal{S}} \|A_{11}^{(k)} B - I\|_F = \|A_{11}^{(k)} B_{11} - I\|_F.$$

Refer to [46,320,131]. Briefly as with most SPAI methods, the use of F-norm decouples the minimization into least squares (LS) problems for individual columns c_j of B_{11}. More precisely, owing to

$$\|A_{11}^{(k)} B_{11} - I\|_F^2 = \sum_{j=1}^{\tau_k} \|A_{11}^{(k)} c_j - e_j\|_2^2,$$

the j-th LS problem is to solve $A_{11}^{(k)} c_j = e_j$ which is not expensive since c_j is sparse. Once B_{11} is found, define an approximation to the true Schur complement $\mathbf{S}^{(k)}$ as

$$S^{(k)} = A_{22}^{(k)} - \overline{A_{21}^{(k)}} B_{11} \overline{A_{12}^{(k)}},$$

and set $A^{(k-1)} = S^{(k)}$. This generates a sequence of matrices $A^{(k)}$.

Now comes the most important step about the new preconditioner. Setting $M^{(0)} = A^{(0)} = S^{(1)}$, on the coarsest level, the fine level preconditioner $M^{(\ell)}$ is defined recursively by

$$M^{(k)} = \begin{bmatrix} B_{11}^{(k)^{-1}} & 0 \\ A_{21}^{(k)} & I \end{bmatrix} \begin{bmatrix} I & B_{11}^{(k)} A_{12}^{(k)} \\ 0 & S^{(k)} \end{bmatrix}, \tag{9.24}$$

where $S^{(k)} \approx M^{(k-1)}$ is an approximation to the true Schur complement $\mathbf{S}^{(k)}$ of $A^{(k)}$. Here $k = 1, 2, \ldots, \ell$. Observe that this preconditioner is defined through the V-cycling pattern recursively using the coarse levels.

To go beyond the V-cycling, we propose a simple residual correction idea. We view the solution $y_k^{[j]}$ of the preconditioning equation (compare to (9.15) and (9.9))

$$M^{(k)} y_k = r_k \tag{9.25}$$

as an approximate solution to the equation

$$T_k y_k = r_k.$$

Then the residual vector is $\bar{r}_k = r_k - T_k y_k^{[j]}$. This calls for a repeated solution $M^{(k)}\delta_j = \bar{r}_k$ and gives the correction and a new approximate solution to (9.25):

$$y_k^{[j+1]} = y_k^{[j]} + \delta_j,$$

for $j = 1, 2, \ldots, \nu_k$. In practice we take $\nu_k = \nu$ for a small ν (say $\nu = 2$ for a W-cycling; see the top plot in Figure 9.4) as our experiments suggest that $\nu \leq 2$ is sufficient to ensure the overall convergence.

Thus an essential feature of this method different from Algorithm 9.3.2 is that every approximated Schur complement matrix needs to be transformed to the next wavelet level in order to admit the matrix splitting (9.23) while inverse transforms are needed to pass coarse level information back to a fine level as illustrated in Figure 9.3. Ideally we may wish to use $A^{(k)} = A_{22}^{(k)} - A_{21}^{(k)} B_{11}^{(k)} A_{12}^{(k)}$ generate the approximate Schur complement but taking $A_{21}^{(k)}$ and $A_{12}^{(k)}$ as full matrices would jeopardize the efficiency of the overall iterative method. We proposed to use band matrices (or thresholding) to approximate these two quantities just as in (9.13) with Algorithm 9.3.2.

To summarize, the solution of the preconditioning equation from the finest level to the coarsest level and back up is illustrated in Figure 9.5 for $k = \ell, \ldots, 2, 1$, where '$\star$' means the same entry and exit point. The general algorithm for solving $M^{(\ell)} y_\ell = r_\ell$ can be stated as follows:

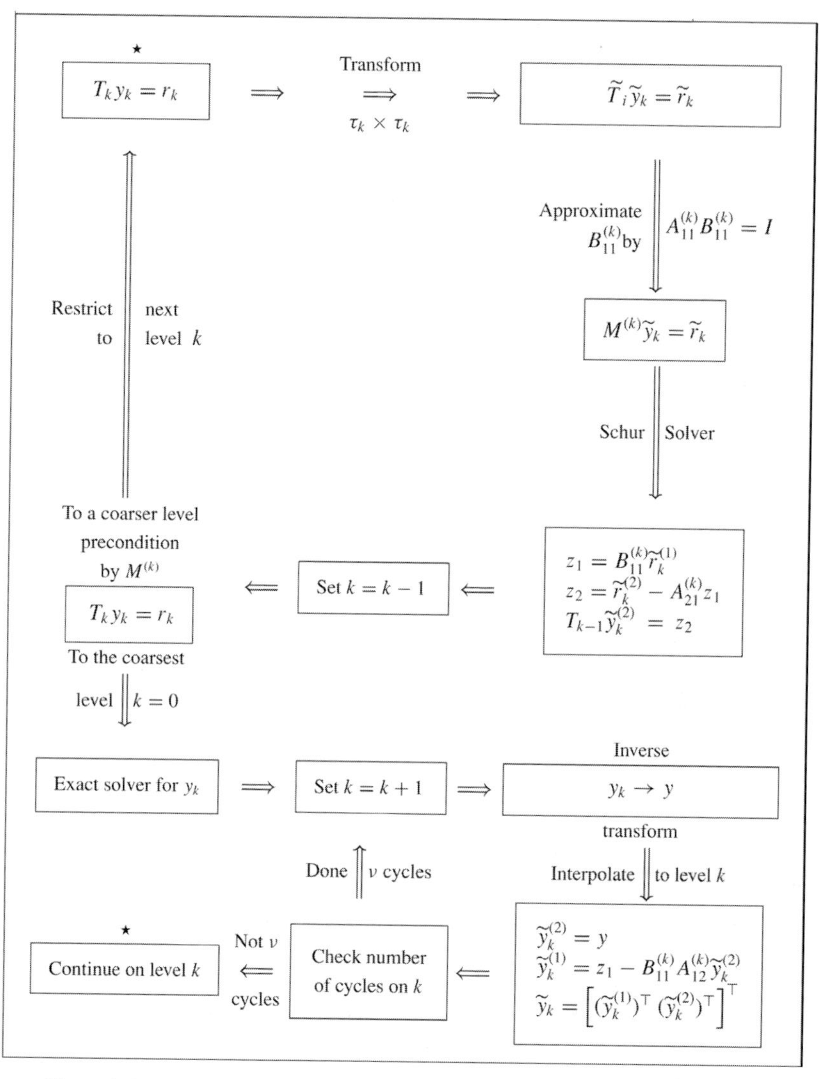

Figure 9.5. Flow chart of Algorithm II (Implementation of the algebraic multilevel iteration preconditioner). Here on a typical fine level k (starting from the finest level ℓ), we illustrate on the top half of the diagram the process of restricting to the coarse level $k - 1$ (up to the coarsest level 0) whilst on a typical coarse level k (starting from ℓ) we show on the bottom half of the diagram the process of interpolating to the fine level $k + 1$ (up to the finest level ℓ). Consult Figure 9.4 for the overall cycling pattern.

Algorithm 9.4.4. (Recursive wavelet Schur II).

Set-up Stage

---

for $k = \ell, \ell - 1, \cdots, 2, 1$

(1) *Apply one level DWT to* T_k *to obtain* $A_{11}^{(k)}, A_{12}^{(k)}, A_{21}^{(k)}, A_{22}^{(k)}$.

(2) *Find the approximate inverse* $B_{11}^{(k)} \approx A_{11}^{(k)^{-1}}$.

(3) *Generate the matrix* $T_{k-1} = A^{(k)} = A_{22}^{(k)} - A_{21}^{(k)} B_{11}^{(k)} A_{12}^{(k)}$.

end

Solution Stage

---

(1) *Set* $k = \ell$ *and start on the finest level.*

(2) *Apply one level DWT to* r_k *and consider* $\tilde{T}_k \tilde{y}_k = \tilde{r}_k$.

(3) **Solve** *the preconditioning equation* $M^{(k)} \tilde{y}_k = \tilde{r}_k$ *by*

Restrict *to the coarse level:*
$$\begin{cases} z_1 = B_{11}^{(k)} \widetilde{r}_k^{(1)} \\ z_2 = \widetilde{r}_k^{(2)} - A_{21}^{(k)} z_1 \\ T_{k-1} \widetilde{y}_k^{(2)} = z_2 \end{cases}$$

(4) **Solve** *for the above third equation at the next level* $T_{k-1} y_{k-1} = r_{k-1}$.

(5) *Set* $k := k - 1$.

(6) *If* $k = 0$ *(on the coarsest level), apply a* **direct solver** *to* $T_k y_k = r_k$ *and proceed with Step 8; otherwise return to Step 2.*

(7) *Set* $k := k + 1$.

(8) **Interpolate** *the coarse level* $k - 1$ *solution to the fine level* k:

$$\widetilde{y}_k^{(2)} = y_{k-1}, \qquad \widetilde{y}_k^{(1)} = z_1 - B_{11}^{(k)} A_{12}^{(k)} \widetilde{y}_k^{(2)} \; i.e.$$

$$\widetilde{y}_k = \begin{bmatrix} \widetilde{y}_k^{(1)} \\ \widetilde{y}_k^{(2)} \end{bmatrix} = \begin{bmatrix} \widetilde{y}_k^{(1)} \\ y_{k-1} \end{bmatrix}.$$

(9) *Apply one level inverse DWT to* $\widetilde{y}_k$ *to obtain* y_k.

(10) *When* $k = \ell$ *(on the finest level), check the residual error – if small enough accept the solution* y_ℓ *and stop the algorithm.*
When $k < \ell$, *check if* ν *cycles have been carried out; if not, find the residual vector and return to Step 2 otherwise continue with Step 7 on level* k.

Remark 9.4.5. It turns out that this algorithm is similar to the algebraic multilevel iteration methods (AMLI) that was developed for a class of symmetric positive definite finite element equations in a hierarchical basis [30,34,469]. In fact, let $\nu_k = \nu$ and then $S^{(k)}$ is implicitly defined by the following

$$S^{(k)} = A^{(k-1)} \left[I - P_\nu \left(M^{(k-1)^{-1}} A^{(k-1)} \right) \right]^{-1},$$

where P_ν is a degree ν polynomial satisfying

$$0 \le P_\nu(t) < 1, \qquad 0 < t \le 1, \qquad P_\nu(0) = 1.$$

As with AMLI, for $\nu = 1$, the valid choice $P_1(t) = 1 - t$ implies $S^{(k)} = M^{(k-1)}$ and gives rise to the V-cycling pattern and For $\nu > 1$, the polynomial $P_\nu(t)$ is chosen to improve the preconditioner; ideally $\kappa \left(M^{(k)^{-1}} A_{11}^{(k)} \right) \approx O(1)$ asymptotically. Refer to these original papers about how to work out the coefficients of $P_\nu(t)$ based on eigenvalue estimates. However we do not use any eigenvalue estimates to construct P_ν.

We also remark that an alternative definition of a recursive preconditioner different from AMLI is the ILUM method as mentioned in Remark 9.3.3, where in a purely algebraic way (using independent sets of the underlying matrix graph) $A_{11}^{(k)}$ is defined as a block diagonal form after a suitable permutation; refer to (7.17). This would give rise to another way of approximating the true Schur complement $\mathbf{S}^{(k)} = A_{22} - A_{21}^{(k)} A_{11}^{(k)^{-1}} A_{12}^{(k)}$. However the sparsity structures of blocks $A_{21}^{(k)}$ and $A_{12}^{(k)}$ will affect the density of nonzeros in matrix $\mathbf{S}^{(k)}$ and an incomplete LU decomposition has to be pursued as in [412,70,418].

9.5 Some analysis and numerical experiments

The two algorithms presented are essentially the normal Schur type preconditioners applied to the wavelet spaces. The beauty of adopting the wavelet idea here lies in avoiding the somewhat difficult task of ensuring the 'top left' block A_{11} is easily invertible – wavelets do this automatically for a large class of operators! The supplied Mfiles follow the usual style of a multigrid (multilevel) implementation.

9.5.1 Complexity analysis

Here we mainly compare the complexity of Algorithms 9.3.2 and 9.4.4 in a full cycle. Note that both algorithms can be used by a main driving iterative solver where each step of iteration will require n^2 flops (one flop refers to 1 multiplication and 1 addition) unless some fast matrix vector multiplication methods are used. One way to reduce this flop count is to use a small threshold so that only a sparse form of $\widetilde{A}$ is stored. Also common to both algorithms are the DWT steps which are not counted here.

To work out flop counts for differing steps of the two algorithms, we list the main steps as follows

	Algorithm 9.3.2	Algorithm 9.4.4
Set-up		Approximate inverses: $\mu^2 n_i^2$
Main step	3 band solves: $3n_i\mu^2$ 4 band-vector multiplications: $8n_i\mu$ 3 off-band-vector multiplications (R_i): $3(n_i - 2\mu)n_i$	2 band-band multiplications: $8n_i\mu^2$ 4 band-vector multiplications: $8n_i\mu$

Therefore an ν-cycling (iteration) across all levels would require these flops (assuming $\nu \leq 3$)

$$F_I = \sum_{i=1}^{\ell-1} \frac{6n^2\nu^i}{2^{2i}} + \frac{3n\mu^2}{2^i} + \frac{2n\mu}{2^i} \approx \frac{6\nu}{4-\nu}n^2,$$

and

$$F_{II} = \sum_{i=1}^{\ell-1} \frac{n^2}{2^{2i}}\mu^2 + 8\sum_{i=1}^{\ell-1} \frac{n\mu^2\nu^2}{2^i} + \frac{n\mu}{2^i} \approx \frac{\mu^2}{3}n^2,$$

after ignoring the low order terms. Therefore we obtain $F_{II}/F_I = \frac{(4-\nu)\mu^2}{9\nu}$; for a typical situation with $\mu = 10$ and $\nu = 2$, $F_{II}/F_I \approx 10$. Thus we expect Algorithm I to be cheaper than II if the same number of iteration steps are recorded. Here by Algorithm I we meant the use of a Richardson iteration (in SOLVE of Algorithm 9.3.2); however if a GMRES iteration is used for preconditioning then the flop count will increase. Of course as is well known, flop count is not always a reliable indicator for execution speed; especially if parallel computing is desired a lot of other factors have to be considered.

Remark 9.5.6. For sparse matrices, all flop counts will be much less as DWT matrices are also sparse. The above complexity analysis is done for a dense matrix case. Even in this case, setting up preconditioners only adds a few equivalent iteration steps to a conventional iteration solver. One can usually observe overall speed-up. For integral operators, the proper implementation is to use the biorthogonal wavelets as trial functions to yield sparse matrices A_i, B_i, C_i directly (for a suitable threshold); in this case a different complexity analysis is needed as all matrices are sparse and experiments have shown that although this approach is optimal a much larger complexity constant (independent of n) is involved. For structured dense matrices where FFT is effective, wavelets may have to be applied implicitly to preserve FFT representation.

9.5.2 An analysis of preconditioner I in Section 9.3

We now show that Algorithm 9.3.2 does not break down for a class of Calderon–Zygmund and pseudo-differential operators [60], or precisely that $\overline{A_i} = Band_\mu(A_i)$ is always invertible. More importantly the bandwidth μ can be very small, independent of the problem size (n).

For our analysis purpose, we assume that all principal submatrices A_i (resulting from matrix $A_\ell = A$) are nonsingular and diagonalizable. Of course this is a much weaker assumption than requiring matrix A being symmetric positive definite. For non-negative integers of μ and m, define the function

$$b(n, m, \mu) = \sum_{k=1}^{n-\mu-1} \frac{2}{(\mu + k)^{m+1}},$$

which is understood in the usual summation convention i.e. if $\mu \geq n - 1$, $b(n, m, \mu) = 0$. We can verify the simple properties: $b(n, m, \mu) \leq (\mu + 1)^{1-m} b(n, 1, \mu)$ if $m \geq 1$, $b(n, m, \mu) < b(\infty, m, \mu)$, and

$$b(n, 1, \mu) < b(\infty, 1, \mu) \leq b(\infty, 1, 0) = \sum_{k=1}^{\infty} 2/k^2 = \pi/3.$$

Therefore if $m > 1$, $\lim_{\mu\to\infty} b(\infty, m, \mu) = 0$ as $\lim_{\mu\to\infty}(\mu + 1)^{1-m} = 0$. Then the following result holds.

Lemma 9.5.7. *Let matrix $O_i = A_i - \overline{A_i}$ with $\overline{A_i}$ from Algorithm 9.3.2 and $c_{m,i} \leq c$ in (9.5) for a generic constant c that does not depend on m, n, μ. Then*

$$\|O_i\|_\infty \leq cb(\tau_i, m, \mu) < cb(\infty, m, \mu).$$

Proof. It suffices to consider the case of $i = 1$, i.e. $\overline{A_1}$. Write $A_1 = (a_{kj})$ for simplicity. Since $\overline{A_1} = Band_\mu(A_1)$ is banded and of dimension τ_1, to estimate the ∞-norm of O_1, take any row index $k \in [1,\ \tau_1]$ with $\mu \geq 2m$ (note that for the first and last μ rows there is only one nonzero sum at the right-hand side of the following line 1 as O_1 is an off band matrix)

$$\begin{aligned}\sum_{j=1}^{\tau_1} |(O_1)_{kj}| &= \sum_{j=1}^{k-\mu-1} |A_{kj}| + \sum_{j=k+\mu+1}^{\tau_1} |A_{kj}| \\ &\le c_{m,1}\left[\sum_{j=1}^{k-\mu-1} \frac{1}{1+|k-j|^{m+1}} + \sum_{j=k+\mu+1}^{\tau_1} \frac{1}{1+|k-j|^{m+1}}\right] \\ &= c_{m,1}\left[\sum_{j=\mu+1}^{k-1} \frac{1}{1+j^{m+1}} + \sum_{j=\mu+1}^{\tau_1-k} \frac{1}{1+j^{m+1}}\right] \\ &\le c_{m,1}\left[\sum_{j=1}^{k-\mu-1} \frac{1}{(\mu+j)^{m+1}} + \sum_{j=1}^{\tau_1-\mu-k} \frac{1}{(\mu+j)^{m+1}}\right] \\ &\le c_{m,1}\left[\sum_{j=1}^{\tau_1-\mu-1} \frac{1}{(\mu+j)^{m+1}} + \sum_{j=1}^{\tau_1-\mu-1} \frac{1}{(\mu+j)^{m+1}}\right] \\ &\le cb(\tau_1, m, \mu).\end{aligned}$$

Therefore

$$\|O_1\|_\infty = \max_{1\le k\le \tau_1} \sum_{j=1}^{\tau_1} |(O_1)_{kj}| \le cb(\tau_1, m, \mu).$$

Using the properties of function b completes the proof. ■

Note that this Lemma suggests $\|O_i\|_\infty$ can be arbitrarily small if μ is large. However it is more useful to quantify how large μ needs to be in order for $\overline{A}_i$ be invertible. This is now stated as follows.

Theorem 9.5.8. *Under the above assumption of A_i, the band matrix $\overline{A}_i$ (used in the preconditioning) is invertible if $m \ge 2$ (i.e. there are 2 or more vanishing moments) and the semi-bandwidth μ of $\overline{A}_i$ satisfies $\mu \ge \mu_{\min}$ where*

$$\mu_{\min} = \left(\frac{c\pi}{3|\lambda_i^s|}\right)^{\frac{1}{m-1}} - 1$$

and λ_i^s denotes the smallest eigenvalue of A_i (in modulus).

Remark 9.5.9. The theorem implies that Algorithm 9.3.2 does not break down. The confirmation in Lemma 9.5.7 of a fast decay in elements of O_i or fast deduction of $\|O_i\|_\infty$ (as μ is large) ensures that a small bandwidth μ is practically sufficient; this justifies the efficient approximation of $\widetilde{T}_i$ by $\overline{T}_i$ in Algorithm 9.3.2. For instance, when $c = 20$ and $\lambda_i^s = 0.1$, $\mu_{\min} = 2.8$ if $m = 5$ and $\mu_{\min} = 13.5$ if $m = 3$. Here the requirement of $m \ge 2$ (number of vanishing moments) is not restrictive. Also as remarked before, the assumption of nonsingularity and diagonalization is assured if $A_\ell = A$ is SPD [30,469].

Proof. Following Lemma 9.5.7, let $\lambda(A_i)$ and $\lambda(\overline{A}_i)$ denote an eigenvalue of the matrix A_i and $\overline{A}_i$ respectively. From the eigenvalue perturbation theorem (for a diagonalizable matrix),

$$\left|\lambda(A_i) - \lambda(\overline{A}_i)\right| \le \|O_i\|_\infty.$$

Using the above inequality with Lemma 9.5.7 yields the sufficient condition

$$(\mu + 1)^{m-1}\frac{\pi}{3}c \le |\lambda_i^s|$$

and further the required result for $\mu_{\min}$. With such a $\mu \ge \mu_{\min}$, the preconditioning matrix $\overline{A}_i$ has only nonzero eigenvalues and hence is invertible. ■

9.5.3 Numerical experiments

The new presented algorithms have recently been tested in [108] with comparisons with other preconditioners; here we only illustrate the same matrices (in Section 4.9) as tested in previous chapters.

We test these supplied Mfiles (especially the first one) implementing Algorithm 9.3.2.

(1) `gmres_nr.m` – Method 1 for implementing Algorithm 9.3.2 with the SOLVE step replaced by a GMRES(k) iteration method on the finest level and a Richardson iteration method for ν steps on all other all levels.
(2) `richa_nr.m` – Method 2 for implementing Algorithm 9.3.2 with the main iteration method iteration method on the finest level and all coarse level correction equations all given by Richardson iterations.

and display the results in Figures 9.6–9.8. As pointed in [108], Algorithm 9.4.4 only works for definite problems (and so it does not work for `matrix1`); this is similar to the behaviour of the AMLI algorithm (Section 7.1).

Clearly the Schur methods in the wavelet space perform better than the single scale methods (of Chapter 4 and 5) as well as the normal wavelet methods (of Chapter 8). Here `matrix3` also involves 'geometric singularities' which affect the decaying rate (8.30) away from the diagonal and so the result is not surprisingly less remarkable (Figure 9.8).

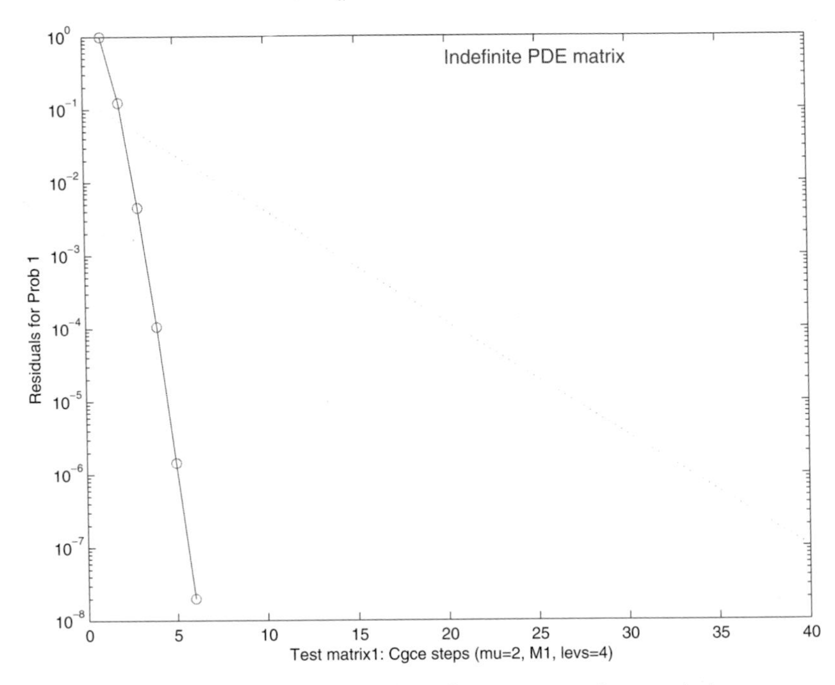

Figure 9.6. Convergence behaviour of `gmres_nr.m` for `matrix1`.

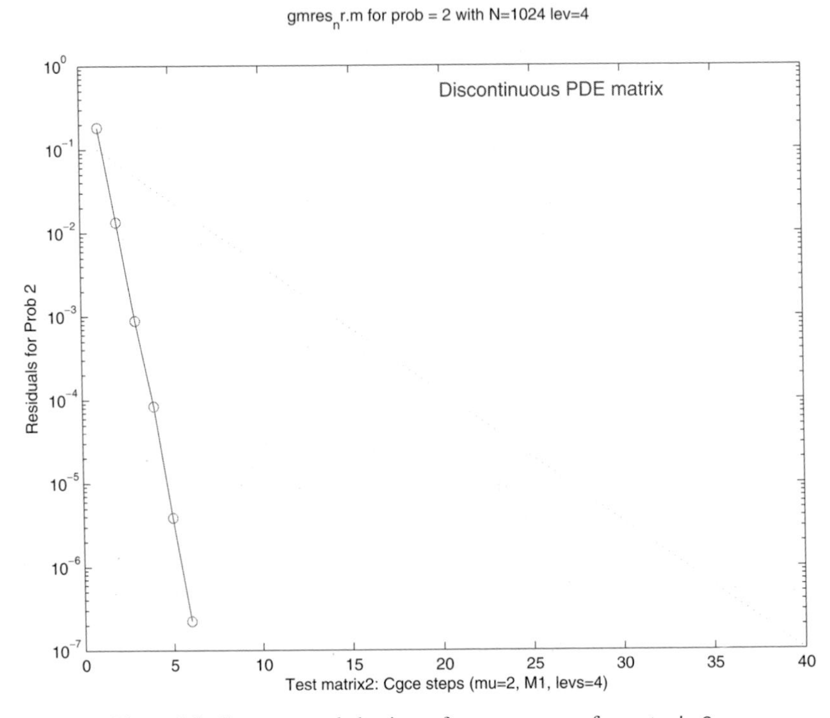

Figure 9.7. Convergence behaviour of `gmres_nr.m` for `matrix2`.

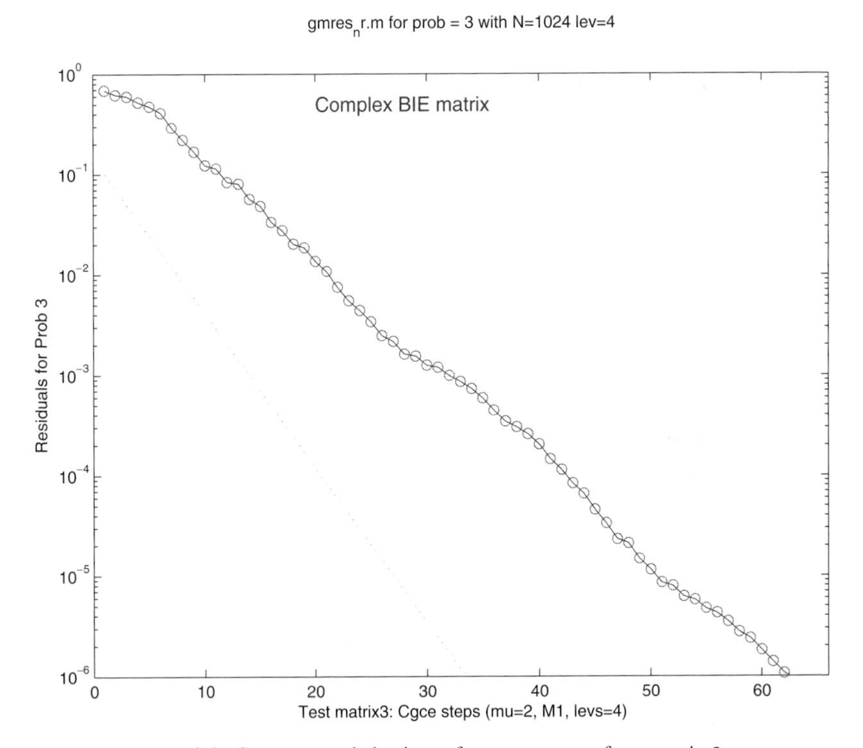

Figure 9.8. Convergence behaviour of gmres_nr.m for matrix3.

9.6 Discussion of the accompanied Mfiles

The literature does not seem to have similar software similar to that described.

This book has supplied these Mfiles for investigating preconditioning-related issues.

[1] gmres_nr.m – Implement Method 1 as described above (of Algorithm 9.3.2).

[2] richa_nr.m – Implement Method 2 as described above (of Algorithm 9.3.2).

[3] run9.m – The drive Mfile for experimenting gmres_nr.m and richa_nr.m.

[4] iwts.m – The sparse version of an inverse DWT (compare to the full matrix version ifwt.m).

10

Implicit wavelet preconditioners [T7]

In fact, advancements in the area (of wavelets) are occurring at such a rate that the very meaning of "wavelet analysis" keeps changing to incorporate new ideas.

BJÖRN JAWERTH and WIM SWELDENS *SIAM Review*, Vol. 36 (1994)

As indicated in Chapter 8, it is often viable to apply a DWT to a sparse linear system to derive a wavelet type preconditioner. In this chapter we consider a new way of obtaining the same wavelet preconditioner *without* applying a DWT. The assumption is that the sparse representation of A in a single scale finite element basis is already available and the corresponding wavelet $\widetilde{A}$ is less sparse than A.

Our idea is to work with this sparse matrix A in order to implicitly compute the representation $\widetilde{A}$ of A and its preconditioner $\widetilde{M}$ in the wavelet basis. Thus the main advantage is that the new strategy removes the costs associated with forming the wavelet matrix $\widetilde{A}$ explicitly and works with a sparse matrix A directly while making full use of the robust preconditioning property of wavelets. The general difficulty of specifying a suitable pattern $\mathcal{S}$ (Chapter 5) is resolved in the new setting. The obtained preconditioners are good sparse approximations to the inverse of A computed by taking advantage of the compression obtained by working in a wavelet basis. In fact, efficient application to both sparse and dense A can be considered as shown in the following.

10.1 Introduction

We consider the fast solution of the linear system (1.1) i.e.

$$Ax = b \tag{10.1}$$

where A is a large $n \times n$ matrix. Such linear systems arise in finite element or finite difference discretizations where typically A is sparse, and in boundary element discretizations where typically A is dense. The previous chapters have discussed various preconditioning techniques to accelerate the GMRES method. The challenge cases are when A is ill conditioned and especially indefinite.

We shall consider improving the inverse type preconditioners (Chapter 5), using the wavelet idea but without applying it explicitly. Recall that system (10.1) may be replaced by the left preconditioned linear system

$$MAx = Mb, \tag{10.2}$$

or the right preconditioned linear system

$$AMy = b, \tag{10.3}$$

or the left and right preconditioned linear system

$$M_2 A M_1 y = M_2 b. \tag{10.4}$$

It is the latter system that we shall be concerned with in this chapter.

The following observations motivate the new preconditioning idea.

(1) **SPAI**. Sparse approximate inverse techniques (Chapter 5) are widely used to produce preconditioners M. The prescribed sparsity pattern $\mathcal{S}$ for M is a detrimental factor in the success of a SPAI approach. The pattern found from applying an adaptive process is generally reliable but the method is expensive while the cheaper PSM approach is practical but not always robust for a large class of problems. Besides, there is no sound theory to support PSM for a small index power.

(2) **Wavelets**. When A is a discretization of a pseudo-differential operator one can view A^{-1} as the discretization of an integral operator whose kernel acts like the Green's function associated with A (as in (8.27)). The inverse A^{-1} is

typically dense but representation in a wavelet basis compresses A^{-1} when the Green's function is smooth away from the diagonal. The compression of the matrix A^{-1} in the wavelet basis allows it to be well approximated by a sparse matrix. Furthermore, A^{-1} has a special *finger pattern* in the wavelet basis and the finger pattern of the large entries is predictable, i.e. S can be specified a priori in the context of SPAI. However designing generally efficient schemes and adaptive quadratures for a wavelet method is not yet a trivial task.

(3) **Least squares method**. The usual procedure of obtaining the F-norm minimization of $\|AM - I\|$ may be generalized to accommodate the related minimization $\|AM - L\|$ for any matrix L. We shall consider the case of $L = W^T$, with W a DWT matrix.

We shall propose to compute a SPAI type preconditioner, making use of the sparsity pattern predictability of a wavelet matrix.

As before, denote representation in the wavelet basis with $\widetilde{\cdot}$ so that, in the wavelet basis, (10.1) becomes

$$\widetilde{A}\widetilde{x} = \widetilde{b}. \tag{10.5}$$

One can approximate $\widetilde{A^{-1}}$ with a finger-patterned sparse approximate inverse of $\widetilde{A}$ because $\widetilde{A^{-1}} = \widetilde{A}^{-1}$.

Remark 10.1.1. The WSPAI preconditioner *Section* 8.3 for (10.5) restricts the pattern of the sparse approximate inverse to banded form or block diagonal. This simplifies the computation but interactions between wavelets at different scales are omitted in the calculation of the preconditioner. This is in line with the theory of Dahmen and Kunoth [168], who show that diagonal preconditioning yields a condition number that is independent of the number of unknowns.

However, as observed by [157], sometimes diagonal preconditioning is not sufficient for efficient iterative solution and it is suggested to compute finger-patterned sparse approximate inverses for (10.5) in the slightly different context where (10.5) is obtained directly from the underlying partial differential equation (PDE) or boundary integral equation (BIE) with discretization using wavelet basis functions. Use of the full finger pattern for the sparse approximate inverse means that all interactions between scales are included in this preconditioner. This makes the preconditioner more robust but can sometimes make the computation of the sparse approximate inverse prohibitively expensive because a small number of dense columns or rows can be present in the finger pattern.

Without using analytical information, one may compute a sparse approximate inverse of $\widetilde{A}$ where the sparsity pattern of the approximate inverse is determined adaptively as in [75]. The adaptive sparse approximate inverse algorithm does not require the entries of $\widetilde{A}$ to be available. and allows second generation wavelets to be used efficiently. As noted before for example in [147], adaptive sparse approximate inverse algorithms can be more expensive to implement than sparse approximate inverse algorithms for which the pattern of the approximate inverse is known in advance. Outside the sparse approximate inverse framework, the preconditioners from Chapter 9 do not require $\widetilde{A}$ to be computed.

The new preconditioner will take advantage of the structure possessed by A^{-1} when it is represented in the wavelet basis. Like Cohen and Masson [157] we incorporate all of the interactions between wavelets at different scales, which is sometimes necessary to produce a robust preconditioner. We describe strategies that overcome the computational difficulties presented by the subsequent sparsity pattern of the preconditioner. Many of our computations involve A rather than $\widetilde{A}$ because our formulation applies the discrete wavelet transform implicitly. When A is sparse it is often sparser than $\widetilde{A}$ and working with A rather than $\widetilde{A}$ reduces the cost of computing the sparse approximate inverse. The cost of computing $\widetilde{A}$ is also removed.

Before proceeding, we clarify the main notation. A typical discrete wavelet transform is applied to the vector $x \in \mathbb{R}^n$ as done in (1.48). For simplicity we assume $n = 2^N$ for some $N \in \mathbb{Z}$, but the theory in this paper applies whenever $n = 2^N p$ for some $N, p \in \mathbb{Z}$. Let $c_0, c_1, \ldots, c_{m-1}$ and $d_0, d_1, \ldots, d_{m-1}$ be the low pass and high pass filter coefficients of the transform. Then the level $L \leq N$ discrete wavelet transform of x is the vector $\widetilde{x} = (\mathbf{s}^L, \mathbf{d}^L, \mathbf{d}^{L-1}, \ldots, \mathbf{d}^1)$, that can be rewritten as

$$\mathbf{s}^k = U_k \mathbf{s}^{k-1}, \qquad \mathbf{s}^0 = x$$

where

$$U_k = \begin{bmatrix} c_0 & c_1 & c_2 & \cdots & c_{m-1} & & & & & \\ & & c_0 & c_1 & c_2 & \cdots & c_{m-1} & & & \\ & & & \vdots & & & & & & \\ c_2 & \cdots & c_{m-1} & & & & & & c_0 & c_1 \end{bmatrix}$$

is an $n/2^k \times n/2^{k-1}$ matrix and

$$\mathbf{d}^k = V_k \mathbf{s}^{k-1}$$

where

$$V_k = \begin{bmatrix} d_0 & d_1 & d_2 & \dots & d_{m-1} & & & & \\ & & d_0 & d_1 & d_2 & \dots & d_{m-1} & & \\ & & & & \vdots & & & & \\ d_2 & \dots & d_{m-1} & & & & & d_0 & d_1 \end{bmatrix}$$

is an $n/2^k \times n/2^{k-1}$ matrix. When $m - 1 > n/2^{k-1}$ the entries wrap around, for example, when $m = 4$ the 1×2 level N matrices are

$$U_N = \begin{bmatrix} c_0 + c_2 & c_1 + c_3 \end{bmatrix},$$
$$V_N = \begin{bmatrix} d_0 + d_2 & d_1 + d_3 \end{bmatrix};$$

alternatively one may prefer to restrict m (or increase n) so that $m - 1 \le n/2^{k-1}$.

The level L DWT can be written

$$\widetilde{x} = Wx = W_L W_{L-1} \dots W_1 x \tag{10.6}$$

with, similar to (1.49),

$$W_k = \left[\begin{array}{c|c} U_k & \\ V_k & 0 \\ \hline 0 & I_{n-n/2^{k-1}} \end{array}\right].$$

The rest of this chapter is structured as follows. In Section 10.2 we review the wavelet sparse approximate inverse preconditioner of [116] as from Section 8.3. In Section 10.3 we introduce our new preconditioner that avoids transforming A and in Section 10.4 we discuss how to implement our new algorithm and how to compute the entries of the wavelet basis vectors, which are required to implement the new method, before commenting on the relative costs of the new preconditioner in relation to the preconditioner of [116]. In Section 10.5 we consider modifications to these preconditioners for the case when A is dense while in Section 10.6 we present some theory on the conditioning of the new preconditioned iteration matrix. In Section 10.8 we present some numerical experiments, before we summarize our supplied Mfile in Section 10.9.

10.2 Wavelet-based sparse approximate inverse

As shown in [157] and Section 8.2, for a wide class of PDE problems, A^{-1} is sparse in the wavelet basis. Noting this we review the preconditioning method of [116].

Let W be as in (10.6). Since $\widetilde{A^{-1}} = WA^{-1}W^T = (WAW^T)^{-1} = \widetilde{A}^{-1}$, Chan, Tang and Wan [116] propose a sparse approximate inverse preconditioner $\widetilde{M}$

for $\widetilde{A}$ computed by solving a minimization problem analogous to minimizing $\|AM - I\|_F$ in Chapter 5 as (8.34) i.e.

$$\begin{aligned} \|AM - I\|_F &= \|WAW^T WMW^T - I\|_F \\ &= \|\widetilde{A}\widetilde{M} - I\|_F \end{aligned} \tag{10.7}$$

Algorithm 8.3.5 implementing this idea made the simplification on the sparsity pattern of $\widetilde{M}$, which in turn restricted its robustness.

As established in [157], for a wide class of pseudo-differential equations, both $\widetilde{A}$ and $\widetilde{A}^{-1}$ have a finger pattern which is not fully used by Algorithm 8.3.5 or [116]. Thus a sparse approximate inverse of $\widetilde{A}$ is justified and a finger pattern for $\widetilde{M}$ can usually be prescribed. Finger-patterned matrices, though sparse, can have a small number of dense columns and rows. As noted by [232] the presence of dense columns in $\widetilde{M}$ require that care be taken in its computation. Reducing the sparsity pattern of $\widetilde{M}$ to block diagonal, as described by [116], is one way to solve this problem but relinquishes robustness because interactions between scales are disregarded.

In this chapter we present a new and cheaper way of using wavelets to give a sparse preconditioner with a predictable pattern but without computing $\widetilde{A}$. Interactions between wavelets at different scales are included.

10.3 An implicit wavelet sparse approximate inverse preconditioner

We describe a new wavelet sparse approximate inverse preconditioner that is closely related to the preconditioner in Algorithm 8.3.5 but the requirement that $\widetilde{A}$ be available is removed and the wavelet transform is applied implicitly. This is facilitated by changing the way in which the sparse approximate inverse is computed. Additionally we use the whole finger pattern for the preconditioner to ensure that all interactions between scales are included.

Observe that

$$\begin{aligned} \|\widetilde{A}\widetilde{M} - I\|_F &= \|WAW^T WMW^T - I\|_F \\ &= \|WAMW^T - I\|_F \\ &= \|WW^T AMW^T - W^T\|_F \\ &= \|A\widehat{M} - W^T\|_F \end{aligned} \tag{10.8}$$

where $\widetilde{M} = WMW^T$ and $\widehat{M} = MW^T = W^T\widetilde{M}$. Here the new idea is that once $\widehat{M}$ is computed, the solution (10.1) becomes that of

$$WA\widehat{M}\,\widetilde{y} = Wb \tag{10.9}$$

which does not require transforming A, b before hand. Note that system (10.9) can be solved by the two-sided solver gmrest_k.m (see Algorithm 3.6.18).

It is not difficult to see that $\widehat{M}$ has a predictable sparsity pattern, because $\widetilde{M}$ has a predictable sparsity pattern. Minimizing $\|A\widehat{M} - W^T\|_F$ reduces to n independent least squares problems in the same way as minimizing $\|AM - I\|_F$. These observations lead to a new algorithm.

Algorithm 10.3.2. (Wavelet SPAI without transforms).

(1) Compute $\widehat{M}$, the minimizer of $\|A\widehat{M} - W^T\|_F$ subject to the prescribed sparsity pattern for $\widehat{M}$.
(2) Solve $WA\widehat{M}\widetilde{y} = Wb$.
(3) Compute $x = \widehat{M}\widetilde{y}$.

Here $\widehat{M}$ is a quasi-approximation to A^{-1} represented from the wavelet basis to the standard, single scale basis. We establish in Theorem 10.6.3 that this matrix possesses a band pattern like that shown in Figure 10.1(b) and which is analogous to the finger pattern shown in Figure 10.1(a). In fact W possesses such a band pattern and our practical experience has shown that the pattern of W provides a suitable pattern for $\widehat{M}$.

In fact we have implicitly applied a one-sided wavelet transform to M, compared with the two-sided wavelet transform of M in Algorithm 8.3.5. In this way we achieve compression and structure in A^{-1} and like the Cohen and Masson [157] preconditioner, our preconditioner includes interactions between

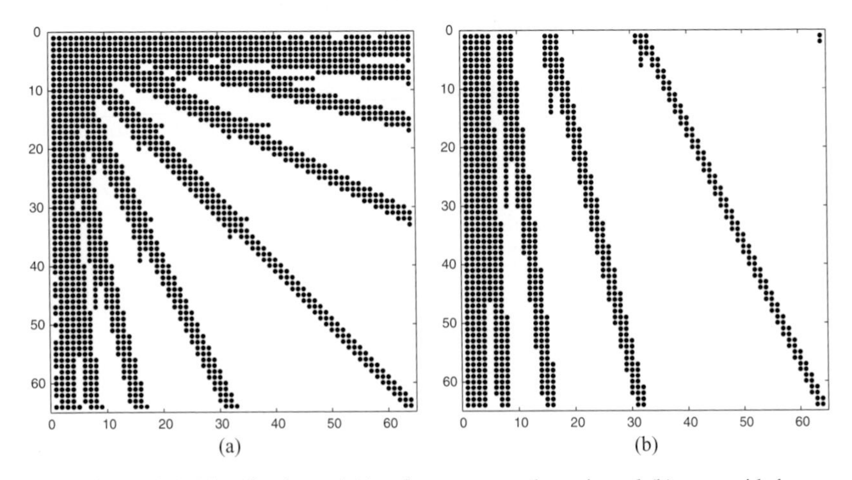

Figure 10.1. Visualizations of (a) a finger patterned matrix and (b) a one-sided finger patterned matrix. (See also Figure 10.2.)

different scales. The one-sided application of the wavelet transform provides less compression than the two-sided application but our practical experience is that the compression achieved is sufficient.

In general the preconditioner $\widehat{M}$ is unsymmetric, even when A is symmetric. Symmetrization techniques such as the one presented for Algorithm 8.3.5 are precluded because the nonsymmetric pattern of $\widehat{M}$ is dictated by the decay of the inverse of A and the properties of the wavelet basis. Though in general it is more expensive to solve unsymmetric systems than symmetric systems, use of this new strategy for symmetric systems is justified because construction of the preconditioner in Algorithm 10.3.2 is cheaper than construction of the preconditioner in Algorithm 8.3.5.

10.4 Implementation details

We now discuss how to solve in practice the least squares problem in Algorithm 10.3.2. Analysis of the least squares problem reveals two important considerations for practical implementation of the algorithm. We show how to address these considerations by modifying slightly how the preconditioner is constructed, and by finding cheaper ways of computing the wavelet matrix W. We finish with some comments about the complexity of computing the matrices $\widetilde{A}$ and W.

Consider the problem of minimizing $\|A\widehat{M} - W^T\|_F$ at Step 1 of Algorithm 10.3.2. Let S_j be the set of indices of the nonzero entries of $\widehat{m}_j$ where $\widehat{m}_j$ is the jth column of $\widehat{M}$. The classical approach to solving the minimization problem is to observe that

$$\begin{aligned}\|A\widehat{m}_j - w_j\|_2 &= \|A(:, S_j)\widehat{m}_j(S_j) - w_j\|_2 \\ &= \|A(T_j, S_j)\widehat{m}_j(S_j) - w_j(T_j)\|_2\end{aligned}$$

where T_j indexes the nonzero rows of $A(:, S_j)$ and w_j is the jth column of W^T. With the reduced QR factorization $A(T_j, S_j) = QR$ we have

$$\|A(T_j, S_j)\widehat{m}_j(S_j) - w_j(T_j)\|_2 = \|QR\widehat{m}_j(S_j) - w_j(T_j)\|_2, \quad (10.10)$$

which can be minimized by solving $R\widehat{m}_j(S_j) = Q^T w_j(T_j)$ by back substitution.

We make two observations. Firstly, the entries $w_j(T_j)$ must be known. Secondly, the principal cost of minimizing (10.10) is that of computing the QR factorization of $A(T_j, S_j)$. This requires $O(|T_j||S_j|^2)$ arithmetic operations, becoming $O(|S_j|^3)$ when A is banded because then $|T_j| \sim |S_j|$. Algorithm 10.3.2

becomes impractical in its basic form when $L = N$ because then $|S_j| = n$ for some columns j. The following modifications address the second observation.

1. Use a wavelet transform of level $L < N$. Note $\max_{j=1,n} |S_j| \approx 2^L$.
2. Compute $\widehat{m}_j$ by minimizing (10.10) whenever $|S_j|/n > \rho$ for some parameter $\rho \in (0, 1)$. Otherwise compute $\widehat{m}_j$ by solving $A\widehat{m}_j = w_j$ using a small number of GMRES steps.
3. Restrict S_j. This approach is used by [116]. We do not use this approach because we aim to include the off diagonal finger pattern into our preconditioner.

To illustrate the issue of sparsity of W^T (as well as the one-sided finger patterns), we show some results on Figure 10.2 for some selected DWT levels. Clearly

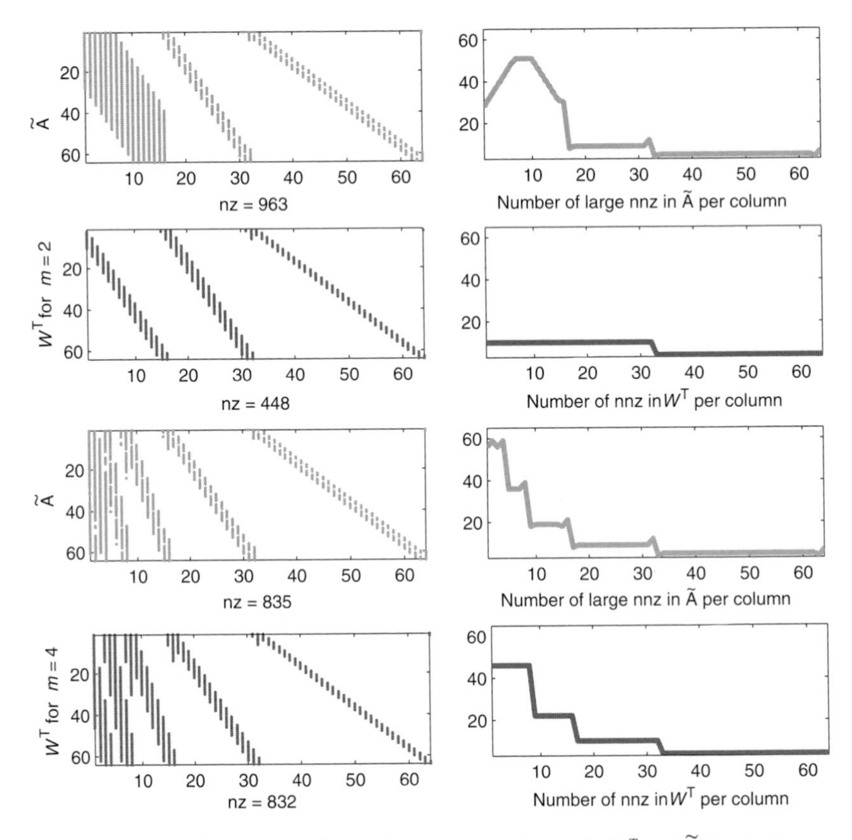

Figure 10.2. Illustration of the density of nonzeros in matrix W^T and $\widetilde{A}$(threshold of 0.08) for A $=$ cz(64, 2) with two different levels. Clearly there is more compression shown in $\widetilde{A}$ and more density in W^T as the level increases.

there is a fine balance between the density of nonzeros per column and the wavelet level used.

The ability to compute the necessary entries of W^T is crucial. Many discrete wavelet transforms are applied level by level, with

$$W^T = W_1^T W_2^T \dots W_{L-1}^T W_L^T.$$

The matrices W_i^T represent single levels of the transform and are often readily available, although W^T may not be. Computation of W^T by multiplying the factors W_i^T is possible because W^T and its factors are sparse.

More generally, the kth column of W^T can be computed by inverse-transforming the kth Euclidean vector e_k using any transform algorithm. Only $O(L)$ wavelets need to be computed in this way because the wavelets at each level are translations of each other – thus only one wavelet at each level must be computed.

Complexity issues. It remains to comment on the relative costs of Algorithms 8.3.5 and 10.3.2. Sparse approximate inverse and setup are considered. In computing the sparse approximate inverse in Algorithm 8.3.5, the restricted pattern for $\widetilde{M}$ yields $|S_j| = c$ for some constant c. However $|T_j|$ is difficult to estimate because it depends on the pattern of $\widetilde{A}$. Nevertheless, one can bound the costs of computing $\widetilde{M}$ by $O(n^2c^2)$. A realistic estimate is $O(nc^2)$ because there are few dense columns of $\widetilde{A}$.

In Algorithm 10.3.2, recall that $|T_j| \sim |S_j|$ when A is banded. The cost of computing $\widehat{M}$ is then bounded by $O(n[\max_{j=1,n} |S_j|]^3)$. Note that $\max_{j=1,n} |S_j| \approx 2^L$, yielding a cost $O(n2^{3L})$. When L is kept constant the cost becomes linear in n.

Setup for Algorithm 8.3.5 involves computing $\widetilde{A}$. Setup for Algorithm 10.3.2 involves computing W. The former requires wavelet forward-transforming $2n$ vectors. The latter requires inverse wavelet transforming $O(L) \le O(\log_2 n)$ vectors. W may be stored and reused for other problems of the same size. We will show that $\widetilde{A}$ can be computed in $O(12mn\alpha)$ arithmetic operations when A has bandwidth α, and W can be computed in $O(4m(m-1)n)$ arithmetic operations. $\widetilde{A}$ and W may be computed naively by multiplication of sparse matrices. This is easily implemented but introduces a computational overhead, although the arithmetic operation estimates will be realized. Computation of $\widetilde{A}$ remains more expensive than that of W.

Algorithm 1.6.18 is the typical pyramid algorithm for the forward discrete wavelet transform. Let $\alpha \in \mathbb{Z}$ be large compared with $m/2$. Then the complexity of transforming a vector with α nonzeros is approximately $O(2m \times 2\alpha[1 - 1/2^L])$ and the transform of the vector contains approximately $2\alpha[1 - 1/2^L]$

nonzeros. It follows that transforming a banded matrix with α nonzeros in each column involves transforming n columns each with α nonzeros and then n rows each with $2\alpha[1 - 1/2^L]$ nonzeros. The complexity of this transformation is approximately

$$O\left(2mn \times \left[2\alpha[1 - 1/2^L] + 2\left(2\alpha[1 - 1/2^L]\right)[1 - 1/2^L]\right]\right) \approx O(12mn\alpha).$$

Algorithm 1.6.19 is the typical pyramid algorithm for the inverse discrete wavelet transform. The complexity of inverse transforming the Euclidean vector whose nonzero entry appears in s^l or d^l is approximately $O(2m(m-1)2^l)$. Thus the complexity of transforming L Euclidean vectors whose nonzeros appear in $s^1, s^2, \ldots, s^L$ respectively is approximately

$$O\left(\sum_{l=1}^{L} 2m(m-1)2^l\right) \approx O(4m(m-1)2^L).$$

When $n = 2^L$ the second estimate becomes $O(4m(m-1)n)$ and we see that computing $\widetilde{A}$ is approximately $12\alpha/4(m-1)$ times more expensive than computing W.

10.5 Dense problems

Although our primary purpose is to precondition sparse problems arising from PDEs, we consider in this section the application of Algorithms 8.3.5 and 10.3.2 to dense problems where A is either fully available, or is partially available, for example, using the fast multipole method; see Section 3.8 and [248,405].

Consider the general problem of finding M that minimizes $\|BM - C\|_F$ subject to a specified sparsity pattern for M. When $B = \widetilde{A}$ and $C = I$ this corresponds to the minimization problem at Step (2) of Algorithm 8.3.5. When $B = A$ and $C = W^T$ this corresponds to the minimization problem at Step 1 of Algorithm 10.3.2.

Let S_j index the non-zero entries of m_j, where m_j is the jth column of M. A generalization of the argument in Section 10.4 shows that the minimization problem reduces to solving n problems of the form $R\,m_j(S_j) = Q^T c_j(T_j)$ where we have the QR factorization $B(T_j, S_j) = QR$, where T_j indexes the nonzero rows of $B(:, S_j)$, and where c_j is the jth column of C.

The principal task in solving each least squares problem is the QR factorization of the $t \times s = |T_j| \times |S_j|$ matrix $B(T_j, S_j)$. This operation has complexity $O(ts^2)$. Note that s depends upon the chosen sparsity pattern for M, and that t, loosely speaking, depends upon the sparsity of B.

When A is dense then $\widetilde{A}$ in Algorithm 8.3.5 is also dense. However, if A is also smooth then $\widetilde{A}$ will have a structure containing many small entries. Computation of a column of the sparse approximate inverse of a dense matrix using the least squares approach has complexity $O(ns^2)$ and for all but the smallest s is too expensive. Computation using the GMRES approach (for example, when the column is dense) is also expensive because it requires repeated matrix vector multiplication with a dense matrix.

We consider two adaptations to the method of computing the sparse approximate inverse. Write $B = S + F$ where S is sparse and F is dense.

1. B is replaced by S in the least squares minimization problems. Thus we minimize $\|SM - C\|_F$ instead of $\|BM - C\|_F$.
2. Dense columns of M are computed by solving $Sm_j = c_j$ instead of $Bm_j = c_j$ using GMRES.

We propose that (2) is always necessary for computing dense columns, but that (1) is required only when $s = |S_j|$ is not small.

In both cases the idea is to let S be a good sparse approximation to B so that M, which approximates S^{-1}, is a good preconditioner for B. A similar approach is taken by Chen [131] who shows that if the continuous operator $\mathcal{S}$ underlying S is bounded and the continuous operator $\mathcal{F}$ underlying F is compact then $\mathcal{S}^{-1}(\mathcal{S} + \mathcal{F}) = \mathcal{I} + \mathcal{S}^{-1}\mathcal{F}$ and $\mathcal{S}^{-1}\mathcal{F}$ is compact. One expects the corresponding matrix $S^{-1}(S + F)$ to have the clustered eigenvalues associated with the discretization of such an operator.

The choice of S and F is crucial. The following strategies are proposed in [94].

1. *Algebraic.* S is obtained from B by discarding entries whose absolute value falls below a specified threshold.
2. *Topological.* When B is obtained by a finite element discretization, S is obtained by discarding those entries b_{ij} of B whose corresponding nodes i and j are not (at some given level) mesh neighbours.
3. *Geometric.* When B is obtained by a finite element discretization, S is obtained by discarding those entries of b_{ij} of B whose corresponding nodes i and j are separated by more than a specified distance.

For Algorithm 8.3.5 we propose Strategy 1 because Strategies 2 and 3 do not directly apply since the entries in $\widetilde{A}$ do not directly relate to connections between nodes in a mesh (unless one works with the sparse graph $G(\widetilde{A})$ obtained algebraically). However, analogous strategies to 2 and 3 can be developed by replacing relationships between nodes with relationships between supports of wavelet basis functions, see for example Hawkins and Chen [271].

For Algorithm 10.3.2 one could apply any of these strategies. In some circumstances details of the mesh may be available but the entries of A are not explicitly known. Then $\widetilde{A}$ cannot easily be computed and our method is particularly useful. Under this context Carpentieri *et al.* [94] propose strategies 2 and 3. This situation arises, for example, in particle simulations or integral equation formulations for Laplace's equation when they are solved using the fast multipole method [248,405]. Here matrix vector multiplication with A can be performed cheaply and to a specified accuracy but A is not computed explicitly. One way to obtain columns of A for the purposes of computing a least squares sparse approximate inverse is to multiply Euclidean vectors by A. This naive approach may be very expensive, because all columns of A will be needed, though they will not be needed or stored simultaneously when the preconditioner is sparse. A better way to compute the sparse approximate inverse in this case might be to use Algorithm 10.3.2 with Adaptation 1, replacing A in the sparse approximate inverse computations by a sparse approximation S to A. The entries of S might be obtained using only near field contributions associated with the singularity.

10.6 Some theoretical results

In this section we justify the banded pattern chosen for $\widehat{M}$ and present a theoretical bound on $\|WA\widehat{M} - I\|_F$. This provides a measure of the conditioning of the iteration matrix $WA\widehat{M}$.

The following theorem establishes that $A^{-1}W$, which we approximate by $\widehat{M}$, has the banded pattern illustrated in Figure 10.1(b). Let $\{\psi_i\}_{i=1}^n$ be a wavelet basis in L_2 with m vanishing moments. Such properties of wavelets are discussed fully by Strang and Nguyen [441]. Let I_i denote the support of ψ_i, and suppose that $\|\psi_i\|_2 = 1$. Let $\{h_i\}_{i=1}^n$ be a single scale basis in L_2, with $\bar{I}_i$ denoting the support of h_i, and with $\|h_i\|_2 = 1$.

Theorem 10.6.3. *Suppose that the Green's function $K(x, y)$ satisfies*

$$\left|\frac{\partial^m K}{\partial x^m}(x, y)\right| \le \frac{C_m}{|x - y|^{m+1}}, \quad \text{for } x \ne y \tag{10.11}$$

for some constant C_m. Let $\widehat{N}$ be the discrete operator of $K(x, y)$ from the wavelet basis $\{\psi_j\}_{j=1}^n$ to the single scale basis $\{h_i\}_{i=1}^n$. Then

$$|\widehat{N}_{ij}| \le b(I_j, \bar{I}_i)$$

with

$$b(I_j, \bar{I}_i) = C|I_j|^{(2m+1)/2} \frac{1}{\text{dist}(I_j, \bar{I}_i)^{m+1}}$$

for some constant C.

Before the proof of Theorem 10.6.3, we remark that such a proof is typical of wavelet compression results where Taylor expansions are combined with applying the vanishing moments' conditions (see [441, Ch.7] and [13]).

Proof. We begin by expanding K as a Taylor series in x about $x_0 = \inf I_j$ to give

$$K(x, y) = K(x_0, y) + (x - x_0)\frac{\partial K}{\partial x}(x_0, y) + \cdots + \frac{(x - x_0)^{m-1}}{(m-1)!}\frac{\partial^{m-1} K}{\partial x^{m-1}}(x_0, y) + R(x, y)$$

where the remainder

$$R(x, y) = \frac{(x - x_0)^m}{m!}\frac{\partial^m K}{\partial x^m}(\xi, y)$$

for some $\xi \in I_j$. All terms in the Taylor series, except for the remainder term, are orthogonal to ψ_j because of the vanishing moments properties of the wavelets.

Now

$$\widehat{N}_{ij} = \int_{\bar{I}_i}\int_{I_j} K(x, y)\,\psi_j(x)\,dx\,h_i(y)\,dy,$$

and

$$\begin{aligned}
\int_{I_j} K(x, y)\,\psi_j(x)\,dx &= \int_{I_j} R(x, y)\,\psi_j(x)\,dx \\
&= \int_{I_j} \frac{(x - x_0)^m}{m!}\frac{\partial^m K}{\partial x^m}(\xi, y)\,\psi_j(x)\,dx \\
&= \frac{\partial^m K}{\partial x^m}(\xi, y)\int_{I_j} \frac{(x - x_0)^m}{m!}\,\psi_j(x)\,dx.
\end{aligned}$$

It follows that

$$\begin{aligned}
&\int_{\bar{I}_i}\int_{I_j} K(x, y)\,\psi_j(x)\,dx\,h_i(y)\,dy \\
&= \int_{\bar{I}_i} \frac{\partial^m K}{\partial x^m}(\xi, y)\int_{I_j} \frac{(x - x_0)^m}{m!}\,\psi_j(x)\,dx\,h_i(y)\,dy \\
&= \left(\int_{I_j} \frac{(x - x_0)^m}{m!}\,\psi_j(x)\,dx\right)\int_{\bar{I}_i} \frac{\partial^m K}{\partial x^m}(\xi, y)\,h_i(y)\,dx.
\end{aligned}$$

The Cauchy–Schwartz inequality gives

$$\begin{aligned}
&\left|\int_{\bar{I}_i} \frac{\partial^m K}{\partial x^m}(\xi, y)\, h_i(y)\, dy\right| \\
&\leq \left(\int_{\bar{I}_i} \frac{\partial^m K}{\partial x^m}(\xi, y)^2\, dy\right)^{1/2} \left(\int_{\bar{I}_i} h_i(y)^2\, dy\right)^{1/2} \\
&\leq \left(\int_{\bar{I}_i} \frac{\partial^m K}{\partial x^m}(\xi, y)^2\, dy\right)^{1/2} \|h_i\|_2 \\
&= \left(\int_{\bar{I}_i} \frac{\partial^m K}{\partial x^m}(\xi, y)^2\, dy\right)^{1/2} \\
&\leq |\bar{I}_i|^{1/2} \sup_{y\in\bar{I}_i} \left|\frac{\partial^m K}{\partial x^m}(\xi, y)\right|.
\end{aligned}$$

A similar argument shows

$$\left|\int_{I_j} \frac{(x-x_0)^m}{m!}\, \psi_j(x)\, dx\right| \leq \frac{|I_j|^{(2m+1)/2}}{m!\,(2m+1)^{1/2}}.$$

Combining these gives

$$\begin{aligned}
\left|\int_{\bar{I}_i}\int_{I_j} K(x, y)\, \psi_j(x)\, dx\; h_i(y)\, dy\right| &\leq \frac{|I_j|^{(2m+1)/2}}{m!\,(2m+1)^{1/2}}\, |\bar{I}_i|^{1/2} \sup_{y\in\bar{I}_i} \left|\frac{\partial^m K}{\partial x^m}(\xi, y)\right| \\
&\leq \frac{|I_j|^{(2m+1)/2}}{m!\,(2m+1)^{1/2}}\, |\bar{I}_i|^{1/2} \frac{C_m}{\operatorname{dist}(I_j, \bar{I}_i)^{m+1}} \\
&\leq C|I_j|^{(2m+1)/2} \frac{1}{\operatorname{dist}(I_j, \bar{I}_i)^{m+1}}
\end{aligned}$$

where

$$C = C_m \frac{|\bar{I}_i|^{1/2}}{m!\,(2m+1)^{1/2}}.$$

Here we make use of the fact that the functions $\{h_i\}$ are a fixed scale basis and so $|\bar{I}_i|$ is constant. ■

Choose $\epsilon > 0$ and let $\mathcal{S} = \{(i, j) : |b(I_j, \bar{I}_i)| > \epsilon\}$. Denote by $\widehat{N}^{\mathcal{S}}$ the matrix obtained by cutting $\widehat{N}$ to the pattern $\mathcal{S}$. Then

$$\begin{aligned}
\max_{(i,j)} |\widehat{N}^{\mathcal{S}}_{ij} - \widehat{N}_{ij}| &\leq \max_{(i,j)\notin\mathcal{S}} |b(I_j, \bar{I}_i)| \\
&\leq \epsilon.
\end{aligned}$$

Theorem 10.6.4. *If we choose $\mathcal{S}$ as the sparsity pattern for $\widehat{M}$ then*

$$\|WA\widehat{M} - I\|_F \leq n\epsilon\|A\|_F.$$

Proof. Denote the jth column of $\widehat{N}^S$ by $\widehat{n}_j^S$. Similarly denote the jth column of $\widehat{M}$ by $\widehat{m}_j$ and the jth column of W by w_j.

Then

$$\begin{aligned}
\|WA\widehat{M} - I\|_F^2 &= \|A\widehat{M} - W^T\|_F^2 \\
&= \sum_{j=1}^{n} \|A\widehat{m}_j - w_j\|_2^2 \\
&\le \sum_{j=1}^{n} \|A\widehat{n}_j^S - w_j\|_2^2 \\
&= \| A\widehat{N}^S - W^T \|_F^2 \\
&\le \|\widehat{N}^S - A^{-1}W^T\|_F^2 \|A\|_F^2 \\
&\le \|\widehat{N}^S - \widehat{N}\|_F^2 \|A\|_F^2 \\
&\le n^2\epsilon^2 \|A\|_F^2.
\end{aligned}$$

■

Remark 10.6.5. These results give a guide to the effectiveness of $\widehat{M}$ for small ϵ. The eigenvalues of the preconditioned matrix will have a desirable distribution when ϵ is small, because they will be clustered close to 1.

Theorem 10.6.4 suggests that one should choose ϵ sufficiently small, so that the pattern $\mathcal{S}$ leads to a sufficiently small bound $\|WA\widehat{M} - I\|_F^2$ for the preconditioner $\widehat{M}$. A similar bound for $\|\widetilde{A}\widetilde{M} - I\|_F$ is given by [116]. In practice, good results can be obtained when ϵ is large.

10.7 Combination with a level-one preconditioner

The idea of an implicit DWT may be combined with the method of [107] to allow the application of a level-one preconditioner to A:

$$Ax = b \quad \Rightarrow \quad D^{-1}Ax = D^{-1}b \quad \Rightarrow \quad D^{-1}A\widetilde{M} = W^T, \tag{10.12}$$

where D^{-1} is some suitable preconditioner that can 'smooth out' A.

Moreover, to avoid forming $D^{-1}A$, the implicit idea may be generalized to yield

$$A\widetilde{M} = DW^T \qquad \text{i.e.} \qquad \min_{\widetilde{M}} \|A\widetilde{M} - \widetilde{D}\|_F, \tag{10.13}$$

where $\widetilde{D} = DW^T$ is the one-sided DWT of D, which will be sparse if D is. On the other hand, regardless of the sparsity of $\widetilde{D}$, the solution of (10.13) will mainly involve A and hence be no more expensive than solving (10.7) or (10.8).

Once $\widetilde{M}$ (whose sparsity is again assured by an one-sided finger pattern; see Figure 10.2) is computed, the solution of (10.1) comes from applying Algorithm 3.6.18 to

$$\widetilde{D}^{-1}A\widetilde{M}\widetilde{y} = \widetilde{D}^{-1}b. \tag{10.14}$$

10.8 Numerical results

The new preconditioning strategy has been tested on a range of sparse and dense linear systems arising from discretization of partial differential equations and boundary integral equations. In [270], experiments were also presented on using other DWTs [9,447] In all cases the preconditioner has been shown to be very useful in reducing the number of GMRES iterations required to solve the linear system.

Here we have used the supplied Mfile `ch0_w2.m` to demonstrate the working of Algorithm 10.3.2 for accelerating GMRES [415,413] iterations. The test matrices from Section 4.9 are again used. In Figures 10.3

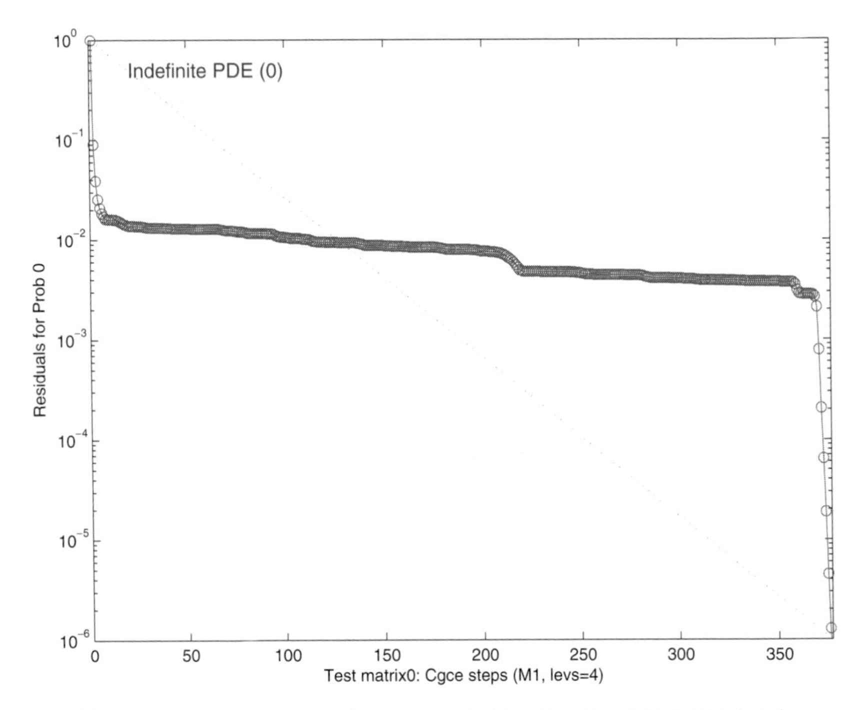

Figure 10.3. Convergence results for `matrix0` by Algorithm 10.3.2. This indefinite matrix is the hardest problem for all preconditioners.

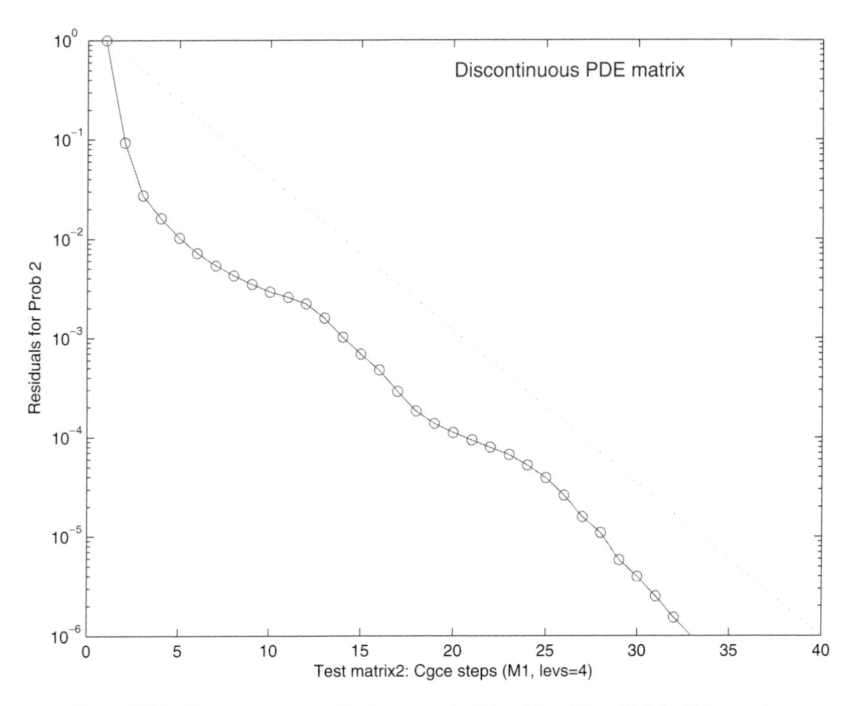

Figure 10.4. Convergence results for matrix2 by Algorithm 10.3.2. This matrix is known to be difficult for the usual application of the wavelet preconditioner [116].

(via ch0_w2(0,4,4)) and 10.4 (via ch0_w2(2,4,6)), we show the convergence results respectively for matrix0 and matrix2. Although the convergence does not appear too fast for matrix0, we did observe convergence while all the previous methods cannot solve this matrix problem. For matrix2, the improvement is also dramatic with the implicit preconditioner.

10.9 Discussion of the supplied Mfile

Combination of advantages of excellent sparsity from a conventional FEM and predictability in large elements of an inverse matrix from a wavelet method is a feasible approach for developing robust preconditioners. The proposed implicit method may be further refined, for example, in better treatment of coarse level-related preconditioners. There does not appear to exist software similar to the method described.

We supply a sample Mfile for experimenting our test matrices:

[1] ch0_w2.m – The main driver file for testing the matrices from Section 4.9. The Mfile may be adapted for testing other matrices.

Remark 10.9.6. ♦ So far we have discussed seven classes of general preconditioning techniques for solving (1.1) using an iterative method. For a given (and less well-known) problem type, one is extremely fortunate if one of the above classes can lead to an efficient solution. For harder problems, specific and problem-dependent techniques may have to be developed. In the next four chapters, we shall consider some useful application problems each offering at least one open problem that deserves further research work. For all numerical methods, parallel computing offers a potentially revolutionizing method to speeding up; this is discussed in Chapter 15.

11

Application I: acoustic scattering modelling

An important class of problems in which significantly higher accuracies are needed relate to low-observable applications, where the quantities of interest are small residuals of large incident fields.

OSCAR P. BRUNO. Fast, high-order, high-frequency integral methods for computational acoustics and electromagnetics. *Lecture Notes in Computational Science and Engineering 31.* Springer-Verlag (2003)

However, wavelet representation of an oscillating matrix appears to be as dense as the original, i.e. oscillatory kernels cannot be handled efficiently by representing them in wavelet bases.

A. AVERBUCH *et al.* On efficient computation of multidimensional oscillatory integrals with local Fourier bases. *Nonlinear Analysis* (2001)

The acoustic scattering modelling provides a typical example of utilizing a boundary element method to derive a dense matrix application as shown in Chapter 1. Such a physical problem is only a simple model of the full wave equations or the Maxell equations from electromagnetism. The challenges are:

(i) the underlying system is dense and non-Hermitian;
(ii) the kernel of a boundary integral operator is highly oscillatory for high wavenumbers, implying that a large linear system must be solved. The oscillation means that the fast multipole method and the fast wavelet methods are not immediately applicable.

This chapter reviews the recent work on using preconditioned iterative solvers for such linear systems arising from acoustic scattering modelling and points out the various challenges for future research work. We consider the following.

Different from (1.53), we shall consider the exterior problem as specified below.

11.1 The boundary integral equations for the Helmholtz equation in $\mathbb{R}^3$ and iterative solution

Whenever the underlying physics allows the use of time-harmonic dependence assumption, the space-dependent part of a function can be elegantly separated and the result is often a much simplified equation. In our case, we shall obtain the Helmholtz equation. This section will consider the fast solution issues.

11.1.1 The time-harmonic wave equation

The linear and dissipative wave equation for the velocity potential $u = u(p, t)$

$$\frac{\partial^2 u}{\partial t^2} + \gamma \frac{\partial u}{\partial t} - c^2 \nabla^2 u = 0 \tag{11.1}$$

that models the propagating acoustic waves reflected and diffracted off a bounded scatterer (e.g. submarine) under some incident field (e.g. sonar transducer), with the use of a time-harmonic dependence assumption

$$u = u(x, y, z, t) = u(p, t) = \phi(p) \exp(-i\omega t),$$

reduces (11.1) to

$$(\nabla^2 + k^2)\phi = 0, \tag{11.2}$$

where $k = \sqrt{\omega^2 + i\gamma}/c$ is the wavenumber, c the speed of sound in a homogeneous isotropic medium (often water) exterior to the scatterer and γ is the damping coefficient. Here the angular frequency $\omega = 2\pi f$ corresponds to the frequency f. Once $\phi(p)$ or $u(p, t)$ is determined, the velocity field and the pressure difference due to the wave disturbance can be computed respectively by $v = \text{grad} u/\rho$ and $p = -(\partial u/\partial t) - \gamma u$ with ρ the density of the medium. Here we are mainly interested in the case of no damping $\gamma = 0$ so $k = \omega/c \geq 0$. Note that, opposite (11.2), the resembling equation $(\nabla^2 - k^2)\phi = 0$ is much easier to solve since it will be more like a Laplace equation.

11.1.2 A unique BIE formulation in $\mathbb{R}^3$

To reformulate (11.2) defined in some infinite domain Ω_+, exterior to the closed surface $\partial\Omega$, into a boundary integral equation defined over $\partial\Omega$ only, we may use the indirect approach with the layer potentials as shown in Section 1.7 or the direct approach with the Green's theorems (1.57) and (1.71); see [16,159]. In any case, we need to specify the boundary conditions. At the finite surface $\partial\Omega$, these may be one of the following

(i) Dirichlet: $\phi(p) = f(p)$; $\quad (f(p) = 0$ for an acoustically soft scatterer)

(ii) Neumann: $\dfrac{\partial\phi}{\partial n}(p) = f(p)$; $(f(p) = 0$ for an acoustically rigid scatterer)

(iii) Impedance: $\dfrac{\partial\phi}{\partial n}(p) + \lambda\phi(p) = f(p)$, (a more realistic assumption on the scatterer)

where $\lambda = i\chi\rho\omega$ with χ the acoustic impedance and $n = n_p$ is the unit outward normal to $\partial\Omega$ at p. The wave potential ϕ is consisted of two components: $\phi(p) = \phi_s(p) + \phi_{inc}(p)$ with $\phi_s(p)$ the scattered and radiated wave and $\phi_{inc}(p)$ the incident field. The commonly used Sommerfeld radiation condition ensures that all waves are outgoing and gradually damped at infinity

$$\begin{cases} \lim_{r\to\infty} r\left\{\dfrac{p}{r}\cdot \text{grad } \phi_s(p) - ik\phi_s(p)\right\} = \lim_{r\to\infty} r\left\{\dfrac{\partial\phi_s(r)}{\partial r} - ik\phi_s(r)\right\} = 0, \\ \lim_{r\to\infty} r\left\{\dfrac{\partial\phi_{inc}(r)}{\partial r} + ik\phi_{inc}(r)\right\} = 0, \end{cases} \tag{11.3}$$

where $p = (x, y, z)$, $r = |p|$ and $p \cdot \text{grad } \phi = r\frac{\partial\phi}{\partial r}$.

An application of Green's second theorem (1.71) leads to [16,159]

$$\int_{\partial\Omega}\left(\phi(q)\frac{\partial G_k(p,q)}{\partial n_q} - G_k(p,q)\frac{\partial\phi(q)}{\partial n_q}\right) dS_q = \begin{cases} \dfrac{1}{2}\phi(p) - \phi_{inc}(p), & p \in \partial\Omega \\ \phi(p) - \phi_{inc}(p), & p \in \Omega_+ \end{cases} \tag{11.4}$$

where the free-space Green's function, or the fundamental solution,

$$G_k(p,q) = \frac{e^{ik|p-q|}}{4\pi|p-q|} = \frac{e^{ikr}}{4\pi r} \tag{11.5}$$

satisfies the Helmholtz equation in the sense of $\nabla^2 G_k(p,q) + k^2 G_k(p,q) = \delta(p-q)$ and n_q is as before the unit outward normal to $\partial\Omega$ at q. However, for any boundary condition over $\partial\Omega$, it is well known that (11.4) does not possess a unique solution for the characteristic (or resonance) wavenumbers, the location of which will depend on the shape of the surface $\partial\Omega$. As $\phi_{\text{inc}}(p)$ is always given (or can be measured), we shall assume $\phi_{\text{inc}}(p) = 0$ to simplify the discussion.

The Burton and Miller method [83] for overcoming the non-uniqueness problem consists of differentiating (11.4) along the normal at p to give

$$\int_{\partial\Omega}\left(\phi(q)\frac{\partial^2 G_k(p,q)}{\partial n_p \partial n_q} - \frac{\partial G_k(p,q)}{\partial n_p}\frac{\partial \phi(q)}{\partial n_q}\right) dS_q = \frac{1}{2}\frac{\partial \phi(p)}{\partial n_p} \tag{11.6}$$

and then taking a linear combination of (11.4) and (11.6) in the form

$$\begin{aligned} -\frac{1}{2}\phi(p) + \int_{\partial\Omega}\phi(q)\left(\frac{\partial G_k(p,q)}{\partial n_q} + \alpha\frac{\partial^2 G_k(p,q)}{\partial n_p \partial n_q}\right) dS_q \\ = \frac{\alpha}{2}\frac{\partial \phi(p)}{\partial n_p} + \int_{\partial\Omega}\frac{\partial \phi(q)}{\partial n_q}\left(G_k(p,q) + \alpha\frac{\partial G_k(p,q)}{\partial n_p}\right) dS_q \end{aligned} \tag{11.7}$$

where α is a coupling constant. It can be shown that provided that the imaginary part of α is non-zero then (11.7) has a unique solution [83] for all real and positive k. However, this formulation has introduced the kernel function $(\partial^2 G_k(p,q))/(\partial n_p \partial n_q)$ which has a $1/(|p-q|^3)$ hyper-singularity. Below we introduce different ways of overcoming this singularity; the use of finite part integration is a separate method [267]. Here, although (11.7) provides a unique formulation for any wavenumber k, one may find that in the literature some authors still choose to experiment on some fixed wavenumber k and carry on using (11.4) or (11.6) alone.

11.1.3 The piecewise constant approximation for collocation

To implement the piecewise constant collocation method, it is possible to use the result from [353]

$$\int_{\partial\Omega}\frac{\partial^2 G_k(p,q)}{\partial n_p \partial n_q} dS_q = k^2 \int_{\partial\Omega} G_k(p,q) n_p \cdot n_q \, dS_q, \tag{11.8}$$

in order to write the hyper-singular integral as

$$\begin{aligned} \int_{\partial\Omega}\phi(q)\frac{\partial^2 G_k(p,q)}{\partial n_p \partial n_q} dS_q = \int_{\partial\Omega}(\phi(q)-\phi(p))\frac{\partial^2 G_k(p,q)}{\partial n_p \partial n_q} dS_q \\ + k^2\phi(p)\int_{\partial\Omega} G_k(p,q) n_p \cdot n_q \, dS_q. \end{aligned} \tag{11.9}$$

Let ϕ and $\partial\phi/\partial n$ be approximated by interpolatory functions of the form

$$\phi(q) = \sum_{j=1}^{m}\phi_j \psi_j(q), \qquad \frac{\partial \phi(q)}{\partial n} = \sum_{j=1}^{m} v_j \psi_j(q) \tag{11.10}$$

where $\{\psi_1, \psi_2, \ldots, \psi_m\}$ are a set of linearly independent basis functions. Substituting (11.10) into (11.7) and (11.9), and applying the collocation method at points $\{p_1, p_2, \ldots, p_m\}$, usually chosen such that $p_i \in S_i$ (defined below), yields

$$\sum_{j=1}^{m} \phi_j \left\{ -\frac{1}{2}\psi_j(p_i) + \int_{\partial\Omega} \left(\psi_j(q)\frac{\partial G_k(p_i,q)}{\partial n_q} + \alpha\Big(\psi_j(q) - \psi_j(p_i)\Big) \times \frac{\partial^2 G_k(p_i,q)}{\partial n_p \partial n_q} \right) dS_q + \psi_j(p_i)\alpha k^2 \int_{\partial\Omega} G_k(p_i,q) n_p \cdot n_q \, dS_q \right\}$$

$$= \sum_{j=1}^{m} v_j \left[\frac{\alpha}{2}\psi_j(p_i) + \int_{\partial\Omega} \psi_j(q) \left(G_k(p_i,q) + \alpha \frac{\partial G_k(p_i,q)}{\partial n_p} \right) dS_q \right]. \tag{11.11}$$

For a piecewise constant approximation, the ψ_j functions are

$$\psi_j(p) = \begin{cases} 1, & p \in S_j \\ 0, & \text{otherwise}, \end{cases} \tag{11.12}$$

where $\{S_1, S_2, \cdots, S_m\}$ is some partition of the surface $\partial\Omega$ into m nonoverlapping elements. For this choice of basis functions then all the integrals appearing in (11.11) are at worst weakly singular, since in this case $\psi_j(p) - \psi_j(q)$ is zero whenever p and q are in the same element. However, if a higher order approximation is used then the term involving the second derivative becomes a Cauchy principal value integral which is difficult to evaluate. Hence, the piecewise constant approximation collocation method has been widely used in practice.

11.1.4 The high-order Galerkin approximation

In [268], we presented an alternative way of using high-order methods via the Galerkin formulation to solve the integral equation (11.7). Essentially our method reformulates the hyper-singular integral equation into a weakly-singular one. As the operator in (11.6) or (11.7) is not Hermitian, we do not have to construct complex basis functions because the resulting linear system will not be Hermitian. Therefore in our Galerkin method, we used the well-known real trial basis functions $\psi_j(p) \in H^{\tau-1/2}(\partial\Omega)$ and construct interpolation functions with complex coefficients as in (11.8).

Taking the inner product of (11.7) with each basis function in turn gives

$$\sum_{j=1}^{m} \phi_j \int_{\partial\Omega} \psi_i(p) \left[\frac{1}{2}\psi_j(p) + \int_{\partial\Omega} \left(\psi_j(q)\frac{\partial G_k(p,q)}{\partial n_q} + \alpha\psi_j(q)\frac{\partial^2 G_k(p,q)}{\partial n_p \partial n_q} \right) dS_q \right] dS_p$$

$$= \sum_{j=1}^{m} v_j \int_{\partial\Omega} \psi_i(p) \left[\frac{\alpha}{2}\psi_j(p) + \int_{\partial\Omega} \psi_j(q) \Big(G_k(p,q)\alpha \frac{\partial G_k(p,q)}{\partial n_p} \Big) dS_q \right] dS_p \tag{11.13}$$

where the last integral on the first line

$$\int_{\partial\Omega}\int_{\partial\Omega}\psi_i(p)\,\psi_j(q)\,\frac{\partial^2 G_k(p,q)}{\partial n_p\,\partial n_q}\,dS_q\,dS_p \tag{11.14}$$

is hyper-singular (to be dealt with) while all other integrals are weakly singular.

To derive a formulation where we only have to evaluate integrals which are at worst weakly singular, the following result is essential:

Lemma 11.1.1. (Harris and Chen [268]).
For piecewise polynomial basis functions $\{\psi_j\}$ of any order τ,

$$\begin{aligned}\int_{\partial\Omega}\int_{\partial\Omega}&\psi_i(p)\,\psi_j(q)\,\frac{\partial^2 G_k(p,q)}{\partial n_p\,\partial n_q}\,dS_q\,dS_p\\ &=\frac{1}{2}\int_{\partial\Omega}\int_{\partial\Omega}[\psi_i(p)-\psi_i(q)]\,[\psi_j(q)-\psi_j(p)]\,\frac{\partial^2 G_k(p,q)}{\partial n_p\,\partial n_q}\,dS_q\,dS_p\\ &\quad+k^2\int_{\partial\Omega}\psi_i(p)\,\psi_j(p)\left[\int_{\partial\Omega}n_p\cdot n_q\;G_k(p,q)\;dS_q\right]dS_p,\end{aligned} \tag{11.15}$$

where all the integrals on the right-hand side are weakly singular.

With this result, we can derive a Galerkin formulation that enables us to use basis functions of any order τ without having to construct special quadrature rules to deal with the hyper-singular integrals. The reformulated form of (11.13) is the following

$$\begin{aligned}\sum_{j=1}^{m}\phi_j\int_{\partial\Omega}\Bigg[&\psi_i(p)\left(-\frac{1}{2}\psi_j(p)+\int_{\partial\Omega}\psi_j(q)\frac{\partial G_k(p,q)}{\partial n_q}dS_q\right)\\ &+\frac{1}{2}\alpha\int_{\partial\Omega}\left((\psi_i(p)-\psi_i(q))\,(\psi_j(q)-\psi_j(p))\,\frac{\partial^2 G_k(p,q)}{\partial n_p\partial n_q}\right)dS_q\\ &+\;\alpha\psi_i(p)\psi_j(p)k^2\int_{\partial\Omega}G_k(p,q)n_p\cdot n_q dS_q\Bigg]dS_p \end{aligned}\tag{11.16}$$

$$=\sum_{j=1}^{m}v_j\int_{\partial\Omega}\psi_i(p)\left[\frac{\alpha}{2}\psi_j(p)+\int_{\partial\Omega}\psi_j(q)\left(G_k(p,q)+\alpha\frac{\partial G_k(p,q)}{\partial n_p}\right)dS_q\right]dS_p.$$

In [268] when iterative solution of (11.16) is developed, we generalize the work of [136] to this case and essentially use the idea of Section 4.7 for constructing a sparse preconditioner. Below our attention will be mainly paid to the collocation method.

11.1.5 The high-order collocation approximation using finite part integration

The finite part integration (in the sense of Hadamard) is essentially to assign a meaning to a singular integral that is otherwise infinity. We now review how to compute a finite part integral, before applying the idea to (11.7).

For a suitably smooth function $f(s)$, consider the problem of evaluating an integral of the form

$$\int_a^b \frac{f(s)}{(s-a)^2}\, ds. \tag{11.17}$$

If $F(s)$ is the anti-derivative of $f(s)/((s-a)^2)$ then the finite part of (11.17) is defined as $F(b)$. In order to approximate (11.17) we need to construct a quadrature rule of the form

$$\int_a^b \frac{f(s)}{(s-a)^2}\, ds = \sum_{j=1}^{m} w_j f(s_j). \tag{11.18}$$

The simplest way of doing this is to use the method of undetermined coefficients, where the quadrature points $s_1, s_2, \ldots, s_m$ take assigned values and then (11.18) is made exact for $f(s) = (s-a)^i, i = 0, 1, \ldots, m-1$. The resulting equations can be written in matrix form as $A\underline{w} = \underline{g}$ where

$$A_{ij} = (s_j - a)^{i-1} \quad \text{and} \quad g_i = \int_a^b (s-a)^{i-3}\, ds, \quad 1 \le i,\ j \le m, \tag{11.19}$$

and, to compute g_i for $i = 1,\ 2$, we can define the following finite part integrals

$$\int_a^b \frac{1}{(s-a)^2}\, ds = -\frac{1}{b-a} \quad \text{and} \quad \int_a^b \frac{1}{s-a}\, ds = \ln(b-a). \tag{11.20}$$

Now consider using finite part integration for (11.7). One way of interpreting the hyper-singular operator is finding out an appropriate change of variables so that *the interested integral is effectively reduced to one-dimensional integral with a hyper-singular integrand and the singularity located at an end point.* Suppose that the surface $\partial\Omega$ is approximated by N nonoverlapping triangular quadratic surface elements $S_1, S_2, \ldots, S_N$. If $p_i, i = 1, \ldots 6$, denote the position vectors of the six nodes used to define a given element, then that element can be mapped into a reference element in the (u, v) plane

$$p(u, v) = \sum_{j=1}^{6} \psi_j(u, v) p_j \qquad 0 \le u \le 1,\ 0 \le v \le 1 - u. \tag{11.21}$$

Now suppose that the singular point corresponds to the point (u_1, v_1) in the (u, v) plane. The reference element is divided into three triangular sub-elements by connecting the point (u_1, v_1) to each of the vertices of the reference triangle. We

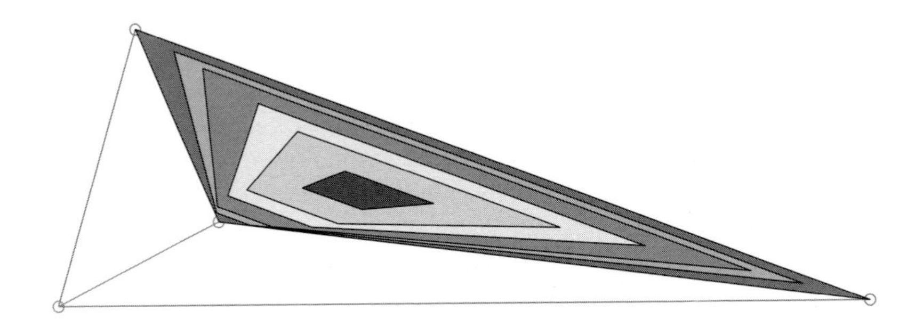

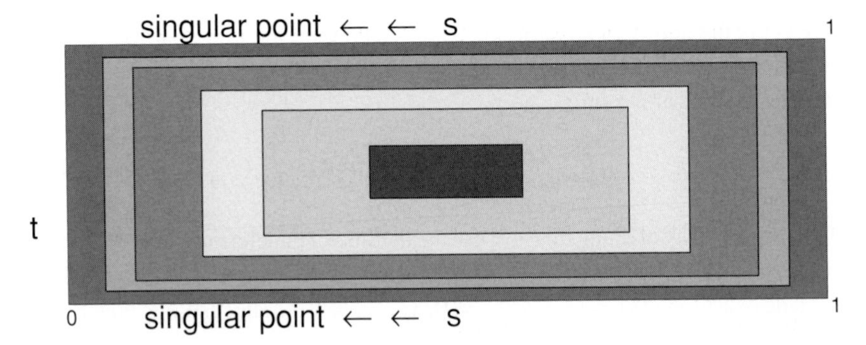

Figure 11.1. Illustration of mapping a triangular domain (top plot in shaded region) to a rectangular domain (the bottom plot) in order to expose the radial singularity away from the mid-point in the triangle.

need to decide on a new coordinate transform in which the singularity is only present in one variable. Since the singularity is in the radial direction (away from point (u_1, v_1)), a suitable transform must be polar like.

Within each sub-element we now propose the following transformation

$$\left.\begin{aligned} u(s,t) &= (1-s)u_1 + stu_2 + s(1-t)u_3 \\ v(s,t) &= (1-s)v_1 + stv_2 + s(1-t)v_3 \end{aligned}\right\} \quad 0 \le s, t \le 1 \tag{11.22}$$

where (u_2, v_2) and (u_3, v_3) are the other vertices of the current sub-triangle. As illustrated in Figure 11.1, clearly, the only way for $(u(s,t), v(s,t)) = (u_1, v_1)$ is for $s = 0$ as these are bi-linear functions of s and t. Further, the mapping (11.21) is bijective as its Jacobian is nonzero for all (u, v) of interest. Hence the only way that p can equal the singular point is when $s = 0$. After some manipulation it is possible to show that $r(s,t) = |p(s,t) - q| = s\tilde{r}(s,t)$ where $\tilde{r}(s,t) \neq 0$ for $0 \le s, t \le 1$. The Jacobian of the transformations (11.21) and (11.22) can be written as

$$J = s\sqrt{D_1^2 + D_2^2 + D_3^2}\,\Big|(u_2 - u_1)(v_3 - v_1) - (u_3 - u_1)(v_2 - v_1)\Big| = J_s s \tag{11.23}$$

where

$$D_1 = \begin{vmatrix} \frac{\partial y}{\partial u} & \frac{\partial z}{\partial v} \\ \frac{\partial y}{\partial v} & \frac{\partial z}{\partial u} \end{vmatrix}, \quad D_2 = \begin{vmatrix} \frac{\partial z}{\partial u} & \frac{\partial x}{\partial v} \\ \frac{\partial z}{\partial v} & \frac{\partial x}{\partial u} \end{vmatrix}, \quad D_3 = \begin{vmatrix} \frac{\partial x}{\partial u} & \frac{\partial y}{\partial v} \\ \frac{\partial x}{\partial v} & \frac{\partial y}{\partial u} \end{vmatrix}. \tag{11.24}$$

Hence, denoting $\partial\Omega_e$ the current sub-element, we can write

$$\int_{\partial\Omega_e} \frac{f(q)}{r^3}\, dS = \int_0^1 \frac{1}{s^2} \left[\int_0^1 \frac{f(q(u(s,t), v(s,t)))}{(\tilde{r}(s,t))^3} J_s\, dt \right] ds. \tag{11.25}$$

We note that the inner integration (with respect to t) is nonsingular and can be approximated by an appropriate quadrature rule. However, the outer integral needs to be interpreted as a Hadamard finite part in a (desirably) single variable s. Thus equation (11.7) is tractable.

11.1.6 A new high-order collocation approximation

In order to develop the collocation method that does not use any finite part integration, we have to tackle the issue of hyper-singularity in (11.7) differently. It turns out that we can devise such an approach that uses the Green theorem to 'undo' the double differentiation $(\partial^2 G_k(p,q))/(\partial n_p \partial n_q)$, as shown in [133].

11.1.6.1 Interpretation of the hyper-singular integral

First we prove the following lemma which will be used to transform the hyper-singular integral to an integral with a weak singularity.

Lemma 11.1.2. *Let* $\mathbf{a} \in \mathbb{R}^3$ *be a constant vector. Then*

$$\begin{aligned}\int_{\partial\Omega} \mathbf{a}\cdot(q-p)\frac{\partial^2 G_k(p,q)}{\partial n_p \partial n_q}\, dS_q = & \int_{\partial\Omega} \mathbf{a}\cdot n_q \frac{\partial G_k(p,q)}{\partial n_p}\, dS_q \\ & - k^2 \mathbf{a}\cdot \int_{\Omega} (q-p)\frac{\partial G_k(p,q)}{\partial n_p}\, dV_q \\ & - \frac{\mathbf{a}\cdot n_p}{2}\end{aligned} \tag{11.26}$$

where Ω *is some interior region in* $\mathbb{R}^3$ *with closed surface* $\partial\Omega$.

Proof. Assume that p lies on smooth part of $\partial\Omega$. That is, n_p is well defined. Let $\partial\Omega_\varepsilon$ be the surface of a sphere centered on p, $\tilde{\Omega}$ be Ω excluding the interior of $\partial\Omega_\varepsilon$ and $\partial\tilde{\Omega}$ denote the surface bounding $\tilde{\Omega}$. We notice that, for $q \in \tilde{\Omega}$, it holds that

$$\nabla_q^2 \frac{\partial G_k(p,q)}{\partial n_p} + k^2 \frac{\partial G_k(p,q)}{\partial n_p} = 0. \tag{11.27}$$

Applying the Green's second theorem

$$\int_{\partial\tilde{\Omega}} \left(\phi_1 \frac{\partial \phi_2}{\partial n_q} - \phi_2 \frac{\partial \phi_2}{\partial n_q} \right) dS_q = \int_{\tilde{\Omega}} \left(\phi_1 \nabla_q^2 \phi_2 - \phi_2 \nabla_q^2 \phi_1 \right) dV_q \tag{11.28}$$

with $\phi_1 = \mathbf{a} \cdot (q - p)$ and $\phi_2 = (\partial G_k(p, q))/\partial n_p$ leads to (taking note of (11.27) in the first integral on the right hand side of (11.28))

$$\begin{aligned} \int_{\partial\tilde{\Omega}} \left[\mathbf{a} \cdot (q - p) \frac{\partial^2 G_k(p, q)}{\partial n_p \partial n_q} - \mathbf{a} \cdot n_q \frac{\partial G_k(p, q)}{\partial n_p} \right] dS_q \\ = \int_{\tilde{\Omega}} -k^2 \mathbf{a} \cdot (q - p) \frac{\partial G_k(p, q)}{\partial n_p} dV_q. \end{aligned} \tag{11.29}$$

The surface integral in (11.29) can be expressed as the sum of integral over the relevant part of the surface $\partial\Omega_\varepsilon$ and the integral over the remaining part of $\partial\tilde{\Omega}$, say $\partial\tilde{\Omega}_\varepsilon$. As $\varepsilon \to 0$, then $\partial\tilde{\Omega}_\varepsilon \to \partial\Omega$, $\tilde{\Omega} \to \Omega$ and

$$\lim_{\varepsilon \to 0} \int_{\partial\Omega_\varepsilon} \left[\mathbf{a} \cdot (q - p) \frac{\partial^2 G_k(p, q)}{\partial n_p \partial n_q} - \mathbf{a} \cdot n_q \frac{\partial G_k(p, q)}{\partial n_p} \right] dS_q = \frac{\mathbf{a} \cdot n_p}{2}. \tag{11.30}$$

Hence we obtain

$$\begin{aligned} \int_{\partial\Omega} \left[\mathbf{a} \cdot (q - p) \frac{\partial^2 G_k(p, q)}{\partial n_p \partial n_q} - \mathbf{a} \cdot n_q \frac{\partial G_k(p, q)}{\partial n_p} \right] dS_q \\ = \int_{\Omega} -k^2 \mathbf{a} \cdot (q - p) \frac{\partial G_k(p, q)}{\partial n_p} dV_q - \frac{\mathbf{a} \cdot n_p}{2} \end{aligned} \tag{11.31}$$

which can be re-arranged to give the desired result so our proof is completed. ■

11.1.6.2 An initial reformulation

We now consider how to make use of Lemma 11.1.2 to reformulate (11.7) into a weakly singular equation, by Green's theorem and introduction of gradient (tangent) variables.

Firstly, we rewrite the underlying hyper-singular integral as

$$\begin{aligned} &\int_{\partial\Omega} \phi(q) \frac{\partial^2 G_k(p, q)}{\partial n_p \partial n_q} dS_q \\ &= \int_{\partial\Omega} \left(\left\{ \phi(q) - \phi(p) - \nabla\phi(p)(q - p) \right\} \frac{\partial^2 G_k(p, q)}{\partial n_p \partial n_q} \right) dS_q \\ &\quad + \phi(p) \int_{\partial\Omega} \frac{\partial^2 G_k(p, q)}{\partial n_p \partial n_q} dS_q + \int_{\partial\Omega} \{ \nabla\phi(p)(q - p) \} \frac{\partial^2 G_k(p, q)}{\partial n_p \partial n_q} dS_q \end{aligned}$$

$$= \int_{\partial\Omega} \left(\{\phi(q) - \phi(p) - \nabla\phi(p)(q-p)\} \frac{\partial^2 G_k(p,q)}{\partial n_p \partial n_q} \right) dS_q$$
$$+ k^2 \phi(p) \int_{\partial\Omega} n_p \cdot n_q G_k(p,q) dS_q$$
$$+ \int_{\partial\Omega} \{\nabla\phi(p) \cdot (q-p)\} \frac{\partial^2 G_k(p,q)}{\partial n_p \partial n_q} dS_q.$$

Note that the first two integrals are weakly singular. It remains to consider reformulating the last singular integral.

To this end, taking $\mathbf{a} = \nabla\phi(p)$ and using Lemma 11.1.2, we obtain

$$\int_{\partial\Omega} \{\nabla\phi(p) \cdot (q-p)\} \frac{\partial^2 G_k(p,q)}{\partial n_p \partial n_q} dS_q$$
$$= \int_{\partial\Omega} \nabla\phi(p) \cdot n_q \frac{\partial G_k(p,q)}{\partial n_p} dS_q$$
$$-k^2 \int_{\Omega} \nabla\phi(p) \cdot (q-p) \frac{\partial G_k(p,q)}{\partial n_p} d\sigma_q - \frac{1}{2} \nabla\phi(p) \cdot n_p. \quad (11.32)$$

and thus the hyper-singular integral is successfully reformulated as follows

$$\int_{\partial\Omega} \phi(q) \frac{\partial^2 G_k(p,q)}{\partial n_p \partial n_q} dS_q$$
$$= \int_{\partial\Omega} \left(\{\phi(q) - \phi(p) - \nabla\phi(p)(q-p)\} \frac{\partial^2 G_k(p,q)}{\partial n_p \partial n_q} \right) dS_q$$
$$+ k^2 \phi(p) \int_{\partial\Omega} n_p \cdot n_q G_k(p,q) dS_q \quad (11.33)$$
$$+ \int_{\partial\Omega} \nabla\phi(p) \cdot n_q \frac{\partial G_k(p,q)}{\partial n_p} dS_q$$
$$- k^2 \int_{\Omega} \nabla\phi(p) \cdot (q-p) \frac{\partial G_k(p,q)}{\partial n_p} d\sigma_q.$$

Substituting (11.33) into (11.7) yields our initial weakly-singular reformulation:

$$-\frac{1}{2}\phi(p)$$
$$+ \int_{\partial\Omega} \left(\phi(q) \frac{\partial G_k(p,q)}{\partial n_q} + \alpha \{\phi(q) - \phi(p) - \nabla\phi(p) \cdot (q-p)\} \frac{\partial^2 G_k(p,q)}{\partial n_p \partial n_q} \right) dS_q$$
$$+ \left[\alpha k^2 \phi(p) \int_{\partial\Omega} n_p \cdot n_q G_k(p,q) dS_q + \alpha \int_{\partial\Omega} \nabla\phi(p) \cdot n_q \frac{\partial G_k(p,q)}{\partial n_p} dS_q \right]$$

$$-\alpha k^2 \int_\Omega \nabla\phi(p)\cdot(q-p)\frac{\partial G_k(p,q)}{\partial n_p}dS_q - \frac{\alpha}{2}\nabla\phi(p)\cdot n_p$$
$$= \left[\frac{\alpha}{2}\frac{\partial\phi(p)}{\partial n_p} + \int_{\partial\Omega}\frac{\partial\phi(q)}{\partial n_q}\left(G_k(p,q)+\alpha\frac{\partial G_k(p,q)}{\partial n_p}\right)dS_q\right]. \quad (11.34)$$

Although all of the integrals appearing on the right-hand side of (11.34) are at worst weakly singular, this formulation is not immediately useful for any subsequent solution. This is because the formulation has introduced a volume integral over the interior of the radiating or scattering object. One possible remedy of this problem is to apply Lemma 11.1.2 to a small closed surface $\partial\Omega_p$ (enclosing the relatively small volume Ω_p) associated with each collocation point p:

$$\int_{\partial\Omega}\phi(q)\frac{\partial^2 G_k(p,q)}{\partial n_p\partial n_q}dS_q = \int_{\partial\Omega\backslash\partial\Omega_p}\phi(q)\frac{\partial^2 G_k(p,q)}{\partial n_p\partial n_q}dS_q + \int_{\partial\Omega_p}\phi(q)\frac{\partial^2 G_k(p,q)}{\partial n_p\partial n_q}dS_q.$$

Note that for a fixed p, the first term on the right-hand side is nonsingular while the second term may be treated by Lemma 11.1.2. Then an equation similar to (11.34) could be derived and can lead to a viable solution approach since integration in a small and local volume is practical. However, instead of pursuing the idea further, we seek another more elegant reformulation replacing (11.34).

11.1.6.3 A new reformulation

We now follow the same idea in Lemma 11.1.2 but consider a specialized case, with a view of combining with a kernel substraction idea later.

Corollary 11.1.3. *Let $\mathbf{a}\in\mathbb{R}^3$ be a constant vector and $G_0(p,q) = 1/(4\pi|p-q|)$ as from (11.5). Then*

$$\int_{\partial\Omega}\mathbf{a}\cdot(q-p)\frac{\partial^2 G_0(p,q)}{\partial n_p\partial n_q}dS_q = \int_{\partial\Omega}\mathbf{a}\cdot n_q\frac{\partial G_0(p,q)}{\partial n_p}dS_q - \frac{\mathbf{a}\cdot n_p}{2}, \quad (11.35)$$

where Ω is some interior region in $\mathbb{R}^3$ with closed surface $\partial\Omega$.

This result is a consequence of Lemma 11.1.2.

Clearly (11.35) does not involve any volume integral but the kernel function G_0 is not yet directly related to (11.5) for the Helmholtz problem. It turns out that another use of the singularity subtraction idea will lead to our desired result.

First, express the hyper-singular integral as

$$\int_{\partial\Omega} \phi(q) \frac{\partial^2 G_k(p,q)}{\partial n_p \partial n_q} \, dS_q = \int_{\partial\Omega} \phi(q) \left(\frac{\partial^2 G_k(p,q)}{\partial n_p \partial n_q} - \frac{\partial^2 G_0(p,q)}{\partial n_p \partial n_q} \right) dS_q + \int_{\partial\Omega} \phi(q) \frac{\partial^2 G_0(p,q)}{\partial n_p \partial n_q} \, dS_q \tag{11.36}$$

where the first integral on the right-hand side of (11.36) is weakly singular and can be evaluated using an appropriate quadrature rule. The second integral on the right hand side of (11.36) can be written as

$$\begin{aligned} \int_{\partial\Omega} \phi(q) \frac{\partial^2 G_0(p,q)}{\partial n_p \partial n_q} \, dS_q = & \int_{\partial\Omega} \Big[\phi(q) - \phi(p) - (q-p) \cdot \nabla\phi(p) \Big] \\ & \times \frac{\partial^2 G_0(p,q)}{\partial n_p \partial n_q} dS_q \\ & + \phi(p) \int_{\partial\Omega} \frac{\partial^2 G_0(p,q)}{\partial n_p \partial n_q} dS_q \\ & + \int_{\partial\Omega} \nabla\phi(p) \cdot (q-p) \frac{\partial^2 G_0(p,q)}{\partial n_p \partial n_q} dS_q \end{aligned} \tag{11.37}$$

where the first integral on the right-hand side is weakly singular, the second integral is zero (using (11.8) with $k = 0$) and the third can be rewritten using (11.35) in Corollary 11.1.3 with $\mathbf{a} = \nabla\phi(p)$ to give

$$\begin{aligned} & \int_{\partial\Omega} \phi(q) \frac{\partial^2 G_0(p,q)}{\partial n_p \partial n_q} \, dS_q \\ & \quad = \int_{\partial\Omega} [\phi(q) - \phi(p) - (q-p) \cdot \nabla\phi(p)] \frac{\partial^2 G_0(p,q)}{\partial n_p \partial n_q} \, dS_q \\ & \quad + \int_{\partial\Omega} \nabla\phi(p) \cdot n_q \frac{\partial G_0(p,q)}{\partial n_p} \, dS_q - \frac{1}{2} \nabla\phi(p) \cdot n_p. \end{aligned} \tag{11.38}$$

Using this result, finally, our new weakly-singular reformulation for (11.7) is the following

$$\begin{aligned} -\frac{1}{2}\phi(p) + & \int_{\partial\Omega} \phi(q) \frac{\partial G_k(p,q)}{\partial n_p} \, dS_q \\ & + \alpha \int_{\partial\Omega} \phi(q) \left(\frac{\partial^2 G_k(p,q)}{\partial n_p \partial n_q} - \frac{\partial^2 G_0(p,q)}{\partial n_p \partial n_q} \right) dS_q \\ & + \alpha \int_{\partial\Omega} \Big[\phi(q) - \phi(p) - (q-p) \cdot \nabla\phi(p) \Big] \frac{\partial^2 G_0(p,q)}{\partial n_p \partial n_q} \, dS_q \end{aligned}$$

$$+\alpha \int_{\partial\Omega} \nabla\phi(p)\cdot n_q \frac{\partial G_0(p,q)}{\partial n_p}\, dS_q - \frac{\alpha}{2}\nabla\phi(p)\cdot n_p \qquad (11.39)$$
$$= \left[\frac{\alpha}{2}\frac{\partial\phi(p)}{\partial n_p} + \int_{\partial\Omega} \frac{\partial\phi(q)}{\partial n_q}\left(G_k(p,q) + \alpha\frac{\partial G_k(p,q)}{\partial n_p}\right) dS_q\right].$$

Evidently all integral operators are at worst weakly-singular and so are more amenable to effective numerical methods – in our case the collocation method.

The above formulation has again introduced the gradient function $\nabla\phi(p)$ (which is unusual in a BEM context) as in the previous subsection. However no volume integrals are involved. To compute the gradient function on the surface, if the surface element is parameterized in terms of the two variables u and v, then

$$\begin{cases} \dfrac{\partial\phi}{\partial u} = \dfrac{\partial\phi}{\partial x}\dfrac{\partial x}{\partial u} + \dfrac{\partial\phi}{\partial y}\dfrac{\partial y}{\partial u} + \dfrac{\partial\phi}{\partial z}\dfrac{\partial z}{\partial u}, \\ \dfrac{\partial\phi}{\partial v} = \dfrac{\partial\phi}{\partial x}\dfrac{\partial x}{\partial v} + \dfrac{\partial\phi}{\partial y}\dfrac{\partial y}{\partial v} + \dfrac{\partial\phi}{\partial z}\dfrac{\partial z}{\partial v}, \\ 0 = \dfrac{\partial\phi}{\partial x}n_x + \dfrac{\partial\phi}{\partial y}n_y + \dfrac{\partial\phi}{\partial z}n_z. \end{cases} \qquad (11.40)$$

where the final equation is obtained by using the property that the surface function does not vary in the direction perpendicular to the surface, that is $\partial\phi(p)/\partial n_p = 0$. The (u, v) coordinates of the current collocation point can be substituted into the system of equations (11.40) which can then be solved to give the gradient function at any (surface) collocation point later.

We shall use the collocation method to solve (11.7) via our new weakly-singular reformulation (11.39). As before, approximate both ϕ and $\partial\phi/\partial n$ by (11.10) where we now assume that the basis functions $\psi_1, \psi_2, \cdots, \psi_m$ can be of any high order (much higher than the commonly-used order 0 for constants), with m collocation points $p_1, p_2, \cdots, p_m$. Then the weakly-singular collocation method for solving (11.39) is

$$\sum_{j=1}^{m} \phi_j \left[-\frac{1}{2}\psi_j(p_i) + \int_{\partial\Omega} \psi_j(q) \frac{\partial G_k(p_i,q)}{\partial n_p}\, dS_q \right.$$
$$+\alpha \int_{\partial\Omega} \psi_j(q) \left(\frac{\partial^2 G_k(p_i,q)}{\partial n_p \partial n_q} - \frac{\partial^2 G_0(p_i,q)}{\partial n_{p_i} \partial n_q} \right) dS_q$$
$$+\alpha \int_{\partial\Omega} \left[\psi_j(q) - \psi_j(p_i) - (q - p_i)\cdot\nabla\psi_j(p_i)\right] \frac{\partial^2 G_0(p,q)}{\partial n_p \partial n_q}\, dS_q$$

$$+\alpha \int_{\partial\Omega} \nabla\psi_j(p_i)\cdot n_q \frac{\partial G_0(p_i,q)}{\partial n_p}\, dS_q - \frac{\alpha}{2}\nabla\psi_j(p_i)\cdot n_p \Bigg] \tag{11.41}$$

$$= \sum_{j=1}^{m} v_j \left[\frac{\alpha}{2}\psi_j(p_i) + \int_{\partial\Omega} \left(G_k(p_i,q) + \alpha \frac{\partial G_k(p_i,q)}{\partial n_p} \right) dS_q \right].$$

11.2 The low wavenumber case of a Helmholtz equation

The previous section has discussed various formulations and feasible numerical methods that would produce a linear system such as (1.1)

$$Ax = b$$

with a dense and non-Hermitian matrix $A \in \mathbb{C}^{n\times n}$.

As with all linear systems, iterative solvers become necessary whenever n is sufficiently large. However, for the Helmholtz equation, there exists a minimum number $n = n_{\min}$ demanded by each wavenumber k to avoid 'aliasing' effects as with the Nyquist principle in signal sampling. One way to find such a $n_{\min}$ is to allow at least $5 - 10$ mesh points in each wavelength[1]

$$\ell = \frac{c}{f} = \frac{2\pi c}{\omega} = \frac{2\pi}{k}. \tag{11.42}$$

Clearly for moderate wavenumbers k (say $k = 1$), such a minimum requirement on n can be easily satisfied – using any larger $n \geq n_{\min}$ would of course lead to higher resolution. *However for large wavenumbers k (say $k = 100$), such a minimum requirement on n may not be satisfied easily; if not, then the final solution may not make sense and then there is no point discussing errors since the solution is qualitatively incorrect.* We shall discuss this case in the next section.

If it is feasible to generate and store matrix A, there is a large selection of fast solvers available.

(1) The conjugate gradient methods of Chapter 3; see [17,472,14,136].
(2) The geometric multigrid methods of Chapter 6; see [259,125,132].
(3) The wavelet transform preconditioners of Chapters 8 and 10; see [130,205, 206].

Related to these techniques is the multilevel preconditioner [5], which may be designed in various ways (Chapter 6).

[1] This concept is quite easy to understand. For instance, consider $y = sin(x)$ in a typical period $[0, 2\pi]$. We do need at least five discrete points (samples) to represent y properly.

If it is not feasible to generate and store matrix A, there is still a choice of two competing fast solvers.

(1) The fast multipole methods of Chapter 3; see [365,395,255,240].
(2) The wavelet discretization methods; see [169,167,272].

11.3 The high wavenumber case of a Helmholtz equation

As remarked following (11.42), if the wavenumber k is large, the immediate consequence is that the number n from discretization must increase correspondingly because the above-mentioned n_{min} is large. From the analysis point of view, for an usual polynomial basis (in a BEM), the numerical error will be bounded by some power of $O(kh)$ where k is the wavenumber and h the largest mesh size. Clearly for small and moderate-sized k, the accuracy of the type $O(h)$ in some power is acceptable but extremely large k the numerical accuracy may be severely reduced.

Moreover, large wavenumbers also can deteriorate the performance of both the fast multipole methods and the wavelet discretization methods, since the usual assumptions by both methods on the smoothness of ϕ become increasingly invalid as ϕ starts to oscillate. This implies that the FMM expansions may need more than an acceptable number of terms and the wavelet matrices will become less sparse.

To design suitable numerical methods, one has to realize that the usual FEM spaces (of piecewise polynomials) are no longer sufficient to approximate oscillating functions. The oscillating functions must enter a new approximating space. In [350,35], the method of the partition of unity method was introduced to the finite element method, which essentially augment the usual spaces by oscillating functions. For solving the Helmholtz equations, the method has been considered by [327,78,121,387] to design the so-called plane waves-based basis functions:

$$\phi(p) = \sum_{j=1}^{n} P_j(\xi, \eta) \sum_{\ell=1}^{m_j} \alpha_j^{\ell} \exp\left(i \mathbf{k}_j^{\ell} \cdot \mathbf{r}\right), \tag{11.43}$$

to replace the usual approximating function (here in $\mathbb{R}^2$)

$$\phi(p) = \sum_{j=1}^{n} \alpha_j P_j(\xi, \eta)$$

where (ξ, η) defines the computational variables, $\mathbf{r} = (x(\xi, \eta), y(\xi, \eta))$ and

$\mathbf{k} = (k_1, k_2)$ to be specified. For simplicity, one may choose

$$\mathbf{k}_j^\ell = (k \cos\theta_{\ell j}, k \cos\theta_{\ell j}), \qquad \theta_{\ell j} = \ell \frac{2\pi}{m_j}, \quad \ell = 1, \ldots, m_j,$$

associated with a uniform distribution of wave (scattering) directions. There exist many other choices one can take in designing hybrid basis functions. It should be remarked that, although (11.43) is locally based, integrating oscillating functions is not a trivial matter. As with the nontrivial treatment of wavelet quadratures, naive integration will undo the benefits gained by the new methods.

Clearly another option is to allow some basis functions to be global and this will lead to less sparse matrices. It is anticipated that much work will be done in this area of high wavenumber modelling and simulation.

11.4 Discussion of software

The standard BEM has been implemented in the useful software.

- 'Boundary elements software' (by Stephen Kirkup [317]):

http : //www.soundsoft.demon.co.uk/tbemia.htm

- The commercial software BEASY

http : //www.beasy.com/index.html

can deal with complicated and industrial geometries.

There do not appear to exist public codes for high wavenumber modelling.

12
Application II: coupled matrix problems

> A clustered [eigenvalue] spectrum often translates in rapid convergence of GMRES.
>
> MICHELE BENZI and GENE H. GOLUB. *SIAM Journal on Matrix Analysis and Applications*, Vol. 24 (2003)

The coupled matrix problems represent a vast class of scientific problems arising from discretization of either systems of PDE's or coupled PDE's and integral equations, among other applications such as the Karush–Kuhn–Tucker (KKT) matrices from nonlinear programming [273,43]. The reader may be aware of the fact that many coupled (nonlinear) systems may be solved by Uzawa type algorithms [92,153,144,115], i.e. all equations are 'artificially' decoupled and solved in turns. A famous example of this strategy is the SIMPLE algorithm widely used in computational fluid dynamics along with finite volume discretization [473,334]. While there is much to do in designing better and more robust preconditioners for a single system such as (1.1), one major challenge in future research will be to solve the coupled problems many of which have only been tackled recently.

This chapter will first review the recent development on a general coupled system and then discuss some specific coupled problems. The latter samples come from a large range of challenging problems including elasticity, particle physics and electromagnetism. We shall discuss the following.

Section 12.1 Generalized saddle point problems
Section 12.2 The Oseen and Stokes saddle point problems
Section 12.3 The mixed finite element method
Section 12.4 Coupled systems from fluid structure interaction
Section 12.5 Elasto-hydrodynamic lubrication modelling
Section 12.6 Discussion of software and a supplied Mfile

12.1 Generalized saddle point problems

We first consider the general framework of a generalized saddle point problem [51,52,342]

$$\mathcal{A}\mathbf{x} = \begin{bmatrix} A & B^T \\ D & C \end{bmatrix} \begin{bmatrix} u \\ p \end{bmatrix} = \begin{bmatrix} f \\ g \end{bmatrix} = \mathbf{b} \tag{12.1}$$

to which the coupled fluid structure interaction is very much related and many other important classes of coupled problems belong [196]. Here $A \in \mathbb{R}^{n\times n}$, $B \in \mathbb{R}^{m\times n}$, $C \in \mathbb{R}^{m\times m}$, $D \in \mathbb{R}^{m\times n}$. It should be remarked that, to develop efficient preconditioners or iterative solvers, we have to put some restrictions on the type of the blocks A, B, C, D otherwise the whole matrix is simply arbitrary and without structures to explore. Although (12.1) defines the general notation for block matrix problems, some well-known problems may be further specified.

(1) Even for the same class of fluid problems (Section 12.2), matrix A may be either symmetric or unsymmetric. However, when A, C are symmetric, $D = B$ leads to a symmetric block matrix. When matrix A is unsymmetric, it may be advantageous to take $D = -B$ (discussed below) as $D = B$ can no longer lead to a symmetric block matrix.
(2) For the coupled fluid structure interaction (Section 12.4), A is symmetric but C is unsymmetric and $D \neq B$. This differs from the fluid problems (Section 12.2).
(3) The coupled matrix problem from elasto-hydrodynamic lubrication (Section 12.5) has no symmetry in its blocks, although each block is discretized from the familiar operators.

The case of an unsymmetric A implies that (12.1) is a challenging problem to precondition. In this case, the usual symmetry from $D = B$ is not useful and hence [52] proposes to replace (12.1) by

$$\mathcal{A}\mathbf{x} = \begin{bmatrix} A & B^T \\ -B & C \end{bmatrix} \begin{bmatrix} u \\ p \end{bmatrix} = \begin{bmatrix} f \\ g \end{bmatrix} = \mathbf{b}. \tag{12.2}$$

For this case of an unsymmetric A, one set of fairly general assumptions is made.

(i) $D = -B$ has full rank.
(ii) $\mathcal{N}(H) \cap \mathcal{N}(B) = \{0\}$.
(iii) $H = \frac{1}{2}(A + A^T)$ is positive semidefinite (i.e. $\lambda(H) \geq 0$).
(iv) C is symmetric positive semidefinite (i.e. including $C = 0$).

These conditions are sufficient to ensure that the coupled matrix $\mathcal{A}$ in (12.1) is nonsingular, $(\mathcal{A}\mathbf{v}, \mathbf{v}) = \mathbf{v}^T\mathcal{A}\mathbf{v} \geq 0$ for all $\mathbf{v} \in \mathbb{R}^{m+n}$ and $\Re(\lambda(\mathcal{A})) \geq 0$.

The above coupled system (12.2), or (12.1), with $\mathcal{A} \in \mathbb{R}^{(m+n)\times(m+n)}$, can be written in a symmetric and skew-symmetric form

$$\mathcal{A} = \mathcal{H} + \mathcal{S}, \qquad \mathcal{H} = \frac{1}{2}\left(\mathcal{A} + \mathcal{A}^T\right), \qquad \mathcal{S} = \frac{1}{2}\left(\mathcal{A} - \mathcal{A}^T\right), \quad (12.3)$$

whereas in the case of $\mathcal{A} \in \mathbb{C}^{(m+n)\times(m+n)}$ the splitting is similar

$$\mathcal{A} = \mathcal{H} + \mathcal{S}, \qquad \mathcal{H} = \frac{1}{2}\left(\mathcal{A} + \mathcal{A}^H\right), \qquad \mathcal{S} = \frac{1}{2}\left(\mathcal{A} - \mathcal{A}^H\right), \quad (12.4)$$

where $\mathcal{H}$ is Hermitian and $\mathcal{S}$ is skew-Hermitian (or $-i\mathcal{S}$ Hermitian). In either case, $\lambda(\mathcal{H}) = \Re(\lambda(\mathcal{H}))$ and $\lambda(\mathcal{S}) = i\Im(\lambda(\mathcal{S}))$ or $\lambda(-i\mathcal{S}) = \Re(\lambda(-i\mathcal{S}))$ as $(-i\mathcal{S})$ is Hermitian. One can show that[1] $\lambda_{\min}(\mathcal{H}) \leq \Re(\lambda(\mathcal{A})) \leq \lambda_{\max}(\mathcal{H})$ and $\lambda_{\min}(-i\mathcal{S}) \leq \Im(\lambda(\mathcal{A})) \leq \lambda_{\max}(-i\mathcal{S})$, where $-i = 1/i$ in the latter relationship is required even for the real case $\mathcal{A} \in \mathbb{R}^{(m+n)\times(m+n)}$. For the real case, the choices of $\mathcal{H}$ and $\mathcal{S}$ are

$$\mathcal{H} = \begin{bmatrix} H & \\ & C \end{bmatrix} \quad \text{and} \quad \mathcal{S} = \begin{bmatrix} S & B^T \\ -B & \end{bmatrix}, \quad (12.5)$$

where $S = \frac{1}{2}(A - A^T)$; in particular $H = A$ and $S = 0$ if A has symmetry.

Then from writing $\mathcal{A}\mathbf{x} = \mathbf{b}$ into

$$(\mathcal{H} + \alpha\mathcal{I}) - (\alpha\mathcal{I} - \mathcal{S})\mathbf{x} = \mathbf{b}, \quad \text{or} \quad (\mathcal{S} + \alpha\mathcal{I}) - (\alpha\mathcal{I} - \mathcal{H})\mathbf{x} = \mathbf{b},$$

the following alternating splitting scheme is proposed for (12.1) [52,38]

$$\begin{cases} (\mathcal{H} + \alpha I)\mathbf{x}^{(k+1/2)} = (\alpha\mathcal{I} - \mathcal{S})\mathbf{x}^{(k)} + \mathbf{b}, \\ (\mathcal{S} + \alpha I)\mathbf{x}^{(k+1)} = (\alpha\mathcal{I} - \mathcal{H})\mathbf{x}^{(k+1/2)} + \mathbf{b}, \end{cases} \quad (12.6)$$

which can also be written as a stationary iteration scheme

$$\mathbf{x}^{(k+1)} = \mathcal{T}_\alpha\mathbf{x}^{(k)} + c \quad (12.7)$$

with

$$\mathcal{T}_\alpha = (\mathcal{S} + \alpha\mathcal{I})^{-1}(\alpha\mathcal{I} - \mathcal{H})(\mathcal{H} + \alpha I)^{-1}(\alpha\mathcal{I} - \mathcal{S})$$

and $c = (\mathcal{S} + \alpha\mathcal{I})^{-1}\left[\mathcal{I} + (\alpha\mathcal{I} - \mathcal{H})(\mathcal{H} + \alpha\mathcal{I})^{-1}\right]\mathbf{b} = \left[\frac{1}{2\alpha}(\mathcal{H} + \alpha\mathcal{I})(\mathcal{S} + \alpha\mathcal{I})\right]^{-1}\mathbf{b}$. Under the above assumptions, it can be shown [52] that $\rho((1-\beta)\mathcal{I} + \beta\mathcal{T}_\alpha) < 1$ for all $\alpha > 0$ and $\beta \in (0, 1)$; $\beta = 1$ is permitted

[1] From $\mathcal{A} = \mathcal{H} + i(-i\mathcal{S})$ for either or complex case and $\lambda_j(\mathcal{A})x_j = \mathcal{A}x_j = \mathcal{H}x_j + i(-i\mathcal{S})x_j$, taking inner products with x_j (or multiplying x^H both sides) gives $\lambda_j(\mathcal{A})x_j^Hx_j = x_j^H\mathcal{H}x_j + ix_j^H(-i\mathcal{S})x_j$ so $\Re(\lambda_j(\mathcal{A})) = x_j^H\mathcal{H}x_j/(x_j^Hx_j)$, $\Im(\lambda_j(\mathcal{A})) = x_j^H(-i\mathcal{S})x_j/(x_j^Hx_j)$. Then the inequalities become apparent.

if $u^T Au > 0$ for all $u \in \mathbb{R}^n$ (i.e. positive real). Here $\alpha > 0$ compensates the semi-indefiniteness of A (or H).

As all stationary iterations imply a matrix splitting, equation (12.7) admit the following splitting (Section 3.3.1)

$$\mathcal{A} = M - N, \qquad \text{with} \quad \begin{cases} M = \dfrac{1}{2\alpha}(\mathcal{H} + \alpha\mathcal{I})(\mathcal{S} + \alpha\mathcal{I}), \\ N = \dfrac{1}{2\alpha}(\mathcal{H} - \alpha\mathcal{I})(\mathcal{S} - \alpha\mathcal{I}). \end{cases} \tag{12.8}$$

As with Section 4.4, if the stationary iteration (12.7) is not converging very fast, we may use a Krylov subspace iterative method with M as a matrix splitting preconditioner, i.e. solve the following instead of (12.7)

$$(\mathcal{I} - \mathcal{T}_\alpha)\mathbf{x} = M^{-1}b, \quad \text{or} \quad M^{-1}\mathcal{A}\mathbf{x} = \mathbf{b}. \tag{12.9}$$

The preconditioning step $Mz = r$ involves solving two substeps

$$\begin{cases} (\mathcal{H} + \alpha\mathcal{I})v = r, \\ (\mathcal{S} + \alpha\mathcal{I})z = 2\alpha v. \end{cases} \tag{12.10}$$

Here the preconditioning step (12.10) can also be solved by a so-called *inner* preconditioned iterative solver for the purpose of solving the *outer* equation (12.9).

12.2 The Oseen and Stokes saddle point problems

The Navier–Stokes equations represent an important set of fluid dynamics equations for the fluid velocity vector $\mathbf{u}$ and the pressure variable p at steady state

$$\begin{cases} -\nu\Delta\mathbf{u} + (\mathbf{u}\cdot\nabla)\mathbf{u} + \nabla p = \mathbf{f}, \\ \nabla\cdot\mathbf{u} = 0, \end{cases} \qquad \left(\begin{array}{c|c} -\nu\Delta + (\mathbf{u}\cdot\nabla) & \nabla \\ \hline \nabla\cdot & \end{array}\right)\begin{pmatrix}\mathbf{u}\\ p\end{pmatrix} = \begin{pmatrix}\mathbf{f}\\ 0\end{pmatrix} \tag{12.11}$$

which is defined in some closed domain $\mathbb{R}^d$ with suitably specified boundary conditions [473,164,341,197]. Here $\nu > 0$ is the viscosity parameter, inversely proportional to two other physical quantities: the Reynolds number Re and the Peclet number μ.

If ν is not too small, one solution method for (12.11) is the following linearization method (called the Oseen equations)

$$\begin{cases} -\nu\Delta\mathbf{u} + (\overline{\mathbf{u}}\cdot\nabla)\mathbf{u} + \nabla p = \mathbf{f}, \\ \nabla\cdot\mathbf{u} = 0, \end{cases} \qquad \left(\begin{array}{c|c} -\nu\Delta + (\overline{\mathbf{u}}\cdot\nabla) & \nabla \\ \hline \nabla\cdot & \end{array}\right)\begin{pmatrix}\mathbf{u}\\ p\end{pmatrix} = \begin{pmatrix}\mathbf{f}\\ 0\end{pmatrix} \tag{12.12}$$

where $\overline{\mathbf{u}}$ is taken to be some old iterate.

If the flow speed is small (i.e. both $\mathbf{u}$ and $\nabla\mathbf{u}$ are small so the dot product is negligible), a simpler fluid model (dropping one term in (12.11)) is the Stokes problem

$$\begin{cases} -\nu\Delta\mathbf{u} + \nabla p = \mathbf{f}, \\ \nabla\cdot\mathbf{u} = 0, \end{cases} \qquad \left(\begin{array}{c|c} -\nu\Delta & \nabla \\ \hline \nabla\cdot & \end{array}\right)\begin{pmatrix}\mathbf{u}\\ p\end{pmatrix} = \begin{pmatrix}\mathbf{f}\\ 0\end{pmatrix}. \tag{12.13}$$

Upon discretization of (12.12) and (12.13), a coupled linear system of type (12.1) is obtained

$$\mathcal{A}\mathbf{x} = \begin{bmatrix} A & B^T \\ B & 0 \end{bmatrix}\begin{bmatrix}\mathbf{u}\\ p\end{bmatrix} = \begin{bmatrix}\mathbf{f}\\ g\end{bmatrix} = \mathbf{b}, \tag{12.14}$$

where A has different symmetry properties for (12.12) and (12.13). Here as a block matrix (7.1), the (2, 2) block is conveniently 0 which means that the Schur complement $S = A_{22} - A_{21}A_{11}^{-1}A_{12} = -BA^{-1}B^T$ is correspondingly simpler than the general case. Two observations can be made of this particular situation.

(i) if B is square and invertible, then $S^{-1} = -B^{-T}AB^{-1}$ does not involve A^{-1}.

(ii) if the block matrix $\mathcal{A}$ is preconditioned by the block diagonal matrix D_1 or a upper triangular matrix D_2 [364,297,485]

$$D_1 = \begin{bmatrix} A & \\ & -S \end{bmatrix}, \qquad D_2 = \begin{bmatrix} A & B^T \\ & -S \end{bmatrix}, \tag{12.15}$$

then the preconditioned matrix $D^{-1}\mathcal{A}$ has only 3 eigenvalues: $1, \frac{1}{2} \pm \frac{\sqrt{5}}{2}$ or $-0.618, 1, 1.618$ while matrix $D_2^{-1}A$ has 2 eigenvalues ± 1, regardless of the size of matrix $\mathcal{A}$. (Note: the sign (-1) is involved. Without it, the results are slightly different but also good – see the supplied Mfile *wathen.m*)

Here the condition in (i) is a strong assumption and cannot be satisfied usually. However, a related idea of making use of the square matrix BB^T (or 'forcing' A^{-1} to commute with B^T in S to make up BB^T) turns out to be feasible [314]. The observation (ii) implies that it will be profitable to approximate A, S for the 'optimal' preconditioner.

From the above observation (i) and its remark, we wish to demonstrate (or to achieve) that approximately $S = -BA^{-1}B^T \approx -(BB^T)A^{-1}$. This latter product does not make sense as matrix dimensions do not match. To gain the preconditioning insight, the clever trick proposed in [314] is the following: re-consider the terms (involving S) in the continuous spaces and propose a related matrix that has the correct dimensions to realize the above thinking (of simplifying and approximating S).

Ignoring boundary conditions, one notes that for (12.12)

$$\begin{array}{lll} A & \sim -\nu\nabla^2 + \mathbf{b}\cdot\nabla & \text{(convection-diffusion)} \\ B & \sim \nabla\cdot & \text{(divergence)} \\ B^T & \sim \nabla & \text{(gradient)} \\ BB^T & \sim \nabla^2 & \text{(Laplacian)} \end{array} \tag{12.16}$$

where $\mathbf{b} = \bar{\mathbf{u}}$ for the Oseen operator and $\mathbf{b} = 0$ for the Stoke's. Thus at the *operator* level, we see that

$$\begin{array}{l} SA = -BA^{-1}B^TA = -BA^{-1}(B^TA) = -BA^{-1}(AB^T) = -BB^T, \\ S = -(BB^T)A^{-1}, \qquad\qquad S^{-1} = -A(BB^T)^{-1}, \end{array} \tag{12.17}$$

since $B^TA = AB^T$. Once this desirable result (of seeing BB^T in S) is visible, it only remains to construct a 'Schur' matrix (approximating the true Schur matrix) accordingly. At the matrix level, matrices B, B^T, A are already available with B, B^T rectangular (with B "short and flat" or more columns than rows). Matrix product AB^T makes sense while B^TA does not (on account of dimension alone). The proposal [314] is to construct another convection–diffusion operator (matrix) in the pressure space, i.e. A_p, that would make the product B^TA_p meaningful. Thus at the matrix level, we expect

$$\begin{array}{l} AB^T \sim B^TA_p \\ \widetilde{S} = -(BB^T)A_p^{-1}, \qquad\qquad \widetilde{S}^{-1} = -A_p(BB^T)^{-1}. \end{array} \tag{12.18}$$

Further, as BB^T is a Laplacian, we should form it directly without multiplying BB^T [314]. Then once the approximated Schur matrix $\widetilde{S}$ is available in a product form, we may use (12.15) to design an effective preconditioner (although the application of the preconditioner may require an inner iterative solver to 'invert' A which is usually done by a multigrid method [396]).

12.3 The mixed finite element method

The mixed finite element method is perhaps the main variational method for solving high-order PDEs, by introducing auxiliary variables. For linear and second-order PDEs, such an approach can lead to first-order systems and less approximations required and hence to more accurate solutions. For nonlinear problems, this approach offers a better linearization. The order reduction idea mimics the commonly used approach (of introducing intermediate derivatives) in dealing with high order ODEs (ordinary differential equations). However, it should be remarked that theoretical analysis of the mixed FEM is more challenging than the traditional FEM [29,164,282,308,355,470,499,197]. For our purpose here, we wish to highlight the resulting block matrix as an interesting preconditioning problem of the type (12.1).

Consider the model PDE in $\mathbb{R}^2$:

$$\nabla \cdot \gamma \nabla u = f \tag{12.19}$$

or

$$\nabla \cdot (\mathbf{D}. * \nabla u) = f \tag{12.20}$$

with $\mathbf{D} = (D_1(x, y), D_2(x, y))$. The mixed FEM introduces $\mathbf{w} = \gamma \nabla u \equiv \mathcal{L}u$ for (12.19) and $\mathbf{w} = \mathbf{D}. * \nabla u \equiv \mathcal{L}u$ for (12.20) so that both model equations become the first order system (with suitable boundary conditions)

$$\begin{cases} \mathbf{w} - \mathcal{L}u = 0, & \text{(auxiliary PDE)} \\ \nabla \mathbf{w} = f, & \text{(main PDE)} \end{cases} \tag{12.21}$$

Further a FEM can be applied to this first-order system, resulting the mixed FEM; what is interesting to note is that the usual FEM spaces H^1 is no longer needed for the resulting bilinear forms.

The final linear system will take a similar form to (12.14)

$$\mathcal{A}\mathbf{x} = \begin{bmatrix} A & B^T \\ D & 0 \end{bmatrix} \begin{bmatrix} \mathbf{w} \\ u \end{bmatrix} = \begin{bmatrix} 0 \\ g \end{bmatrix} = \mathbf{b}, \tag{12.22}$$

where $D = B$ for special cases. Then the preconditioning techniques discussed in the previous section will apply.

Remark 12.3.1. For system of PDEs, other auxiliary variables may be defined. For instance, to solve the Navier–Stokes system (12.11) using the streamline diffusion formulation [308] in $\mathbb{R}^2$, one would introduce the new variable

$$\omega = \frac{\partial u_2}{\partial x_1} - \frac{\partial u_1}{\partial x_2} \tag{12.23}$$

which can be combined with the second equation $\nabla \cdot \mathbf{u} = 0$ to replace the second-order term

$$-\nu \Delta \mathbf{u} = -\nu(\Delta u_1, \Delta u_2) = \nu \text{ rot } \omega,$$

where $\text{rot } \omega = \text{curl } \omega = (\frac{\partial \omega}{\partial x_2}, -\frac{\partial \omega}{\partial x_1})$. The two variable $(\mathbf{u}, p)$ system becomes the three variable $(\mathbf{u}, \omega, p)$ system that does not possess second order derivatives.

12.4 Coupled systems from fluid structure interaction

The fluid and structure interaction problem is only one simple modelling situation of a larger topic of investigating dynamics of an elastic solid structure interacting with its surrounding medium. The particular model that is of interest to us [137,291,16] consists of a vibrating elastic structure and an acoustic field exterior to this structure.

The behaviour of the elastic structure under the influence of the applied and acoustic fields is modelled by linear elasticity. Using a finite element method, either through discretizing the structural differential equation from beam theory or by energy conservation of the structure body, a structural equation can be written. As in [16] we take the latter approach to derive a linear system, which is of the form

$$(K - \omega^2 M)\underline{q} = \underline{f}, \tag{12.24}$$

where K is the stiffness matrix, M the mass matrix, $\underline{q}$ the displacement, and $\underline{f} = \underline{f}^k + \underline{f}^\phi$ the total load due to the applied forces $(\underline{f}^k)$ and the fluid pressure $(\underline{f}^\phi)$. Here ω is the time harmonic frequency for the kinetic energy of the structure and the stiffness matrix contains a large Lame constant λ. In the usual discretization scheme, we approximate the structure surface using piecewise quadratic surfaces defined over triangular elements. Further, prism-shaped finite elements are formed by projection from these surfaces to the centre of the structure and we define piecewise quadratic interpolation functions on the triangular faces and piecewise linear functions on the rectangular faces to approximate $\underline{q}$. Overall, the matrices in (12.24) are real, symmetric, and sparse.

For the corresponding acoustic field, as in Section 11.1, the use of single frequency harmonic waves of the form $\Phi(p, t) = \phi(p)e^{-i\omega t}$, where $\Phi(p, t)$ is the excess pressure at the point p at time t, reduces the linear wave equation (governing Φ) to a Helmholtz equation [159,16]

$$\nabla^2\phi(p) + k^2\phi(p) = 0, \tag{12.25}$$

where $k = \omega/c$ is the acoustic wave number, c being the wave speed. To ensure that all waves are outgoing at infinity, we use the Sommerfeld radiation condition ($r = |p - q|$ with $p, q \in \mathbb{R}^3$)

$$\lim_{r\to\infty} r\left\{\frac{\partial\phi(r)}{\partial r} - ik\phi(r)\right\} = 0.$$

Formulating the Helmholtz equation in equivalent boundary integral equation form over the surface of the structure gives

$$\int_S \left\{ \phi(q) \frac{\partial G_k(p,q)}{\partial n_q} - G_k(p,q) \frac{\partial \phi(q)}{\partial n_q} \right\} dS_q = \frac{1}{2}\phi(p),$$

where $G_k(p,q) = (e^{ik|p-q|})/(4\pi |p-q|)$ is the free-space Green's function for the Helmholtz equation. To avoid problems of nonexistence and nonuniqueness of the solutions of the integral equation at the natural frequencies of the structure, we use the Burton and Miller formulation,

$$\left(-\frac{1}{2}\mathcal{I} + \mathcal{M}_k + \alpha \mathcal{N}_k\right)\phi = \left[\mathcal{L}_k + \alpha\left(\frac{1}{2}\mathcal{I} + \mathcal{M}_k^T\right)\right]\frac{\partial \phi}{\partial n}, \tag{12.26}$$

where $\mathcal{L}_k$ and $\mathcal{M}_k$ are the single- and double-layer Helmholtz potential operators, $\mathcal{M}_k^T$ and $\mathcal{N}_k$ are their normal derivatives, and α is a coupling parameter whose imaginary part must be nonzero. The above integral equation is discretized using the boundary element (BE) method. We use the collocation method, with triangular elements for the surface of the structure and piecewise constant interpolation functions for the solution, which discretizes (12.26) as a linear system of the form

$$R\underline{\phi} = i\omega\rho B\underline{v} - \underline{c}, \tag{12.27}$$

where ρ is the fluid density, $\underline{\phi}$ is the pressure, $\underline{v}$ the velocity, and $\underline{c}$ the incident wave pressure. The matrices in this system are dense and complex (non-Hermitian).

For the coupled system, we require that the fluid particle velocity be continuous at the surface of the structure to couple the BE and FE systems. The load due to the fluid pressure, $\underline{f}^\phi$, can be written in terms of the pressure potential $\underline{\phi}$ as

$$\underline{f}^\phi = -L\underline{\phi} \qquad \text{with} \qquad \underline{f}^\phi = \underline{f} - \underline{f}^k,$$

where L is a matrix derived from the basis functions used in the BE and FE analyzes. Also, the velocity, $\underline{v}$, can be written in terms of the displacement, $\underline{q}$, as

$$\underline{v} = -i\omega L'\underline{q},$$

where L' is a matrix derived from the basis functions used in the FE analysis.

Then the coupled problem is to solve, simultaneously, the equations

$$\begin{cases} R\underline{\phi} = i\omega\rho B\underline{v} - \underline{c}, \\ (K - \omega^2 M)\underline{q} = \underline{f}^k - L\underline{\phi}. \end{cases} \tag{12.28}$$

This gives the partitioned system for $\underline{\phi}$ and $\underline{q}$

$$\begin{bmatrix} R & -\omega^2\rho BL' \\ & \\ L & K - \omega^2 M \end{bmatrix} \begin{bmatrix} \underline{\phi} \\ \\ \underline{q} \end{bmatrix} = \begin{bmatrix} -\underline{c} \\ \\ \underline{f}^k \end{bmatrix},$$

which will be written as the generic equation

$$\mathcal{A}\mathbf{x} = \mathbf{b}. \tag{12.29}$$

Diagrammatically, the block structure of the matrix $\mathcal{A}$ is

$$\left[\begin{array}{l|l} \text{BE} & \text{BEC} \\ \text{Boundary element block} & \text{Coupling strip} \\ \text{(small, dense, complex)} & \text{(full, complex)} \\ \hline \text{FEC} & \text{FE} \\ \text{Coupling strip} & \text{Finite element block} \\ \text{(sparse, real)} & \text{(sparse, real,} \\ & \quad\text{symmetric)} \end{array}\right]. \tag{12.30}$$

The preconditioner proposed in [291] is a block diagonal preconditioner taking the following form

$$M = \begin{bmatrix} M_1 & \\ & M_2 \end{bmatrix} = \left[\begin{array}{l|l} \text{BEINV} & \\ \text{SPAI approximate} & \qquad 0 \\ \text{inverse for BE block} & \\ \hline & \text{FEINV} \\ \qquad 0 & \text{SPAI approximate} \\ & \text{inverse for FE block} \end{array}\right] \tag{12.31}$$

from solving (see (12.30))

$$\min_{M_1} \|\text{BE}M_1 - I\|_F \quad \text{and} \quad \min_{M_2} \|\text{FE}M_2 - I\|_F.$$

Here the a priori patterns for M_1, M_2 are based on the patterns of the original blocks, with sparsification using a dropping strategy based on the global mean of the absolute values of the entries. It should be remarked that, without preconditioning, iterative methods such as GMRES will not converge as the coupled system (12.29) is indefinite.

12.5 Elasto-hydrodynamic lubrication modelling

The elasto-hydrodynamic lubrication modelling is an interesting and multidisciplinary research subject [186] involving solid state mechanics, fluid mechanics and chemistry. Mathematically this is another example of coupled differential and integral equations. The coupling will translate to a coupled linear system of dense and sparse matrix blocks, making an interesting problem for testing iterative solvers and preconditioning.

This section briefly reviews the model equations, the discretized systems and some preconditioning techniques involving the FWT.

12.5.1 Coupled isothermal equations

The steady-state contact problem may be described by the Reynolds equation for the lubricating film thickness H (which represents a simplified Navier–Stokes equation) coupled with the equation for the elastic deformation of the surrounding surfaces:

$$\begin{cases} \dfrac{d}{dx}\left(\dfrac{\rho H^3}{\eta}\dfrac{\partial p}{\partial x}\right) - \lambda\dfrac{d}{dx}(\rho H) = 0, \\ H - H_0 - \dfrac{x^2}{2} + \dfrac{1}{\pi}\displaystyle\int_{-\infty}^{\infty} \ln\left|x - x'\right| p(x')dx' = 0, \\ \dfrac{2}{\pi}\displaystyle\int_{-\infty}^{\infty} p(x')dx' - 1 = 0, \end{cases} \tag{12.32}$$

where the three equations are solved for the three dependent unknowns:

$H = H(x)$	the film thickness between two rollers (typically cylinders);
H_0	the film thickness at the point of lubricant separation;
$p = p(x)$	the pressure of the lubricant;

in some interval $[-L, L]$ which approximates the interval $(-\infty, \infty)$. The parameter λ is determined by several known quantities including the applied load w which is linked to the level of difficulties in solving the overall equations. The density function $\rho = \rho(p, \theta)$ is a known nonlinear function of pressure $p = p(x)$ and the ambient temperature θ [186]. Here L is assumed to be large enough to ensure that $p(\pm L) = 0$. In [210] and [207], following the earlier work of [372,421], we considered the line and point contact problems of two deforming cylinders as the sounding surfaces. In this isothermal case, the ambient temperature in the fluid is assumed to be constant while in the thermal case the ambient temperature will be a variable that must be solved with the main equations.

Therefore, for the isothermal case where θ is given, the nonlinear equation (12.32) effectively has two unknown quantities $p(x),\ H_0$. We shall be primarily interested in exposing the structure of the matrices of the linearized and discretized equations.

Consider the finite difference method, with a uniform mesh $x_j,\ j = 0, 1, \ldots, n+1$ with n internal mesh points and mesh size $h = \Delta x = 2L/(n+1)$. The boundary conditions for the pressure $p(x)$ are $p_0 = p(x_0) = p(x_{n+1}) = p_{n+1} = 0$, implying that the true unknowns will be

$$p(x_1),\ \ p(x_2),\ \ \cdots,\ \ p(x_n).$$

Thus equation (12.32) can be discretized to obtain a nonlinear system of algebraic equations, that must be solved by some nonlinear solvers. In details, the following nonlinear system is derived

$$\begin{cases} \dfrac{\rho(x_j^+)H^3(x_j^+)}{\eta(x_j^+)}(p(x_{j+1} - p(x_j)) - \dfrac{\rho(x_j^-)H^3(x_j^-)}{\eta(x_j^-)}(p(x_j - p(x_{j-1})) \\ \qquad - \dfrac{\lambda}{2}\Big[3\rho(x_j)H(x_j) - 4\rho(x_{j-1})H(x_{j-1}) + \rho(x_{j-2})H(x_{j-2})\Big] = 0, \\ H(x_j) = H_0 + \dfrac{x_j^2}{2} - \displaystyle\sum_{k=1}^{n+1} \dfrac{p(x_k) + p(x_{k-1})}{2\pi}\ell_{jk}, \\ \displaystyle\sum_{k=1}^{n+1} [p(x_k) + p(x_{k-1})] - \dfrac{\pi}{\Delta x} = 0, \end{cases}$$

where $x_{1} = x_0,\ x_j^{\pm} = x_j \pm \Delta x/2$, and

$$\ell_{jk} = \begin{cases} (x_k - x_j)\big[\ln(x_k - x_j) - 1\big] - (x_{k-1} - x_j)\big[\ln(x_{k-1} - x_j) - 1\big], & k - 1 > j, \\ (x_j - x_{k-1})\big[\ln(x_j - x_{k-1}) - 1\big] - (x_j - x_k)\big[\ln(x_j - x_k) - 1\big], & k < j, \\ (x_k - x_j)\big[\ln(x_k - x_j) - 1\big], & k - 1 = j, \\ (x_j - x_{k-1})\big[\ln(x_j - x_{k-1}) - 1\big], & k = j. \end{cases}$$

To solve the above system by a Newton–Raphson method, we have to solve a linear system involving the dense Jacobian matrix J. The logarithmic kernel in (12.32) defines a smooth integral operator similar to the well-known single-layer potential for the Laplace equation (1.73). Thus it is feasible to represent it in some (biorthogonal) wavelet space to obtain a nearly sparse Jacobian matrix $\widetilde{J}$. This optimal implementation is not yet attempted.

In [210], we took instead the easier option of applying a DWT (Section 1.6.2) and found the centring DWT algorithm of Section 8.4, i.e. [130] (denoted by

DWTPer in [210]) particularly useful in transforming J and subsequently selecting a (non-diagonal) sparse preconditioner. In [205], the idea was further refined by detecting the nonsmooth blocks in J and proposing a banded arrow type sparse preconditioner while an efficient way of computing the wavelet preconditioner based on partially transforming the Jacobian matrix J was presented in [206]. Based on the idea of a banded arrow preconditioner and a refined permutation [204] (see Chapter 8), an improved banded arrow preconditioner can be designed [204].

12.5.2 Coupled thermal equations

We now discuss the thermal case where the local mean temperature θ is a function and must be solved together with the equations, (12.32), for the pressure p and the film thickness at separation H_0. The new equations including the energy equation, as used by [421,207], will be the following set

$$\begin{cases} \dfrac{d}{dx}\left(\dfrac{\rho H^3}{\eta}\dfrac{\partial p}{\partial x}\right) - \lambda \dfrac{d}{dx}(\rho H) = 0, \\ H - H_0 - \dfrac{x^2}{2} + \dfrac{1}{\pi}\displaystyle\int_{-\infty}^{\infty} \ln\left|x - x'\right| p(x')dx' = 0, \\ \dfrac{2}{\pi}\displaystyle\int_{-\infty}^{\infty} p(x')dx' - 1 = 0, \\ \rho\left[u_m \dfrac{\partial \theta}{\partial x} + \dfrac{\theta - \theta_b}{H}(u_m - u_b)\dfrac{\partial H}{\partial x}\right] = \dfrac{3k(\theta_a + \theta_b - 2\theta)}{2B^2H^2} \\ \qquad + \left[(\beta_e\theta - 2B\mu)u_m + B\mu u_e\right]\dfrac{\partial p}{\partial x} + \dfrac{B\mu k}{3}\eta\Gamma_m^2 \end{cases} \tag{12.33}$$

where θ_a, θ_b are the surface temperatures and u_a, u_b the speeds of the rollers (the two contacting objects).

When the FDM is applied to (12.33), within each step of a homotopic parameter continuation [376], a linear system of $2(n+1)$ equations in $2(n+1)$ unknowns

$$p_1, \ p_2, \cdots, p_n; \ H_0; \ \theta_1, \ \theta_2, \cdots, \theta_{n+1},$$

are generated, which may be denoted by

$$\left[\begin{array}{c|cc} D & C & b_1 \\ \hline B & A & b_2 \\ c_1^T & c_2^T & d \end{array}\right] \mathbf{x} = \mathbf{b}. \tag{12.34}$$

Here the structure of the coefficient matrix can be explored [421,207]:

D — a lower Hessenberg matrix (invertible)
C — a dense matrix
B — a singular matrix
A — a dense matrix (invertible).

In particular, the Schur complement method (using the partition in (12.34)) was found beneficial and the main Schur equation can be solved by the GMRES method in the DWT space using a banded matrix preconditioner based on the DWT (centering DWT) of the dense A_{22} matrix, in the notation of (7.1),

$$A_{22} = \begin{bmatrix} A & b_2 \\ c_2^T & d \end{bmatrix}.$$

Remark 12.5.2. As an individual integral or differential operator can be preconditioned effectively, a block matrix problem requires a careful exploration of all operators (blocks). The Schur complements method is frequently a natural choice. The idea of using operator relations to find a more computable Schur matrix (or approximation) [314] in Section 12.2 may be used more widely.

12.6 Discussion of software and a supplied Mfile

There has been a lot of attention towards developing block matrix algorithms. Many existing software have a block version. Here we list of a few examples.

(1) SuperLU (by James Demmel, John Gilbert and Xiaoye Li) is a general purpose library for the direct solution of large, sparse, nonsymmetric systems of linear equations.

http://crd.lbl.gov/~xiaoye/SuperLU/

(2) IFISS (Incompressible flow and iterative solver software by David Silvester and Howard Elman) aims to solve several types of coupled fluid problems, using the preconditioned conjugate gradient and multigrid methods.

http://www.ma.umist.ac.uk/djs/software.html

(3) BILUM (A software package of multi-level block ILU preconditioning techniques for solving general sparse linear systems by Youcef Saad and Jun Zhang).

http : //www.cs.uky.edu/~jzhang/bilum.html

The supplied Mfile `wathen.m` is simply to demonstrate the effectiveness of the block diagonal preconditioning (as a guide to approximation of Schur complements).

13

Application III: image restoration and inverse problems

An inverse problem assumes a direct problem that is a well-posed problem of mathematical physics. In other words, if we know completely a "physical device," we have a classical mathematical description of this device including uniqueness, stability and existence of a solution of the corresponding mathematical problem.

VICTOR ISAKOV. *Inverse Problems for Partial Differential Equations*. Springer-Verlag (1998)

Image restoration is historically one of the oldest concerns in image processing and is still a necessary preprocessing step for many applications.

GILLES AUBERT and PIERRE KORNPROBST. *Mathematical Problems in Image Processing*. Springer-Verlag (2002)

However, for the time being it is worthwhile recalling the remark of Lanczos: "A lack of information cannot be remedied by any mathematical trickery." Hence in order to determine what we mean by a solution it will be necessary to introduce "nonstandard" information that reflects the physical situation we are trying to model.

DAVID COLTON and RAINER KRESS. *Integral Equation Methods in Scattering Theory*. Wiley (1983)

The research of inverse problems has become increasingly popular for two reasons:

(i) there is an urgent need to understand these problems and find adequate solution methods; and
(ii) the underlying mathematics is intriguingly nonlinear and is naturally posed as a challenge to mathematicians and engineers alike.

It is customary for an introduction to inverse problems of boundary value problems to discuss the somewhat unhelpful terms of 'ill-posed problems' or 'improperly-posed problems'. Indeed without introducing constraints, all inverse problems do not admit meaningful solutions. The suitable constraints, often based on a knowledge derived from solving the corresponding forward problems (i.e. the physical and modelling situation), can ensure uniqueness to an inverse problem.

Image restoration represents a useful topic in the large class of inverse problems. Our exposition here will mainly highlight the various challenges and issues in developing fast iterative solvers and suitable preconditioners. The energy minimization-related PDE models are mainly discussed; there exist other physics-motivated PDE models [10,11,26,154,419]. We shall first discuss the modelling equations and then focus on numerical solution techniques. This chapter will present the following.

Section 13.1 Image restoration models and discretizations
Section 13.2 Fixed point iteration method
Section 13.3 The primal-dual method
Section 13.4 Explicit time marching schemes
Section 13.5 Nonlinear multigrids for optimization
Section 13.6 The level set method and other image problems
Section 13.7 Numerical experiments
Section 13.8 Guide to software and the use of supplied Mfiles

13.1 Image restoration models and discretizations

We denote by $u = u(x, y)$ the true image and $z = z(x, y)$ the observed image, both defined in the bounded, open and rectangular domain $\Omega = [a, b] \times [c, d] \subset \mathbb{R}^2$ [230,18,388]. Without loss of generality, we may take $\Omega = [0, b] \times [0, d]$. The observed image z has been contaminated in the data collection stage, due to either environmental influences or technical limitations.

The purpose of image restoration is to recover u as much as we can using a degradation model

$$Ku - z = \eta_0, \qquad \text{or} \qquad Ku = z + \eta_0, \tag{13.1}$$

where η_0 is a Gaussian white noise and K is a known linear degradation operator; for deblurring problems K is often a convolution operator and for denoising problems K is simply the identity operator. (It should be remarked that if the degradation is relatively small, filtering techniques can 'recover' the images; one may even use wavelets [193]. However, the problem that concerns us here is the harder case where filtering cannot recover the image.)

For the general deblurring problem, operator K takes the convolution form

$$(Ku)(x,y) = \int_\Omega k(x,x';y,y')u\,d\Omega = \int_0^d \int_0^b k(x,x';y,y')u(x',y')dx'dy' \tag{13.2}$$

where the kernel k is usually called the *point spread function* (PSF), that measures the level of degradation in the observed image z. The case of k being spatially variant (i.e. general) implies a computationally intensive problem (at least in storage). In many cases, fortunately, the PSF acts uniformly so we may assume that it is *spatially invariant*

$$k(x,x';y,y') = k(x-x',y-y'). \tag{13.3}$$

Then we may write (13.2) in the convolution '$*$' notation

$$(Ku)(x,y) = K * u. \tag{13.4}$$

The PSF is often a smooth function, given analytically e.g.

$$k(s,t) = \frac{1}{4\pi\sigma} \exp\left(-(s^2+t^2)/(4\sigma)\right)$$

for a Gaussian blur with some small σ (say $\sigma = 0.01$), so it is safe to assume that operator K is a compact linear operator in a normed space. If the white noise η_0 has the mean of 0 and the standard deviation of η, solving (13.1) for u amounts to seeking the solution of

$$\|Ku - z\|_2^2 = \eta^2. \tag{13.5}$$

13.1.1 Solution of a first kind equation

To see that such a problem does not have a stable solution, simply assume $\eta = 0$ for a moment and solve

$$Ku = z. \tag{13.6}$$

Using the Picard theorem (see [160, Theorem 4.8, p. 92]), one can write the exact solution u to the first kind equation (13.6) as an infinite series in terms of singular values μ_n and singular functions ϕ_n, g_n of K:

$$u = \sum_{n=1}^{\infty} \frac{1}{\mu_n}(z, g_n)\phi_n.$$

As $\mu_n \to 0$ as $n \to \infty$ (also true for eigenvalues of K), solving (13.6) numerically (amounting to certain series truncation) is extremely unstable as small perturbation of z (e.g. with some noise η) will destroy any quality in the solution.

Clearly image restoration is an inverse problem that may not have a unique or rather meaningful solution. As remarked before, as with all inverse problems, some extra constraint or information must be given to ensure the solvability of (13.1). The typical constraint as regularity condition for u is imposed on the solution space in order to turn the underlying problem to a well posed one [476].

13.1.2 Regularity condition: total variation and bounded variation

To restrict the solution to (13.1) and (13.5), one may consider the L_2 space by requiring

$$\min_u \|u\|_2^2 = \min_u \int_\Omega |u|^2 d\Omega$$

or

$$\min_u \|\nabla u \cdot \nabla u\|_2^2 = \min_u \int_\Omega |\nabla u \cdot \nabla u|^2 d\Omega.$$

These requirements are sufficient to define a unique formulation. It turns out [64,377] that these extra conditions are only useful in recovering images with smooth features, as they do not allow jumps or discontinuities which are common in general images.

A better constraint in solving (13.1) and (13.5) to use is the total variation semi-norm (the TV norm [408])

$$\min_u TV(u), \qquad TV(u) = \int_\Omega |\nabla u| d\Omega = \int_\Omega \sqrt{u_x^2 + u_y^2} d\Omega, \quad (13.7)$$

with $|\nabla u| = \|\nabla u\|_2$, which is equivalent to the more rigorous definition of the norm of bounded variation for the distributional gradient Du of a nondifferentiable function u [26,223,99]

$$\min_u BV(u), \qquad BV(u) = \sup_{\phi} \left\{ \int_\Omega u \mathrm{div} \phi d\Omega \mid \phi \in C_0^1(\Omega), \|\phi\|_\infty \le 1 \right\}. \tag{13.8}$$

For vectors (as a discretized function) e.g. $u = (u_1, \ldots, u_n)^T \in \mathbb{R}^n$, the discrete norm is much easier to understand

$$TV(u) = BV(u) = \sum_{j=1}^{n-1} \left| u_{j+1} - u_j \right|.$$

13.1.3 Tikhonov regularization

The solution of the above regularity minimization with the main equation (13.5) as a constraint may be proceeded by the method of Lagrange multipliers. This was the approach first adopted by Tikhonov in the early 1960s for solving inverse problems (see [26,453] and the references therein). In this regularization framework, the Tikhonov technique proceeds as

$$\min_u J(u), \qquad J(u) = \alpha R(u) + \frac{1}{2} \|Ku - z\|_2^2, \tag{13.9}$$

where the regularization functional $R(u)$ is selected as the TV-norm [408,110]

$$R(u) = TV(u) = \int_\Omega |\nabla u| dxdy = \int_\Omega \sqrt{u_x^2 + u_y^2} dxdy. \tag{13.10}$$

Here the parameter α, usually given a priori, represents a tradeoff between the quality of the solution and the fit to the observed data. However once a solver is available for a fixed α, one may use the modular solver in [64] to solve the fully constrained problem to identify the correct parameter α.

Thus the overall image restoration problem is modelled by the following minimization formulation [10,64,110,408,476]

$$\min_u J(u), \qquad J(u) = \int_\Omega \left[\alpha \sqrt{u_x^2 + u_y^2} + \frac{1}{2} (Ku - z)^H (Ku - z) \right] dxdy \tag{13.11}$$

or

$$\min_u \left(\alpha \left\| \nabla u \right\|_{L_1} + \frac{1}{2} \left\| Ku - z \right\|_{L_2} \right). \tag{13.12}$$

The theoretical solution to problem (13.11) is given by the Euler–Lagrange[1] equation (assuming homogeneous Neumann boundary conditions)

$$\alpha \nabla \cdot \left(\frac{\nabla u}{|\nabla u|} \right) - K^H K u = -K^H z, \tag{13.13}$$

where K^H is the adjoint operator of K. Notice that the resulting nonlinear PDE has an interesting coefficient that may have a zero denominator so the equation is not well defined at such points (corresponding to flat regions of the solution). Consider any closed curve (along edges of features in u) in Ω. One can observe that the terms

$$\mathbf{n} = \frac{\nabla u}{|\nabla u|}, \qquad \mathbf{k} = \nabla \cdot \left(\frac{\nabla u}{|\nabla u|} \right) = \nabla \cdot \mathbf{n} \tag{13.14}$$

respectively, define the unit (outward) normal vector and the curvature of this curve. These quantities are heavily used in the level set method [379,377]. For 1D problems, $\Omega = [0, 1]$, the above equation reduces to

$$\alpha \frac{d}{dx} \left(\left| \frac{du}{dx} \right|^{-1} \frac{du}{dx} \right) - K^H K u = -K^H z. \tag{13.15}$$

A commonly adopted idea to overcome this apparent difficulty was to introduce (yet) another parameter β to (13.11) and (13.13) so the new and better-defined Euler–Lagrange equation becomes

$$\alpha \nabla \cdot \left(\frac{\nabla u}{\sqrt{|\nabla u|^2 + \beta}} \right) - K^H K u = -K^H z, \tag{13.16}$$

where in theory $u = u_\beta(x, y)$ differs from u in (13.13). Observe that when $\beta = 0$ equation (13.16) reduces to the previous (13.13); moreover as $\beta \to 0$, we have $u_\beta \to u$ as shown in [3].

13.1.4 Discretizations

The observed image z of size $m \times n$ may be thought of as an image function z restricted onto an *uniform* and finest grid with mesh size $h = h_1 = b/m = h_2 = d/n$;

[1] Recall that the Euler–Lagrange solution to the functional minimization problem

$$\min_u J(u), \qquad J(u) = \int_\Omega F(x, y, u, u_x, u_y) dx dy$$

is the following

$$\frac{\partial F}{\partial u} - \frac{\partial}{\partial x}\left(\frac{\partial F}{\partial u_x} \right) - \frac{\partial}{\partial y}\left(\frac{\partial F}{\partial u_y} \right) = 0.$$

here each pixel is located in the centre of a mesh box. Usually the FDM (finite difference method) is used to discretize the differential operator while the Nyström method is used to discretize the integral operator. For readers' benefit, we give a brief summary below (as most research papers do not present full details often due to space limitation).

Assume we have a lexicographical ordering for the underlying mesh centre points (image pixels). Denote the forward and backward difference operators by (see (13.63))

$$
\begin{aligned}
A_k^T u = A_{ij}^T u &= \begin{bmatrix} u_{i+1,j} - u_{ij} \\ u_{i,j+1} - u_{ij} \end{bmatrix} = \begin{bmatrix} u_{k+1} - u_k \\ u_{k+m} - u_k \end{bmatrix}, \\
A_k u = A_{ij} u &= \begin{bmatrix} u_{ij} - u_{i,j-1} \\ u_{ij} - u_{i-1,j} \end{bmatrix} = \begin{bmatrix} u_k - u_{k-m} \\ u_k - u_{k-1} \end{bmatrix},
\end{aligned}
\tag{13.17}
$$

where we have shown the cases of both a single index $k = (j-1)m + i$ and double indices (i, j) for u. At image boundaries, the differences are adjusted accordingly to accommodate the Neumann's boundary condition. With the above notation, the differential operator in (13.16) will be approximated by

$$
\nabla \cdot \Big(\frac{\nabla u}{\sqrt{|\nabla u|^2 + \beta}} \Big) = \frac{1}{h^2} A_k \Big(\frac{A_k^T u}{\sqrt{|A_k^T u|^2 + h^2 \beta}} \Big), \tag{13.18}
$$

where the term $h^2 \beta$ is still denoted by β in practice.

Now we consider the discretization of the integral operator K. Let $\bar{x}_j = jh$, $\bar{y}_k = kh$ for $j = 0, 1, \ldots, m$; $k = 0, 1, \ldots, n$ be the mesh points for the image domain Ω with $\bar{x}_0 = 0, \bar{x}_m = b$ and $\bar{y}_0 = 0, \bar{y}_n = d$. Then the image pixel (j, k) has the coordinate (x_j, y_k) with $x_j = (j - 1/2)h$, $y_k = (k-1)h$ for $j = 1, \ldots, m$; $k = 1, \ldots, n$. Therefore our two-dimensional integral in (13.16) for a fixed pixel $x = x_i$, $y = y_r$ may be evaluated by the mid-point quadrature rule

$$
(Ku)(x_i, y_r) = \int_\Omega k(x_i - x', y_r - y') u(x', y') dx' dy' \tag{13.19}
$$

$$
\approx h^2 \sum_{k=1}^{n} \sum_{j=1}^{m} k(x_i - x_j, y_r - y_k) u(x_j, y_k) \; = \; Ku, \tag{13.20}
$$

where K of size $mn \times mn$ is a block Toeplitz matrix with Toeplitz blocks

(BTTB)

$$K = \begin{bmatrix} \bullet & \leftarrow & \leftarrow & \leftarrow & \dashv \\ \downarrow & K_0 & K_{-1} & \cdots & K_{1-n} \\ \downarrow & K_1 & K_0 & \cdots & K_{2-n} \\ \downarrow & K_2 & K_1 & \cdots & K_{3-n} \\ \downarrow & \vdots & \vdots & \ddots & \vdots \\ \downarrow & K_{n-1} & K_{n-2} & \cdots & K_0 \end{bmatrix}_{n\times n} \tag{13.21}$$

where for $|j| \leq n-1$

$$K_j = \begin{bmatrix} (0,j) & (-1,j) & \cdots & (1-m,j) \\ (1,j) & (0,j) & \cdots & (2-m,j) \\ \vdots & \vdots & \ddots & \vdots \\ (m-2,j) & (m-3,j) & \cdots & (-1,j) \\ (m-1,j) & (m-2,j) & \cdots & (0,j) \end{bmatrix}$$

$$= \begin{bmatrix} k(0,jh) & k(-h,jh) & \cdots & k((1-m)h,jh) \\ k(h,jh) & k(0h,jh) & \cdots & k((2-m)h,jh) \\ \vdots & \vdots & \ddots & \vdots \\ k((m-2)h,jh) & k((m-3)h,jh) & \cdots & k(-h,jh) \\ k((m-1)h,jh) & k((m-2)h,jh) & \cdots & k(0,jh) \end{bmatrix}$$

with the index pair (i, j) denoting the value $k(ih, jh)$. Clearly the second index j in the index pair (i, j) is also the block index j for K_j, which is of size $m \times m$. Hence, the discretized equation takes the form

$$\alpha\mathcal{N}(u)u - K^H K u = -K^H z, \tag{13.22}$$

or (sometimes seen in the literature)

$$-\alpha\mathcal{N}(u)u + K^H K u = K^H z,$$

where $\mathcal{N}$ is defined by (13.18).

We shall approximate a BTTB matrix by a BCCB (block circulant matrix with circulant blocks) of the same size $mn \times mn$

$$C = \begin{bmatrix} \top & C_0 & C_{n-1} & \cdots & C_2 & C_1 \\ \downarrow & C_1 & C_0 & \cdots & C_3 & C_2 \\ \downarrow & C_2 & C_1 & \cdots & C_4 & C_3 \\ \downarrow & \vdots & \vdots & \ddots & \vdots & \\ \downarrow & C_{n-1} & C_{n-2} & \cdots & C_1 & C_0 \end{bmatrix}_{n\times n} \tag{13.23}$$

where each C_j is a circulant matrix of size $m \times m$. Refer to [307].

To manipulate matrix K, it is necessary to discuss the following definitions; note that both conversions can be done in the MATLAB command `reshape`.

Definition 13.1.1. (2D FFT-related operations). *A matrix $A \in \mathbb{R}^{m\times n}$ can be uniquely mapped to the vector $a \in \mathbb{R}^{mn}$ and vice versa by (actions of packing and unpacking)*

$$a = \mathbf{Vector}(A) = \begin{bmatrix} A_{11} \cdots A_{m1} \ldots A_{1j} \cdots A_{mj} \ldots A_{1n} \cdots A_{mn} \end{bmatrix}, \quad (13.24)$$

$$A = \mathbf{Matrix}(a) = \begin{bmatrix} a_1 & \cdots & a_{(j-1)m+1} & \cdots & a_{(n-1)m+1} \\ a_2 & \cdots & a_{(j-1)m+2} & \cdots & a_{(n-1)m+2} \\ \vdots & \vdots & \vdots & \vdots & \vdots \\ a_m & \cdots & a_{(j-1)m+m} & \cdots & a_{(n-1)m+m} \end{bmatrix}. \quad (13.25)$$

Remark 13.1.2. At this point, three remarks are in order. Firstly the dense matrix K is too large to be stored for a typical image sized 1024×1024 (as the matrix of size $1024^2 \times 1024^2$ has $1099511627776 > 10^{12}$ entries!). Once its root (the first column of K) is stored, for a given vector v, computing $w = Kv$ can be done using FFT [103,465] without storing K explicitly as a matrix (as shown below). Secondly the PSF is frequently given as a matrix (a discrete kernel array representing the kernel function k evaluated only at $(\pm ih, \pm h)$ as required by a fixed pixel):

$$PSF = \text{kernel} = \begin{bmatrix} k_{11} & k_{12} & \cdots & k_{1n} \\ k_{21} & k_{22} & \cdots & k_{2n} \\ \vdots & \vdots & \ddots & \vdots \\ k_{m1} & k_{m2} & \cdots & k_{mn} \end{bmatrix}_{m\times n} = \begin{bmatrix} K_{11} & K_{12} \\ K_{21} & K_{22} \end{bmatrix} \quad (13.26)$$

i.e. k not available as a function. Here in the partitioned matrix of 4 blocks of size $\ell_m \times \ell_n$, with $\ell_m = m/2$ and $\ell_n = n/2$, the entry $K_{22}(1,1) = k(0,0)$ contains the center of the point source and $K_{22}(1,2) = k(h,0)$, $K_{22}(2,1) = k(0,h)$ and so on. Finally the Definition 13.1.1 should remind the reader that the mapping can be used to apply 'vector-norms' to matrices; see [280] for the S-norm.

Lemma 13.1.3. (Diagonalization of a BCCB matrix). *Let C be a BCCB matrix with the root matrix $c \in \mathbb{R}^{m\times n}$, as in (13.21), and $\mathbf{F}_\tau$ denote the 1D discrete Fourier matrix of size τ as in (1.38). Then matrix C can diagonalized*

$$FCF^{-1} = diag(\widetilde{c}), \qquad \widetilde{c} = F\mathbf{Vector}(c) \quad (13.27)$$

where $F = \mathbf{F}_n \otimes \mathbf{F}_m$ *is the 2D FFT matrix (refer to (14.3.3)) and we have used* F^{-1} *instead of the usual notation* F^H *(to avoid adding a constant, since* $F^H F = mn$ *and* $\mathbf{F}_n^H \mathbf{F}_n = n$ *– see § 1.6). Thus the matrix-vector product* $w = Cv$, $v \in \mathbb{R}^{mn}$, *can be implemented efficiently as*

$$
\begin{aligned}
w &= C\,\mathbf{Vector}(v) = F^{-1}\, diag(\widetilde{c}) F\,\mathbf{Vector}(v) \\
&= \underbrace{F^{-1} \quad \underbrace{F\,\mathbf{Vector}(c)}_{\textit{Forward 2D FFT}} \quad .* \quad \underbrace{F\,\mathbf{Vector}(v)}_{\textit{Forward 2D FFT}}}_{\textit{Inverse 2D FFT}} .
\end{aligned}
\tag{13.28}
$$

Note that the 2D FFT transforms are implemented in MATLAB in matrix form i.e. both input data and output are in matrix form. Thus (13.28), after setting $v = \mathbf{Matrix}(v)$, will be done by

```
>>  c_tilde = fft2(c);
>>  v_tilde = fft2(v);
>>          w = ifft2( c_tilde .* v_tilde ) ;
>>          w = reshape( real(w), m*n, 1);
                                   % Output the vector form
```

To address how to generate $w = Kv$ using such PSF information i.e. to make the link of (13.26) to the overall matrix K, the latter may be conveniently expressed in structural form

$$
\left[\begin{array}{c|c|c}
\begin{array}{cccc}
(0,0) & (-1,0)\cdots & (1-m,0) \\
(1,0) & (0,0)\cdots & (2-m,0) \\
(2,0) & (1,0)\cdots & (3-m,0) \\
\ddots & \ddots\cdots & \ddots \\
(m-1,0) & (m-2,0)\cdots & (0,0)
\end{array} & \cdots & \begin{array}{ccc}
(0,(1-n)) & (-1,(1-n))\cdots & (1-m,(1-n)) \\
(1,(1-n)) & (0,(1-n))\cdots & (2-m,(1-n)) \\
(2,(1-n)) & (1,(1-n))\cdots & (3-m,(1-n)) \\
\ddots & \ddots\cdots & \ddots \\
(m-1,(1-n)) & (m-2,(1-n))\cdots & (0,(1-n))
\end{array} \\
\hline
\begin{array}{ccc}
(0,1) & (-1,1)\cdots & (1-m,1) \\
(1,1) & (0,1)\cdots & (2-m,1) \\
(2,1) & (1,1)\cdots & (3-m,1) \\
\ddots & \ddots\cdots & \ddots \\
(m-1,1) & (m-2,1)\cdots & (0,1)
\end{array} & \cdots & \begin{array}{ccc}
(0,(2-n)) & (-1,(2-n))\cdots & (1-m,(2-n)) \\
(1,(2-n)) & (0,(2-n))\cdots & (2-m,(2-n)) \\
(2,(2-n)) & (1,(2-n))\cdots & (3-m,(2-n)) \\
\ddots & \ddots\cdots & \ddots \\
(m-1,(2-n)) & (m-2,(2-n))\cdots & (0,(2-n))
\end{array} \\
\hline
\ddots & \cdots & \ddots \\
\hline
\begin{array}{ccc}
(0,n-1) & (-1,n-1)\cdots & (1-m,n-1) \\
(1,n-1) & (0,n-1)\cdots & (2-m,n-1) \\
(2,n-1) & (1,n-1)\cdots & (3-m,n-1) \\
\ddots & \ddots\cdots & \ddots \\
(m-1,n-1) & (m-2,n-1)\cdots & (0,n-1)
\end{array} & \cdots & \begin{array}{ccc}
(0,0) & (-1,0)\cdots & (1-m,0) \\
(1,0) & (0,0)\cdots & (2-m,0) \\
(2,0) & (1,0)\cdots & (3-m,0) \\
\ddots & \ddots\cdots & \ddots \\
(m-1,0) & (m-2,0)\cdots & (0,0)
\end{array}
\end{array}\right].
$$

As with a Toeplitz matrix (§ 2.5.2), matrix K is determined precisely by the collection of all of its first columns (from the top right corner to the bottom right):

$$\left[\begin{array}{c|c|c|c|c|c|c}
(1-m,1-n) & (1-m,2-n) & \cdots & (1-m,0) & \cdots & (1-m,n-2) & (1-m,n-1)\\
(2-m,1-n) & (2-m,2-n) & \cdots & (2-m,0) & \cdots & (2-m,n-2) & (2-m,n-1)\\
\vdots & \vdots & \vdots & \vdots & \vdots & \vdots & \vdots\\
(0,1-n) & (0,2-n) & \cdots & (0,0) & \cdots & (0,n-2) & (0,n-1)\\
\vdots & \vdots & \vdots & \vdots & \vdots & \vdots & \vdots\\
(m-2,1-n) & (m-2,2-n) & \cdots & (m-2,0) & \cdots & (m-2,n-2) & (m-2,n-1)\\
(m-1,1-n) & (m-1,2-n) & \cdots & (m-1,0) & \cdots & (m-1,n-2) & (m-1,n-1)
\end{array}\right] \tag{13.29}$$

which, when used in the column form, provides the root vector for (13.21). In fact, the PSF (13.26) is often given (along with the data z), in the following form and with the meaning (to serve (13.29))

$$PSF = \begin{bmatrix} K_{11} & K_{12} \\ K_{21} & K_{22} \end{bmatrix}$$

$$= \left[\begin{array}{c|c|c|c}
k((1-m)h,(1-n)h) & k((1-m)h,(2-n)h) & \cdots & k((1-m)h,(n-1)h)\\
k((2-m)h,(1-n)h) & k((2-m)h,(2-n)h) & \cdots & k((2-m)h,(n-1)h)\\
\vdots & \vdots & \vdots & \vdots\\
k(0,(1-n)h) & k(0,(2-n)h) & \cdots & k(0,(n-1)h)\\
\vdots & \vdots & \vdots & \vdots\\
k((m-2)h,(1-n)h) & k((m-2)h,(2-n)h) & \cdots & k((m-2)h,(n-1)h)\\
k((m-1)h,(1-n)h) & k((m-1)h,(2-n)h) & \cdots & k((m-1)h,(n-1)h)
\end{array}\right]. \tag{13.30}$$

Here we assume that the PSF matrix is of size $(2m-1) \times (2n-1)$; if a smaller $m \times n$ matrix given instead, we have to pack rows and columns of zeros to enclose the smaller matrix in order to extend it to a new PSF matrix of size $(2m-1) \times (2n-1)$, e.g. the following illustrates how to extend a given smaller 4×4 PSF matrix to a required 7×7 (watch the centring position of $K_{22}(1,1)$)

$$\left[\begin{array}{cc|cc}
k_{11} & k_{12} & k_{13} & k_{14}\\
k_{21} & k_{22} & k_{23} & k_{24}\\ \hline
k_{31} & k_{32} & k_{33} & k_{34}\\
k_{41} & k_{42} & k_{43} & k_{44}
\end{array}\right] \Rightarrow \left[\begin{array}{ccc|cccc}
0 & 0 & 0 & 0 & 0 & 0 & 0\\
0 & k_{11} & k_{12} & k_{13} & k_{14} & 0 & 0\\
0 & k_{21} & k_{22} & k_{23} & k_{24} & 0 & 0\\ \hline
0 & k_{31} & k_{32} & k_{33} & k_{34} & 0 & 0\\
0 & k_{41} & k_{42} & k_{43} & k_{44} & 0 & 0\\
0 & 0 & 0 & 0 & 0 & 0 & 0\\
0 & 0 & 0 & 0 & 0 & 0 & 0
\end{array}\right]. \tag{13.31}$$

13.1.5 Computation of $w = Kv$ by FFT

In the nonblock version of FFT (Section 2.5.2), we have discussed how to compute $w = Kv$ type product for a Toeplitz matrix via a circulant matrix. Here we briefly address the block version. (It should be noted that all discussions for computing $w = Kv$ with the root matrix R for K apply to $w = K^H v$, if the new root matrix R^T is used.) First we show the computational details in an example (that can be reproduced by the Mfile `ch13.m`). Consider the following PSF kernel matrix for K (intended for some 3×3 image), and the given vector v

$$R = \left[\begin{array}{c|c|ccc} \mathbf{1} & 3 & 7 & 3 & 8 \\ \mathbf{5} & 6 & 2 & 3 & 2 \\ \mathbf{2} & 8 & 8 & 8 & 9 \\ \mathbf{3} & 9 & 1 & 9 & 5 \\ \mathbf{9} & 7 & 9 & 7 & 4 \end{array}\right], \qquad \mathbf{Matrix}(v) = \begin{bmatrix} 1 & 3 & 5 \\ 2 & 3 & 1 \\ 1 & 1 & 2 \end{bmatrix}.$$

Here is the 'wrong' and direct method to compute $w = Kv$ using K:

$$w = Kv = \left[\begin{array}{ccc|ccc|ccc} 8 & 2 & 7 & 8 & 6 & 3 & \mathbf{2} & \mathbf{5} & \mathbf{1} \\ 1 & 8 & 2 & 9 & 8 & 6 & \mathbf{3} & 2 & 5 \\ 9 & 1 & 8 & 7 & 9 & 8 & \mathbf{9} & 3 & 2 \\ \hline 8 & 3 & 3 & 8 & 2 & 7 & 8 & 6 & 3 \\ 9 & 8 & 3 & 1 & 8 & 2 & 9 & 8 & 6 \\ 7 & 9 & 8 & 9 & 1 & 8 & 7 & 9 & 8 \\ \hline 9 & 2 & 8 & 8 & 3 & 3 & 8 & 2 & 7 \\ 5 & 9 & 2 & 9 & 8 & 3 & 1 & 8 & 2 \\ 4 & 5 & 9 & 7 & 9 & 8 & 9 & 1 & 8 \end{array}\right] \begin{bmatrix} 1 \\ 2 \\ 1 \\ 3 \\ 3 \\ 1 \\ 5 \\ 1 \\ 2 \end{bmatrix} = \begin{bmatrix} 81 \\ 103 \\ 127 \\ 106 \\ 122 \\ 131 \\ 113 \\ 96 \\ 141 \end{bmatrix}. \tag{13.32}$$

The plan is to embed the above $mn \times mn$ BTTB matrix to a larger $4mn \times 4mn$ BCCB matrix, as was done in the 1D case (Section 2.5.2), which in turn will be similarly applied by the FFT technique. Once we extend v to

$$v_{ext} = \mathbf{Vector}\left(\begin{bmatrix} \mathbf{Matrix}(v) & 0_{m\times n} \\ 0_{m\times n} & 0_{m\times n} \end{bmatrix} \right) = \mathbf{Vector}\left(\left[\begin{array}{ccc|ccc} 1 & 3 & 5 & 0 & 0 & 0 \\ 2 & 3 & 1 & 0 & 0 & 0 \\ 1 & 1 & 2 & 0 & 0 & 0 \\ \hline 0 & 0 & 0 & 0 & 0 & 0 \\ 0 & 0 & 0 & 0 & 0 & 0 \\ 0 & 0 & 0 & 0 & 0 & 0 \end{array}\right] \right),$$

the method using an extended BCCB matrix C to compute $\widetilde{v} = Cv_{ext}$ goes like in the following Table 13.1, where the boxed elements of size 3 are related to $w = Kv$ and the rest is to ensure that C is a BCCB (see Section 2.5.2). From

Table 13.1. *Embedding of a BTTB matrix into a large BCCB matrix (see* `ch13.m`*).*

$$
\left[\begin{array}{cccccccccccccccccccccccccccccccccccc}
8&2&7&0&9&1&8&6&3&0&7&9&2&5&1&0&9&3&0&0&0&0&0&0&9&2&8&0&4&5&8&3&3&0&7&9\\
1&8&2&7&0&9&9&8&6&3&0&7&3&2&5&1&0&9&0&0&0&0&0&0&5&9&2&8&0&4&9&8&3&3&0&7\\
9&1&8&2&7&0&7&9&8&6&3&0&9&3&2&5&1&0&0&0&0&0&0&0&4&5&9&2&8&0&7&9&8&3&3&0\\
0&9&1&8&2&7&0&7&9&8&6&3&0&9&3&2&5&1&0&0&0&0&0&0&0&4&5&9&2&8&0&7&9&8&3&3\\
7&0&9&1&8&2&3&0&7&9&8&6&1&0&9&3&2&5&0&0&0&0&0&0&8&0&4&5&9&2&3&0&7&9&8&3\\
2&7&0&9&1&8&6&3&0&7&9&8&5&1&0&9&3&2&0&0&0&0&0&0&2&8&0&4&5&9&3&3&0&7&9&8\\
8&3&3&0&7&9&8&2&7&0&9&1&8&6&3&0&7&9&2&5&1&0&9&3&0&0&0&0&0&0&9&2&8&0&4&5\\
9&8&3&3&0&7&1&8&2&7&0&9&9&8&6&3&0&7&3&2&5&1&0&9&0&0&0&0&0&0&5&9&2&8&0&4\\
7&9&8&3&3&0&9&1&8&2&7&0&7&9&8&6&3&0&9&3&2&5&1&0&0&0&0&0&0&0&4&5&9&2&8&0\\
0&7&9&8&3&3&0&9&1&8&2&7&0&7&9&8&6&3&0&9&3&2&5&1&0&0&0&0&0&0&0&4&5&9&2&8\\
3&0&7&9&8&3&7&0&9&1&8&2&3&0&7&9&8&6&1&0&9&3&2&5&0&0&0&0&0&0&8&0&4&5&9&2\\
3&3&0&7&9&8&2&7&0&9&1&8&6&3&0&7&9&8&5&1&0&9&3&2&0&0&0&0&0&0&2&8&0&4&5&9\\
9&2&8&0&4&5&8&3&3&0&7&9&8&2&7&0&9&1&8&6&3&0&7&9&2&5&1&0&9&3&0&0&0&0&0&0\\
5&9&2&8&0&4&9&8&3&3&0&7&1&8&2&7&0&9&9&8&6&3&0&7&3&2&5&1&0&9&0&0&0&0&0&0\\
4&5&9&2&8&0&7&9&8&3&3&0&9&1&8&2&7&0&7&9&8&6&3&0&9&3&2&5&1&0&0&0&0&0&0&0\\
0&4&5&9&2&8&0&7&9&8&3&3&0&9&1&8&2&7&0&7&9&8&6&3&0&9&3&2&5&1&0&0&0&0&0&0\\
8&0&4&5&9&2&3&0&7&9&8&3&7&0&9&1&8&2&3&0&7&9&8&6&1&0&9&3&2&5&0&0&0&0&0&0\\
2&8&0&4&5&9&3&3&0&7&9&8&2&7&0&9&1&8&6&3&0&7&9&8&5&1&0&9&3&2&0&0&0&0&0&0\\
0&0&0&0&0&0&9&2&8&0&4&5&8&3&3&0&7&9&8&2&7&0&9&1&8&6&3&0&7&9&2&5&1&0&9&3\\
0&0&0&0&0&0&5&9&2&8&0&4&9&8&3&3&0&7&1&8&2&7&0&9&9&8&6&3&0&7&3&2&5&1&0&9\\
0&0&0&0&0&0&4&5&9&2&8&0&7&9&8&3&3&0&9&1&8&2&7&0&7&9&8&6&3&0&9&3&2&5&1&0\\
0&0&0&0&0&0&0&4&5&9&2&8&0&7&9&8&3&3&0&9&1&8&2&7&0&7&9&8&6&3&0&9&3&2&5&1\\
0&0&0&0&0&0&8&0&4&5&9&2&3&0&7&9&8&3&7&0&9&1&8&2&3&0&7&9&8&6&1&0&9&3&2&5\\
0&0&0&0&0&0&2&8&0&4&5&9&3&3&0&7&9&8&2&7&0&9&1&8&6&3&0&7&9&8&5&1&0&9&3&2\\
2&5&1&0&9&3&0&0&0&0&0&0&9&2&8&0&4&5&8&3&3&0&7&9&8&2&7&0&9&1&8&6&3&0&7&9\\
3&2&5&1&0&9&0&0&0&0&0&0&5&9&2&8&0&4&9&8&3&3&0&7&1&8&2&7&0&9&9&8&6&3&0&7\\
9&3&2&5&1&0&0&0&0&0&0&0&4&5&9&2&8&0&7&9&8&3&3&0&9&1&8&2&7&0&7&9&8&6&3&0\\
0&9&3&2&5&1&0&0&0&0&0&0&0&4&5&9&2&8&0&7&9&8&3&3&0&9&1&8&2&7&0&7&9&8&6&3\\
1&0&9&3&2&5&0&0&0&0&0&0&8&0&4&5&9&2&3&0&7&9&8&3&7&0&9&1&8&2&3&0&7&9&8&6\\
5&1&0&9&3&2&0&0&0&0&0&0&2&8&0&4&5&9&3&3&0&7&9&8&2&7&0&9&1&8&6&3&0&7&9&8\\
8&6&3&0&7&9&2&5&1&0&9&3&0&0&0&0&0&0&9&2&8&0&4&5&8&3&3&0&7&9&8&2&7&0&9&1\\
9&8&6&3&0&7&3&2&5&1&0&9&0&0&0&0&0&0&5&9&2&8&0&4&9&8&3&3&0&7&1&8&2&7&0&9\\
7&9&8&6&3&0&9&3&2&5&1&0&0&0&0&0&0&0&4&5&9&2&8&0&7&9&8&3&3&0&9&1&8&2&7&0\\
0&7&9&8&6&3&0&9&3&2&5&1&0&0&0&0&0&0&0&4&5&9&2&8&0&7&9&8&3&3&0&9&1&8&2&7\\
3&0&7&9&8&6&1&0&9&3&2&5&0&0&0&0&0&0&8&0&4&5&9&2&3&0&7&9&8&3&7&0&9&1&8&2\\
6&3&0&7&9&8&5&1&0&9&3&2&0&0&0&0&0&0&2&8&0&4&5&9&3&3&0&7&9&8&2&7&0&9&1&8
\end{array}\right]
\left[\begin{array}{c}
1\\2\\1\\0\\0\\0\\3\\3\\1\\0\\0\\0\\5\\1\\2\\0\\0\\0\\0\\0\\0\\0\\0\\0\\0\\0\\0\\0\\0\\0\\0\\0\\0\\0\\0\\0
\end{array}\right]
=
\left[\begin{array}{c}
81\\103\\127\\64\\55\\69\\106\\122\\131\\76\\69\\69\\113\\96\\141\\54\\81\\53\\90\\103\\96\\42\\57\\48\\76\\50\\60\\35\\58\\25\\45\\51\\71\\53\\22\\30
\end{array}\right]
\tag{13.33}
$$

Lemma 13.1.3, the above product can be efficiently computed by 2D FFT:

```
>>  v = reshape( v, m, n);  % Input data in matrix form
>>  v_ext = reshape( real( ifft2( fft2(c) .* fft2(v) ) ), m*n, 1)
```

which should produce an v_{ext} identical to the product in (13.33).

It only remains to highlight the construction of the root for matrix C based on the PSF matrix R above. Clearly one observes that the root matrix R_C for C

in (13.33) is related to the block permuted form of an extended R:

$$R_C = \left[\begin{array}{ccc|c|cc} 8 & 8 & 9 & 0 & 2 & 8 \\ 1 & 9 & 5 & 0 & 3 & 9 \\ 9 & 7 & 4 & 0 & 9 & 7 \\ \hline 0 & 0 & 0 & 0 & 0 & 0 \\ \hline 7 & 3 & 8 & 0 & 1 & 3 \\ 2 & 3 & 2 & 0 & 5 & 6 \end{array}\right] = \left[\begin{array}{c|c} R_{22} & R_{21} \\ \hline R_{12} & R_{11} \end{array}\right], \quad \left[\begin{array}{c|cc|ccc} 0 & 0 & 0 & 0 & 0 & 0 \\ \hline 0 & 1 & 3 & 7 & 3 & 8 \\ 0 & 5 & 6 & 2 & 3 & 2 \\ \hline 0 & 2 & 8 & 8 & 8 & 9 \\ 0 & 3 & 9 & 1 & 9 & 5 \\ 0 & 9 & 7 & 9 & 7 & 4 \end{array}\right] = \left[\begin{array}{c|c} R_{11} & R_{12} \\ \hline R_{21} & R_{22} \end{array}\right]. \tag{13.34}$$

In fact, the idea and formulae shown from this example apply to the general case where the PSF matrix R such as (13.30) is given to define K of size $mn \times mn$. The general procedure to compute a BTTB matrix via a BCCB is as follows.

Algorithm 13.1.4. (Computation of a BTTB matrix product).

Assume a PSF root matrix R_1 for a BTTB matrix K is given. For $v \in \mathbb{R}^{m\times n}$, to compute $\mathbf{Vector}(w) = K\mathbf{Vector}(v)$,

(1) If R_1 is of size $(2m-1)\times(2n-1)$, extend it to R_2 of size $2m \times 2n$ as in (13.31):

$$R_1 = \begin{bmatrix} K_{11} & K_{12} \\ K_{21} & K_{22} \end{bmatrix} \Rightarrow R_2 = \begin{bmatrix} 0 & 0_{1\times(n-1)} & 0_{1\times n} \\ 0_{(n-1)\times 1} & K_{11} & K_{12} \\ 0_{n\times 1} & K_{21} & K_{22} \end{bmatrix}.$$

(2) If R_1 is of size $m \times n$ (the same as an observed image), extend it to a $2m \times 2n$ root matrix R_2 by embedding it to the centre of a large matrix (similar to (13.31)):

$$R_1 = \begin{bmatrix} K_{11} & K_{12} \\ K_{21} & K_{22} \end{bmatrix} \Rightarrow R_2 = \begin{bmatrix} 0_{m_1\times n_1} & 0_{m_1\times n_1} & 0_{m_1\times n_1} & 0_{m_1\times n_1} \\ 0_{m_1\times n_1} & K_{11} & K_{12} & 0_{m_1\times n_1} \\ 0_{n\times 1} & K_{21} & K_{22} & 0_{m_1\times n_1} \\ 0_{m_1\times n_1} & 0_{m_1\times n_1} & 0_{m_1\times n_1} & 0_{m_1\times n_1} \end{bmatrix},$$

where $m_1 = m/2$ and $n_1 = n/2$.

(3) Block permute matrix R_2 to get the root matrix R_C for $C_{mn\times mn}$:

$$R_2 = \begin{bmatrix} R_{11} & R_{12} \\ R_{21} & R_{22} \end{bmatrix} \Rightarrow R_C = \begin{bmatrix} R_{22} & R_{21} \\ R_{12} & R_{11} \end{bmatrix}.$$

(This step may be completed by the MATLAB® command `fftshift`*.)*

(4) Compute the 2D FFT of R_C, $\widetilde{R_C} = \texttt{fft2}(R_C)$;

(5) Compute the 2D FFT of the input data v, $\widetilde{v} = \texttt{fft2}(v)$;

(6) Compute the inverse 2D FFT of the product, $w = \texttt{ifft2}(\widetilde{R_C} \;.*\; \widetilde{v})$.

The overall algorithm may be experimented by using the Mfile `ch13.m`. If the PSF matrix R_1 in the above algorithm is symmetric, then $w = K^H v = K v$; otherwise one has to set $R1 = R_1^H$ before using the Algorithm similarly. Also note that if R_1 is solely consisted of values of the kernel, one should set $R_1 = h^2 R_1$ in view of (13.20).

So far we have completely described the discretization process. It remains to discuss the preconditioning and fast solver issues.

Iterative methods for image restoration. As shown, the underlying operator is a summation of a nonlinear differential operator ('sparse' when discretized) and an integral operator ('dense' when discretized). Such a problem poses a significant challenge in computational mathematics due to the need of using FFT to present the dense matrix (or avoiding storage of this latter matrix).

The existing work on the topic fall into the following three categories as described separately. There ought be more work appearing in the near future.

13.2 Fixed point iteration method

This was done in a number of papers including [3,101,105,474,477,478,475, 476]. Once the coefficients in $\mathcal{N}(u)$ of (13.22) are 'freezed' at each outer nonlinear step, various iterative solver techniques have been considered for the inner solution [477,478,101,105,102,100,335]. This is a linearization technique, commonly used in treating nonlinear PDEs (refer to (6.13) and (12.12)), by repeatedly solving for the new u:

$$\alpha \nabla \cdot \left(\frac{\nabla u}{\sqrt{|\nabla \bar{u}|^2 + \beta}} \right) - K^H K u = -K^H z, \tag{13.35}$$

or for short

$$\alpha \mathcal{L} u - K^H K u = -K^H z \tag{13.36}$$

where $\bar{u}$ is assigned the previous iterate u and is initially set $\bar{u} = z$. More precisely, we solve a lagged diffusion problem for u^{k+1} until $u^{k+1} - u^k$ is small

$$\alpha \nabla \cdot \left(\frac{\nabla u^{k+1}}{\sqrt{|\nabla u^k|^2 + \beta}} \right) - K^H K u^{k+1} = -K^H z. \tag{13.37}$$

Although the method converges, the convergence (for the nonlinear iterations) may not be always fast. There exists a large literature on this topic, mainly due to wide interest in developing fast iterative solvers for the above linearized equations. As stated, since K is a convolution operator, the challenge is to solve the resulting linear system without forming the discretized matrix of $K^H K$ (mimicking the capability of the fast multipole method) [477,478,101,105,102, 100,335]. Below we summarize two contrasting ideas (of operator splitting type) for preconditioning the linear 'summation' operator, before we mention the potential for a third possibility.

13.2.1 A differential operator-based preconditioner

To make use of the sparsity of operator $\mathcal{L}$ (and various fast solvers associated with $\mathcal{L}$), the first idea is to approximate $K^H K$ by some sparse matrix with compatible sparsity pattern to $\mathcal{L}$.

The technique proposed by [477,478] uses $\bar{A} = \alpha\mathcal{L} + \gamma I$ to precondition (13.36) in the simple version and the full preconditioner is

$$M = \frac{1}{\gamma}\left(\gamma I + \bar{K}^H \bar{K}\right)^{-1/2} (\alpha\mathcal{L} + \gamma I)\left(\gamma I + \bar{K}^H \bar{K}\right)^{-1/2} \tag{13.38}$$

where $(\bar{K}^H \bar{K} + \gamma I)^{-1/2}$ denotes a BTTB matrix with the root matrix $(\bar{R})_{jk} = 1/\sqrt{|R_{jk}|^2 + \gamma}$ with R the root matrix for K.

13.2.2 An integral operator-based preconditioner

A related but different preconditioner was attempted by several researchers [103,113,101,104]. This is a dense circulant (or BCCB) matrix preconditioner. The idea of approximating a BTTB by a BCCB matrix follows from the 1D case (see [439] and Section 4.6) while the idea of of approximation of a sparse matrix by a BCCB matrix generalizes the 1D case (see [106] and (4.12)).

Denote by $c(X)$ the circulant approximation of a matrix X, as done in (4.11). Thus, with A_{ij} of size $m \times m$, a block matrix

$$A = \begin{bmatrix} A_{11} & A_{12} & \cdots & A_{1n} \\ A_{21} & A_{22} & \cdots & A_{2n} \\ \vdots & \vdots & \ddots & \vdots \\ A_{n1} & A_{n2} & \cdots & A_{nn} \end{bmatrix} \tag{13.39}$$

will admit the so-called level-one preconditioner

$$c(A) = \begin{bmatrix} c(A_{11}) & c(A_{12}) & \cdots & c(A_{1n}) \\ c(A_{21}) & c(A_{22}) & \cdots & c(A_{2n}) \\ \vdots & \vdots & \ddots & \vdots \\ c(A_{n1}) & c(A_{n2}) & \cdots & c(A_{nn}) \end{bmatrix}. \tag{13.40}$$

As each block entry can be diagonalized by the FFT matrix $\mathbf{F}_m$

$$\Lambda = (I_n \otimes \mathbf{F}_m) c(A) (I_n \otimes \mathbf{F}_m)^{-1} = \begin{bmatrix} d_{11} & d_{12} & \cdots & d_{1n} \\ d_{21} & d_{22} & \cdots & d_{2n} \\ \vdots & \vdots & \ddots & \vdots \\ d_{n1} & d_{n2} & \cdots & d_{nn} \end{bmatrix}. \quad (13.41)$$

where the diagonal matrix $d_{ij} = \mathbf{F}_m c(A_{ij}) \mathbf{F}_m^{-1}$, the level-one preconditioner is thus denoted by

$$c(A) = (I_n \otimes \mathbf{F}_m)^{-1} \Lambda (I_n \otimes \mathbf{F}_m).$$

We note in passing that, from (4.11), $d_{ij} = \text{diag}(\mathbf{F}_m A_{ij} \mathbf{F}_m^{-1})$. Here Λ is a block matrix with diagonal blocks which can be solved by permuting all diagonals to the main diagonal blocks (see Lemma 4.1.1):

$$P \Lambda P^T = \text{diag}(D_1, \cdots, D_m), \qquad \Lambda = P^T \text{diag}(D_1, \cdots, D_m) P. \quad (13.42)$$

More specifically, Px groups all first elements of x's block vectors and then second elements etc e.g. for $n = 4$ and $m = 3$,

$$P \begin{bmatrix} 3 & 2 & 4 | 1 & 12 & 5 | 11 & 10 & 7 | 0 & 8 & 6 \end{bmatrix}^T = \begin{bmatrix} 3 & 1 & 11 & 0 | 2 & 12 & 10 & 8 | 4 & 5 & 7 & 6 \end{bmatrix}^T.$$

Therefore, for a general A, the level-one preconditioner is formally

$$c(A) = (I_n \otimes \mathbf{F}_m)^{-1} P^T \text{diag}(D_1, \cdots, D_m) P (I_n \otimes \mathbf{F}_m). \quad (13.43)$$

For the main linear system (13.36), the level-one preconditioner takes the form $M_1 = \alpha c(L) + c(K)^H c(K)$ or

$$M_1 = (I_n \otimes \mathbf{F}_m)^{-1} P^T \text{diag}(\alpha E_1 + D_1^2, \cdots, \alpha E_m + D_m^2) P (I_n \otimes \mathbf{F}_m), \quad (13.44)$$

where E_j denotes the diagonal block j for matrix $\mathcal{L}$.

The involvement of dense blocks D_j may be removed if we approximate each D_j by its circulant counterpart $c(D_j)$ by (4.11). This is the idea [103] of the level 2 preconditioner

$$\begin{aligned} M_2 = (I_n \otimes \mathbf{F}_m)^{-1} P^T & \text{diag}(\alpha c(E_1) + c(D_1)^2, \cdots, \alpha c(E_m) + c(D_m)^2) \\ & \times P (I_n \otimes \mathbf{F}_m). \end{aligned} \quad (13.45)$$

As $\hat{D}_j = \mathbf{F}_n c(D_j) \mathbf{F}_n^{-1}$ and $\hat{E}_j = \mathbf{F}_n c(E_j) \mathbf{F}_n^{-1}$ are diagonal matrices of size

$n \times n$, the level 2 preconditioner to (13.36) can be formally written as

$$\begin{aligned} M_2 &= (I_n \otimes \mathbf{F}_m)^{-1} P^T \operatorname{diag}(\alpha c(E_1) + c(D_1)^2, \cdots, \alpha c(E_m) + c(D_m)^2) \\ &\quad \times P(I_n \otimes \mathbf{F}_m) \\ &= (I_n \otimes \mathbf{F}_m)^{-1} P^T (I_m \otimes \mathbf{F}_n)^{-1} \\ &\quad \operatorname{diag}(\alpha \hat{E}_1) + \hat{D}_1)^2, \cdots, \alpha \hat{E}_m) + \hat{D}_m)^2)(I_m \otimes \mathbf{F}_n) P(I_n \otimes \mathbf{F}_m) \\ &= (\mathbf{F}_n \otimes \mathbf{F}_m)^{-1} \operatorname{diag}(\alpha \hat{E}_1) + \hat{D}_1)^2, \cdots, \alpha \hat{E}_m) + \hat{D}_m)^2)(\mathbf{F}_n \otimes \mathbf{F}_m). \end{aligned} \tag{13.46}$$

Here note that $(\mathbf{F}_n \otimes \mathbf{F}_m) = (I_m \otimes \mathbf{F}_n)P(I_n \otimes \mathbf{F}_m)$ and also the matrix-vector $\hat{v} = (\mathbf{F}_n \otimes \mathbf{F}_m)v = \texttt{fft2}(\mathbf{Matrix}(v))$.

13.2.3 The use of DWT for a mixed preconditioner

The effectiveness of the above two integral operator-based preconditioners hinges on the approximation (4.11). Here it might be difficult to improve on the approximation of $K^H K$ by a BCCB matrix $C^H C$. However the approximation of the sparse matrix $\mathcal{L}$ by a dense BCCB matrix $c(\mathcal{L})$ may be done differently. Our idea below is first to transform $\mathcal{L}$ and $C^H C$ using a DWT and then to design a sparse preconditioner in the wavelet space. To proceed, we must consider the detailed structure of a circulant as well as a BCCB matrix under a DWT.

Lemma 13.2.5. (DWT for a circulant matrix). *Let C be a circulant matrix, as defined by (4.10), with the root vector $c = [c_0\ c_1 \ldots c_{n-1}]^T$ and a DWT as defined by (1.48). Then the one level DWT, $\widetilde{C}$, of C is a block 2×2 matrix of four circulant matrices of equal size $n/2 \times 2$:*

$$\widetilde{C} = WCW^T = \begin{bmatrix} D & T \\ S & U \end{bmatrix}.$$

Moreover the roots of these four circulant matrices are uniquely determined by the DWT of the first two row vectors of C.

Proof. This is by direct construction. For the circulant matrix C as in (4.10), the most important observation is that C is also circulant in blocks of size 2×2! (In fact, C is a BCCB matrix of any block size 2^τ).

Let the first two rows (block row 1) of $C_1 = CW^T$ be denoted by

$$\begin{array}{cccccc|cccccc} d_1 & d_2 & d_3 & d_4 & \cdots & d_{n/2-1}\ d_{n/2} & s_1 & s_2 & s_3 & s_4 & \cdots & s_{n/2-1}\ s_{n/2} \\ t_1 & t_2 & t_3 & t_4 & \cdots & t_{n/2-1}\ t_{n/2} & u_1 & u_2 & u_3 & u_4 & \cdots & u_{n/2-1}\ u_{n/2}. \end{array}$$

Then one observes that C_1 is consisted of two (left and right) block circulant matrices of block size 2×1:

$$\left[\begin{array}{ccccccc|ccccccc} d_1 & d_2 & d_3 & d_4 & \cdots & d_{n/2-1} & d_{n/2} & s_1 & s_2 & s_3 & s_4 & \cdots & s_{n/2-1} & s_{n/2} \\ t_1 & t_2 & t_3 & t_4 & \cdots & t_{n/2-1} & t_{n/2} & u_1 & u_2 & u_3 & u_4 & \cdots & u_{n/2-1} & u_{n/2} \\ d_{n/2} & d_1 & d_2 & d_3 & \cdots & d_{n/2-2} & d_{n/2-1} & s_{n/2} & s_1 & s_2 & s_3 & \cdots & s_{n/2-2} & s_{n/2-1} \\ t_{n/2} & t_1 & t_2 & t_3 & \cdots & t_{n/2-2} & t_{n/2-1} & u_{n/2} & u_1 & u_2 & u_3 & \cdots & u_{n/2-2} & u_{n/2-1} \\ \cdots & \cdots & \cdots & \cdots & \cdots & \cdots & \cdots & \cdots & \cdots & \cdots & \cdots & \cdots & \cdots & \cdots \\ d_2 & d_3 & d_4 & d_5 & \cdots & d_{n/2} & d_1 & s_2 & s_3 & s_4 & s_5 & \cdots & s_{n/2} & s_1 \\ t_2 & t_3 & t_4 & t_5 & \cdots & t_{n/2} & t_1 & u_2 & u_3 & u_4 & u_5 & \cdots & u_{n/2} & u_1 \end{array}\right].$$

Clearly $\widetilde{C} = WC_1 = WCW^T$ will be circulant in four blocks. To generate $\widetilde{C}$, we again need to transform the first vector of each $n \times n/2$ block of C_1

$$\begin{array}{c} d_1\ t_1\ d_{n/2}\ t_{n/2}\ \cdots d_2\ t_2 \\ s_1\ u_1\ s_{n/2}\ u_{n/2} \cdots s_2\ u_2 \end{array}$$

respectively to the following (for simplicity, we re-use the same notation)

$$\begin{array}{c|c} d_1\ d_2\ d_3\ d_4 \cdots d_{n/2-1}\ d_{n/2} & s_1\ \ s_2\ s_3\ s_4\ \cdots s_{n/2-1}\ \ s_{n/2} \\ \hline t_1\ \ t_2\ \ t_3\ \ t_4\ \ \cdots t_{n/2-1}\ \ t_{n/2} & u_1\ u_2\ u_3\ u_4 \cdots u_{n/2-1}\ u_{n/2}. \end{array}$$

This completes the proof. ■

To apply the DWT to (13.36), we have to consider the case of a BCCB matrix. The block version of a DWT is $W_{2D} = W_n \otimes W_n = (W \otimes W)_{n^2 \times n^2}$; similarly we can generalize to $W_{2D} = W_m \otimes W_n$. To use Lemma 13.2.5, we need the decomposition of the tensor-product [465] (see Section 14.3.1)

$$W_{2D} = (I_n \otimes W)P(I_n \otimes W) = (W \otimes I_n)(I_n \otimes W) = (I_n \otimes W)(W \otimes I_n) \tag{13.47}$$

where P is as defined in (13.46).

Lemma 13.2.6. (DWT for a BCCB matrix). *Let $C_{n^2 \times n^2}$ be a BCCB matrix, as defined by (13.23), with the root matrix $c = [c_0\ c_1\ \ldots\ c_{n-1}]^T$ and a DWT as defined by (13.47). Then the one-level DWT, $\widetilde{C}$, of C is a block 2×2 circulant matrix with 4 circulant blocks of size $n/2 \times n/2$. Moreover the roots of these circulant matrices are uniquely determined by the DWT of $8n$ vectors of C.*

Proof. Firstly consider $W_{2D} = (I_n \otimes W)(W \otimes I_n)$ and the block level transform first i.e. $C_1 = (W \otimes I_n)C(W \otimes I_n)^T$. From Lemma 13.2.5, C_1 is a block 2×2 circulant matrix, determined by transforming all first columns of the first two

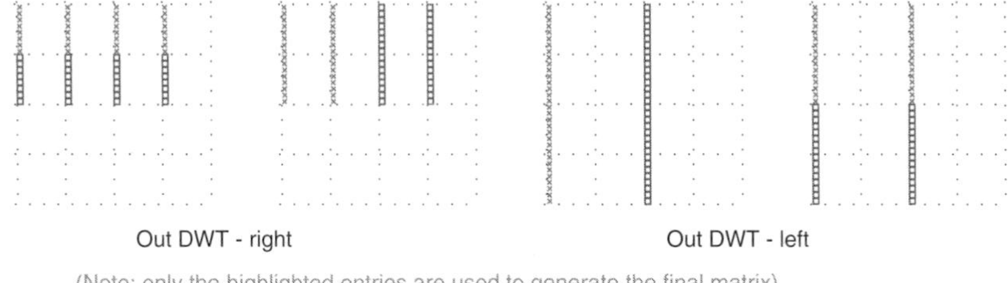

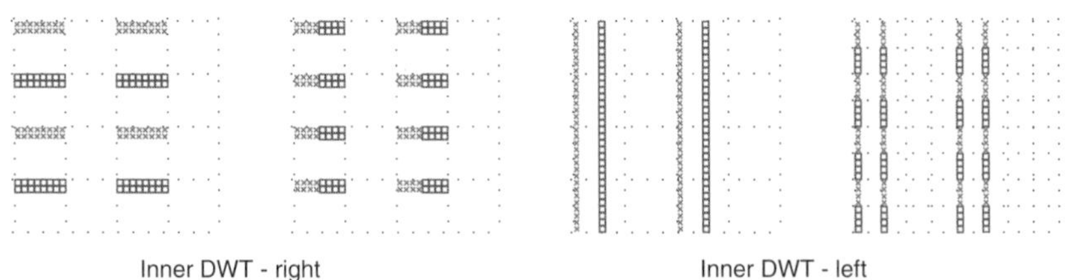

Figure 13.1. Graphical illustration of the proof of Lemma 13.2.6. The top plots show the outer block level transform $C_1 = (W \otimes I_n)C(W \otimes I_n)^T$ while the bottom plots show the inner block transform $\widetilde{C} = (I_n \otimes W)C_1(I_n \otimes W)^T = W_{2D}CW_{2D}^T$.

block rows of C i.e.

$$\begin{array}{lllll} c_0 & c_{n-1} & \cdots & c_2 & c_1 \\ c_1 & c_0 & \cdots & c_3 & c_2. \end{array}$$

Secondly consider the transform within each block i.e. $\widetilde{C} = (I_n \otimes W)C_1(I_n \otimes W)^T$. Again using Lemma 13.2.5 we see that for each block, transforming the first two rows is sufficient. The overall data access and minimal transformations can be illustrated by Figure 13.1, where only 'highlighted' positions are transformed. Overall, $8n$ vectors are transformed from left and right transformations. Thus the proof is complete. ■

For a BCCB matrix, $\widetilde{C}$ is a block circulant matrix with four circulant blocks (BC4CB, i.e. not directly a BCCB matrix unlike the 1D case with Lemma 13.2.5) but the result of the outer block level transform is BCCB. The former BC4CB matrix requires a discussion of how to compute a matrix–vector product with it and the latter fact will be explored to resolve the potential difficulty associated with using multiple levels.

Firstly, we consider the matrix–vector product. We claim that a BC4CB matrix G (as in the last plot of Figure 13.1) can be permuted to four BCCB matrices. Define $P_{n^2 \times n^2} = \Pi^T \otimes I$ as the (block) odd-even permutation matrix with Π^T the FFT permutation matrix from §1.6: $\Pi^T = I([1\ 3\ \ldots n-1\ 2\ 4 \ldots\ n], :)$.

Then matrix $G_1 = PGP^T$ is a 2×2 block matrix with BCCB blocks so $y = Gx = P^T G_1(Px)$ can be evaluated by four separate 2D FFT operations. For clarity, consider the following BC4CB matrix

$$G = \left[\begin{array}{cc|cc|cc} A_1 & B_1 & A_3 & B_3 & A_2 & B_2 \\ C_1 & D_1 & C_3 & D_3 & C_2 & D_2 \\ \hline A_2 & B_2 & A_1 & B_1 & A_3 & B_3 \\ C_2 & D_2 & C_1 & D_1 & C_3 & D_3 \\ \hline A_3 & B_3 & A_2 & B_2 & A_1 & B_1 \\ C_3 & D_3 & C_2 & D_2 & C_1 & D_1 \end{array}\right], \quad PGP^T = \left[\begin{array}{ccc|ccc} A_1 & A_3 & A_2 & B_1 & B_3 & B_2 \\ A_2 & A_1 & A_3 & B_2 & B_1 & B_3 \\ A_3 & A_2 & A_1 & B_3 & B_2 & B_1 \\ \hline C_1 & C_3 & C_2 & D_1 & D_3 & D_2 \\ C_2 & C_1 & C_3 & D_2 & D_1 & D_3 \\ C_3 & C_2 & C_1 & D_3 & D_2 & D_1 \end{array}\right]$$

$$= \left[\begin{array}{c|c} A & B \\ \hline C & D \end{array}\right],$$

where all matrices A_j, B_j, C_j, D_j are circulant and clearly A, B, C, D are BCCB. Therefore

$$y = Gx = P^T G_1 Px = P^T \begin{bmatrix} Ax_{odd} + Bx_{even} \\ Cx_{odd} + Dx_{even} \end{bmatrix}, \quad \begin{bmatrix} x_{odd} \\ x_{even} \end{bmatrix} = Px,$$

which is amenable to `fft2` operations.

Secondly we attempt to separate the outer and inner block transforms. Similar to (9.4), our one-step DWT matrix $W_{2D} = (I_n \otimes W)(W \otimes I_n)$ may be written as

$$W_{2D} = \begin{bmatrix} I_{n/2} \otimes W & \\ & I_{n/2} \otimes W \end{bmatrix} \begin{bmatrix} Q_n \otimes I_n \\ P_n \otimes I_n \end{bmatrix}, \quad W = \begin{bmatrix} Q_n \\ P_n \end{bmatrix}, \tag{13.48}$$

where P_n (corresponding to the 'sums' in a DWT) and Q_n are applied in the outer block DWT as in Lemma 13.2.6 while the inner block DWT is kept separate and applied later.

Finally a preconditioner similar to Section 9.3 (or the work of [60] and [108]) may be proposed as follows. Let

$$T_0 x_0 = b_0 \tag{13.49}$$

denote the equation $(\alpha\mathcal{L} + M)x_0 = b_0$ approximating (13.36) (with M representing the usual BCCB term $C^T C$). Then the use of W_{2D} will convert (13.49) to

$$\widetilde{T_0}\widetilde{x_0} = \widetilde{b_0}, \qquad \widetilde{T_0} = \left[\begin{array}{c|c} A_1 & B_1 \\ \hline C_1 & T_1 \end{array}\right], \tag{13.50}$$

where A_1, B_1, C_1 are sparse matrices plus BC4BC matrices by Lemma 13.2.6. Here the trick is not to form T_1 explicitly (involving a BC4CB matrix):

$$T_1 = (I_{n/2} \otimes W)\widehat{T_1}(I_{n/2} \otimes W)^T,$$

where $\widehat{T_1}$ is a sparse matrix plus a BCCB matrix resulting from the out block DWT. Thus a fine level matrix T_0 involving the sparse matrix $\mathcal{L}_0$ and a BCCB matrix M_0 is directly linked to the coarser level matrix $\widehat{T}_1$ involving the sparse matrix $\widetilde{\mathcal{L}}_1$ and a BCCB matrix M_1 – a ready and repetitive setting for using the wavelet Schur type preconditioner (Section 9.3).

Remark 13.2.7. So far we have mainly discussed the FFT approach for K. Other related transforms such as the DCT [101] and DST [105] have also been considered in this context. See [104].

There is still much scope left to search for a robust preconditioner for preconditioning the linear equation (13.36). However, the overall efficiency can be restricted by the nonlinearity in the main equation (13.22). Below we review some methods for solving (13.22) by different (or no) linearization.

13.3 Explicit time marching schemes

The idea explored by [408,378] is to turn the nonlinear PDE (13.22) into a parabolic equation before using an explicit Euler method to march in time to convergence. The original idea in [408], refined in [378], aims to solve the following parabolic PDE until a steady state has been reached

$$u_t = |\nabla u|\Big[\alpha\nabla\cdot\Big(\frac{\nabla u}{\sqrt{|\nabla u|^2+\beta}}\Big) - K^*Ku + K^*z\Big]. \tag{13.51}$$

The explicit time marching schemes become the following

$$\frac{u^{k+1}-u^k}{\Delta t} = |\nabla u^k|\Big[\alpha\nabla\cdot\Big(\frac{\nabla u^k}{\sqrt{|\nabla u^k|^2+\beta}}\Big) - K^*Ku^k + K^*z\Big]. \tag{13.52}$$

As remarked in [390], for linear problems, this type of ideas represents a kind of relaxation schemes. The drawback may be that the artificial time step Δt must be small due to stability requirement.

13.4 The Primal-dual method

This is probably the most reliable method for solving (13.22) in many ways. It was proposed in [110] and discussed also in [112,64,377]. The method solves for both the primal and dual variables together in order to achieve faster convergence with the Newton method (and a constrained optimization with the dual variable). As discussed in [64], the Newton method for equation (13.16) or (13.13) leads to very slow or no convergence because z is often not a sufficiently close initial guess for u and the operator is highly nonlinear. However introducing the dual variable

$$\omega = \frac{\nabla u}{|\nabla u|} \tag{13.53}$$

to (13.13) appears to have made the combined system

$$\begin{cases} -\alpha \nabla \cdot \omega + K^* K u = K^* z, \\ \omega |\nabla u| - \nabla u = 0. \end{cases}$$

in two variables (u, ω) more amenable to Newton iterations as the new system is nearly 'linear' in the two variables (not so linear as a single variable u after elimination of ω). The idea resembles that of a mixed finite element method Section 12.3. Note that ω is constrained in each iteration step so the overall algorithm needs some care in any implementation.

Owing to the definition (13.53) using $|\nabla u|$ explicitly, the formulation will not be rigorous for non-differentiable functions u. In this case, the general derivation of the primal-dual algorithm starts from the alternative norm (13.8) below. As we aim to arrive at a numerical scheme, we shall consider the discrete optimization (without using the the Euler–Lagrange equation (13.13)) directly from (13.9) and with the dual norm (to replace (13.8)) [110]

$$\|v\|_2 = \max_{\|\omega\|_2 \le 1} v^T \omega, \qquad v \in \mathbb{R}^n, \ \forall n. \tag{13.54}$$

Denote the discretized problem of (13.9) by, with h^2 due to 2D integrals,

$$\min_u \alpha \sum_{k=1}^{mn} h^2 \|\frac{1}{h} A_k^T u\| + \frac{h^2}{2} \|Ku - z\|_2^2$$

i.e. setting $\alpha = \alpha / h$ (and dividing by h^2)

$$\min_u \alpha \sum_{k=1}^{mn} \|A_k^T u\| + \frac{1}{2} \|Ku - z\|_2^2 \tag{13.55}$$

where A_k^T and K are as defined in (13.18).

Now using (13.54) for each k i.e.

$$\|A_k^T u\| = \max_{\|\mathbf{x}_k\| \le 1} (A_k^T u)^T \mathbf{x}_k \tag{13.56}$$

with $\mathbf{x}_k \in \mathbb{R}^2$, the above minimization problem becomes

$$\min_u \alpha \sum_{k=1}^{mn} \max_{\|\mathbf{x}_k\| \le 1} u^T A_k \mathbf{x}_k + \frac{\|Ku - z\|_2^2}{2} = \min_u \max_{\|\mathbf{x}_k\| \le 1} \alpha u^T A \mathbf{x} + \frac{\|Ku - z\|_2^2}{2} \tag{13.57}$$

where $u \in \mathbb{R}^{mn}$ is the usual unknown, $\mathbf{x}$ is the new dual variable with

$$\begin{aligned}&\mathbf{x} = (\mathbf{x}_1, \mathbf{x}_2, \ldots, \mathbf{x}_{mn})^T \in \mathbb{R}^{2mn},\\ &\mathbf{x}_k \in \mathbb{R}^2 \qquad \text{and}\\ &A = [A_1, A_2, \ldots, A_{mn}] \in \mathbb{R}^{mn \times 2mn} \qquad \text{as}\\ &A_k^T \in \mathbb{R}^{2 \times mn}.\end{aligned}$$

As the object functional in (13.57) is optimized in a bounded domain, convex in one variable (u) and concave in the other ($\mathbf{x}$), therefore the operations min and max can be interchanged [404] to give

$$\max_{\|\mathbf{x}_k\| \le 1} \min_{u} \alpha u^T A\mathbf{x} + \frac{1}{2}\|Ku - z\|_2^2. \tag{13.58}$$

The solution to the inner min problem of (13.58) is given by

$$\alpha A\mathbf{x} + K^H(Ku - z) = 0. \tag{13.59}$$

The second equation comes from the solution $\omega = v/\|v\|_2$ to (13.54) or specifically the solution of (13.56)

$$\mathbf{x}_k = \frac{A_k^T u}{\|A_k^T u\|_2}, \quad \|A_k^T u\|_2 \mathbf{x}_k = A_k^T u, \quad \sqrt{\|A_k^T u\|_2^2 + \beta}\,\mathbf{x}_k = A_k^T u, \tag{13.60}$$

where adding a small β is to ensure that $\mathbf{x}_k$ always has a unique solution. To put (13.60) to a compact form, we define a new block matrix [110]

$$E = \operatorname{diag}(I_2\eta_1, \; I_2\eta_2, \; \cdots, \; I_2\eta_{mn})$$

where $\eta_k = \sqrt{\|A_k^T u\|_2^2 + \beta}$. Then (13.60) together with (13.59) defines the primal-dual method of Chan *et al.* [110] as follows

$$\begin{cases} E\mathbf{x} - A^T u = 0,\\ \alpha A\mathbf{x} + K^H(Ku - z) = 0,\\ \|\mathbf{x}_k\|_2 \le 1, \end{cases} \tag{13.61}$$

which is a nonlinear system for ($\mathbf{x}$, u), though 'less' nonlinear than (13.22). The constraint in (13.61) will be treated separately.

To solve the nonlinear system of the first two equations in (13.61), the Newton method may be used to yield

$$\begin{bmatrix} E & -FA^T \\ \alpha A & K^H K \end{bmatrix} \begin{bmatrix} \Delta\mathbf{x} \\ \Delta u \end{bmatrix} = -\begin{bmatrix} E\mathbf{x} - A^T u \\ \alpha A\mathbf{x} + K^H(Ku - z) \end{bmatrix}, \tag{13.62}$$

where

$$F = \operatorname{diag}\left(I_2 - \frac{\mathbf{x}_1 u^T A_1}{\eta_1}, \; I_2 - \frac{\mathbf{x}_2 u^T A_2}{\eta_2}, \; \cdots, \; I_2 - \frac{\mathbf{x}_{mn} u^T A_{mn}}{\eta_{mn}} \right),$$

since $\mathbf{x}_k \in \mathbb{R}^{2\times 1}$, $u^T \in \mathbb{R}^{1\times mn}$, $A_k \in \mathbb{R}^{mn\times 2}$,

$$\begin{aligned} \frac{\partial \eta_k}{\partial u_{m+k}} &= (u_{m+k} - u_k)/\eta_k, \\ \frac{\partial \eta_k}{\partial u_{k+1}} &= (u_{k+1} - u_k)/\eta_k, \\ \frac{\partial \eta_k}{\partial u_k} &= (2u_k - u_{m+k} - u_{k+1})/\eta_k, \\ A_k^T &= \begin{bmatrix} \cdot & \overbrace{-1}^{(k)} & 1 & \cdot & 0 & \cdot \\ \cdot & -1 & \cdot & & 1 & \cdot \end{bmatrix}. \end{aligned} \tag{13.63}$$

Viewing (13.62) as a block 2×2 matrix, since it is feasible to invert $E = A_{11}$, the Schur complement method for (13.62) (see Chapter 7) gives

$$\begin{cases} S\Delta u = -(\alpha A E^{-1} A^T u + K^H(Ku - z)) \\ \quad \Delta \mathbf{x} = -\mathbf{x} + E^{-1} A^T u + E^{-1} F A^T \Delta u, \end{cases} \tag{13.64}$$

where the Schur matrix $S = A_{22} - A_{21} A_{11}^{-1} A_{12} = \alpha A E^{-1} F A^T + K^H K$ resembles the summation operator (13.36) in the case of fixed iterations. Therefore the preconditioners developed in Section 13.2 are applicable to the iterative solution of (13.64) and (13.62).

In [110], the symmetrized matrix $\widehat{S} = (S + S^H)/2$ is used to replace S in (13.64) and the preconditioned conjugate gradient method (Section 3.4) is used. Other iterative approaches have not yet been tried. Finally we remark that the constraint in (13.61) is imposed by the backtracking procedure (at the end of each Newton step)

$$\mathbf{x} = \mathbf{x} + s\Delta\mathbf{x}, \qquad y = y + s\Delta y, \tag{13.65}$$

with $s = \rho \sup\{\tau \mid \|\mathbf{x}_k + \tau \Delta \mathbf{x}_k\|_2 < 1\}$ for some $\rho \in (0, 1)$ and assuming initially $\|\mathbf{x}_k\|_2 < 1$.

13.5 Nonlinear multigrids for optimization

There are many other methods one can consider, for example, using the Krylov subspace methods [88,219] or wavelets methods [367,368]. Here we briefly discuss the fast solver issues [109] via the nonlinear multigrid approach and, as this is on-going work, we only give a short introduction. The multilevel subspace correction framework was considered in [449] and the references therein for convex optimization problems.

An alternative to variational PDEs via the Euler–Lagrange equations is to first discretize the object functional in (13.11) and then carry out discrete optimization. Although we have discussed various multilevel methods for operator equations in Chapter 6, there are very few papers devoted to developing multilevel methods for optimization problems, specially for non-differentiable optimization to which the problem (13.11) belongs.

♦ **Differentiable optimization.** The main (and perhaps so far the only early) source of references of a working multigrid algorithm for this problem appears to be [448], which considered the following differentiable optimization problem

$$\min_{\alpha^h} E^h(u^h, \alpha^h) \quad s.t. \quad L^h(u^h, \alpha^h) = f^h, \tag{13.66}$$

where h refers to a fine grid for the objective functional $E(u, \alpha)$ and the constraint $L(u, \alpha) = f$ with u the solution vector and α the unknown scalar parameter (actually [448] dealt with multiple constraints and considered α to be a parameter vector). *For an optimization problem, unlike an operator equation, it is no longer obvious how to define coarse grid problems.*

As both E and f are differentiable, the necessary conditions for minimizing (13.66) are

$$\begin{cases} L^h(u^h, \alpha^h) = f^h, \\ L_u^h(u^h, \alpha^h)\lambda + E_u^h = 0, \\ L_\alpha^h(u^h, \alpha^h)\lambda + E_\alpha^h = 0, \end{cases} \tag{13.67}$$

where there are $n+2$ equations for $n+2$ unknowns: $\alpha, \lambda \in \mathbb{R}$ and $u^h \in \mathbb{R}^n$. Here a naive choice of a coarse grid equation for (13.66) could be

$$\min_{\alpha^H} E^H(u^H, \alpha^H) \quad s.t. \quad L^H(u^H, \alpha^H) = f^H, \tag{13.68}$$

which, unfortunately, may not help the solution of (13.66). It turns out that the correct coarse grid equation for (13.66) is the following

$$\min_{\alpha^H} E^H(u^H, \alpha^H) - < g_1^H, u^H > - < g_2^H, \alpha^H > \quad s.t. \quad L^H(u^H, \alpha^H) = f^H, \tag{13.69}$$

where f^H, g_1^H, g_2^H are the restricted residual quantities while u^H, α^H are the usual restricted solution quantities for the necessary conditions

$$\begin{cases} L^H(u^H, \alpha^H) = f^H, \\ L_u^H(u^H, \alpha^H)\lambda^H + E_u^H = g_1^H, \\ L_\alpha^H(u^H, \alpha^H)\lambda^H + E_\alpha^H = g_2^H. \end{cases} \tag{13.70}$$

Clearly the trick is to connect the coarse grids to fine grids via the intermediate operator equations (13.67) – the necessary (or first-order) conditions!

♦ **Non-differentiable optimization.** This concerns with our problem from (13.11). We restrict ourselves to the denoising case of (13.11) by taking $K = I$ in the following discussion. In the 1D case, the discretization of

$$J(u) = \alpha \int_a^b |\frac{du}{dx}|dx + \frac{1}{2}\int_a^b (u-z)^2 dx$$

leads to the functional (with $u = [u_1, u_2, \ldots, u_n]^T$)

$$\begin{aligned} J(u) &= \alpha \sum_{k=1}^{n-1} h|\frac{1}{h} A_k^T u\| + \frac{1}{2}\sum_{k=1}^{n} h(u_k - z_k)^2 \\ &= h\left[\frac{\alpha}{h}\sum_{k=1}^{n-1} |A_k^T u\| + \frac{1}{2}\|u-z\|_2^2\right], \end{aligned}$$

which may be simply written as (on setting $\alpha = \alpha/h$)

$$J(u) = \alpha \sum_{k=1}^{n-1} |A_k^T u| + \frac{1}{2}\|u-z\|_2^2. \tag{13.71}$$

In the 2D case, the discretization of (13.11) leads to the functional (with $u = [u_1, u_2, \ldots, u_{mn}]^T$)

$$\begin{aligned} J(u) &= \alpha \sum_{k=1}^{mn} h^2|\frac{1}{h} A_k^T u| + \frac{1}{2}\sum_{k=1}^{mn} h^2(u_k - z_k)^2 \\ &= h^2\left[\frac{\alpha}{h}\sum_{k=1}^{mn} |A_k^T u| + \frac{1}{2}\|u-z\|_2^2\right], \end{aligned}$$

which is similarly written as (on setting $\alpha = \alpha/h$)

$$J(u) = \alpha \sum_{k=1}^{mn} |A_k^T u| + \frac{1}{2}\|u-z\|_2^2. \tag{13.72}$$

Here the operator A_k^T is as defined in (13.63) with a minor adjustment in 1D:

$$A_k^T = \left[\ \cdot \ \underbrace{-1}_{(k)}\ \ 1\ \ \cdot\ \ 0\ \ \cdot\ \right].$$

One proposal is to convert the above nondifferentiable optimization $\min_u J(u)$ to a differentiable optimization by adding a small parameter $\beta > 0$

$$\begin{cases} J_\varepsilon(u) = \alpha \sum_{k=1}^{n-1} \sqrt{|A_k^T u|^2 + \beta} + \frac{1}{2}\|u - z\|_2^2, & \text{the 1D case} \\ J_\varepsilon(u) = \alpha \sum_{k=1}^{mn} \sqrt{|A_k^T u|^2 + \beta} + \frac{1}{2}\|u - z\|_2^2, & \text{the 2D case.} \end{cases} \tag{13.73}$$

This will ensure that a coarse grid construction can be done as in (13.69). It is also possible to avoid this parameter β in alternative formulation, if the idea from backtracking is used in the interpolation stage similar to (13.65). Other AMG-based multilevel approaches may also be developed.

13.6 The level set method and other image problems

The above discussed image restoration problem is only part of the long list of problems from imaging science or computer vision in general [26]. Below we shall briefly discuss another problem, image segmentation, from the list and show how it can be tackled by the increasingly popular and powerful tool of level set methods in the context of variational PDEs [379,377]. The variational PDE models provide a unified framework for solving a large class of problems; see also [118,114,338,115].

♦ **The level set function and interfaces.** Let an interested interface Γ (enclosing the domain D) be a closed curve in $\mathbb{R}^2$ (or a surface in $\mathbb{R}^3$ and open or multiple closed curves can also be treated), and formally represented by $x_2 = f(x_1)$. In practice, f may not be a simple and closed formula. Here both D and Γ are contained in the image domain Ω for the given and observed image z (as Section 13.1). Traditionally, to evolve or locate this curve Γ in a practical context, some complicated parametrization (and adaptive meshing) has to be done.

The idea proposed by [379] is to define a function $\phi(\mathbf{x})$, $\mathbf{x} = (x_1, x_2) \in \mathbb{R}^2$, in the space $\mathbb{R}^3$ of one dimension higher so that the zero level curve of $\phi(\mathbf{x})$ defines Γ precisely. Figure 13.2 shows the simple curve $x_1^2 + x_2^2 = 1$ in $\mathbb{R}^2$ is the interface of $x_3 = \phi(\mathbf{x}) = 1 - \sqrt{x_1^2 + x_2^2}$ with the plane $z = 0$ in $\mathbb{R}^3$. Such a function $\phi(\mathbf{x})$ is called the *level set function* for the interface Γ. There are many advantages to evolve $\phi(\mathbf{x})$ to track Γ: the topology of Γ can be flexible, merging and breaking are automatic, and the computational grid (not moving anyway) is regular.

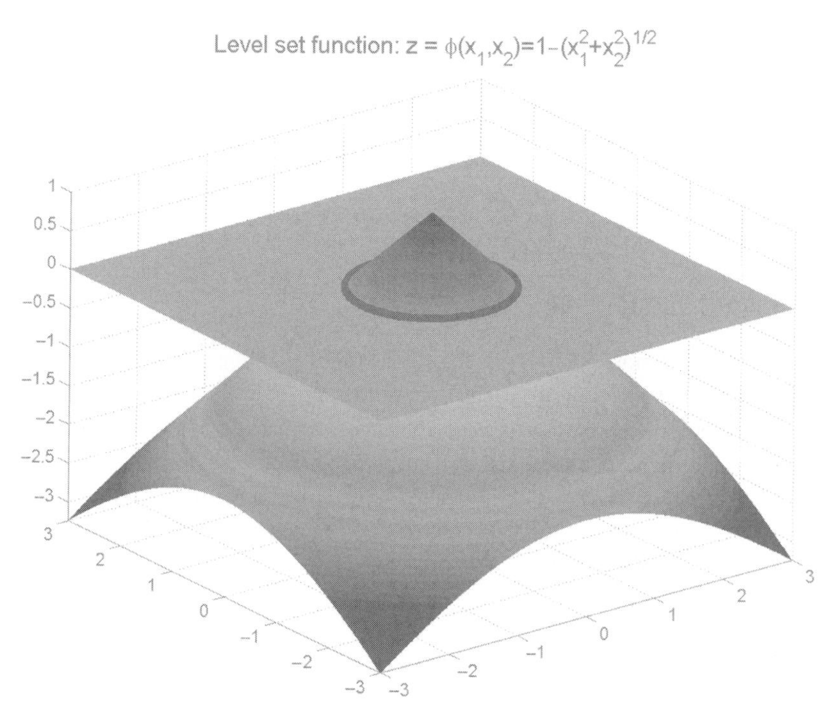

Figure 13.2. Graphical illustration of the level set function $\phi(\mathbf{x})$ and $\phi(\mathbf{x}) = 0$ in $\mathbb{R}^3$.

If $\Gamma = \Gamma(t)$ is represented by the level curve of $\phi(\mathbf{x}, t) = 0$, assuming ϕ is positive in D^- (inside Γ) and negative in D^+ (outside Γ) as illustrated in Figure 13.2, then computing the total differentiation $D\phi/Dt$ of $\phi(\mathbf{x}(t), t) = 0$ gives

$$\frac{\partial \phi}{\partial t} + \nabla_{\mathbf{x}}\phi \cdot (\frac{dx_1(t)}{dt}, \frac{dx_2(t)}{dt}) = \frac{\partial \phi}{\partial t} + V \cdot \nabla_{\mathbf{x}} = 0, \qquad (13.74)$$

where $V = (dx_1(t)/dt, dx_2(t)/dt)$ is the velocity of the evolving curves at time t, V provides the driving force for the evolution equation (or the Hamilton–Jacobi equation [379]) and it will couple with other equations for a given problem [26,377].

♦ **The Heaviside function and the Delta function.** The equation (13.74) alone is not very exciting. The power of a level set method lies in the use of $\phi(\mathbf{x})$ to reformulate an underlying problem. This is where the Heaviside function $H(x)$ and and the Delta function $\delta(x)$ and their regularized variants (note $H_\varepsilon(x) = \frac{1}{2}(1 - (x^3/2\varepsilon^3) + (3x/2\varepsilon))$ for $|x| \le \varepsilon$ is used in [26])

$$H(x) = \begin{cases} 1, \; x \ge 0, \\ 0, \; x < 0, \end{cases} \qquad H_\varepsilon(x) = \begin{cases} 1, & x > \varepsilon, \\ \frac{1}{2}(1 + \frac{x}{\varepsilon} + \frac{\sin(\pi x/\varepsilon)}{\pi}), & |x| \le \varepsilon, \\ 0, & x < -\varepsilon, \end{cases}$$

$$\delta(x) = H'(x) = \begin{cases} \infty, \; x = 0, \\ 0, \;\; x \ne 0, \end{cases} \quad \delta_\varepsilon(x) = H'_\varepsilon(x) = \begin{cases} \frac{1+\cos(\pi x/\varepsilon)}{2\varepsilon}, & |x| \le \varepsilon, \\ 0, & |x| > \varepsilon, \end{cases} \tag{13.75}$$

are useful, where one notes that $\delta'_\varepsilon(x) \approx 0$. The purpose of image segmentation is to locate Γ or more precisely to classify an observed image z into meaningful segments which have sharp variations in the image intensity across the boundaries Γ of such segments (image objects). For a binary image (of essentially two classes), the level set method will attempt to identify the desirable level set function ϕ such that $\phi(\mathbf{x}) = 0$ locates the boundaries of the image. Mathematically speaking, with ϕ found, the underlying image function is approximated by two piecewise constants u_1, u_2

$$u = u(\mathbf{x}) = u_1 H(\phi) + u_2(1 - H(\phi)), \tag{13.76}$$

where $u = u_1$ inside Γ i.e. $\phi > 0$ and $H(\phi) = 1$, and $u = u_2$ outside Γ i.e. $\phi < 0$ and $1 - H(\phi) = 1$. In general, one can consider a solution with p constants $u_1, u_2, \ldots, u_p$ with more level set functions; if n level set functions are used, up to $p = 2^n$ constants can be found and, as each ϕ function may represent more than 1 constant, $n = 2$ (and $p = 4$) is theoretically sufficient to cover all constants and solution regions [377,114] using the four colour theorem in $\mathbb{R}^2$.

It is a necessity to introduce the size measures of $\Gamma(t)$, $D^+(t)$ respectively as, noting $D^+ = \{\mathbf{x} \mid \phi > 0\}$, $D^- = \{\mathbf{x} \mid \phi < 0\}$ and $D^+ \bigcup D^- \bigcup \Gamma = \Omega$,

$$\begin{aligned} |\Gamma(t)| = H^{d-1}(\Gamma) &= \int_\Omega |\nabla H(\phi(\mathbf{x}))| d\mathbf{x} \\ &= \int_{D^+} |\nabla H(\phi(\mathbf{x}))| d\mathbf{x} = \int_\Omega \delta(\phi) |\nabla \phi(\mathbf{x})| d\mathbf{x}, \end{aligned} \tag{13.77}$$

$$|D^+(t)| = L(D^+) = \int_\Omega H(\phi(\mathbf{x})) d\mathbf{x} \tag{13.78}$$

in addition to the normal vector and the curvature as in (13.14)

$$\mathbf{n} = \frac{\nabla \phi}{|\nabla \phi|}, \qquad \mathbf{k} = \nabla \cdot \left(\frac{\nabla \phi}{|\nabla \phi|} \right) = \nabla \cdot \mathbf{n}.$$

Here $H^{d-1}(\Gamma)$ denotes the more general Hausdorff measure for a $(d-1)$-manifold Γ in the $\mathbb{R}^d$ space [26]. The above formulae assume that $\Gamma(t)$ and ϕ remain Lipschitz boundaries and function respectively. Refer to [377,114].

♦ **Segmentation by active contours.** The image segmentation problem was earlier studied in [363] in the mathematical framework of a variational approach based on minimizing an energy functional. Although the theory was attractive, the resulting algorithm was complicated to implement (before the days of level set methods). Using the level set idea, Chan and Vese [117] proposed a related but different variational method that is easier to implement and much refinement and generalization have been done ever since (see [377,114] and the references therein).

For the piecewise approximation u in (13.76) to the given image z, the Chan–Vese model minimizes the following functional

$$
\begin{aligned}
F(u,\phi) &= \underbrace{\mu \int_\Omega \delta(\phi)|\nabla\phi| d\mathbf{x} + \nu \int_\Omega H(\phi) d\mathbf{x}}_{\text{regularity requirement}} + \underbrace{\lambda \int_\Omega |u-z|^2 d\mathbf{x}}_{\text{data fidelity}}, \\
i.e.\ F(u_1,u_2,\phi) &= \mu \int_\Omega \delta(\phi)|\nabla\phi| d\mathbf{x} + \nu \int_\Omega H(\phi) d\mathbf{x} \\
&\quad + \lambda_1 \int_{D^+} |z-u_1|^2 d\mathbf{x} + \lambda_2 \int_{D^-} |z-u_2|^2 d\mathbf{x} \\
&= \mu \int_\Omega \delta(\phi)|\nabla\phi| d\mathbf{x} + \nu \int_\Omega H(\phi) d\mathbf{x} \\
&\quad + \lambda_1 \int_\Omega |z-u_1|^2 H(\phi) d\mathbf{x} + \lambda_2 \int_\Omega |z-u_2|^2 (1-H(\phi)) d\mathbf{x},
\end{aligned}
\tag{13.79}
$$

where μ, ν, λ_1, λ_2 are four nonnegative (Lagrangian) parameters. To derive the Euler–Lagrange equations for (13.79), we first approximate (13.79) using the regularized functions

$$
\begin{aligned}
F(u_1,u_2,\phi) &= \mu \int_\Omega \delta_\varepsilon(\phi)|\nabla\phi| d\mathbf{x} + \nu \int_\Omega H_\varepsilon(\phi) d\mathbf{x} \\
&\quad + \lambda_1 \int_\Omega |z-u_1|^2 H_\varepsilon(\phi) d\mathbf{x} + \lambda_2 \int_\Omega |z-u_2|^2 (1-H_\varepsilon(\phi)) d\mathbf{x}.
\end{aligned}
\tag{13.80}
$$

Thus the first-order conditions and Euler–Lagrange equations for (13.79) are

$$
\begin{cases}
u_1 = \dfrac{\int_\Omega z(\mathbf{x}) H(\phi) d\mathbf{x}}{\int_\Omega H(\phi) d\mathbf{x}}, \qquad u_2 = \dfrac{\int_\Omega z(\mathbf{x})(1-H(\phi)) d\mathbf{x}}{\int_\Omega (1-H(\phi)) d\mathbf{x}}, \\
\delta_\varepsilon(\phi)\left[\mu \nabla \cdot \dfrac{\nabla\phi}{|\nabla\phi|} - \nu - \lambda_1 |z-u_1|^2 + \lambda_2 |z-u_2|^2\right] = 0
\end{cases}
\tag{13.81}
$$

where we have applied $\delta'_\varepsilon(\phi) \approx 0$. Clearly from (13.81), $u_1 = mean(z)$ in D^+ and $u_2 = mean(z)$ in D^-. The solution of (13.81) is as challenging as (13.22) for the restoration model. We remark that the usual name for the method, '*active contours without edges*', does not imply that the method cannot detect edges; in fact, the method can detect sharp edges and discontinuities so the name is merely to differentiate itself from classical methods where explicit edge detectors based on gradients are used.

♦ **Fast solver issues.** There appears to exist no work on the fast solution of (13.81) unlike (13.22), partly because the model was quite recent or more probably because the default solver by time-marching schemes (Section 13.3), in line with the idea of level set methods for tracking evolution curves, is widely used. Therefore the widely accepted solution strategy is to let $\phi(\mathbf{x}) = \phi(\mathbf{x}, t)$ and solve the following (instead of (13.81))

$$\frac{\partial \phi}{\partial t} = \delta_\varepsilon(\phi)\left[\mu \nabla \cdot \frac{\nabla \phi}{|\nabla \phi|} - \nu - \lambda_1 |z - u_1|^2 + \lambda_2 |z - u_2|^2\right], \qquad (13.82)$$

coupled with the updates of $u_1 = mean(z)$ in D^+ and $u_2 = mean(z)$ in D^- after each step of a new ϕ in a time-marching numerical scheme. In fact, based on the work of [379], some researchers have applied a re-scaling to (13.82), by replacing $\delta_\varepsilon(\phi)$ by $|\nabla \phi|$, and hence proposed to solve [377,26]

$$\frac{\partial \phi}{\partial t} = |\nabla \phi|\left[\mu \nabla \cdot \frac{\nabla \phi}{|\nabla \phi|} - \nu - \lambda_1 |z - u_1|^2 + \lambda_2 |z - u_2|^2\right]. \qquad (13.83)$$

13.7 Numerical experiments

To show a flavour of the type of problems considered in this chapter, we only give one simple denoising example using the CGM [110] method. Many examples, illustrations and comparisons can be found in [11,110,3,97]. The fast solvers issues are not yet fully addressed, as research work is on-going. In Figure 13.3, the left plot shows a test image with noise in it and the right plot shows the restored image using the primal-dual method after 50 Newton iterations (with $\beta = 10^{-10}$ and $\alpha = 50$). Clearly the processing has made the image much more identifiable by people who know the Mathematics building in Liverpool.

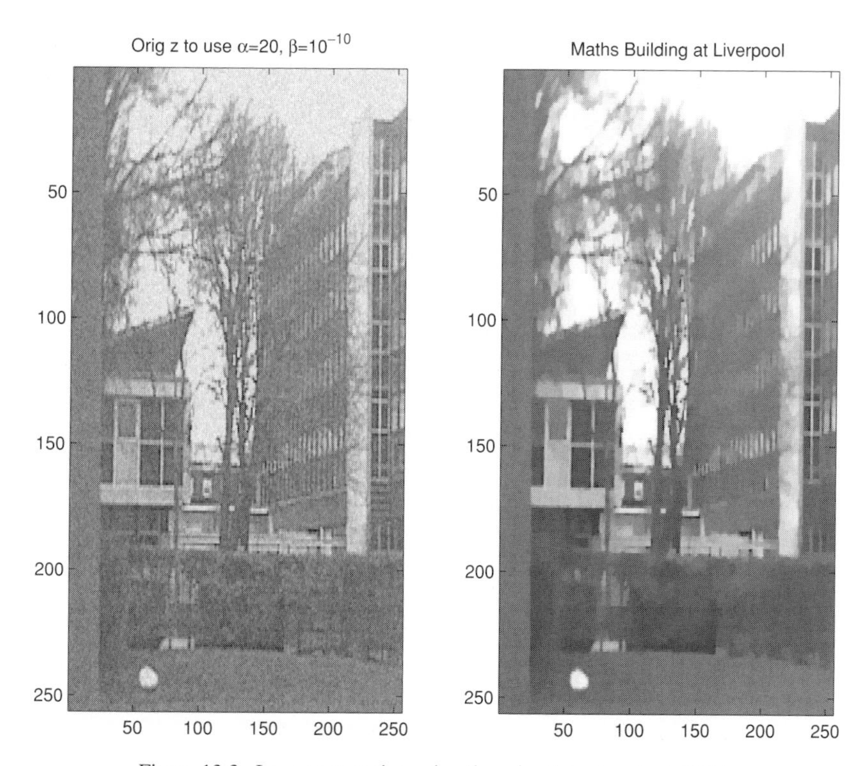

Figure 13.3. Image restoration using the primal-dual method [110].

13.8 Guide to software and the supplied Mfiles

For many topics in image restoration, the suite of Mfiles developed in association with [476,401] is very useful for study and development

http : //www.math.montana.edu/~vogel/Software/deconv/

and another software that may be used to run examples is the imagetool package (which is written in C with a MATLAB interface):

http : //gata.matapl.uv.es/~mulet/imagetool/

The following simple Mfiles are supplied for this chapter for readers.

[1] `BCCB.m` – Generate a BCCB matrix C from a root matrix R;
[2] `BTTB.m` – Generate a BTTB matrix T from a root matrix R for either a symmetric or unsymmetric T;
[3] `ch13.m` – Compute $w = Kv$ using three methods: BTTB, BCCB and FFT2.

It should be remarked that, to load in an image `file.jpg` onto MATLAB and store it in a matrix A, do the following

```
>> A = imread('file','jpg');
>> A = double(A(:,:,1));  % to convert to double, ready for
                            processing A ...
>> imagesc(A);            % to display an image stored in A.
>> colormap(gray)         % to set to clear B/W image
```

14

Application IV: voltage stability in electrical power systems

> As an example of what is often called a complex system, the power grid is made up of many components whose complex interactions are not effectively computable. Accordingly, some scientists have found it more useful to study the power grid's macroscopic behaviour than to dissect individual events.
>
> SARA ROBINSON. The power grid as complex system. *SIAM News,* Vol. 36 (2003)

The electrical power network is a real life necessity all over the world; however, delivering the power supply stably while allowing various demand pattern changes and adjustments is an enormous challenge. Mathematically speaking, the beauty of such networks lies in their providing a challenging set of nonlinear differential-algebraic equations (DAEs) in the transient case and a set of nonlinear algebraic equations in the equilibrium case [438,89,292].

This chapter introduces the equilibrium equations, discusses some recent fast nonlinear methods for computing the fold bifurcation parameter and finally highlights the open challenge arisen from computing the Hopf bifurcation parameter, where one must solve a new system of size $O(n^2)$ for an original problem of size n. We shall consider the following.

Section 14.1 The model equations
Section 14.2 Fold bifurcation and arc-length continuation
Section 14.3 Hopf bifurcation and solutions
Section 14.4 Preconditioning issues
Section 14.5 Discussion of software and the supplied Mfiles

14.1 The model equations

The power flow problem involves the calculation of voltages at all nodes of an alternating current network when subject to a specified loading condition and the power and voltage constraints that are applied to the system. The essential physical laws are the KCL and KVL (Kirchhoff current and voltage laws) coupled with power balance (note that the complex power $S = P + iQ$ is linked to the voltage V and current I via $S = VI^*$, where $*$ denotes a complex conjugate). The main equations can be found in applied mathematics books such as [439,28] as well as in most power system analysis books, for example [224,239,325].

Here we give a brief description to assist readers. Assume that we are dealing with an electrical power system of $m + 1$ nodes (called buses): $j = 1, 2, \ldots, m, m + 1$ with the last one, $m + 1$, used as the reference bus. In nodal analysis, the novelty lies in converting (somewhat complicated) system control quantities into equivalent quantities in terms of admittance (the reciprocals of impedance) and equivalent circuits. At bus j, write into polar form $V_j = v_j \exp(i\delta_j)$ with v_j and δ_j denoting the voltage magnitude and phase angle. Similarly between any two connecting buses j and k, write the admittance as $Y_{jk} = y_{jk} \exp(i\theta_{jk})$.

Then at bus j, letting the combined active and reactive power (due to other buses) equal to the net injected active and reactive power yields the power flow equations as follows

$$S_j = V_j I_j^*, \tag{14.1}$$

that is,

$$\begin{aligned} P_j + iQ_j &= V_j \sum_{k=1}^{m+1} (Y_{jk} V_k)^* \\ &= v_j \sum_{k=1}^{m+1} y_{jk} v_k \exp(\delta_j - \delta_k - \theta_{jk}). \end{aligned} \tag{14.2}$$

Therefore, we can write the nonlinear equations for Q_j, v_j, δ_j as follows (P_j is usually known as discussed below):

$$\begin{cases} P_j = v_j \sum_{k=1}^{m+1} y_{jk} v_k \cos(\delta_j - \delta_k - \theta_{jk}) \\ Q_j = v_j \sum_{k=1}^{m+1} y_{jk} v_k \sin(\delta_j - \delta_k - \theta_{jk}). \end{cases} \tag{14.3}$$

At a network equilibrium, the net injected power $S_j = P_j + iQ_j$ is equal to the difference of the generation power $P_{G_j} + iQ_{G_j}$ and the (user consumed) load power $P_{L_j} + iQ_{L_j}$.

In iterating the power equations or in a power disturbance, the two powers (S_j and the difference) are different and hence we have the term *power mismatch*, referring to (for bus j)

$$\begin{cases} \Delta P_j = P_{G_j} - P_{L_j} - P_j = P_{G_j} - P_{L_j} - v_j \sum_{k=1}^{m+1} y_{jk} v_k \cos(\delta_j - \delta_k - \theta_{jk}) \\ \Delta Q_j = Q_{G_j} - Q_{L_j} - Q_j = Q_{G_j} - Q_{L_j} - v_j \sum_{k=1}^{m+1} y_{jk} v_k \sin(\delta_j - \delta_k - \theta_{jk}). \end{cases} \tag{14.4}$$

In normal circumstances, the power equations are then simply $\Delta P_j = \Delta Q_j = 0$ that are sometimes called the mismatch equations (a confusing usage!).

The precise number of equations to be solved depends upon the bus type:

0 ▶ the slack (reference) bus $j = m + 1$ with $v_j = 1$ and $\delta_j = 0$ and no equation is needed;
1 ▶ a PQ (load) bus at $j = 1, \ldots, m_1$ where two power equations for v_j and δ_j are needed;
2 ▶ a PV (voltage control) bus at $j = m_1 + 1, \ldots, m$ where v_j is given and only one power equation is needed for δ_j.

Here m_1 is the total number of PQ buses (which are assumed to be the *first* m_1 buses out of the network). Then the power equations for the network can be written as follows

$$\begin{bmatrix} \Delta P_1 \\ \vdots \\ \Delta P_m \\ \Delta Q_1 \\ \vdots \\ \Delta Q_{m_1} \end{bmatrix} = \begin{bmatrix} P_{G_1} - P_{L_1} - P_1 \\ \vdots \\ P_{G_m} - P_{L_m} - P_m \\ Q_{G_1} - Q_{L_1} - Q_1 \\ \vdots \\ Q_{G_{m_1}} - Q_{L_{m_1}} - Q_{m_1} \end{bmatrix} = \mathbf{0}. \tag{14.5}$$

Let $n = m + m_1$ be the total number of unknowns and let $\mathbf{x} \in \mathbb{R}^n$ denote the system state variables i.e.

$$\mathbf{x} = [x_1 \ \cdots \ x_n]^\top = [\delta_1 \ \delta_2 \ \ldots \ \delta_{m_1} \ \ldots \ \delta_m \ v_1 \ v_2 \ \cdots v_{m_1}]^\top.$$

Define $\mathbf{f} : \mathbb{R}^n \rightarrow \mathbb{R}^n$ as follows

$$\mathbf{f}(\mathbf{x}) = \begin{bmatrix} f_1(\mathbf{x}) \\ \vdots \\ f_m(\mathbf{x}) \\ f_{m+1}(\mathbf{x}) \\ \vdots \\ f_n(\mathbf{x}) \end{bmatrix} = \begin{bmatrix} \Delta P_1(\mathbf{x}) \\ \vdots \\ \Delta P_m(\mathbf{x}) \\ \Delta Q_1(\mathbf{x}) \\ \vdots \\ \Delta Q_{m_1}(\mathbf{x}) \end{bmatrix}. \tag{14.6}$$

In the above setting of nonlinear equations, the load can be a varying parameter. When the network is operating at a stationary equilibrium, we use Newton's method to solve the power flow problem to determine the voltages of the entire network [454]. Assume the current state vector is $\mathbf{x}_0$ associated with the present load $[P_{L_{10}}\ P_{L_{20}} \ldots P_{L_{m0}}\ P_{L_{m+1,0}}]$.

Voltage collapse occurs in power systems as a result of a sequence of events that accompany a loss of stability where a change in system conditions causes a progressive and uncontrollable drop in voltage in significant parts of a power system [195,89]. The main factor causing this collapse has been shown to be the depletion of reactive load power on the power system. Mathematically, voltage collapse is associated with fold bifurcations resulting from a loss of stability in the parameterized nonlinear equations that describes the static power system [90]. Over the last few years many articles and papers have been written on the subject [89].

To describe the load increase in terms of a varying parameter λ, define the new load as

$$\begin{cases} P_{L_j} = P_{L_{j0}} + \lambda\alpha_j \\ Q_{L_j} = Q_{L_{j0}} + \lambda\beta_j \end{cases} \tag{14.7}$$

where P_{L_j} and Q_{L_j} are the new real and reactive loads increased after the initial state at $\lambda_0 = 0$; and α_j and β_j describe the load increase pattern at each bus j for the real and reactive loads respectively. Then we may write the parameterized power flow equations at bus j as follows:

$$\begin{cases} \Delta\mathcal{P}_j = P_{G_j} - (P_{L_{j0}} + \lambda\alpha_j) - P_j = 0 \\ \Delta\mathcal{Q}_j = Q_{G_j} - (Q_{L_{j0}} + \lambda\beta_j) - Q_j = 0 \end{cases} \tag{14.8}$$

where P_j and Q_j are as defined in equations (14.3). Combining equation (14.8) with (14.4) we obtain

$$\begin{cases} \Delta\mathcal{P}_j = \Delta P_j - \lambda\alpha_j = 0 \\ \Delta\mathcal{Q}_j = \Delta Q_j - \lambda\beta_j = 0. \end{cases} \tag{14.9}$$

Now we define $\mathbf{f}(.,.) : \mathbb{R}^n \times \mathbb{R} \to \mathbb{R}^n$ as follows

$$\mathbf{f}(\mathbf{x},\lambda) = \begin{bmatrix} \Delta\mathcal{P}_1(\mathbf{x},\lambda) \\ \vdots \\ \Delta\mathcal{P}_m(\mathbf{x},\lambda) \\ \Delta\mathcal{Q}_1(\mathbf{x},\lambda) \\ \vdots \\ \Delta\mathcal{Q}_{m_1}(\mathbf{x},\lambda) \end{bmatrix} = \begin{bmatrix} \Delta P_1(\mathbf{x}) - \lambda\alpha_1 \\ \vdots \\ \Delta P_m(\mathbf{x}) - \lambda\alpha_m \\ \Delta Q_1(\mathbf{x}) - \lambda\beta_1 \\ \vdots \\ \Delta Q_{m_1}(\mathbf{x}) - \lambda\beta_{m_1} \end{bmatrix} = \mathbf{0} \tag{14.10}$$

and combining with (14.6) we obtain the our main system as a special case of (14.14)

$$\mathbf{f}(\mathbf{x},\lambda) = \mathbf{f}(\mathbf{x}) - \lambda\mathbf{b} = 0, \tag{14.11}$$

where we only allow fixed power changes to distribute the total system load change represented by λ. Here the constant vector $\mathbf{b} \in \mathbb{R}^n$ denotes the system load pattern i.e. $\mathbf{b} = [\alpha_1 \ldots \alpha_m\ \beta_1 \ldots \beta_{m_1}]^\top$ and is such that $\sum_{k=1}^n b_k = 1$. Here we also consider the special case

$$\mathbf{b} = \sum_{k=1}^{s} w_{l_k}\mathbf{e}_{l_k} \qquad \text{with } w_{l_k} = 1/s \tag{14.12}$$

where $\mathbf{e}_{l_k}$ is the l_kth column of $I_{n\times n}$ with $m+1 \le l_k \le n$ i.e. we only consider variations of reactive load for a single bus or a selection of any s load buses.

Finally, to get familiarized with the unified notation, we expand the Jacobian equation in terms of the power quantities

$$\begin{bmatrix} \frac{\partial\Delta P_1}{\partial\delta_1} & \cdots & \frac{\partial\Delta P_1}{\partial\delta_m} & \frac{\partial\Delta P_1}{\partial v_1} & \cdots & \frac{\partial\Delta P_1}{\partial v_{m1}} \\ \vdots & & \vdots & \vdots & & \vdots \\ \frac{\partial\Delta P_m}{\partial\delta_1} & \cdots & \frac{\partial\Delta P_m}{\partial\delta_m} & \frac{\partial\Delta P_m}{\partial v_1} & \cdots & \frac{\partial\Delta P_m}{\partial v_{m1}} \\ \frac{\partial\Delta Q_1}{\partial\delta_1} & \cdots & \frac{\partial\Delta Q_1}{\partial\delta_m} & \frac{\partial\Delta Q_1}{\partial v_1} & \cdots & \frac{\partial\Delta Q_1}{\partial v_{m1}} \\ \vdots & & \vdots & \vdots & & \vdots \\ \frac{\partial\Delta Q_{m_1}}{\partial\delta_1} & \cdots & \frac{\partial\Delta Q_{m_1}}{\partial\delta_m} & \frac{\partial\Delta Q_{m_1}}{\partial v_1} & \cdots & \frac{\partial\Delta Q_{m_1}}{\partial v_{m1}} \end{bmatrix} \begin{bmatrix} \Delta\delta_1(\mathbf{x}) \\ \vdots \\ \Delta\delta_m(\mathbf{x}) \\ \Delta v_1(\mathbf{x}) \\ \vdots \\ \Delta v_{m_1}(\mathbf{x}) \end{bmatrix} = - \begin{bmatrix} \Delta P_1(\mathbf{x}) \\ \vdots \\ \Delta P_m(\mathbf{x}) \\ \Delta Q_1(\mathbf{x}) \\ \vdots \\ \Delta Q_{m_1}(\mathbf{x}) \end{bmatrix}. \tag{14.13}$$

We wish to remark that the real life application can be more difficult than our normal continuity assumption on $\mathbf{f}$ [139].

To show the performance of various methods, we use the following IEEE nine-bus system which is in IEEE Common Data Format; see [295]. This system is shown in Figure 14.1 and has nine buses: $m_1 = 6$ load buses (Type 1: bus

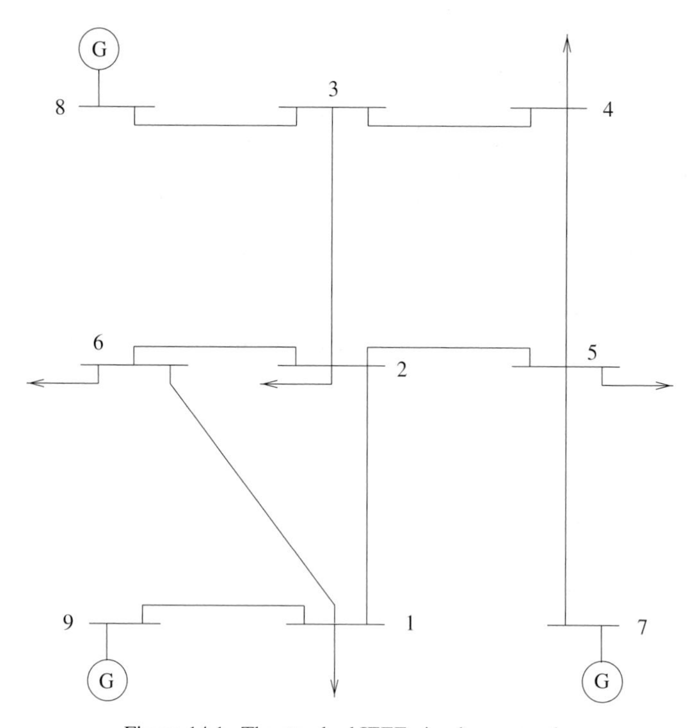

Figure 14.1. The standard IEEE nine-bus example.

$1,\ 2,\ldots,6$), $m = 8 - m_1 = 2$ generators (Type 2: bus 7, 8) and 1 reference generator bus (Type 2: bus 9).

14.2 Fold bifurcation and arc-length continuation

The model introduced above fits into a large class of nonlinear systems of equations, whose solution is addressed here.

In the mathematical literature, the term 'fold bifurcation' is often called the turning point because the solution turns at a fold point [183,425,434]. Adopting the usual notation, denote a nonlinear system of equations by

$$\mathbf{f}(\mathbf{x}, \lambda) = \mathbf{0} \tag{14.14}$$

where $\mathbf{x} \in \mathbb{R}^n$ is the state vector, $\lambda \in \mathbb{R}$ is the bifurcation parameter and $\mathbf{f} : \mathbb{R}^n \times \mathbb{R} \to \mathbb{R}^n$ with $\mathbf{f}(\mathbf{x}) = [f_1(\mathbf{x}) \ \ldots \ f_n(\mathbf{x})]^\top$. Assume that the current state variable is $\mathbf{x} = \mathbf{x}_0$ corresponding to the parameter λ_0.

The convergence problem with direct bifurcation methods is well known (due to the Jacobian matrix being singular) and one usually uses these methods only for accurately locating the fold point $(\mathbf{x}^*, \lambda^*)$ given a good estimate. Ideally, we want a method that has no difficulties near or passing round a fold point. With it we can numerically trace the solution path by generating a sequence of points satisfying a chosen tolerance criterion. This is reasonable since there is nothing wrong geometrically with the curve but the parameter λ is not the right parameter to use.

One such method that overcomes the convergence problem completely is the continuation method as extensively discussed in the literature [6,183,425,434]. These kinds of methods together with suitable monitoring steps for locating the fold bifurcation (e.g. computing the minimum singular value) are very reliable in most cases. In this section we briefly discuss the popular pseudo-arc length continuation method which is due to Keller [315].

Let s denote the parameter describing the solution path of (14.11). Suppose we have a solution $(\mathbf{x}_0, \lambda_0)$ of (14.11), then the Keller's method consists of solving the following equations for $(\mathbf{x}, \lambda) = (\mathbf{x}_1, \lambda_1)$ close to $(\mathbf{x}_0, \lambda_0)$

$$\mathbf{G}(\mathbf{Y}, s) = \begin{bmatrix} \mathbf{f}(\mathbf{x}, \lambda) \\ \hat{\mathbf{x}}_0^\top(\mathbf{x} - \mathbf{x}_0) + \hat{\lambda}_0(\lambda - \lambda_0) - \Delta s \end{bmatrix} = \mathbf{0} \tag{14.15}$$

where $\mathbf{Y} = (\mathbf{x}^\top, \lambda)^\top \in \mathbb{R}^{n+1}$ and $\mathbf{G} : \mathbb{R}^{n+2} \to \mathbb{R}^{n+1}$ and $(\hat{\mathbf{x}}_0, \hat{\lambda}_0)$ is the normalized tangent vector at $(\mathbf{x}_0, \lambda_0)$.

This system is solvable from the (Schur complement like) ABCD lemma [315,434] and can be solved by using a predictor–corrector procedure as follows.

- **Predictor** (Euler's method)

$$\begin{bmatrix} \mathbf{x}^{(1)} \\ \lambda^{(1)} \end{bmatrix} = \begin{bmatrix} \mathbf{x}_0 \\ \lambda_0 \end{bmatrix} + \Delta s \begin{bmatrix} \hat{\mathbf{x}}_0 \\ \hat{\lambda}_0 \end{bmatrix}$$

- **Corrector** (Newton's method)
  for $\ell \geq 1$ iterate

$$\begin{bmatrix} (\mathbf{f}_\mathbf{x})^{(\ell)} & (\mathbf{f}_\lambda)^{(\ell)} \\ \hat{\mathbf{x}}_0^\top & \hat{\lambda}_0 \end{bmatrix} \begin{bmatrix} \Delta\mathbf{x}^{(\ell)} \\ \Delta\lambda^{(\ell)} \end{bmatrix} = -\begin{bmatrix} \mathbf{f}(\mathbf{x}^{(\ell)}, \lambda^{(\ell)}) \\ \hat{\mathbf{x}}_0^\top(\mathbf{x}^{(\ell)} - \mathbf{x}_0) + \hat{\lambda}_0(\lambda^{(\ell)} - \lambda_0) - \Delta s \end{bmatrix} \tag{14.16}$$

  with updates

$$\begin{bmatrix} \mathbf{x}^{(\ell+1)} \\ \lambda^{(\ell+1)} \end{bmatrix} = \begin{bmatrix} \mathbf{x}^{(\ell)} \\ \lambda^{(\ell)} \end{bmatrix} + \begin{bmatrix} \Delta\mathbf{x}^{(\ell)} \\ \Delta\lambda^{(\ell)} \end{bmatrix}. \tag{14.17}$$

  On convergence of the corrector step, we obtain $(\mathbf{x}_1, \lambda_1)$. Then starting at $(\mathbf{x}_1, \lambda_1)$ to find $(\mathbf{x}, \lambda) = (\mathbf{x}_2, \lambda_2)$, the next tangent vector is computed from (note $\mathbf{f}_\mathbf{x}^1 = \mathbf{f}_\mathbf{x}(\mathbf{x}_1, \lambda_1)$ and $\mathbf{f}_\lambda^1 = \mathbf{f}_\lambda(\mathbf{x}_1, \lambda_1)$)

$$\begin{bmatrix} \mathbf{f}_\mathbf{x}^1 & \mathbf{f}_\lambda^1 \\ \hat{\mathbf{x}}_0^\top & \hat{\lambda}_0 \end{bmatrix} \begin{bmatrix} \hat{\mathbf{x}}_1 \\ \hat{\lambda}_1 \end{bmatrix} = \begin{bmatrix} 0 \\ 1 \end{bmatrix}. \tag{14.18}$$

It is interesting to observe that the predictor step is simply the initial Newton step because $(\mathbf{x}_0, \lambda_0)$ is a point on the solution curve.

However, there are at least *two* reasons why a continuation method is not the answer to the practical power control problem:

(1) a continuation method (either arc-length or pseudo-arc length-based) is usually expensive because many path-following steps (including monitoring steps) must be carried out to reach a bifurcation point λ^*; and
(2) the exact bifurcation point $\mathbf{x}^*$ is not always needed and only the parameter location λ^* is practically required in power system analysis.

Test functions, as smooth functions $\tau : \mathbb{R}^{n+1} \to \mathbb{R}$ defined along the solution path of $\mathbf{f}(\mathbf{x}, \lambda) = \mathbf{0}$, are necessary tools for detecting bifurcation points during the process of continuation [425]. A test function satisfies $\tau(\mathbf{x}^*, \lambda^*) = 0$ when $(\mathbf{x}, \lambda) = (\mathbf{x}^*, \lambda^*)$ is a bifurcation point.

The obviously qualified test function $\tau(\mathbf{x}, \lambda) = \det(\mathbf{J}(\mathbf{x}, \lambda))$ suffers from scaling problems [131,233,235,62]. To alleviate these problems, the bordering method sets up an $(n + 1)$-dimensional bordered system to compute the test function τ

$$\begin{bmatrix} \mathbf{J}(\mathbf{x}, \lambda) & \mathbf{d} \\ \mathbf{g}^\top & 0 \end{bmatrix} \begin{pmatrix} \mathbf{w} \\ \tau \end{pmatrix} = \mathbf{M} \begin{pmatrix} \mathbf{w} \\ \tau \end{pmatrix} = \begin{pmatrix} \mathbf{0} \\ 1 \end{pmatrix}, \tag{14.19}$$

where $\mathbf{d}, \mathbf{g} \in \mathbb{R}^n$ are chosen such that the matrix $\mathbf{M}$ is nonsingular for all regular solutions of $\mathbf{f}(\mathbf{x}, \lambda) = \mathbf{0}$. Then by Cramer's rule, such a τ is clearly a scaled Jacobian (satisfying $\tau(\mathbf{x}^*, \lambda^*) = 0$)

$$\tau(\mathbf{x}, \lambda) = \frac{\det(\mathbf{J})}{\det(\mathbf{M})}. \tag{14.20}$$

A related and mathematically equivalent test function is the tangent test function due to Abbott–Seydel [2,1,424]

$$\tau = \mathbf{e}_k^\top \mathbf{B}^{-1} \mathbf{J} \mathbf{e}_k. \tag{14.21}$$

Here, $\mathbf{B}$ is defined to be the matrix $\mathbf{J} = \mathbf{f}_\mathbf{x}$ with column k replaced with $\mathbf{b} = \mathbf{f}_\lambda$, i.e.

$$\mathbf{B} = \mathbf{J}(\mathbf{I}_n - \mathbf{e}_k \mathbf{e}_k^\top) + \mathbf{b}\mathbf{e}_k^\top, \tag{14.22}$$

Table 14.1. *Comparison of the performance index (using the Abbott–Seydel test function τ) with the continuation power flow solution ('exact index'). 'CPF' stands for continuation power flow and 'C steps' for 'CPF steps'. Note that only two CPF steps are needed to obtain the predicted index using τ.*

Varied bus	Exact index	CPF	C steps	Predicted index using τ
4	5.257974	5.2539	55	5.3251
5	2.454823	2.4305	25	2.4531
6	5.814020	5.7499	58	5.8133
7	3.201959	3.1859	34	3.1984
8	5.245256	5.2247	54	5.2968
9	2.342621	2.3287	25	2.3514

where $\mathbf{I}_n$ is the $n \times n$ identity matrix, and $\mathbf{e}_k$ is the k-th unit vector. In [142], the Abbott–Seydel test function (14.21) was reformulated into

$$\tau_k(\mathbf{x}, \lambda) = \frac{1}{\mathbf{e}_k^\top (d\mathbf{x}/d\lambda)}. \tag{14.23}$$

Furthermore, a formal link was made to the $(n+1)$ augmented system in (14.19) and also to the TVI method [90,175] that has been in use by the power engineering community for many years. Refer to [142,292].

As remarked, a mathematical as well as engineering challenge is to be able to predict λ^* without having to compute many nonlinear stepping steps. We now consider how to adapt the test function $\tau(\lambda)$ in (14.23) to yield an approximate performance index, namely, an approximation to parameter λ^*.

The essence of this work is to analyse the special analytical behaviour of functions $det(J)$, $det(B)$ and hence τ_k. It turns out that this is possible for power systems (14.11) and a reliable approximation to λ^* can be computed from the Newton-like formula at the current solution point (x_0, λ_0)

$$\lambda^* = \lambda_0 - \frac{1}{2} \frac{\tau(\lambda_0)}{d\tau/d\lambda(\lambda_0)}, \tag{14.24}$$

where a finite difference formula may be used to compute

$$\frac{d\tau}{d\lambda} = \frac{\tau(\lambda + \delta\lambda) - \tau(\lambda)}{\delta\lambda}, \tag{14.25}$$

for suitable $\delta\lambda > 0$. We remark that the reliability of (14.24) is only established well after it has been used for general test functions and found unreliable for some test functions [145,424,427].

We now present some numerical results to illustrate the effectiveness of the the performance index (i.e. the predicted fold parameter λ^*). The IEEE nine-bus system (Figure 14.1) consisting of three generators, six load buses and nine lines is used to illustrate the application of the performance index. Equation (14.23) is implemented by computing two successive Newton solutions at $\lambda = \lambda_0$ and $\lambda = \lambda_0 + \delta\lambda$ for a suitably small $\delta\lambda$. $d\tau/d\lambda$ is then then found via the finite difference described in (14.25). A performance index is then evaluated using (14.24). For simplicity, we set $\lambda_0 = 0$. Table 14.1 shows, for single bus variations, the appropriate fold bifurcation parameter λ^* evaluated using the tangent test function method and the pseudo-arc-length continuation method with a fixed step length of 0.1. These results are compared with the exact λ^*. Observe that the test function (14.23) provides a very good prediction for λ^* at a cost of two Newton steps.

In summary, formula (14.24) defines a fast nonlinear method, based on the Abbott–Seydel test function (14.21), for location of the bifurcation parameter λ^*. This method only requires two successive power solutions. We shall discuss how to speed up the linear system solvers within each nonlinear power solution for large networks.

14.3 Hopf bifurcation and solutions

The dynamic behaviour of an electrical power system may be modelled by the time-dependent system of ordinary differential equations

$$\frac{d\mathbf{x}}{dt} = \mathbf{f}(\mathbf{x}, \lambda) \tag{14.26}$$

where $\mathbf{x} \in \mathbb{R}^n$ is the vector of the usual state variables (i.e. voltage magnitude and phase angle), $\lambda \in \mathbb{R}$ is the bifurcation parameter, and $\mathbf{f} : \mathbb{R}^n \times \mathbb{R} \to \mathbb{R}^n$ is a sufficiently smooth function. A stability analysis of system (14.26) is achieved by examining the steady state solutions of the nonlinear equations

$$\mathbf{f}(\mathbf{x}, \lambda) = \mathbf{0}. \tag{14.27}$$

Given an initial solution $(\mathbf{x}_0, \lambda_0)$, the implicit function theorem [315] guarantees that (14.27) can be solved for each $\lambda \in \mathbf{B}(r, \lambda_0)$ for some $r > 0$, where $\mathbf{B}$ denotes an interval centred at λ_0 with radius r.

The stability of a stationary solution of (14.27) is indicated by the eigenvalues of the Jacobian matrix $J = \mathbf{f}_{\mathbf{x}}(\mathbf{x}, \lambda)$ evaluated at the stationary solution. All eigenvalues need to lie in the left half complex plane; that is, have negative real part. When one or more eigenvalues crosses over from the left half plane to the right, upon varying the parameter λ, the real part of these eigenvalues becomes positive and the associated stationary solution loses stability. The onset of instability, called a bifurcation point, corresponds to the critical value of λ for which the eigenvalues lie precisely on the imaginary axis. We already know that when one real eigenvalue crosses over from the left half plane to the right, then at the critical point we have a fold bifurcation point which is characterized by J being singular. Such bifurcations are easy to detect during a numerical continuation using test function methods of [425,235,142,140,141] which check sign changes for detecting a critical point.

When a pair of complex conjugate eigenvalues cross over from the left half complex plane to the right, the critical point is called a Hopf bifurcation point. These bifurcations are much more difficult to detect since J is nonsingular at the Hopf point and constructing test functions that change sign relies in general on computing the appropriate dominant eigenvalue which does not give rise to a robust method. We will discuss this in further detail later but first we give a formal definition for the Hopf bifurcation.

Definition 14.3.1. *A solution point* $(\mathbf{x}_0, \lambda_0)$ *of equation (14.27) is called a Hopf point if the Jacobian* $J = \mathbf{f}_\mathbf{x}(\mathbf{x}_0, \lambda_0)$ *has a conjugate pair of pure imaginary eigenvalues* $\mu(\lambda_0) = \pm i\beta_0$, $\beta_0 > 0$, *with* $\dfrac{dRe(\mu_0)}{d\lambda} \neq 0$.

At a Hopf bifurcation point, either stable limit cycles are created from the unstable stationary solutions called a *supercritical Hopf bifurcation*, or unstable limit cycles are created from the stable stationary solutions which is called a *subcritical Hopf bifurcation*.

In our context, we are interested in constructing appropriate test functions for the detection of such points, in particular, our problem of interest is in computing a security index for the onset of the subcritical Hopf bifurcation. Before that we review some of the classical methods used to detect and compute Hopf bifurcations in general.

Direct methods for computing Hopf bifurcations. Recall that to compute a fold bifurcation of $\mathbf{f}(\mathbf{x}, \lambda) = 0$, we set up the extended system of the form

$$\begin{pmatrix} \mathbf{f}(\mathbf{x}, \lambda) \\ \mathbf{f}_\mathbf{x}(\mathbf{x}, \lambda)\phi \\ \phi^\top\phi - 1 \end{pmatrix} = \mathbf{0},$$

where ϕ is the null eigenvector (see for example [425,235]). We can set up similar systems to directly compute a Hopf bifurcation. These have been discussed in [235,407,251] and take the form

$$G(y) = \begin{pmatrix} \mathbf{f}(\mathbf{x}, \lambda) \\ \mathbf{f_x}(\mathbf{x}, \lambda)u + \beta v \\ \mathbf{f_x}(\mathbf{x}, \lambda)v - \beta u \\ N_1(u, v) \\ N_2(u, v) \end{pmatrix} = \mathbf{0}, \tag{14.28}$$

where $G : \mathbb{R}^{3n+2} \to \mathbb{R}^{3n+2}$ and $y = (\mathbf{x}^\top, \lambda, u^\top, v^\top, \beta)^\top$ are the unknowns. Here $N_1(u, v)$ and $N_2(u, v)$ are two normalizing equations required to ensure that we obtain a system of $3n + 2$ equations with $3n + 2$ unknowns.

System (14.28) can be obtained by considering the eigenvalue problem

$$J\phi = \mu\phi \tag{14.29}$$

where $J = \mathbf{f_x}(\mathbf{x}, \lambda)$. Assuming the eigenvalue $\mu = \alpha + i\beta$ with corresponding eigenvector $\phi = u + iv$ and so equation (14.29) becomes

$$J(u + iv) = (\alpha + i\beta)(u + iv).$$

Separating the real and imaginary parts and noting that at a Hopf bifurcation $\alpha = 0$, we can write

$$\begin{cases} Ju + \beta v = 0 \\ Jv - \beta u = 0. \end{cases} \tag{14.30}$$

We consider two normalizations. The first due to Griewank and Reddien in [251] is given by

$$\begin{cases} N_1 = c^\top u \\ N_2 = c^\top v - 1 \end{cases} \tag{14.31}$$

for some suitable vector $c \in \mathbb{R}^n$. The choice of vector c needs only to satisfy $c^\top u = 0$ and $c^\top v = 1$ at a Hopf point $(\mathbf{x}_H, \lambda_H)$. The system (14.28) thus becomes

$$G(y) = \begin{pmatrix} \mathbf{f}(\mathbf{x}, \lambda) \\ \mathbf{f_x}(\mathbf{x}, \lambda)u + \beta v \\ \mathbf{f_x}(\mathbf{x}, \lambda)v - \beta u \\ c^\top u \\ c^\top v - 1 \end{pmatrix} = \mathbf{0}. \tag{14.32}$$

It was shown in [251] that the Jacobian G_y of (14.32) is nonsingular at a Hopf point. We remark that this choice of normalization allows the solution $y_0 = (\mathbf{x}_0^\top, \lambda_0, u_0^\top, \mathbf{0}^\top, 0)^\top$ which corresponds to a fold point. Therefore it is possible that one computes a fold point rather than a Hopf point if the solution path of $\mathbf{f}(\mathbf{x}, \lambda) = 0$ has both points close to each other, as noted in [251,407,434].

The second normalization was introduced by Roose and Hlavacek [407] given by

$$\begin{cases} N_1 = c^\top u \\ N_2 = u^\top u - 1 \end{cases} \tag{14.33}$$

and so (14.28) becomes

$$G(y) = \begin{pmatrix} \mathbf{f}(\mathbf{x}, \lambda) \\ \mathbf{f}_\mathbf{x}(\mathbf{x}, \lambda)u + \beta v \\ \mathbf{f}_\mathbf{x}(\mathbf{x}, \lambda)v - \beta u \\ c^\top u \\ u^\top u - 1 \end{pmatrix} = \mathbf{0}. \tag{14.34}$$

Again, with this normalization, it can be shown that G_y is nonsingular [407]. More importantly, the choice of normalization ensures that a solution of system (14.32) cannot be a fold point provided that the vector c is not orthogonal to the solution u corresponding to $i\beta$.

Remark 14.3.2. The second and third equations of system (14.28) can be written in the form

$$\mathbf{f}_\mathbf{x}^2(\mathbf{x}, \lambda)u + \beta^2 u.$$

This simplifies equation (14.28) to a reduced $2n + 2$ system. The same reduction holds for equation (14.34) on applying the above relation [407]:

$$G(y) = \begin{pmatrix} \mathbf{f}(\mathbf{x}, \lambda) \\ (\mathbf{f}_\mathbf{x}^2(\mathbf{x}, \lambda) + \beta^2 I_n)u \\ c^\top u \\ u^\top u - 1 \end{pmatrix} = \mathbf{0}, \tag{14.35}$$

where I_n denotes the $n \times n$ identity matrix and $y = (\mathbf{x}^\top, \lambda, u^\top, \beta)^\top$. Notice here that the matrix $\mathbf{f}_\mathbf{x}(\mathbf{x}, \lambda)^2 + \beta^2 I_n$ will have a rank defect of 2.

System (14.34) or (14.35) can be solved efficiently to compute a Hopf bifurcation using the Newton method *provided that* the initial solution is relatively close to the Hopf bifurcation. However, this is not usually the case since in general we are interested in problems where the Hopf point may be some distance away. Therefore we will use a continuation method [315,425,234] first on $\mathbf{f}(\mathbf{x}, \lambda) = 0$ to detect the approximate location of the Hopf point. We review these detection methods next.

Test function methods for detecting Hopf bifurcations. We assume that at the initial λ the steady-state solutions of $\mathbf{f}(\mathbf{x}, \lambda) = 0$ are stable; namely that, all eigenvalues of the system Jacobian $J = \mathbf{f}_\mathbf{x}(\mathbf{x}, \lambda)$ have negative real part. Most of the methods available use a continuation process to compute the steady-state solutions for varying λ. At each solution point $(\mathbf{x}, \lambda)$ we wish to find whether the real part of a complex pair of eigenvalues cross the imaginary axis. The simplest approach would be to estimate all the eigenvalues to establish a change of stability. If the dimension of the Jacobian $J = \mathbf{f}_\mathbf{x}$ is small then this would be feasible. However, for larger dimensions of the Jacobian matrix this certainly would not be practical. Another approach was to compute some dominant eigenvalues (ones with smallest absolute real part) where the 'some' depends on the nonlinearity of the problem. Most of the early research was on accelerating the computation of these eigenvalues especially for large systems [425,237,369,213]. For example, in [213], the authors developed a method to estimate the two dominant eigenvalues by performing the Arnoldi algorithm on a modified Cayley transform of these eigenvalues to accelerate convergence. Although these techniques are efficient when applicable, they are generally not useful for many problems, especially when the Hopf bifurcation may be some distance away. This is so since in general the eigenvalues have a nonlinear behaviour which means that determining which ones are dominant is difficult. Therefore we turn our attention to constructing test functions for the detection of Hopf bifurcations.

Recall that a test function has the property that it has a zero value at the required bifurcation point [425,142]. Such test functions depend on the determinant of the Jacobian J and are simple to construct for detecting fold points during a continuation process since a change of sign signals that a fold has been passed. However, constructing test functions for detecting Hopf bifurcations is not so simple since the Jacobian matrix is nonsingular at a Hopf point. Nevertheless, test functions do exist to detect Hopf bifurcations; as we show next, the trick is to construct a new and more desirable Jacobian matrix.

Before we present a more robust method, let us consider a simple test function based on the Hopf definition. Suppose that the Jacobian matrix J has a conjugate pair of eigenvalues and so we can write the eigenvalue problem:

$$J\phi = \mu\phi \tag{14.36}$$

where $\mu = \alpha + i\beta$ is an eigenvalue of J with corresponding eigenvector $\phi = u + iv$. As we stated in Section 14.3, at a Hopf point, we are interested in $\alpha = 0$ and so equation (14.36) can be written as:

$$(J - i\beta I_n)\phi = 0. \tag{14.37}$$

Thus at a Hopf point we have $\det(J - i\beta I_n) = 0$. This suggests that during a continuation process of $\mathbf{f}(\mathbf{x}, \lambda) = 0$ a change of sign will occur to the determinant $\det(J - i\beta I_n)$ if a Hopf point is passed. Therefore appropriate Hopf test functions can be defined similar to those defined for detecting folds [235,425]. For example, the test function (or its scaled version)

$$\tau_H = \det(J - i\beta I_n) \tag{14.38}$$

can be monitored during a continuation process to detect a Hopf point. Notice here that in general $\tau_H \in \mathcal{C}$ and that the computations involve complex arithmetic. Moreover, in practical applications, the test function described above (and others similar [356,357]) requires identifying one of the eigenvalues that will eventually cross the imaginary axis (which can be difficult), known as the critical eigenvalue. Many researchers have assumed that the dominant eigenvalue is the critical one [357,7] but this may not always be the case especially when the Hopf point is some distance away from the current solution point.

To overcome these difficulties we require a method for detecting Hopf bifurcations that does not depend on any direct computation of eigenvalues of J. Essentially this method constructs a new (and real) Jacobian matrix, based on this available J, that exhibits a single zero eigenvalue whenever the underlying J has a pair of pure imaginary eigenvalues. This gives rise to an elegant way of detecting a Hopf bifurcation by checking the singularity of the new Jacobian matrix just as in the fold case.

Hopf Test functions using bialternate products. We review a method which is due to Stephanos [435] in 1900 and later Fuller [212] in 1968, and more recently has become the approach to Hopf points [235,62,254,236]. The method is known as the bialternate matrix product or biproduct. First we go through some preliminaries (see [235,329,45]) and present a method (tensor sum) as a first version of biproduct.

14.3.1 The tensor product of matrices

Definition 14.3.3. (Tensor product). *Let A and B be $n \times n$ matrices with elements (a_{ij}) and (b_{kl}) respectively, $1 \le i, j, k, l \le n$. Then the tensor product of A and B, denoted $A \otimes B$, is an $n^2 \times n^2$ matrix with elements defined by*

$$(A \otimes B)_{rs} = a_{ij} b_{kl} \tag{14.39}$$

where

$$r \equiv (i-1)n + k, \qquad s \equiv (j-1)n + l.$$

Consider the case where A and B are 2×2 matrices, then

$$A \otimes B = \begin{pmatrix} a_{11}B & a_{12}B \\ a_{21}B & a_{22}B \end{pmatrix} = \begin{pmatrix} a_{11}b_{11} & a_{11}b_{12} & a_{12}b_{11} & a_{12}b_{12} \\ a_{11}b_{21} & a_{11}b_{22} & a_{12}b_{21} & a_{12}b_{22} \\ a_{21}b_{11} & a_{21}b_{12} & a_{22}b_{11} & a_{22}b_{12} \\ a_{21}b_{21} & a_{21}b_{22} & a_{22}b_{21} & a_{22}b_{22} \end{pmatrix}.$$

Note that this tensor product notation is extensively used in (1.41), (13.47) and in studying the FFT framework [465]. The properties of tensor products which follow from the definition are as follows. Let A, B, C, D be $n \times n$ matrices. Then:

1. if $\alpha, \beta \in C^n$, $\alpha A \otimes \beta B = \alpha\beta(A \otimes B)$;
2. $A \otimes (B + C) = A \otimes B + A \otimes C$;
3. $(A + B) \otimes C = A \otimes C + B \otimes C$;
4. $(A \otimes B)(C \otimes D) = (AC) \otimes (BD)$;
5. $A \otimes B = (A \otimes I_n)(I_n \otimes B)$;
6. $\det(A \otimes B) = (\det A)^n (\det B)^n$;
7. $I_n \otimes I_n = I_{n^2}$;
8. if A and B are nonsingular, then $(A \otimes B)^{-1} = A^{-1} \otimes B^{-1}$;
9. if A and B are triangular, so is $A \otimes B$. In particular, if A and B are diagonal, so is $A \otimes B$.

The following important result concerns the eigenvalues and eigenvectors:

Theorem 14.3.4. *If ν_i and μ_j are the eigenvalues of the matrices A and B respectively, the eigenvalues of the matrix $A \otimes B$ are $\nu_i \mu_j$ for all $1 \leq i, j \leq n$. The associated eigenvectors have the form*

$$\omega_{ij} = \begin{pmatrix} \phi_i^{[1]}\psi_j & \phi_i^{[2]}\psi_j & \cdots & \phi_i^{[n]}\psi_j \end{pmatrix}^T \tag{14.40}$$

where ϕ_i and ψ_j are the eigenvectors of A and B respectively.

Proof. As an illustration of the result, we prove the theorem for the case of $n = 2$ along the lines of [45]. Let A and B have distinct eigenvalues ν_i and μ_j where $1 \leq i, j \leq 2$ with corresponding eigenvectors $\phi_i = (\phi_i^{[1]}, \phi_i^{[2]})^\top$ and $\psi_j = (\psi_j^{[1]}, \psi_j^{[2]})^\top$ respectively. Consider the case $i = j = 1$, then we can write

$$A\phi_1 = \nu_1\phi_1, \qquad B\psi_1 = \mu_1\psi_1.$$

Expand both systems

$$a_{11}\phi_1^{[1]} + a_{12}\phi_1^{[2]} = \nu_1\phi_1^{[1]} \tag{14.41}$$

$$a_{21}\phi_1^{[1]} + a_{22}\phi_1^{[2]} = \nu_1\phi_1^{[2]} \tag{14.42}$$

$$b_{11}\psi_1^{[1]} + b_{12}\psi_1^{[2]} = \mu_1\psi_1^{[1]} \tag{14.43}$$

$$b_{21}\psi_1^{[1]} + b_{22}\psi_1^{[2]} = \mu_1\psi_1^{[2]}. \tag{14.44}$$

Multiply both equations (14.41) and (14.42) by equation (14.43)

$$\begin{aligned}
&a_{11}b_{11}\phi_1^{[1]}\psi_1^{[1]} + a_{11}b_{12}\phi_1^{[1]}\psi_1^{[2]} + a_{12}b_{11}\phi_1^{[2]}\psi_1^{[1]} + a_{12}b_{12}\phi_1^{[2]}\psi_1^{[2]} \\
&\quad = \nu_1\mu_1\phi_1^{[1]}\psi_1^{[1]} \\
&a_{21}b_{11}\phi_1^{[1]}\psi_1^{[1]} + a_{21}b_{12}\phi_1^{[1]}\psi_1^{[2]} + a_{22}b_{11}\phi_1^{[2]}\psi_1^{[1]} + a_{22}b_{12}\phi_1^{[2]}\psi_1^{[2]} \\
&\quad = \nu_1\mu_1\phi_1^{[2]}\psi_1^{[1]}
\end{aligned}$$

and, similarly, multiply equations (14.43) and (14.44) by equation (14.42)

$$\begin{aligned}
&a_{21}b_{11}\phi_1^{[1]}\psi_1^{[1]} + a_{21}b_{12}\phi_1^{[1]}\psi_1^{[2]} + a_{22}b_{11}\phi_1^{[2]}\psi_1^{[1]} + a_{22}b_{12}\phi_1^{[2]}\psi_1^{[2]} \\
&\quad = \nu_1\mu_1\phi_1^{[1]}\psi_1^{[2]} \\
&a_{21}b_{21}\phi_1^{[1]}\psi_1^{[1]} + a_{21}b_{22}\phi_1^{[1]}\psi_1^{[2]} + a_{22}b_{21}\phi_1^{[2]}\psi_1^{[1]} + a_{22}b_{22}\phi_1^{[2]}\psi_1^{[2]} \\
&\quad = \nu_1\mu_1\phi_1^{[2]}\psi_1^{[2]}.
\end{aligned}$$

Finally, gather and re-arrange these equations

$$\begin{aligned}
&a_{11}b_{11}\phi_1^{[1]}\psi_1^{[1]} + a_{11}b_{12}\phi_1^{[1]}\psi_1^{[2]} + a_{12}b_{11}\phi_1^{[2]}\psi_1^{[1]} + a_{12}b_{12}\phi_1^{[2]}\psi_1^{[2]} \\
&\quad = \nu_1\mu_1\phi_1^{[1]}\psi_1^{[1]} \\
&a_{11}b_{21}\phi_1^{[1]}\psi_1^{[1]} + a_{11}b_{22}\phi_1^{[1]}\psi_1^{[2]} + a_{12}b_{21}\phi_1^{[2]}\psi_1^{[1]} + a_{12}b_{22}\phi_1^{[2]}\psi_1^{[2]} \\
&\quad = \nu_1\mu_1\phi_1^{[1]}\psi_1^{[2]} \\
&a_{21}b_{11}\phi_1^{[1]}\psi_1^{[1]} + a_{21}b_{12}\phi_1^{[1]}\psi_1^{[2]} + a_{22}b_{11}\phi_1^{[2]}\psi_1^{[1]} + a_{22}b_{12}\phi_1^{[2]}\psi_1^{[2]} \\
&\quad = \nu_1\mu_1\phi_1^{[1]}\psi_1^{[2]} \\
&a_{21}b_{21}\phi_1^{[1]}\psi_1^{[1]} + a_{21}b_{22}\phi_1^{[1]}\psi_1^{[2]} + a_{22}b_{21}\phi_1^{[2]}\psi_1^{[1]} + a_{22}b_{22}\phi_1^{[2]}\psi_1^{[2]} \\
&\quad = \nu_1\mu_1\phi_1^{[1]}\psi_1^{[2]}.
\end{aligned}$$

Writing these equations in matrix form

$$\begin{pmatrix} a_{11}b_{11} & a_{11}b_{12} & a_{12}b_{11} & a_{12}b_{12} \\ a_{11}b_{21} & a_{11}b_{22} & a_{12}b_{21} & a_{12}b_{22} \\ a_{21}b_{11} & a_{21}b_{12} & a_{22}b_{11} & a_{22}b_{12} \\ a_{21}b_{22} & a_{21}b_{22} & a_{22}b_{21} & a_{22}b_{22} \end{pmatrix} \begin{pmatrix} \phi_1^{[1]}\psi_1^{[1]} \\ \phi_1^{[1]}\psi_1^{[2]} \\ \phi_1^{[2]}\psi_1^{[1]} \\ \phi_1^{[2]}\psi_1^{[2]} \end{pmatrix} = \nu_1\mu_1 \begin{pmatrix} \phi_1^{[1]}\psi_1^{[1]} \\ \phi_1^{[1]}\psi_1^{[2]} \\ \phi_1^{[2]}\psi_1^{[1]} \\ \phi_1^{[2]}\psi_1^{[2]} \end{pmatrix}, \tag{14.45}$$

and in more compact form, we obtain that

$$(A \otimes B)(\phi_1 \otimes \psi_1) = \nu_1\mu_1(\phi_1 \otimes \psi_1). \tag{14.46}$$

It follows that $\nu_1\mu_1$ is an eigenvalue of $A \otimes B$ with the associated eigenvector $\phi_1 \otimes \psi_1$. In a similar way, we see that the remaining eigenvalues of $A \otimes B$ are $\nu_1\mu_2$, $\nu_2\mu_1$ and $\nu_2\mu_2$ with associated eigenvectors $\phi_1 \otimes \psi_2$, $\phi_2 \otimes \psi_1$ and $\phi_4 \otimes \psi_4$. ■

The next result concerns the sum of tensor products[1]

Theorem 14.3.5. *Let A and B be $n \times n$ matrices defined as before with eigenvalues ν_i and μ_j respectively, $1 \leq i, j \leq n$. Then the eigenvalues of the matrix*

$$G = A \otimes B + B \otimes A \tag{14.47}$$

are given by the n^2 pairwise sums $\nu_i\mu_j + \nu_j\mu_i$ with the associated eigenvectors $\phi_i \otimes \psi_j$ for $1 \leq i, j \leq n$.

More general discussion and results can be found in Stephanos [435], Fuller [212], Govaerts [235] and [329]. The above theorem has a very important consequence:

Corollary 14.3.6. *The eigenvalues of the matrix*

$$D = A \otimes I_n + I_n \otimes A \tag{14.48}$$

are the n^2 values $\nu_i + \nu_j$ with associated eigenvectors $\phi_i \otimes \phi_j$ where $1 \leq i, j \leq n$.

Here matrix D in (14.48) is known as the tensor sum of the matrix A with itself. Corollary 14.3.6 shows that if any pair of eigenvalues of A sum to zero then its tensor sum is singular i.e. $\det(D) = 0$. Additionally, since each pair of distinct eigenvalue sums occurs twice and a stable system has all eigenvalues with negative real parts, the tensor sum will have a rank-2 deficiency providing of course that the matrix A is nonsingular.

This result lays the foundation for our first method of detecting a Hopf bifurcation. Consider the system Jacobian of $J = \mathbf{f_x}$ of equation (14.26). Recall that a Hopf bifurcation occurs when a pair of complex conjugate eigenvalues are purely imaginary. For the tensor sum of J defined by

$$D = J \otimes I_n + I_n \otimes J,$$

[1] For two general matrices A, B, their eigenvalues $\lambda(A)$, $\lambda(B)$ may not be related to $\lambda(A + B)$. For precise conditions and more special cases, refer to the work of Prof. T. Tao, UCLA, USA.

we have $\det(D) = 0$ with rank-2 deficiency at a Hopf bifurcation. This points a way of Hopf detection. Although with this method it is possible to detect a Hopf point during a continuation process, it would not be convenient or straightforward to use because the determinant of the tensor sum D does not change sign after a Hopf point is passed. Therefore, the application of the tensor sum in its current form is considered impractical. In the next section we describe a method to overcome these problems. But first we give a numerical example which illustrates the properties of the tensor sum.

Example 14.3.7. *Consider a matrix that mimics matrix J at a Hopf point*

$$A = \begin{pmatrix} -7 & 9 & 2 \\ 0 & -3i & 4 \\ 0 & 0 & 3i \end{pmatrix}.$$

Clearly A has eigenvalues $\{-7, -3i, 3i\}$ and is nonsingular. The tensor sum of A is given by

$$D = A \otimes I_n + I_n \otimes A$$
$$= \begin{pmatrix} -14 & 9 & 2 & 9 & 0 & 0 & 2 & 0 & 0 \\ 0 & -7-3i & 4 & 0 & 9 & 0 & 0 & 2 & 0 \\ 0 & 0 & -7+3i & 0 & 0 & 9 & 0 & 0 & 2 \\ 0 & 0 & 0 & -7-3i & 9 & 2 & 4 & 0 & 0 \\ 0 & 0 & 0 & 0 & -6i & 4 & 0 & 4 & 0 \\ 0 & 0 & 0 & 0 & 0 & 0 & 0 & 0 & 4 \\ 0 & 0 & 0 & 0 & 0 & 0 & -7+3i & 9 & 2 \\ 0 & 0 & 0 & 0 & 0 & 0 & 0 & 0 & 4 \\ 0 & 0 & 0 & 0 & 0 & 0 & 0 & 0 & 6i \end{pmatrix}$$

and D has eigenvalues $\{-14, -7-3i, -7+3i, -7-3i, -6i, 0, -7+3i, 0, 6i\}$ as predicted by Corollary 14.3.6. Here D is singular and has rank-2 deficiency. Notice that all the eigenvalues of D are pairwise sums of the eigenvalues of A and also that D is triangular so the structure of A is preserved.

14.3.2 The biproduct of matrices

One of the features of the matrix $G = A \otimes B + B \otimes A$ as we showed in the previous section is that each distinct eigenvalue occurs *twice*. We also remarked on how this creates a problem when trying to detect a singularity for the matrix $D = J \otimes I_n + I_n \otimes J$. What is needed is a new matrix that has only a single zero eigenvalue (instead of a double) when J has a pair of pure imaginary eigenvalues.

The method we now describe defines a new matrix that meets the above requirement and in addition it is more efficient to form than G since the matrix size is smaller [235,254]. Like D (see Corollary 14.3.6) the eigenvalues of the new matrix are possible sums of eigenvalues of J but unlike D each eigenvalue of J only contributes once. Although the new biproduct matrix can be viewed as derived (restricted or refined) from the tensor sum matrix G via eigenspace decomposition [235], we present the more computationally efficient formula defined originally by Stephanos [435]:

Definition 14.3.8. *Let A and B be $n \times n$ matrices with elements (a_{ij}) and (b_{ij}) respectively, $1 \le i, j \le n$. Set $m = \dfrac{n(n-1)}{2}$. Then the biproduct of A and B denoted $A \odot B$, is an $m \times m$ matrix whose rows are labelled (p, q) for $(p = 2, 3, \ldots, n; q = 1, 2, \ldots, p-1)$ and whose columns are labelled (r, s) for $(r = 2, 3, \ldots, n; s = 1, 2, \ldots, r-1)$ and with elements given by*

$$(A \odot B)_{\alpha,\beta} = (A \odot B)_{(p,q)(r,s)} = \frac{1}{2}\left(\begin{vmatrix} b_{qs} & b_{qr} \\ a_{ps} & a_{pr} \end{vmatrix} + \begin{vmatrix} a_{qs} & a_{qr} \\ b_{ps} & b_{pr} \end{vmatrix} \right). \tag{14.49}$$

Here the integer pairs (p, q) and (r, s), corresponding to the (α, β) entry and representing positions of a strictly lower triangular matrix, are ordered lexicographically, as illustrated by the Mfile `hopf.m`. That is,

$$\begin{aligned}
(p, q) = (2, 1) \text{ corresponds to } & \quad \alpha = 1 \\
(p, q) = (3, 1) \text{ corresponds to } & \quad \alpha = 2 \\
(p, q) = (3, 2) \text{ corresponds to } & \quad \alpha = 3 \\
(p, q) \text{ with } p > q \ge 2 \text{ in general corresponds to } & \quad \alpha = \tfrac{(p-2)(p-1)}{2} + q.
\end{aligned}$$

To work out (p, q), given α, we first find the largest integer that is less than or equal to the positive root of $f(p) = (p-2)(p-1)/2 - \alpha$ as done in `hopf.m`. One can also visualize how $A \odot B$ captures the interaction of entries of A, B from the following diagram for entry $(A \odot B)_{(p,q)(r,s)}$ — row p of A with row q of B, and row q of A with row p of B in their r, s columns:

$$A = \begin{bmatrix} & \vdots & & \vdots & \\ \cdots & a_{qs} & \cdots & a_{qr} & \cdots \\ & \vdots & \ddots & \vdots & \\ \cdots & \underline{a_{ps}} & \cdots & \underline{a_{pr}} & \cdots \\ & \vdots & & \vdots & \end{bmatrix}, \qquad B = \begin{bmatrix} & \vdots & & \vdots & \\ \cdots & \underline{b_{qs}} & \cdots & \underline{b_{qr}} & \cdots \\ & \vdots & \ddots & \vdots & \\ \cdots & b_{ps} & \cdots & b_{pr} & \cdots \\ & \vdots & & \vdots & \end{bmatrix},$$

where the underlined entries and the others interact separately from (14.49). Biproduct matrices (sharing the similar properties to the tensor products) have the additional commutative property: let A, B and C be $n \times n$ matrices, then

1. $A \odot B = B \odot A$;
2. if $\alpha, \beta \in C^n$, $\alpha A \odot \beta B = \alpha\beta(A \odot B)$;
3. $A \odot (B + C) = A \odot B + A \odot C$;
4. $(A \odot B)^\top = A^\top B^\top$;
5. if A and B are nonsingular, then $(A \odot B)^{-1} = A^{-1} \odot B^{-1}$;
6. if A and B are triangular then so is $A \odot B$, in particular, if A and B are diagonal then so $A \odot B$.

Therefore the special matrix corresponding to the tensor sum D takes the simpler form $2A \odot I_n$ and is known as the biproduct of A. From (14.49) we can write

$$(2A \odot I_n)_{(p,q)(r,s)} = \begin{vmatrix} a_{pr} & a_{ps} \\ \delta_{qr} & \delta_{qs} \end{vmatrix} + \begin{vmatrix} \delta_{pr} & \delta_{ps} \\ a_{qr} & a_{qs} \end{vmatrix} \tag{14.50}$$

where

$$\delta_{ij} = \begin{cases} 1, & i = j \\ 0, & i \neq j \end{cases}$$

which can be explicitly written as

$$(2A \odot I_n)_{(p,q)(r,s)} = \begin{cases} -a_{ps}, & \text{if} \quad r = q, \\ a_{pr}, & \text{if} \quad r \neq p \quad \text{and} \quad s = q, \\ a_{pp} + a_{qq}, & \text{if} \quad r = p \quad \text{and} \quad s = q, \\ a_{qs}, & \text{if} \quad r = p \quad \text{and} \quad s \neq q, \\ -a_{qr}, & \text{if} \quad s = p, \\ 0, & \text{otherwise.} \end{cases} \tag{14.51}$$

Notice that the entries of this matrix, coming directly from the matrix A are relatively easy to compute. As an example the biproduct of a general 4×4 matrix A has dimension 6×6 and is given by

$$(2A \odot I_3) = \left(\begin{array}{ccc|ccc} a_{11} + a_{22} & a_{23} & -a_{13} & a_{24} & -a_{14} & 0 \\ a_{32} & a_{11} + a_{33} & a_{12} & a_{34} & 0 & -a_{14} \\ -a_{31} & a_{21} & a_{22} + a_{33} & 0 & a_{34} & -a_{24} \\ \hline a_{42} & a_{43} & 0 & a_{11} + a_{44} & a_{12} & a_{13} \\ -a_{41} & 0 & a_{43} & a_{21} & a_{22} + a_{44} & a_{23} \\ 0 & -a_{41} & -a_{42} & a_{31} & a_{32} & a_{33} + a_{44} \end{array}\right).$$

As remarked, further details on (14.47) and (14.49) can be found in [235]. The main result on eigenvalues of biproducts is the following [435]:

Theorem 14.3.9. *Let A be an $n \times n$ matrix with eigenvalues $\{\nu_i\}_{1 \le i \le n}$. Then the matrix $A \odot I_n + I_n \odot A = 2A \odot I_n$ has eigenvalues $\{\nu_i + \nu_j\}_{1 \le j < i \le n}$.*

The supplied Mfile `bprod.m` may be tried to illustrate this main result.

Example 14.3.10. *We compute the biproduct of A given in Example 14.3.7 ($n = 3$):*

$$2A \odot I_3 = \begin{pmatrix} -7-3i & 4 & -2 \\ 0 & -7+3i & 9 \\ 0 & 0 & 0 \end{pmatrix}.$$

Here the eigenvalues of $2A \odot I_3$ are $\{-7-3i, -3i, 0\}$. As predicted by Theorem 14.3.9 there are no eigenvalue duplicates in terms of contributions from eigenvalues of A i.e. each pairwise sum of distinct eigenvalues of A occurs only once. The matrix $2A \odot I_3$ is singular but this time with the desirable rank-1 deficiency. Notice again that the biproduct of A has preserved the sparsity structure of A.

14.3.3 Hopf test functions

We now return to the dynamic problem (14.26) and consider how to make use of the above results. The $m \times m$ biproduct matrix of the system Jacobian $J = \mathbf{f_x}$ of (14.26) is $2J \odot I_n$. Then at a Hopf bifurcation we will have $\det(2J \odot I_n) = 0$ with $rank(2J \odot I_n) = m - 1$ because Theorem 14.3.9 predicts that $(2J \odot I_n)$ will have a single zero eigenvalue i.e. $i\beta - i\beta = 0$. Thus $\det(2J \odot I_n)$ will change sign when a Hopf point is passed during a continuation process of equation (14.27). Therefore we can define the test function

$$\tau_H = \det(2J \odot I_n) \tag{14.52}$$

which can be monitored during a continuation process to detect a Hopf bifurcation. As is known, without numerical scaling, a determinant is not the right quantity to measure singularity. Thus, to alleviate from scaling problems that this test function can suffer from, we now propose a test function using a bordering method, namely the framework set for fold point detection [235,142,62,315]. This consists of setting up in this case the $(m + 1)$-dimensional system.

$$\begin{pmatrix} 2J \odot I_n & \mathbf{d} \\ \mathbf{g}^\top & 0 \end{pmatrix} \begin{pmatrix} \mathbf{w} \\ \tau \end{pmatrix} = \begin{pmatrix} \mathbf{0} \\ 1 \end{pmatrix} \tag{14.53}$$

where $\mathbf{d}, \mathbf{g} \in \mathbb{R}^n$ are chosen such that the matrix

$$\mathbf{M} = \begin{pmatrix} 2J \odot I_n & \mathbf{d} \\ \mathbf{g}^\top & 0 \end{pmatrix} \tag{14.54}$$

is non-singular for all regular solutions of $\mathbf{f}(\mathbf{x}, \lambda) = \mathbf{0}$.

The simplest choice for vectors $\mathbf{d}$ and $\mathbf{g}$ are the unit vectors $\mathbf{e}_l$ and $\mathbf{e}_k$ respectively such that matrix $\mathbf{M}$ is nonsingular. The bordered system (14.53) thus becomes

$$\begin{pmatrix} 2J \odot I_n & \mathbf{e}_l \\ \mathbf{e}_k^\top & 0 \end{pmatrix} \begin{pmatrix} \mathbf{w} \\ \tau \end{pmatrix} = \begin{pmatrix} \mathbf{0} \\ 1 \end{pmatrix}. \tag{14.55}$$

This linear system has the dimension of $O(n^2) = (n-1)n/2 + 1 = m + 1$, the solution of which represent a major computational challenge!

By Cramer's rule the last equation is reduced to the Seydel test function [425,426] for the fold case:

$$\tau_H = \frac{\det(2J \odot I_n)}{\det(\mathbf{M})} = \frac{\det(2J \odot I_n)}{\det(2J \odot I_n)_{lk}}, \tag{14.56}$$

where $(2J \odot I_n)_{lk}$ is the matrix $(2J \odot I_n)$ with row l replaced by $\mathbf{e}_l$ and column k replaced by $\mathbf{e}_k$. As we only found a suitable scaling, at a Hopf point $(\mathbf{x}_H, \lambda_H)$, we clearly have $\tau_H = 0$. This test function can be utilized to effectively detect a Hopf bifurcation point.

However this test function can also be used to efficiently compute a Hopf bifurcation point $(\mathbf{x}_H, \lambda_H)$ of equation (14.26) by setting up the $(m+1)$-dimensional augmented system of equations

$$\begin{pmatrix} \mathbf{f}(\mathbf{x}, \lambda) \\ \tau_H(\mathbf{x}, \lambda) \end{pmatrix} = \mathbf{0}. \tag{14.57}$$

This will be a direct solution method (as usual requiring the initial point be close to the Hopf point for convergence) just as in the fold case [425].

Remark 14.3.11. As noted, the new biproduct matrix $2J \odot I_n$ poses a computational challenge for us. This is because dimension has been increased from n to $m = (n(n-1))/2$. If J is large, sparse matrix techniques must be used. Fortunately if J is sparse, $2J \odot I_n$ is equally sparse as was illustrated in Example 14.3.10. In [254], it was observed that $2J \odot I_n$ can be band structured and sparse and this observation can be exploited when implementing a Hopf continuation process. Another possible method is exploiting the structural preserving properties of $2J \odot I_n$. For example if the matrix A could be reduced to tridiagonal then $2J \odot I_n$ would be block tridiagonal which can be exploited to speed computations. Clearly there is a need to develop fast solvers. All these issues are open for further research.

Hopf bifurcation index. We have presented a comprehensive review of the methods suitable for detecting Hopf bifurcation. The next question for us is whether we can define an index that will give some kind of measure as to how close we are to a Hopf bifurcation: in particular, a subcritical one that occurs before a fold bifurcation since this signals the onset of instability. This is important for applications to voltage stability in power systems, and can help avoid a voltage collapse. The answer to that is yes, since the distance in the parameter space from the current stationary solution to the first detected (subcritical) Hopf bifurcation will give a measure to instability. This is similar to the situation of designing indices for measuring distances to fold bifurcations in power system applications; see for example [182].

Therefore the distance to a Hopf bifurcation for a positive or negative direction in the parameter space (Hopf index) can be defined by

$$\iota_\lambda = \lambda_H - \lambda_0 \tag{14.58}$$

where λ_0 is the current (stable) stationary point and λ_H is the first occurrence of a Hopf bifurcation. Often such a bifurcation occurs before a fold bifurcation (if the latter exists!). Overall, a suitable algorithm for computing the Hopf stability index ι_λ can be summarized as follows:

Algorithm 14.3.12. (Hopf stability index).

(1) Solve $\mathbf{f}(\mathbf{x}, \lambda) = 0$ *by the Newton method to obtain an initial solution* $(\mathbf{x}_0, \lambda_0)$*;*

(2) Compute the test function (14.56)

$$\tau_H(\mathbf{x}_0, \lambda_0) = \frac{\det(2J \odot I_n)}{\det(\mathbf{M})} = \frac{\det(2J \odot I_n)}{\det(2J \odot I_n)_{lk}}.$$

For $j \geq 1$

(3) Compute a solution $(\mathbf{x}_j, \lambda_j)$ *of* $\mathbf{f}(\mathbf{x}, \lambda) = 0$ *using pseudo-arclength continuation;*

(4) Compute the test function (14.56)

$$\tau_H(\mathbf{x}_j, \lambda_j) = \frac{\det(2J \odot I_n)}{\det(\mathbf{M})} = \frac{\det(2J \odot I_n)}{\det(2J \odot I_n)_{lk}},$$

(5) if $\tau_{H_j}\tau_{H_{j-1}} < 0$ *set* $\lambda_H = \lambda_j$ *then the Hopf point is found and exit to Step (6);*

otherwise return to Step (3) with next j*;*

(6) Compute the Hopf stability index (14.58)

$$\iota_\lambda = \lambda_H - \lambda_0.$$

Remark 14.3.13. This proposal of a test function is only the very first step towards a fast solution to locate the Hopf point λ_H. There are two further open challenges:

(i) how to solve the large linear system (14.55) of size $O(n^2) = (n-1)n/2 + 1 = m + 1$ by exploring its sparsity pattern in order to find τ_H; and
(ii) how to examine the analytical behaviour of τ_H with respect to λ so that an index method can be proposed to substitute the expensive continuation method.

Some simple cases and preliminary studies were in [254,235] but a general fast method is not yet available. In [293], we considered applying the Hopf index algorithm for studying some model power systems (without addressing the fast solver issues). Refer also to [146,181].

14.4 Preconditioning issues

The numerical solution of nonlinear systems provides a rich source of matrices for developing fast solvers. In an electrical power system as with most other cases of engineering importance, such a nonlinear system arises from modelling a physical process at state equilibrium and this implies that the system Jacobian matrix (in the sense of linear stability) must have all of its eigenvalues on the left half of the complex plane, i.e. the real parts of such eigenvalues are negative. The only exception is when the a subcritical Hopf bifurcation (eigenvalues touching the imaginary axis) is passed and some pairs of eigenvalues may pass to the right half place before returning back to the left. Therefore, matrices arising from a Jacobian context are almost always indefinite and hence are difficult to work with an iterative solver. Suitable preconditioning is essential.

For the linear systems involved in the fold bifurcation study Section 14.2, we have found [294] the SPAI approach combined with a deflation method is adequate in speeding up the GMRES method. For solving the Hopf systems Section 14.3, work is still in progress.

14.5 Discussion of software and the supplied Mfiles

There are two main sources of useful software that should be mentioned. Firstly, for numerical data and power system related software, we refer to

```
http://www.power.uwaterloo.ca
```

Secondly, for general pseudo-arc length continuation for nonlinear bifurcation systems, we refer to the well-known AUTO software due to E. Doedel:

ftp://ftp.cs.concordia.ca/pub/doedel/auto/auto.ps.gz

The MATLAB® command for computing $A \otimes B$ is `C = kron(A,B)`. We have supplied two Mfiles.

[1] `hopf.m` – To illustrate the definition of the biproduct matrix $A \odot B$.
[2] `bprod.m` – To compute and examine the biproduct matrix $A \odot B$.

15

Parallel computing by examples

Parallelism has sometimes been viewed as a rare and exotic subarea of computing, interesting but of little relevance to the average programmer. A study of trends in applications, computer architecture, and networking shows that this view is no longer tenable. Parallelism is becoming ubiquitous, and parallel computing is becoming central to the programming enterprise.

IAN FOSTER. *Designing and Building Parallel Programs*. Addison-Wesley (1995)

I rather kill myself than debug a MPI program.

ANONYMOUS

Parallel computing represents a major research direction for the future and offers the best and often the only solution to large-scale computational problems in today's technology. A book on fast solvers is incomplete without a discussion of this important topic. However, any incomplete description of the subject is of no use and there are already too many books available. Nevertheless, the author believes that too much emphasis has been put from a computer scientist's view (on parallelization) so that a beginner may feel either confused with various warnings and jargons of new phrases or intimated by the complexity of some published programs (algorithms) of well-known methods. Hence we choose to give complete details for a few selected examples that fall into the category of 'embarrassingly parallelizable' methods.

Therefore the purpose of this chapter is to convey two simple messages.

(i) Parallel computing is relatively simple to implement, so all readers should gain certain experience by implementing some algorithms.
(ii) Many parallel inefficiencies may well be due to the nonexistence of reliable and parallel algorithms. Serious imbalance between sequential time and

communication time should be addressed by future research and further hardware improvement.

Thus we emphasize on expositions and easy algorithms. The reader is encouraged to implement their own parallel algorithms and understand possible research gaps in algorithm development. We attempt to use concrete and meaningful examples to achieve this purpose. For each example, motivating (i) is our main purpose and hence the programs given may not be yet optimal even if (ii) is not an issue. The following outline is planned.

Section 15.1 A brief introduction to parallel computing and MPI
Section 15.2 Some commonly used MPI routines
Section 15.3 Example 1 of a parallel series summation
Section 15.4 Example 2 of a parallel power method
Section 15.5 Example 3 of a parallel direct method
Section 15.6 Discussion of software and the supplied MPI programs

As with other chapters, all illustrating computer codes (in MPI[1] Fortran) are available from this book's web page (see Preface on page XV). For general references on MPI, see [20,362,252].

15.1 A brief introduction to parallel computing and MPI

There are two main reasons for using a parallel computer.

(1) A sequential program may need too much memory: the memory (storage) issue.
(2) A sequential program may take too long to run: the timing (cpu) issue.

The two issues are related with the first one easier to settle (often automatically achieved) and the second one harder to resolve optimally. It must be remarked that a sequential program is not always parallelizable; if one uses brute force to parallelize a code, the underlying mathematics may not be correct.

Assuming parallelization is feasible, it remains to address how to distribute the computing tasks among processors to achieve a minimal time for the overall execution. The main factors to consider are the following.

(i) **Communication cost**. As each individual communication requires a fixed amount of startup time (or the latency) as well as the normal time expected for transferring a required length of data (depending on the communication

[1] MPI stands for 'Message Passing Interface'. See the official home page: http://www.mpi.org/. There exist other message passing libraries; see [433,184].

speed or the bandwidth), the general advice is that one should aim to send a 'small number' of large data sets rather than a 'large number' of small data sets.

(ii) **Load balancing**. This is the technical aspect but also the easy concept for readers to accept – we need to keep all processors equally busy or to distribute the processor subtasks fairly.

Before we show some simple examples to illustrate the working of MPI programs, we assume that MPI packages have been properly set up on your system, whether it is a shared memory mode (where communication is fast and all processors share a usually large memory) or a distributed memory model (where each processor has its own memory and data). For instance, on a Unix (or Linux) operating system, the typical way to compile a MPI Fortran 77 program **myfile.f** is

```
Unix>   f77 myfile.f -o myfile -lmpi
```

and to launch the compiled program **myfile** using four processors

```
Unix>   mpirun -np 4  myfile
```

If some additional library is needed, such as NAG[2] or IMSL[3], one would compile using `f77 myfile.f -o myfile -lnag -lmpi` or if IMSL is desired `f77 myfile.f -o myfile -limsl -lmpi`

A MPI program implements the so-called SPMD (single program multiple data) model. If p processors are used, all processors execute the same piece of code but the processors are distinguished only by their own

identity number (or called the rank) – an integer ID between 0 and $p - 1$,

which is used to command a particular processor to operate own data and operations. On a network of (physically) different computers, the alternative identifier may be

the processor name (or called the host name) – a character string that denotes its true name (e.g. uxb.liv.ac.uk or cam007)

which may not be unique if a computer has more than one processor. One reason for the popularity of MPI programming, as one finds out below, is the minimal amount work that one requires to do to set up the syntax – apart from one line for the library header, one only needs four MPI calls

[2] See `http://www.nag.co.uk` [3]See `http://www.imsl.org`

```
MPI_INIT        ---  The initialization routine
MPI_COMM_RANK   ---  Obtain the assigned identity number: j
MPI_COMM_SIZE   ---  Find out the total number of processors: p
MPI_FINALIZE    ---  Signal the end of a MPI program
```

to obtain a MPI code! Here $0 \le j \le p - 1$ for the identity number j. Also the MPI commands offer the transparency between Fortran and C versions.

15.2 Some commonly used MPI routines

The MPI programming library offers a large selection of parallel routines for possible needs of an experienced user. The selection may be too large for a beginner. Here we summarize and highlight a few commonly used ones only, hoping a reader can read about other advanced or adventurous routines after familiarizing these easy ones. As with all IT technology, the rule of thumb is that the answer is (or will be) out there if you dare to ask the question: *May I do this task this way?* We now provide tabular forms of some selected routines for easy reference, with further illustrations to be found in the concrete examples below. We shall use this generic notation for easy typesetting.

ie	Scalar integer type for displaying possible error indicator
er	Another scalar integer type for displaying error codes
cm	The internal communicator MPI_COMM_WORLD
st	The internal status object, declared initially as st(MPI_STATUS_SIZE)
vs	The data as a **s**ending quantity Note the tuple (vs, ls, types)
ls	The data length of a **s**ending quantity
types	The data type declaration for a **s**ending quantity
vr	The data as a **r**eceiving quantity Note the tuple (vr, lr, typer)
lr	The data length of a **r**eceiving quantity
typer	The data type declaration for a **r**eceiving quantity
PID	The active processor identity number ($0 \le \text{PID} \le p - 1$)
Tag	The data tag (an assigned integer $0 \le \text{Tag} \le 2^{32} - 1$)
Req	An integer quantity used as the alias name for an execution step
sas	The sized-p starting address vector for the sending quantity vs
sar	The sized-p starting address vector for the receiving quantity vr (Here sas and sar refer to address increments so the first entry contains 0.)

vls	The variable lengths vector for all sending quantities in vs
vlr	The variable lengths vector for all receiving quantities in vr
OPS	Alias for one of these commonly-used operations of scalars or arrays e.g. MPI_SUM, MPI_PROD, MPI_MAX, MPI_MIN, MPI_MAXLOC, MPI_MINLOC) or a user-defined function via MPI_OP_CREATE.

Here vr and vs cannot refer to the same quantity in most cases. For types or typer, the usual Fortran data types are supported and, as a parameter, these are used as

Integer	MPI_INTEGER
Real	MPI_REAL
Double precision	MPI_DOUBLE_PRECISION (or MPI_REAL8)
Sized 2 integers	MPI_2INTEGER (for MPI_MAXLOC or MPI_MINLOC)
Sized 2 reals	MPI_2REAL (for MPI_MAXLOC or MPI_MINLOC)
Sized 2 double precisions	MPI_2DOUBLE_PRECISION (for MPI_MAXLOC or MPI_MINLOC)

♦ **(1) The minimal set of starting up and ending routines.**

MPI_INIT(ie)	The initialization routine
MPI_COMM_RANK(cm,j,ie)	Obtain the assigned identity number: j
MPI_COMM_SIZE(cm,p,ie)	Find out the total number of processors: p
MPI_FINALIZE(ie)	Signal the end of a MPI program
MPI_GET_PROCESSOR_NAME(name,er,ie)	Get the host processor name, (a character string)
MPI_ABORT(cm,er,ie)	Abort a MPI program from inside
MPI_ANY_SOURCE	Wild card for not specifying which PID is used.
MPI_ANY_TAG	Wild card for not checking which tag is specified
MPI_PROC_NULL	The use of a null processor PID to disable a step.

♦ **(2) The set of point communication routines.**

MPI_SEND(vs,ls,types, PID,Tag, cm,st,ie)

Send the data (vs,ls,types) with Tag from the current processor to PID.

MPI_RECV(vr,lr,typer, PID,Tag, cm,st,ie)
Receive the data (vr,lr,typer) with Tag on the current processor from PID.

MPI_SENDRECV(vs,ls,types, PID1,Tag1, vr,lr,typer, PID2,Tag2, cm,st,ie)
Send the data (vs,ls,types) with Tag1 to PID1
and Receive the data (vr,lr,typer) with Tag1 from PID2.

MPI_ISEND(vs,ls,types, PID,Tag, cm,Req,ie)
Send the data (vs,ls,types) with Tag from the current processor to PID.
MPI_IRECV(vr,lr,typer, PID,Tag, cm,Req,ie)
Receive the data (vr,lr,typer) with Tag from processor PID.
MPI_WAIT(Req, st,ie)
Waiting point for completing (the recent Req) either ISEND or IRECV.

♦ **(3) The set of collective communication routines.**
MPI_BCAST(vs,ls,types, PID, cm,ie)
Broadcast the data (vs,ls,types) from processor PID to all others.
MPI_IBCAST(vs,ls,types, PID, cm,Req,ie)
Broadcast the data (vs,ls,types) from processor PID to all others
(need to use MPI_WAIT(Req, er,ie) to complete).
MPI_BARRIER(cm,ie)
Waiting point for all processors to reach for synchronization.

MPI_GATHER(vs,ls,types, vr,lr,typer, PID, cm,ie)
Receive the sent data (vs,ls,types) from each processor by PID
and pack them to the received data (vr,lr,typer) in **rank order**.
MPI_SCATTER(vs,ls,types, vr,lr,typer, PID, cm,ie)
All processors to receive data (vr,lr,typer) from the sent
data (vs,ls,types) by PID in **rank order**
MPI_GATHERV(vs,ls,types, vr,vlr,sar,typer, PID, cm,ie)
Receive the sent data (vs,ls,types) from each processor j by PID
and pack it at the starting address sar(j + 1) in (vr,vlr(j),typer).
MPI_SCATTERV(vs,vls,sas,types, vr,lr,typer, PID, cm,ie)
All processors to receive data (vr,lr,typer) from the sent data
(vs,vls(j),types) by PID starting at sas(j + 1) for processor j.

MPI_ALLGATHER(vs,ls,types, vr,lr,typer, cm,ie)
All to receive the data (vs,ls,types) from other processors and
pack them to the received data (vr,lr,typer) in **rank order**.
MPI_ALLSCATTER(vs,ls,types, vr,lr,typer, cm,ie)
All processors to receive data (vr,lr,typer) from the sent
data (vs,ls,types) by others in **rank order**

MPI_ALLGATHERV(vs,ls,types, vr,vlr,sar,typer, cm,ie)
All to receive the sent data (vs,ls,types) from other processors
and pack it using the starting vector sar(j) in (vr,vlr(j),typer).
(clearly, there is no need for MPI_ALLSCATTERV to exist).

♦ **(4) The set of collective operation routines.**

MPI_REDUCE(vs,vr,ls,types, OPS, PID, cm,ie)
Processor PID to receive the data in vr from some OPS operation
with all data (vs,ls,types).
(when OPS = MPI_MAXLOC or MPI_MINLOC, ls = 2).

MPI_ALLREDUCE(vs,vr,ls,types, OPS, cm,ie)
All processors to receive the data in vr from some OPS operation
with all data (vs,ls,types).
(again when OPS = MPI_MAXLOC or MPI_MINLOC, ls = 2).

♦ **(5) The timing routines.**

T1 = MPI_WTIME() Takes the wall clock time so the consumed CPU will be T = MPI_WTIME() − T1.

ETIME(T1) Takes the Unix system clock time so the consumed CPU will be T = T2 − T1 if ETIME(T2) is called later. (On other systems,use the normal measure for CPU, e.g. one uses $DCLOCK@(T1)$ with the Salford[4] compiler on Windows[5] systems.)

15.3 Example 1 of a parallel series summation

We now show the first example of a sample MPI program in Fortran 77; the program listing is available from this book's web page (see Preface).

Our first example attempts to compute the series

$$S = \sum_{k=1}^{N} \frac{(-1)^{k+1} \sin k}{k} \tag{15.1}$$

for some large N using p processors. The alternating series is known to converge since the general term goes to zero (though slowly) as $k \to \infty$. Here we

[4] See http://www.salford.co.uk for their product details.
[5] See http://www.microsoft.com/.

allow the number of terms N and the number of processors p to be arbitrary integers.

♦ **Preparation for parallel computing.** Firstly, we assign the subtask for each processor. The number of terms to be summed on processor j (for $0 \le j \le p-1$) will be

$$N_p = \left[\frac{N}{p}\right] \tag{15.2}$$

where $[\cdot]$ denotes the integer part of a quantity and N_p should be adjusted for one processor (say the last one) so that an arbitrary N can be accommodated. The overall assignment may be illustrated by

N_p terms for 0	N_p terms for 1	$\cdots$	N_p terms for $p-1$

Secondly, we clarify the formula used on processor $j = 0, 1, \ldots, p-1$:

$$S_j = \sum_{k=k_1}^{k_2} \frac{(-1)^{k+1} \sin k}{k}, \quad k_1 = jN_p + 1, k_2 = (j+1)N_p. \tag{15.3}$$

If $N_p p \neq N$, to accommodate an arbitrary N, we have to set $k_2 = N$ on the last processor (so possibly processor $(p-1)$ does slightly more work).

Finally, we highlight the quantities that must be communicated and also the corresponding MPI commands that may be used.

(1) Let N be input on processor 0 and be broadcast to others. We need `MPI_BCAST`. (Alternatively all processors can read N from a pre-defined file).

Option 1

(2) The individual partial sums S_j will have to be collected together to form the final sum for the series S. We can use `MPI_REDUCE` with the task option of `MPI_SUM`. The individual CPU timing t_j can be collected together using `MPI_REDUCE` with the task option of `MPI_MAXLOC` or `MPI_MAX`.

Option 2

(3) We can alternatively avoid the collective command `MPI_REDUCE` by using the more one-to-one communication commands to collect information on S_j and t_j. These will be `MPI_ISEND`, `MPI_SEND`, `MPI_IRECV`, `MPI_RECV`.

♦ **The sample program and the test results.** A sample program (available as **cup1.f**) using these MPI routine calls has been written to illustrate the above strategy:

```
    call MPI_INIT              ! The minimal call
    call MPI_COMM_RANK         ! The minimal call
    call MPI_COMM_SIZE         ! The minimal call
 CALL MPI_GET_PROCESSOR_NAME ! The optional use for processor
                               name
    call MPI_bcast             ! Broadcast N
    call MPI_barrier           ! Set a waiting point for bcast
 CALL MPI_REDUCE              != Collect the final results
    call MPI_IRECV             ! Test alternative communications
    call MPI_RECV              !
    call MPI_ISEND             !
    call MPI_SEND              !
    call MPI_WAIT             != Used by IRECV/ISEND
 CALL MPI_FINALIZE             ! The minimal call
```

as shown in Tables 15.1 and 15.2.

Running the program for $N = 8000000$ terms, the answer we get is

$$S = 0.500000050141.$$

On a network of Unix clusters, we have observed the following CPU timings

Processors p	CPU timing T_p (seconds)	Ratio with $p = 1$ case T_1/T_p
1	10.7	1.0
4	3.2	3.3
8	1.6	6.7
16	1.1	9.7

Clearly the optimal value p for the 'ratio' is not yet achieved (even though there is only 1 broadcasting step).

Note that the web version of **cup1.f** has more comments, some of which are left out in Tables 15.1 and 15.2 to typeset more easily.

15.4 Example 2 of a parallel power method

Given a matrix $A \in \mathbb{R}^{n\times n}$ that has a largest single eigenvalue $\lambda_a(A)$, the power method can be used to find λ_1 effectively. Recall that a power method may be described by the following algorithm

Table 15.1. *Part 1 of the sample program* **cup1.f** *available from the book web page.*

```
      PROGRAM CUP1    !-------! MPI SAMPLE PROGRAM 1
      IMPLICIT NONE           ! (Section 16.1.2)
      include 'mpif.h'        ! This is necessary
      INTEGER MY_ID, ierr, MASTER, NPROC,NLAST,NDIV,
     +        I,INFO,SN, N, K2,K1,K,REQUEST
     +        status(MPI_STATUS_SIZE)
      DOUBLE PRECISION SUM_all, TT1,TT2, WHO(2),
     +        WHO_ALL(2), SUM, SUM1
      CHARACTER*256 MY_NAME
         CALL MPI_INIT(ierr)
         CALL MPI_COMM_RANK(MPI_COMM_WORLD, MY_ID,ierr)
         CALL MPI_COMM_SIZE(MPI_COMM_WORLD, NPROC,ierr)
      NLAST = NPROC - 1 !  Proc ID for the last proc
      MASTER = 0        !  Master can be any proc>1
      TT1 = MPI_WTIME()
c---------------------------------------------------
      IF (MY_ID .EQ. 0) THEN ! Main Proc
      PRINT*,'Type in N (required number of terms)?'
      N = 8000000
      CALL MPI_GET_PROCESSOR_NAME(MY_NAME,ierr,INFO)
       PRINT'(A)','============cup1.f==========='
       PRINT*,'Number of processes started: ',NPROC
       PRINT*,'Main processor is: ID =',MY_ID,' on ',
     +         MY_NAME(1:13)
      ENDIF
      CALL MPI_barrier(MPI_COMM_WORLD, ierr)
      IF (MY_ID .NE. 0) THEN
      CALL MPI_GET_PROCESSOR_NAME(my_name,ierr,INFO)
      PRINT*,' and ID =',MY_ID,' on ', MY_NAME(1:13)
      ENDIF
c---------------------------------------------------
c Let other procs know about N (from MY_ID=0)
      CALL MPI_bcast(N,1,MPI_INTEGER, 0,
     +         MPI_COMM_WORLD, ierr)
      CALL MPI_barrier(MPI_COMM_WORLD, ierr)
      NDIV = N / NPROC    ! Subtask for each processor
      K1 = MY_ID*NDIV+1   ! Start of series
      K2 = (MY_ID+1)*NDIV ! End of the series
      IF (NDIV*NPROC.LT.N .and. MY_ID.eq.NLAST) K2=N
              SN = -1  !! Sign for Even Terms
      IF (MOD((K1+1),2 ).eq.1 ) SN = 1   !! Odd terms
```

```
c---------------------------------Even Terms (local)
        SUM1 = 0.0D0
      DO K = K1+1, K2, 2
     SUM1 = SUM1 + SIN(DBLE(K))/DBLE(K)
   ENDDO
     SUM1 = SUM1*SN
       SN = -SN
c----------------------------------Odd  Terms (local)
      SUM = 0.0D0
      DO K = K1, K2, 2
         SUM = SUM + SIN(DBLE(K))/DBLE(K)
      ENDDO
         SUM = SUM1 + SUM*SN  !-Partial Sum on My_ID
```

Table 15.2. *Part* 2 *of the program* **cup1.f** *as available from the book web page.*

```
c--Collect partial sums to get the answer on MASTER
     CALL MPI_reduce(sum,sum_all,1,MPI_DOUBLE_PRECISION
    +    ,MPI_SUM, MASTER, MPI_COMM_WORLD, ierr)
c---------------------STOP here to check CPU-----------
     TT2 = MPI_WTIME() - TT1 !reduce to compute max CPU
      who(1)=TT2
      who(2)=MY_ID
      CALL MPI_REDUCE(who,who_all,1,
    +   MPI_2DOUBLE_PRECISION,
    +   MPI_MAXLOC, MASTER, MPI_COMM_WORLD, ierr)
      IF (MY_ID .EQ. MASTER) then
      WRITE(*,'(2X,''The series summing up to '',I9,
    + '' terms ='', G18.12)') N, sum_all
      WRITE(*,'(2X,''The Parallel CPU = '',G12.3,
    + '' on Proc '',I2
    + '' [Method 1]'')') who_all(1),INT(who_all(2))
      ENDIF
c== Alternative if MPI_REDUCE is not used ======|
     IF (MY_ID .EQ. 0) then
     WRITE(*,'(/''The partial sum from P'',I2,'' is'',
    +   G12.3,'' using cpu:'',G10.2)') my_id,sum,TT2
            TT1 = TT2
        SUM_All = SUM
     DO I = 1, NLAST
        call MPI_IRECV(TT2, 1, MPI_DOUBLE_PRECISION,
    +         MPI_ANY_SOURCE,
```

```
+          I, MPI_COMM_WORLD, Request, ierr)
  call MPI_RECV(SUM, 1, MPI_DOUBLE_PRECISION,
+          MPI_ANY_SOURCE,
+          NPROC+I, MPI_COMM_WORLD, status, ierr)
  call MPI_WAIT(Request, status,ierr)
          TT1 = MAX(TT1,TT2)
      SUM_All = SUM_All + SUM
  WRITE(*,'(''The partial sum from P'',I2,'' is'',
+     G12.3,'' using cpu:'',G10.2)') I,sum,TT2
  ENDDO
  WRITE(*,'(2X,''The series summing up to '',I9,
+ '' terms ='',G18.12,/''The Parallel CPU = '',
+ G12.3, '' NPROC ='',I3, '' [Method 2]'',/)')
+ N, sum_all, TT1, NPROC
c-----------------------------------------------------
  ELSE ! MY_ID > 0 other procs
     call MPI_ISEND(TT2, 1, MPI_DOUBLE_PRECISION,
+    MASTER,MY_ID, MPI_COMM_WORLD,Request, ierr)
     call MPI_SEND(SUM, 1, MPI_DOUBLE_PRECISION,
+    MASTER,NPROC+my_id, MPI_COMM_WORLD, ierr)
     call MPI_WAIT(Request, status,ierr)
  ENDIF
c== Alternative if REDUCE is not used ======|
     CALL MPI_barrier(MPI_COMM_WORLD, ierr)
     CALL MPI_FINALIZE(ierr)
  STOP
  END
```

Algorithm 15.4.1. (The power method).

(1) *Generate the random vector z of size n and set $k = 0$, $\mu = 1$, $\mu_0 = 0$;*
while $|\mu - \mu_0| > TOL$ *or* $k \le 1$
(2) *Compute $y = Az/\mu$ and let $\mu_0 = \mu$;*
(3) *Find the largest component $\mu = y_m$ such that $|y_m| = \max_j |y_j|$;*
(4) *Set $z = y/\mu$ and $k = k + 1$.*
end while
(5) *Accept μ as the approximation to the largest eigenvalue of A and y the eigenvector.*

We intend to take test data directly from a Harwell–Boeing (HB) data format (Appendix B). In the supplied files **cup2.f** and **cup3.f**, we have adopted a simple version similar to routines from Sparskit (see [414]) for reading a HB matrix.

We now consider how to parallelize the power method or rather how the two supplied Mfiles are designed. As we intend to read an existing matrix, it is convenient for all processors to keep the same matrix so it only remains to partition the matrix for multiplication purpose.

Remark that the data partition will involve the notation identical to (15.2) and (15.3) (note again we need to reset N_p and k_2 on processor $p-1$ if $N_p p \neq n$) i.e.

$$N_p = \left[\frac{N}{p}\right], \quad k_1 = jN_p + 1, \quad k_2 = (j+1)N_p, \tag{15.4}$$

where $N = n$ is the dimension here, p is the number of processors and j is the identity number for processor j ($0 \leq j \leq p-1$).

♦ **Column partition of a sparse matrix.** Let A be partitioned in blocks of columns as illustrated below

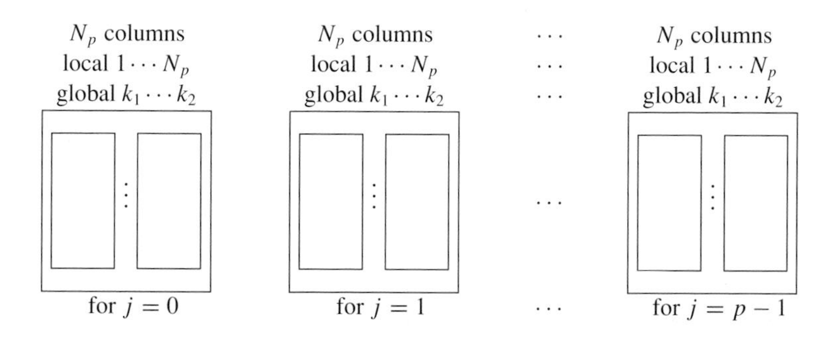

This partition is quite natural as the sequential data is stored in columns. However, matrix vector products may not be formed conveniently.

Although processor j is only in charge of N_p columns, the main parallel step (for Algorithm 15.4.1), $y = Az/\mu$, produces a full length vector $y^{(j)}$ of size n. These vectors must be added together by communication steps. Clearly there exists a fine balance between the number of blocks (or p processors) and the number of columns within each block (or N_p) to be considered for optimal performance. It is not hard to imagine that we may not gain much by using $p = n$ processors (so $N_p = 1$ and communication is dominant).

To test this data partition, we have written **cup2.f** for implementing the power Algorithm 15.4.1. It takes any input matrix in the HB format (Appendix B).

♦ **Row partition of a sparse matrix.** To facilitate matrix vector products, we consider the following row partition of a sparse matrix.

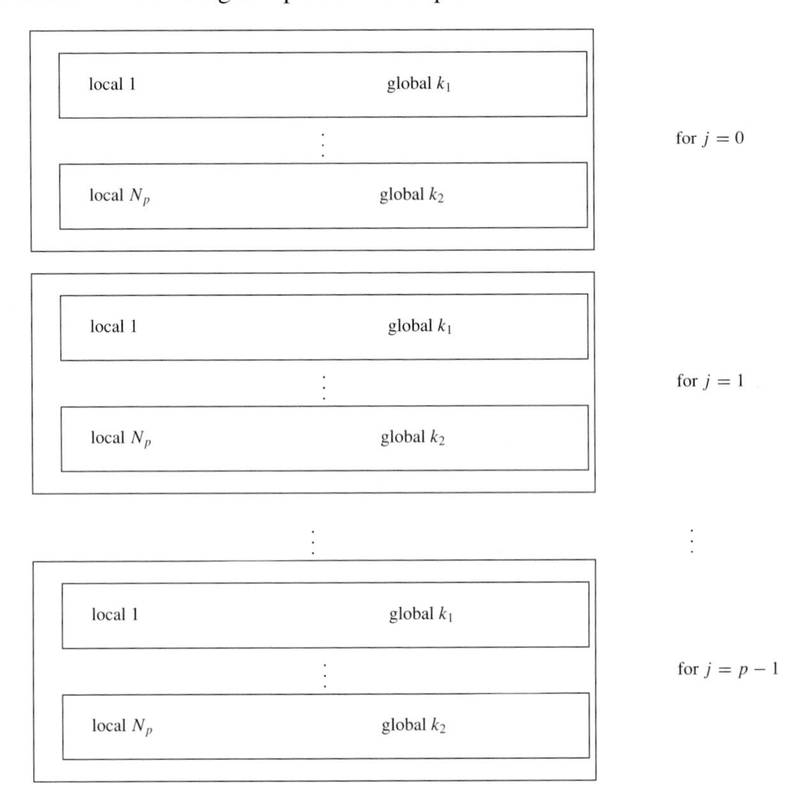

In our context, the matrix A is stored in HB format in columns so we have to convert a column based storage into a row based storage. This is illustrated in our sample program **cup3.f** (for implementing the power Algorithm 15.4.1) where `COL2RW` does the conversion.

In summary, the sample programs **cup2.f** and **cup3.f** used these MPI routines

```
MPI_INIT               ! The minimal call
MPI_COMM_RANK          ! The minimal call
MPI_COMM_SIZE          ! The minimal call
MPI_GET_PROCESSOR_NAME ! The optional use for processor name
MPI_bcast              ! Broadcast
MPI_barrier            ! Set a waiting point for bcast
MPI_ALLREDUCE          ! Compute across processors
MPI_ALLGATHERV         ! Combine vectors of varying lengths
```

```
MPI_REDUCE              ! Compute across processors
MPI_ABORT               !
MPI_WTIME               ! Obtain the cpu timing
MPI_FINALIZE            ! The minimal call
```

To demonstrate the performance of **cup2.f**, we have taken the input matrix as `pores_3.rua` from the Matrix Market collection (see [347] and Appendix E). With $p = 3$ processors, the result is the following

```
 ===================cup2.f====================
 Number of processes started:   3 ulgsmp1 1
 Matrix loaded from HB file : pores_3.rua
 ... Pointers read ok :   533
 ... Reading row indices for nnz :   3474
 ... Reading row indices done
 ... Reading A starts for nnz =  3474 FORMAT: (4D20.10)
 ... Reading A values done
 Matrix Size  =      532 X   532
 Matrix Title = 1UNSYMMETRIC MATRIX FROM PORES
 Market Title = PORES 3      GUESOL Symbol :
  Market Type = RUA  (error =0)
  CPU after read =    0.18E-01
Max eigenvalue=  -149867.499378 found by cup2.f after 1065
   iterations
Entire : CPU = 0.27   P0 using 3 Processors (Err=0.10E-05)
Parallel CPU = 0.25   P0 using 3 Processors (Err=0.10E-05)
My CPU = 0.27 0.25 on P0 using 3 Processors (Err=0.10E-05)
My CPU = 0.27 0.25 on P1 using 3 Processors (Err=0.10E-05)
My CPU = 0.27 0.25 on P2 using 3 Processors (Err=0.10E-05)
```

Clearly, the load balance is excellent since the underlying power method is easily parallelizable.

15.5 Example 3 of a parallel direct method

The sequential Purcell method as mentioned in Section 2.3.3 is a direct solution method for (1.1). Two variants of the GJ method, the Gauss–Huard method [288,163,178,283] and the Purcell method [392], are of special interest because both require a flop count comparable to the Gaussian elimination method. The pivoting strategy used in these variants mimics the GJ method with row pivoting the reliability of which was established in [176]. Parallel algorithms based on the Gauss–Huard method were developed in [284,177,178]. The Purcell method is also known as a row projection method [47,54,49].

Without pivoting, the Gauss–Huard method [288,163] and the Purcell method [392], are identical. So they offer no major advantages over the GE, apart from the fact that they are of the GJ type (not the GE type). However, with partial row pivoting, the Purcell method has been shown to perform more robustly than the GE with the usual partial pivoting [134,143].

As a further example of parallel computing, we illustrate a parallel algorithm for the Purcell method that was taken from [143]. Our parallel algorithm based on the Purcell method will address the issue of pivoting and load balancing. This topic of a parallel Purcell method, to our knowledge, is not widely known in the literature (see [47]). We encourage the reader to try out this interesting method.

To proceed, rewrite the linear system $Ax = b$ in the new augmented notation

$$Ax = b \qquad \Rightarrow \qquad \begin{bmatrix} A & \Big| & -b \end{bmatrix} \begin{bmatrix} x \\ 1 \end{bmatrix} = CV = 0, \tag{15.5}$$

where $A \in \mathbb{R}^{n\times n}$, $C \in \mathbb{R}^{n\times(n+1)}$, $x, b \in \mathbb{R}^n$, $V \in \mathbb{R}^{n+1}$. Denote the ith row vector of C by C_i^T i.e. $C_i = [a_{i1}\ a_{i2} \ldots a_{in}\ -b_i]^\top$. Then a vector V, with $V_{n+1} = 1$, is said to be the solution of the system if V is orthogonal to all C_i, namely

$$C_i^\top V = 0, \qquad \text{for } i = 1, 2, \ldots, n. \tag{15.6}$$

Clearly this method, looking for V in $\mathbb{R}^{n+1}$, is based on the orthogonality of vectors. As the final subspace $\mathbb{R}^{n+1}\backslash\mathtt{span}\,(C_1, \ldots, C_n)$ is only of dimension 1, one may imagine that the task of finding x can be done.

We remark that both the Gauss–Huard method and the Purcell method share the advantage that at step k of elimination, rows $j = k + 2, \ldots, n$ are neither required nor need to be updated. Therefore both methods can be considered as flexible elimination methods [328]. That is, the direct solution process can be combined with the elimination and coefficients-forming (rows of A) processes to achieve better performance. For some applications such as the solution of boundary element equations [328], this flexible elimination aspect of methods is potentially useful but remains to be fully explored.

15.5.1 The Purcell method

The above orthogonality condition (15.6) will be satisfied in a step-by-step manner. The order in which such conditions are met can be arbitrary [328]. Here we assume that constraints $C_1, C_2, \ldots, C_n$ are eliminated in turn, although any permutation of this order is allowed. Let $\mathcal{C} = \mathbf{C}^{(n)} = \{C_1, \ C_2, \ldots, \ C_n\}$. For $i = 1, 2, \ldots, n-1$, define the set

$$\mathbf{C}^{(i)} = \{C_1, \ C_2, \cdots, \ C_i\}.$$

Clearly we have $\mathbf{C}^{(i)} = \mathbf{C}^{(i-1)} \bigcup C_i$ with $\mathbf{C}^{(0)} = \{\emptyset\}$ empty. Similarly for $i = 1, 2, \ldots, n$, define $\mathbf{R}^{(i)}$ as the subspace, of dimension i, of the vector space $\mathbf{R}^{n+1}$ which consists of vectors orthogonal to $\mathbf{C}^{(n+1-i)}$. Let $\mathbf{R}^{(n+1)} = \mathbf{R}^{n+1}$. We shall use matrix $\mathbf{V}^{(i)}$ to denote the basis vectors of $\mathbf{R}^{(i)}$. That is,

$$\mathbf{V}^{(i)} = \left[\begin{array}{c|c|c|c} V_1^{(i)} & V_2^{(i)} & \cdots & V_i^{(i)} \end{array} \right]_{(n+1)\times i}$$

and therefore (note the solution vector $V \in \mathbf{R}^{(1)}$)

$$\mathbf{R}^{(i)} = \mathrm{span}\left(V_1^{(i)}, \ V_2^{(i)}, \cdots, \ V_i^{(i)}\right).$$

The basis for the $(n+1)$-dimensional space $\mathbf{R}^{(n+1)}$ may be chosen as the natural basis, i.e.

$$\mathbf{V}^{(n+1)} = \left[V_1^{(n+1)} \cdots V_{n+1}^{(n+1)}\right] = \left[[1\ 0 \cdots 0]^\top, \cdots, \ [0 \cdots 0\ 1]^\top\right]. \quad (15.7)$$

We are now ready to state the Purcell method. The objective is to reduce the large solution space $\mathbf{R}^{(n+1)}$ of dimension $(n+1)$ to the final solution subspace $\mathbf{R}^{(1)}$ of dimension 1.

Starting from this known space $\mathbf{R}^{(n+1)}$, at step $(n+1-i)$ for each i from n to 1, subspace $\mathbf{R}^{(i)}$ can be constructed by performing linear combinations of a chosen vector (the pivot) from the basis $\mathbf{V}^{(i+1)}$ with the remaining vectors, subject to the condition that the resulting vectors are orthogonal to C_{n+1-i}. More specifically, for $C_{n+1-i} \in \mathbf{C}^{(n+1-i)}$, the main construction involves the following (for $i = n, \ldots, 1$ and $k = 1, \ldots, i$)

$$V_k^{(i)} := \alpha_k V_{s(n+1-i)}^{(i+1)} + V_{m(k)}^{(i+1)}, \quad C_{n+1-i}^\top V_k^{(i)} = 0 \qquad i.e. \quad (15.8)$$

$$\alpha_k = -\frac{C_{n+1-i}^\top V_{m(k)}^{(i+1)}}{C_{n+1-i}^\top V_{s(n+1-i)}^{(i+1)}} \quad (15.9)$$

where $1 \le s(n+1-i), m(k) \le i+1, \ \ s(n+1-i) \ne m(k)$. Here the pivot index $s(n+1-i)$ is so selected that $|\alpha_k| \le 1$ i.e. the denominator on the right-hand side of (15.9) is the largest. Once the final subspace $\mathbf{R}^{(1)}$, of dimension 1, is found, its basis vector $V_1^{(1)}$ is orthogonal to every vector in $\mathbf{C}^{(n)} = \mathcal{C}$. Thus this vector gives rise to the solution of the system $Ax = b$.

We observe that, by construction, the vector $V_k^{(i)}$ is orthogonal to each vector of $\mathbf{C}^{(n+1-i)} \subset \mathcal{C}$. Pivoting by the above choice of $s(n+1-i)$ and $m(k)$ leads to a more reliable method than the Gaussian, GJ and the Gauss–Huard [288] methods as illustrated shortly. For the pivoted version, the Purcell method as described can reduce to the Gauss–Huard method [288,283] if we restrict the choice of $s(n+1-i)$ and impose the condition that $1 \le s(n+1-i) \le n+1-i$. Recall that for a Purcell method we have $1 \le s(n+1-i) \le n+2-i$. This seemingly simple restriction that distinguishes the Purcell method from the Gauss–Huard method turns out to be a vital condition, which makes the former a better method. However, the unpivoted version of the Purcell method with the choice of $s(n+1-i) = 1$ and $m(k) = k+1$ is not useful.

Finally from $\mathbf{R}^{(i+1)} = \mathbf{R}^{(i)} \bigoplus \mathtt{span}\,(C_{n+1-i})$, we can summarize the Purcell method in terms of subspace decomposition. We can derive the following

$$
\begin{aligned}
\mathbf{R}^{n+1} = \mathbf{R}^{(n+1)} &= \mathbf{R}^{(n)} \bigoplus \mathtt{span}\,(C_1) \\
&= \mathbf{R}^{(n-1)} \bigoplus \mathtt{span}\,(C_1,\ C_2) \\
&\ \ \vdots \\
&= \mathbf{R}^{(j)} \bigoplus \mathtt{span}\left(C_1,\ C_2, \cdots, C_{n+1-j}\right) \\
&\ \ \vdots \\
&= \mathbf{R}^{(2)} \bigoplus \mathtt{span}\,(C_1, \cdots,\ C_{n-1}) \\
&= \mathbf{R}^{(1)} \bigoplus \mathtt{span}\,(C_1, \cdots,\ C_n) \ = \ \mathbf{R}^{(1)} \bigoplus \mathit{range}(\mathcal{C}).
\end{aligned}
$$

Here the idea of finding the solution subspace $\mathbf{R}^{(1)}$, of dimension 1, out of the initial large space $\mathbb{R}^{n+1}$ is by squeezing out C_j's by orthogonal decomposition. Before we discuss parallel algorithms, we give some examples.

Example 15.5.2. (Solution of a 4×4 system by the Purcell method). *To illustrate the sequential method, we now consider the following example*

$$
\begin{bmatrix} 5 & 1 & 2 & 1 \\ 2 & 10 & 3 & 1 \\ 1 & 4 & 8 & 2 \\ 6 & 2 & 4 & 20 \end{bmatrix}
\begin{bmatrix} x_1 \\ x_2 \\ x_3 \\ x_4 \end{bmatrix}
=
\begin{bmatrix} 17 \\ 35 \\ 41 \\ 102 \end{bmatrix},
\qquad
\begin{bmatrix} C_1^\top \\ C_2^\top \\ C_3^\top \\ C_4^\top \end{bmatrix}
=
\begin{bmatrix} 5 & 1 & 2 & 1 & -17 \\ 2 & 10 & 3 & 1 & -35 \\ 1 & 4 & 8 & 2 & -41 \\ 6 & 2 & 4 & 20 & -102 \end{bmatrix}.
$$

Step 1, $i = n = 4$: *pivot* $s(1) = 5$, $m(k) = 1, 2, 3, 4$

$$C_1^\top \mathbf{V}^{(i+1)} = \begin{bmatrix} 5 & 1 & 2 & 1 & -17 \end{bmatrix}$$

$$\mathbf{V}^{(i)} = \mathbf{V}^{(i+1)} \begin{bmatrix} 1 & & & \\ & 1 & & \\ & & 1 & \\ & & & 1 \\ \alpha_1 & \alpha_2 & \alpha_3 & \alpha_4 \end{bmatrix} = \begin{bmatrix} 1 & 0 & 0 & 0 \\ 0 & 1 & 0 & 0 \\ 0 & 0 & 1 & 0 \\ 0 & 0 & 0 & 1 \\ 0.2941 & 0.0588 & 0.1176 & 0.0588 \end{bmatrix}.$$

Step 2, $i = n - 1 = 3$: *pivot* $s(2) = 1$, $m(k) = 2, 3, 4$

$$C_2^\top \mathbf{V}^{(i+1)} = \begin{bmatrix} -8.2941 & 7.9412 & -1.1176 & -1.0588 \end{bmatrix}$$

$$\mathbf{V}^{(i)} = \mathbf{V}^{(i+1)} \begin{bmatrix} \alpha_1 & \alpha_2 & \alpha_3 \\ 1 & & \\ & 1 & \\ & & 1 \end{bmatrix} = \begin{bmatrix} 0.9574 & -0.1348 & -0.1277 \\ 1 & 0 & 0 \\ 0 & 1 & 0 \\ 0 & 0 & 1 \\ 0.3404 & 0.0780 & 0.0213 \end{bmatrix}.$$

Step 3, $i = n - 2 = 2$: *pivot* $s(3) = 1$, $m(k) = 2, 3$

$$C_3^\top \mathbf{V}^{(i+1)} = \begin{bmatrix} -9 & 4.6667 & 1 \end{bmatrix}$$

$$\mathbf{V}^{(i)} = \mathbf{V}^{(i+1)} \begin{bmatrix} \alpha_1 & \alpha_2 \\ 1 & \\ & 1 \end{bmatrix} = \begin{bmatrix} 0.3617 & -0.0213 \\ 0.5185 & 0.1111 \\ 1 & 0 \\ 0 & 1 \\ 0.2545 & 0.0591 \end{bmatrix}.$$

Step 4, $i = n - 3 = 1$: *pivot* $s(4) = 1$, $m(k) = 2$

$$C_4^\top \mathbf{V}^{(i+1)} = \begin{bmatrix} -18.7549 & 14.0662 \end{bmatrix}$$

$$\mathbf{V}^{(i)} = \mathbf{V}^{(i+1)} \begin{bmatrix} \alpha_1 \\ 1 \end{bmatrix} = \begin{bmatrix} 0.25 \\ 0.50 \\ 0.75 \\ 1.00 \\ 0.25 \end{bmatrix}, \qquad V = \frac{V_1^{(1)}}{[V_1^{(1)}]_{n+1}} = \frac{\mathbf{V}^{(i)}}{0.25} = \begin{bmatrix} 1.0 \\ 2.0 \\ 3.0 \\ 4.0 \\ 1.0 \end{bmatrix}.$$

15.5.2 Comparison of the Purcell method with other related methods

We now use four related examples from [281,280,229] to demonstrate that the Purcell method has better stability properties. These examples are often used to test growth factors of the Gaussian method. Define the usual growth factor

[229] by (for $i, k = 1, 2, \ldots, n$ and $j = 1, 2, \ldots, n, n+1$)

$$\rho = \max_{i,j,k} \frac{|A_{ij}^{(k)}|}{\|A\|_\infty},$$

where we assume that $A_{i,n+1} = -b_i$; see also [280,281]. For the Purcell method we measure the growth of the components of all $V_k^{(i)}$.

(1) **Example** 1, with $\mu = 0.4$: $[A]_{ij} = [\mathcal{A}_1]_{ij} = \begin{cases} 1, & \text{if } i = j \text{ or } j = n, \\ -\mu, & \text{if } i > j, \\ 0, & \text{otherwise.} \end{cases}$

(2) **Example** 2, transpose of $\mathcal{A}_1$: $A = \mathcal{A}_1^\top$.

(3) **Example** 3 (with $\mu = 1$ in $\mathcal{A}_1$): $[A]_{ij} = [\mathcal{A}_3]_{ij} = \begin{cases} 1, & \text{if } i = j \text{ or } j = n, \\ -1, & \text{if } i > j, \\ 0, & \text{otherwise.} \end{cases}$

(4) **Example** 4, transpose of $\mathcal{A}_3$: $A = \mathcal{A}_3^\top$.

Table 15.3 shows results of solving the linear system (15.5) for the above four examples using

Gaussian (c) — the Gaussian elimination method with complete pivoting,
Gaussian (p) — the Gaussian elimination method with partial pivoting,
Huard [288] — the Gauss–Huard method with partial pivoting,
Purcell [392] — the Purcell method with partial pivoting.

The exact solution is chosen as $x_i^* = 1$ that defines the right-hand side b. In the table, "$\bowtie$" indicates failure by a numerical method and "$\frown$" shows a loss of accuracy due to large growth factors. Note an accuracy is considered acceptable if it is less than $10^2\epsilon \approx 10^{-13}$ with ϵ associated with double precision.

The results clearly demonstrate that the performance of the Purcell method is close to Gaussian (c), and much better than Gaussian (p) and Huard [288]. Moreover, the partial pivoting used by the Purcell method is inexpensive (unlike Gaussian (c)), making it a serious candidate for wider applications and more research and development.

15.5.3 Parallelization of the Purcell method

Denote by p the number of parallel processors that are accessible, and set $n_j = n/p$ for $j = 0, 1, \ldots, p-1$. Then step 1 can be simultaneously be carried out by all processors as no real computing is involved. First we locate the maximum entry $m = s(1)$ in row 1, C_1^T, of matrix C. Then assign n_j columns of unit vectors to basis vectors v_k, in $\mathbf{V}^{(n+1)}$, that are given $v_k(m) = \alpha_k = -C_1(k)/C_1(m)$. For

Table 15.3. *Comparison of the Purcell method with other direct methods.*

Problem	Size n	Method	Growth ρ	Accuracy $\|x - x^*\|_2$	Failure
	30	Gaussian (c)	1.4	3.8×10^{-14}	
		Gaussian (p)	1.7×10^{4}	6.0×10^{-11}	$\frown$
		Huard [288]	1.1×10^{-1}	3.6×10^{-14}	
1		Purcell [392]	1.3×10^{-1}	3.9×10^{-14}	
	60	Gaussian (c)	1.4	1.3×10^{-13}	
		Gaussian (p)	4.2×10^{8}	2.8×10^{-6}	$\bowtie$
		Huard [288]	5.6×10^{-2}	1.2×10^{-13}	
		Purcell [392]	6.6×10^{-2}	1.2×10^{-13}	
	30	Gaussian (c)	1.4	4.6×10^{-14}	
		Gaussian (p)	1.4	5.0×10^{-14}	
		Huard [288]	5.8×10^{2}	5.9×10^{-11}	$\frown$
2		Purcell [392]	1.2×10^{-1}	3.9×10^{-14}	
	60	Gaussian (c)	1.4	1.6×10^{-13}	
		Gaussian (p)	1.4	1.9×10^{-14}	
		Huard [288]	7.0×10^{6}	2.4×10^{-4}	$\bowtie$
		Purcell [392]	5.8×10^{-2}	1.6×10^{-13}	
	30	Gaussian (c)	2.0	1.1×10^{-14}	
		Gaussian (p)	5.4×10^{8}	3.0×10^{-7}	$\bowtie$
		Huard [288]	6.7×10^{-2}	2.6×10^{-14}	
3		Purcell [392]	8.8×10^{-2}	3.8×10^{-14}	
	60	Gaussian (c)	2.0	5.5×10^{-14}	
		Gaussian (p)	5.8×10^{17}	1.5×10^{2}	$\bowtie$
		Huard [288]	3.4×10^{-2}	1.2×10^{-13}	
		Purcell [392]	4.3×10^{-2}	1.5×10^{-13}	
	30	Gaussian (c)	2.0	5.8×10^{-15}	
		Gaussian (p)	2.0	5.8×10^{-15}	
		Huard [288]	1.8×10^{7}	2.6×10^{-7}	$\bowtie$
4		Purcell [392]	1.3×10^{-1}	2.7×10^{-14}	
	60	Gaussian (c)	2.0	7.9×10^{-14}	
		Gaussian (p)	2.0	7.9×10^{-14}	
		Huard [288]	9.6×10^{15}	2.8×10^{1}	$\bowtie$
		Purcell [392]	6.5×10^{-2}	1.5×10^{-13}	

simplicity, we can assume that n is divisible by p; if not we can let the last processor take the extra vectors i.e. $n_p = n_1 + (n - pn_1)$.

Parallel generation of rows of matrix C. For many applications (e.g. BEM), rows of a matrix can be formed independently. An easy example shown in programs **cup4.f** and **cup5.f** is to use a parallel counter *my_count* which is allowed to take the value 1 every p steps, when we let all processors compute (fetch) a row of matrix C. To ensure that only one processor broadcasts its row content to all others when needed, we also set another parallel counter *my_role* and check whether *my_role=0*; such a counter value corresponds to the processor identity number *my_first*. For example, with $p = 4$ processors, the counters may be organized as follows for the first few steps

row i	processor j	*my_role*	*my_first*	*my_count*
2	0	0	0	1
	1	1	0	1
	2	2	0	1
	3	3	0	1
3	0	3	1	2
	1	0	1	2
	2	1	1	2
	3	2	1	2
4	0	2	2	3
	1	3	2	3
	2	0	2	3
	3	1	2	3
5	0	1	3	4
	1	2	3	4
	2	3	3	4
	3	0	3	4

Clearly *my_first* tracks down the identity of the processor that has *my_role*= 0.

Storage. The amount of storage required on processor i is $(n+1) \times n_i$, corresponding to n_i column vectors. Specifically, it is appropriate to assume that after step 1 the following matrix of size $(n+1) \times n_i$ is stored on processor i

$$\mathbf{V} = \left[V_1 \,\middle|\, V_2 \,\middle|\, \cdots \,\middle|\, V_{n_i} \right]_{(n+1)\times n_i}$$

which corresponds (globally) to the vectors

$$\left[V^{(n)}_{t(1,i)} \,\middle|\, V^{(n)}_{t(2,i)} \,\middle|\, \cdots \,\middle|\, V^{(n)}_{t(n_i,i)} \right]$$

where (note $t(n_p, p) = (p-1)n_p + n_p = n$)

$$t(k,i) = (i-1)n_i + k \qquad \text{for } k = 1, 2, \cdots, n_i.$$

That is, the subspace $\mathbf{R}^{(n)}$ is split into smaller subspaces in p processors. For each step $j = 1, 2, \ldots, n$, let the jth pivoting vector $V_{s(j)}^{(n+1-j)}$ reside in processor p_j; here $p_j \in \{0, 1, \ldots, p-1\}$.

Once data are distributed as above, the pivoting vector $V_{s(j)}^{(n+1-j)}$ at a subsequent step j has to be broadcast from processor p_j to all other processors; we denote such a step by 'bcast' for `MPI_BCAST`. In implementation it is important to note that $V_{s(j)}^{(n+1-j)}$ has at most j nonzero positions (e.g. 1 nonzero at step $j = 1$ and 3 nonzeros at step $j = 3$). One other observation is that the processor that holds the pivot vector (i.e. uses 'bcast') will have one less vector to work with; this fact may be called dimension reduction for simplicity. Therefore an ideal load balancing will be achieved if p_j (the pivoting processor) takes all values from the set $\{0, 1, \ldots, p-1\}$ (in any order) every p steps. The dimension reduction in the active processor will lead to load unbalancing.

♦ **Method I.** Our first parallel method is illustrated in program **cup4.f**, where the global pivoting is restricted in search columns in order to achieve a 'perfect' load balancing. In details, every p steps, no processors are allowed to host the pivot for more than once; hence each processor hosts the pivot exactly once in any order.

As it turns out, such an idea is quite easy to implement with the command `MPI_ALLREDUCE` by modifying a control variable; compare for the maximum value of

$$\begin{cases} \text{local maximum pivot product } C_i^T v_1, & \text{if } n_j \geq [n/p] - [(i-2)/p], \\ \text{the value -1}, & \text{if } n_j < [n/p] - [(i-2)/p], \end{cases}$$

where, $i \geq 2$ denotes the row i for matrix C, n_j is the total number of working vectors left on processor j and note that $[(i-2)/p]$ advances by 1 for every p steps. As an illustration, we list the case of $p = 4$ processors, the active dimensions in the first 4 steps may be look like this

row i	processor 0	processor 1	processor 2	processor 3
2	$n_0 - 1$	n_1	n_2	n_3
3	$n_0 - 1$	n_1	n_2	$n_3 - 1$
4	$n_0 - 1$	$n_1 - 1$	n_2	$n_3 - 1$
5	$n_0 - 1$	$n_1 - 1$	$n_2 - 1$	$n_3 - 1$

As the reader may find out from running **cup4.f** with the built in data, the method is adequate in terms of parallel performance (or scalability) but is not suitable for two of the problems (giving extremely inaccurate solutions) where pivoting is essential.

♦ **Method II.** This algorithm attempts the global (column) pivoting strategy, which is not somehow compatible with our data partition. At each step, define the active processor *my_act* as the one that keeps the pivoting vector. One way to balance the load (in terms of dimension reduction) is for other processors in turn to send a basis vector to this active processor, which has just reduced its dimension n_j. Thus all processors are forced to reduce their dimension at the same speed and consequently the work load is balanced. This is our program **cup5.f**.

We need `MPI_REDUCE` with the option `MPI_MAXLOC` to find out the pivot and the pivoting processor number. To achieve load balance by shifting vectors between processors, our idea is to set up a parallel cyclic counter my_role that takes values from $\{0, 1, 2, \ldots, p-1\}$ cyclically; whenever $my\_role = 0$ we expect the number of vectors V_j's to be reduced by one and if the underlying processor is not the pivot processor, we propose to shift a vector V_1 from it to the pivoting processor. This forces an even dimension reduction across processors.

Initially set $my\_role = i$ on processor i i.e. we anticipate at step j, $p_j = j$ otherwise we shift a vector from j to the pivoting processor p_j. Therefore this achieves the aim of load balance while maintaining the global row pivoting, as depicted in Figure 15.1. For instance, assume $n = 16$ with $p = 4$. If the active processors for the first five steps of $i \geq 2$ are $my\_act = 3,\ 1,\ 2,\ 2,\ 0$, then the load shifting is as shown in Table 15.4. It is possible to adapt the idea further for specific problems with some predictable pivoting patterns.

Finally we remark on the extra effort made in **cup5.f** in allowing both n and p to be arbitrarily integers, by using the counters *my_turn* and *my_tid* to control fetching row i's and by focusing the counters *my_role* and *my_first* on the task of load shifting. In this general case, $n \neq [n/p]p$ so we have to defer the normal shifting process until the last processor $p-1$ shifts out all of its extra $nc\_left = n - [n/p]p$ vectors allocated. We use this feature to highlight what can be done to deal with general dimensions.

In summary, the sample programs **cup4.f** and **cup5.f** used these MPI routines

```
MPI_INIT                 ! The minimal call
MPI_COMM_RANK            ! The minimal call
```

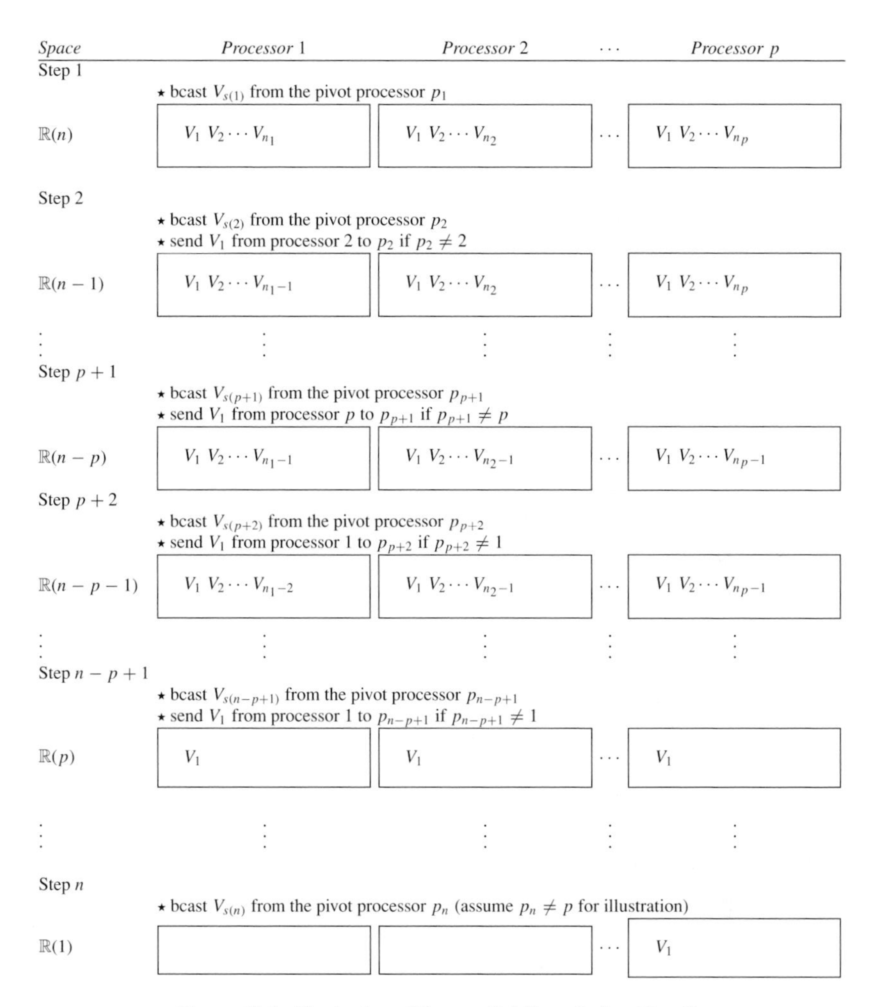

Figure 15.1. Illustration of the parallel Purcell algorithm II.

```
MPI_COMM_SIZE           ! The minimal call
MPI_GET_PROCESSOR_NAME  ! The optional use for processor name
MPI_bcast               ! Broadcast
MPI_barrier             ! Set a waiting point for bcast
MPI_ALLREDUCE           ! Compute across processors
MPI_SEND                !
MPI_RECV                !
MPI_WTIME               ! Obtain the cpu timing
MPI_FINALIZE            ! The minimal call
```

Table 15.4. *Illustration of load shifting according to positioning of the active processor my_act* $= 3,\ 1,\ 2,\ 2,\ 0,$ *in relation to the preset cyclic counter* my_role. *Both the pivoting vector and the shifted vector are shown in* **bold** *font.*

Row i	Processor 0	Processor 1	Processor 2	Processor 3
2 start ⊩	$V_1^{(0)}\ V_2^{(0)}\ V_3^{(0)}\ V_4^{(0)}$	$V_1^{(1)}\ V_2^{(1)}\ V_3^{(1)}\ V_4^{(1)}$	$V_1^{(2)}\ V_2^{(2)}\ V_3^{(2)}\ V_4^{(2)}$	$\mathbf{V_1^{(3)}}\ V_2^{(3)}\ V_3^{(3)}\ V_4^{(3)}$
update ⊩	$V_1^{(0)}\ V_2^{(0)}\ V_3^{(0)}\ V_4^{(0)}$	$V_1^{(1)}\ V_2^{(1)}\ V_3^{(1)}\ V_4^{(1)}$	$V_1^{(2)}\ V_2^{(2)}\ V_3^{(2)}\ V_4^{(2)}$	$V_1^{(3)}\ V_2^{(3)}\ V_3^{(3)}$
my_role ⊩	0	1	2	3
shift ⊨	$V_2^{(0)}\ V_3^{(0)}\ V_4^{(0)}$	$V_1^{(1)}\ V_2^{(1)}\ V_3^{(1)}\ V_4^{(1)}$	$V_1^{(2)}\ V_2^{(2)}\ V_3^{(2)}\ V_4^{(2)}$	$V_1^{(3)}\ V_2^{(3)}\ V_3^{(3)}\ \mathbf{V_1^{(0)}}$
3 start ⊩	$V_1^{(0)}\ V_2^{(0)}\ V_3^{(0)}$	$\mathbf{V_1^{(1)}}\ V_2^{(1)}\ V_3^{(1)}\ V_4^{(1)}$	$V_1^{(2)}\ V_2^{(2)}\ V_3^{(2)}\ V_4^{(2)}$	$V_1^{(3)}\ V_2^{(3)}\ V_3^{(3)}\ V_4^{(3)}$
update ⊩	$V_1^{(0)}\ V_2^{(0)}\ V_3^{(0)}$	$V_1^{(1)}\ V_2^{(1)}\ V_3^{(1)}$	$V_1^{(2)}\ V_2^{(2)}\ V_3^{(2)}\ V_4^{(2)}$	$V_1^{(3)}\ V_2^{(3)}\ V_3^{(3)}\ V_4^{(3)}$
my_role ⊩	3	0	1	2
4 start ⊩	$V_1^{(0)}\ V_2^{(0)}\ V_3^{(0)}$	$V_1^{(1)}\ V_2^{(1)}\ V_3^{(1)}$	$\mathbf{V_1^{(2)}}\ V_2^{(2)}\ V_3^{(2)}\ V_4^{(2)}$	$V_1^{(3)}\ V_2^{(3)}\ V_3^{(3)}\ V_4^{(3)}$
update ⊩	$V_1^{(0)}\ V_2^{(0)}\ V_3^{(0)}$	$V_1^{(1)}\ V_2^{(1)}\ V_3^{(1)}$	$V_1^{(2)}\ V_2^{(2)}\ V_3^{(2)}$	$V_1^{(3)}\ V_2^{(3)}\ V_3^{(3)}\ V_4^{(3)}$
my_role ⊩	2	3	0	1
5 start ⊩	$V_1^{(0)}\ V_2^{(0)}\ V_3^{(0)}$	$V_1^{(1)}\ V_2^{(1)}\ V_3^{(1)}$	$\mathbf{V_1^{(2)}}\ V_2^{(2)}\ V_3^{(2)}$	$V_1^{(3)}\ V_2^{(3)}\ V_3^{(3)}\ V_4^{(3)}$
update ⊩	$V_1^{(0)}\ V_2^{(0)}\ V_3^{(0)}$	$V_1^{(1)}\ V_2^{(1)}\ V_3^{(1)}$	$V_1^{(2)}\ V_2^{(2)}$	$V_1^{(3)}\ V_2^{(3)}\ V_3^{(3)}\ V_4^{(3)}$
my_role ⊩	1	2	3	0
shift ⊨	$V_1^{(0)}\ V_2^{(0)}\ V_3^{(0)}$	$V_1^{(1)}\ V_2^{(1)}\ V_3^{(1)}$	$V_1^{(2)}\ V_2^{(2)}\ \mathbf{V_1^{(3)}}$	$V_2^{(3)}\ V_3^{(3)}\ V_4^{(3)}$
6 start ⊩	$\mathbf{V_1^{(0)}}\ V_2^{(0)}\ V_3^{(0)}$	$V_1^{(1)}\ V_2^{(1)}\ V_3^{(1)}$	$V_1^{(2)}\ V_2^{(2)}\ V_3^{(2)}$	$V_1^{(3)}\ V_2^{(3)}\ V_3^{(3)}$
update ⊩	$V_1^{(0)}\ V_2^{(0)}$	$V_1^{(1)}\ V_2^{(1)}\ V_3^{(1)}$	$V_1^{(2)}\ V_2^{(2)}\ V_3^{(2)}$	$V_1^{(3)}\ V_2^{(3)}\ V_3^{(3)}$
my_role ⊩	0	1	2	3

15.5.4 Numerical examples

As shown earlier (in Section 15.5.1), the Purcell method in its sequential form appears to produce a consistently small growth factor in comparison to other well-known direct elimination methods, making it a potential winner of all practical direct methods. Our test examples will be the four simple matrices shown in Section 15.5.2; for BEM examples, see [328].

Below we demonstrate that our new algorithm II performs better than the parallel Gauss–Jordan method with global (partial) column pivoting as in [178,283], in terms of reliability (accuracy expected of a direct method) and scalability (as the number of processors increases). The tests were carried out on a SGI IP25 machine with 14 processors and have also been verified on a Sun-sparc work station cluster.

Reliability and accuracy test. Here we test the four tough (but simple) problems as in Section 15.5.2 to see how accurate the different algorithms with our pivoting strategies are; each problem is solved with $n = 1024,\ 2048$ equations. Table 15.5 shows the details of our experiments, where ⊞ denotes the problem number, △ denotes an acceptable accuracy is achieved for all n and ▼ denotes a solution failure or inaccuracy. By accuracy, we mean the machine precision because these are direct methods. Clearly one can observe that only the Purcell algorithm (from Method II) is very reliable. We remark that Method I is not so reliable for two of the problems.

Scalability test. We now test the scalability of the parallel Purcell method II as proposed. This is presented in Table 15.6. One observes that the scalability of the proposed parallel method is generally good.

Table 15.5. *Comparison of accuracy of parallel direct methods on SGI IP25.*

	Method GJ				Purcell		
⊞	$p = 1$	2	4	8	2	4	8
1	▼	▼	▼	▼	△	△	△
2	▼	▼	▼	▼	△	△	△
3	▼	▼	▼	▼	△	△	△
4	▼	▼	▼	▼	△	△	△
5	△	△	△	△	△	△	△

Table 15.6. *Performance of parallel direct algorithms on SGI IP25.*

Algorithm used	Processors p	Problem size n	IP25 CPU (s)	Efficiency over $p = 1$
GJ	1	1024	244	
		2048	1868	
	2	1024	122	2
		2048	935	2
	4	1024	61	4
		2048	469	4
	8	1024	31	7.9
		2048	238	7.8
PC	2	1024	69	2.0
		2048	544	2.0
	4	1024	35	4.0
		2048	277	3.9
	8	1024	20	6.9
		2048	143	7.5

Once a reader becomes familiar with the functionality of basic MPI routines, there are 'endless' algorithms to be attempted. For instance, in [209], we considered how to parallelize a DWT. Parallelizing SPAI type preconditioners [148,149,253] will be another interesting task to complete. The multigrid methods (Chapter 6) may also be parallelized [21,185,346].

15.6 Discussion of software and the supplied MPI Fortran files

Parallel computing, also refered to as the high performance computing, is a fast expanding subject. There exist ample examples of MPI programs, that are available often along with sequential software (as remarked, not all of the parallel programs are easy to follow). Here we highlight three of such resources for information.

(1) The ScaLAPACK (Scalable LAPACK):

http : //www.netlib.org/scalapack/

(2) The PLAPACK (parallel LAPACK):

http : //www.cs.utexas.edu/users/plapack/

(3) The BlockSolve package:

http : //www−unix.mcs.anl.gov/sumaa3d/BlockSolve/index.html
http : //www.netlib.org/misc/blocksolve.tgz

This book has supplied some more gentle programs for first attempts of MPI programming.

[1] `cup1.f` – Example 1 of a parallel series summation.
[2] `cup2.f` and `cup3.f` – Example 2 of a parallel power method using respectively the column and row partition.
[3] `cup4.f` and `cup5.f` – Example 3 of a parallel direct (Purcell) method, where pivoting is done differently in each program.

Appendix A: a brief guide to linear algebra

> Information in science and mathematics is often organized into rows and columns to form rectangular arrays, called "matrices" (plural of "matrix"). Matrices are often tables of numerical data that arise from physical observations, but they also occur in various mathematical contexts.
>
> HOWARD ANTON. *Elementary Linear Algebra.* Wiley (1973 1st edn, 2000 8th edn).

To be able to read or work on matrix computing, a reader must have completed a course on linear algebra. The inclusion of this Appendix A is to review some selected topics from basic linear algebra for reference purposes.

A.1 Linear independence

Let $v_j \in \mathbb{R}^n$ for $j = 1, \ldots, r$ with $r \le n$. The set of r vectors $v_1, v_2, \ldots, v_r$ is said to be **linearly independent** if the only solution of the equation

$$k_1 v_1 + k_2 v_2 + \cdots + k_r v_r = 0 \tag{A.1}$$

is $k_1 = k_2 = k_3 = \ldots = k_r = 0$. Otherwise, the set of vectors is said to be **linearly dependent**.

An interesting fact is that equation (A.1) can be put into matrix vector form

$$A\mathbf{k} = 0, \quad \text{with} \quad A = [v_1\ v_2\ \cdots\ v_r]_{n \times r}, \ \mathbf{k} = [k_1\, k_2\, \cdots\, k_r]^T \in \mathbb{R}^r. \tag{A.2}$$

Let the subspace $V = \mathtt{span}(v_1, v_2, \ldots, v_n)$ and the matrix $A = [v_1, v_2, \ldots, v_n]$. Then $y \in V$ is the same as

$$y = Ax, \qquad \text{for} \ \ x \in \mathbb{R}^n. \tag{A.3}$$

A.2 Range and null spaces

Given a matrix $A \in \mathbb{R}^{m \times n}$, the range space and the null space are defined by

$$\begin{aligned} \text{Range}(A) &= \{y \mid y = Ax \in \mathbb{R}^m, \ x \in \mathbb{R}^n\} \\ \text{Null}(A) &= \{x \mid y = Ax = 0 \in \mathbb{R}^m, \ x \in \mathbb{R}^n\} \end{aligned} \tag{A.4}$$

with the dimension of Range(A) called the *rank* of A and the dimension of Null(A) called the *nullity* of A. The important connection is that

$$rank(A) + nullity(A) = n.$$

If $m = n$, we have the direct sum: $\mathbb{R}^n = \text{Range}(A) \oplus \text{Null}(A)$. Note also that $\text{Range}(A) \neq \text{Range}(A^T)$ unless A is symmetric, but $rank(A) = rank(A^T)$ for any $A_{m \times n}$.

A.3 Orthogonal vectors and matrices

Two vectors $p, q \in \mathbb{R}^n$ are called **orthogonal** to each other if

$$p^T q = (p, q) = \sum_{j=1}^{n} p_j q_j = 0.$$

Mutually orthogonal vectors $p_1, p_2, \ldots, p_r$, if normalized with $(p_j, p_j) = 1$, are called **orthonormal** i.e. they satisfy

$$p_i^T p_j = (p_i, p_j) = \delta_{ij} = \begin{cases} 1, & i = j, \\ 0, & i \neq j. \end{cases} \tag{A.5}$$

An **orthogonal matrix** $A_{n \times r}$ has orthonormal column vectors

$$A = [p_1 \ p_2 \ \cdots \ p_r].$$

From (A.5), we see that

$$A^T A = I_{r \times r}.$$

When $r = n$, we have the usual result $A^T A = A A^T = I$.

A.4 Eigenvalues, symmetric matrices and diagonalization

For $A \in \mathbb{R}^{n \times n}$, its eigenvalue λ and eigenvector x are defined by

$$Ax = \lambda x \tag{A.6}$$

where $\lambda \in \mathbb{C}$, $x \in \mathbb{C}^n$. At most, there are n such pairs (λ_j, x_j) for a general A. If

- either all n eigenvalues λ_j's are distinct,
- or all n eigenvectors x_j's are linearly independent,

then matrix A can be diagonalized by its eigenvector matrix X

$$X^{-1}AX = \text{diag}(\lambda_1, \cdots, \lambda_n), \qquad X = [x_1\ x_2\ \cdots\ x_n]. \tag{A.7}$$

The above conditions imply that the only case where (A.7) does not hold is when some λ_j's are the same and there are not enough corresponding eigenvectors (linearly independent). This case is referred to as A having defective eigenvalues; that is, at least one eigenvalue has an algebraic multiplicity (multiple eigenvalues) larger than its geometric multiplicity (eigenvector count). Then the alternative to (A.7) will be the Jordan decomposition.

When A is symmetric, all λ_j's are real and X will be invertible; if all columns are normalized, X is orthogonal. (Likewise a Hermitian A will lead to an unitary X). Once diagonalization is established, there are various applications.

A.5 Determinants and Cramer's rule

The determinant $|A|$ or $det(A)$ is normally computed by its row-wise (or columnwise) expansion: for any k

$$|A| = a_{k1}A_{k1} + a_{k2}A_{k2} + \cdots + a_{kn}A_{kn} = \sum_{j=1}^{n} a_{kj}A_{kj} = \sum_{j=1}^{n} (-1)^{k+j} a_{kj} M_{kj} \tag{A.8}$$

where $M_{kj} = |A(r, r)|$, with $r = [1\ \cdots\ k-1\ k+1\ \cdots\ n]$, is the minor of A at entry a_{kj} and $A_{kj} = (-1)^{k+j} M_{kj}$ is the cofactor at a_{kj}. Here the MATLAB notation $A(r, r)$ for submatrix extraction is used.

Formula (A.8) is only a special case of the Laplace theorem for determinant expansion. In general, let vector $\mathbf{k} = [k_1, k_2, \ldots, k_\tau]$ be a subset of size τ (ordered sequentially) taken from the index set $\mathbf{N} = [1, 2, \ldots, n]$; e.g. $\mathbf{k} = [2\,4]$ with $\tau = 2$. Consider the determinant $|A(\mathbf{k}, \mathbf{j})|$ with $\mathbf{j} = [j_1, j_2, \ldots, j_\tau]$ and $1 \le j_1 < j_2 < \ldots < j_\tau \le n$. There are $t = \begin{pmatrix} \tau \\ n \end{pmatrix} = C_n^\tau = \dfrac{n!}{\tau!(n-\tau)!}$ such determinants; name them as $M_1, M_2, \ldots, M_t$. For each determinant $|A(\mathbf{k}, \mathbf{j})|$, define its 'cofactor' by $|A(\hat{\mathbf{k}}, \hat{\mathbf{j}})|(-1)^{k_1+\cdots+k_\tau+j_1+\cdots+j_\tau}$ with $\hat{\mathbf{k}} = \mathbf{N}\backslash\mathbf{k}$ and $\hat{\mathbf{j}} = \mathbf{N}\backslash\mathbf{j}$. Denote such cofactors by $A_1, A_2, \ldots, A_t$. Then the Laplace theorem states that

$$|A| = M_1A_1 + M_2A_2 + \cdots + M_tA_t \tag{A.9}$$

which reduces to (A.8) if we choose $\tau = 1$ and $\mathbf{k} = k$.

Once determinants are defined, for a square linear system $Ax = b$ with $rank(A) = n$, denote $B_j = [a_1 \ldots, a_{j-1}, b, a_{j+1}, \ldots, a_n]$ the matrix A with its jth column replaced by vector b. Then the solution by Cramer's rule is

$$x_j = \frac{|B_j|}{|A|}, \qquad j = 1, 2, \ldots, n,$$

which may be derived from using $x = A^{-1}b$ and considering $|A^{-1}B_j|$. Replacing $b = e_k$ (the kth unit vector) in turns gives rise to the solution of $Az_k = e_k$

$$z_k = \frac{1}{|A|}\begin{pmatrix} |B_1^k| \\ |B_2^k| \\ \vdots \\ |B_n^k| \end{pmatrix}, Z = [z_1, z_2, \cdots, z_n] = \frac{1}{|A|}\begin{bmatrix} |B_1^1| & |B_1^2| & \cdots & |B_1^n| \\ |B_2^1| & |B_2^2| & \cdots & |B_2^n| \\ \vdots & \vdots & \cdots & \vdots \\ |B_n^1| & |B_n^2| & \cdots & |B_n^n| \end{bmatrix}, \tag{A.10}$$

where B_j^k is the matrix of A with its jth column replaced by e_k so $|B_j^k| = A_{kj}$ is the cofactor for entry (k, j) and hence $Z = A^{-1}$. Therefore the last matrix in (A.10) is precisely the adjoint matrix.

A.6 The Jordan decomposition

Whenever (A.7) does not hold for a matrix A, we have to use the Jordan decomposition which exists for any A. Note that in (A.7), X is complex. The same is true with a Jordan decomposition: for any matrix $A \in \mathbb{C}^{n\times n}$, there exists a nonsingular matrix $C \in \mathbb{C}^{n\times n}$ such that

$$\begin{aligned} C^{-1}AC = J &= \operatorname{diag}(B_1, B_2, \cdots, B_d), \\ B_i &= \operatorname{diag}(\underbrace{\lambda_i, \cdots, \lambda_i,}_{p_i - 1 \text{ terms}}\ J_i), \end{aligned} \qquad J_i = \begin{bmatrix} \lambda_i & 1 & & \\ & \lambda_i & \ddots & \\ & & \ddots & 1 \\ & & & \lambda_i \end{bmatrix}_{q_i\times q_i} \tag{A.11}$$

where d is the total number of distinct eigenvalues and the eigenvalue λ_i has p_i linearly independent eigenvectors (i.e. p_i is the geometric multiplicity for λ_i and λ_i has $q_i - 1$ deficient eigenvectors). Here $p_i + q_i - 1$ is the algebraic multiplicity for λ_i. As C has n independent column vectors, so

$$\sum_{i=1}^{d}(p_i + q_i - 1) = n.$$

Assuming that p_i, q_i are known, to find C, we identify its column vectors from

$$C = [c_1 \; c_2 \; \cdots \; c_n]$$

and (A.11)

$$\begin{array}{ll} Ac_j = \lambda_i c_j, & \text{if } c_j \text{ belongs to the set of independent eigenvectors} \\ Ac_j = c_{j-1} + \lambda_i c_j, & \text{if } c_j \text{ belongs to the set of deficient eigenvectors} \end{array} \tag{A.12}$$

where the 'deficient' eigenvectors (usually called the generalized eigenvectors for λ_i) will have to be replaced and computed while the independent eigenvectors can be assumed to be known. Therefore all c_j can be found from the above linear systems. Practically, the generalized eigenvectors c_j for λ_i can be found from solving the least-squares problem (by the QR method of §2.4 and (5.25))

$$(A-\lambda_i I)c_j = c_{j-1}, \qquad j = s_i + 1, \ldots, s_i + q_i - 1, \quad s_i = \sum_{k=1}^{i-1}(p_k + q_k - 1) + p_i,$$

which has the property

$$(A - \lambda_i I)^{j-s_i+1} c_j = (A - \lambda_i I)c_{s_i} = 0, \qquad j = s_i + 1, \ldots, s_i + q_i - 1.$$

That is, the generalized eigenvectors are simply the null vectors of powers of matrix $(A - \lambda_i I)$; see [508]. Consider the following 4×4 example, where the matrix has a single eigenvalue $\lambda_1 = 4$ with algebraic multiplicity $p_1 + q_1 - 1 = 4$ and geometric multiplicity $p_1 = 1$ i.e. there exists only one eigenvector $x_1 = [0.5\,0.5\,0.5\,0.5]^T$. Using the QR method (see the Mfile `my_jordan.m`), we find that

$$A = \begin{bmatrix} 3 & 1 & 0 & 0 \\ -1 & 4 & 1 & 0 \\ -1 & 0 & 4 & 1 \\ -1 & 0 & 0 & 5 \end{bmatrix}, \quad X^{-1}AX = \begin{bmatrix} 4 & 1 & & \\ & 4 & 1 & \\ & & 4 & 1 \\ & & & 4 \end{bmatrix},$$

$$X = \begin{bmatrix} 0.5 & -0.5 & 0 & 0.5 \\ 0.5 & 0 & -0.5 & 0.5 \\ 0.5 & 0 & 4 & 0 \\ 0.5 & 0 & 0 & 0.5 \end{bmatrix}.$$

Here x_2, x_3, x_4 are $q_1 - 1 = 3$ deficient eigenvectors. One can verify that $(A - 4I)x_1 = (A - 4I)^2 x_2 = (A - 4I)^3 x_3 = (A - 4I)^4 x_4 = 0$ and $(A - 4I)x_j = x_{j-1}$ etc.

A.7 The Schur and related decompositions

The Jordan decomposition provides a similarity transform to simplify A. The transform is not unitary. The Schur decomposition supplies us with an unitary transform to simplify A, not to a block diagonal matrix such as the Jordan matrix J but to a triangular matrix whose diagonals are the eigenvalues of A. Related decompositions aim for the diagonal matrix and two simultaneous matrices.

Schur. For any matrix $A \in \mathbb{C}^{n\times n}$, there exists an unitary matrix $U \in \mathbb{C}^{n\times n}$ that can transform A to an upper triangular form, with $\lambda_j \in \mathbb{C}$,

$$U^H AU = T, \qquad \text{diag}(T) = \text{diag}(\lambda_1, \lambda_2, \cdots, \lambda_n). \tag{A.13}$$

Note that most columns of U, u_j's for $j > 1$, are in general not eigenvectors of A; the first column is an eigenvector as $Au_1 = T_{11}u_1$ from equating $AU = UT$. If A has n independent eigenvectors, the Schur decomposition is equivalent to orthogonalizing them to make up an orthonormal basis via the Gram–Schmidt process.

For a real matrix $A \in \mathbb{R}^{n\times n}$, there exists an orthogonal matrix $U \in \mathbb{R}^{n\times n}$ that can transform A to an upper triangular form with diagonal blocks,

$$U^H AU = T, \quad \text{diag}(T) = \text{diag}(b_1, b_2, \cdots, b_d), \quad b_j \in \mathbb{R} \text{ or } b_j \in \mathbb{R}^{2\times 2}, \tag{A.14}$$

where the 2×2 blocks correspond to a complex pair of eigenvalues. On MATLAB®, the default setting is to give a real Schur decomposition, unless specified otherwise

```
>> rand('state',0),  A=rand(4) % set the seed
>> a = eig(A)'                  % 2.3230, 0.0914+/-0.4586i, 0.2275
>> b = schur(A)                 % b1,b2,b3 with b2=2 x 2
>> c = schur(A,'complex') %diag = 2.3230, 0.0914+/-0.4586i, 0.2275
```

Related decompositions. For $A \in \mathbb{C}^{n\times n}$, the singular value decomposition is

$$U^H AV = D, \quad D = \text{diag}(\sigma_1, \sigma_2, \cdots, \sigma_n), \tag{A.15}$$

where U, V are unitary matrices, and $\sigma_j \geq 0$ is the jth singular value of A. For two matrices $A, B \in \mathbb{C}^{n\times n}$, there exist Schur type decompositions

$$U^H AV = T_1, \quad U^H BV = T_2, \quad \text{diag}(T_1)./\ \text{diag}(T_2) = \text{diag}(g_1, g_2, \cdots, g_n), \tag{A.16}$$

where T_1, T_2 are upper triangular, './' is the usual pointwise division and $g_j \in \mathbb{C}$ is the jth generalized eigenvalue as defined by [329,280]

$$Ax_j = g_j Bx_j. \tag{A.17}$$

Similarly to the Schur case, if A is real, U, V can be real [329].

Appendix B: the Harwell–Boeing (HB) data format

The matrix market [347,188] contains a rich source of test matrices from various applications, many of which are challenging test examples (other sources can be found from Appendix E). The Harwell–Boeing data format is one of the two text formats used to store matrices. To help the reader to understand this sparse storage, we give a brief introduction.

Let A be sparse. For simplicity, we first consider the sample matrix (as from the supplied Mfile `size6.m`):

$$A = \begin{bmatrix} -8.4622 & 0 & 0.3046 & 0 & 0 & 0 \\ 0 & 0 & 0.1897 & 0.6979 & 0 & 0.2897 \\ 0.2026 & 0.8318 & 0 & 0 & 0 & 0 \\ 0 & 0 & 0 & 0 & 0.6449 & 0.5341 \\ 0.8381 & 0.7095 & 0.3028 & 0.8537 & 0 & 0 \\ 0.0196 & 0.4289 & 0.5417 & 0 & 0 & 0 \end{bmatrix}.$$

For this matrix, we have prepared a sample data set `sample.rua` which keeps this A in HB format:

```
Sample HB data (see size6.m - the Mfile)   sample.rua          } Title
            1            2            8            14       0  } No of lines
rua                      6            6            16          } Type/Rows/Cols/NNZ
       (7I3)        (9I3)        (2D20.13)       (4D20.14)     } Listed Data Formats
  1  5  8 12 14 15 17                                          } Column pointers 7I3
  1  3  5  6  3  5  6  1                                       } Row indices
  2  5  6  2  5  4  2  4                                       } (Format 8I3)
-8.4622141782399D+00 2.0264735764999D-01                       }
 8.3811844505199D-01 1.9639513864800D-02                       } The 16 non-zeros
 8.3179601761001D-01 7.0947139270301D-01                       }
 4.2889236534100D-01 3.0461736686901D-01                       } of matrix A
 1.8965374754701D-01 3.0276440077701D-01                       }
 5.4167385389801D-01 6.9789848186001D-01                       } (Format 2D20.13)
 8.5365513066301D-01 6.4491038419400D-01                       }
 2.8972589585600D-01 5.3407901762700D-01                       }
```

The best way to read a matrix of this type is to use the utility routines from the 'official site' http : //math.nist.gov/MatrixMarket/formats.html or from the Sparskit [414]

http : //www−users.cs.umn.edu/∼saad/software/SPARSKIT/sparskit.html

A simple example can be found in **cup2.f** in Section 15.4.

Appendix C: a brief guide to MATLAB®

There are many books written on MATLAB® at both tutorial and advanced levels e.g. [135,279]; see also the official site

http : //www.mathworks.com,
http : //www.mathworks.com/support/books/

This short appendix is to illustrate some MATLAB notation and highlight the excellent graphics facilities. Further elementary introductions may be found from

http : //www.liv.ac.uk/maths/ETC/matbook/ (Chen *et al.* [135])
http : //www.math.uiowa.edu/~atkinson/m72_master.html
(Atkinson and Han [25])

C.1 Vectors and matrices

In mathematics, we use either the round braces or the square brackets to denote a vector or matrix. MATLAB sticks to the square brackets for entering a matrix e.g.

```
>> x = [ 1 2 3 ]       % for a row vector of size 3
>> y = [ 1 2 3 4:6 ]' % for a column vector of size 6
>> A = [ 1 2 3; 4 5 6;
         7 8 9         % for entering a 4 x 3 matrix
         0 1 2]        % (note the use of ';' and end-of-line)
>> B = [ 1 2 3 4 ...
         5 6; 7 8 ... % for entering a 2 x 6 matrix
         9 0 1 2]      % (note the use of ... for continuation)
>> z = 0:0.001:2*pi;  % An uniform vector of stepsize h=0.001
>> rand('state',0); V=rand(3,5);  % A random matrix of 3 x 5
```

Once a matrix (or vector) is entered, the power of MATLAB really starts

```
>> A = round( rand(6)*9 )      % Generate a 6 x 6 random matrix
>> [X D] = eig(A)              % Compute the eigen-system
>> [U D V] = svd(A)            % Compute  A=U*D*V' or U'*A*V=D
>> b = A(:,1)                  % Extract the first column of A
>> x = A \ b                   % Solve Ax = b
>> [v j]=max(abs(x)),xm=x(j)% Locate the largest component
>> s1  = eigs(A,1,0)           % The smallest 1 eigenvalue near 0
>> s2  = svds(A,2,1)           % Smallest 2 singular values near 1
```

Data extraction via an index vector is very convenient. Below we show a few examples using submatrices or building larger blocks

```
>> row = randperm(6)            % Set up an index vector 'row'
>> T = toeplitz(row)            % Create a Toepliz matrix
>> add = 6:-1:2                 % index list
>> new = [row 0 row(add)]       % New extended row vector
>> C = gallery('circul',new); % Create a larger circulant matrix
>> mat_pr0(C,6,1,0)             % Observe T is embedded in C
>> T = C(1:6,1:6)               % Extract T out of C directly
```

where `mat_pr0.m` is a supplied Mfile.

C.2 Visualization of functions

Plotting is really easy with MATLAB, although complex plots may take a lot of practice. First, we plot a peanut-shaped curve.

```
>> t = 0:0.01:2*pi;
>> r = sqrt( 100*cos(4*t) + sqrt(11000 - 10000*sin(4*t).^2) );
>> x = 4.1*r.*cos(t);  y = 4.1*r.*sin(t);
>> H = plot(x,y,'r');  set(H,'linewidth',6)
```

Second, we plot selected curves over a sphere:

```
>>   p=0:0.001:2*pi;      z=cos(p);  % Vertical
>> for t=0:pi/8:2*pi %-----------------------
>>   x=sin(p)*cos(t*2);  y=sin(p)*sin(t*2);
>>   h=plot3(x,y,z,'b'); hold on
>>   set(h,'linewidth',2)
>> end
>>   t=0:0.001:2*pi;
>> for p=0:pi/10:2*pi %-----------------------
>>   x=sin(p)*cos(t*2); y=sin(p)*sin(t*2);
```

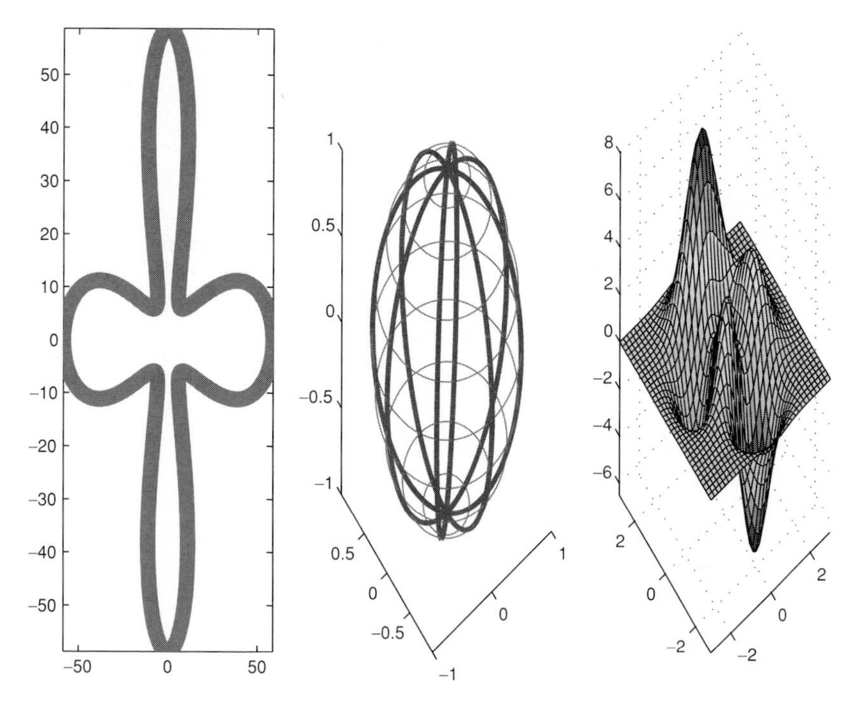

Figure C.1. MATLAB plots of functions.

```
>>   z=cos(p).*ones(size(x));      % Horizontal
>>   plot3(x,y,z,'r'); hold on
>> end
>>   hold off
```

Finally, we show a surface plot:

```
>> [x y z]=peaks(35);
>> surf(x,y,z), axis tight
```

The three plots are shown in Figure C.1. In addition, we should mention the 'non-numerical' and 'easy' plotting commands.

```
>> ezplot('x^2-4*y^2=1')
>> ezsurf('sqrt(1-x^16-y^18)')
```

Since MATLAB version 6, a user can annotate a figure by adding texts and arrows interactively. From version 7, more conveniently, the annotated results can be saved as a Mfile, with a view to reproduce the same plot later.

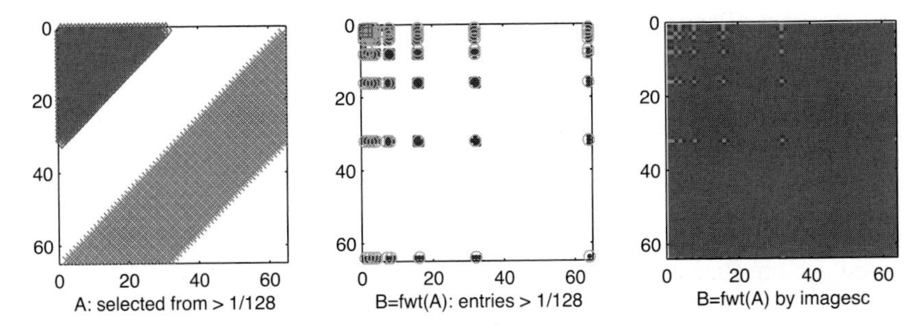

Figure C.2. MATLAB plots of matrices.

C.3 Visualization of sparse matrices

Visualization of matrices is necessary especially for large matrices, because it is not possible to print out all entries. The sparse matrices are dealt with very efficiently in MATLAB with commands like `A=sparse(I,J,V)`, `speye(n)`, `A=spdiags([e -2*e e], -1:1, n, n)` and associated operations!

The command `spy` is the most useful to visualize a matrix and also relevant are the plotting routines such as `mesh`. A somewhat surprising choice may be the command `imagesc`, usually for displaying images. If necessary, use different colours, symbols and superimposition as illustrated below and with results shown in Figure C.2.

```
>>   n = 64;  A=hilb(n);
>> subplot(131)
>>   spy(A>1/n/2,'b.'),hold on
>>   spy(A>2/n/3,'mx'),hold on
>>   spy(A>1/n,'go'),  hold on
>>   spy(A>2/n,'rd'),  hold on
>>   xlabel(['Hilbert matrix A: entries > 1/' num2str(n*2)])
>> subplot(132)
>>   B = fwt(A);
>>   spy(abs(B)>1/n/2,'b.'),hold on
>>   spy(abs(B)>0.1,'go'),  hold on
>>   spy(abs(B)>0.5,'rs'),  hold on
>>   xlabel(['Hilbert B=fwt(A): entries > 1/' num2str(n*2)])
>> subplot(133)
>>   imagesc(B)
>>   xlabel('Hilbert B=fwt(A) displayed by imagesc')
>>   axis equal; axis([0 n 0 n])
```

C.4 The functional Mfile and string evaluations

The functional Mfile is the subroutine (or sub-program) facility offered by MATLAB. It is allowed to contain its own functional Mfiles within the same file – a convenient option.

Any reader with experience of a programming language should find Mfiles easy to use and understand, for example, regarding passing a list of input and output parameters. We only need to point out two useful facilities.

(1) `global` allows all Mfiles that declare the same members of global variables (of any type) to access the latest content. (This resembles *COMMON* statements in Fortran).
(2) `eval` of a string that may be generated by `sprintf` can trigger any normal commands in MATLAB. The facility is specially useful for generating a variable number of matrix or vector names, depending on other control variables. For instance, given an $n \times n$ matrix A, illustrating LU factorization will require saving all intermediate matrices $L_1, L_2, \ldots, L_{n-1}$, all depending on n. That is to say, if $n = 3$, we only need L_1, L_2 and there is no need to generate L_3, L_4 etc. This idea has been used in various parts of the book notably in Chapter 6 when programming multilevel methods.

To be brief, we give an example of using a functional Mfile and the above facilities to automatically generate a certain number of subplots (depending on the input parameter). These plots illustrate the Harr wavelets from a fixed coarse level to a used-specified fine level:

```
%ap2f.m - Functional Mfile for Harr wavelets -- lev >= 0
%Usage:  figs = ap2f(lev)  %lev can be -2, -1, 0, 1, 2, 3, etc


function out=ap2f(lev)
  if nargin<1, help ap2f, return, end
  x=(-10:0.01:10)'; figure
  global total num color
    total = lev+3; num = 0; color=1;
    out = total;
for j=-2:lev
    num = num + 1;
 for k_shift=-10*2^j : 10*2^j
  my_plot(x,j,k_shift)
 end
    if j==-2,title(['Levels =' num2str(out)]),end
end
    set(gca,'color','y') % Mark the finest level
```

```
return

%_ _ _ _ _ _ _ _ _ _ _ _ _ _ _ _ _ _ _
function my_plot(x,j_scale,k_shift)
  global total num color
  str = sprintf('subplot(%d,1,%d)',total,num); eval(str)
  hold on
Scale = 2^(j_scale/2);
y = 2^j_scale * x - k_shift; yy=zeros(size(y));
yy = (y>=0 \& y<1/2) - ...
     (y<1 \& y>=1/2) + ...
     (y<0 } y>1) * 0;
 p=plot(x,yy*Scale);
 if color==1, set(p,'color','r'); color=2; else
              set(p,'color','b'); color=1; end
 hold off
return
```

The slight long Mfile is supplied as `ap2f.m` and the result from `out=ap2f(0)` is shown in Figure C.3.

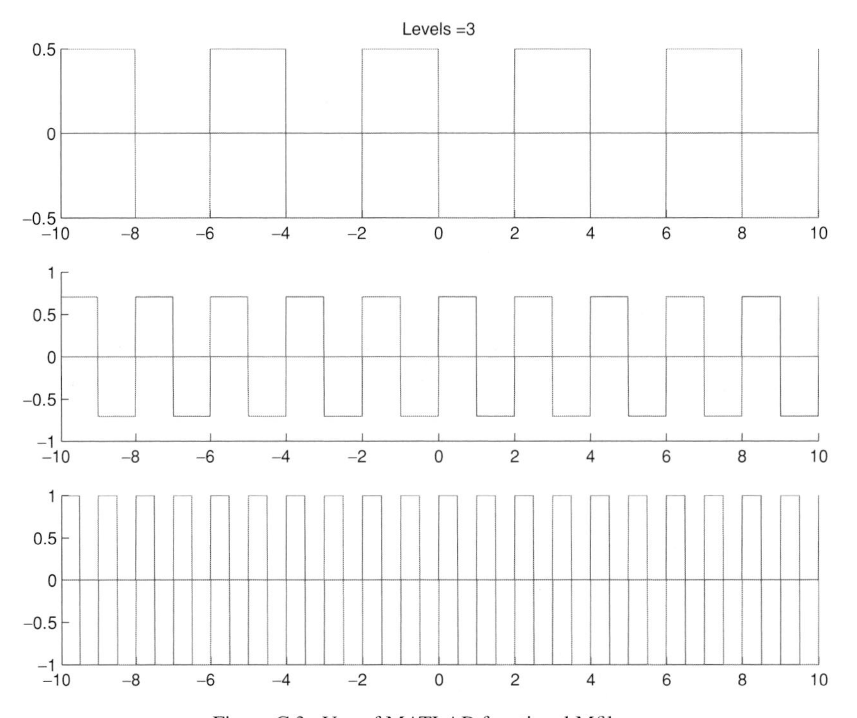

Figure C.3. Use of MATLAB functional Mfiles.

C.5 Interfacing MATLAB® with Fortran or C

MATLAB is a useful research and development tool. As such, the reader may find a need either to experiment on a specific matrix (or other data) from a different platform (e.g. Fortran or C) to MATLAB or to output results from MATLAB to be further tested by another platform. This is a topic not widely discussed. We give a brief summary.

(1) **MATLAB output of a dense matrix to text files**. As high level programming languages often require a formatted data input, one can use the MATLAB command `fprintf` to format data as in the following (assume a real $n \times n$ matrix A exists):

```
>> str = sprintf('fid=fopen(''mpta_%d.txt'',''w'');',n);
                                                    eval(str)
>> fprintf(fid, '%d %% dimension n for n x n\n',n);
>> for j=1:n      % by columns
>>   fprintf(fid, '%20.12e %20.12e %20.12e %20.12e\n',
                                                    A(:,j));
>> end
>> fclose(fid);  % File mpta_64.txt ready (if n=64)
```

(2) **MATLAB output of a sparse matrix to text files**. Suppose A is a real and sparse $n \times n$ matrix. We wish to save it as an ascii text file in the Harwell–Boeing format Section 15.4. Fortunately, such Mfiles exist; download from

http : //math.nist.gov/MatrixMarket/formats.html

which also keeps Mfiles that read such data onto MATLAB.

(3) **Simple text files loadable by MATLAB**. This is a simple and neat way to load data onto MATLAB, but the file can only contain one matrix having the *same* number of columns per row! Therefore for large matrices, one has to think about folding rows correctly and recovering them later. Otherwise, the command to import the data file `mpta.txt` is

```
>> load mpta.txt              % Internal name => mpta     or
>> importdata mpta.txt        % Internal name => ans
```

(4) **Generating Mfiles for MATLAB**. This is a robust and yet basic method of loading multiple data sets onto MATLAB. Simply generate a text file with data formatted as in Section C.1 for manual data entry. For example, with

Fortran,

```
...
   STR='mpta_gen.m'
   OPEN(24, FILE=STR)
   WRITE(24,'('' n ='',I5)') N
   WRITE(24,'('' A = [ % Start of a matrix'')') ! By rows
   DO I=1,N
   WRITE(24,'(4E22.10, ''...'')') (A(I,J),J=1,N)
   ENDDO
   WRITE(24,'( / )')
   WRITE(24,'('' ];          % End of a matrix'')') ! By rows
   WRITE(24,'('' b = [ % Start of a vector'')') ! By rows
   WRITE(24,'(4E22.10, ''...'')') (B(J),J=1,N)
   WRITE(24,'('' ]; b=b''''; % End of a vector'')') ! By rows
   WRITE(24,'('' whos A b '')')
   CLOSE(24)
...
```

(5) **MAT and MEX files on MATLAB**. These two facilities are provided by MATLAB for advanced data and programs integration. Firstly MAT files are normally generated by the MATLAB command `save FILE A b` for saving matrix A and vector b to FILE.MAT, which can be loaded in next time when one types `load FILE`. However, such MAT files can be read by Fortran and C directly and likewise one can generate MAT files directly from Fortran and C for MATLAB use.

Secondly, MEX files are subroutines produced from Fortran or C source code and can be used by MATLAB like Mfiles! In this way, our own (efficient) codes can be used by simple commands within MATLAB. Vice versa, Fortran or C sources are allowed to call the MATLAB engine library – in this latter case, the original motivation for inventing MATLAB to relieve users of working with Fortran routines (in using LINPACK[1]) is completely lost. This does not suggest that there will not be needs for MAT and MEX files for some readers (and developers).

However, to use either facility, one may have to learn quite a lot about external interfaces (or application program interface - API) and this is worthwhile if the effort can help many people to access an otherwise difficult code. (The dilemma may be that if an average reader must really know all these, again there might be less of a need for linking the interfaces in the first place). The two facilities will surely improve in the near future.

[1] See `http://www.netlib.org/lapack/` for LAPACK – the modern version of early LINPACK and EISPACK.

Finally, it should be remarked that there exists a C++ translation compiler to turn MATLAB Mfiles (or scripts) to C++ codes, which can then be run on a standalone basis. This might be useful if a MATLAB development code is to be converted. Check the MATLAB website.

C.6 Debugging an Mfile

It is inevitable that an experienced MATLAB user will soon write long Mfiles. MATLAB version 7 has provided a useful command `mlint` which can suggest modifications and pinpoint fatal errors:

```
>>  mlint('file')      % to check the Mfile file.m
>>  mlintrpt('file')   % in graphical user interface (GUI) mode
```

However it cannot do the algorithms for us, e.g. it will not pick out $\mathtt{A} = 3 + \mathtt{x}$; as an error (which will be found out at the running stage) when x is not defined.

C.7 Running a MATLAB® script as a batch job

Running a large job at background or off-peak periods is a standard practice on Unix (or Linux) systems. Although few readers run large MATLAB jobs, the increasing computer power makes it feasible for MATLAB to solve bigger problems.

To run `file.m` as a batch job, use the usual commands **at hh.mm** (e.g. **at 11:55pm**) or **batch** using

```
matlab < file.m
```

where graphics is displayed in the device defined by DISPLAY (check **echo $DISPLAY**). If there is graphics involved in `file.m` and there is no display facility during a batch run, then use

```
matlab -display /dev/null < file.m
```

C.8 Symbolic computing

Symbolic computing is another useful tool for a researcher as well as a graduate student or similar. Maple is a well-known symbolic computing package, that has been included in MATLAB (from version 5) as a symbolic toolbox. With it, symbolic computation can be done with ease. Interestingly with `syms`, the adoption of the MATLAB notation for some commands such as y = (x + 5) ∧ 6 makes the usage easier than within a standalone Maple where one must type y := (x + 5) ∧ 6; including ':' and ';'; the difference is more remarkable with matrix input and operations. However for other commands such as `int` for integration, the MATLAB adaption is not as flexible and yet some Maple commands share names with the normal MATLAB commands as *sum, plot, roots*; in these cases, we must call the Maple kernel explicitly, e.g. `maple('sum(k ∧ 2, k = 1..n)')` or use the newly defined names such as `symsum, poly2sym, ezplot.`

Here we display only a couple of short examples in a MATLAB session (given as `ap2m.m`):

```
>> syms A w x y z  % Declare variables (allowed to be variables)
>> A = [ -5  2  4
          2 -8  2   % Note the syntax is in MATLAB
          4  2 -5 ] % (Maple itself would need the "array"
                       command)
>> P3 = det(x*eye(3) - A)   % Result = x^3+18*x^2+81*x
>> cof = sym2poly(P3)       % Result = 1    18    81     0
>> A2=double(A)             % Convert back to normal matrix
>> c2 = poly(A2)            % Result = 1.0  18.0  81.0   0 (usual)
>> c2 = poly2sym(c2)        % Result = x^3+18*x^2+81*x
>> ezplot(c2,[-12,2])       % PLOT the cubic polynomial
>> Integral=int(c2,x,-12,2)% Result = -434 (integrate the cubic)
>> p =taylor(1/(1-x)^2,x,3)% Result = 1+2*x+3*x^2
>> w = cos(x*y*z)           % Define another function
>> a = diff(w,x,2)+diff(w,y,2) ...    % Compute 3D Laplacian
                  +diff(w,z,2)
>> a = factor(a)   % a = -cos(x*y*z)*(y^2*z^2+x^2*z^2+x^2*y^2)
>> latex(a)        % Convert to LaTeX for word processing
>> mhelp('dsolve') % Invoke the help system for doing "dsolve"
```

Owing to the MATLAB adaption of commands, not all Maple script files can be run with the toolbox, e.g. the usual Maple command `read('file.txt');` may not be run by using

```
>> maple('read('file.txt')')  %% NEED the full Maple package
```

Appendix D: list of supplied M-files and programs

Most supplied Mfiles for MATLAB® experiments and illustrations have been explained in the text and at the end of the concerned chapters. The summary given here as a single and brief list is for reference purposes. As shown in the preface, the Mfiles supplied can be downloaded from the internet

http://www.cambridge.org/9780521838283
http://www.liv.ac.uk/maths/ETC/mpta

Chapter 1 . . . .

- hess_as.m
- exafft16.m
- mygrid.m
- mat_prt.m
- mat_prt4.m
- fwt.m
- ifwt.m
- fft_fwt.m

Chapter 2 . . . .

- circ_toep.m
- ge_all.m
- gh_all.m
- givens.m
- houses.m
- mgs.m
- gj_ek.m
- g_e.m
- g_h.m
- g_j.m
- mk_ek.m
- nsh.m
- waz_fox.m
- waz_zol.m
- waz_all.m

Chapter 3 . . . .

- ch3_fmm.m
- gmrest_k.m
- gmres_c.m
- gmres_k.m
- index2.m
- intera.m
- iter3.m

Chapter 4 . . . .

- banda.m

- bandb.m
- bandg.m
- circ_pre.m
- detect.m
- ilu_0.m
- ilu_t.m
- matrix0.mat
- matrix1.mat
- matrix2.mat
- matrix3.mat
- multic.m
- schts.m
- t_band.m
- t_dete.m

Chapter 5 ....
- chebg.m
- cheb_fun.m
- def_no.m
- run_ai.m
- spai2.m
- waz_t.m

Chapter 6 ....
- cf_split.m
- ch6_gs.m
- ch6_mg2.m
- lap_lab.m
- mgm_2d.m
- mgm_2f.m
- mgm_2s.m

Chapter 7 ....
- hb1.m

Chapter 8 ....
- ch8.m
- cz.m
- fwts.m
- iperm0.m
- iperm2.m
- perm0.m
- perm2.m
- run8.m
- spyc.m

Chapter 9 ....
- gmres_nr.m
- iwts.m
- richa_nr.m
- run9.m

Chapter 10 ....
- ch0_w2.m

Chapter 13 ....
- bccb.m
- bttb.m
- ch13.m

Chapter 14 ....
- bprod.m
- hopf.m

Chapter 15 ....
- cup1.f
- cup2.f
- cup3.f
- cup23.in
- cup4.f
- cup5.f

Appendices *A–E* ....
- ap2f.m
- ap2m.m
- my_jordan.m
- mat_pr0.m
- size6.m
- size6.rua

Appendix E: list of selected scientific resources on Internet

In this Appendix we list a small and selected sample of internet sites that contain specific topics and useful software relating to matrix preconditioning, fast solvers and scientific computing in general. Any list of this kind cannot be complete because there are simply too sites to be found and listed (unfortunately Internet addresses can be time sensitive).

E.1 Freely available software and data

Here the commonly used word 'freely' should better be replaced by 'generously'. Indeed, only some selected sites are listed below some of which might require a simple online registration which is generously reasonable.

⟨1► Survey of freely available software for linear algebra on the web (Jack Dongarra):

http://www.netlib.org/utk/people/JackDongarra/la-sw.html

⟨2► Overview of Iterative Linear System Solver Packages (Victor Eijkhout):

http://www.netlib.org/utk/papers/iterative-survey/

⟨3► NETLIB (Collection of mathematical software, papers, and databases) (Chief editors: Jack Dongarra and Eric Grosse)

http : //www.netlib.org/

⟨4► TOMS (ACM Transactions on Mathematical Software) (Association for Computing):

http : //math.nist.gov/toms/

⟨5▶ Matrix Market:

http : //math.nist.gov/MatrixMarket/formats.html

⟨6▶ University of Florida Sparse Matrix Collection (Tim Davis):

http://www.cise.ufl.edu/research/sparse/matrices/

⟨7▶ BPKIT (Block Preconditioning Toolkit) (Edmond Chow):

http://www-users.cs.umn.edu/∼chow/bpkit.html/

⟨8▶ PETSc (Portable, Extensible Toolkit for Scientific Computation): (PETSc team):

http://www-unix.mcs.anl.gov/petsc/petsc-2/

⟨9▶ SPARSLAB (Michele Benzi and Miroslav Tuma):

http://www.cs.cas.cz/∼tuma/sparslab.html

⟨10▶ SPAI (SParse Approximate Inverse) (Stephen Barnard and Marcus Grote):

http://www.sam.math.ethz.ch/∼grote/spai/

⟨11▶ SAINV (Stabilized Approximate Inverse) (Robert Bridson):

http://www.cs.ubc.ca/∼rbridson/download/ainvInC.tar.bz2

⟨12▶ The Sparskit (Youcef Saad):

http : //www − users.cs.umn.edu/∼saad/software/SPARSKIT/
sparskit.html

⟨13▶ The BILUM package (Jun Zhang and Youcef Saad):

http : //cs.engr.uky.edu/∼jzhang/bilum.html

⟨14▶ FMM Toolbox (The MadMax Optics, Inc):

http : //www.madmaxoptics.com/

⟨15▶ Meshless Methods (Ching-Shyan Chen):

http://www.neveda.edu/∼chen/computer_code.html

⟨16▶ NSPCG package (Solving Large Sparse Linear Systems by Various Iterative Methods) (Thomas Oppe, Wayne Joubert and David Kincaid):

http : //www.ma.utexas.edu/CNA/NSPCG/

⟨17► The MATLAB® file exchange: (Various contributors)

http : //www.mathworks.com/matlabcentral/fileexchange/

E.2 Other software sources

Some of the following software may already be licensed and available in a reader's institution.

1 ► NAG (The Numerical Algorithms Group Ltd): comprehensive Fortran and C libraries

http : //www.nag.co.uk/

2 ► IMSL (Visual Numerics, Inc): comprehensive Fortran and C libraries

http : //www.vni.com/products/imsl/

3 ► AMG (Algebraic MGM by Fraunhofer-SCAI)

http : //www.scai.fraunhofer.de/samg.htm

4 ► FEMLAB (Finite Element Modelling Laboratory, COMSOL Ltd)

http : //www.uk.comsol.com/

5 ► HSL (Harwell Subroutine Library)

http : //www.cse.clrc.ac.uk/nag/hsl/contents.shtml

E.3 Useful software associated with books

There are many mathematics and scientific computing related books, that provide associated software – the main MATLAB site (at www.mathworks.com) keeps an up-to-date list.

---

http : //www.mathworks.com/support/books/

Below we highlight a few of such books.

⟨1► Templates for the Solution of Linear Systems: Building Blocks for Iterative Methods, 2nd ed in [41]: Mfiles again from

http : //www.mathworks.com/support/books/

⟨2► Templates for the Solution of AEP in [36]: software from

http : //www.cs.ucdavis.edu/∼bai/ET/sa_alg_list_book.html

⟨3► Matrix Computation Toolbox [280] (Nicholas Higham):

http://www.ma.man.ac.uk/∼higham/mctoolbox/

⟨4► Spectral Methods in MATLAB [458] (Nicholas Lloyd Trefethen):

http : //web.comlab.ox.ac.uk/oucl/work/nick.trefethen/spectral.html

⟨5► IFISS (Incompressible Flow and Iterative Solver Software) [197] (David Silvester and Howard Elman):

http : //www.ma.umist.ac.uk/djs/software.html
http : //www.ma.umist.ac.uk/djs/vers.html

⟨6► Numerical Mathematics and Computing (5th ed) (Ward Cheney and David Kincaid, 2004, Brooks/Cole):

http : //www.ma.utexas.edu/CNA/NMC5/sample.html
ftp : //ftp.ma.utexas.edu/pub/kincaid−cheney/

⟨7► Numerical Computing with MATLAB (Cleve Moler, 2004, SIAM publications, USA):

http : //www.mathworks.com/moler/ncmfilelist.html

⟨8► Numerical Recipes Software [390] (Cambridge University Press):

http : //www.nr.com/

⟨9► Mathematical software (Curt Vogel [476]):

http : //www.math.montana.edu/∼vogel/

⟨10► Boundary elements software (Stephen Kirkup [317]):

http : //www.soundsoft.demon.co.uk/tbemia.htm

E.4 Specialized subjects, sites and interest groups

1 ► nanet – for community of numerical analysts and other researchers:

http : //www.netlib.org/na−net/

2 ▶ mgnet – for community of multigrid and multilevel related researchers:

http : //www.mgnet.org/

3 ▶ ddm net – for community of multidomain and multilevel related researchers:

http : //www.ddm.org/

4 ▶ waveletnet – for community of multiresolution researchers:

http : //www.wavelet.org/

5 ▶ at-net – for community of approximation theory researchers:

http://www.uni−giessen.de/www−Numerische−Mathematik/at−net/

6 ▶ mathscinet – excellent sources to find research from various results:

http : //www.ams.org/mathscinet

7 ▶ google specializes in everything (but the search list is often too lengthy):

http : //www.google.com

References

1. Abbott J. P. (1977). Numerical continuation methods for nonlinear equations and bifurcation problems. Ph.D. thesis, Australian National University.
2. Abbott J. P. (1978). An efficient algorithm for the determination of certain bifurcation points. *J. Comput. Appl. Math*, **4**, 19–27.
3. Acar R. and Vogel C. R. (1994). Analysis of total variation penalty method for ill-posed problems. *Inverse Probs.*, **10**, 1217–29.
4. Adams L. M., Leveque R. J. and Young D. M. (1988). Analysis of the SOR iteration for the nine point Laplacian. *SIAM J. Numer. Anal.*, **25**, 1156–80.
5. Ainsworth M. and McLean W. (2003). Multilevel diagonal scaling preconditioners for boundary element equations on locally refined meshes. *Numer. Math.*, **93**, 387–413.
6. Ajjarapu V. and Christy C. (1992). The continuation power flow: a tool for steady state voltage stability analysis. *IEEE Trans. Power Systems*, **7**, 416–23.
7. Ajjarapu V. and Lee B. (1992). Bifurcation theory and its application to non-linear dynamical phenomena in an electrical power system. *IEEE Trans. Power Systems*, **7**, 424–31.
8. Aksoylou B., Bond S. and Holst M. (2003). An odyssey into local refinement and multilevel preconditioning III: implementation and numerical experiments. *SIAM J. Sci. Comput.*, **25**, 478–98. (See also the software from `http://www.fetk.org`).
9. Alpert B., Beylkin G., Coifman R. and Rokhlin V. (1993). Wavelet-like bases for the fast solution of second-kind integral equations. *SIAM J. Sci. Comput.*, **14**, 159–84.
10. Alvarez L., Lions P. L. and Morel J. M. (1992). Image selective smoothing and edge detection by nonlinear diffusion II. *SIAM J. Numer. Anal.*, **29**, 845–66.
11. Alvarez L. and Morel J. M. (1994). Formulation and computational aspects of image analysis. *Acta Numer.*, 1994, 1–59.
12. Amini S. (1999). On boundary integral operators for the Laplace and the Helmholtz equations and their discretisations. *J. Engng. Anal. Bound. Elements*, **23**, 327–37.
13. Amini S., Chadwick E. A., Nixon X. P. and Pistillo S. (2001). Multiwavelet solution of boundary integral equations – application to the radiosity problem, *Proc. UK BIE* **3**, Brighton University Press, pp. 99–108.

14. Amini S. and Chen K. (1989). Conjugate gradient method for the second kind integral equations – applications to the exterior acoustic problem. *J. Engng. Anal. Boundary Elements*, **6**, 72–7.
15. Amini S., Chen K. and Harris P. J. (1990). Iterative solution of boundary element equations for the exterior acoustic problem. *ASME J. Vib. Acoust.*, **112**, 257–62.
16. Amini S., Harris P. J. and Wilton D. T. (1992). Coupled boundary and finite element methods for the solution of the dynamic fluid-structure interaction problem. In: *Lecture Note in Engineering, 77*, eds. C. A. Brebbia and S. A. Orszag. Springer-Verlag, London.
17. Amini S. and Maines N. (1998). Preconditioned Krylov subspace methods for boundary element solution of the Helmholtz equation. *Int. J. Numer. Meth. Engrg*, **41**, 875–98.
18. Andrews H. C. and Hunt B. R. (1977). *Digital Image Restoration*, Prentice-Hall.
19. Anselone P. M. (1971). *Collectively Compact Operator Approximation Theory*. Prentice-Hall.
20. Aoyama Y. and Nakano J. (1999). *RS/6000 SP: Practical MPI Programming*, IBM Corp., from `http://www.redbooks.ibm.com/redbooks/SG245380.html`.
21. Ashby S. F. and Falgout R. D. (1996). A parallel multigrid preconditioned conjugate gradient algorithm for groundwater flow simulations. *Nucl. Sci. Engng.*, **124**, 145–59.
22. Atkinson K. E. (1973). Iterative variants of the Nystrom method for the numerical solution of integral equations. *Numer. Math.*, **22**, 17–31.
23. Atkinson K. E. (1989). *Introduction to Numerical Analysis*. Wiley.
24. Atkinson K. E. (1997). *The Numerical Solution of Integral Equations of the Second Kind*. Cambridge University Press. (See also the first edition published by SIAM in 1976).
25. Atkinson K. E. and Han W. M. (2004). *Elementary Numerical Analysis*, 3rd edn. Wiley.
26. Aubert G. and Kornprobst P. (2002). *Mathematical Problems in Image Processing*. Springer.
27. Axelsson O. (1985). A survey of preconditioned iterative methods for linear systems of algebraic equations. *BIT*, **25**, 166–87.
28. Axelsson O. (1994). *Iterative Solution Methods*. Cambridge University Press.
29. Axelsson O. and Barker V. A. (1984). *Finite Element Solution of Boundary Values Problems: Theory and Computation*. Academic Press. (2nd edn reprinted by SIAM in 2001).
30. Axelsson O. and Neytcheva M. G. (1994). Algebraic multilevel iteration method for Stieljes matrices. *Numer. Lin. Alg. Appl.*, **1**, 213–36.
31. Axelsson O. and Neytcheva M. G. (1994). *The Algebraic Multilevel Iteration Methods – Theory and Applications*. Preprint.
32. Axelsson O. and Padiy A. (1999). On the additive version of the algebraic multilevel iteration method for anisotropic elliptic problems. *SIAM J. Sci. Comput.*, **20**, 1807–30.
33. Axelsson O. and Vassilevski P. S. (1989). A survey of multilevel preconditioned iterative methods. *BIT*, **29**, 769–93.

34. Axelsson O. and Vassilevski P. S. (1989). Algebraic multilevel precondtioning methods I. *Numer. Math.*, **56**, 157–77. (Also Part II, *SIAM J. Num. Anal.*, **27**, 1569–90, 1990).
35. Babuska I. and Melenk J. M. (1997). The partition of unity method. *Int. J. Numer. Meth. Eng.*, **40**, 727–58.
36. Bai Z. J., Chen T. Y., Day D., Demmel J. W., Dongarra J., Edelman A., Ericsson T., Freund R., Gu M., Kagstrom B., Knyazev A., Koev P., Kowalski T., Lehoucq R., Li R. C., Li X. Y., Lippert R., Maschoff K., Meerbergen K., Morgan R. B., Ruhe A., Saad Y., Sleijpen G., Sorensen D. and van der Vorst H. A. (2000). *Templates for the Solution of Algebraic Eigenvalue Problems: a Practical Guide*. SIAM Publications. See also http://www.cs.ucdavis.edu/~bai/ET/contents.html
37. Bai Z. Z., Duff I. S. and Wathen A. J. (2001). A class of incomplete orthogonal factorization methods I. *BIT*, **41**, 53–70.
38. Bai Z. Z., Golub G. H. and Ng M. K. (2003). On successive-over-relaxation acceleration of the Hermitian and skew-Hermitian splitting iterations. *SIAM J. Matr. Anal. Appl.*, **24**, 603–26. (See also SCCM 01-06 from the web [340]).
39. Bai Z. Z., Golub G. H. and Pan J. Y. (2004). Preconditioned Hermitian and Skew-Hermitian Splitting Methods for non-Hermitian positive semidefinite linear systems. Numer. Math., 98, pp. 1–32. (Also as SCCM-02-12 from the web [340].)
40. Barakat K. and Webb J. P. (2004). A clustering algorithm for multilevel fast multipole methods. *IEEE Trans. Magnetics*, **40**, 1072–5.
41. Barrett R., Berry M. W., Chan T. F., Demmel J. W., Donato J., Dongarra J., Eijkhout V., Pozo R., Romine C. and van der Vorst H. A. (1993). *Templates for the Solution of Linear Systems: Building Blocks for Iterative Methods*. SIAM Publications.
42. Barulli M. and Evans D. J. (1996). Implicit Gauss–Jordan scheme for the solution of linear systems. *Para. Algor. Applics.*, **10**, 145–59.
43. Battermann A. and Sachs E. W. (2001). Block preconditioner for KKT systems in PDE-governed optimal control problems. In: *Workshop on Fast Solutions of Discretized Optimization Problems*, eds. R. H. W. Hoppe, K.-H. Hoffmann, and V. Schulz, pp. 1–18. Birkhäuser.
44. Beauwens R. and Quenon L. (1976). Existence criteria for partial matrix factorization in iterative methods. *SIAM J. Numer. Anal.*, **13**, 615–43.
45. Bellman R. (1997). Introduction to Matrix Analysis. vol. 19 of *Classics in Applied Mathematics*, 2nd edn, SIAM.
46. Benson M. W. and Frederickson P. O. (1982). Iterative solution of large sparse linear systems arising in certain multidimensional approximation problems. *Utilitas Math.*, **22**, 127–40.
47. Benzi M. (2001). *Who was E. Purcell? NA Digest*, 2001 (4), http://www.netlib.org/na-net
48. Benzi M. (2002). Preconditioning techniques for large linear systems: a survey. *J. Comput. Phys.*, **182**, 418–77.
49. Benzi M. (2004). A direct projection method for Markov chains. *Lin. Alg. Appl.*, **386**, 27–49.
50. Benzi M., Cullum J. K. and Tuma M. (2000). Robust approximate inverse preconditioning for the conjugate gradient method. *SIAM J. Sci. Comput.*, **22**, 1318–32.

51. Benzi M., Gander M. J. and Golub G. H. (2003). Optimization of the Hermitian and skew-Hermitian splitting iteration for saddle-point problems. *BIT*, **43**, 1–19. (See also SCCM 03-06 from the web [340]).
52. Benzi M. and Golub G. H. (2003). A preconditioner for generalised saddle point problems. *SIAM J. Matr. Anal. Appl.*, 26, pp. 20–41. (See also SCCM 02-14 from the web [340]).
53. Benzi M., Haws J. C. and Tuma M. (2000). Preconditioning highly indefinite and nonsymmetric matrices. *SIAM J. Sci. Comput.*, **22**, 1333–53.
54. Benzi M. and Meyer C. D. (1995). A direct projection method for sparse linear systems. *SIAM J. Sci. Comput.*, **16**, 1159–76.
55. Benzi M. and Tuma M. (1998). A sparse approximate inverse preconditioner for nonsymmetric linear systems. *SIAM J. Sci. Comput.*, **19**, 968–94.
56. Benzi M. and Tuma M. (1999). A comparative study of sparse approximate inverse preconditioners. *Appl. Numer. Math.*, **30**, 305–40.
57. Benzi M. and Tuma M. (2000). Ordering for factorised sparse approximate inverses. *SIAM J. Sci. Comput.*, **21**, 1851–68.
58. Bergamaschi L., Pini G. and Sartoretto F. (2003). Computational experience with sequential and parallel, preconditioned Jacobi–Davidson for large, sparse symmetric matrices. *J. Comput. Phys.*, **188**, 318–31.
59. Beylkin G. (1993). Wavelets and fast numerical algorithms. Lecture Notes for short course, AMS-93 AMS, *Proceedings of Symposia in Applied Mathematics*, v.47, 89–117.
60. Beylkin G., Coifman R. and Rokhlin V. (1991). Fast wavelet transforms and numerical algorithms I. *Comm. Pure Appl. Math.*, **44**, 141–83.
61. Beylkin G. and Cramer R. (2002). A multiresolution approach to regularization of singular operators and fast summation. *SIAM J. Sci. Comput.*, **24**, 81–117.
62. Beyn W. J., Champneys A., Doedel E., Govaerts W., Sandstede B. and Kuznetsov Yu A.(2002). Numerical continuation and computation of normal forms. In: *Handbook of Dynamical Systems II: Towards Applications*, eds. B. Fiedler, G. Iooss, and N. Coppell. Elsevier Science.
63. Bjorck A. (1996). *Numerical Methods for Least Squares Problems*. SIAM Publications.
64. Blomgren P., Chan T. F., Mulet P., Vese L. and Wan W. L. (2000). Variational PDE models and methods for image processing. In: *Research Notes in Mathematics*, 420, pp. 43–67, Chapman & Hall/CRC.
65. Board J. A. (1997). Introduction to 'a fast algorithm for particle simulations'. *J. Comput. Phys.*, **135**, 279.
66. Bollhöfer M. (2003). A robust and efficient ILU that incorporates the growth of the inverse triangular factors. *SIAM J. Sci. Comput.*, **25**, 86–103.
67. Bollhöfer M. and Saad Y. (2004). Multilevel preconditioners constructed from inverse–based ILUs. University of Minnesota, UMSI report 2004/75.
68. Bond D. and Vavasis S. (1994). *Fast Wavelet Transforms for Matrices arising from Boundary Element Methods*. Computer Science Research Report TR-174, Cornell University.
69. Borm S., Grasedyck L. and Hackbusch W. (2003). Introduction to hierarchical matrices with applications. *J. Engng. Anal. Bound. Elements*, **27**, 405–22.

70. Botta E. F. F. and Wubs F. W. (1999). Matrix renumbering ILU: An effective algebraic multilevel ILU preconditioner for sparse matrices. *SIAM J. Matr. Anal. Appl.*, **20**, 1007–26.
71. Bramble J. H., Pasciak J. E. and Xu J. C. (1990). Parallel multilevel preconditioners. *Math. Comp.*, **55**, 1–12.
72. Brandt A. (1977). Multilevel adaptive solutions to boundary value problems. *Math. Comp.*, **31**, 333–90.
73. Brandt A. (2000). General highly accurate algebraic coarsening. *Elec. Trans. Num. Anal.*, **10**, 1–20; from `http://etna.mcs.kent.edu`.
74. Brezinski C. (1997). *Projection Methods for Systems of Equations*. North-Holland.
75. Bridson R. and Tang W. P. (2001). Multiresolution approximate inverse preconditioners. *SIAM J. Sci. Comput.*, **23**, 463–79.
76. Briggs W. L., Henson V. E. and McCormick S. F. (2000). *A Multigrid Tutorial*, 2nd edn. SIAM Publications.
77. Brown P. N and Walker H. F. (1997). GMRES on (nearly) singular systems. *SIAM J. Matr. Anal. Appl.*, **18**, 37–51.
78. Bruno O. (2003). Fast, high-oder, high-frequency integral methods for computational acoustics and electromagnetics. In: *Topics in Computational Wave Propagation – Direct and Inverse Problems. Lecture notes in Computational Science and Engineering* 31, eds. M. Ainsworth, *et al.* Springer-Verlag.
79. Buhmann M. (2003). *Radial Basis Functions: Theory and Implementations*. Cambridge University Press.
80. Burden R. J. and Faries J. D. (1981). *Numerical Analysis*, 2nd edn. PWS-Kent.
81. Burrage K. and Erhel J. (1998). On the performance of various adaptive preconditioned GMRES strategies. *Numer. Lin. Alg. Applics.*, **5**, 101–21.
82. Burton A. J. (1976). *Numerical Solution of Acoustic Radiation Problems*. NPL report OC5/535. National Physical Laboratory.
83. Burton A. J. and Miller G. F. (1971). The application of integral equation methods to numerical solution of some exterior BVP. *Proc. Roy. Soc. Lond.*, **A323**, 201–10.
84. Buzbee B. L., Dorr F. W., George J. A. and Golub G. H. (1971). The direct solution of the discrete Poisson equation on irregular domains. *SIAM J. Numer. Anal.*, **8**, 722–36.
85. Caf D. and Evans D. J. (1998). A preconditioning strategy for banded circulant and Toeplitz systems. *Int. J. Comput. Math.*, **69**, 283–94.
86. Cai X. C. (1994). A family of overlapping Schwarz algorithms for nonsymmetric and indefinite elliptic problems. In: *Domain-based Parallelism and Problem Decomposition Methods in Computational Science and Engineering*, eds. D. E. Keyes, Y. Saad and D. G. Truhlar, Ch. 1. SIAM Publications.
87. Cai X. C. and Widlund O. B. (1992). Domain decomposition algorithms for indefinite elliptic problems. *SIAM J. Sci. Stat. Comp.*, **13**, 243–58.
88. Calvetti D., Lewis B. and Reichel L. (2001). Krylov subspace iterative methods for nonsymmetric discrete ill-posed problems in image restoration. In: *Advanced Signal Processing Algorithms, Architectures, and Implementations X*, ed. F. T. Luk. Proceedings of the Society of Photo-Optical Instrumentation Engineers (SPIE 01), 4116. The International Society for Optical Engineering, Bellingham, WA, USA.

89. Canizares C. A. (2002). Voltage stability indices. In: *Voltage Stability Assessment, Procedures and Guides*, IEEE/PES PSS Subcommittee Special Publication; see http://www.power.uwaterloo.ca
90. Canizares C. A., de Souza A. Z. and Quintana V. H. (1996). Comparison of performance indices for detection of proximity to voltage collapse. *IEEE Trans. Power Systems*, **11**, 1441–50.
91. Canuto C., Hussaini M. Y., Quarteroni A. and Zang T. A. (1988). *Spectral Methods in Fluid Dynamics*. Springer-Verlag.
92. Cao Z. H. (2004). Fast Uzawa algorithms for solving non-symmetric stabilized saddle point problems. *Numer. Lin. Alg. Applics.*, **11**, 1–24.
93. Carpentieri B., Duff I. S. and Giraud L. (2000). *Experiments with Sparse Preconditioners of Dense Problems from Electromagnetic Applications*. CERFACS TR/PA/00/04, France; http://www.cerfacs.fr/algor/reports/2000/TR_PA_00_04.ps.gz.
94. Carpentieri B., Duff I. S. and Giraud L. (2000). Some sparse pattern selection strategies for robust Frobenius norm minimization preconditioners in electromagnetism. *Numer. Lin. Alg. Appl.*, **7**, 667–85. (also RAL-TR-2000-009, http://www.numerical.rl.ac.uk/reports/)
95. Carpentieri B., Duff I. S. and Giraud L. (2003). A class of spectral two-level preconditioners. *SIAM J. Sci. Comput.*, **25**, 749–65. (See also CERFACS TR/PA/02/55, France. http://www.cerfacs.fr/algor/reports/2002/TR_PA_02_55.ps.gz).
96. Carpentieri B., Duff I. S., Giraud L. and Sylvand G. (2003). Combining fast multipole techniques and approximate inverse preconditioner for large electromagnetism calculations. (RAL-TR-2003-024, http://www.numerical.rl.ac.uk/reports/)
97. Carter J. L. (2002). *Dual Method for Total Variation-based Image Restoration*. CAM report 02-13, UCLA, USA; see http://www.math.ucla.edu/applied/cam/index.html
98. Cerdan J., Marin J. and Martinez A. (2002). Polynomial preconditioners based on factorized sparse approximate inverses. *Appl. Math. Comput.*, **133**, 171–86.
99. Chambolle A. (2000). *Inverse Problems in Image Processing and Image Segmentation*, ICTP Lecture Notes Series, Vol.II, `http://www.ictp.trieste.it`.
100. Chan R. H., Chan T. F. and Wan W. L. (1997). Multigrid for differential convolution problems arising from image processing. In: *Proc. Sci. Comput. Workshop*, eds. R. Chan, T. F. Chan and G. H. Golub. Springer-Verlag. See also CAM report 97-20, UCLA, USA.
101. Chan R. H., Chan T. F. and Wong C. K (1995). Cosine transform based preconditioners for total variation deblurring. *IEEE Trans. Image Proc.*, **8**, 1472–8. (See also the web page in [112] for CAM report 95-23).
102. Chan R. H., Chang Q. S. and Sun H. W. (1998). Multigrid method for ill-conditioned symmetric Toeplitz systems. *SIAM J. Sci. Comput.*, **19**, 516–29.
103. Chan R. H., Nagy J. and Plemmons R. J. (1994). Circulant preconditioned Toeplitz least squares iterations. *SIAM J. Matrix Anal. Appl.*, **15**, 80–97.
104. Chan R. H. and Ng M. K. (1996). Conjugate gradient methods for Toeplitz systems. *SIAM Rev.*, **38**, 427–82.
105. Chan R. H. and Wong C. K. (1997). Sine transform based preconditioners for elliptic problems. *Num. Linear Algebra Applic.*, **4**, 351–68.

106. Chan T. F. (1988). An optimal circulant preconditioner for Toeplitz systems. *SIAM J. Sci. Stat. Comput.*, **9**, 766–71.
107. Chan T. F. and Chen K. (2000). *Two-stage Preconditioners using Wavelet Band Splitting and Sparse Approximation*. UCLA CAM report CAM00-26, Dept of Mathematics, UCLA, USA.
108. Chan T. F. and Chen K. (2002). On two variants of an algebraic wavelet preconditioner. *SIAM J. Sci. Comput.*, **24**, 260–83.
109. Chan T. F. and Chen K. (2005). On nonlinear multigrid solvers for the primal total variation formulation. *J. Numer. Algor.*
110. Chan T. F., Golub G. H. and Mulet P. (1999). A nonlinear primal dual method for total variation based image restoration. *SIAM J. Sci. Comput.*, **20**, 1964–77.
111. Chan T. F. and Matthew T. P. (1994). Domain decomposition algorithms. In: *Acta Numerica 1994*, ed. A. Iserles, pp. 61–143. Cambridge University Press.
112. Chan T. F. and Mulet P. (1996). *Iterative Methods for Total Variation Restoration*. CAM report 96-38, UCLA, USA; see http://www.math.ucla.edu/applied/cam/index.html
113. Chan T. F. and Olkin J. (1994). Circulant Preconditioners for Toeplitz-block matrices. *J. Numer. Algorithms*, **6**, 89–101.
114. Chan T. F., Shen J. H. and Vese L. (2004). *Variational PDE Models in Image Processing*. UCLA CAM report CAM02-61, Dept of Mathematics, UCLA, USA. (See [112] for the web address).
115. Chan T. F. and Tai X. C. (2003). Level set and total variation regularization for elliptic inverse problems with discontinuous coefficients. *J. Comput. Phys.*, **193**, 40–66.
116. Chan T. F., Tang W. P. and Wan W. L. (1997). Wavelet sparse approximate inverse preconditioners. *BIT*, **37**, 644–60.
117. Chan T. F. and Vese L. (1999). An active contour model without edges. In: *Scale-Space Theories in Computer Vision*, Lecture Notes in Comput. Sci. 1682, pp. 141–51.
118. Chan T. F. and Vese L. (2000). Image Segmentation using Level Sets and the Piecewise-constant Mumford–Shah model. UCLA CAM report CAM00-14, Dept of Mathematics, UCLA, USA. (See [112] for the web address).
119. Chan T. F. and Wan W. L. (1999). *Robust Multigrid Methods for Elliptic Linear Systems.* SCCM report 99-08 from `http://www.sccm-stanford.edu/wrap/pub-tech.html`.
120. Chan T. F. and Wan W. L. (2000). *Robust Multigrid Methods for Nonsmooth Coefficient Elliptic Linear Systems*. *J. Comput. Appl. Math.*, **123**, 323–52.
121. Chandler-Wilde S. N., Langdon S. and Ritter L. (2004). A high wavenumber boundary element method for an acoustic scattering problem. *Proc. Roy. Soc. Lon.*, Ser. **A362**, 647–71.
122. Chen C. S., Hon Y. C. and Schaback R. (2005). *Radial Basis Functions with Scientific Computation*. Monograph to appear. (See 'Reconstruction of Multivariate Functions from Scattered Data' by R. Schaback, preprint, 1997, http://www.num.math.uni-goettingen.de/schaback/research/papers/rbfbook.ps and also 'Scattered Data Approximation' by H. Wendland. Cambridge University Press, 2005.)
123. Chen D. and Toledo S. (2003). Vaidyas preconditioners: implementation and experimental study. *Electron. Trans. Numer. Anal.*, **16**, 30–49.

124. Chen K. (1986). *Multigrid algorithms with linear and nonlinear applications.* Master's thesis, Department of Mathematics, University of Manchester, UK.
125. Chen K. (1990). *Analysis of iterative methods for solutions of boundary integral equations with applications to the Helmholtz problem*, Ph.D. thesis, Dept of Mathematics & Statistics, University of Plymouth, Plymouth, UK.
126. Chen K. (1991). Conjugate gradient methods for the solution of boundary integral equations on a piecewise smooth boundary. *J. Comput. Phys.*, **97**, 127–43.
127. Chen K. (1994). Efficient iterative solution of linear systems from discretizing singular integral equations. *Elec. Trans. Numer. Anal.*, **2**, 76–91.
128. Chen K. (1998). On a class of preconditioning methods for dense linear systems from boundary elements. *SIAM J. Sci. Comput.*, **20**, 684–98.
129. Chen K. (1998). On a new preconditioning algorithm for iterative solution of generalized boundary element systems. J. Dalian University of Technology, **49**, 1–18.
130. Chen K. (1999). Discrete wavelet transforms accelerated sparse preconditioners for dense boundary element systems. *Elec. Trans. Num. Anal.*, **8**, 138–53.
131. Chen K. (2001). An analysis of sparse approximate inverse preconditioners for boundary integral equations. *SIAM J. Matr. Anal. Appl.*, **22** 1058–78.
132. Chen K. and Amini S. (1993). Numerical analysis of boundary integral solution of the Helmholtz equation in domains with non-smooth boundaries. *IMA J. Numer. Anal.*, **13**, 43–66.
133. Chen K., Cheng J. and Harris P. J. (2005). An efficient weakly-singular reformulation of the Burton–Miller method for solving the exterior Helmholtz problem in three dimensions. *Proc. Roy. Soc. Lond., Series A.*
134. Chen K. and Evans D. J. (2000). An efficient variant of Gauss–Jordan type algorithms for direct and parallel solution of dense linear systems. *Int. J. Comp. Math.*, **76**, 387–410.
135. Chen K., Giblin P. J. and Irving I. R. (1999). *Mathematical Exploration with Matlab.* Cambridge University Press.
136. Chen K. and Harris P. J. (2001). Efficient preconditioners for iterative solution of the boundary element equations for the three-dimensional Helmholtz equation. *J. Appl. Num. Math.*, **36**, 475–89.
137. Chen K. and Harris P. J. (2001). On preconditioned iterative solution of the coupled block linear system arising from the three-dimensional fluid-structure interaction problem. In: *Proc. 3rd UK BIE Meth. Conf.*, Brighton University Press.
138. Chen K., Hawkins S. C. and Hughes M. D. (2003). Effective sparse preconditioners of the two-level deflation type for fast solution of the Helmholtz equation and related problems. In: *Proc. UK BIE* 4, pp. 147–56, Salford University Press.
139. Chen K., Hussein A., Bradley M. and Wan H. B. (2003). A performance-index guided continuation method for fast computation of saddle-node bifurcation in power systems. *IEEE Trans. Power Systems*, **18**, 753–60.
140. Chen K., Hussein A. and Wan H. B. (2001). On a class of new and practical performance indexes for approximation of fold bifurcations of nonlinear power flow equations. *J. Comput. Appl. Math.*, **140**, 119–41.
141. Chen K., Hussein A. and Wan H. B. (2001). An analysis of Seydel's test function methods for nonlinear power flow equations. *Int. J. Comput. Math.*, **78**, 451–70.
142. Chen K., Hussein A. and Wan H. B. (2002). On adapting test function methods for fast detection of fold bifurcations in power systems. *Int. J. Bifurc. Chaos*, **12**, 179–85.

143. Chen K. and Lai C. H. (2002). Parallel algorithms of the Purcell method for direct solution of linear systems. *J. Parallel Comput.*, **28**, 1275–91.
144. Chen X. J. (1998). Global and superlinear convergence of inexact Uzawa methods for saddle point problems with nondifferentiable mappings. *SIAM J. Numer. Anal.*, **35**, 1130–48.
145. Chiang H. D. and Jean-Jumeau R. (1995). Toward a practical performance index for predicting voltage collapse in electric power systems. *IEEE Trans. Power Systems*, **10**, 584–92.
146. Chiang H. D. and Liu C. W. (1993). Chaos in a simple power system. *IEEE Trans. Power Systems*, **8**, 1407–17.
147. Chow E. (2000). A priori sparsity patterns for parallel sparse inverse preconditioners. *SIAM J. Sci. Compt.*, **21**, 1804–22.
148. Chow E. (2001). Parallel implementation and practical use of sparse approximate inverse preconditioners with a priori sparsity patterns. *Int. J. High Perf. Comput. Appl.*, **15**, 56–74.
149. Chow E. and Saad Y. (1997). Parallel approximate inverse preconditioners. In: *Proc. 8th SIAM Conference on Parallel Processing for Scientific Computing*, Minneapolis, MN, March 1997, pp. 14–17 (http://www.llnl.gov/CASC/people/chow/pubs/history.ps).
150. Chow E. and Saad Y. (1998). Approximate inverse preconditioners via sparse-sparse iterations. *SIAM J. Sci. Comput.*, **19**, 995–1023.
151. Christensen O. (2001). Frames, Riesz bases and discrete Gabor/wavelet expansions. *Bull. AMS*, **38**, 273–91.
152. Chui C. K. (1992). *An Introduction to Wavelets*. Academic Press.
153. Ciarlet P. G. (1988). *Introduction to Numerical Linear Algebra and Optimisation*. Cambridge University Press.
154. Cochran W. K., Plemmons R. J. and Torgersen T. C. (2000). Exploiting Toeplitz structure in atmospheric image reconstruction. *Contemp. Math.*, **280**, 177–89. (See also http://www.wfu.edu/~plemmons)
155. Cohen A., Dahmen W. and DeVore R. (2001). Adaptive wavelet methods for elliptic operator equations – convergence rates. *Math. Comp.*, **70**, 27–75.
156. Cohen A., Daubechies I. and Feauveau J. (1992). Biorthogonal bases of compactly supported wavelets. *Comm. Pure Appl. Math.*, **45**, 485–560.
157. Cohen A. and Masson R. (1999). Wavelet methods for second-order elliptic problems, preconditioning, and adaptivity. *SIAM J. Sci. Comput.*, **21**, 1006–26.
158. Cohen A. and Masson R. (2000). Wavelet adaptive method for second order elliptic problems: boundary conditions and domain decomposition. *Numer. Math.*, **86**, 193–238.
159. Colton D. and Kress R. (1983). *Integral Equation Methods in Scattering Theory*. Wiley.
160. Colton D. and Kress R. (1998). *Inverse Acoustic and Electromagnetic Scattering Theory*. 2nd edn. Springer-Verlag.
161. Concus P. and Golub G. H. (1976). *A Generalized Conjugate Gradient Method for Non-symmetric Systems of Linear Systems*. Stanford CS report STAN-CS-76-535 (Presently available from the web [340]).
162. Cosgrove J. D. F., Diaz J. C. and Griewank A. (1992). Approximate inverse preconditioning for sparse linear systems. *Int. J. Comput. Math.*, **44**, 91–110.

163. Cosnard M., Robert Y. and Trystram D. (1987). Parallel solution of dense linear systems using diagonalization methods. *Int. J. Comput. Math.*, **22**, 249–70.
164. Cuvelier C, Segal A. and van Steenhoven A. (1986). *Finite Element Methods and the Navier–Stokes*. D. Reidel Publishing Co.
165. Dahlquist G. and Bjorck A. (2004). *Numerical Mathematics and Scientific Computation*. SIAM Publications.
166. Dahmen W. (1997). Wavelet and multiscale methods for operator equations. *Acta Numerica*, **6**, 55–228 (Cambridge University Press).
167. Dahmen W., Harbrecht H. and Schneider R. (2002). *Compression Techniques for Boundary Integral Equations – Optimal Complexity Estimates*. Preprint SFB393/02-06, University of Chemnitz, Germany. `http://www.tu-chemnitz.de/sfb393`.
168. Dahmen W. and Kunoth A. (1992). Multilevel preconditioning. *Numer. Math.*, **63**, 315–44.
169. Dahmen W., Prossdorf S. and Schneider R. (1993). Wavelet approximation methods for periodic pseudo-differential equations II – fast solution and matrix compression. *Adv. Comp. Math.*, **1**, 259–335.
170. Dahmen W. and Stevenson R. (1999). Element by element construction of wavelets satisfying stability and moment conditions. *SIAM J. Numer. Anal.*, **37**, 319–25.
171. Darve E. and Have P. (2004). Efficient fast multipole method for low-frequency scattering. *J. Comp. Phys.*, **197**, 341–63.
172. Daubechies I. (1988). Orthonormal bases of compactly supported wavelets. *Comm. Pure Appl. Math.*, **41**, 909–96.
173. Daubechies I. (1992). *Ten Lectures on Wavelets*. SIAM Publications.
174. Davey K. and Bounds S. (1998). A generalized SOR method for dense linear systems of boundary element equations. *SIAM J. Sci. Comput.* **19**, 953–67.
175. de Souza A. Z., Canizares C. A. and Quintana V. H. (1997). New techniques to speed up voltage collapse computations using tangent vectors. *IEEE Trans. Power Systems*, **12**, 1380–7.
176. Dekker T. J. and Hoffmann W. (1989). Rehabilitation of the Gauss–Jordan algorithm. *Numer. Math.*, **54**, 591–9.
177. Dekker T. J., Hoffmann W. and Prota K. (1994). Parallel algorithms for solving large linear systems. *J. Comput. Appl. Math.*, **50**, 221–32.
178. Dekker T. J., Hoffmann W. and Prota K. (1997). Stability of the Gauss–Huard algorithm with partial pivoting. *Computing*, **58**, 225–44.
179. Demko S., Moss W. F. and Smith P. W. (1984). Decay rates for inverse of band matrices. *Math. Comp.*, **43**, 491–9.
180. Demmel J. W. (1997). *Applied Numerical Linear Algebra*. SIAM Publications.
181. Dobson I., Chiang H. D., Thorp J. S. and Fekih-Ahmed L. (1998). A model for voltage collapse in power systems. In: *Proc. of the 27th IEEE Conference on Decision and Control*, Austin, TX, Dec. 1988, 2104–9.
182. Dobson I. and Lu L. (1993). New methods for computing a closest saddle node bifurcation and worst case load power margin for voltage collapse. *IEEE Trans. Power Systems*, **8**, 905–913.
183. Doedel E. (1997). *Lecture Notes on Numerical Analysis of Bifurcation Problems*. Hamburg, March 1997. (See `ftp://ftp.cs. concordia.ca/pub/doedel/doc/montreal.ps.Z`).

184. Dongarra J., Duff I. S., Sorensen D. C. and van der Vorst H. A. (1998). *Numerical Linear Algebra on High-Performance Computers*. SIAM Publications.
185. Douglas C. C., Haase G. and Langer U. (2003). *A Tutorial on Elliptic PDE Solvers and Their Parallelization*. SIAM Publications.
186. Dowson D. and Higginson G. R. (1977). *Elastohydrodynamic Lubrication*. Pergamon Press.
187. Dubois D., Greenbaum A. and Rodrigue G. (1979). Approximating the inverse of a matrix for use in iterative algorithms on vector processors. *Computing*, **22**, 257–68.
188. Duff I. S. (1992). *The HB Exchange Format User's Guide*, from `ftp://ftp.cerfacs.fr/pub/harwell_boeing/userguide.ps.Z`.
189. Duff I. S., Erisman A. M. and Reid J. K. (1986). *Direct Methods for Sparse Matrices*. Clarendon Press (2nd edn to appear).
190. Duff I. S. and Koster J. (1999). The design and use of algorithms for permuting large entries to the diagonal of sparse matrices. *SIAM J. Matr. Anal. Appl.*, **20**, 889–901. (also RAL-TR-1997-059, http://www.numerical.rl.ac.uk/reports/)
191. Duff I. S. and Koster J. (2001). On algorithms for permuting large entries to the diagonal of a sparse matrix. *SIAM J. Matr. Anal. Appl.*, **22**, 973–96. (also RAL-TR-1999-030, http://www.numerical.rl.ac.uk/reports/)
192. Duff I. S. and Meurant G. A. (1989). The effect of ordering on preconditioned conjugate gradients. *BIT*, **29**, 635–57.
193. Egiazarian K. O., Astola J. T., Atourian S. M. and Gevorkian D. Z. (1997). Combining the discrete wavelet transforms and nearest neighbour filters for image restoration. In: *Proc. 1997 IEEE Workshop on Nonlinear Signal and Image Processing*, preprint from http://www.ecn.purdue.edu/NSIP/tp15.ps
194. Eiermann M. (1993). Field of values and iterative methods. *Lin. Alg. Applics.*, **180**, 167–97.
195. Ekwue A. O., Wan H. B., Cheng D. T. Y. and Song Y. H. (1998). Voltage stability analysis on the NGC systems. *Elec. Power Sys. Res.*, **47**, 173–80.
196. Elman H. C., Silvester D. and Wathen A. J. (2002). Performance and analysis of saddle point preconditioners for the discrete steady-state Navier–Stokes equations. *Numer. Math.*, **90**, 665-88.
197. Elman H. C., Silvester D. and Wathen A. J. (2005). *Finite Elements and Fast Iterative Solvers*. Oxford University Press.
198. Embree M. (1999). *Convergence of Krylov subspace methods for non-normal matrices*. Ph.D thesis, Oxford University Computing Laboratory, UK.
199. Erhel J., Burrage K. and Pohl B. (1996). Restarted GMRES preconditioned by deflation. *J. Comput. Appl. Math*, **69**, 303–18.
200. Evans D. J. (1968). The use of pre-conditioning in iterative methods for solving linear systems with symmetric positive definite matrices. *J. Inst. Math. Appl.*, **4**, 295–303.
201. Evans D. J. (1973). The analysis and application of sparse matrix algorithms in the finite element method. In: *Proc. the Mathematics of Finite Elements and its Applications* (*MAFELAP'72*), ed. J. R. Whiteman, pp. 427–47. Academic Press.
202. Evans D. J. (1994). *Preconditioning Iterative Methods*. Gordon and Breach Science Publishers.

203. Evans D. M. W. (1987). An improved digit-reversal permutation algorithms for the fast Fourier and Hartley transforms. *IEEE Trans. Acous. Speech Sig. Proc.*, **35**, 1120–5. (updates in *IEEE Trans. ASSP*, **37**, 1288–91 (1989)).
204. Ford J. M. (2003). An improved DWT preconditioner for dense matrix problems, *SIAM J. Matr. Anal. Applics.*, **25**, 642–61.
205. Ford J. M. and Chen K. (2001). Wavelet-based preconditioners for dense matrices with non-smooth local features. *BIT (Q. J. Numer. Math.)*, **41**, 282–307.
206. Ford J. M. and Chen K. (2001). An algorithm for accelerated computation of DWTPer-based band preconditioners. *J. Numer. Algor.*, **26**, 167–72.
207. Ford J. M. and Chen K. (2002). Speeding up the solution of thermal elastohydrodynamic lubrication problems. *Int. J. Numer. Methods Engng*, **53**, 2305–10.
208. Ford J. M., Chen K. and Evans D. J. (2003). On a recursive Schur preconditioner for iterative solution of a class of dense matrix problems. *Int. J. Comput. Math.*, **80**, 105–122.
209. Ford J. M., Chen K. and Ford N. J. (2003). Parallel Algorithms of fast wavelet transforms. *J. Parallel Algor. Applics.*, **18**, 155–69.
210. Ford J. M., Chen K. and Scales L. E. (2000). A new wavelet transform preconditioner for iterative solution of elastohydrodynamic lubrication problems. *Int. J. Comput. Math.*, **75**, 497–513.
211. Fox L. (1964). *An Introduction to Numerical Linear Algebra*. Oxford University Press.
212. Fuller A. T. (1968). Conditions for a matrix to have only characteristic roots with negative real parts. *J. Math. Anal. Appl.*, **23**, 71–98.
213. Garrat T. J., Moore G. and Spence A. (1991). Two methods for the numerical detection of Hopf bifurcations. *Int. Series Numer. Math.*, **97**, 129–33.
214. George J. A. (1973). Nested dissection of a regular finite element mesh. *SIAM J. Num. Anal.*, **10**, 345–63.
215. George J. A. and Liu J. W. (1981). *Computer Solution of Large Sparse Positive Definite Systems*. Prentice Hall.
216. Geurts R. J., van Buuren R. and Lu H. (2000). Application of polynomial preconditioners to conservation laws. *J. Engng. Math*, **38**, 403–26.
217. Gibson J. B., Zhang K., Chen K., Chynoweth S. and Manke C. W. (1999). Simulation of colloidal-polymer systems using dissipative particle dynamics. *J. Molec. Simul.*, **23**, 1–41.
218. Gill P. E., Murray W. and Wright M. H. (1989). *Numerical Linear Algebra and Optimization*. Addison Wesley.
219. Gill P. E. and Murray W., Ponceleon D. B. and Saunders M. A. (1992). Preconditioners for indefinite systems arising in optimization. *SIAM J. Matrix Anal. Appl.*, **13**, 292–311.
220. Gimbutas Z. and Rokhlin V. (2002). A generalized fast multipole method for nonoscillatory kernels. *SIAM J. Sci. Comput.*, **24**, 796–817.
221. Gines D., Beylkin G. and Dunn J. (1998). LU factorization of non-standard forms and direct multiresolution solvers. *Appl. Comput. Harmonic Anal.*, **5**, 156–201.
222. Giroire J. and Nedelec J. C. (1978). Numerical solution of an exterior Neumann problem using a double layer potential. *Math. Comp.*, **32**, 973–90.
223. Giusi E. (1984). *Minimal Surfaces and Functions of Bounded Variation*. Birkhauser.

224. Glover J. D. and Sarma M. (1994). *Power System Analysis and Design*. PWS-ITP.
225. Gohberg I., Hanke M. and Koltracht I. (1994). Fast preconditioned conjugate gradient algorithms for Wiener–Hopf integral equations. *SIAM J. Numer. Anal.*, **31**, 429–43.
226. Golberg R. R. (1965). *Fourier Transforms*. Cambridge University Press.
227. Goldstine H. (1972). *The Computer: from Pascal to van Neumann*. Princeton University Press.
228. Golub G. H. and O'Leary D. P. (1989). Some history of the conjugate gradient and Lanczos methods: 1948–1976. *SIAM Rev*. **31**, 50.
229. Golub G. H. and van Loan C. F. (1996). *Matrix Computations*, 3rd edn. Johns Hopkins University Press.
230. Gonzalez R. C. and Woods R. E. (1993). *Digital Image Processing*. Addison-Wesley.
231. Gottlieb D. and Orszag S. A. (1977). *Numerical Analysis of Spectral Methods*. SIAM Publications.
232. Gould N. I. M. and Scott J. A. (1998). Sparse approximate-inverse preconditioners using norm-minimization techniques. *SIAM J. Sci. Comput.*, **19**, 605–25. (also RAL-TR-1995-026, http://www.numerical.rl.ac.uk/reports/)
233. Govaerts W. J. F. (1995). Bordered matrices and singularities of large nonlinear systems. *Int. J. Bifurc. Chaos*, **5**, 243–50.
234. Govaerts W. J. F. (2000). Numerical bifurcation analysis for ODEs. *J. Comp. Appl. Math.*, **125**, 57–68.
235. Govaerts W. J. F. (2000). *Numerical Methods for Bifurcations of Dynamical Equilibria*. SIAM Publications.
236. Govaerts W. J. F. and Sijnave B. (1999). Matrix manifolds and the Jordan structure of the bialternate matrix product. *Lin. Algebra Appl.*, **292**, 245–66.
237. Govaerts W. J. F. and Spence A. (1996). Detection of Hopf points by counting sectors in the complex plane. *Numer. Math.*, **75**, 43–58.
238. Graham I. G., Hackbusch W. and Sauter S. A. (2000). Hybrid Galerkin boundary elements: theory and implementation. *Numer. Math.*, **86**, 139–72.
239. Grainger J. J. and Stevenson W. D. (1994). *Power System Analysis*. McGraw-Hill.
240. Grama A., Kumar V. and Sameh A. (1999). Parallel hierarchical solvers and preconditioners for boundary element methods. *SIAM J. Sci. Comput.*, **20**, 337–58.
241. Gray R. M. (2002). *Toeplitz and Circulant Matrices: a review*. Tutorial document from http://ee.stanford.edu/~gray/toeplitz.pdf (69 pages; draft 1 from 1971).
242. Graybill F. A. (1969). *Introduction to Matrices with Applications in Statistics*. Wadsworth.
243. Greenbaum A. (1997). *Iterative Methods for Solving Linear Systems*. SIAM Publications.
244. Greenbaum A., Ptak V. and Strakos Z. (1996). Any convergence curve is possible for GMRES. *SIAM Matrix Anal. Appl.*, **17**, 465–70.
245. Greenbaum A. and Trefethen L. N. (1994). GMRES/CR and Arnoldi/ Lanczos as matrix approximation problems. *SIAM J. Sci. Comput.*, **15**, 359–68.
246. Greengard L. (1988). *The Rapid Evaluation of Potential Fields in Particle Systems*, The MIT Press.
247. Greengard L. and Huang J. F. (2002). A new version of the fast multipole method for screened coulomb interactions in three dimensions. *J. Comp. Phys.*, **180**, 642–58.

248. Greengard L. and Rokhlin V. (1987). A fast algorithm for particle simulations. *J. Comp. Phys.*, **73**, 325–48. (See also the reprint from *J. Comp. Phys.*, **135**, 280–92, 1997).
249. Greengard L. and Rokhlin V. (1997). A new version of the fast multipole method for the Laplace equation in three dimensions. *Acta Numerica* 1997, (ed. A. Iserles), pp. 229–69.
250. Grenander U. and Szego G. (1984). *Toeplitz Forms and Their Applications*, 2nd edn., Chelsea, New York.
251. Griewank A. and Reddien G. (1983). The calculation of Hopf points by a direct method. *IMA J. Numer. Anal.*, **3**, 295–303.
252. Gropp W., Lusk E. and Skjellum A. (1999). *Using MPI*, 2nd edn., MIT Press.
253. Grote M. and Huckle T. (1997). Parallel preconditioning with sparse approximate inverses. *SIAM J. Sci. Comput.*, **18**, 838–53.
254. Guckenheimer J., Meyers M. and Sturmfels B. (1997). Computing Hopf bifurcations I. *SIAM J. Numer. Anal.*, **31**, 1–21.
255. Guillaume P., Huard A. and Le Calvez C. (2002). A multipole approach for preconditioners. In: *Lecture Notes in Computer Science*, LNCS 2330, pp. 364–73. Springer-Verlag.
256. Gustafson K. E. and Rao D. K. M. (1997). *Numerical Range: the Fields of Values of Linear Operators and Matrices*. Springer-Verlag.
257. Gutknecht M. H. and Röllin S. (2002). The Chebyshev iteration revisited. *Parallel Comput.*, **28**, 263–83.
258. Hackbusch W. (1982). Multigrid convergence theory. In: [263], 177–219.
259. Hackbusch W. (1985). *Multigrid Methods and Applications*. Springer-Verlag.
260. Hackbusch W. (1994). *Iterative Solution of Large Sparse Systems*. Springer-Verlag.
261. Hackbusch W. (1995). *The Integral Equation Method*. Birkhauser Verlag.
262. Hackbusch W. and Nowak Z. P. (1989). On the fast matrix multiplication in the boundary element method by panel clustering. *Numer. Math.*, **54**, 463–91.
263. Hackbusch W. and Trottenberg U. (1982). *Multigrid methods, Proc. of First European MGM conference, Lecture Notes in Mathematics 960*, Springer-Verlag.
264. Hageman A. L. and Young D. M. (1981). *Applied Iterative Methods*. Academic Press. (See also its earlier version: D. M. Young. *Iterative Solution of Large Linear systems*, Academic Press, New York, 1971).
265. Hansen P. C. (1998). *Rank-deficient and Discrete Ill-posed Problems: Numerical Aspects of Linear Inversion*. SIAM Publications.
266. Harbrecht H. (2003). *Wavelet-based Fast Solution of Boundary Integral Equations*. SFB report 03-07 from `http://webdoc.sub.gwdg.de/ebbok/e/2003/tu-chemnitz/sfb393`.
267. Harris P. J. (1992). A boundary element method for the Helmholtz equation using finite part integration. *Comput. Methods Appl. Mech. Engrg.*, **95**, 331–42.
268. Harris P. J. and Chen K. (2003). On efficient preconditioners for iterative solution of a Galerkin boundary element equation for the three dimensional exterior Helmholtz problem. *J. Comput. Appl. Math.*, **156**, 303–18.
269. Harten A. (1994). *Multiresolution Representation and Numerical Algorithms: a Brief Review*. CAM report 94-12, Dept of Mathematics, UCLA, USA.
270. Hawkins S. C. and Chen K. (2005). An implicit wavelet sparse approximate inverse preconditioner. *SIAM J. Sci. Comput.*, in press.

271. Hawkins S. C. and Chen K. (2004). New wavelet preconditioner for solving boundary integral equations over nonsmooth boundaries. *Int. J. Comput. Math.*, **81**, 353–60.
272. Hawkins S. C., Chen K. and Harris P. J. (2004). A new wavelet preconditioner for solving boundary integral equations over closed surfaces. To appear, *International Journal of Wavelets, Multiresolution and Information Processing*.
273. Haws J. C. and Meyer C. D. (2004). Preconditioning KKT systems. *Numer. Lin. Alg. Applics.*, to appear.
274. Hemker P. W. and Schippers H. (1981). Multigrid methods for the solution of Fredholm integral equations of the second kind. *Math. Comp.*, **36**, 215–32.
275. Henrici P. (1962). Bounds for iterates, inverses, spectral variation and fields of values of non-normal matrices. *Numer. Math.*, **4**, 24–40.
276. Henson V. E. (1999). *An Algebraic Multigrid Tutorial*, lecture slides, from http://www.casc.gov/CASC/people/henson, 101 pp.
277. Hess J. L. and Smith A. M. O. (1967). Calculations of potential flow about arbitrary bodies. In *Progress in Aeronautical Sciences*, (ed. D. Kucheman), **8**, 1–138. Pergamon Press.
278. Hestenes M. R. and Stiefel E. L. (1952). Methods of conjugate gradients for solving linear systems. *J. Res. Natl. Bur. Stand.*, **49**, 409–36.
279. Higham D. J. and Higham N. J. (2000). *MATLAB Guide*. SIAM Publications.
280. Higham N. J. (2002). *Accuracy and Stability of Numerical Algorithms*, 2nd edn. SIAM Publications. [1st edn, 1996].
281. Higham N. J. and Higham D. J. (1989). Large growth factors in Gaussian elimination with partial pivoting. *SIAM J. Matrix Anal. Appl.*, **10**, 155–64.
282. Hinton E. and Owen D. R. J. (1979). *An Introduction to Finite Element Computations*. Pineridge Press.
283. Hoffmann W. (1998). The Gauss–Huard algorithm and LU factorization. *Lin. Alg. Appl.*, **275–276**, 281–6.
284. Hoffmann W., Potma K. and Pronk G. (1994). Solving dense linear systems by Gauss–Huard's method on a distributed memory system. *Future Gen. Comp. Sys.*, **10**, 321–5.
285. Holm H., Maischak M. and Stephan E. P. (1995). *The hp-Version of the Boundary Element Method for Helmholtz Screen Problems*, Institute For Applied Mathematics report, IFAM 7, University of Hannover, Germany. (Available via ftp from ftp.ifam.uni-hannover.de/pub/preprints)
286. Hotelling H. (1943). Some new results in matrix calculation. *Ann. Math. Stats.*, **14**, 1–14.
287. Householder A. S. (1964). *The Theory of Matrices in Numerical Analysis*. Blaisdell.
288. Huard P. (1979). La methode du simplexe sans inverse explicite, *E. D. F. Bull. de la Direction des Etudes at des Recherches Serie C2*, **2**, 79–98.
289. Huckle T. (1993). Some aspects of circulant preconditioners. *SIAM J. Sci. Comput.*, **14**, 531–41.
290. Huckle T. (1999). Approximate sparsity patterns for the inverse of a matrix and preconditioning. *Appl. Numer. Math.*, **30**, 291–303.
291. Hughes M. D. and Chen K. (2004). Effective sparse preconditioners for 3D fluid structure interaction. *Int. J. Comput. Math.*, **81**, 583–94.

292. Hussein A. (2003). Fast computational methods for power flow bifurcation problems. Ph.D. thesis, University of Liverpool.
293. Hussein A. and Chen K. (2003). On efficient methods for detecting Hopf bifurcation with applications to power system instability prediction. *Int. J. Bifurc. Chaos*, **13**, 1247–62.
294. Hussein A. and Chen K. (2004). Fast computational algorithms for locating fold points for the power flow equations. *J. Comput. Appl. Math.*, **165**, 419–30.
295. IEEE (1973). Working Group on a Common Format, 'Common Format for Exchange of solved load flow data'. *IEEE Trans. Power Appa. Sys*, **92**, (November 1973), 1916–1925.
296. Il'in V. P. (1992). *Iterative Incomplete Factorization Methods*. World Scientific.
297. Ipsen I. C. F. (2001). A note on preconditioning nonsymmetric matrices. *SIAM J. Sci. Comput.*, **23**, 1050–1.
298. Isaacson E. and Keller H. B. (1966). *Analysis of Numerical Methods*. John Wiley. (See also paperback by Dover publications, 1994).
299. Iserles A. (1996). *A First Course in the Numerical Analysis of Differential Equations*. Cambridge University Press.
300. James G., Burley D., Dyke P., Searl J., Steele N. and Wright J. (1993). *Advanced Modern Engineering Mathematics*. Addison-Wesley.
301. Jaswon M. A. and Symm G. T. (1977). *Integral Equation Methods in Potential Theory and Electrostatics*. Academic Press.
302. Jawerth B. and Sweldens W. (1994). An overview of wavelet based multiresolution analysis. *SIAM Rev.*, **36**, 377–412.
303. Jennings A. (1977). Influence of the eigenvalue spectrum on the convergence rate of the conjugate gradient method. *J. Inst. Math. Applics.*, **20**, 61–72.
304. Jennings A. and Malik G. M. (1978). The solution of sparse linear equations by the conjugate gradient method. *Int. J. Numer. Meth. Engng.*, **12**, 141–58.
305. Jia Z. X. (2002). The refined harmonic Arnoldi method and an implicitly restarted refined algorithm for computing interior eigenpairs of large matrices. *Appl. Numer. Math.*, **42**, 489–512.
306. Jiang Z. B. and Yang F. X. (1992). *Numerical Solution of Partial Differential Equations in Exterior Domains*. Tianjin University Press.
307. Jin X. Q. (2003). *Developments and Applications of Block Toeplitz Iterative Solvers*. Kluwer Academic Publishers.
308. Johnson C. (1987). *Numerical Solution of Partial Differential Equations by the Finite Element Method*. Cambridge University Press.
309. Johnson O., Micchelli C. and Paul G. (1983). Polynomial preconditioning for conjugate gradient calculation. *SIAM J. Numer. Anal.*, **20**, 362–76.
310. Kailath T. and Sayed A. H. (1999). *Fast Reliable Algorithms for Matrices with Structure*. SIAM Publications.
311. Kang N., Zhang J. and Carlson E. S. (2004). Performance of ILU preconditioning techniques in simulating anisotropic diffusion in the human brain. *Future Generation Computer Sys.*, **20**, 687–98.
312. Kaporin I. E. (1992). Two-level explicit preconditioning of the conjugate gradient method. *Diff. Eqn.*, **28**, 329–39.
313. Karp A. H. (1996). Bit reversal on uniprocessors. *SIAM Rev.*, **38**, 1–26.

314. Kay D., Loghin D. and Wathen A. J. (2002). A preconditioner for the steady-state Navier–Stokes equation. *SIAM J. Sci. Comput.*, **24**, 237–56.
315. Keller H. B. (1987). *Numerical Methods in Bifurcation Problems*. Springer-Verlag.
316. Kelley C. T. (1995). *Iterative Methods for Solving Linear and Nonlinear Equations*. SIAM Publications.
317. Kirkup S. (1998). *The BEM in Acoustics*, Integrated Sound Software. See also http://www.soundsoft.demon.co.uk/tbemia/bookweb.htm
318. Klawonn A. and Starke G. (1999). Block triangular preconditioners for nonsymmetric saddle point problems: field of values analysis. *Numer. Math.*, **81**, 577–94.
319. Kolotilina L. Y., Nikishin A. A. and Yeremin A. Y. (1999). Factorized sparse approximate inverse preconditioners IV: simple approaches to rising efficiency. *Numer. Lin. Alg. Applics.*, **6**, 515–31.
320. Kolotilina L. Y. and Yeremin A. Y. (1986). On a family of two-level preconditionings of the incomplete block factorization type. *Soviet J. Numer. Anal. Math. Model.*, **1**, 293–320.
321. Kolotilina L. Y. and Yeremin A. Y. (1993). Factorized sparse approximate inverse preconditioners I – Theory. *SIAM J. Matr. Anal. Appl.*, **14**, 45–58.
322. Koschinski C. (1999). *Properties of approximate inverse and adaptive control concepts for preconditioning*. Internal Report 74/99, Ph.D. thesis, University of Karlsrule, Germany; www.ubka.uni-karlsruhe.de/indexer-vvv/1999/mathematik/4.
323. Kress R. (1989). *Linear Integral Equations*. Springer-Verlag.
324. Kreszig E. (1999). *Advanced Engineering Mathematics*. 8th edn., John Wiley.
325. Kundur P. (1994). *Power System Stability and Control*. McGraw-Hill.
326. Kuo C. C. J. and Levy B. C. (1989). A two-level four colour SOR method. *SIAM J. Num. Anal.*, **26**, 129–51.
327. Laghrouche O. and Bettess P. (2000). Short wave modelling using special finite elements. *J. Comput. Acoustics*, **8**, 189–210.
328. Lai C. H. and Chen K. (1998). Solutions of boundary element equations by a flexible elimination process. *Contemp. Math.*, **218**, 311–17.
329. Lancaster P. and Tismenetsky M. (1985). *The Theory of Matrices: with Applications*, 2nd edn, Academic Press.
330. Leem K. H., Oliveira S. and Stewart D. E. (2003). Algebraic Multigrid (AMG) for saddle point systems from meshfree discretizations. *Num. Lin. Alg. Applics.*, **11**, 293–308.
331. Lehoucq R. B. and Sorensen D. C. (1996). Deflation techniques for an implicitly restarted Arnoldi iteration. *SIAM J. Matrix Anal. Applics.*, **17**, 789–821,
332. Leuze R. (1989). Independent set orderings for parallel matrix factorizations by Gaussian elimination. *Para. Comput.*, **10**, 177–91.
333. Lewis J. G., Peyton B. W. and Pothen A. (1989). A fast algorithm for reordering sparse matrices for parallel factorizations. *SIAM J. Sci. Stat. Comput.*, **6**, 1146–73.
334. Li C. G. and Vuik K. (2004). Eigenvalue analysis of the SIMPLE preconditioning for incompressible flow. *Numer. Lin. Alg. Applics.*, **11**, 511–23.
335. Li Y. Y. and Santosa F. (1996). A computational algorithm for minimizing total variation in image restoration. *IEEE Trans. Image Proc.*, **5**, 987–95.

336. LI Z. Z., Saad Y. and Sosonkina M. (2003). PARMS: a parallel version of the algebraic recursive multilevel solver. *Num. Lin. Alg. Applics.*, **10**, 485–509.
337. Liang Y., Weston J. and Szularz M. (2002). Generalized least-squares polynomial preconditioners for symmetric indefinite linear system. *Parallel Comput.*, **28**, 323–41.
338. Lie J., Lysakera M. and Tai X. C. (2003). *A Variant of the Level Set Method and Applications to Image Segmentation*, UCLA CAM report CAM03-50, Dept of Mathematics, UCLA, USA. (See [112] for the web address).
339. Lin Z. C. (1982). The numerical solution of the Helmholtz equation using integral equations. Ph.D. thesis, Dept of Mathematics, University of Iowa.
340. Livne O. E. and Golub G. H. (2003). *Scaling by Binormalization*. SCCM report 03-12 from `http://www.sccm-stanford.edu/ wrap/pub-tech.html.` and *J. Numer. Alg.*, **35**, 94–120, January 2004.
341. Loghin D. and Wathen A. J. (2002). Schur complement preconditioners for the Navier–Stokes equations. *Int. J. Num. Methods Fluids*, **40**, 403–12.
342. Loghin D. and Wathen A. J. (2004). Analysis of preconditioners for saddle-point problems. *SIAM J. Sci. Comput.*, **25**, 2029–49.
343. Lu T., Shih T. M. and Liem C. B. (1992). *Domain Decomposition Methods – New Techniques for Numerical Solution of Partial Differential Equations*. Science Press, Beijing, China.
344. Mallat S. (1988). Multiresolution representation and wavelets. Ph.D. thesis, University of Pennsylvania, USA. (See also Multiresolution approximations and wavelet orthonormal bases of $L_2(\mathbb{R})$. *Trans. Am. Math. Soc.*, **315**, 69–87, 1989).
345. Mandel J. (1990). On block diagonal and Schur complement preconditioning. *Numer. Math.*, **58**, 79–93.
346. Margenov S., Wasniewski J. and Yalamov P. (2001). *Large-Scale Scientific Computing*. Lecture Notes in Computer Science 2179, Springer-Verlag.
347. Matrix market (2003). A repository of matrix test data, `http://math.nist.gov/MatrixMarket/.`
348. McCormick S. F. (1992). *Multilevel Projection Methods for Partial Differential Equations*. SIAM Publications.
349. Meijerink J. A. and van der Vorst H. A. (1977). An iterative solution method for linear systems of which the coefficient matrix is a symmetric M-matrix. *Math. Comput.*, **31**, 148–62.
350. Melenk J. M. and Babuska I. (1996). The partition of unity finite element method – basic theory and applications. *Comp. Meth. Appl. Mech. Eng.*, **136**, 289–314.
351. Meurant G. A. (1992). A review on the inverse of symmetric tridiagonal and block tridiagonal matrices. *SIAM J. Matr. Anal. Appl.*, **13**, 707–28.
352. Meurant G. A. (1999). *Computer Solution of Large Linear Systems*. North-Holland-Elsevier, 753 pp.
353. Meyer W. L., Bell W. A., Zinn B. T. and Stallybrass M. P. (1978). Boundary integral solution of three dimensional acoustic radiation problems. *J. Sound Vib.*, **59**, 245–62.
354. Mitchell R. and Griffiths D. F. (1980). *The Finite Difference Method in Partial Differential Equations*. John Wiley & Sons.
355. Mitchell R. and Wait R. (1985). *Finite Element Analysis and Applications*. John Wiley & Sons.

356. Mithulananthan N., Canizares C. A. and Reeve J. (1999). Hopf bifurcation control in power systems using power system stabilizers and static var compensators. In: *Proceedings of the North American Power Symposium (NAPS)*, pp. 155–63, California.
357. Mithulananthan N., Canizares C. A. and Reeve J. (2000). Indices to detect Hopf bifurcations in power systems. In: *Proceedings of the North American Power Symposium (NAPS)*, pp. 1232–9, Waterloo, ON.
358. Moret I. (1997). A note on the superlinear convergence of GMRES. *SIAM J. Numer. Anal.*, **34**, 513–16.
359. Morgan R. B. (1996). On restarting the Arnoldi method for large nonsymmetric eigenvalue problems. *Math. Comp.*, **65**, 1213–30.
360. Morgan R. B. (2000). Preconditioning eigenvalues and some comparison of solvers. *J. Comp. Appl. Math.*, **123**, 101–16.
361. Morgan R. B. (2002). GMRES with deflated restarting. *SIAM J. Sci. Comput.*, **24**, 20–37.
362. MPI (2000). *The Message Passing Interface Standard.* Argonne National Laboratory, `http://www.mcs.anl.gov/mpi/index.html`.
363. Mumford D. and Shah J. (1989). Optimal approximation by piecewise smooth functions and associated variational problems. *Comm. Pure Appl. Math.*, **42**, 477–685.
364. Murphy M. F., Golub G. H. and Wathen A. J. (2000). A note on preconditioning for indefinite linear systems. *SIAM J. Sci. Comput.*, **21**, 1969–72.
365. Nabors K., Korsmeyer F. T., Leighton F. T. and White J. (1994). Preconditioned, adaptive, multipole-accelerated iterative methods for three-dimensional first-kind integral equations of potential theory. *SIAM J. Sci. Comput.*, **15**, 713–35.
366. Nachtigal N. M., Reddy S. C. and Trefethen L. N. (1992). How fast are nonsymmetric matrix iterations? *SIAM J. Matr. Anal. Appl.*, **13**, 778–95.
367. Neelamani R. (1999). Wavelet-based deconvolution for ill-conditioned systems. M.Sc. thesis, Rice University, Houston, USA.
368. Neelamani R., Hyeokho C. and Baraniuk R. (2004). *ForWaRD:* Fourier-wavelet regularized deconvolution for ill-conditioned systems. *IEEE Trans. Image Proc.*, **52**, 418–33.
369. Neubert R. (1993). Predictor–corrector techniques for detecting Hopf points. *Int. J. Bifurc. Chaos*, **3**, 1311–18.
370. Neytcheva M. G. (1996). *Algebraic Multilevel Iteration Preconditioning Technique – a Matlab Implementation*. The document (28 pages) along with the software *amli* from `http://www.netlib.org/linalg/amli.tgz`.
371. Notay Y. (2002). Robust parameter-free algbraic multilevel preconditioning. *Numer. Lin. Alg. Applics.*, **9**, 409–28.
372. Nurgat E., Berzins M. and Scales L. E. (1999). Solving EHL problems using iterative, multigrid, and homotopy methods. *Trans. ASME*, **121**, 28–34.
373. Olkin J. (1986). Linear and nonlinear deconvolution problems. Ph.D. thesis, Rice University, Houston, USA.
374. Oosterlee C. W. and Washio T. (1998). An evaluation of parallel multigrid as a solver and a preconditioner for singularly perturbed problems. *SIAM J. Sci. Comput.*, **19**, 87–110.

375. Oppe T. C., Joubert W. D. and Kincaid D. R. (1988). *NSPCG User's Guide Version 1.0 – A Package for Solving Large Sparse Linear Systems by Various Iterative Methods*, available from `http://www.ma.utexas. edu/CNA/NSPCG/`
376. Ortega J. M. and Rheinboldt W. C. (1970). *Iterative Solution of Nonlinear Equations in Several Variables*, Academic Press. (Reprinted in 2nd edn as *SIAM Classics in Applied Mathematics*, 30, 2001).
377. Osher S. and Fedkiw R. (2003). *Level Set Methods and Dynamic Implicit Surfaces*. Springer.
378. Osher S. and Marquina A. (2000). Explicit algorithms for a new time dependent model based on level set motion for nonlinear deblurring and noise removal. *SIAM J. Sci. Comput.*, **22**, 387–405.
379. Osher S. and Sethian J. (1988). Fronts propagating with curvature dependent speed: algorithms based on Hamilton–Jacobi formulations. *J. Comp. Phys.*, **79**, 12–49.
380. Paige C. C. and Saunders M. A. (1975). Solution of sparse indefinite systems of linear equations. *SIAM J. Numer. Anal.* **12**, 617–29.
381. Paige C. C. and Saunders M. A. (1982). LSQR: an algorithm for sparse linear equations and sparse least squares. *ACM Trans. Math. Softw.*, **8**, 43–71.
382. Parks M. L., de Sturler E., Mackey G., Johnson D. D. and Maiti S. (2004). *Recycling Krylov Subspaces for Sequences of Linear Systems*. Dept of Computer Science, University of Illinois at UC, Technical Report UIUC DCS-R-2004-2421.
383. Parlett B. N. (1980). *The Symmetric Eigenvalue Problem*. Prentice-Hall.
384. Parter S. V. and Wong S. P. (1991). Preconditioning second-order elliptic operators: condition numbers and the distribution of the singular values. *J. Sci. Comput.*, **6**, 129–157.
385. Partridge P. W., Brebbia C. A. and Wrobel L. C. (1992). *The Dual Reciprocity Boundary Element Method*. Computational Mechanics Publications.
386. Peaceman D. W. and Rachford H. H. (1955). The numerical solution of parabolic and elliptic differential equations. *J. Soc. Ind. Appl. Math. (SIAM)*, **3**, 28–41.
387. Perry-Debain E., Trevelyan J. and Bettess P. (2003). Plane wave interpolation in direct collocation boundary element method for radiation and wave scattering: numerical aspects and applications. *J. Sound and Vib.*, **261**, 839–58.
388. Petrou M. and Bosdogianni P. (1999). *Image Processing: The Fundamentals*. John Wiley & Sons.
389. Pickering W. M. and Harley P. J. (1993). FFT solution of the Robbins problem. *IMA J. Numer. Anal.*, **13**, 215–33.
390. Press W. H., Teukolsky S. A., Vetterlin W. T. and Flannery B. P. (1992). *Numerical Recipes in C*. Cambridge University Press. [Also available *Numerical Recipes in Fortran*].
391. Proshurowski W. and Widlund O. B. (1976). On the numerical solution of Helmholtz equation by the capacitance matrix method. *Math. Comp.*, **30**, 433–68.
392. Purcell E. W. (1953). The vector method of solving simultaneous linear systems. *J. Maths. Phys.*, **32**, 180–3.
393. Puschel M. and Moura J. M. (2003). The algebraic approach to the discrete cosine and sine transforms and their fast algorithms. *SIAM J. Sci. Comput.*, **32**, 1280–1316.
394. Raghavan P., Teranishi K. and Ng E. (2003). A latency tolerant hybrid sparse solver using incomplete Cholesky factorization. *Numer. Lin. Alg. Applics.*, **10**, 541–60.

395. Rahola J. (1996). Diagonal forms of the translation operators in the multipole algorithm for scattering problems. *BIT*, **36**, 333–58.
396. Ramage A. (1999). A multigrid preconditioner for stabilised discretizations of advection-diffusion problems. *J. Comput. Appl. Math.*, **101**, 187–203.
397. Rathsfeld A. and Schneider R. (2000). *On a Quadrature Algorithm for the Piecewise Linear Wavelet Collocation Applied to Boundary Integral Equations*. Preprint SFB393/00-15. Univeristy of Chemnitz, Germany. `http://www.tu-chemnitz.de/sfb393`.
398. Reid J. K. (1971). On the method of conjugate gradients for the solution of large sparse systems of linear equations. In: *Large Sparse Sets of Linear Equations*, ed. J. K. Reid, pp. 231. Academic Press.
399. Reusken A. (1994). Multigrid with matrix-dependent transfer operators for convection-diffusion problems. In: *Seventh Int. Symp. DDMs for PDEs*, eds P. W. Hemker and P. Wesseling. Birkhauser.
400. Richtmyer R. D. and Morton K. W. (1967). *Difference Methods for Initial-value Problems*, 2nd edn. Wiley.
401. Riley K. L. (1999). Two-level preconditioners for regularized ill-posed problems. Ph.D. thesis, Montana State University.
402. Rius J. M. and De Porrata-Doria R. (1995). New FFT bit-reversal algorithm. *IEEE Trans. Sig. Proc.*, **43**, 991–4.
403. Rivlin T. J. (1974). *The Chebyshev Polynomials*. Wiley.
404. Rockafellar R. T. (1970). *Convex Analysis*. Princeton University Press.
405. Rokhlin V. (1985). Rapid solution of integral equations of classical potential theory. *J. Comp. Phys.*, **60**, 187–207.
406. Rokhlin V. (1990). Rapid solution of integral equations of scattering theory in two dimensions. *J. Comp. Phys.*, **86**, 414–39.
407. Roose D. and Hlavacek V. (1985). A direct method for the computation of Hopf bifurcation points. *SIAM J. Appl. Math.*, **45**, 879–94.
408. Rudin L. I., Osher S. and Fatemi E. (1992). Nonlinear total variation based noise removal algorithms. *Physica D*, **60**, 259–68.
409. Ruge J. W. and Stuben K. (1987). Algebraic multigrid. In: *Multigrid Methods*, ed. S. F. McCormick. SIAM Publications.
410. Saad Y. (1985). Practical use of polynomial preconditionings for the conjugate gradient method. *SIAM J. Sci. Statist. Comput.*, **6**, 865–81.
411. Saad Y. (1988). Preconditioning techniques for nonsymmetric and indefinite linear system. *J. Comput. Appl. Math.*, **24**, 89–105.
412. Saad Y. (1996). ILUM: a multi-elimination ILU preconditioner for general sparse matrices. *SIAM J. Sci. Comput.*, **17**, 830–47.
413. Saad Y. (1996). *Iterative Methods for Sparse Linear Systems*. PWS. (2nd edn by SIAM Publications, 2003).
414. Saad Y. (1999). *The Sparskit package* (a basic tool-kit for sparse matrix computations). from `http://www-users.cs.umn.edu/saad/software/SPARSKIT/sparskit.html`.
415. Saad Y. and Schultz M. H. (1986). GMRES: A generalized minimal residual algorithm for solving nonsymmetric linear systems. *SIAM J. Sci. Stat. Comput.*, **7**, 856–69.

416. Saad Y. and van der Vorst H. A. (2000). Iterative solution of linear systems in the 20th century. *J. Comput. Appl. Math.*, **123**, 1–33.
417. Saad Y. and Zhang J. (1999). BILUM: Block versions of multielimination and multilevel ILU preconditioner for general sparse linear systems. *SIAM J. Sci. Comput.*, **20**, 2103–21.
418. Saad Y. and Zhang J. (2001). Enhanced multi-level block ILU preconditioning strategies for general sparse linear systems. *J. Comput. Appl. Math.*, **130**, 99–118. See also (1998) http://www.cs.engr.uky.edu/~jzhang/pub/PREPRINT/enhanced.ps.gz
419. Sapiro G. (2001). *Geometrical Differential Equations and Image Analysis*, Cambridge University Press.
420. Schippers H. (1985). Multigrid methods for boundary integral equations. *Numer. Math.*, **46**, 351–63.
421. Schlijper A. G., Scales L. E. and Rycroft J. E. (1996). Current tools and techniques for EHL modeling. *Tribology Int.*, **29**, 669–73.
422. Schneider R., Levin P. L. and Spasojevic M. (1996). Multiscale compression of BEM equations for electrostatic systems. *IEEE Trans. Dielectrics Elec. Insulation*, **3**, 482–93.
423. Schulz G. (1933). Iterative calculation of inverse matrix (iterative berechnung der reziproken matrix). *Z. Angew. Math. Mech.*, **13**, 57–9.
424. Seydel R. (1979). Numerical computation of branch points in non-linear equations. *Numer. Math.*, **33**, 339–52.
425. Seydel R. (1994). *Practical Bifurcation and Stability Analysis – from Equilibrium to Chaos*, Vol 5 of *Interdisciplinary Applied Mathematics*, Springer-Verlag, 2nd edn.
426. Seydel R. (1997). On a class of bifurcation test functions. *Chaos, Solitons and Fractals*, **8**, 851–5.
427. Seydel R. (2001). Assessing voltage collapse. *Latin American Applied Research*, **31**, 171–6.
428. Shanazari K. and Chen K. (2003). On an overlapping DDM algorithm for the dual reciprocity method. *J. Numer. Alg.*, **32**, 275–86.
429. Shewchuk J. R. (1994). *An Introduction to the Conjugate Gradient Method without the Agonizing Pain*. Available from Carnegie Mellon University, Pittsburgh, USA, `http://www.cs.cmu.edu/~jrs/ jrspapers.html`, 58pp.
430. Sloan I. H. (1976). Improvement by iteration for compact operator equation. *Math. Comp.*, **30**, 758–64.
431. Sloan I. H. (1992). Error analysis of boundary integral equations. *Acta Numerica*, ed. A. Iserles (Cambridge University Press), **1**, 287–339.
432. Smith B., Gropp W. and Borjstad P. (1996). *Domain Decomposition Methods*. Cambridge University Press.
433. Soverance C. and Dowd K. (1998). *High Performance Computing*, 2nd edn., O'Reilly Media, Inc.
434. Spence A. and Graham I. G. (1999). Numerical methods for bifurcation problems. In: *Leicester 1998 EPSRC NA Summer School, The Graduate Student's Guide to Numerical Analysis '98*, eds. M. Ainsworth, J. Levesley, M. Marletta and W. A. Light. Springer Series in Computational Mathematics, Springer-Verlag.

435. Stephanos C. (1900). Sur une extension du calcul des substitutions lineaires, *J. Math. Pure Appl.*, **6**, 73–128.
436. Stevenson R. (1997). A robust hierarchical basis preconditioner on general meshes. *Numer. Math.*, **78**, 269–303.
437. Stewart G. W. (1998). *Matrix Algorithms I: Basic Decompositions*. SIAM Publications. (See also *Matrix Algorithms II: Eigensystems*, 2001).
438. Strang G. (1986). *An Introduction to Applied Mathematics*, Wellesley-Cambridge Press.
439. Strang G. (1986). A proposal for Toeplitz matrix calculations. *Stud. Appl. Math.*, **74**, 171–6.
440. Strang G. (1989). Wavelets and Dilation Equations: a Brief Introduction. *SIAM Rev.*, **31**, 613–27.
441. Strang G. and Nguyen T. (1996). *Wavelets and Filter Banks*. Wellesley-Cambridge Press.
442. Stuben K. (2000). *An Introduction to Algebraic Multigrid*, in Appendix A of [460], pp. 413–532. Also appeared as GMD report 70 from `http://www.gmd.de` and `http://publica.fhg.de/english/index.htm`.
443. Stuben K. and Trottenberg U. (1982). *Multigrid methods: fundamental algorithms, model problem analysis and applications*, pp. 1–176, in [263].
444. Süli E. and Mayers D. (2003). *An Introduction to Numerical Analysis*. Cambridge University Press.
445. Sun X. B. and Pitsiansi N. P. (2001). A matrix version of the fast multipole method. *SIAM Rev.*, **43**, 289–300.
446. Sweldens W. (1994). The construction and application of wavelets in numerical analysis. Ph.D. thesis, Catholic University of Leuven, Belgium (on Web).
447. Sweldens W. (1997). The lifting scheme: a construction of second generation wavelets. *SIAM J. Math. Anal.*, **29**, 511–46.
448. Ta'asan S. (2001). *Multigrid one-shot methods and design startegy*. Lecture note 4 of *Von-Karmen Institute Lectures*, `http://www.math.cmu.edu/~shlomo/VKI-Lectures/ lecture4`, 18pp.
449. Tai X. C. and Xu J. C. (2001). Global and uniform convergence of subspace correction methods for some convex optimization problems. *Math. Comp.*, **237**, 105–24.
450. Le Tallec P. (1994). Domain decomposition methods in computational mechanics. *Comp. Mech. Adv.*, **2**, 121–220.
451. Tang W. P. (1999). Towards an effective sparse approximate inverse preconditioner. *SIAM J. Matr. Anal. Appl.*, **20**, 970–86.
452. Tang W. P. and Wan W. L. (2000). Sparse approximate inverse smoother for multigrid. *SIAM J. Matr. Anal. Appl.*, **21**, 1236–52.
453. Tikhonov A. N. and Arsenin V. Y. (1977). *Solutions of Ill-posed Problems*. Winston and Sons.
454. Tinney W. F. and Hart C. E. (1967). Power flow solution by Newton's method. *IEEE Trans. Power Apparatus Syst.*, **pas-86** (11), November 1967.
455. Titchmarsh E. C. (1948). *Introduction to the Theory of Fourier Integrals*, 2nd edn. Oxford University Press.
456. Trefethen L. N. (1990). *Approximation theory and numerical linear algebra*. In: *Algorithms for Approximation II*, eds. J. C. Mason and M. G. Cox. Chapman and Hall.

457. Trefethen L. N. (1999). *Spectra and pseudospectra: the behaviour of non-normal matrices*. In: *The Graduate Student's Guide to Numerical Analysis'98*, eds. M. Ainsworth, J. Levesley and M. Marletta, pp. 217–50. Springer-Verlag.
458. Trefethen L. N. (2000). *Spectral Methods in MATLAB*. SIAM Publications. (See also a related but different book: http://web.comlab.ox.ac.uk/oucl/work/nick.trefethen/pdetext.html).
459. Trefethen L. N. and Bau D. III (1997). *Numerical Linear Algebra*. SIAM Publications.
460. Trottenberg U., Oosterlee C. W and Schuller A. (2000). *Multigrid*. Academic Press.
461. Turing A. M. (1948). Rounding-off errors in matrix processes. *Q. J. Mech. Appl. Math*. **1**, 287–308.
462. Tyrtyshnikov E. (1992). Optimal and superoptimal circulant preconditioners. *SIAM J. Matr. Anal. Appl*., **13**, 459–73.
463. van der Sluis A. and van der Vorst H. A. (1986). The rate of convergence of conjugate gradients. *Numer. Math*., **48**, 543–60.
464. van der Vorst H. A. (2003). *Iterative Krylov Methods for Large Linear Systems*. Cambridge University Press.
465. van Loan C. F. (1992). *Computational Frameworks of the FFT*. SIAM Publications.
466. Vanek P., Mandel J. and Brezina M. (1996). Algebraic multigrid on unstructured meshes. *Computing*, **56**, 179–96.
467. Varga R. (2000). *Matrix Iteration Analysis*, 2nd edn, Springer. (See also the 1st edn, Prentice-Hall, 1962).
468. Vassilevski P. S. (1989). *Nearly Optimal Iterative Methods for Solving Finite Element Elliptic Equations based on the Multilevel Splitting of the Matrix*. Report 1989-09, Inst of Scientific Computation, University of Wyoming, Laramie, WY, USA. (See also Hybrid V-cycle algebraic multilevel preocnditioners. *Math. Comp*., **58**, 489–512, 1992).
469. Vassilevski P. S. (1997). On two ways of stabilizing the hierarchical basis multilevel methods. *SIAM Rev*., **39**, 18–53.
470. Vassilevski P. S. and Lazarov R. D. (1996). Preconditioning mixed finite element saddle-point elliptic problems. *Num. Lin. Alg. Applics*., **3**, 1–20.
471. Vassilevski P. S. and Wang J. P. (1998). Stabilizing the hierarchical basis by approximate wavelets II: implementation and numerical results. *SIAM J. Sci. Comput*., **20**, 490–514. (See also Part I: theory, *Num. Lin. Alg. Appl*., **4**, 103–126, 1997).
472. Vavasis S. (1992). Preconditioning for boundary integral equations. *SIAM J. Matr. Anal. Appl*., **13**, 905–25.
473. Versteeg H. K. and Malalasekera W. (1995). *An Introduction to Computational Fluid Dynamics – the Finite Volume Method*. Longman Scientific & Technical.
474. Vogel C. R. (1995). A multigrid method for total variation-based image denoising. In: *Computation and Control IV*, 20, *Progress in Systems and Control Theory*, eds. K. Bowers and J. Lund. Birkhauser.
475. Vogel C. R. (1999). Negative results for multilevel preconditioners in image deblurring. In: *Scale-space Theories in Computer Vision*, eds. M. Nielson *et al*, pp. 292–304. Springer-Verlag.
476. Vogel C. R. (2002). *Computational Methods for Inverse Problems*. SIAM Publications.

477. Vogel C. R. and Oman M. (1996). Iterative methods for total variation denoising. *SIAM J. Sci. Comput.*, **17**, 227–37.
478. Vogel C. R. and Oman M. (1998). Fast, robust total variation-based reconstruction of noisy, blurry images. *IEEE Trans. Image Proc.*, **7**, 813–24.
479. von Petersdorff T. and Schwab C. (1996). Wavelet approximations of the first kind integral equation on polygons. *Numer. Math.*, **74**, 479–516.
480. Wagner C. (1999). An Introduction to Algebraic Multigrid. Course notes at University of Heidelberg, ver 1.1, from `http://www.iwr.uni-heidelberg.de/iwr/techsim/chris`, 127pp.
481. Walker J. S. (2002). *Fourier and Wavelet Analysis*. CRC Press.
482. Walnut D. F. (2001). *An Introduction to Wavelet Analysis*. Birkhauser Verlag.
483. Wan W. L. (1998). Scalable and multilevel iterative methods. Ph.D. thesis, CAM report 98-29, Dept of Mathematics, UCLA, USA.
484. Wang Z. and Hunt B. (1985). The discrete W transform. *Appl. Math. Comput.*, **16**, 19–48.
485. Wei Y. M. and Zhang N. M. (2004). Further note on constraint preconditioning for nonsymmetric indefinite matrices. *Appl. Math. Comp.*, **152**, 43–6.
486. Wei Y. M. and Wu H. B. (2000). Convergence properties of Krylov subspace methods for singular linear systems with arbitrary index. *J. Comput. Appl. Math.*, **114**, 305–18.
487. Wendlend W. L. (1980). On Galerkin collocation methods for integral equations of elliptic boundary value problems. In: *Numerical Treatment of Integral Equations*, eds. J Albrecht and L. Collatz, pp. 244–75, ISNM 53. Birkhauser.
488. Wendlend W. L. (1990). Boundary element methods for elliptic problems. In: Mathematical Theory of Finite and Boundary Element Methods, eds. H. Schatz, V. Thomee and W. L. Wendlend, pp. 219–76. Birkhauser Verlag.
489. Wesseling P. (1982). A robust and efficient multigrid method. In: [263, p. 614].
490. Wesseling P. (1992). *An Introduction to Multigrid Methods*, Wiley (from `http://www.mgnet.org/mgnet-books-wesseling.html`).
491. Widlund O. B. (1978). A Lanczos method for a class of nonsymmetric systems of linear systems. *SIAM J. Num. Anal.*, **15**, 801–12.
492. Wilkinson J. H. (1965). *The Algebraic Eigenvalue Problem*. Oxford University Press.
493. Winther R. (1980). Some superlinear convergence results for the conjugate gradient method. *SIAM J. Num. Anal.*, **17**, 14–17.
494. Xu J. C. (1992). Iterative methods by space decomposition and subspace correction. *SIAM Rev.*, **34**, 581–613.
495. Xu J. C. (1999). *An Introduction to Multilevel Methods*. In: Leicester 1998 EPSRC NA Summer School, *The Graduate Student's Guide to Numerical Analysis '98*, eds M. Ainsworth, J. Levesley, M. Marletta and W. A. Light, Springer Series in Computational Mathematics. Springer-Verlag.
496. Xu J. C. and Cai X. C. (1992). A preconditioned GMRES method for nonsymmetric or indefinite problems. *Math. Comp.*, **59**, 311–19.
497. Yan Y. (1994). Sparse preconditioned iterative methods for dense linear systems. *SIAM J. Sci. Comput.*, **15**, 1190–200.
498. Yang U. M. (2004). On the use of relaxation parameters in hybrid smoothers. *Numer. Lin. Alg. Applics.*, **11**, 155–72.

499. Ying L. A. (1988). *Lecture Notes in Finite Elements*, Beijing University Press.
500. Ying L. X., Biros G. and Zorin D. (2004). A kernel-independent adaptive fast multipole algorithm in two and three dimensions. *J. Comput. Phys.*, **196**, 591–626.
501. Young D. M. (1971). *Iterative Solution of Large Linear Systems*. Academic Press.
502. Yserentant H. (1986). On the multilevel splitting of finite element spaces. *Numer. Math.*, **49**, 379–412.
503. Yserentant H. (1993). Old and new convergence proofs for multigrid methods. *Acta Numerica*, **2**, 285–326.
504. Yu D. Y. (1993). *Natural Boundary Elements and Applications*. Science Press, China.
505. Yuan Y. X. and Sun W. Y. (1997). *Theory and Methods of Optimization*. Series of Doctoral Training. Science Press, China.
506. Zhang J. (2000). On preconditioning Schur complement and Schur complement preconditioning. *Electron. Trans. Numer. Anal.*, **10**, 115–30.
507. Zhang S. L. (2001). *Int. Conf. on Numerical Optimization and Numerical Linear Algebra*. Dun-Huang, China. Private communication.
508. Zhang Z. C. (1992). *Linear Algebra and its Applications*, University lecture notes series. Northwestern Industrial University Press, Xian, China.
509. Zhou H. M. (2000). *Wavelet transforms and PDE techniques in image compression*. Ph.D. Thesis, CAM report 00-21, Dept of Mathematics, UCLA, USA.
510. Zollenkopf K. (1971). Bi-factorisation – basic computational algorithm and programming technique. In: *Large Sparse Sets of Linear Equations*, ed. J. K. Reid, pp. 75–97. Academic Press.

Author Index

Subject Index